Lecture Notes in Mathematics

A. Carboni M. C. Pedicchio G. Rosolini (Eds.)

Category Theory

Proceedings of the International Conference
held in Como, Italy, July 22-28, 1990

Springer-Verlag
Berlin Heidelberg New York
London Paris Tokyo
Hong Kong Barcelona
Budapest

Editors

Aurelio Carboni
Dipartimento di Matematica
Università di Milano
Via C. Saldini 50
20133 Milano, Italy

Maria Cristina Pedicchio
Dipartimento di Scienze Matematiche
Università di Trieste
Piazzale Europa 1
34100 Trieste, Italy

Guiseppe Rosolini
Dipartimento di Matematica
Università di Parma
43100 Parma, Italy

Mathematics Subject Classification (1991): 18-06

ISBN 3-540-54706-1 Springer-Verlag Berlin Heidelberg New York
ISBN 0-387-54706-1 Springer-Verlag New York Berlin Heidelberg

Typesetting: Camera ready by author
Printing and binding: Druckhaus Beltz, Hemsbach/Bergstr.
46/3140-543210 - Printed on acid-free paper

PREFACE

The Conference "Category Theory '90" was held in Como (Italy) from July 22 to July 28, 1990. The organizing committee was composed of:
A. Carboni, University of Milan, Italy
M.C. Pedicchio, University of Trieste, Italy
G. Rosolini, University of Parma, Italy.
The scientific committee was composed of:
J. Adamek (Prague), J. Bénabou (Paris), A. Heller (New York), P.T. Johnstone (Cambridge), G.M. Kelly (Sydney), F.W. Lawvere (Buffalo).
125 persons took part to the Conference and 46 of them gave a talk. The editors would like to express their gratitude to the organisms that supported the meeting:
"Consiglio Nazionale delle Ricerche"
"Gruppi Nazionali 40% di Topologia e Logica"
Universities of Milan, Parma and Trieste,
as well as to all referees for their competent and prompt collaboration and to Springer for publishing the Proceedings.

The volume is divided in two parts: the first one contains research papers and is preceded by an introductory article by F.W. Lawvere; the second part consists of a monograph by A. Joyal and R. Street on Tannaka duality and quantum groups. It is intended to be an introduction to this rapidly growing subject and is addressed to a wider audience than category theorists; the paper is not only expository, but also contains new results and new proofs of classical results.

TABLE OF CONTENTS

PART I

PART II

SOME THOUGHTS ON THE FUTURE OF CATEGORY THEORY

F. William Lawvere
Department of Mathematics, S.U.N.Y. at Buffalo
Buffalo, NY 142140

The Como meeting was something of a milestone, coming as it did just twenty-five years after the first international meeting on category theory held at La Jolla, California in 1965. The work of Kan, Grothendieck, and others had greatly intensified the elaboration and application of the subject in the ten years prior to La Jolla, and enormous development has continued uninterruptedly since. I have been asked, as a participant at both meetings, to speculate on how at least some of the threads of the subject might develop in the immediate future. The threads I have selected now were only dimly visible then, for when J. L. Verdier described topos theory on the beach at La Jolla, most of us were slow to grasp its significance.

The crystallized philosophical discoveries which still propel our subject include the idea that a category of objects of thought is not specified until one has specified the category of maps which transform these objects into one another and by means of which they can be compared and distinguished. Thus, for applications of mathematics, to objectify is to mapify. Quite non-trivial in fact is also the idea that there must be definite domains and definite codomains and that there must be identity maps; even today there are many who think one could usefully "generalize" by omitting those requirements, sometimes on grounds of dislike for the "stasis" they think they imply. However, in modern Greek "stasis" means "bus-stop"; how useless an intricate network of speeding buses would be without them, and how disembodied would be processes without states. In fact category theory is the first to capture in reproducible form an incessant contradiction in mathematical practice: we must, more than in any other science, hold a given object quite precisely in order to construct, calculate, and deduce; yet we must also constantly transform it into other objects. These precepts, together with the powerful guide to look for and use adjoints in all categories large and small because they are the form of most constructions and deductions and many calculations and estimates, have guided us in our work in all the varied fields of mathematics. Most of us have struggled to explicitly introduce these principles also into our teaching, and those who have persisted find that this explicit use of the unity and cohesiveness of mathematics sparks the many particular processes whereby ignorance becomes knowledge, in learning just as it does in investigating. The need to teach, to explain and to respond to students' probing is often the genesis of problems taken up in "pure" research.

Though much remains to be done, it seems to some that we (that is, the community of category theorists with our ties to all the fields of pure and applied mathematics) have reached a unique position with regard to philosophy. I concentrate here on an outline of what is intended as a positive mathematical program. The history of possible philosophical objections to it will be treated elsewhere. Suffice it to suggest that Möbius, Hamilton, Grassman, Maxwell, etc. would not be among the naysayers. At least we can hope that sober application, of category theory to the ancient philosophical categories, will not only clarify both but also renew respect for serious thought, through solid examples approaching adequacy to their concept.

This attempt, by an admirer of rational mechanics, to include objective logic among the tools for arriving at a more accurate conception of space, will, I hope, not be dismissed by confusing it with objective idealism. The general science of the development of scientific ideas has a big overlap with category theory. That general science does not claim that scientific ideas are self-generating nor does it depend on faith for the acceptance of its own conclusions, as idealism would.

In the first section I start from the opposition between connected and separable objects to propose the tentative clarification, by a certain disjoint pair of classes of categories, of the conceptions of Being and Becoming respectively; how the one class arises from the other is the content of some resulting mathematical conjectures. In the second section, a specific mathematical formulation of the principle "unity-and-identity-of-opposites" is described in hopes of clarifying dimensionality in general and infinitesimals in particular, with again some mathematical conjectures aiming at further clarification. In the third section it is urged that certain pathologies "commonplace" since 1861/1890 need not be included in a more accurate conception of space and that both more physically-realistic models of computers as well as a more "objective" approach to Diophantine problems are already emerging from certain fascinating calculations.

I. In the remarkable paper [QDC] (Quotients of Decidable Objects, Cambridge) a certain epsilon difference between classes of toposes is mentioned. This epsilon is in a way the victory of geometry over narrow logicism and is what [QDB] (Qualitative Distinctions, Boulder) is groping to clarify. There were actually two kinds of mathematical examples which around 1960 forced the qualitative generalization of the previous notion of sheaf: for a particular algebraic space, the replacement of mere open subspaces by objects unramified over it, and for the category of all analytic spaces, the enlargement to a much nicer *category* within which those *particular* spaces which happen to be "nicer" according to some previously-achieved definition could be styled "representable". This raised the question: given a space in a topos of the second kind, what is the reasonable topos of pseudo-classical sheaves on it? Taking only a few of the simplest ideas from those circulated by the dozen or so who have strenuously worked on this problem, one arrives at the suggestion to study QD *sub*toposes of the topos of all spaces over a given space. Let's recall what QD means.

While the term "decidable" has subjective connotations which are a powerful guide to certain investigations, and "separable" is well-established in commutative algebra, in geometry "unramified" has an objective history; here I may use "SUD-object" to remind myself of the essential identity of all three. For brevity I'll use the term "neat" for objects which are *both* SUD and connected. There is a reflection from any locally connected topos to a topos in which every object is a colimit of neat objects; on the other hand, there is no coreflection (like Booleanization). In the cases where the reflection map is local, we have a start on the investigation of QD subtoposes. Note that "locally connected" and "sum" are relative to a base topos, which itself is quite special but (as in the Galois base of algebraic geometry) is sometimes much better *not* the topos of abstract sets; it would be fortunate if the hypothesis that the base itself be QD turned out to be sufficient . As usual, if I say "set", we should imagine an object of the base topos.

To clarify the above considerations, generalize to distributive categories and seek philosophical guidance. Even though the determination of which maps are epimorphisms is the more profound question studied with Grothendieck topologies, it takes place within a topos of the following kind. Call a small category C "extensive" if it has finite coproducts which yield an equivalence C/A+B = C/A x C/B and C/0 = 1 (this seems a minimum requirement on an op-fibration to conform with the notion of "family" and with Grassmann's "combinatorics of continuous magnitudes"); for example, "the" homotopy category or the category of spaces of dimension at most 4.

Then the topos G(C) of all those presheaves X on C for which merely X(A+B) = X(A) x X(B) & X(0) = 1 contains any conceivable theory of intensive quantities, either cohomological or function-theoretic, as an (algebra) *object*.

Further subtoposes of G(C) and G(C)/X based on "flatness" or the need to classify structures satisfying existential axioms are obviously of great importance, but passed over here to get to the main point. Those extensive categories which also have finite limits are called distributive, as discussed in the paper on their Burnside rigs in this volume [ECD].

A general category of Being, particular categories of Becoming: this is a suggested philosophical guide for sorting the two original kinds of toposes and what they have become. The unity and cohesiveness of Being provides the basis for Becoming, and the historicity and controlled variability of Becoming produces new Being from old. The unity and cohesiveness of space suggests the following condition on a category: "Every object can be included in a connected object". This axiom is not true in many toposes, for example, in the Weil (or infinitesimal) topos. If the category is not a topos, the axiom should perhaps be strengthened to say that every object is an equalizer of maps between connected objects; if the category is a topos, the axiom can be sufficiently checked on the one object 1 + 1, but implies that any object is functorially included in a contractible object (i.e. C for which all C^X are connected). This axiom for a category of Being will be paired with another one.

Here is a dialogue which suggests how the unity-and-cohesiveness axiom may be used: Suppose you claim that the surface of the earth and the point called the sky have "nothing to do with each other", whereas I claim they must. As a first step I consider the contractible container C of E+S which (though simple) may then become the basis for a more concrete connection, such as a scheme C' for a system of airlines and airports. . . . .[By the way, the unity-and-cohesiveness axiom can sometimes be demonstrated without invocation of power sets by using properties of the four adjoint functors: components, discrete, points, codiscrete; namely, if points map surjectively to components and if each discrete space maps injectively to the codiscrete space it negates, and finally, if nonempty codiscrete spaces are connected, then we need the injectivity of pushouts of injections.]

The connected objects and the unramified objects are "orthogonal" in the sense that any two maps $C \rightrightarrows U$ are either totally equal or nowhere equal. Hence the subcategory of neat objects in a given distributive category is always a category in which every map is an epimorphism, i.e. a [QDC] site. The orthogonal axiom, that "every object X can be covered by a SUD object U", is proposed as a characterization of a category of Becoming. Why? Considering the maps between objects in the site as control processes or deformations, a figure $U \longrightarrow X$ in a sheaf X may be considered to be a state of X, and a composite $U' \xrightarrow{t} U \xrightarrow{x} X$ to be the state x' which x "becomes" under the process t. Dialogue: I think that if $x_1 t = x_2 t$ now, surely $x_1 = x_2$ originally. You say no, there are many dissipative systems X. But no, I reply, you forgot to maintain enough information about history in your definition of present state; if you correct this neglect, you will obtain an epic $hX \longrightarrow X$ where hX satisfies my original injectivity-of-becoming claim. I resolve this dialogue iff my category satisfies the above "QD" axiom (a really ineradicable dissipation would require another sort of site, possibly with "relaxation" idempotents in it.) Note that a fundamental process of analysis, where a neighborhood becomes a smaller neighborhood $U' \subset U$, inducing a section of any sheaf to become its restriction, is of this kind.

[In a distributive category, an object U is SUD iff for any two maps $A \rightrightarrows U$ the equalizer E is a coproduct summand: A = E+E'. $E' \rightarrow A$ has the property that for any map $T \longrightarrow E'$, if the composites

$$T \longrightarrow E' \longrightarrow A \rightrightarrows U$$

are equal, then T = 0. Thus the requirement that all objects in a distributive category be SUD could be extended to merely extensive categories by demanding that every pair of maps have such a pair E, E'.]

For any distributive category C and any space X in G(C), the locally distributive site of SUD objects in C/X determines a QD subtopos P(X) of G(C)/X which is an approximation to "the particular category of Becoming which X is". Of course we have plucked X from its environment, so P(X) by itself is a too-clean abstraction from which to recover X; however, the composite $P(X) \longrightarrow G(C)/X \longrightarrow G(C)$, which we may call $\mathcal{O}_X$, retains the ties: for any R in G(C), $\mathcal{O}_X{}^*(R)$ is the pseudo-classical sheaf of intensive quantities of type R. $\mathcal{O}_X$ as a classifying map shows that P(X) is canonically given the additional structure of a sheaf of C-algebras ("without idempotents"). Note that I still have not succeeded to describe this in a site-invariant manner starting from a given pair of toposes $\mathcal{X}$, $\mathcal{S}$ satisfying suitable axioms, with the nature of $\mathcal{X}$ itself determining the corresponding refined version of the fiber P(X). I hope that the above clarifies the problem and that the several efforts in this direction will combine to solve it.

The normalization P(1) = $\mathcal{S}$ and the QD reflection suggest that a suitable axiom on $\mathcal{X}$ might just be that its QD reflection map is local (which is similar to the possible dual axiom that $\neg\neg$ is essential). This strong localness tie persists when 1 is generalized to a discrete space. However, for X of higher dimension, the extra essentialness adjoint of the refined $P(X) \rightrightarrows \mathcal{X}/X$, while (remarkably) product-preserving, is not exact, and $\mathcal{X}/X \longrightarrow (\mathcal{X}/X)_{QD}$ may not be local; the image of the composite may be a significant topos. We'll return in the next section to the meaning of "higher dimension."

There are many distributive categories which satisfy both axioms: in that case every object X is the image $U \twoheadrightarrow X \hookrightarrow C$ of a map from a SUD object to a connected object. For example, consider the topos of quivers (i.e. irreflexive graphs). However, they don't satisfy the further requirement on a general category of Being that "the product of connected spaces is connected". For example, if A is the connected quiver with a single arrow, then $A^2 = A+2D$ where D is the naked-dot quiver.

The condition, that a category of Being should not only be cohesively unifying but also have its connected objects closed under finite product, justifies Hurewicz's definition $[X,Y] = \pi_0(Y^X)$ of the homotopy category, expressing a definite kind of qualitative aspect of spaces. Such a category of Being cannot be simultaneously also a pure category of Becoming. For, in that case, the neat objects would be subobjects of 1, the topos thus localic; but a product-preserving cocontinuous functor on a localic topos is always left exact, hence preserves any equalizer $2 \longrightarrow I \rightrightarrows I'$ of connected objects representing 2, so an inconsistency would be reached by taking π_0 as the functor.

II. The intuitive idea that any one-dimensional connected group must be abelian could probably be proved in any suitable topos. We know what "connected" means, but what is "one-dimensional" for an object?

It seems that a significant portion of algebraic geometry and differential geometry does not depend so much on the particular algebraic theory used to construct models for it but is of a more fundamental conceptual nature. "One-dimensional", like "connected", is actually a philosophical concept, related to the minimal Hegelian level of figures which must be considered within an arbitrary space in order to determine that space's connectedness.

By a level in a category of Being, I mean a ("downward") functor from it to a smaller category which has both left and right adjoints which are full inclusions. Such a pair of categories and triple of functors is a unity-and-identity-of-opposites (UIO) in the sense that the big category unites the two opposite subcategories which in themselves are identical with the smaller category. One can picture the big category as a (horizontal) cylinder, some objects of which lie on the identical right or left ends. The two ends are opposite not only because we picture them so, but for the intrinsic reason of adjointness; every object in the category lies on a unique horizontal thread, two objects lying on the same thread iff the downward functor assigns to them isomorphic objects in the smaller (or lower) category. All is determined by the one functor. If the big category is a topos, the right-hand end will automatically be a subtopos, but the "identical" left hand end will usually not be. To say that a particular object belongs to the level has two sharply opposed meanings: we may say that it is a sheaf for the level if it belongs to the right-hand end, but that it negates a sheaf for the level if it belongs to the left hand end. The two idempotent adjoint endofunctors of the big category obtained by composing the three are called the coskeleton (right) and skeleton (left) functors for the level; the skeleton and coskeleton of any given space (object) in Being provide a kind of interval, graspable at this level, within which the possibly more complex space being studied must lie. The basic starting example of all this is that where the downward functor is the unique one to the terminal category; then the whole big category of Being constitutes just one thread, the unique sheaf being the terminal object ("pure Being") and its negative being the initial object ("non Being"); the two opposed subcategories are singletons in this case.

Within a given category of Being, consider the partially-ordered class of all levels within it whose adjointness is enriched over a given base topos which is a category of Becoming and which itself has the structure of a level, to be thought of as the lowest non-trivial level; assume that both the initial object as well as the terminal object are sheaves for this base level; the general sheaves for this base level are commonly called "codiscrete" or "chaotic" objects within the big category of Being, and the subtopos of them may be called "pure Becoming". The negative objects for this level are commonly called "discrete" and the subcategory of them deserves to be styled "non Becoming". The objects of the base category (which is identical with the two opposite subcategories of pure Becoming and non Becoming when the inclusion functors are neglected) can just be called "sets". However, this base topos, although we have restricted it to be QD, is not necessarily the category of abstract sets; part of the philosophical content of the work of Galois is that, for the Being of algebraic geometry over a non-algebraically closed base field, a much more accurate picture is achieved if the base is taken to be a well-determined Boolean topos of more-subtly Becoming sets which is not of the purest abstract kind where the axiom of choice would hold. The base in fact seems in examples to be determined by the given category of Being itself, either as the latter's QD reflection with the extra localness condition supplying the right adjoint pure Becoming inclusion, or else (for example simplicial sets) as the double-negation sheaves with the extra essentialness condition supplying the left adjoint inclusion (in the latter case it is in Hegelian fashion always the smallest level for which both 0,1 are sheaves). Within the class of all levels over the base (of course it is a set in fact if the category of Being is a topos), the base itself is often further distinguished by having a still further left adjoint to its discrete inclusion, this extra functor therefore assigning to every space in Being its set of components. The downward (non faithful) functor itself we regard, of course, as assigning to any space its set of points.

The relation between the trivial level and the base level above it is only the first case of a possible strong relation between two levels which (hoping not to do too great an injustice to Hegel) I will call Aufhebung relative to the given category of Being: this is the relation between a lower level and a higher level whereby the first level is not only included (on the left and equivalently on the right, or simply that the longer downward functor factors across the shorter one) in the higher, but moreover that the longer left adjoint inclusion factors across the shorter right adjoint inclusion; equivalently, the higher coskeleton functor fixes both the skeleta and the coskeleta in the sense of the lower level. A very simple picture (not involving toposes) involves taking the basic downward functor to be any given map from a seven-element totally ordered set onto a three-element one, which just amounts to a partition of the big set into three non-empty closed intervals; an arbitrary intermediate level is simply a finer partition of each of these coarse intervals into finer subintervals, but an intermediate level is an Aufhebung of the lower one iff the following more stringent condition is satisfied: among the subintervals within each coarse interval, the left-most one is a singleton. In this simple example, as in some but not all examples involving toposes, every level has a smallest aufgehobenen level over it, which could reasonably be called "the" Aufhebung of it.

Unity-and-identity-of-opposites, the Aufhebung relation between two such within a given unity: this is a second proposed philosophical guide. It is not limited to distributive categories, nor is the dual case of an inclusion which has both left and right adjoint retractions without interest; that dual relation holds for example between graded modules and chain complexes, and it is the image of the canonical map between the opposites which defines the homology functor.

Having described a basic framework, we can now return to the question of the intrinsic meaning of "one-dimensionality" of an object within such a framework. The basic idea is simply to identify dimensions with levels and then try to determine what the general dimensions are in particular examples. More precisely, a space may be said to have (less than or equal to) the dimension grasped by a given level if it belongs to the negative (left adjoint inclusion) incarnation of that level. Thus a zero-dimensional space is just a discrete one (there are several answers, not gone into here, to the objection which general topologists may raise to that) and dimension one is the Aufhebung of dimension zero. Because of the special feature of dimension zero of having a components functor to it (usually there is no analogue of that functor in higher dimensions), the definition of dimension one is equivalent to the quite plausible condition: the smallest dimension such that the set of components of an arbitrary space is the same as the set of of components of the skeleton at that dimension of the space, or more pictorially: if two points of any space can be connected by anything, then they can be connected by a curve. Here of course by "curve" we mean any figure in (i.e. map to) the given space whose domain is one-dimensional.

Continuing, two-dimensional spaces should be those negating a subtopos which itself contains both the one-dimensional spaces and the identical-but-opposite sheaves which the one-dimensional spaces negate.

If by "function" we mean a map with one-dimensional codomain, then any function naturally defined along each of the surfaces in an arbitrary space uniquely extends to a smooth function on the space itself. That "surfaces" might even be replaced by "curves" is the basis of recent interesting work on infinite-dimensional differentiation (as it was the basis of the very first work 250 years ago on that subject); the possibility of using curves as test figures may not be the result of the somewhat restrictive category of spaces considered, but rather of a more refined property of the basic codomains of functions, such as the line and circle. These are not only one-dimensional but even belong also to the Aufhebung of a still smaller level, since they are retracts of map-spaces of infinitesimal spaces.

The infinitesimal spaces, which contain the base topos in its non-Becoming aspect, are a crucial step toward determinate Becoming, but fall short of having among themselves enough connected objects, i.e. they do not in themselves constitute fully a "category of cohesive unifying Being." In examples the four adjoint functors relating their topos to the base topos coalesce into two (by the theorem that a finite-dimensional local algebra has a unique section of its residue field) and the infinitesimal spaces may well negate the largest essential subtopos of the ambient one which has that property. This level may be called "dimension $\in$"; calling the levels (i.e. the subtoposes essential over the base) "dimensions" does not imply that they are linearly ordered nor that the Aufhebung process touches each of them. The infinitesimal spaces provide (in many ways) a good example of a non trivial unity-and-identity-of-opposites inside the ambient topos of Being: explicitly recognizing the *two* inclusions, as spaces which could be called infinitesimal and formal spaces respectively, may help clarify the confusing but powerful interplay between these two classes which are opposite but in themselves identical. The calculation of the $\in$-skeleton and $\in$-coskeleton, of a space which is neither, needs to be carried out, and also the calculation of the Aufhebung of dimension $\in$.

The idea behind the identification of the levels in a category of Being with dimensions is that a higher level is a more determinate general Becoming, that is, it contains spaces having in them possibly-more-varied information for determining processes. Thus one conjectures that dimX only depends on the category P(X) of particular Becoming associated to X (and not on the important structure sheaf which recalls for the little category the big environment in which it was born). In other words, if we have an equivalence of categories $P(X) \equiv P(Y)$, then X,Y should belong to the same class of UIO levels within the category of Being in which they are objects. Suitable hypotheses to make this conjecture true should begin to clarify the relationships between the two suggested philosophical guides.

III. Why does the epsilon difference leave room for the triumph of geometry over narrow logicism? What might have led Grothendieck to propose his (still unpublished) program for "tame topology", wherein he arrived at roughly the same real analytic spaces as logicians working on a Tarski problem of "decidability"? It seems that all attempts to characterize continuity in a purely intensive logical way, such as the frame algebra, leave another kind of room in spite of their profound contribution to calculations - room for the obviously non-physical space-filling curves and nowhere-differentiable functions. Though we have been led to believe that this subjectively-generated Raum shows that our basic intuition of space is unreliable, still we have not been shown anything in the real world which could more than provisionally be modeled as a discrete infinity. Rather than such speculations about the unreliability of knowledge, it seems that still more serious work is needed, marshalling all the achievements of subjective logic *and* objective logic, of geometry *and* algebra to hone still more realistic models of continuous space. As several have suggested, a guide is to consider figures, i.e. maps to the space, as fundamental in determining it, with intensive quantities (i.e. maps from the space) being derived by naturality rather than the other way around; this does not metaphysically mean of course that the nature of the domains of the figures is not derived by a careful ascending interplay between all four of the mentioned subjects.

The above considerations are related, as suggested in a Como discussion, with old problems such as Fermat's. Diophantus probably considered natural numbers not in the abstract way which we habitually now do, but as born from actual objects. While the method of formally adjoining negatives and ascending to powerful cohomological calculations etc. leads to many results, we should not forget the objects themselves. Just as realizing cohomology classes by vector bundles via K-theory permitted powerful interplay between those calculations and directed manipulations of the objects by actual maps, so a similar possibility is opened by the Burnside rig of a distributive category, wherein polynomial equations satisfied by objects are revealed as specific structure on the objects themselves. For example, the equation $x = 1 + 2x$ arises from an object with a point and operated on by a 2-generator monoid, with an additional inverse map. But even the dangerous $x = 1+x$ does not lead to unbridled infinity. While the study of linear equations on distributive categories is packed with surprising subtleties, higher-degree equations are also approachable with, for example, homogeneity retaining some of its usual properties when interpreted in this more demanding functorial manner. I was surprised to note that an isomorphism $x = 1 + x^2$ (leading to complex numbers as Euler characteristics if they don't collapse) always induces an isomorphism $x^7=x$.

The rough similarity between Grothendieck's tame spaces and the finitely sub-analytic [FSA] objects considered in logic is in fact a major difference; the same sort of difference exists between real algebraic geometry and semi-algebraic sets, as well as between the ordinary continuous PL category and the polyhedral category of negative sets. All of these are quite different from categories in which countable coproduct decompositions are common. The difference within each of the three pairs mentioned may all exemplify a single general process which I'll call Aristotelian intervention; some such analysis seems also basic to attempts to hone a more physically realistic model for the programming of "digital" computers. Aristotle pointed out that the continuum is divisible but not infinitely divided. One can break a stick. Repeating that and all which it led to for 40,000 years has created a lot of indispensable chairs, buildings, etc. but has not changed the fundamental continuity of space; neither will billions of times dividing possible current-levels into "yes and no".

Given a map of coherent toposes, there is not only the usual induced topology but there are also topologies in which only the inverted maps between *coherent* objects are taken as covers. For example, to construct the generic solution of the equation $x = 1+3x+2x^2$, consider first the classifying topos for the free algebraic theory with one constant, three unary, and two binary operations; the free algebra on no generators determines a point of this topos. If we were to consider the full induced topology, our topos would collapse to that of discrete sets; a coherently induced topos however might not only satisfy the equation but have a non-trivial Burnside ring, whereas just because it lacks the metaphysical "minimality of the fixed point", it may provide a more physical model of potential lists and trees. We don't yet know which presentations of rigs can be Burnside-realized from distributive categories, because the very concreteness of the non-isomorphisms in the latter may give rise also to unexpected isomorphisms. We also don't know whether finitely subanalytic sets can be obtained via such a uniform procedure from some sort of continuously tame ones. Another possibility would be to take not the base topos but the infinitesimal one as the domain of the topos map which is used to induce the Aristotelian intervention - is it possible that even after such an explosion, functions could still have a well-defined derivative at every point? Certainly the resulting sites need not be Boolean; for example, consider a half-open interval x: it should satisfy $x = 2x$ but its endpoint is not a coproduct summand.

Naturally, models like the polyhedra constructed from below from real space are more satisfying than those constructed from above by classifying abstract algebras, but as usual the goal is to be approached by pushing hard from both sides.

If the general program proposed above is correct at least in rough outline, it would serve both the advancement and the dissemination of the subject to have it clearly worked out. As clearly formulated in Grassmann's introduction, only a good philosophical preamble can orient the student toward what kind of applications of a purely mathematical development he should look out for; that theory of pedagogy is at least as deserving of trial as the pragmatist theory of teaching only skills, which as we have seen did not achieve its goal.

In spite of temporary setbacks of all kinds, the many-sided and passionate advance of category theory has been on the whole remarkably steady. On the basis of all that work many questions of both fundamental and applied nature are now becoming clear, thus the future of our science is bright.

Bibliography

[QDC] Johnstone, Peter, "Quotients of Decidable Objects in a Topos" in *Math. Proc. Camb. Phil. Soc.* **93**, (1983), 409-419.

[QDB] Lawvere, F. William, "Qualitative Distinctions Between Some Toposes of Generalized Graphs" in *Contemporary Mathematics* **92**, (1989), 261-299.

[ECD] Schanuel, Stephen, "Negative Sets have Euler Characteristic and Dimension", (this volume).

[FSA] Van den Dries, Lou, "A Generalization of the Tarski-Seidenberg Theorem and Some Nondefinability Results" *Bull. Am. Math. Soc.* **15**, (1986), 189-193.

This paper is in final form and will not be published elswhere.

WHAT ARE LOCALLY GENERATED CATEGORIES?

JIŘÍ ADÁMEK AND JIŘÍ ROSICKÝ

ABSTRACT. The answer to the question in the title, as given by Gabriel and Ulmer, is well enough known: precisely the locally presentable categories. We prove that the same result holds for mono-locally generated categories, a concept we introduce since we find it more "natural" than that of locally generated categories.

Recall from [GU] that a category $\mathcal{K}$ is called locally λ-presentable iff it is cocomplete, and has, for some regular cardinal λ, a set of λ-presentable objects such that every $\mathcal{K}$-object is a λ-direct colimit of objects lying in that set. And $\mathcal{K}$ is called locally λ-generated iff

(a) $\mathcal{K}$ is cocomplete
(b) $\mathcal{K}$ has for some regular cardinal λ a set $\mathcal{A}$ of λ-generated objects (recall that an object A is λ-presentable (λ-generated) iff its hom-functor $\hom(A, -)$ preserves λ-direct colimits (λ-direct colimits of monomorphisms)) such that every $\mathcal{K}$-object is a λ-direct colimit of objects lying in that set
(c) each $\mathcal{A}$-object has only a set of strong quotients.

The condition (c) is quite unpleasant, particularly in view of the fact that each locally presentable category is co-wellpowered (as proved directly in [GU] and very elegantly in [MP]). Now, Gabriel and Ulmer prove that a category is locally generated iff it is locally presentable (with unequal λ's, in general). This leads to the following

Open problem 1. Is each category, satisfying (a) and (b) above, locally presentable? That is, can (c) be omitted?

Remark 1. (i) This problem can be stated in a more suggestive form: Is each full reflective subcategory of a locally presentable category $\mathcal{L}$, closed in $\mathcal{L}$ under λ-direct colimits of monomorphisms, locally presentable?

In fact, to see that the affirmative answer to this question is equivalent to the affirmative answer to Problem 1, let $\mathcal{K}$ be a category satisfying (a), (b) above. The closure $\bar{\mathcal{A}}$ of $\mathcal{A}$ under all colimits of less than λ objects is dense in $\mathcal{K}$ and consists of λ-generated objects (see [GU]). Thus, the Yoneda embedding $E : \mathcal{K} \to Set^{\bar{\mathcal{A}}^{op}}$, $EK = \hom(-, K)$ is full and faithfull. Since $\mathcal{K}$ is cocomplete, E is a right adjoint (see [GU]) and since $A \in \bar{\mathcal{A}}$ is λ-generated, E is easily seen to preserve λ-direct colimits of monomorphisms. Hence $\mathcal{K}$ is (isomorphic to) a full reflective subcategory of a locally presentable category which is closed under λ-direct colimits of monomorphisms.

Conversely, let $\mathcal{K}$ be a full reflective subcategory of a locally presentable category $\mathcal{L}$ which is closed in $\mathcal{L}$ under λ-direct colimits of monomorphisms. Then $\mathcal{K}$ is cocomplete and, taking λ such that $\mathcal{L}$ is locally λ-presentable, reflections of λ-presentable objects of $\mathcal{L}$ form the desired set $\mathcal{A}$ from (b).

(ii) The answer is either affirmative, or it depends on set theory because, as we have proved in [RTA], under Vopěnka's principle every limit-closed full subcategory of a locally presentable category is locally presentable.

Thus, an absolute negative answer to the above question would imply that Vopěnka's principle is wrong (and thus imply the non-existence of huge cardinals, see [ART]).

(iii) The value of the concept of locally generated category lies in the fact that λ-direct unions of subobjects are more natural and more often used (e.g. in model theory) than general λ-direct colimits. The trouble with the concept of Gabriel and Ulmer is that, whereas λ-direct colimits of monomorphisms appear in the condition put upon $\mathcal{A}$-objects, they do not appear in the representation of objects of $\mathcal{K}$ as colimits of $\mathcal{A}$-objects. We believe that the natural condition is the following: each object is a λ-direct colimit of a diagram of $\mathcal{A}$-objects and monomorphisms. The aim of this note is to prove that by changing (b) in this manner, (c) can be omitted.

Since in model theory one works with submodels (=strong subobjects of Σ-structures) rather than just subobjects, whereas in algebra subobjects are sometimes more natural than strong subobjects, we formulate our result relatively w.r.t. a class $\mathcal{M}$ of monomorphisms.

By a *factorization system* in a category $\mathcal{K}$ we mean the classical concept of J. Isbell: compositive classes $\mathcal{E} \subseteq \mathrm{Epi}(\mathcal{K})$ and $\mathcal{M} \subseteq \mathrm{Mono}(\mathcal{K})$ with $\mathcal{E} \cap \mathcal{M} = \mathrm{Iso}(\mathcal{K})$ such that each morphism has an essentially unique $(\mathcal{E}, \mathcal{M})$-factorization.

Definition. *Let $\mathcal{M}$ be a class of monomorphisms in a category $\mathcal{K}$, and let λ be a regular cardinal.*

(1) *An object A of $\mathcal{K}$ is called λ-generated w.r.t $\mathcal{M}$ provided that* $\hom(A, -)$ *preserves colimits of λ-direct diagrams of $\mathcal{M}$-monomorphisms.*

(2) *$\mathcal{K}$ is said to be $\mathcal{M}$-locally generated if*
   (a) *$\mathcal{K}$ is cocomplete, and*
   (b) *$\mathcal{K}$ has, for some regular cardinal λ, a set $\mathcal{A}$ of λ-generated objects w.r.t. $\mathcal{M}$ such that every $\mathcal{K}$-object is a λ-direct colimit of a diagram of $\mathcal{A}$-objects and $\mathcal{M}$-morphisms.*

Theorem 1. *For each category $\mathcal{K}$ equivalent are:*

(i) *$\mathcal{K}$ is locally presentable,*
(ii) *$\mathcal{K}$ is mono-locally generated,*
(iii) *$\mathcal{K}$ is $\mathcal{M}$-locally generated for some factorization system $(\mathcal{E}, \mathcal{M})$.*

Remark 2. We will actually prove that whenever $\mathcal{K}$ has a factorization system $(\mathcal{E}_0, \mathcal{M}_0)$ such that $\mathcal{K}$ is $\mathcal{M}_0$-locally generated, then $\mathcal{K}$ is $\mathcal{M}$-locally generated for *each* factorization system $(\mathcal{E}, \mathcal{M})$.

Proof If $\mathcal{K}$ is $\mathcal{M}$-locally generated, then the set $\mathcal{A}$ in the above definition is easily seen to be dense. Therefore $\mathcal{K}$ is equivalent to a full reflective subcategory of $Set^{\mathcal{A}^{op}}$ which is closed under λ-direct colimits of $\mathcal{M}$-morphisms (via the Yoneda embedding, cf. Remark 1). It follows that $\mathcal{K}$ is complete and wellpowered - thus, it has (strong epi, mono)-factorizations. Therefore, we see that given $(\mathcal{E}_0, \mathcal{M}_0)$ as in Remark 2, then *(ii)* holds. The implication $(ii) \to (iii)$ is evident. Furthermore, (i) implies that $(\mathcal{E}_0, \mathcal{M}_0)$ in Remark 2 exists: if $\mathcal{K}$ is locally λ-presentable then the (essentially small) collection $\mathcal{A}$ of all $\mathcal{E}$-quotients of all λ-presentable objects satisfies (b) above.

Thus, it is sufficient to prove that $(iii) \to (i)$. By the above observation, since $\mathcal{L} = Set^{\mathcal{A}^{op}}$ is locally ω-presentable, it is only necessary to prove the following: Given a locally ω-presentable category $\mathcal{L}$ and its full, reflective subcategory $\mathcal{K}$ closed under λ-direct colimits of $\mathcal{M}$-morphisms where $(\mathcal{E}, \mathcal{M})$ is a factorization system of $\mathcal{K}$, then $\mathcal{K}$ is locally presentable.

(1) We first observe that the set $\mathcal{A}$ of (b) above represents all λ-generated objects w.r.t. $\mathcal{M}$. Given a λ-generated A, let D be a λ-direct diagram of $\mathcal{M}$-morphisms with the colimit $(D_i \xrightarrow{d_i} A)_{i \in I}$. Since it is the λ-direct colimit of monomorphisms in $\mathcal{L}$ and $\mathcal{L}$ is locally ω-presentable, any d_i is a monomorphism. Since A is λ-generated w.r.t. $\mathcal{M}$, id_A factors through some d_{i_0}. Thus d_{i_0} is a split epimorphism and, as a monomorphism, it is an isomorphism.

(2) Let P denote the collection of all reflection arrows of $\mathcal{L}$-objects in $\mathcal{K}$ which are either finite colimits in $\mathcal{L}$ of diagrams in $\mathcal{A}$ or codomains of multiple pushouts in $\mathcal{L}$ of $\mathcal{E}$-morphisms with a domain in $\mathcal{A}$.

Observe that P is essentially small: the case of finite colimits is clear, for the multiple pushout use the fact that each $\mathcal{E}$-morphism with a domain λ-generated w.r.t. $\mathcal{M}$ has a codomain λ-generated w.r.t. $\mathcal{M}$, and apply (1).

Thus, the class $P^\perp$ of all $\mathcal{L}$-objects orthogonal to any arrow from P is a locally presentable category (see [GU]). The choice of P guarantees that $\mathcal{K}$ is closed under finite colimits and $\mathcal{E}$-cointersections in $P^\perp$. Besides, $\mathcal{K}$ is reflective and, since any domain and codomain of any P-morphism is λ-generated w.r.t. $\mathcal{M}$, $\mathcal{K}$ is closed under λ-direct colimits of $\mathcal{M}$-morphisms in $P^\perp$. Consider a morphism $f : K \to L$ such that $K \in \mathcal{A}$ and $L \in P^\perp$. Let $e_f : K \to H_f$ be the cointersection in $\mathcal{K}$ of all $\mathcal{E}$-morphisms $K \to K', K' \in \mathcal{K}$ through which f factors. Since $\mathcal{K}$ is closed in $P^\perp$ under $\mathcal{E}$-cointersections, there is $m_f : H_f \to L$ such that $f = m_f \cdot e_f$. Since $\mathcal{E}$ is closed under $\mathcal{E}$-cointersections, e_f belongs to $\mathcal{E}$. We will show that m_f is $\mathcal{E}$-extremal, i.e., given $e : H_f \to H'_f$ in $\mathcal{E}$ such that m_f factors through e, then e is an isomorphism. In fact, given $m_f = m'_f \cdot e$, $e \in \mathcal{E}$, then f factors through $e \cdot e_f \in \mathcal{E}$, and the definition of e_f guarantees that $e \cdot e_f$ factors through e_f - thus, e is an isomorphism.

Observe the following "functoriality" of the above factorization. Given $f' : K' \to L, K' \in \mathcal{A}$, then for each $h : K \to K'$ with $f = f' \cdot h$ there exists an $\mathcal{M}$-morphism $h^* : H_f \to H_{f'}$ with $m_f = m_{f'} \cdot h^*$. In fact, let us form a pushout (in $\mathcal{K}$ or $P^\perp$) of h and e_f:

$$\begin{array}{ccc} K & \xrightarrow{e_f} & H_f \\ {\scriptstyle h}\downarrow & & \downarrow{\scriptstyle \tilde{h}} \\ K' & \xrightarrow[\tilde{e}_f]{} & \tilde{H} \end{array}$$

Since $m_f \cdot e_f = f' \cdot h$, there is $q : \tilde{H} \to L$ such that $q \cdot \tilde{h} = m_f$ and $g \cdot \tilde{e}_f = f'$. Since $\tilde{e}_f \in \mathcal{E}$, $e_{f'} = d \cdot \tilde{e}_f$ for some $d : \tilde{H} \to H_{f'}$. Then $h^* = d \cdot \tilde{h}$ fulfils $m_{f'} \cdot h^* \cdot e_f = m_{f'} \cdot d \cdot \tilde{e}_f \cdot h = f' \cdot h = m_f \cdot e_f$, thus, $m_{f'} \cdot h^* = m_f$. To prove that h^* is an $\mathcal{M}$-morphism, factor it as $h^* = m \cdot e_0, e_0 \in \mathcal{E}$, then f factors through the $\mathcal{E}$-morphism $e_0 \cdot e_f$, hence, e_0 is an isomorphism.

(3) $\mathcal{K}$ is locally presentable because for the (essentially small) collection Q of all reflection arrows of λ-presentable objects of $\mathcal{L}$ in $\mathcal{K}$ we will prove that $\mathcal{K} = (P \cup Q)^{\perp}$. In fact, $\mathcal{K} \subseteq (P \cup Q)^{\perp}$, and to prove the reverse inclusion, consider any $L \in P^{\perp} \cap Q^{\perp}$. Since $\mathcal{L}$ is locally ω-presentable, we have a λ-direct colimit $(D_i \xrightarrow{d_i} L)_{i \in I}$ with D_i λ-presentable in $\mathcal{L}$. For each i form the reflection $r_i : D_i \to D'_i$ in $\mathcal{K}$. Since $r_i \in Q$, d_i factors through r_i, say, $d_i = d'_i \cdot r$. It is easy to see that $D'_i \in \mathcal{A}$ and we can take the factorization $d'_i = m_i \cdot e_i$ from (2), $m_i = D''_i \to L$. By the functoriality, $i \leq j$ implies that there is a morphism $d''_{ij} : D''_i \to D''_j$ with $m_i = m_j \cdot d''_{ij}$. Each d''_{ij} is an $\mathcal{M}$-morphism because given an $(\mathcal{E}, \mathcal{M})$-factorization $d''_{ij} = m \cdot e$, then m_i $(= m_j \cdot m \cdot e)$ factors through e, however, m_i is $\mathcal{E}$-extremal, thus e is an isomorphism. Thus, we have a λ-direct diagram D'' of $\mathcal{M}$-morphisms whose colimit obviously is $(m_i : D''_i \to L)$. Since $\mathcal{K}$ is closed in $\mathcal{L}$ under such colimits, we get $L \in \mathcal{K}$.

Remark 3. In the proof of Theorem 1 we made use of ideas presented in [GU].

Let us call a category $\mathcal{M}$-*accessible* if it has λ-direct colimits of $\mathcal{M}$-morphisms and satisfies the condition (b) in the above definition of $\mathcal{M}$-locally generated category. (The concept of accessible category [MP] then precisely corresponds to $\mathcal{M}$ = all morphisms.) By Theorem 1., each cocomplete $\mathcal{M}$-accessible category is locally presentable, we now prove the following related result:

Theorem 2. *A category $\mathcal{K}$ is locally presentable iff it is complete and $\mathcal{M}$-accessible.*

Proof As in the proof of Theorem 1 we have a Yoneda embedding $E : \mathcal{K} \to \mathcal{L} = Set^{\mathcal{A}^{op}}$ preserving λ-direct colimits of $\mathcal{M}$-morphisms. Moreover, since E preserves and reflects limits, we can consider $\mathcal{K}$ as an actual full subcategory of $\mathcal{L}$ closed under limits and λ-direct colimits of $\mathcal{M}$- morphisms. We will prove that $\mathcal{K}$ is a reflective subcategory of $\mathcal{L}$. Then $\mathcal{K}$ is cocomplete, and we can apply Theorem 1.

Let $\widetilde{\mathcal{M}}$ denote the class of all monomorphisms $m : L \to K$ in $\mathcal{L}$ such that in each commutative square

(*)
$$\begin{array}{ccc} A & \xrightarrow{\;e\;} & B \\ {\scriptstyle p}\downarrow & & \downarrow{\scriptstyle q} \\ L & \xrightarrow[\;m\;]{} & K \end{array}$$

with $A, B \in \mathcal{K}$ and $e \in \mathcal{E}$ there exists a diagonal $d : B \to L$ with $p = d \cdot e$ and $q = m \cdot d$. We have $\mathcal{M} = \widetilde{\mathcal{M}} \cap \mathcal{K}$. A monomorphism $m : L \to K$ in $\mathcal{L}$ will be called λ-pure w.r.t. $\mathcal{M}$ (shortly pure) if in each commutative square (*) with A λ-generated in $\mathcal{L}$, B λ-generated in $\mathcal{K}$ w.r.t. $\mathcal{M}$ and $q \in \widetilde{\mathcal{M}}$ (e is an arbitrary $\mathcal{K}$-morphism) there exists $d : B \to L$ which belongs to $\widetilde{\mathcal{M}}$ and $p = d \cdot e$ (no condition on $m \cdot d$!). We will prove that $\mathcal{K}$ is closed in $\mathcal{L}$ under pure subobjects.

Let $m : L \to K$ be pure and $K \in \mathcal{K}$. Since $\mathcal{L}$ is locally ω-presentable, there is a λ-direct diagram D of λ-presentable objects in $\mathcal{L}$ with the colimit $(D_i \xrightarrow{p_i} L)_{i \in I}$. For any $i \in I$, since D_i is λ-generated, there is $q_i : \bar{D}_i \to K$ from $\mathcal{M}$, with $\bar{D}_i \in \mathcal{A}$, such that $m \cdot p_i$ factorizes through q_i:

$$\begin{array}{ccc} D_i & \xrightarrow{e_i} & \bar{D}_i \\ {\scriptstyle p_i}\downarrow & & \downarrow{\scriptstyle q_i} \\ L & \xrightarrow[m]{} & K \end{array}$$

Since m is pure, there is $d_i : \bar{D}_i \to L$ with $d_i \cdot e_i = p_i$ and $d_i \in \widetilde{\mathcal{M}}$. Denote by $\bar{D}$ the diagram whose objects are $\bar{D}_i, i \in I$ and whose morphisms are $f : \bar{D}_i \to \bar{D}_j$ such that $d_j \cdot f = d_i$. Since $d_i \in \widetilde{\mathcal{M}}$, any such f belongs to $\mathcal{M}$. Since any $A \in \mathcal{A}$ is ω-presentable (thus, λ-presentable) in $\mathcal{L} = Set^{\mathcal{A}^{op}}$, $\bar{D}$ is λ-direct. Since L is the colimit of $\bar{D}$ (via $d_i : \bar{D}_i \to L$), $L \in \mathcal{K}$.

For each morphism $f : X \to K$ in $\mathcal{L}$ with $K \in \mathcal{K}$ we will construct a pure monomorphism $f^* : X^* \to K$ through which f factors as follows: we define a chain $m_i : L_i \to K$ ($i \leq \lambda$) of monomorphisms and put $f^* = m_\lambda$, where

m_0 is a monomorphism such that $f = m_0 \cdot e_0$ for some epimorphism e_0 in $\mathcal{L}$,

$m_i = \bigcup\limits_{j<i} m_j$ for each limit ordinal i,

given m_i, we define m_{i+1} as follows: consider a (small!) set of representatives of all spans

$$\begin{array}{ccc} Y_t & \xrightarrow{n_t} & X_t \\ {\scriptstyle p_t}\downarrow & & \\ L_i & & \end{array}$$

such that Y_t is λ-generated in $\mathcal{L}$, X_t λ-generated w.r.t. $\mathcal{M}$ in $\mathcal{K}$ and there is $q_t : X_t \to K$ from $\widetilde{\mathcal{M}}$ with $q_t \cdot n_t = m_i \cdot p_t$. Then $m_{i+1} = m_i \cup \bigcup\limits_{t \in T} q_t$. For $f^* = m_\lambda$, we clearly have a factorization of f through f^*. We have to prove that f^* is pure. Consider a square (*) in the above definition of pure monomorphism $m = m_\lambda$. Since A is λ-generated, p factorizes as $p' : A \to L_i$ followed by the inclusion $L_i \to L_\lambda$ for some $i < \lambda$. The span (p', e) is, due to q, isomorphic to one of the spans (p_t, n_t) used to define m_{i+1}. Since $q_t : X_t \to K$ given by (p_t, n_t) factorizes through m_λ via an $\widetilde{\mathcal{M}}$-morphism (since $\widetilde{\mathcal{M}}$ is right cancelative: $m_\lambda \cdot v \in \widetilde{\mathcal{M}}$ implies $v \in \widetilde{\mathcal{M}}$), there is an $\widetilde{\mathcal{M}}$-morphism $d : B \to L_\lambda$ such that $d \cdot e = p$. Thus m_λ is pure.

It is clear from the construction that, taking all $f : X \to K, K \in \mathcal{K}$, we get only a set of L'_f's. This yields the solution set condition, hence the reflectivity of $\mathcal{K}$.

Remark 4. The proof of Theorem 2 is analogous to the proof of the following result of [AR]: each full subcategory of a locally presentable category closed under limits and λ-direct colimits is reflective. Theorem 2 would follow from a generalization of that result to λ-direct colimits of $\mathcal{M}$-morphisms. However, we even don't know whether this is true for monomorphisms:

Open problem 2. Is each full subcategory of a locally presentable category $\mathcal{L}$, closed in $\mathcal{L}$ under limits and λ-direct colimits of monomorphisms, reflective?

References

[AHS] J. Adámek, H. Herrlich and G. Strecker,, *Abstract and concrete categories*, J. Wiley-Interscience Publ. New York, 1990.

[AR] J. Adámek and J. Rosický, *Reflections in locally presentable categories*, Arch. Math. (Brno) **25** (1989), 89-94.

[ART] J. Adámek, J. Rosický and V. Trnková, *Are all limit-closed subcategories of locally presentable categories reflective?*, Proc. Categ. Conf. Louvain-La-Neuve (1987); Lecture Notes in Math. **1348** (1988), 1-18, Springer- Verlag, Berlin.

[GU] P. Gabriel and F. Ulmer, *Lokal präsentierbare Kategorien*, Lect. Notes in Math. **221** (1971), Springer-Verlag, Berlin.

[MP] M. Makkai and R. Paré, *Accessible categories: The foundations of categorical model theory*, Contemporary Mathematics **104** (1989), Amer. Math. Society, Providence.

[RTA] J. Rosický, V. Trnková and J. Adámek, *Unexpected properties of locally presentable categories*, Alg. Univ. 27 (1990), 153-170.

Technical University, Suchbátarova 2, 166 27 Prague, Czechoslovakia

Masaryk University, Janáčkovo nám. 2a, 662 95 Brno, Czechoslovakia

SOME REMARKS ON FREE MONOIDS IN A TOPOS

by

JEAN BENABOU

20, Rue Lacepede, 75005 Paris

Introduction

Let **E** be a topos with natural numbers object. We denote by Mon(**E**) the category of monoids of **E**, by $U : \text{Mon}(\mathbf{E}) \to \mathbf{E}$ the forgetful functor, by $F : \mathbf{E} \to \text{Mon}(\mathbf{E})$ its left adjoint and by $M : \mathbf{E} \to \mathbf{E}$ the composite UF.

Since F is a left adjoint it preserves all colimits. However, at least when **E** is the topos **E**ns of sets, we know that it preserves also some limits, e.g. pullbacks, and this is not expected for a left adjoint. Moreover F has also some right exactness properties that it "should not" have, e.g. if $f : X \to Y$ is surjective, left adjointness implies that $Ff : FX \to FY$ is an epi in the category of monoids, but such epis need not be surjective, and yet we know that Ff is surjective.

Thus it is natural to ask the following questions: Why does F have such properties? Are there any other "unexpected" properties satisfied by F? Do such properties hold in any topos with NNO?

The object of this paper is to answer these questions, and others, and to show that the answers provide a better understanding, even when $\mathbf{E} = \mathbf{E}ns$, and an effective tool to prove very simply new results. In particular we have separated as much as possible the exacteness properties which hold in the general context of §1, from those which depend on the specific properties of the generic finite cardinal, i.e. the object $\kappa : K \to N$ of $\mathbf{E}/N$ defined as follows:

$K = \{(i, n) \in N \times N \mid i < r\}$ and κ is the map $(i, n) \mapsto n$.

(We assume a little familiarity with the internal language of toposes, with the usual abuses.)

Apart from the clarification, this separation permits to reduce the inductive proofs to a strict minimum, mainly about the generic cardinal. We feel that too many such proofs are usually given which amount to, but in fact hide, important properties of κ. These properties explain why many diagrams, which should normally commute by naturality, turn out to be pullbacks.

Finally, although in many cases (e.g. 2.2) the proofs would have been more precise or more general, fibered categories have not been used.

1. Basic remarks

For each $I \in \mathbf{E}$ and $\alpha : A \to I \in \mathbf{E}/I$ we denote by $G^\alpha : \mathbf{E} \to \mathbf{E}/I$ and $M^\alpha : \mathbf{E} \to \mathbf{E}$ the following composite functors:

$$\mathbf{E} \xrightarrow{I^*} \mathbf{E}/I \xrightarrow{(\ )^\alpha} \mathbf{E}/I \qquad \text{and} \qquad \mathbf{E} \xrightarrow{I^*} \mathbf{E}/I \xrightarrow{(\ )^\alpha} \mathbf{E}/I \xrightarrow{\coprod_I} \mathbf{E},$$

where $(\ )^\alpha$ is the exponential by α in $\mathbf{E}/I$.

1.1. Since I^* and $(\ )^\alpha$ have left adjoints $\coprod_I$ and $\alpha \times (\ )$ so does G^α, hence M^α preserves all the limits which are preserved by $\coprod_I$, in particular pullbacks and equalizers.

1.2. Since I^* and $\coprod_I$ have right adjoints $\prod_I$ and I^*, thus preserve all colimits, M^α preserves all the colimits preserved by $(\ )^\alpha$. In particular if α is internally projective in $\mathbf{E}/I$, the functor M^α preserves epis.

1.3. If $u : J \to I \in \mathbf{E}$ and $\beta = u^*(\alpha) : B \to J \in \mathbf{E}/J$ we can construct a diagram:

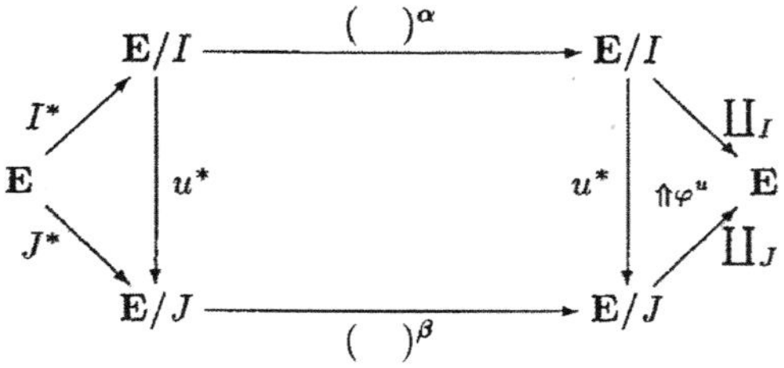

where, up to a unique isomorphism, the left triangle commutes trivially, the square commutes because u^* is logical, and the natural transformation $\varphi^u : \coprod_J u^* \to \coprod_I$ associates to each $\xi : X \to I \in \mathbf{E}/I$ the map $Y \to X$ given by the following pullback

$$\begin{array}{ccc} Y & \longrightarrow & X \\ {\scriptstyle\eta}\downarrow & & \downarrow{\scriptstyle\xi} \\ J & \xrightarrow[u]{} & I \end{array}$$

Hence we have: $u^* \cdot G^\alpha \xrightarrow{\sim} G^\beta$, and we get a natural transformation $\varphi^{u\alpha} : M^\beta \to M^\alpha$ defined as the composite:

$$M^\beta = \coprod_J \cdot G^\beta \xrightarrow{\sim} \coprod_J u^* G^\alpha \longrightarrow \coprod_I G^\alpha = M^\alpha.$$

We note moreover that all the categories in the previous diagram have finite limits, and all the functors preserve pullbacks.

1.4. If $\mathbf{C}$ and $\mathbf{D}$ are categories with pullbacks, F and $F' : \mathbf{C} \to \mathbf{D}$ are pullback preserving functors, a natural transformation $\varphi : F \to F'$ is *cartesian* if for each

$f : X \to Y \in \mathbf{C}$ the following diagram is a pullback:

$$\begin{array}{ccc} FX & \xrightarrow{Ff} & FY \\ {\scriptstyle \varphi X}\downarrow & & \downarrow{\scriptstyle \varphi Y} \\ F'X & \xrightarrow[F'f]{} & F'Y \end{array}$$

Clearly if $\mathbf{C}$ has a terminal object it suffices to verify the pullback property for $Y = 1$. We note also that such transformations are stable by composition, and more generally by pasting along pullback preserving functors.

The transformation φ^u is obviously cartesian, hence so is $\varphi^{u\alpha}$.

For each $I \in \mathbf{E}$ let T_I be the constant functor $\mathbf{E} \longrightarrow \mathbf{1} \xrightarrow{I} \mathbf{E}$. We denote by $\tau : 1_{\mathbf{E}} \to T_I$ the unique natural transformation, and by $\lambda^\alpha : M^\alpha \to T_I$ the transformation which associates to each $X \in \mathbf{E}$ the structural map $\lambda^\alpha_X : M^\alpha X \to I$ of $G^\alpha X \in \mathbf{E}/I$.

It is clear that $M^\alpha 1 = I$ and $\lambda^\alpha_X = M^\alpha \tau_X$ i.e. $\lambda^\alpha = M^\alpha \tau$.

1.5. We say that α is *skeletal* if for all $J \in \mathbf{E}$ and maps $u_0, u_1 : J \rightrightarrows I$, if $u_0{}^*(\alpha)$ and $u_1{}^*(\alpha)$ are isomorphic in $\mathbf{E}/J$ then $u_0 = u_1$. For example $t : 1 \to \Omega$ is skeletal. If we think of α as an I-indexed family (A_i) of objects of $\mathbf{E}$, this means that if there exists an iso $A_{i_0} \xrightarrow{\sim} A_{i_1}$ then $i_0 = i_1$.

This property can be characterized internally as follows. By pulling back α along the projections $\pi_0, \pi_1 : I \times I \rightrightarrows I$ we get two objects $\alpha_0{}' : A_0{}' \to I \times I$ and $\alpha_1{}' : A_1{}' \to I \times I$ of $\mathbf{E}/I \times I$. Let $\delta = (\delta_0, \delta_1) : \mathrm{Iso}(\alpha_0{}', \alpha_1{}') \to I \times I$ denote the object of isos of $\alpha_0{}'$ and $\alpha_1{}'$ in $\mathbf{E}/I \times I$. It is characterized as follows: there exists an isomorphism $\sigma_0 : \delta_0{}^*(\alpha) \xrightarrow{\sim} \delta_1{}^*(\alpha) \in \mathbf{E}/\mathrm{Iso}(\alpha_0{}', \alpha_1{}')$ called generic, such that for every map $u = (u_0, u_1) : J \to I \times I$ and every isomorphism $\sigma : u_0{}^*(\alpha) \to u_1{}^*(\alpha) \in \mathbf{E}/J$, there exists a unique map $v : J \to \mathrm{Iso}(\alpha_0{}', \alpha_1{}')$ such that $u = \delta v$ and $\sigma = v^*(\sigma_0)$.

With these notations we have: α is skeletal iff $\delta_0 = \delta_1$.

This means of course that the full subcategory of $\mathbf{E}$ generated by the family α is skeletal, which justifies our terminology.

Clearly if α is skeletal, a map $u : J \to I$ is characterized by the isomorphism class of $u^*(\alpha)$ in $\mathbf{E}/J$.

1.6. Let $\alpha' : A_0{}' + A_1{}' = A' \to I \times I$ be the coproduct $\alpha_0{}' + \alpha_1{}'$ in $\mathbf{E}/I \times I$.

For each object $X \in \mathbf{E}$ we have: $M^\alpha X \times M^\alpha X \xrightarrow{\sim} M^{\alpha'} X$.

Proof. We observe first that: $(I \times I)^* X = \pi_i{}^*(I^* X)$ and $\alpha_i{}' = \pi_i{}^*(\alpha)$ $(i = 0, 1)$, hence, since $\pi_i{}^*$ is logical: $\left[(I \times I)^* X\right]^{\alpha_i{}'} = \pi_i{}^*\left[(I^* X)^\alpha\right] = \pi_i{}^* G^\alpha X$. But $G^{\alpha'} X = \left[(I \times I)^* X\right]^{\alpha_0' + \alpha_1'} = \left[(I \times I)^* X\right]^{\alpha_0'} \times \left[(I \times I)^* X\right]^{\alpha_1'} = \pi_0{}^* G^\alpha X \times \pi_1{}^* G^\alpha X$, where the product is taken in $\mathbf{E}/I \times I$.

But if $\sigma_0 : S_0 \to I$ and $\sigma_1 : S_1 \to I$ are objects of $\mathbf{E}/I$, in $\mathbf{E}/I \times I$ the product $\pi_0{}^*\sigma_0 \times \pi_1{}^*\sigma_1$ is $\sigma_0 \times \sigma_1 : S_0 \times S_1 \to I \times I$, i.e., $\coprod_I \sigma_0 \times \coprod_I \sigma_1 = \coprod_{I \times I} \pi_0{}^*\sigma_0 \times \pi_1{}^*\sigma_1$.

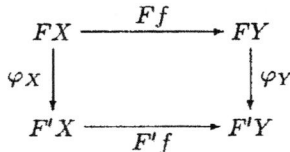

2. Exactness properties

We take now $I = N$ and $\alpha = \kappa : K \to N$, and we abbreviate G^κ, M^κ, φ^κ, λ^κ and $\varphi^{u\kappa}$ by G, M, φ, λ and φ^u. It is clear that $M = UF$.

Since Mon(**E**) is monadic over **E**, we know, for instance from [M.L.], how to deduce exactness properties of F and U from exactness properties of M.

From (1.1) and (1.2) we get that M preserves all limits, internal or external, preserved by $\coprod_N$ and all colimits preserved by $(\ \ )^\kappa$, in particular epis since a finite cardinal is internally projective.

LEMMA 2.1. If $[n]$ is a finite cardinal the functor $(\ \ )^{[n]} : \mathbf{E} \to \mathbf{E}$ preserves (internal) filtered colimits.

Proof. For the notations see [P.J.]. Let $\pi : F \to C$ be a discrete opfibration where C is filtered, and S be the subobject of N defined by:

$S = \{n \in N \mid \text{the canonical map } \lambda_n : \varinjlim F^{[n]} \to (\varinjlim F)^{[n]} \text{ is an iso}\}$.

We have $O \in S$ since $\varinjlim 1 = 1$ because C is filtered, and $(\varinjlim F)^O = 1$.

Suppose $n \in S$. Then $F^{[n+1]} \simeq F^{[n]} \times F$, hence $\varinjlim F^{[n+1]} \simeq \varinjlim F^{[n]} \times \varinjlim F$ since $\varinjlim$ is left exact. Since $n \in S$, λ_n is an iso and $\lambda_{n+1} \simeq \lambda_n \times \mathrm{id}(\varinjlim F)$ is an iso. Hence $n + 1 \in S$ and $S = N$.

REMARKS. (i) Even with the internal language, the proof, as it stands, is very sketchy. One should work in $\mathbf{E}/N$, or better, state the left exactness of $\varinjlim$ for the fibration $\mathbf{E}^2 \to \mathbf{E}$.

(ii) The lemma is a special case of the following result, which does not suppose the existence of N; if I is Kuratowski finite and decidable the functor $(\ \ )^I$ preserves filtered colimits. It will be proved elsewhere.

COROLLARY 2.2. The functors G and M preserve filtered colimits.

Let us call $\mathbf{E}/N$ the category of *graded objects* of **E**. The functor G associates to each $X \in \mathbf{E}$ the graded object $\lambda_X : MX \to N$, where λ_X is the length map. It gives more information than M or F since it permits induction on length, and it behaves much better because it has a left adjoint and preserves filtered colimits, and we know the importance of this flatness condition, which is also an essential feature of models of finitary theories.

The left adjoint L of G is the composite $\mathbf{E}/I \xrightarrow{\kappa \times (\ )} \mathbf{E}/I \xrightarrow{\coprod_I} \mathbf{E}$; thus, if $\eta : Y \to N$ is a graded object, $L\eta$ is given by the pullback:

$$\begin{array}{ccc} L\eta & \longrightarrow & K \\ \downarrow & & \downarrow \kappa \\ Y & \xrightarrow[\eta]{} & N \end{array}$$

In the case $\mathbf{E} = \mathbf{E}ns$, this gives: If $\eta = (Y_n)_{n \in \mathbf{N}}$, then $L\eta = \coprod_n nY_n$.

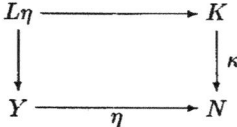

3. Some properties of the generic finite cardinal

Let $\alpha = (d_0, d_1) : \mathrm{Mono}([k_0], [k_1]) \to N \times N$ denote the object $\mathrm{Mono}(\pi_0{}^*\kappa, \pi_1{}^*\kappa)$ of $\mathbf{E}/N \times N$.

PROPOSITION 3.1. The image of d is the order relation of N, i.e. the following formula: $p \leq q \iff \exists f \in \mathrm{Mono}([k_0], [k_1])\ (d_0 f = p \wedge d_1 f = q)$, is true in $N \times N$.

Proof. Let $S = \{q \in N \mid \forall\, p \in N\ \exists f \in \mathrm{Mono}([k_0], [k_1])\ (d_0 f = p \wedge d_1 f = q) \Longrightarrow p \leq q\}$.
Clearly $O \in S$ since $O^*(\kappa) = O$.
Suppose $q \in S$ and $f : [p] \to [q+1]$ is a mono. If $p = 0$ then $p \leq q+1$. If $p > 0$ then $p = n+1$ and $[p] = [n] \coprod \{n\}$. Let $\tau : [q+1] \to [q+1]$ be the transposition which permutes f_n and q. Then τf induces a mono $[n] \to [q]$. Since $q \in S$ we have $n \leq q$ hence $p = n+1 \leq q+1$. (We have used dichotomy and decidability in N and in κ).

COROLLARY 3.2. The generic cardinal κ is skeletal.

Thus by (1.5) we know that any map $u : J \to N$ is characterized by $u^*\kappa = \beta \in \mathbf{E}/J$. But by (1.3) such a map defines a cartesian natural transformation $\varphi^u : M^\beta \to M$. Thus all we have to do is to compute β, M^β and φ^u for significant u's. Moreover for any u, since $\beta = u^*\kappa$ is a finite cardinal in $\mathbf{E}/J$ we know that M^β preserves epis and filtered colimits, and of course all the limits preserved by $\coprod_J$.

In the following examples, to have coherent notations, we make the following conventions: all the monoids are noted additively, with unit 0. Moreover, if S is a monoid, the map $U\epsilon_S$ is denoted by $\sum : MS \to S$. For each $X \in \mathbf{E}$ the maps τ_X, λ_X and φ^u_X are abbreviated by τ, λ and φ^u.

3.3. It is well known that the following diagrams are pullbacks

$$\begin{array}{ccc} 0 & \longrightarrow & K \\ \downarrow & & \downarrow{\scriptstyle \kappa} \\ 1 & \xrightarrow[0]{} & N \end{array} \qquad \begin{array}{ccc} 1 & \longrightarrow & K \\ \downarrow & & \downarrow{\scriptstyle \kappa} \\ 1 & \xrightarrow[1]{} & N \end{array} \qquad \begin{array}{ccc} B & \longrightarrow & K \\ {\scriptstyle \pi_0^*\kappa + \pi_1^*\kappa}\downarrow & & \downarrow{\scriptstyle \kappa} \\ N \times N & \xrightarrow[+]{} & N \end{array}$$

Since $M^0 = T_1$, $M^1 = \mathrm{id}$ and $M^{(\pi_0^*\kappa + \pi_1^*\kappa)} = M \times M$ by (1.6), for each $f : X \to Y \in \mathbf{E}$ we have pullbacks:

$$\begin{array}{ccc} 1 & \xrightarrow{0} & MX \\ \downarrow & & \downarrow{\scriptstyle Mf} \\ 1 & \xrightarrow[0]{} & MY \end{array} \qquad \begin{array}{ccc} X & \xrightarrow{\eta} & MX \\ {\scriptstyle f}\downarrow & & \downarrow{\scriptstyle Mf} \\ Y & \xrightarrow[\eta]{} & MY \end{array} \qquad \begin{array}{ccc} MX \times MX & \xrightarrow{+} & MX \\ {\scriptstyle Mf \times Mf}\downarrow & & \downarrow{\scriptstyle Mf} \\ MY \times MY & \xrightarrow[+]{} & MY \end{array}$$

3.4. We say that a morphism of monoids $h : R' \to R$ is *cartesian* if the two squares of the following diagram are pullbacks.

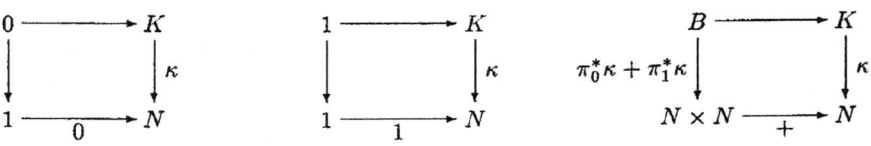

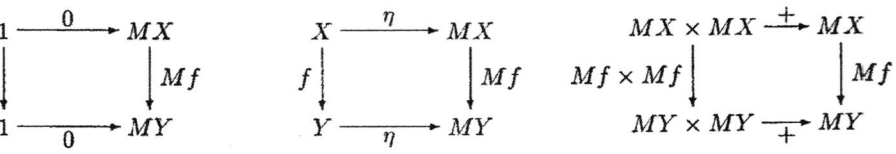

$$\begin{array}{ccccc}
1 & \xrightarrow{0} & R' & \xleftarrow{+} & R' \times R' \\
\downarrow & & \downarrow h & & \downarrow h \times h \\
1 & \xrightarrow[0]{} & R & \xleftarrow[+]{} & R \times R
\end{array}$$

The Mf's are cartesian for all f, but also, e.g., $t : 1 \to \Omega$. This is a very strong condition which permits to lift properties of R to R'. We examine a special case. For all X, $M\tau = \lambda : MX \to N$ is cartesian. This means that the following formulas where u, v, w denote "words", i.e., variables of type MX, and p, q denote numbers, are true: $p + q = \lambda w \Leftrightarrow \exists!(u, v)\ (u + v = w \wedge \lambda u = p \wedge \lambda v = q)$ and $fw = 0 \Leftrightarrow w = 0$.
From this we derive immediately, by lifting well known properties of N:

(i) $u + v = u + v' \Rightarrow v = v'$
Take $w = u + v = u + v'$, $p = \lambda u$, $q = \lambda v$, $q' = \lambda v'$ and use $p + q = p + q' \Rightarrow q = q'$.
Dually: $u + v = u' + v \Rightarrow u = u'$.

(ii) $u + v = 0 \Rightarrow u = 0 \wedge v = 0$: since $p + q = 0 \Rightarrow p = 0 \wedge q = 0$, and λ is cartesian.

(iii) The preorder relation, which can be defined for any monoid:
$u \leq v \Leftrightarrow \exists u'(u + u' = v)$ is an order relation, and for each v, the segment $\downarrow v$ is isomorphic by λ to the segment $\downarrow \lambda v$, hence is a finite cardinal.

(iv) Let pred : $N \to N$ denote the predecessor function. It can be lifted to a predecessor function $MX \to MX$ defined by: pred(w) is the unique v such that $v \leq w$ and $\lambda v =$ predλw. We can of course iterate the process, and define the map: $N \times MX \to MX$; $(n, w) \mapsto \text{pred}^n(w)$, and we have: $\text{pred}^n(w) = 0$ iff $\lambda w \leq n$. Thus MX is a tree.

Because of the importance of the monoid Ω, with "addition" $\wedge$ we must describe the map $\epsilon_\Omega : F\Omega \to \Omega$. It is determined by its underlying map, which we denote by $\wedge : M\Omega \to \Omega$ hence by the subobject of $M\Omega$ classified by $\wedge$.

PROPOSITION 3.3. The following diagram is a pullback:

$$\begin{array}{ccc}
N = M1 & \longrightarrow & 1 \\
Mt \downarrow & & \downarrow t \\
M\Omega & \xrightarrow[\wedge]{} & \Omega
\end{array}$$

Proof. Since $1 \to \Omega$ is cartesian, the result is an obvious consequence of the following lemma:

LEMMA 3.4. If $h : R' \to R$ is cartesian the following diagram is a pullback.

$$\begin{array}{ccc}
MR' & \xrightarrow{\Sigma} & R' \\
Mh \downarrow & & \downarrow h \\
MR & \xrightarrow[\Sigma]{} & R
\end{array}$$

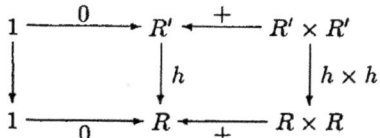

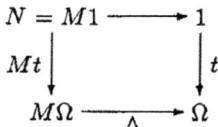

Proof. Let S be the subobject of N defined by the formula:

$$\forall (w,\, x') \in MR \times R'(\lambda w = n \wedge \Sigma w = hx' \rightarrow \exists ! w' \in MR'(Mhw' = w \wedge \Sigma w' = x'))$$

(i) $0 \in S$. Suppose $\lambda w = 0 \wedge \Sigma w = hx'$. Since $\lambda w = 0$ and Mh preserves the length the only w' such that $Mhw' = w$ is 0 but $\Sigma w = 0 = hx'$ and h cartesian $\rightarrow x' = 0 = \Sigma 0$

(ii) suppose $n \in S$, $\lambda w = n + 1 \wedge \Sigma w = hx'$. Let $v = \mathrm{pred} w$, and $y = \Sigma v$. There is a unique $z \in R$ such that $w = v + z$; hence $\Sigma w = y + z = h(x')$. By cartesianness, there is a unique $(y',\, z') \in R' \times R'$ such that $y' + z' = x'$, $hz' = z$ and $h(y') = y = \Sigma v$. But $\lambda v = n \in S$ and $\Sigma v = hy'$ hence there is a unique $v' \in MR$ such that $Mhv' = v$ and $\Sigma v' = y'$. It is then obvious that $v' + z'$ is the unique w' such that $Mhw' = w \wedge \Sigma w' = x'$.
Thus $n + 1 \in S$ and $S = N$.

COROLLARY 3.5. The natural transformations η, ϵ and μ are cartesian.

Proof. We know already that η is cartesian. For μ it follows from the lemma applied in the case $h = Mf$, since we know that Mf is a cartesian morphism. Since $\mu = U\epsilon$ and U reflect pullbacks we get that ϵ is cartesian.

REMARK 3.6. The corollary could also have been obtained from the following property of κ which we have as an exercise: In the following diagram, where $ev : K \underset{N}{\times} M(N) \rightarrow N$ denotes the evaluation deduced from the adjunction $\alpha \times (\ \ ) \dashv (\ \ )^{\alpha}$ in $\mathbf{E}/N$, the two squares are pullbacks:

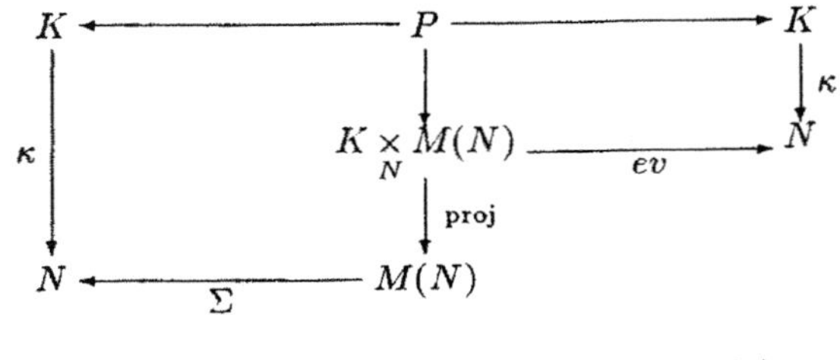

We spell out the meaning, when $\mathbf{E} = \mathbf{E}ns$. For all $n \in \mathbf{N}$ and $(p_0, \ldots, p_{n-1}) \in \mathbf{N}^n$ we have: $[\Sigma p_i] \simeq \coprod [p_i]$ where $[p] = \{0, \ldots, p-1\}$.
This contains as special cases the three diagrams of (3.1). In the same vein, the cartesianness of μ expresses the following property of free monoids. If w is a word of length p, and $p_0, \ldots, p_{n-1} \in \mathbf{N}$ are such that $\Sigma p_i = p$, there is a unique way of decomposing w as the concatenation $w_0 \ldots w_{n-1}$ with w_i of length p_i.

We state briefly some facts which are immediate consequences of the left exactness properties of M hence of F.

3.9. For all X_0, $X_1 \in \mathbf{E}$ we have:

(i) the following diagram is a pullback of monoids

$$\begin{array}{ccc} F(X_0 \times X_1) & \xrightarrow{F\pi_0} & FX_0 \\ {\scriptstyle F\pi_1}\big\downarrow & & \big\downarrow{\scriptstyle \lambda} \\ FX_1 & \xrightarrow[\lambda]{} & N \end{array}$$

(ii) $M(X_0 \times X_1)$ is a complemented subobject of $MX_0 \times MX_1$

Proof. (i) M and hence F preserve pullbacks and $N = F1$. (ii) follows from (i) since the pullback is the inverse image of the diagonal $N \rightarrowtail N \times N$, and N is decidable.

3.10. If $\underline{C}$ is an internal category in **E**, then $F\underline{C}$ is an internal category in Mon**E**, because the definition of categories involves only pullbacks, and F preserves them.

3.11. If $R \rightarrowtail X \times X$ is an order (resp. equivalence) relation on X, then FX is an ordered monoid (resp. FR is a congruence on FX).

4. Decidability properties

PROPOSITION 4.1. If $Y \rightarrowtail X$ is a complemented mono of **E**, so is $MY \rightarrowtail MX$

Proof. Let $X \xrightarrow{f} 1+1 \overset{i}{\rightarrowtail} \Omega$ denote the characteristic map of Y.
From the following diagram where both squares are pullbacks:

$$\begin{array}{ccccc} Y & \longrightarrow & 1 & \longrightarrow & 1 \\ \big\downarrow & & \big\downarrow{\scriptstyle s} & & \big\downarrow{\scriptstyle t} \\ X & \xrightarrow[f]{} & 1+1 & \xrightarrow[i]{} & \Omega \end{array} ,$$

we get the diagram

$$\begin{array}{ccccccc} MY & \xrightarrow{\lambda} & N = M1 & \xrightarrow{\text{id}} & N & \longrightarrow & 1 \\ \big\downarrow & & \big\downarrow{\scriptstyle Ms} & & \big\downarrow{\scriptstyle Mt} & & \big\downarrow{\scriptstyle t} \\ MX & \xrightarrow[Mf]{} & M(1+1) & \xrightarrow[Mi]{} & M\Omega & \xrightarrow[\wedge]{} & \Omega \end{array} ,$$

where the first two squares are pullbacks since M preserves pullbacks, and the third is a pullback by (3.5). Hence $\wedge \circ M_i \circ Mf$ is the characteristic map of $MY \rightarrowtail MX$. But $1+1$ is a submonoid of Ω, hence $\wedge_0 M_i$ factors through $1+1$.

COROLLARY 4.2. If X is decidable so is MX.

Proof. Let $\delta : X \rightarrowtail X \times X$ and $\delta' : MX \rightarrowtail MX \times MX$ denote the diagonals, and $i : M(X \times X) \rightarrowtail MX \times MX$ be the inclusion. By (3.9) and (4.1) both i and $M\delta$ are complemented, but clearly $\delta' = i \circ M\delta$.

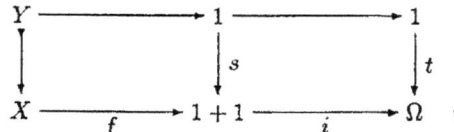

5. The free monoid on N

The free monoid on X may seem a fairly trivial notion, and indeed it is if X is a "set", i.e. an object of $\mathbf{E}$. However if X is equipped with an interesting or complex structure, this structure lifts to MX, moreover "glorified" by the monoid structure on MX and the exactness properties of M become essential to understand this lifting.

Let us give an example which, even when $\mathbf{E} = \mathbf{E}ns$, deserves some attention, namely $X = N$. Thus we want to understand which extra properties $M(N)$ inherits from N. We shall use only elementary properties of the triple (M, η, μ), but the cartesianness properties seen previously will play a central role.

First, $M(N)$ is equipped with maps

$$N = M(1) \xrightarrow{M(\eta_1)} MM(1) = M(N) \underset{M(\tau) = \lambda}{\overset{\mu_1 = \Sigma_N}{\rightrightarrows}} M(1) = N$$

where $\tau : N \to 1$. Since $\mathbf{M}$ is a triple $\mu_1 \cdot M(\eta_1) = \mathrm{id}$; since $\tau\eta_1 = \mathrm{id}$ we have also $M(\tau)M(\eta_1) = \mathrm{id}$, hence we have a reflexive graph. But all the maps are homomorphisms of monoids, hence it is a monoidal graph.

In the following diagram

$$\begin{array}{ccccc}
M^3(1) = M^2(N) & & \xrightarrow{\Sigma_{M(N)} = \mu_{M(N)} = \mu_{M^2 1}} & & M^2(1) = M(N) \\
 & \searrow^{M\mu_N = M\mu M1} & & & \\
M^2(\tau) = M(\lambda) \Big\downarrow & & M^2(1) = M(N) & & \Big\downarrow \lambda = M(\tau) \\
M^2(1) = M(N) & & \xrightarrow[\Sigma_N = \mu_1]{} & & M(1) = N
\end{array}$$

the exterior square is a pullback by (3.6) and the map M_{μ_N} defines a multiplication on the graph. The properties of triples imply that this multiplication is associative and admits $M(\eta_1)$ as identity, hence $M(N)$ is the object of arrows of an internal category C, having N as object of objects. Moreover C is strictly monoidal since all the maps are homomorphisms of monoids.

Since N is decidable, it is clear that all the objects in the previous diagrams are decidable, and all the monos are complemented.

It remains to say what this category is. We shall state the result without proof. The generic cardinal κ is totally ordered in $\mathbf{E}/N$. Let D be the category having N as object of objects, and as object of maps the object $\delta = (\delta_0, \delta_1) : \underline{\mathrm{Ord}}(\pi_0{}^*\kappa, \pi_1{}^*\kappa) \longrightarrow N \times N$ of order preserving maps of $\pi_0{}^*\kappa$ into $\pi_1{}^*\kappa$ in $\mathbf{E}/N \times N$ (here again fibrations would be handy!).

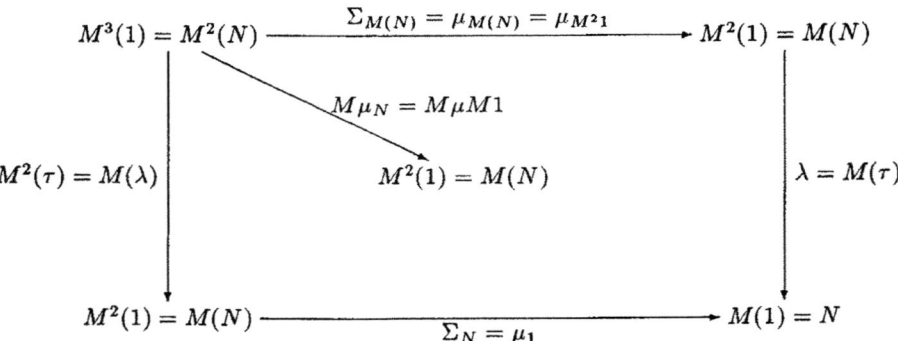

In the case $\mathbf{E} = \mathbf{E}ns$, this category is "the Δ of Mac Lane" [M.L], i.e. "the Δ of topologists" augmented by 0.

THEOREM 1. *$M(N)$ equipped with the monoidal structure just described is isomorphic to D.*

REMARK 5.1. We have clearly $M(0) = 1$; $M(1) = N$ thus $M(N) = M^3(0)$, is already a very complicated object. This process, in any topos with N.N.O. can be internally iterated over "$n \in N$". More precisely the endofunctor $M : \mathbf{E} \to \mathbf{E}$ has a "first fixed point" which has been described in many talks I gave. For details one can consult [B].

BIBLIOGRAPHY

[B] BENABOU, Seminaire 1989-90 & 90-91 (Available in preprints).

[M.L] MAC LANE, Categories for the working mathematician, Springer-Verlag, 1971.

[P.J] PETER JOHNSTONE, Topos Theory, Academic Press, 1977.

This paper is in final form and will not be published elsewhere.

A generic sheaf representation for rings

Francis Borceux* and Gilberte Van den bossche

Dpt. de Mathématique – Université de Louvain
2, chemin du Cyclotron – 1348 Louvain-la-Neuve (Belgium)

0 Introduction

A sheaf representation theorem for a ring **R** consists in defining a topological space $Sp(\mathbf{R})$ (generally called a "spectrum" of the ring) and a sheaf of rings **F** on $Sp(\mathbf{R})$ whose global section are in bijection with the elements of **R**. Most often, the sheaf **F** is presented as a ringed space over $Sp(\mathbf{R})$

$$\mathcal{F} \to Sp(\mathbf{R}).$$

The representation is considered to be interesting for a given ring **R** when the stalks of its ringed space have sufficiently rich properties : local rings, integral domains, fields, and so on ... Very often, a sheaf representation theorem holds for a wide class of rings, but gives pertinent results just for a restricted class of rings.

On a given spectrum $Sp(\mathbf{R})$ of the ring **R**, there are generally several sheaves representing **R**. As previously noticed by Hofmann [3] and Mulvey [5], one of those sheaves can be defined in a rather "canonical" way, and in particular, the stalks of this structural sheaf are quotients of the original ring.

A sheaf can also be defined in terms of generators and relations. The generators are some elements $x \in \mathbf{F}(U)$ for some open subsets U of $Sp(\mathbf{R})$. The relation which exists between two elements $x \in \mathbf{F}(U)$ and $y \in \mathbf{F}(V)$ is the knowledge of the biggest open subset W where $x_{|W} = y_{|W}$. A subset $\mathbf{E} \subseteq \prod \mathbf{F}(U)$ (where U runs through the open subsets of $Sp(\mathbf{R})$) is a set of generators for **F** when every element $x \in \mathbf{F}(U)$ is completely characterized by its relations with the various generators. This yields the definition of a $Sp(\mathbf{R})$-set as being a set **E** together with an "equality"

$$[\cdot = \cdot] : \mathbf{E} \times \mathbf{E} \to \mathcal{O}(Sp(\mathbf{R}))$$

satisfying the two axioms

$$[x = y] = [y = x] \text{ and } [x = y] \wedge [y = z] \leq [x = z].$$

Given another sheaf $\mathbf{F}'$ determined by a $Sp(\mathbf{R})$-set $\mathbf{E}'$, a morphism $\varphi : \mathbf{F} \to \mathbf{F}'$ is completely defined if, for every generator $x \in \mathbf{E}$, we know the precise relations between

*Research supported by NATO grant CRG 900959

$\varphi(x)$ and the various generators of $\mathbf{F}'$. This yields the definition of a morphism $f : \mathbf{E} \rightarrow \mathbf{E}'$ between $Sp(\mathbf{R})$-sets as being a mapping written

$$(f\cdot = \cdot) : \mathbf{E} \times \mathbf{E}' \rightarrow \mathcal{O}(Sp(\mathbf{R}))$$

which satisfies the following properties

$$\begin{aligned} [x = x'] \wedge (fx' = y) &\leq (fx = y) \\ (fx = y) \wedge [y = y'] &\leq (fx = y') \\ (fx = y) \wedge (fx = y') &\leq [y = y'] \\ [x = x] &\leq \bigvee_y (fx = y). \end{aligned}$$

If we think of $(fx = y)$ as the "truth-value of $f(x) = y$", the two first axioms are just compatibility rules with the equality. The third axiom indicates that f is single-valued and the last one the fact it is everywhere defined. Indeed

$$fx = y_1 \text{ or } fx = y_2 \text{ or } fx = y_3 \ldots$$

should be understood as

$$\exists y\, fx = y.$$

What has been said for the locale $\mathcal{O}(Sp(\mathbf{R}))$ can be repeated for every locale Ω and we get in this way the definition of the category of Ω-sets, which turns out to be equivalent to the category of sheaves on Ω [2].

It is a matter of fact that, in most spectral constructions for rings, the spectrum is a sober space, thus completely determined by its locale of open subsets. And this locale, most often, turns out to be a quotient of the lattice of all ideals

$$q : Id(\mathbf{R}) \rightarrow \mathcal{O}(Sp(\mathbf{R}))$$

$Id(\mathbf{R})$ is a complete lattice, but generally not a locale. But $Id(\mathbf{R})$ is provided with a binary multiplication (the usual multiplication of ideals) which distributes over $\bigvee = +$ in each variable. And in most spectral constructions the relation

$$q(I \cdot J) = q(I) \wedge q(J)$$

holds and forces the quotient of $Id(\mathbf{R})$ to be a locale.

Our aim is to prove a "generic" sheaf representation theorem directly on the lattice $Id(\mathbf{R})$ of all ideals. To do this, we first define the category of sheaves on the lattice $Id(\mathbf{R})$ using a description of sheaves in terms of generators and relations. An $Id(\mathbf{R})$-set is thus a set equipped with a suitable equality

$$[\cdot = \cdot] : \mathbf{E} \times \mathbf{E} \rightarrow Id(\mathbf{R})$$

and a morphism $f : \mathbf{E} \rightarrow \mathbf{E}'$ of $Id(\mathbf{R})$-sets is a suitable mapping

$$(f\cdot = \cdot) : \mathbf{E} \times \mathbf{E}' \rightarrow Id(\mathbf{R}).$$

This category turns out to have a terminal object **1** and by a "global element" of an $Id(\mathbf{R})$-set **E**, we mean a morphism $\mathbf{1} \to \mathbf{E}$.

We then construct our generic sheaf $\mathcal{R}$ on $Id(\mathbf{R})$ by taking the set **R** itself equipped with an equality

$$[\cdot = \cdot] : \mathbf{R} \times \mathbf{R} \to Id(\mathbf{R}).$$

In this first approach of the problem, we limit our attention to the case of commutative unital rings and define the equality by

$$[r = s] = \text{Ann } (r - s)$$

where "Ann" stands for annihilator. This gives a representation of the ring **R** in the usual sense that the global elements of $\mathcal{R}$ are in bijection with the elements of **R**. Moreover, given one of the classical constructions on a spectrum of **R** obtained via a quotient

$$q : Id(\mathbf{R}) \to \mathcal{O}(Sp(\mathbf{R}))$$

the composite

$$\mathbf{R} \times \mathbf{R} \xrightarrow{[\cdot=\cdot]} Id(\mathbf{R}) \xrightarrow{q} \mathcal{O}(Sp(\mathbf{R}))$$

is just the corresponding classical sheaf representation of the ring.

1 Quantales and their sheaves

Definition 1.1. *A (commutative non idempotent) quantale is a complete lattice Q equipped with a binary multiplication*

$$\& : Q \times Q \to Q$$

satisfying the following axioms

$$\begin{array}{llcll}
(Q1) & a\&b & = & b\&a & \\
(Q2) & a\&(\bigvee_{i\in I} b_i) & = & \bigvee_{i\in I}(a\&b_i) & \text{for any set } I \\
(Q3) & a\&(b\&c) & = & (a\&b)\&c & \\
(Q4) & a\&1 & = & a &
\end{array}$$

where 1 stands for the top element of Q.

Note that it is immediate from the definition that if $a \leq a'$ then $a\&b \leq a'\&b$ and consequently that $a\&b \leq a$.

Joyal and Tierney [4] did prove that a quantale is a locale (with $\& = \wedge$) precisely when its multiplication is idempotent.

Example 1.2. A basic example of a quantale is given by the lattice of ideals of a commutative unital ring **R** with the usual multiplication of ideals. We will denote this quantale by Id **R**.

Definition 1.3. *Let Q be a quantale. A Q-set is a pair* $(\mathbf{E}, [\cdot = \cdot])$ *where* $\mathbf{E}$ *is a set and* $[\cdot = \cdot]$ *is a mapping*

$$[\cdot = \cdot] : \mathbf{E} \times \mathbf{E} \to Q$$

satisfying

$$\begin{array}{llcl} (S1) & [a = b] & = & [b = a] \\ (S2) & \bigvee_{b \in \mathbf{E}} [a = b] \& [b = c] & = & [a = c] \\ (S3) & \bigvee_{b \in \mathbf{E}} [a = b] \& [b = b] & = & \bigvee_{b \in \mathbf{E}} [a = b] \end{array}$$

Example 1.4. The quantale Q itself provided with the equality $[a = b] = a \& b$ becomes a Q-set.

In the case of Ω-sets, with Ω a locale, the equalities (S1), (S2) and (S3) are consequences of the inequalities given in the introduction, by the idempotency of $\wedge$. But these inequalities are not sufficient to obtain a good workeable category for quantales. For example, in the basic example Id $\mathbf{R}$ of ideals of a ring, the lack of idempotency of ideals does not allow one to write $IJJJ = IJ$ i.e. $[I = J] \& [J = J] = [I = J]$. But one has

$$\bigvee_J IJJJ = \bigvee_J IJ$$

since both sups are obtained by putting $J = R$.

In particular $\bigvee_{a'} [a = a']$ must be thought of as the level at which a is defined, while $[a = a]$ is a lower level.

Definition 1.5. *Given a quantale Q and two Q-sets* $\mathbf{E}$, $\mathbf{E}'$, *a morphism* $f : \mathbf{E} \to \mathbf{E}'$ *is a mapping, written*

$$(f \cdot = \cdot) : \mathbf{E} \times \mathbf{E}' \to Q$$

satisfying

$$\begin{array}{llcl} (M1) & \bigvee_{x' \in \mathbf{E}} [x = x'] \& (fx' = y) & = & (fx = y) \\ (M2) & \bigvee_{y' \in \mathbf{E}'} (fx = y') \& [y' = y] & = & (fx = y) \\ (M3) & (fx = y) \& (fx = y') & \leq & [y = y'] \\ (M4) & [x = x'] & \leq & \bigvee_{y \in \mathbf{E}'} (fx = y) \& (fx' = y) \end{array}$$

Again, axioms (M3) and (M4) can be read intuitively as f is single-valued and everywhere defined. The reason why compatibility rules with the equalities (M1) and (M2) should take this form and not a weaker one (with inequalities) becomes clear when one tries to make a category out of Q-sets.

The reader should notice that a morphism from $\mathbf{E}$ to $\mathbf{E}'$ is, in general, by no means an actual function from the set $\mathbf{E}$ to the set $\mathbf{E}'$. So writing $[fx = x']$ with $x \in \mathbf{E}$, $x' \in \mathbf{E}'$ and $[\cdot = \cdot]$ the equality in $\mathbf{E}$ does not make much sense. We use curly brackets to avoid any such confusion.

Proposition 1.6. *A morphism of Q-sets* $f : \mathbf{E} \to \mathbf{E}'$ *satisfies the following properties*

$$\begin{array}{llcl} (M5) & \bigvee_{y \in \mathbf{E}'} (fx = y] & = & \bigvee_{x' \in \mathbf{E}} [x = x'] \\ (M6) & \bigvee_{x \in \mathbf{E}} [x = x] \& (fx = y) & = & \bigvee_{x \in \mathbf{E}} (fx = y) \\ (M7) & \bigvee_{y \in \mathbf{E}'} (fx = y) \& [y = y] & = & \bigvee_{y \in \mathbf{E}'} (fx = y) \\ (M8) & \bigvee_{x, x' \in \mathbf{E}} [x = x'] & \leq & \bigvee_{y, y' \in \mathbf{E}'} [y = y'] \end{array}$$

Proof : Rather straightforward calculations using the various axioms. For example, the non trivial inequality in (M6) is obtained from (M1) and (S3) in the following manner :

$$\begin{aligned}\bigvee_x (fx = y) &= \bigvee_x (\bigvee_{x'} [x = x'] \& (fx' = y)) \\ &= \bigvee_{x,x'} [x = x] \& [x = x'] \& (fx' = y) \\ &= \bigvee_x [x = x] \& (fx = y).\end{aligned}$$

■

Property (M5) says something as sensible as "the level at which $f(x)$ is defined is the level at which x is defined". (M8) says that in order to have a morphism from $\mathbf{E}$ to $\mathbf{E}'$, the level at which $\mathbf{E}$ is inhabited should be lower than the level at which $\mathbf{E}'$ is.

We now proceed to make a category out of Q-sets. As morphisms are defined in a distributorlike fashion (they are not distributors since Q-sets are not Q-categories) so should it be for their composition. Indeed

Proposition 1.7. *Let $f : \mathbf{E} \to \mathbf{E}'$ and $g : \mathbf{E}' \to \mathbf{E}''$ be morphisms of Q-sets. The formula*

$$(gfx = z) = \bigvee_{y \in \mathbf{E}'} (fx = y) \& (gy = z)$$

defines a morphism of Q-sets $gf : \mathbf{E} \to \mathbf{E}''$.

Proof : To verify that gf has properties (M1), (M2) and (M3) is straightforward computation. In checking (M4) the necessity of axiom (M2) becomes apparent :

$$\begin{aligned}&\bigvee_{z \in \mathbf{E}''} (gfx = z) \& (gfx' = z) = \\ &= \bigvee_{z \in \mathbf{E}'', y, y' \in \mathbf{E}'} (fx = y) \& (gy = z) \& (fx' = y') \& (gy' = z) \\ &\geq \bigvee_{y,y'} [y = y'] \& (fx = y) \& (fx' = y') \\ &= \bigvee_y (fx = y) \& (fx' = y) \\ &\geq [x = x'].\end{aligned}$$

■

Proposition 1.8. *Q-sets and their morphisms constitute a category.*

Proof : It is routine to check that the composition defined above is associative. The identity on a Q-set $\mathbf{E}$ is given by

$$(id\, x = x') = [x = x']$$

which is easily seen to be a morphism. To prove that id is a unit for composition we use (M1) and (M2). ■

Proposition 1.9. *The singleton $\{*\}$ provided with the equality $[* = *] = 1$ is a terminal object in the category of Q-sets. For any Q-set* $\mathbf{E}$*, the unique morphism* $f : \mathbf{E} \to \mathbf{1}$ *is given by*

$$(fx = *) = \bigvee_{x' \in \mathbf{E}} [x = x'].$$

Proof : $(\{*\}, [* = *] = 1)$ is clearly a Q-set. Let $f : \mathbf{E} \to \{*\}$ be a morphism of Q-sets, i.e. a mapping

$$\mathbf{E} \times \{*\} \to Q \; ; \; x \mapsto (fx = *)$$

satisfying in particular property (M5) :

$$\bigvee_{*}(fa = *) = \bigvee_{a'}[a = a'].$$

Hence, if there is such a morphism f it must be given by

$$(fa = *) = \bigvee_{a'}[a = a']$$

Now, all that remains to be done, is to prove that this formula defines a morphism of Q-sets. (M2) and (M3) are trivial. It is very easy to check (M1) and (M4). ■

Isomorphic Q-sets come in many sizes : Q itself, seen as a Q-set as in Example 1.4, is isomorphic to $\mathbf{1}$. More generally

Proposition 1.10. *Given any Q-set* $\mathbf{E}$*, the set* $\mathbf{E} \times Q$ *equipped with the equality*

$$[(a,p) = (b,q)] = [a = b]\&p\&q$$

is a Q-set isomorphic to $\mathbf{E}$.

Proof : Define $f : \mathbf{E} \to \mathbf{E} \times Q$ by

$$(fx = (x', q)) = [x = x']\&q$$

and define $g : \mathbf{E} \times Q \to \mathbf{E}$ by

$$(g(x,q) = x') = [x = x']\&q.$$

It is straightforward, but long and tedious, to verify that $\mathbf{E} \times Q$ is a Q-set, f and g are morphisms of Q-sets and finally that fg and gf are identities. ■

By a "global element" or a "global section" of a Q-set $\mathbf{E}$, we mean as usual a morphism from the terminal object to $\mathbf{E}$. We will denote the global section functor by Γ.

Proposition 1.11. *The global section functor*

$$\Gamma : Q\text{-}Set \to Set \; ; \; \mathbf{E} \mapsto \{f : 1 \to \mathbf{E} \mid f \text{ morphism of } Q\text{-sets}\}$$

has a left adjoint Δ.

Proof : Let A be any set. The "constant" Q-set ΔA is the set A equipped with the Q-valued equality given by

$$\begin{aligned}[a = a'] &= 1 \quad \text{if } a = a' \\ &= 0 \quad \text{if } a \neq a'\end{aligned}$$

Let $\mathbf{E}$ be a Q-set and $f : \Delta A \to \mathbf{E}$ a morphism of Q-sets. For each element $a \in A$, the mapping

$$\{*\} \times \mathbf{E} \to Q \; ; \; x \mapsto (fa = x)$$

defines a morphism

$$\overline{f}_a : \mathbf{1} \to \mathbf{E}$$

Hence we obtain a mapping

$$\overline{f} : A \to \Gamma(\mathbf{E})$$

Conversely, given any map $\varphi : A \to \Gamma(\mathbf{E})$, the mapping

$$\Delta A \times \mathbf{E} \to Q \; ; \; (a, x) \mapsto (\varphi(a)(*) = x)$$

defines a morphism of Q-sets

$$\varphi^0 : \Delta A \to \mathbf{E}.$$

Now

$$\begin{aligned} (\overline{f}^0 a = x) &= (\overline{f}_a * = x) = (fa = x) \\ (\overline{\varphi^0_a} * = x) &= (\varphi^0(a) = x) = (\varphi(a)(*) = x). \end{aligned}$$

This yields a bijection

$$\mathrm{Hom}_{Q-\mathrm{Set}}(\Delta A, \mathbf{E}) \to \mathrm{Hom}_{\mathrm{Set}}(A, \Gamma(E)).$$

Verification of the naturality of these bijections is left to the reader who hates himself. (Δ is defined on mappings between sets in the only possible way :
$(\Delta fa = b) = 1$ if $fa = b$
$= 0$ if $fa \neq b$.) ■

A more detailled study of the properties of the category of Q-sets requires too much space to be included in this paper whose main aim is a representation theorem for rings. They will be treated in a later paper. We conclude part 1 by analysing briefly what happens in a change of base quantale.

Definition 1.12. *A morphism of quantales is a mapping*

$$f : Q \to Q'$$

satisfying the following properties

$$\begin{aligned} (1) \quad f(\textstyle\bigvee_{i \in I} a_i) &= \textstyle\bigvee_{i \in I} f(a_i) \\ (2) \quad f(a \& b) &= fa \& fb \\ (3) \quad f(1) &= 1 \end{aligned}$$

Proposition 1.13. *A morphism of quantales $f : Q \to Q'$ induces a functor*

$$f^* : Q\text{-}Set \to Q'\text{-}Set.$$

Proof : Let $\mathbf{E}$ be a Q-set with equality $[a = b]$ taking value in Q. It is easily verified that the equality $< \cdot = \cdot >$ defined by

$$< a = b >= f[a = b]$$

makes $\mathbf{E}$ into a Q'-set. For any morphism of Q-sets $g : \mathbf{E} \to \mathbf{E}'$, the composite

$$\mathbf{E} \times \mathbf{E}' \xrightarrow{(g \cdot = \cdot)} Q \xrightarrow{f} Q'$$

defines a morphism $\overline{g}$ between the corresponding Q'-sets. And it is again routine to check that f^* so defined is indeed a functor. ∎

2 Sheaf representation of rings

In this section $\mathbf{R}$ will denote an arbitrary commutative unital ring (as well as its underlying set) and Id $(\mathbf{R})$ its quantale of ideals. For any element $r \in \mathbf{R}$, we will denote by Ann (r) the annihilator of r, i.e. the ideal $\{x \in \mathbf{R} \mid xr = 0\}$. It is easy to see that

$$\text{Ann}(r - s) \cdot \text{Ann } (s - t) \leq \text{Ann } (r - t).$$

From this and the fact that Ann $(0) = \mathbf{R}$ one verifies immediately

Proposition 2.1. *The set* $\mathbf{R}$ *equipped with the equality given by*

$$[r = s] = \textit{Ann } (r - s)$$

is a Id $(\mathbf{R})$*-set, denoted by* $\mathcal{R}$*.*

It is just as easy to verify

Proposition 2.2. *For any* $r_0 \in \mathbf{R}$*, the mapping*

$$\{*\} \times \mathbf{R} \to \textit{Id } (\mathbf{R}) \; ; \; r \mapsto \textit{Ann } (r - r_0)$$

defines a global element

$$\rho_{r_0} : \mathbf{1} \to \mathcal{R}.$$

Theorem 2.3. *The mapping*

$$\mathbf{R} \to \Gamma(\mathcal{R}) \; ; \; r_0 \mapsto \rho_{r_0}$$

is a bijection of the elements of the ring $\mathbf{R}$ *on the global elements of the Id* $(\mathbf{R})$*-set* $\mathcal{R}$*.*

Proof : The mapping is injective since for $r_0 \neq r_1$

$$(\rho_{r_0}(*) = r_0) = \text{Ann } (r_0 - r_0) = \mathbf{R} \neq (\rho_{r_1}(*) = r_0) = \text{Ann } (r_1 - r_0) \not\ni 1.$$

Surjectivity is more difficult to prove. Let $\rho : \mathbf{1} \to \mathcal{R}$ be a global element. By the last axiom (M4) for a morphism of Id $(\mathbf{R})$-sets, we obtain

$$\mathbf{R} = +_{r \in \mathbf{R}} (\rho * = r) \cdot (\rho * = r)$$

This implies the existence of finitely many elements r_i such that

$$(\rho * = r_1) \cdot (\rho * = r_1) + \ldots + (\rho * = r_n) \cdot (\rho * = r_n) = \mathbf{R}$$

Since $(\rho * = r) \cdot (\rho * = r) \leq (\rho * = r)$, we also obtain

$$(\rho * = r_1) + \ldots + (\rho * = r_n) = \mathbf{R}.$$

Therefore, there exist $\varepsilon_i \in (\rho * = r_i)$ such that

$$\varepsilon_1 + \ldots + \varepsilon_n = 1. \tag{1}$$

Now take

$$r_0 = \varepsilon_1 r_1 + \ldots + \varepsilon_n r_n. \tag{2}$$

We will prove that $\rho = \rho_{r_0}$. It suffices to prove that $(\rho * = r_0) = \mathbf{R}$ since ρ satisfies (M2) and (M3). Thus, for each $r \in \mathbf{R}$

$$(\rho * = r) \geq (\rho * = r_0) \cdot \text{Ann } (r - r_0)$$

and

$$(\rho * = r) \cdot (\rho * = r_0) \leq \text{Ann } (r - r_0).$$

Now, for any $i \in \{1, \ldots, n\}$ and $t \in (\rho * = r_i)$, using (1) and (2) we obtain

$$(r_0 - r_i)t = \varepsilon_1 t(r_1 - r_i) + \ldots + \varepsilon_n t(r_n - r_i).$$

But $\varepsilon_j t \in (\rho * = r_j) \cdot (\rho * = r_i) \leq \text{Ann } (r_j - r_i)$. Hence

$$(r_0 - r_i)t = 0 \text{ for all } t \in (\rho * = r_i)$$

and

$$(\rho * = r_i) \leq \text{Ann } (r_0 - r_i).$$

Using again axiom (M2) for morphisms of Id $(\mathbf{R})$-sets, we get

$$\begin{aligned}
(\rho * = r_0) &= +_{r \in \mathbf{R}} (\rho * = r) \cdot \text{Ann } (r - r_0) \\
&\geq (\rho * = r_1) \cdot \text{Ann } (r_1 - r_0) + \ldots + (\rho * = r_n) \cdot \text{ Ann } (r_n - r_0) \\
&\geq (\rho * = r_1) \cdot (\rho * = r_1) + \ldots + (\rho * = r_n) \cdot (\rho * = r_n) \\
&= \mathbf{R}
\end{aligned}$$

■

Observe that in the Id $(\mathbf{R})$-set $\mathcal{R}$, $[r = r] = \mathbf{R}$ for all $r \in \mathbf{R}$. This allows one to show that in any change of base Id $(\mathbf{R}) \to Q$, the full strength of a morphism of quantales is not needed to present $\mathbf{R}$ as a Q-set. More precisely

Proposition 2.4. *Let Q be a quantale and $q : Id\ (\mathbf{R}) \to Q$ a functor (i.e. a morphism of posets) preserving the multiplication. Then the composite*

$$\mathbf{R} \times \mathbf{R} \xrightarrow{[\cdot=\cdot]} Id\ (\mathbf{R}) \xrightarrow{q} Q$$

provides $\mathbf{R}$ with a Q-valued equality, making it into a Q-set denoted by $\mathcal{R}_q$.

Choosing for q the various localic quotients producing the various classical spectra of $\mathbf{R}$, we obtain the corresponding structural sheaves.

In fact, any time such a quotient is sufficiently compatible with $0, 1, \bigvee$ and &, the Q-set obtained by composing with q is again a representation of $\mathbf{R}$ on Q. A precise set of conditions to be satisfied is given in the following

Theorem 2.5. *Consider a commutative unital ring $\mathbf{R}$, a quantale Q and a surjective morphism of posets $q : Id\ (\mathbf{R}) \to Q$. If the folowing conditions are satisfied*

1. *q preserves the multiplication ;*
2. *q has a left or a right adjoint i ;*
3. *q reflects 1 ;*
4. *$iq(0) = 0$;*
5. *$i(u) \cdot i(v) \leq i(u\&v)$*

then the composite

$$\mathbf{R} \times \mathbf{R} \xrightarrow{[\cdot=\cdot]} Id\ (\mathbf{R}) \xrightarrow{q} Q$$

is a representation of the ring $\mathbf{R}$ over the quantale Q.

Proof : Sups and multiplication in Id $(\mathbf{R})$ will be denoted as before by $+$ and $\cdot$; in Q they will be denoted by $\bigvee$ and &. First observe that

$$\begin{aligned} u\&v &= q(iu \cdot iv) \\ \bigvee_k u_k &= qi(\bigvee_k u_k) \geq q(+_k iu_k) \geq \bigvee_k qiu_k = \bigvee_k u_k \end{aligned}$$

thus

$$\bigvee_k u_k = q(+_k iu_k).$$

It is again easy to verify that for any $r_0 \in \mathbf{R}$, the mapping

$$\mathbf{R} \to Q\ ;\ r \mapsto q\ \mathrm{Ann}(r - r_0)$$

defines a global element in Q-Set

$$\tau_{r_0} : \mathbf{1} \to \mathcal{R}_q.$$

And, since q reflects the top element, $\tau_{r_0} \neq \tau_{r_1}$ as soon as $r_0 \neq r_1$. Consider $\tau : \mathbf{1} \to \mathcal{R}_q$ a global element of $\mathcal{R}_q$. Axiom (M4) applied to τ then gives

$$\begin{aligned} 1 &= \bigvee_{r \in \mathbf{R}} (\tau * = r)\&(\tau * = r) = \bigvee_{r \in \mathbf{R}} (\tau * = r) \\ &= q(+_{r \in \mathbf{R}} i((\tau * = r)\&(\tau * = r))) = q(+_{r \in \mathbf{R}} i((\tau * = r))) \end{aligned}$$

From the reflection of the top element by q, we then can deduce that there exist finitely many elements $r_i \in \mathbf{R}$ such that

$$\begin{aligned} \mathbf{R} &= i((\tau * = r_1) \& (\tau * = r_1)) + \ldots + i((\tau * = r_n) \& (\tau * = r_n)) \\ &= i((\tau * = r_1)) + \ldots + i((\tau * = r_n)). \end{aligned}$$

Keeping in mind the proof of theorem 2.3, notice that the first equality will be no help. But

$$\begin{aligned} & q(+_{k=1}^{n} i(\tau * = r_k) \cdot i(\tau * = r_k)) \\ & \qquad \geq \bigvee_{k=1}^{n} q(i(\tau * = r_k) \cdot i(\tau * = r_k)) \\ & \qquad = \bigvee_{k=1}^{n} (\tau * = r_k) \& (\tau * = r_k) \\ & \qquad = q(+_{k=1}^{n} i((\tau * = r_k) \& (\tau * = r_k))) \\ & \qquad = q(\mathbf{R}) \\ & \qquad = 1. \end{aligned}$$

This allows us to write

$$\mathbf{R} = i(\tau * = r_1) \cdot i(\tau * = r_1) + \ldots + i(\tau * = r_n) \cdot i(\tau * = r_n).$$

We now proceed as before. The unit $1 \in \mathbf{R}$ may be written

$$1 = \varepsilon_1 + \ldots + \varepsilon_n \quad \text{with } \varepsilon_k \in i(\tau * = r_k)$$

and we define

$$r_0 = \varepsilon_1 r_1 + \ldots + \varepsilon_n r_n.$$

The rest of the proof, but for one point, goes as in theorem 2.3. The only difficulty resides in showing that for each k, $(\tau * = r_k) \leq q \operatorname{Ann}(r_0 - r_k)$. This is where condition 4 and 5 come in (each of them is always true in one adjunction case but not necessarily in the other). For each $t \in i(\tau * = r_k)$ we obtain

$$(r_0 - r_k)t = \varepsilon_1 t(r_1 - r_k) + \ldots + \varepsilon_n t(r_n - r_k),$$

and

$$\begin{aligned} \varepsilon_j t(r_j - r_k) \;\in\; & i(\tau * = r_j) \cdot i(\tau * = r_k) \cdot (r_j - r_k)\mathbf{R} \\ & \leq i((\tau * = r_j) \& (\tau * = r_k)) \cdot (r_j - r_k)\mathbf{R} \\ & \leq iq \operatorname{Ann}(r_j - r_k) \cdot (r_j - r_k)\mathbf{R} \end{aligned}$$

If i is a left adjoint to q

$$iq \operatorname{Ann}(r_j - r_k) \leq \operatorname{Ann}(r_j - r_k)$$

and $(r_0 - r_k)t = 0$.
If i is a right adjoint

$$\begin{aligned} & iq \operatorname{Ann}(r_j - r_k) \cdot (r_j - r_k)\mathbf{R} \\ & \leq iq \operatorname{Ann}(r_j - r_k) \cdot iq((r_j - r_k)\mathbf{R}) \\ & \leq iq(\operatorname{Ann}(r_j - r_k) \cdot (r_j - r_k)\mathbf{R}) \\ & = iq(0) \\ & = 0 \end{aligned}$$

Therefore

$$\begin{aligned} i(\tau * = r_k) &\leq \operatorname{Ann}(r_0 - r_k) \\ (\tau * = r_k) &\leq q \operatorname{Ann}(r_0 - r_k). \end{aligned}$$

■

When i is a left adjoint, a more direct proof can be made by observing that condition 5 forces i to be a morphism of quantales. The equality on $\mathbf{R}$

$$\mathbf{R} \times \mathbf{R} \xrightarrow{[\cdot = \cdot]} \operatorname{Id}(\mathbf{R})$$

can also be seen as a morphism of $\operatorname{Id}(\mathbf{R})$-sets

$$i^*(\mathcal{R}_q) \to \mathcal{R}.$$

Composing this morphism with $i^*(\tau) : \mathbf{1} = i^*(\mathbf{1}) \to i^*(\mathcal{R}_q)$ yields a global element ρ_{r_0} of $\mathcal{R}$. One verifies that $\tau = \tau_{r_0}$.

3 Some examples

3.1 $Sp(\mathbf{R})$ = the prime spectrum of the ring $\mathbf{R}$

The locale of open subsets of $Sp(\mathbf{R})$ is isomorphic to the locale of radical ideals of $\mathbf{R}$. We recall that an ideal $I \subseteq \mathbf{R}$ is said to be radical if

$$(\forall n > 0)(\forall r \in \mathbf{R})(r^n \in I \Rightarrow r \in I).$$

This is equivalent to saying that I is the intersection of all prime ideals containing I. The corresponding quotient map

$$q : \operatorname{Id}(\mathbf{R}) \to \operatorname{Rad} \operatorname{Id}(\mathbf{R}) \cong \mathcal{O}(Sp(\mathbf{R}))$$

is given by

$$q(I) = \sqrt{I} = \cap\{P \mid P \text{ prime ideal} ; I \subseteq P\}.$$

This quotient map admits the inclusion i as a right adjoint. The corresponding structural sheaf can thus be presented as the set $\mathbf{R}$ provided with the equality

$$[r = s] = \sqrt{\operatorname{Ann}(r - s)}.$$

It represents the ring R as soon as R has a zero nilradical, i.e. when $iq(0) = 0$. This result, but in terms of etale spaces and global sections, can be found in [3].

We should remark that theorem 2.5, applied to this case, is rather trivial since in any semi-prime ring every $\text{Ann}(r)$ is already a radical ideal. (Or one could say that this is a total demystification of the prime spectrum representation.)

3.2 $Sp(\mathbf{R})$ = the maximal spectrum of the ring R

Analogous situation with "prime" replaced by "maximal".

3.3 $Sp(\mathbf{R})$ = the pure (neat) spectrum of the ring R

The locale of open subsets is isomorphic to the locale of pure ideals of R [1]. We recall that an ideal $I \subseteq \mathbf{R}$ is said to be pure if

$$(\forall r \in I)(\exists \varepsilon \in I)(r\varepsilon = r).$$

The quotient map

$$q: \text{Id}(\mathbf{R}) \to \text{Pure Id}(\mathbf{R})$$

applies an ideal on its pure part (i.e. the biggest pure ideal contained in it). This time, the inclusion i of pure ideals in all ideals is a left adjoint to q. The corresponding structural sheaf is the set R equipped with the equality

$$[r = s] = \text{Pure part of Ann}(r - s).$$

It always represents the ring R.

References

[1] F. BORCEUX and G. VAN DEN BOSSCHE, Algebra in a localic topos with applications to ring theory, Springer LNM 1038, 1983.

[2] D. HIGGS, Injectivity in the topos of complete Heyting valued sets, Can. J. Math. Vol. 36 n° 3 (1984), 550-568.

[3] K.H. HOFMANN, Representations of algebras by continuous sections, Bull. Amer. Math. Soc. 78 (1972), 291-373.

[4] A. JOYAL and M. TIERNEY, An extension of the Galois theory of Grothendieck, Memoir of the AMS 309, 1984.

[5] C. MULVEY, Representations of rings and modules, Springer LNM 753, 1980, 542-587.

This paper is in final form and will not be published elsewhere.

NORMALIZATION EQUIVALENCE,
KERNEL EQUIVALENCE AND AFFINE CATEGORIES

Dominique Bourn
Fac. de Mathématiques, Université de Picardie
33 rue St Leu, 80039 Amiens France.

In a recent paper [4], A. Carboni gave an interesting characterization of the categories of affine spaces, i. e. slices of additive categories, by means of a "modularity" condition, relating coproducts and pullbacks, which is a categorical version of the modularity condition for lattices, in the same way as the distributive categories are the categorical version of the distributive lattices. At the end of his paper, he shows that, for any modular category $\mathbb{E}$, the forgetful functor $\mathrm{Grd}\,\mathbb{E} \to \mathrm{Grph}\,\mathbb{E}$ from internal groupoïds to internal reflexive graphs in $\mathbb{E}$ is an equivalence of categories. F. W. Lawvere asked whether this equivalence could be part of a characterization of modular or of affine categories.

P.J. Johnstone, studying internal Mal'cev operations in [6], pointed out that the classical category of affine modules satisfies the above equivalence, but is not modular.

But for a left exact additive category $\mathbb{A}$, there is a much more specific equivalence, namely the normalization equivalence $N : \mathrm{Grd}\,\mathbb{A} \to C^1\mathbb{A}$ from internal groupoids to complexes of length 1. This equivalence still holds in any slice of additive categories and therefore in any modular category. Whence a natural question : would the normalization equivalence characterize the modular categories ?

But it is easy to check that the normalization equivalence is the extension to categories of algebras of the kernel equivalence $K : \mathrm{Spl}\,\mathbb{A} \to \mathbb{A}/0$ associating to each split epimorphism its kernel and consequently follows from it. This kernel equivalence synthesizes the two first results the most often produced in a book on additive categories, namely : a) the (split) short five lemma which means that K is conservative, b) the decomposition of the domain of a split epimorphism into the direct sum of its kernel and its codomain, which means that K has a left adjoint right inverse. Now the kernel equivalence still holds not only in any modular category (slice of an additive category) but also in any coslice of an additive category. So, behind our rather anecdotal initial question, whose answer is no according to this last remark, lies the problem of the understanding of what is exactly the heart of additivity.

It is shown here, that for a left exact category $\mathbb{E}$ with an initial objet and O-valued sums, the kernel equivalence is equivalent to the following condition (called the essentially affine condition) : for any commutative square of split epimorphisms :

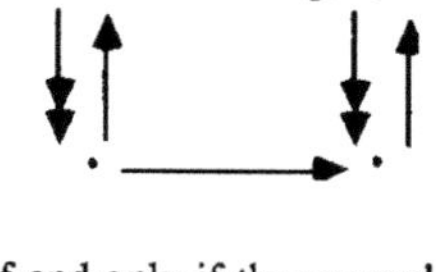

the downward square is a pullback if and only if the upward square is a pushout. Now a left exact category is additive if and only if it essentially affine and pointed $(0 = 1)$. It is modular if and only if it is essentially affine and its terminal object satisfies a condition of modularity. It is equivalent to a coslice of additive category if and only if it is essentially affine and its initial object satisfies a certain condition of comodularity.

On the other hand, it is well known that the short five lemma still holds in the category Grp of (non abelian) abstract groups. So it seems rather relevant to study the property following which the functor K is conservative or equivalently the property following which the following condition (called the "protomodularity" condition) does hold : given two commutative squares between split epimorphisms :

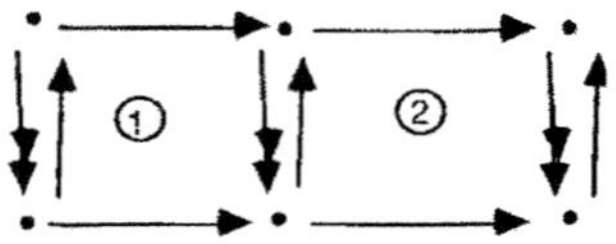

if the squares ① and ① + ② are pullbacks, then the square ② is a pullback. Indeed this condition spreads some light on the categorical aspect of the exactness of the category Grp : in any protomodular category $\mathbb{E}$, the monos are characterized by their kernels, the internal categories are always internal groupoïds and, when moreover $\mathbb{E}$ is Barr-exact, the pullbacks along a regular epimorphism are always pushouts.

It is not unexpected that, when $\mathbb{E}$ is left exact, the category Grp $\mathbb{E}$ of internal groups in $\mathbb{E}$ is protomodular. What is more surprising is that not only Grp $\mathbb{E}$ but also any fiber of the fibration $(\ )_0 : \mathrm{Grd}\,\mathbb{E} \to \mathbb{E}$ associating to each internal groupoïd its object of objects is again protomodular.

0. - Modular categories

A. Carboni defines [4] a modular category $\mathbb{E}$ as a left exact category with finite sums and such that :

① for each slice category $\mathbb{E}/U$ and each arrow $f : X \to Z$ in $\mathbb{E}/U$ the canonical map :

$$\begin{pmatrix}(i_X, f)\\ i_Y \times Z\end{pmatrix} : X + (Y \times Z) \to (X + Y) \times Z$$

is invertible for each object Y in $\mathbb{E}/U$.

② for each arrow $f : X \to Y$ in $\mathbb{E}$ the following commutative square is a pullback :

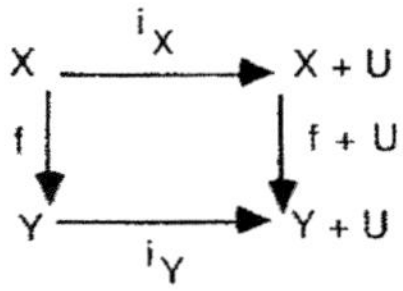

Then he shows that all possible modular categories are the slices of additive categories with kernels.

Remark 1. The axiom 1 says that the category $\mathbb{C} = \mathbb{E} / U$ which admits finite sums and products satisfies a certain condition (the modularity condition) which is, in fact, autodual in $\mathbb{C}$, i. e. which holds in $\mathbb{C}$ if and only if it holds in $\mathbb{C}^{op}$. But obviously the axiom 1 is not autodual in $\mathbb{E}$.

Remark 2. Categorically speaking, the axiom 1 is not very intuitive. Let us make it explicit : when the square in the left hand diagram is a pullback, then the right hand square is a pullback :

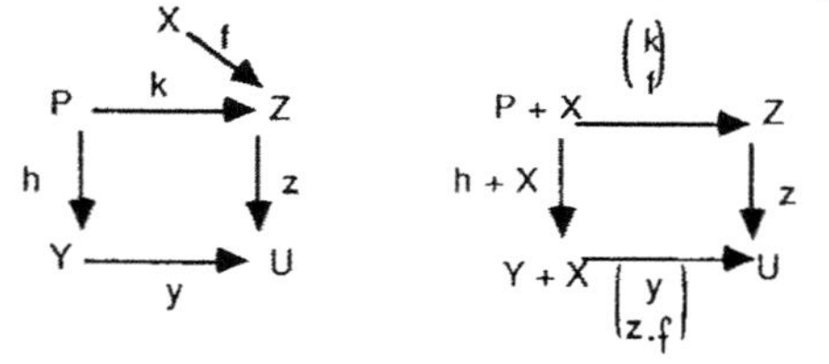

Actually there is an equivalent way of stating axiom 1 which is more connected to the axiom 2 :

Proposition 1 : In any left exact category $\mathbb{E}$ with finite sums, the axiom 1 is equivalent to the axiom 1', for each arrow $l : S \rightarrow T$, then each decomposition of the arrow $l + U : S + U \xrightarrow{m} K \xrightarrow{n} T + U$, is such that K is isomorphic to J + U, where J is given by the pullback (*) :

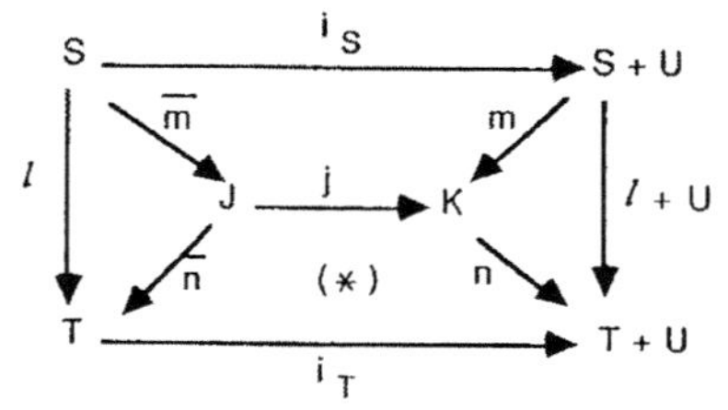

with, furthermore, $m = \bar{m} + U$ and $n = \bar{n} + U$, up to isomorphism.

Proof. Apply the axiom 1 to the left hand square, then the right hand square is a pullback and K isomorphic to J + U :

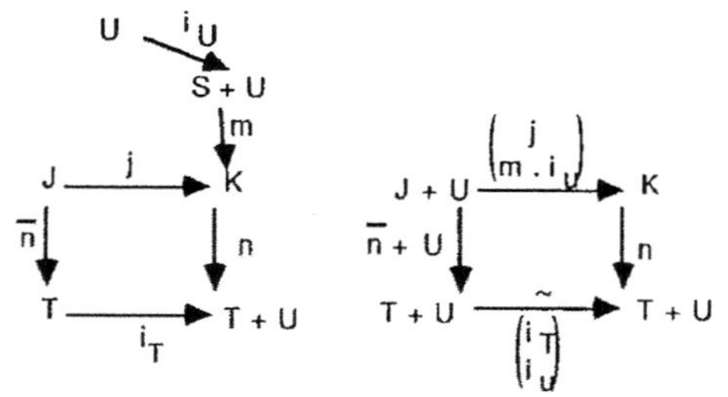

Conversely, let us suppose that the left hand square is a pullback :

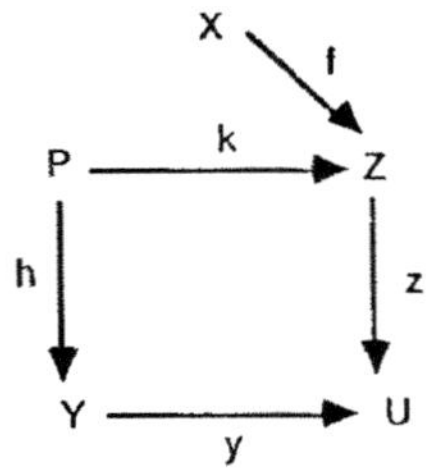

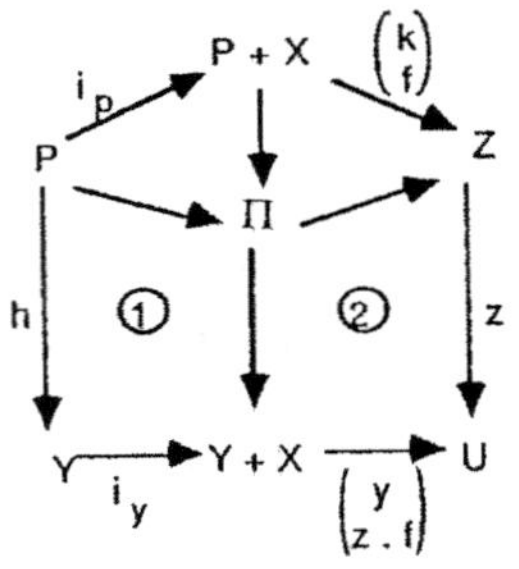

then the whole right hand square (being equal to the left hand one) is a pullback. Thus, if the square ② is a pullback, the square ① is again a pullback. Furthermore there is an arrow : $P + X \to \Pi$ determining a decomposition of the arrow $h + X$ (actually this arrow is exactly the one involved in Carboni's axiom 1).

Following the axiom 1', this arrow is invertible. ■

It is the right place here to introduce the following definition which we shall need later on :

<u>Definition 1</u> : An object U, in any category $\mathbb{E}$, will be called modular when the functor "sum with U" is defined and satisfies the previous axioms 1' and 2.

<u>Remark 3</u> : <u>The kernel equivalence</u> :

Let $f : Y \to X$ be a given epimorphism split by a given morphism s. Considering now the following left hand diagram with the lower square (*) a pullback, the axiom 1' implies that the arrow in the middle of the right hand diagram is invertible :

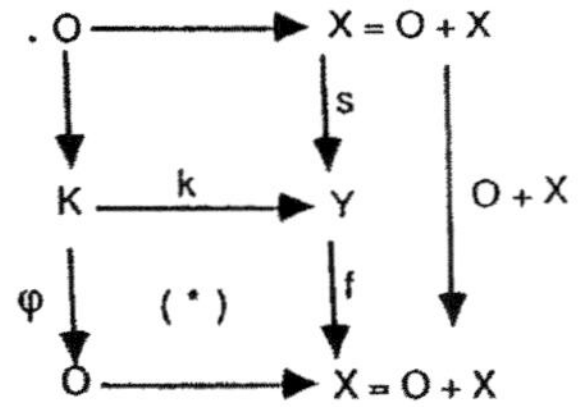

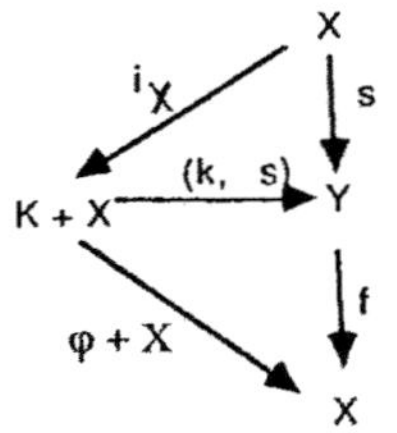

In other words a modular category satisfies the key result of an additive category with kernels following which the domain of a split epimorpism is canonically isomorphic to the sum of its codomain with its kernel.

Let us be a little more precise : let $\mathbb{E}$ be any category. Let us denote by Pt $\mathbb{E}$ the category whose objects are the split epimorphisms of $\mathbb{E}$ with a given splitting and whose morphims are the commutative squares between such data. If $\mathbb{E}$ admits pullbacks of split epimorphisms along any map (we shall say : $\mathbb{E}$ admits split pullbacks) and an initial objet O, then there is a functor **K** (the kernel functor) :

$$K : \mathrm{Pt}\,\mathbb{E} \to \mathbb{E}/0 \times \mathbb{E}$$

defined by $K(f, s) = (\varphi : K \to O, X)$ where φ is given by the pullback (*).

If, moreover, for any O-valued objet K (i.e. an object such that there exists a map $K \to O$) the functor "sum with K" is defined in $\mathbb{E}$ (we shall say that E admits O-valued sums), then the functor **K** has a left adjoint **L**, defined by $L(\varphi : K \to O, X) = (\varphi + X, i_X)$.

Our previous remark means exactly that the axiom 1' implies that the natural transformation $L \cdot K \to 1$ is an isomorphism.

The axiom 2 implies that the natural transformation $1 \to K \cdot L$ is an isomorphism. Thus we have the following result :

Proposition 2 : : In any modular category, the kernel functor is an equivalence of categories.

11] The normalization functor

Let $\mathbb{E}$ now be any category with split pullbacks and let Grd $\mathbb{E}$ denote the category of internal groupoids in $\mathbb{E}$. If moreover $\mathbb{E}$ has an initial object, we shall denote by $C^1\mathbb{E}$ the category whose objects (called 1-complexes) are the spans [2] of the type :

$$O \longleftarrow Y \longrightarrow X$$

and morphisms the morphisms between such spans. The normalization functor $N : \mathrm{Grd}\,\mathbb{E} \to C^1\,\mathbb{E}$ associates to each internal groupoid

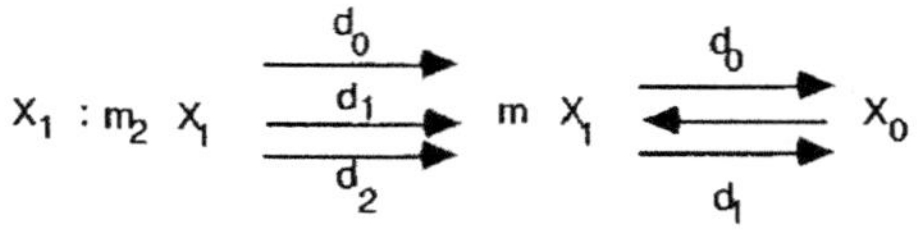

the following 1-complex where the square (*) is a pullback :

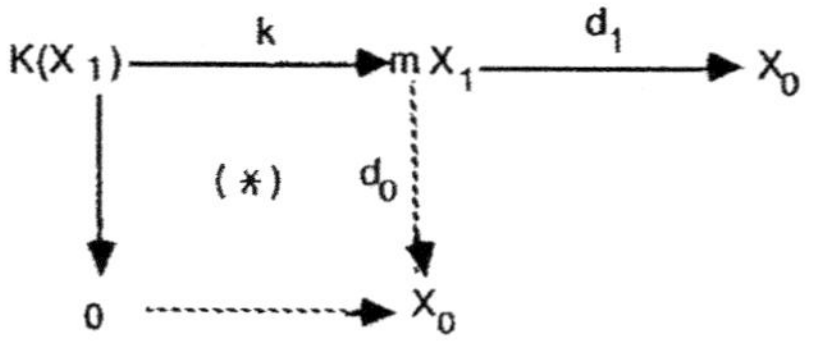

When $\mathbb{E} = \mathbb{A}$ is an addidive category with kernel this functor is well known to be an equivalence of categories [5, 7] . The same result holds for any slice of such categories (i.e. a category of the type $\mathbb{A}/X$ for any object X of $\mathbb{A}$) and consequently for any modular categories.

One of our aim is to investigate the following question : Would this property characterize the modular categories ?

2] The kernel functor and the normalization functor.

We shall show first that, when $\mathbb{E}$ has O-valued sums, the normalization functor **N** is actually an extension to categories of algebras of the kernel functor **K**.

Let us recall indeed, that, when $\mathbb{E}$ has split pullbacks, the category Grd $\mathbb{E}$ is monadic above the category Pt $\mathbb{E}$ [3]. The endofunctor T of this monad is given by $T(f, s) = (p_0, s_0)$ where p_0 is the projection of the following pullback and s_0 is the diagonal map : $Y \rightarrow Y \times_X Y$:

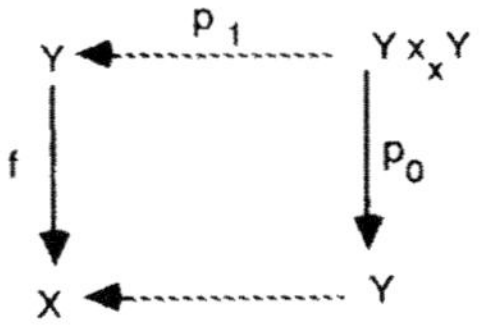

the unit and the multiplication being given by the following diagram :

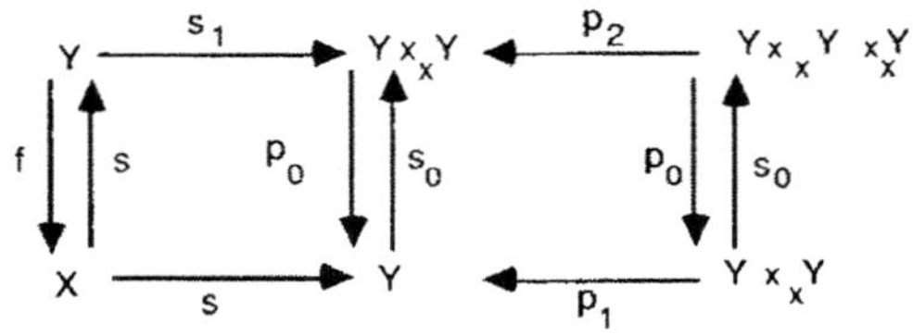

On the other hand, when $\mathbb{E}$ has an initial object O and admits O-valued sums, the category $C^1\,\mathbb{E}$ is monadic above $\mathbb{E}/0 \times \mathbb{E}$.

Indeed, we have the following monad on $\mathbb{E}/0 \times \mathbb{E}$, defined by :

$T'(\varphi : K \rightarrow O, X) = (\varphi : K \rightarrow O , K + X)$

$\lambda'(\varphi : K \rightarrow O, X) = (1_K, i_X)$

$\mu'(\varphi : K \rightarrow O, X) = (1_K, s_X)$

where i_X and s_X are the usual map :

$$X \xrightarrow{i_X} K + X \xleftarrow{s_X} K + K + X .$$

Now, the algebras of this monad are given by the maps :

$(h, \alpha) : (\varphi : K \rightarrow O , K + X) \longrightarrow (\varphi : K \rightarrow O, X)$

satisfying the usual axioms. Necessarily $h = 1_K$ and α is an algebra for the monad $(K + -)$ on $\mathbb{E}$ and consequently is entirely determined by a map $x : K \rightarrow X$. Therefore an algebra of T' is just a span : $0 \xleftarrow{\varphi} K \xrightarrow{x} X$, that is an object of $C^1\,\mathbb{E}$. Furthermore there is a natural transformation $r : T' . \mathbf{K} \rightarrow \mathbf{K} . T$ which determines a morphism of monad :

$$r(f, s) = (1_K, \begin{pmatrix} k \\ s \end{pmatrix}) : (\varphi : K \rightarrow O , K + X) \rightarrow (\varphi : K \rightarrow O, Y).$$

Proposition 3. When IE has O-valued sums , the normalization functor N is the extension to these categories of algebras of the kernel functor **K** .

Proof. Straigtforward. ■

When IE is modular, the axiom 1' implies that the natural transformation r is a natural isomorphism.

On the other hand, we saw that the functor **K** is then an equivalence. Therefore, the normalization equivalence is, for the modular categories, a consequence of the kernel equivalence.

3] The fibration of "pointed objects" in IE.

Actually, as soon as the category IE admits split pullbacks the category Pt IE is fibered over IE along the forgetfull functor p : Pt IE $\rightarrow$ IE defined by $p(f, s) = d_1(f)$. We shall call this fibration p the fibration of pointed objects in IE. [So, what is denoted by Pt IE in Carboni's paper, is here only the fiber above the terminal object 1 and will be denoted by Pt IE [1]].

On the other hand the category IE / 0 x IE is trivially fibered over IE by the canonical projection π . Furthermore the kernel functor **K** is clearly cartesian between these two fibrations (this is the obvious translation of the fact that the cartesian squares in Pt IE (i.e. the split pullbacks) have invertible images in IE / 0).

So, when IE is modular, **K** determines a cartesian equivalence of fibrations. But all the fibers of the fibration π being isomorphic to IE / 0, all the fibers of the fibration p are consequently equivalent, more exactly each change of base functor of the fibration p is an equivalence of categories (Let us call trivial such a kind of fibration).

In order to understand better this point, let us study more closely this fibration of internal points. Let us first establish the following fact :

Proposition 4. Let IE be a category with split pullbacks. If IE admits pushouts of split monos (we shall say that IE has split pushouts), then the change of base functors of the fibration p have a left adjoint (i.e. p is also a cofibration). Conversely if p is a cofibration and IE admits finite products, then the category IE has split pushouts.

Proof. Let h be a morphism : X $\rightarrow$ X' in IE . We must show that the change of base functor h* along h has a left adjoint h ! . When IE has split pushouts, then h ! (f, s) is given by the following pushout :

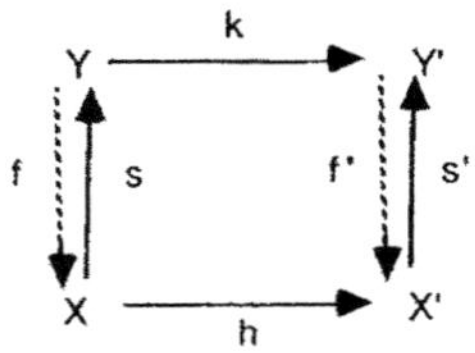

Conversely let us suppose that, for every h, the functor h* has a left adjoint h ! and let us assume that $\mathbb{E}$ has finite products. Let us denote by (f', s') the pair h !(f, s). We must show that the following square is a pushout.

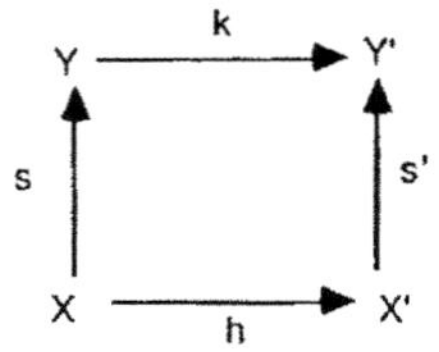

So, let $\bar{k}$ and $\bar{s}$ be two maps with the same codomain $\bar{Y}$ satisfying $\bar{k} . s = \bar{s} . h$.

Existence of a factorization.

The following diagram of split epimorphisms :

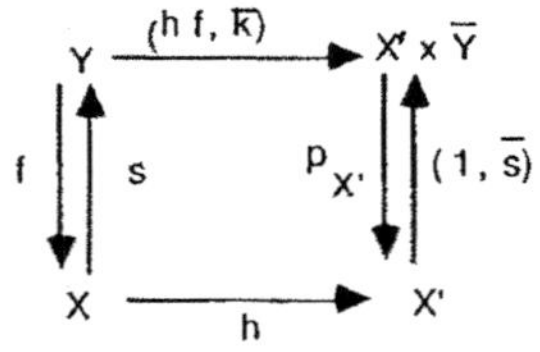

determines, by adjunction, a unique morphism $(\varphi, \psi) : Y' \to X' \times \bar{Y}$ such that $p_{X'} . (\varphi, \psi) = f'$, $(\varphi, \psi) \cdot s' = (1, \bar{s})$ and $(\varphi, \psi) . k = (h f, \bar{k})$. Whence $\varphi = f'$, $\psi . s' = \bar{s}$, $\psi . k = \bar{k}$.

Unicity of the factorization

We must show that the pair (k, s') is jointly epic.

So, let $l_1, l_2 : Y' \to \bar{Y}$ be such that $l_1 . k = l_2 . k \,(= \lambda)$ and $l_1 . s' = l_2 . s' \,(= \sigma)$.

The following diagram of split epimorphisms :

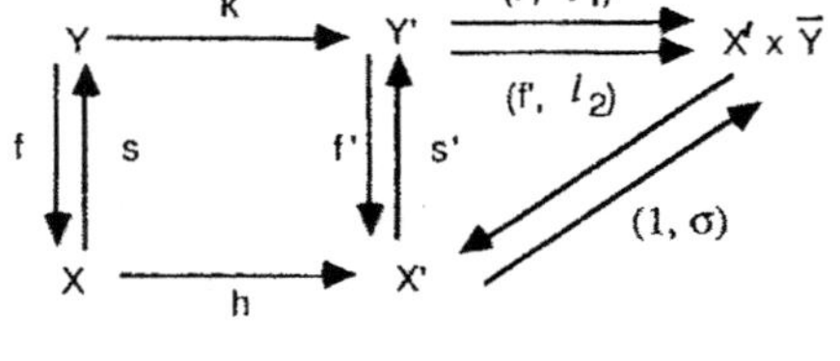

is such that the two upper composites are equal. That means that the adjoint maps of (f', l_1) and

(f', l_2) in the fiber above X are equal (they are both equal to $(f, \lambda) : Y \longrightarrow X \times \overline{Y}$) . Consequently (f', l_1) and (f', l_2) are equal and so l_1 and l_2. ■

4] <u>The essentially affine categories</u>.

Now to assume that the previous change of base functor h* is an equivalence of categories is to assume the following <u>essentially affine condition</u> :

In any commutative square of split epimorphisms :

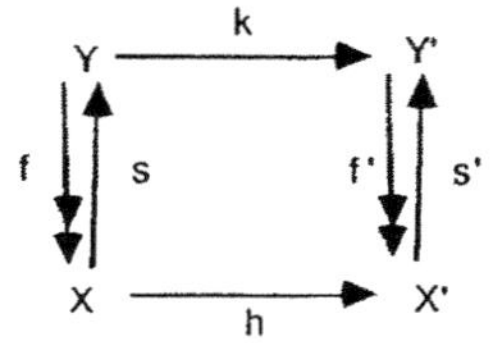

the downward square is a pullback if and only if the upward square is a pushout.

<u>Definition 2</u>. We shall call essentially affine a category $\mathbb{E}$ with split pullbacks and pushouts such that the essentially affine condition holds or equivalently such that the fibration p of pointed objects is trivial.

<u>Remark 4</u>. Clearly, this definition is autodual.

<u>Example 1</u>. When $\mathbb{E}$ is modular, $\mathbb{E}$ has split pullbacks and finite products. Moreover the fibration p is trivial because of the kernel equivalence. Finally, following the proposition 4, $\mathbb{E}$ has split puschouts. So the modular categories are essentially affine.

The previous terminology is due to the following result :

<u>Proposition 5</u>. If $\mathbb{E}$ is essentially affine, the fibration p is additive (i.e. each fiber and each change of base functor is additive).

<u>Proof</u>. For each object X of $\mathbb{E}$, the fiber of p above $\mathbb{E}$ has a terminal object $(1_X, 1_X)$ which is also an initial object (i.e. the fiber is pointed). Furthermore, $\mathbb{E}$ having split pullbacks, each fiber admits finite products and each change of base functor preserves them.

Let us show now that the products in the fibers are sums as well. Given two pointed objects (f, s) and (f', s') above X, their products in the fiber (left hand square) determines by splitting a pullback (right hand square) which is actually a morphism in Pt IE :

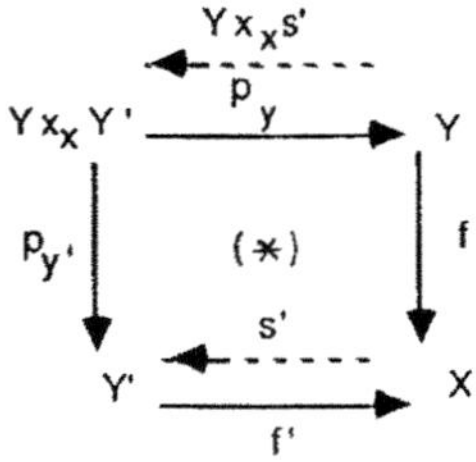

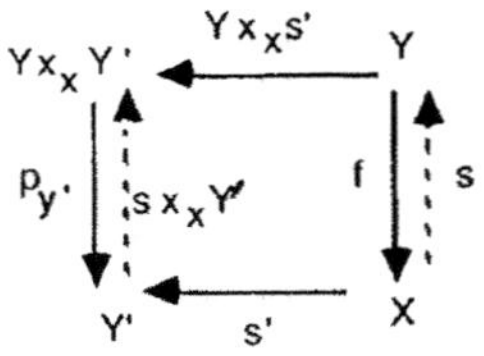

Consequently, IE being essentially affine, the following diagram is a pushout in IE, that is a sum in the fiber :

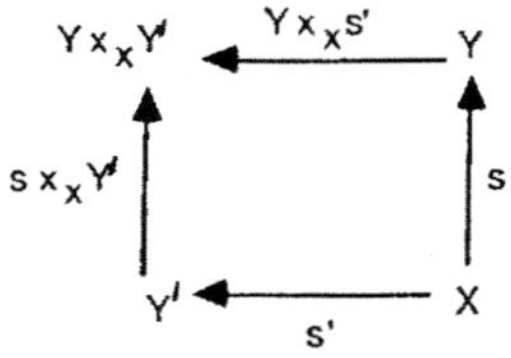

For each (f, s), the codiagonal map $\sigma_Y : Y \times_X Y \to Y$ determines clearly a commutative monoid in the fiber. We must show that this monoid is a group. That will be the case if and only if the following left hand square is a pullback :

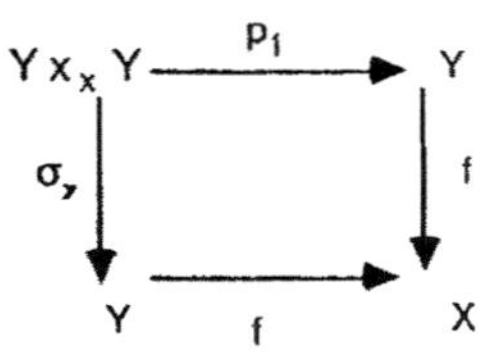

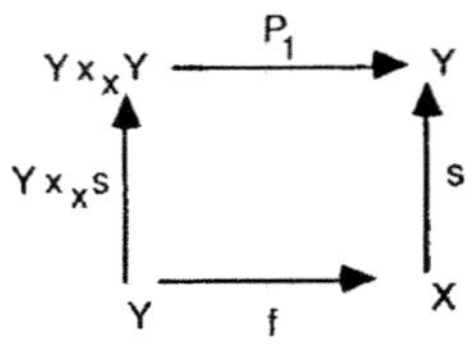

But the right hand square is a splitting of it and is a pushout since the square (*) is a pullback. So, following the essentially affine condition, the left hand square is a pullback. This natural abelian group structure for each split epi (f, s) determines a canonical structure of additive category on the fiber of p above X. Clearly this construction is preserved by change of base. ■

Remark 5. This is an extended and fibered version of the Carboni's result following which, when IE is modular, the category Pt IE [1] = 1 / IE is additive.

Remark 6. Let us suppose that IE is essentially affine. If IE has an initial object 0, then IE / 0 is nothing but Pt IE [0], the fiber of p above 0 (which, so, is additive). The kernel functor is defined and is an equivalence of categories, and therefore the normalization equivalence does hold in IE :
N : Grd IE $\to$ C^1 IE .

Conversely, if IE has split pullbacks, an initial object and admits 0-valued sums, then the kernel functor is defined and has a left adjoint L. If the kernel equivalence holds, the fibration p is trivial and it is easy to check that IE has split pushouts. Then IE is essentially affine.

Thus, in the presence of split pullbacks and 0-valued sums, the kernel equivalence is equivalent to the fact that $\mathbb{E}$ is essentially affine.

Finally, thanks to a dual of the proposition 4, it is also possible to show that if $\mathbb{E}$ admits kernels of split epis and sums, then the kernel equivalence implies that $\mathbb{E}$ is essentially affine.

Remark 7. If $\mathbb{E}$ is essentially affine and has a terminal object 1, there is a dual version of the kernel equivalence, namely that the following functor is an equivalence :

$$\mathrm{Cok} : \mathrm{Pt}\,\mathbb{E} \to (1 / \mathbb{E}) \times \mathbb{E}$$

where the first component is the pushout along the terminal map and the second one is just $d_1(f)$. In particular, the codomain of each split mono is the product of its domain and its cokernel.

Remark 8. If $\mathbb{E}$ is pointed (i.e. the unique map $0 \to 1$ is invertible) and essentially affine, then $\mathbb{E}$ is itself additive since $\mathbb{E} \cong \mathbb{E}/1 \cong \mathbb{E}/0$. Conversely there is a characterization of additive categories with kernels of split epimorphisms.

Corollary. The three following conditions are equivalent

1) $\mathbb{A}$ is additive with kernels of split epis.
2) $\mathbb{A}$ is pointed and essentially affine.
3) a) $\mathbb{A}$ is pointed, has sums and kernels of split epis
   b) the kernel equivalence holds.

Proof. That 1) implies 3) is well known. That 3a) and 3b) implies 2) is our last observation in the remark 6. That 2) implies 1) is just the remark 8. ■

Definition 3. A functor between two essentially affine categories will be said essentially affine when it preserves split pullbacks.

Corollary. A functor between two additive categories is additive if and only if it is essentially affine and it preserves the null object.

Example 2. Let $\mathbb{E}$ be a category with split pushouts and split pullbacks. Let $F : \mathbb{E} \to \mathbb{E}'$ be a functor preserving the two types of squares. If furthermore F is conservative, it reflects such types of squares. Thus, if $\mathbb{E}'$ is essentially affine, then $\mathbb{E}$ is essentially affine.

Now, for any object X of $\mathbb{E}$, the forgetful functor $\mathbb{E} / X \to \mathbb{E}$ fulfils these properties, in the same way as the forgetful functor $X / \mathbb{E} \to \mathbb{E}$. Therefore the essentially affine categories are stable by slices and coslices. In particular, if $\mathbb{A}$ is additive with kernels of split epis, not only $\mathbb{A} / X$ but also $X / \mathbb{A}$ is essentially affine.

Remark 9. The normalization equivalence does not characterize the modular categories. Indeed $X / \mathbb{A}$ ($\mathbb{A}$ additive with kernels of split epis) has an initial object and is essentially affine, then the normalization equivalence holds (remark 6). But $X / \mathbb{A}$ is not modular since the axiom 2 of Carboni does not hold any more, in general, in $X / \mathbb{A}$.

Example 3. When $\mathbb{E}$ is essentially affine, then Pt $\mathbb{E}$ and Grd $\mathbb{E}$ are essentially affine.

5] Essentially affine and modular categories.

If $\mathbb{E}$ is modular, then $\mathbb{E}$ is essentially affine but the converse (remark 9) is not true. What kind of additional axiom has to satisfy an essentially affine category $\mathbb{E}$ to be modular ? A careful inspection of the proof of the Carboni's theorem will give us the answer.

Proposition 6. A category $\mathbb{E}$ is modular, if and only if :

1) $\mathbb{E}$ is essentially affine

2) $\mathbb{E}$ is left exact and has its terminal object modular (see definition 1).

Proof. When $\mathbb{E}$ is modular, $\mathbb{E}$ is essentially affine (example 1) and any object is modular (proposition 1).

Conversely the existence of the sum with 1 determines the following adjunction, where U is monadic :

$$1 / \mathbb{E} \underset{F}{\overset{U}{\rightleftarrows}} \mathbb{E}$$

with $U(x : 1 \to X) = X$ and $F(X) = (i_1 : 1 \to X + 1)$.

The comonad yielded on $1 / \mathbb{E}$ is described by the following diagram :

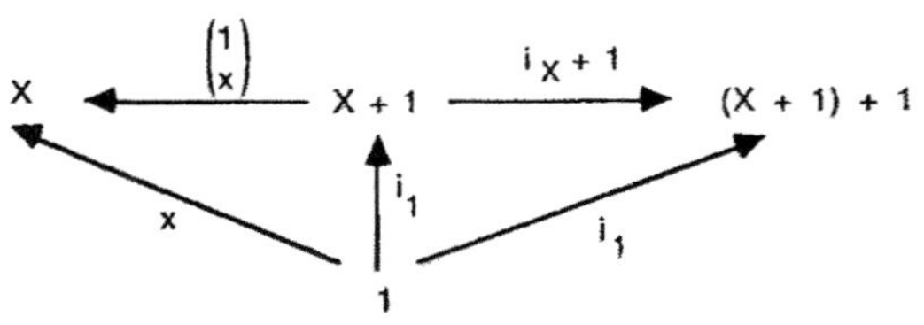

If, moreover, $\mathbb{E}$ is essentially affine, let us show that this comonad on $1 / \mathbb{E}$ is simply the comonad generated by the product in $1 / \mathbb{E}$ with $(i_1 : 1 \to 1 + 1)$. Indeed the left hand square being a pushout in $\mathbb{E}$, the right hand square is a pullback in $\mathbb{E}$ and a product in $1 / \mathbb{E}$:

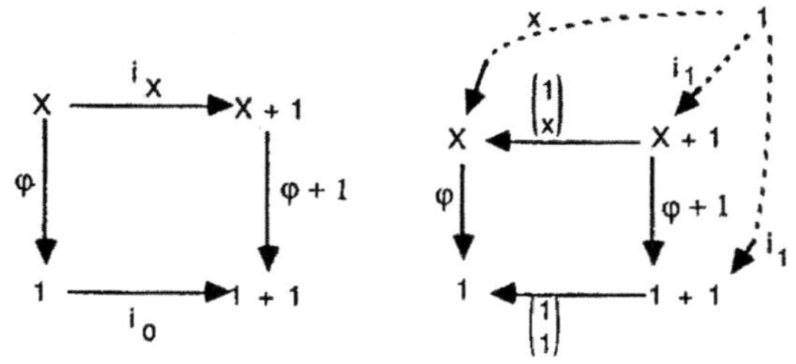

Consequently, the category of coalgebras of this comonad is the slice category $(1 / \mathbb{E}) / i_1$ of the additive category $1 / \mathbb{E}$.

Let us show now that, when $\mathbb{E}$ is left exact and 1 a modular object in $\mathbb{E}$, then the functor F is comonadic. Thus $\mathbb{E}$ will be equivalent to $(1 / \mathbb{E}) / i_1$ and so modular.

1) The following square is a pullback, hence i_X is the kernel of i_{X+1} and $i_X + 1$:

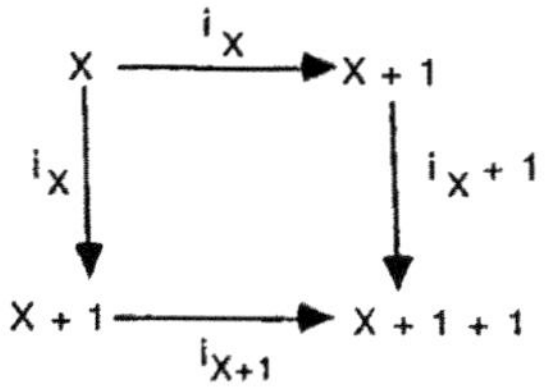

2) If f and g : X ⇉ Y are two maps such that f + 1 and g + 1 have a kernel K, then the left hand pullback (*) determines the kernel of f and g :

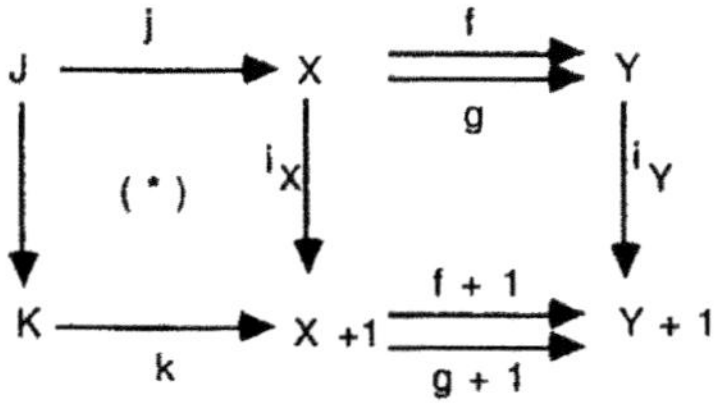

3) The functor (- + 1) preserves these kernels : indeed the morphism $i_1 : 1 \to X + 1$ equalizes f + 1 and g + 1 and so has a factorization $i : 1 \to K$. Now let us consider the following diagram :

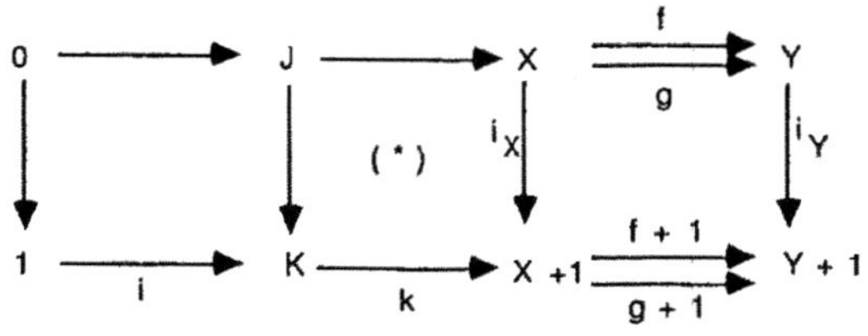

Let us denote by $\alpha_X : 0 \to X$ the initial map. Then $k \cdot i = i_1 = \alpha_X + 1$. The square (*) being a pullback, K is isomorphic to J + 1 by the axiom 1' of the modularity of 1.

4) The functor (- + 1) reflects the kernels :

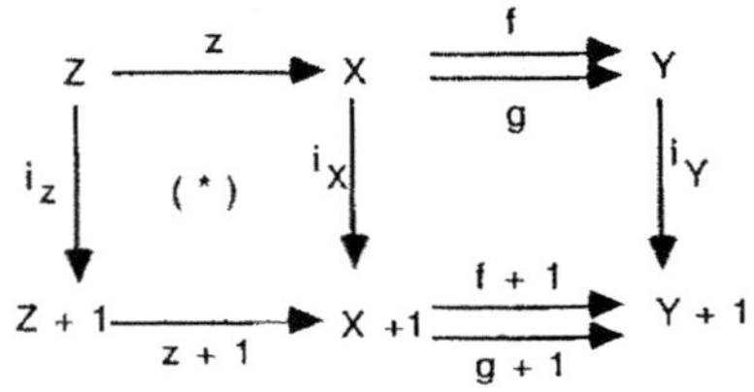

If z + 1 is the kernel of f + 1 and g + 1, then, the square (*) being a pullback by the axiom 2 of the modularity of 1, z is the kernel of f and g . ■

Remark 10. This is a pleasant circumstance where, at the same time, U is monadic and F comonadic.

Remark 11. It is possible to characterize the coslices of additive categories with cokernels as the essentially affine categories with pushouts and an initial object 0 which is comodular (i.e. satisfying the dual conditions of a modular object). The dual of Carboni's theorem would have only characterized the dual of such coslices.

6] Protomodular categories.

The kernel functor K is an equivalence if and only if (as any equivalence)

1) it is conservative

2) its left adjoint L is a right inverse.

When $\mathbb{A}$ is additive, the condition 1 is known as the short five lemma for split exact sequences, the condition 2 means that

$$0 \to A \to A \oplus B \to B \to 0$$

is a split exact sequence.

Now, the short five lemma still holds for non abelian groups and consequently, the kernel functor, for the category Grp of groups, is conservative

$$K : \text{Pt Grp} \to \text{Grp} \times \text{Grp}$$

That implies that the change of base functors of the fibration p of pointed objects are again conservative. So it seems relevant to study such a situation.

Definition 4. Let $\mathbb{E}$ be a category with split pullbacks. It will be called protomodular when the fibration p : Pt $\mathbb{E} \to \mathbb{E}$ of pointed objects has its change of base functor conservative.

Examples. 1) Of course, the essentially affine categories are protomodular.

2) Following our initial remark, the category Grp is protomodular.

3) Let F : $\mathbb{E} \to \mathbb{E}'$ be a functor which preserves split pullbacks and is conservative, then if $\mathbb{E}'$ is protomodular and $\mathbb{E}$ has split pullbacks then $\mathbb{E}$ is protomodular.

4) Thus protomodularity is stable by slices and coslices.

5) If $\mathbb{E}$ has split pullbacks, then the category Grp $\mathbb{E}$ of internal groups in $\mathbb{E}$ has split pullbacks. Furthermore the Yoneda embedding Y : Grp $\mathbb{E} \to \mathrm{Grp}^{\mathbb{E}^{op}}$ is conservative and left exact. Therefore Grp $\mathbb{E}$ is protomomular.

6) What is more surprising is the following fact.

Let $\mathbb{E}$ be a left exact category and $(\ )_0$ be the forgetful functor from the category Grd $\mathbb{E}$ of internal groupoïds in $\mathbb{E}$ to the category $\mathbb{E}$:

$$(\ )_0 : \mathrm{Grd}\,\mathbb{E} \to \mathbb{E}$$

It is a fibration. The fibers of this fibration are protomodular, generalizing the property of Grp $\mathbb{E}$ which is the fiber above the terminal object.

The proof is given by checking first the result in the set theoretical context and then by using the Yoneda embedding.

7) The dual of the category of sets is protomodular.

Let us state now an equivalent definition.

Proposition 7. A category with split pullbacks is protomodular if and only if, when the squares ① and ① + ② are pullbacks in the following diagram, then the square ② is a pullback :

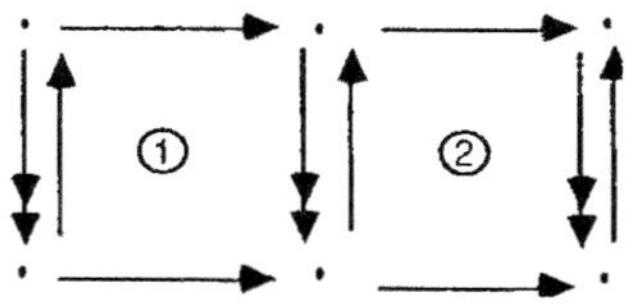

Proof. This condition implies protomodularity : take this kind of diagram with the lower map of the square ② an identity. Then the square ② being a pullback, the upper map is an isomorphism. Conversely, take the pullback ②' of the right hand split epimorphism along the lower map of ② :

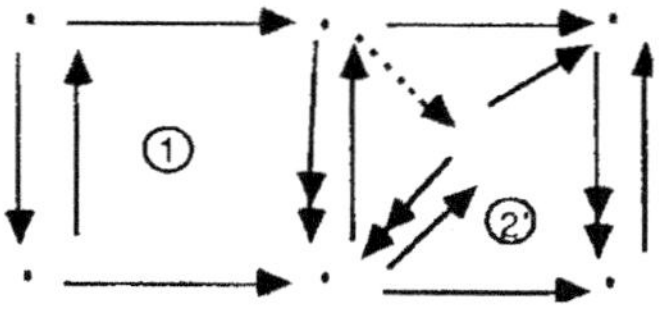

it produces the dotted arrow, which, by the protomodularity condition is invertible. So ② is a pullback. ■

In fact we have a slightly more general result :

Proposition 8. When $\mathbb{E}$ is protomodular, if the following squares ① and ① + ② are pullbacks, then ② is a pullback :

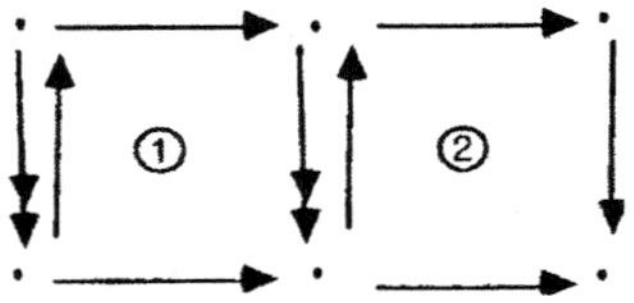

(no splitting condition for the right hand map).

Proof. Take the pullback ②' :

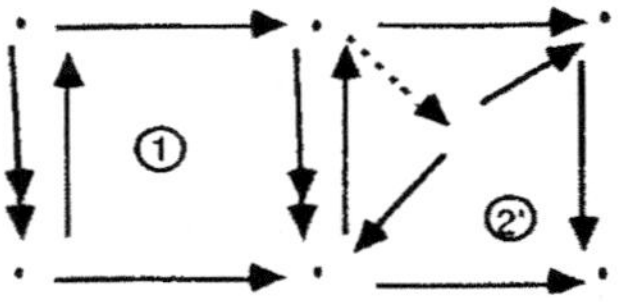

it produces the dotted arrow which extends the splitting from the first edge to the second edge of the middle triangle. Then, apply protomodularity. ■

The main interest of the protomodularity condition is to allow the recovery of some aspects of the exactness of the category Grp of groups.

1) The characterization of the monos.

Proposition 9. The pullbacks reflect monomorphisms.

Proof. Let us suppose that the following square is a pullback with f' a mono :

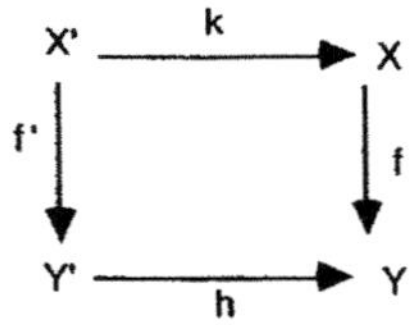

In the left hand diagram, all the squares are pullbacks :

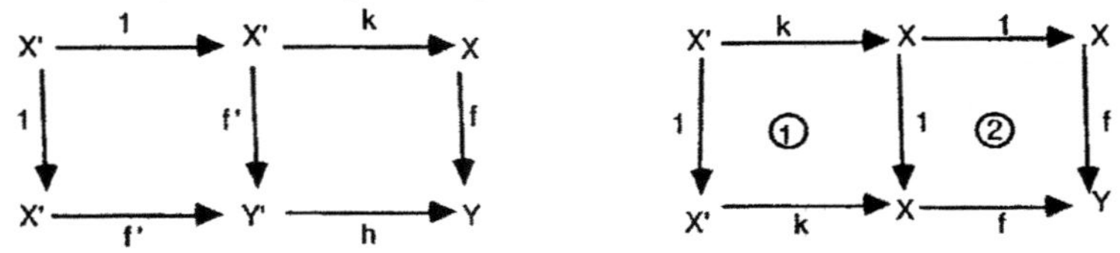

So the square ① + ② is a pullback. Now, ① is a split pullback and following proposition 8, ② is a pullback and f a mono. ■

Corollary 1. A morphism $f : X \to Y$ is a mono, if and only if there is a morphism $h : Y' \to Y$ such that the pullback of f along h is an isomorphism.

Proof. If f is a mono, take $h = f$. Conversely apply proposition 9. ■

Corollary 2. If $\mathbb{E}$ has an initial object, f is a mono if and only if $\ker f = 0$.

2) The internal categories are internal groupoïds.

Proposition 10. When $\mathbb{E}$ is protomodular the internal categories are internal groupoïds.

Proof. Let X_1 be an internal category :

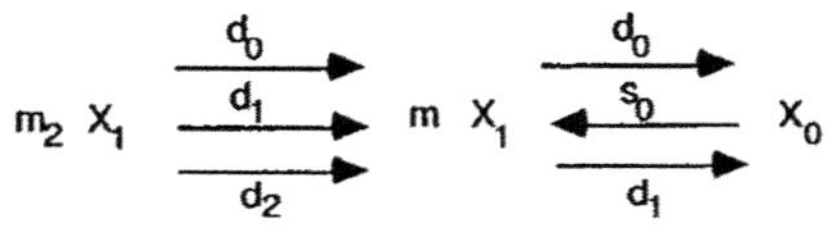

1) In the following diagram, the squares ① + ② and ② are pullbacks so ① is a pullback

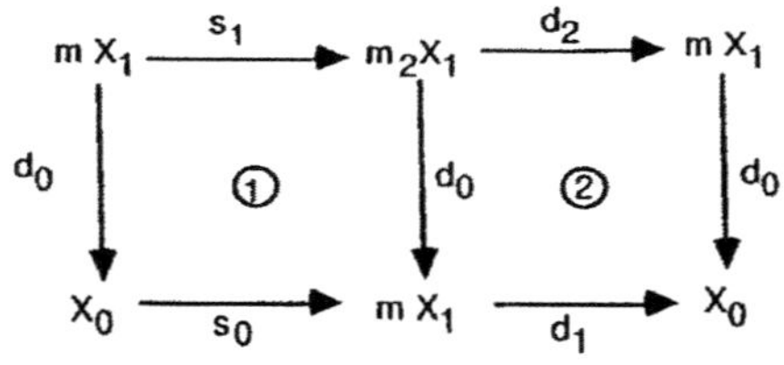

2) the following squares ① and ① + ② are pullbacks, then ② is a pullback and X_1 is a groupoïd (see [3], th 2, corollary 1)

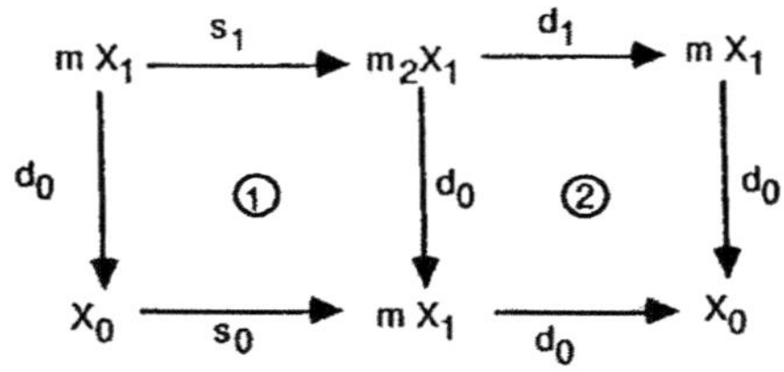

■

Remark. The normalization functor N is also conservative, the kernel functor being conservative by definition.

3) The algebra of exact sequences.

Proposition 11. When $\mathbb{E}$ is left exact and protomodular, then when the following diagram is a pullback, the pair (k, s) is jointly epic :

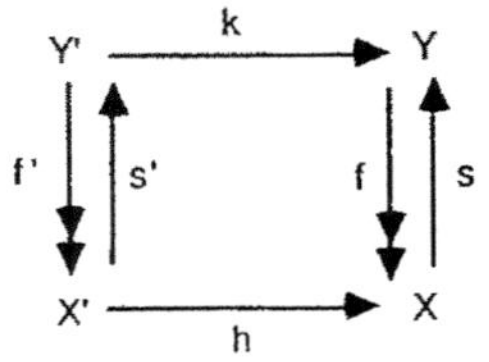

Proof. Let $t, t' : Y \rightrightarrows Z$ be two maps such that $t \cdot k = t' \cdot k$, $t \cdot s = t' \cdot s$ $(= \sigma)$. Then we have the following diagram of split epic :

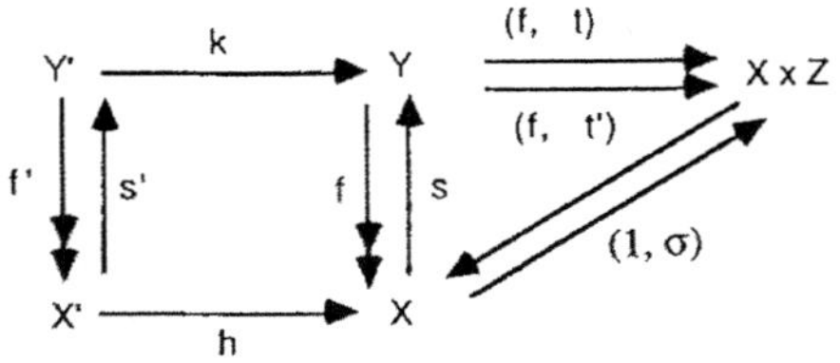

Now $t\,k = t'\,k$ implies that $h^*(f, t) = h^*(f, t') : Y' \to X' \times Z$. But h^* is conservative and preserves kernels, so h^* is faithful. So (f, t) is equal to (f, t') and t equal to t'. ■

Proposition 12. When $\mathbb{E}$ is left exact and protomodular, if the following square is a pullback, it is also a pushout :

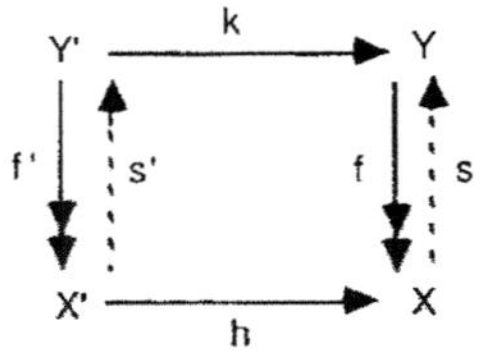

Proof. Let $m : Y \to Z$ and $l : X' \to Z$ such that $m\,k = l \cdot f'$.

1) Then $m \cdot s \cdot f = m$. Indeed.

$m \cdot s \cdot f \cdot k = m \cdot s \cdot h \cdot f' = m \cdot k \cdot s' \cdot f' = l' \cdot f' \cdot s' \cdot f' = l \cdot f' = m \cdot k$

$m \cdot s \cdot f \cdot s = m \cdot s$

2) So $m \cdot s$ is the wanted factorization :

$m \cdot s \cdot f = m$

$m \cdot s \cdot h = m \cdot k \cdot s' = l \cdot f' \cdot s' = l$.

The uniqueness is a consequence of the fact that f is epic. ■

Corollary. If $\mathbb{E}$ has an initial object and f is a split epi then the sequence

$0 \longrightarrow \ker f \longrightarrow Y \overset{f}{\twoheadrightarrow} X \longrightarrow 0$ is exact i.e. the following square is biexact :

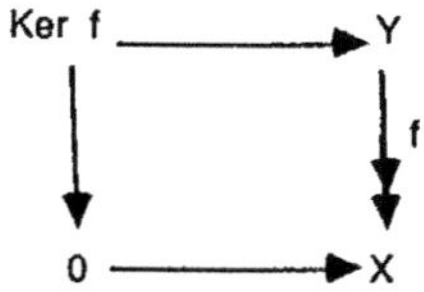

When, moreover, $\mathbb{E}$ is supposed to be Barr exact [1], that is such that every internal equivalence relation has a universal quotient, then we can extend the previous result to any regular epimorphism, through the two following steps :

Proposition 13. When $\mathbb{E}$ is left exact, Barr-exact and protomodular, then if the following squares ① and ① + ② of regular epimorphisms are pullbacks, then ② is a pullback :

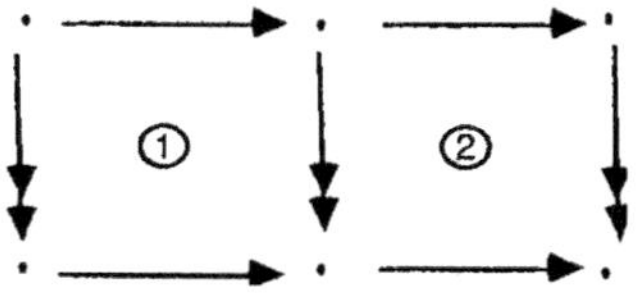

Proof. Take the associated equivalence relations (kernel pairs)

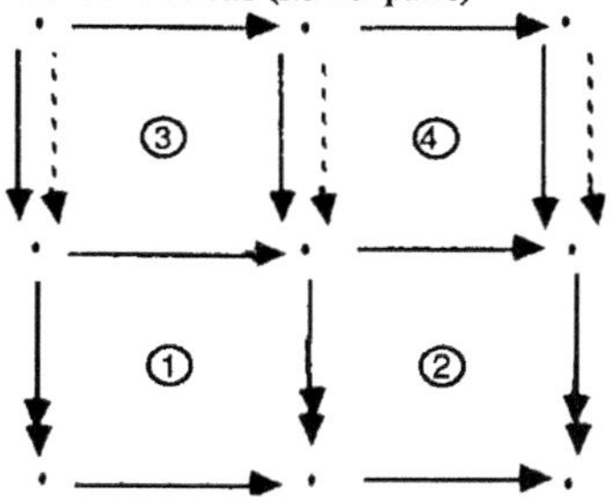

Then ① being a pullback, ③ is a pullback. For the same reason ③ + ④ is a pullback. So ③ and ③ + ④ are split pullbacks, then ④ is a pullback and, following [1] p. 73 , ② is a pullback. ■

Proposition 14. If the following square is a pullback of regular epimorphisms, then it is also a pushout :

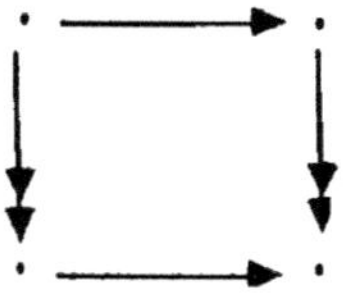

Proof. Extend the diagram by the kernel pairs :

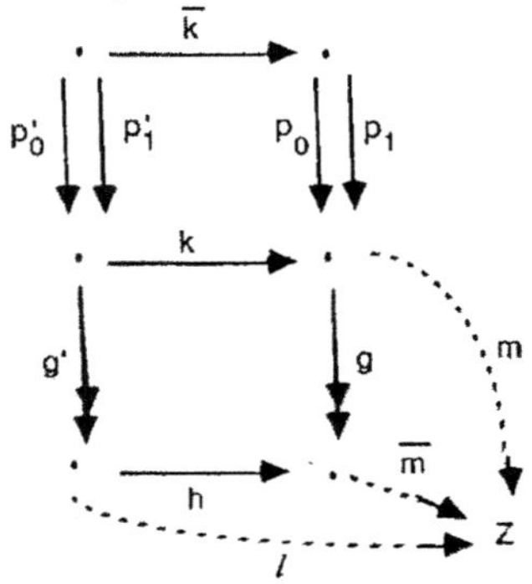

If m and l are such that $m \cdot k = l \cdot g'$, then $m \cdot p_0 = m \cdot p_1$ since (the pair $(\bar{k}, \delta)$ being jointly epic) :

1) $m \cdot p_0 \cdot \bar{k} = m \cdot k \cdot p'_0 = l \cdot g' \cdot p'_0 = l \cdot g' \cdot p'_1 = m \cdot k \cdot p'_1 = m \cdot p_1 \cdot \bar{k}$

2) $m \cdot p_0 \cdot \delta = m = m \cdot p_1 \cdot \delta$

whence a map $\bar{m}$: such that $\bar{m} \cdot g = m$. Moreover $\bar{m} \cdot h = l$ if and only if $\bar{m} \cdot h \cdot g' = l \cdot g'$.
But $\bar{m} \cdot h \cdot g' = \bar{m} \cdot g \cdot k = m \cdot k = l \cdot q'$. ■

Corollary. If $\mathbb{E}$ has an initial object and f is a regular epi, then the sequence
$0 \longrightarrow \ker f \longrightarrow Y \overset{f}{\twoheadrightarrow} X \longrightarrow 0$ is exact.

1] M. Barr — Exact categories, in : M. Barr, P.A. Grillet, D.H. van Osdol : Exact categories and categories of sheaves. Springer L. N. in Math. 236 (1971) 1-120.

2] J. Benabou — Introduction to bicategories . Rep. Midw. Cat. Sem. I. Springer L. N. in Math. 47 (1967).

3] D. Bourn — The shift functor and the comprehensive factorization for internal groupoïds. Cah. Top. Géom. Diff. 28 (3) (1987) 197-226.

4] A. Carboni — Categories of affine spaces. J. Pure Appl. Algebra 61 (1989) 243-250.

5] A. Dold - D. Puppe — Homologie nicht additiver Funktoren. Ann. Inst. Fourier 11 (1961) 201-212.

6] P.T. Johnstone — Affine categories and naturally Mal'cev categories. J. Pure Appl. Algebra 61 (1989), 251-256.

7] D.H. Kan — Functor involving C.S.S. Complexes. Trans. A.M.S. 87 (1958), 330-346.

8] S. Mac Lane — Categories for the working mathematician. Springer Berlin (1971).

This paper is in final form and will not be published elsewhere.

COMPUTING QUOTIENTS OF ACTIONS OF A FREE CATEGORY

S. Carmody*
R. F. C. Walters*
Department of Pure Mathematics,
University of Sydney
Sydney, NSW.

In [1] we have described a procedure for computing left Kan extensions (Chapter 10 of [6]) of functors into **Set** which generalizes the traditional Todd-Coxeter procedure (described in [2],[4],[5]) for coset enumeration. In this paper we describe and prove the correctness of a key sub-procedure – the Quotient Procedure – which computes quotients of actions of a free category. The quotient procedure, being highly recursive, is the most subtle part of the left Kan extension procedure. In fact published proofs (as in [3],[7]) of the Todd-Coxeter procedure appear to lack a proof of the correctness of the quotient procedure for this special case.

A further subtlety arises from the fact that infinite mathematical structures must, for computational purposes, be handled in terms of finite presentations. More specifically, we consider a finite directed graph G and $\mathcal{F}G$, the free category on G, and define the notion of a partial description of a functor $\mathcal{F}G \to \mathbf{Set}$, which we call a *presentation*. Given a presentation we describe its *completion* to a functor $\overline{P}\colon \mathcal{F}G \to \mathbf{Set}$. We then consider a finite system of *equations* S between the elements of a presentation P and develop the notion of the *quotient* P/S which is the natural one to take so that $\overline{P}/S$ corresponds to the usual quotient of the action by the congruence generated by S. The key result justifying computing with finite presentations in this paper, given in THEOREM 1, is that $\overline{P}/S$ is isomorphic to $\overline{P/S}$. The typical situation then is that P is finite, whilst $\overline{P}$ is infinite, but in the procedure to compute a quotient of $\overline{P}$ we need only work with a quotient of P and hence a finite amount of data.

The ideas developed in this paper are very naturally motivated through graphical interpretations and to this end we include several diagrams by way of illustrative example.

In sections 1 and 2 we treat presentations of functors and quotients of presentations. The quotient procedure itself is described in section 3 and analysed in section 4. We finish in section 5 with a sketch of the application of this procedure to the calculation of left Kan extensions.

§1 Presentations of Functors

We first make precise the idea of a *presentation of a functor*, which we shall often refer to simply as a *presentation*.

---

* The authors gratefully acknowledge the support of the Australian Research Council.

Definition 1.1 **A presentation of a functor**, P, on a graph G consists of

$$\text{sets} \quad PA \qquad \text{for all } A \in G$$

$$\text{and partial functions } Pg \subseteq PA \times PB \quad \text{for all } g: A \to B \text{ in } G.$$

Notation: If $(x, y) \in Pg$ we write $x \xrightarrow{g} y$ in P.

Given a presentation P on a graph G, if $x \xrightarrow{g} y$ in P for some y, we say $Pg(x)$ is **defined** and that $Pg(x) = y$; otherwise it is **undefined**.

The following diagram gives an example of a graph G (later illustrations will be based on the same graph) and a presentation P on G:

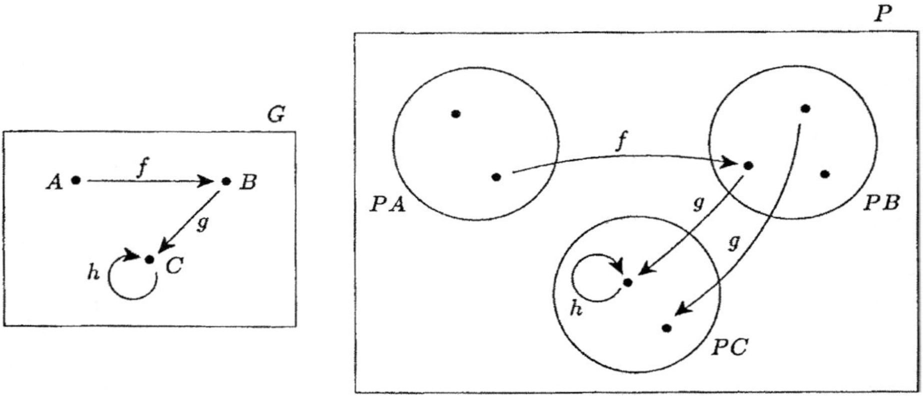

Definition 1.2 A presentation P on a graph G is called **complete** if for each $g: A \to B$ in G, the relations Pg are in fact *functions*.

Note: In this case P can can be considered as a functor $\mathcal{F}G \to \mathbf{Set}$ and in speaking of isomorphisms of complete presentations, we will mean isomorphisms of the corresponding functors.

Remark In the material that follows, given a presentation P, we will frequently be considering families of equivalence relations $\sim_A$ on PA indexed by $A \in G$. By a convenient abuse of terminology, we shall sometimes refer to $\sim$ as an equivalence relation on P, and by an abuse of notation, we shall often drop the subscript from $\sim_A$.

We now turn to the question of obtaining a functor (complete presentation) from any given presentation P. We take two approaches, which turn out to be equivalent. The first, $\overline{P}$, has an obvious connection with the coend formula for left Kan extensions, which will be discussed further in §5, while the second, $P + P^\infty$, suggests itself very naturally from the diagrammatic representations of presentations, and is also convenient to work with as it does not involve a quotient by an equivalence relation.

DEFINITION 1.3 Given a presentation P on a graph G, we define a *complete* presentation $\overline{P}$ — the **completion** of P — as follows:
for each $A \in G$, define

$$\overline{P}A = \Big(\sum_{B \in \mathcal{F}G} \mathcal{F}G(B, A) \times PB\Big)\Big/\sim_A$$

where $\sim$ is the smallest equivalence relation such that:

- for $g: A \to B$ in G, f, $f': A \to B$ in $\mathcal{F}G$, $x, x' \in PA$, $y \in PB$,
  (i) $x \xrightarrow{g} y$ in $P \Rightarrow (g, x) \sim (1_B, y)$
  (ii) $(f, x) \sim (f', x') \Rightarrow (hf, x) \sim (hf', x')$ whenever $\operatorname{dom} h = B$ in G.

We denote the equivalence class of (f, x) by $[f, x]$.

- for each $g: A \to B$ in G, define

$$\overline{P}g : \overline{P}A \to \overline{P}B : [h, x] \longmapsto [gh, x]$$

which is well-defined by (ii).

DEFINITION 1.4 Given a presentation P on a graph G, we define a complete presentation $P + P^\infty$ as follows:

- for each $A \in G$,

$$P^\infty A = \left\{(g_n \ldots g_1, x) \in \sum_{B \in \mathcal{F}G} \mathcal{F}G(B, A) \times PB \;\middle|\; g_i \text{ in } G,\ Pg_1(x) \text{ undefined}\right\}$$

We denote $(g_n \ldots g_1, x)$, with $Pg_1(x)$ not defined, by $g_n \ldots g_1 x$. Now define

$$(P + P^\infty)A = PA + P^\infty A.$$

- for each $h: A \to B$ in G, define

$$(P + P^\infty)h : (P + P^\infty)A \to (P + P^\infty)B$$

by

$$x \longmapsto \begin{cases} Ph(x) & \text{if defined} \\ hx & \text{otherwise} \end{cases}$$

$$g_n \ldots g_1 x \longmapsto hg_n \ldots g_1 x.$$

The following proposition makes the connection between these two completions:

PROPOSITION 1.1 Given a presentation P on a graph G,

$$\overline{P} \cong P + P^{\infty}$$

Proof For each $A \in G$, define $\phi_A: (P + P^{\infty})A \to \overline{P}A$ by

$$x \longmapsto [1_A, x]$$
$$g_n \dots g_1 x \longmapsto [g_n \dots g_1, x].$$

Now by (1.3) (i) and (ii) we can express any element y of $\overline{P}A$ in the form $[g_n \dots g_1, x]$ where either $Pg_1(x)$ is not defined or $g_n \dots g_1 = 1_A$. In the first case, $y = \phi_A([g_n \dots g_1 x])$ and in the second, $y = \phi_A(x)$. Thus ϕ_A is surjective.

To show that ϕ_A is injective, note that two elements u, $v \in \overline{P}A$ are equivalent iff there is a chain

$$u = u_1 \sim u_2 \sim \dots \sim u_n = v$$

where each step of the chain is of the form

$$u_i = (g_n \dots g_1 g, x) \sim (g_n \dots g_1, y) = u_{i+1}$$

where $x \xrightarrow{g} y$ in P, or the reverse of this step.

We may assume that throughout such a chain, the length of the path in the first component of each term is either strictly increasing or strictly decreasing, since the only possible change from decreasing to increasing occurs as

$$(g_n \dots g_1 g, x) \sim (g_n \dots g_1, y) \sim (g_n \dots g_1 g, x)$$

where $x \xrightarrow{g} y$ in P since Pg is a partial function. Such a pair of steps may be removed from the chain. The same applies to changes from increasing to decreasing.

Thus $(1, x) \sim (1, y)$ implies $x = y$ and also if $(g_n \dots g_1, x) \sim (h_m \dots h_1, y)$ with $Pg_1(x)$ undefined then $n \leq m$. Thus if $Pg_1(x)$ and $Ph_1(y)$ are both undefined, it clearly follows that $m = n$ and $g_i = h_i$ for $i = 1, \dots, n$ and $x = y$. Hence the ϕ_A are injective and thus isomorphisms.

We now check naturality of ϕ.

For each $h: A \to B$ in G,

$$\begin{aligned} \overline{P}h\phi_A(x) &= \overline{P}h[1_A, x] \\ &= [h, x], \end{aligned}$$

while

$$\begin{aligned} \phi_B(P + P^{\infty})h(x) &= \begin{cases} \phi_B Ph(x) & \text{if } Ph(x) \text{ defined} \\ \phi_B(hx) & \text{otherwise} \end{cases} \\ &= \begin{cases} [1_A, Ph(x)] = [h, x] & \text{if } Ph(x) \text{ defined} \\ [h, x] & \text{otherwise} \end{cases} \end{aligned}$$

and

$$\begin{aligned} \overline{P}h\phi_A(g_n \dots g_1 x) &= \overline{P}h[g_n \dots g_1, x] \\ &= [hg_n \dots g_1, x] \end{aligned}$$

while

$$\phi_B(P+P^\infty)h(g_n\dots g_1x) = \phi_B hg_n\dots g_1x$$
$$= [hg_n\dots g_1, x].$$

□

We can now illustrate $P+P^\infty$ and hence $\overline{P}$ with an example. We think of P^∞ as adding necessary extra elements to P so as to ensure it is complete. The dotted lines indicate the original P, and the points outside P constitute P^∞.

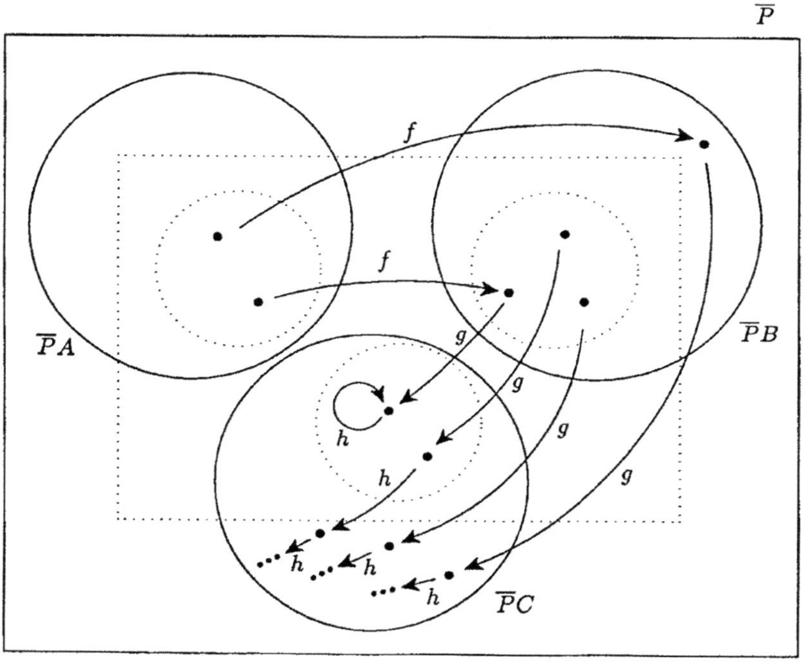

The first simple property of the completion process is the following:

Proposition 1.2 Given a presentation P on a graph G,

$$P \text{ complete } \Rightarrow \overline{P} \cong P.$$

Proof

$$P \text{ complete} \Rightarrow P^\infty A = \emptyset \qquad \text{for all } A \in G$$
$$\Rightarrow P + P^\infty = P$$

□

§2 Quotients of Presentations

DEFINITION 2.1 Given a presentation P on a graph G and a family of equivalence relations $\sim_A$ on PA for each $A \in G$, we say $\sim$ is **closed** under P if for all g in G and x, $x' \in PA$,

$$x \sim x' \Rightarrow Pg(x) \sim Pg(x') \quad \text{when both } Pg(x) \text{ and } Pg(x') \text{ are defined.}$$

DEFINITION 2.2 A presentation with **equations** consists of a presentation P on a graph G and for each $A \in G$ sets $SA \subseteq (PA)^2$. We refer to these sets collectively as the *equations* S.

Notation: If $(x, x') \in SA$, we write x — x' in S.

DEFINITION 2.3 Given a presentation P on a graph G with equations S, we define the **quotient presentation** P/S as follows:

• for all $A \in G$, define

$$(P/S)A = PA/\sim_A$$

where $\sim$ is the smallest equivalence relations on P such that

$$x \text{ — } x' \Rightarrow x \sim x'$$

and $\sim$ is closed under P.

• for all $g: A \to B$ in G, define

$$(P/S)g = \{([x], [y]) \mid x \xrightarrow{g} y \text{ in } P\}$$

where [ ] denotes equivalence class under $\sim$.

We call $\sim$ the **quotient equivalence relation** induced by S on P.

We illustrate the quotient process diagrammatically. The equations S are indicated by a solid line, other induced equivalences by a dotted line.

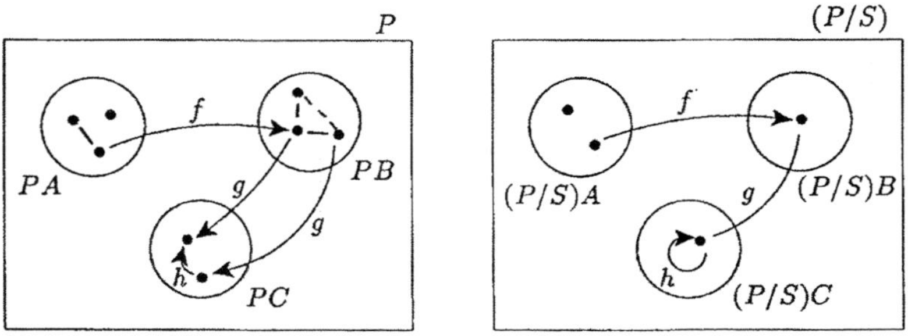

PROPOSITION 2.1 Given a presentation P on a graph G and equations S, if P is complete then (P/S) is complete.

Proof Since the quotient equivalence relation $\sim$ is closed under P,

$$x \sim x' \Rightarrow Pg(x) \sim Pg(x')$$

Hence for each $g: A \to B$ in G,

$$(P/S)g : (P/S)A \to (P/S)B : [x] \longmapsto [Pg(x)]$$

is a well-defined *function*. □

Our first main result justifies computing with finite presentations.

THEOREM 1 Given a presentation P on a graph G and equations S,

$$\overline{(P/S)} \cong (\overline{P}/S).$$

Proof We first construct a closed equivalence relation on $\overline{P}$.

Let [ ] denote equivalence class under the quotient equivalence relation induced on P by S. We now describe a subdivision of P^∞:

- for each $A \in G$, let

$$QA = \{g_n \dots g_1 x \in P^\infty A \mid (P/S)g_n \dots (P/S)g_1[x] \text{ defined}\}$$

$$Q^\infty A = \{g_n \dots g_1 x \in P^\infty A \mid (P/S)g_n \dots (P/S)g_1[x] \textit{ not} \text{ defined}\}.$$

We now partition the sets $\overline{P}A$ as follows:

- for each $A \in G$, $y \in PA$, let

$$[y]^e = [y] \cup \{g_n \dots g_1 x \in QA \mid (P/S)g_n \dots (P/S)g_1[x] = [y]\}$$

which partitions $PA + QA$ and now we partition $Q^\infty A$:
Given $g_n \dots g_1 x \in Q^\infty A$, if

$$[y] = (P/S)g_i \dots (P/S)g_1[x] \text{ is defined,}$$

but

$$(P/S)g_{i+1}[y] \text{ is } \textit{not} \text{ defined}$$

we place $g_n \dots g_1 x$ in a set denoted

$$g_n \dots g_{i+1}[y]^e.$$

This clearly partitions each set $\overline{P}A$. We let $\approx$ denote the equivalence relation corresponding to this partition. We claim this equivalence relation is closed under $\overline{P}$. First we consider the effect of $\overline{P}g$ on elements of $[y]^e \subseteq \overline{P}A$ where $g: A \to B$. This breaks into two cases:

(i) $(P/S)g[y]$ defined, then if $z \in [y]$

(a) if $Pg(z)$ defined then

$$\overline{P}g(z) = Pg(z) \in (P/S)g[y] \subseteq ((P/S)g[y])^e$$

(b) if $Pg(z)$ not defined then $\overline{P}g(z) \in QA$ and

$$\overline{P}g(z) \in ((P/S)g[y])^e,$$

whereas if $z \in [y]^e \setminus [y]$, then clearly

$$\overline{P}g(z) \in ((P/S)g[y])^e.$$

(ii) $(P/S)g[y]$ *not* defined, then clearly for any $z \in [y]^e$,

$$\overline{P}g(z) \in g[y]^e.$$

Now considering $g_n \ldots g_1 x \in Q^\infty A$ in $g_n \ldots g_{i+1}[y]^e$, clearly

$$\overline{P}g(g_n \ldots g_1 x) = gg_n \ldots g_1 x \in gg_n \ldots g_{i+1}[y]^e.$$

Thus $\approx$ is indeed closed under $\overline{P}$. It is clear elements equivalent under $\approx$ are also equivalent under the closed equivalence relation induced by S on $\overline{P}$ and hence

$$(\overline{P}/\approx) \cong (\overline{P}/S).$$

There is, however an obvious isomorphism

$$\phi_A : (\overline{P}A/\approx) \to \overline{(P/S)}A$$

given by

$$g_n \ldots g_1[y]^e \longmapsto g_n \ldots g_1[y]$$

which is clearly natural. Thus

$$\overline{(P/S)} \cong (\overline{P}/S).$$

□

§3 The Quotient Procedure

We now describe a procedure to calculate quotients of the form $\overline{P}/S$. Given a presentation P on a graph G and equations S, the procedure specifies a way to modify both P and S which preserves $\overline{P/S}$ and hence $\overline{P}/S$ by THEOREM 1. We begin with an equation $u — v$ in S with $u \neq v$ and then the modification is carried out in two steps, producing P_1, S_1 and then P_2, S_2. The procedure terminates when every equation has the form $u — u$. When the graph G and the sets PA are finite, the procedure terminates after a finite number of steps. If we commence the procedure with P and S, terminating with Q and T, then clearly $\overline{Q/T} \cong \overline{Q}$. Further, since the procedure preserves $\overline{P/S}$ (proved in §4), it follows that $\overline{P/S} \cong \overline{Q/T}$. Therefore $\overline{P}/S \cong \overline{Q}$ and the desired $\overline{P}/S$ has effectively been calculated. We leave the analysis to section 4, and now simply describe the procedure.

The Procedure

Begin with presentation P, equations S with u — $v \in PC$ and $u \neq v$.

Step 1

For all $A \in G$ and g in G, define

$$P_1A = PA$$
$$P_1g(x) = (Pg(x))^\circ \quad \text{if } Pg(x) \text{ defined}$$

where

$$(\ )^\circ : PA \to P_1A : a \longmapsto \begin{cases} a & \text{if } a \neq v \\ u & \text{if } a = v. \end{cases}$$

Also

$$S_1A = \begin{cases} S^\circ C \cup TC \cup \{(u,v)\} & \text{if } A = C \\ S^\circ A \cup TA & \text{otherwise} \end{cases}$$

where

$$S^\circ A = \{(a^\circ, b^\circ) \mid (a,b) \in SA\}$$

and

$$TA = \{(Pg(u)^\circ, Pg(v)^\circ) \mid \text{if both defined for } g: C \to A \text{ in } G\}.$$

Step 2

For all $A \in G$ and g in G, define

$$P_2A = \begin{cases} P_1C \setminus \{v\} & \text{if } A = C \\ P_1A & \text{otherwise} \end{cases}$$
$$P_2g = P_1g|_{P_2A}$$

and

$$S_2A = \begin{cases} S_1C \setminus \{(u,v)\} & \text{if } A = C \\ S_1A & \text{otherwise.} \end{cases}$$

We can illustrate this modification process graphically. If we begin with the presentation P and equations S as:

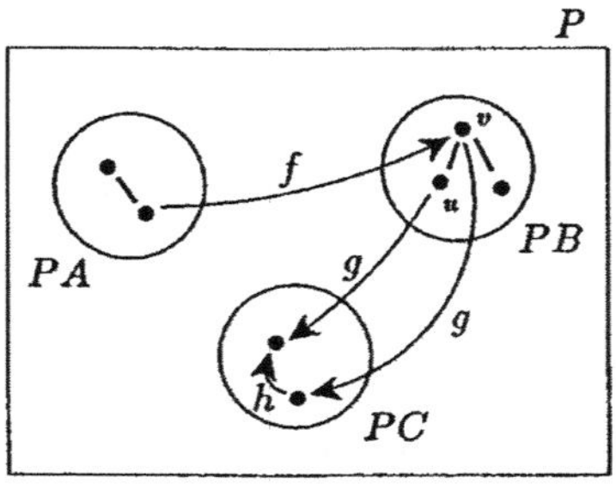

Then the modified presentations P_1 and P_2 can then be illustrated as:

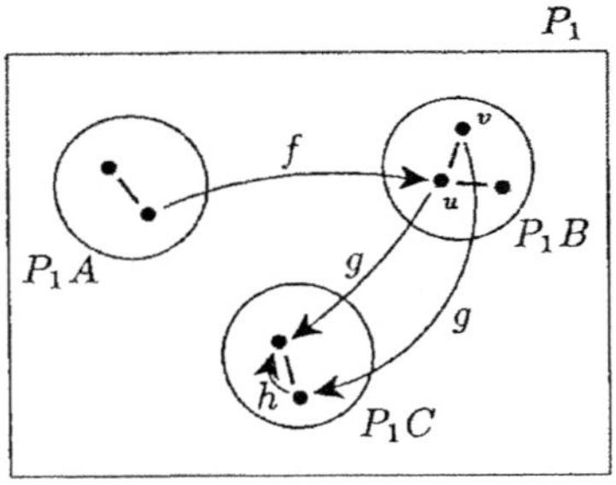

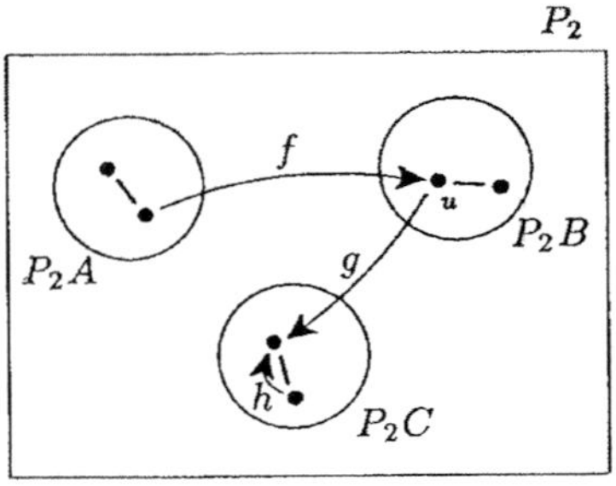

§4 Analysis of the Procedure

The following two Lemmas show that both steps in the procedure preserve the quotient (P/S) at least in the case when P (and hence of course (P/S) by PROPOSITION 2.1) is complete.

LEMMA 4.1 Let P be a complete presentation on a graph G, with equations S. Given $u — v$ in S where $u,\ v \in PC$, with $u \neq v$, if the complete presentation P' and equations S' are defined as in **Step 1** of the quotient procedure then

$$(P/S) \cong (P'/S').$$

Proof Let $\sim_A$ denote the quotient equivalence relation E_{SA} induced by S on PA and $\sim'_A$ the quotient equivalence relation $E_{S'A}$ induced by S' on $P'A$ for each $A \in G$. Note that $u — v$ in $S \Rightarrow$

$$\begin{aligned} \text{for all } a \quad & a^\circ \sim a \\ \text{and} \quad & a^\circ \sim' a. \end{aligned} \qquad (1)$$

Hence, using transitivity,

$$\begin{aligned} a \sim b &\iff a^\circ \sim b^\circ \\ a \sim' b &\iff a^\circ \sim' b^\circ. \end{aligned} \qquad (2)$$

Now

$$\begin{aligned} a — b \text{ in } S &\Rightarrow a \sim b \\ &\Rightarrow a^\circ \sim b^\circ. \end{aligned}$$

Hence $S^\circ A \subseteq E_{SA}$.
Further, for $a,\ b \in PA$,

$$\begin{aligned} a \sim b &\Rightarrow Pg(a) \sim Pg(b) && \text{whenever } \operatorname{dom} g = A \text{ in } G \\ &\Rightarrow Pg(a)^\circ \sim Pg(b)^\circ && \text{by (2)} \\ &\Rightarrow P'g(a) \sim P'g(b). \end{aligned}$$

Thus $\sim$ is closed under P' and in particular, $P'g(u) \sim P'g(v)$ so $TA \subseteq E_{SA}$ for all $A \in G$.
Of course $(u,v) \in E_{SC}$.
Thus $S'A \subseteq E_{SA}$ for all $A \in G$.
Since $\sim$ is closed under P', but $\sim'$ is the smallest such equivalence relation,

$$E_{S'A} \subseteq E_{SA} \quad \text{for all } A \in G$$

Conversely,

$$\begin{aligned} a \text{ — } b \text{ in } S &\Rightarrow a^\circ \text{ — } b^\circ \text{ in } S^\circ \\ &\Rightarrow a^\circ \text{ — } b^\circ \text{ in } S' \\ &\Rightarrow a^\circ \sim' b^\circ \\ &\Rightarrow a \sim' b \qquad \text{by (2)} \end{aligned}$$

Hence $SA \subseteq E_{S'A}$.
Further, for $a, b \in P'A$

$$\begin{aligned} a \sim' b &\Rightarrow P'g(a) \sim' P'g(b) \qquad \text{whenever } \operatorname{dom} g = A \text{ in } G \\ &\Rightarrow Pg(a)^\circ \sim' Pg(b)^\circ \\ &\Rightarrow Pg(a) \sim' Pg(b) \qquad \text{by (2)} \end{aligned}$$

so $\sim'$ is closed under P, thus $E_{SA} \subseteq E_{S'A}$ for all $A \in G$.
Hence

$$E_{SA} = E_{S'A} \quad \text{for all } A \in G.$$

Thus we have isomorphisms

$$\phi_A : (P/S)A \longrightarrow (P'/S')A : [x] \longmapsto [x].$$

To ensure naturality of ϕ we require

$$(P/S)g[x] = (P'/S')g[x] \qquad \text{for all } g \text{ in } G$$

But

$$\begin{aligned} (P/S)g[x] &= [Pg(x)] \\ &= [(Pg(x))^\circ] \\ &= [P'g(x)] \\ &= (P'/S')g[x] \end{aligned}$$

□

DEFINITION 4.1 Let P be a complete presentation on a graph G, with equations S such that u — v in S, where $u, v \in PC$ say, and $u \neq v$. We say that v is **removable** via u if

(i) for all g in G with domain C,

$$Pg(u) \text{ — } Pg(v) \text{ in S}$$

(ii) for all h in G with codomain C,

$$v \notin \operatorname{im} Ph$$

(iii) for all $x — y$ in S with $(x, y) \neq (u, v)$,

$$x, y \neq v.$$

LEMMA 4.2 Given a presentation P on a graph G and coincidences S with v removable via u with u, $v \in PC$, if we define a complete presentation P' and coincidences S' as follows:
for all $A \in G$,

$$P'A = \begin{cases} PC \setminus \{v\} & \text{if } A = C \\ PA & \text{otherwise} \end{cases}$$

for all g in G,

$$P'g = Pg|_{P'A}$$

and

$$S'A = \begin{cases} SC \setminus \{(u, v)\} & \text{if } A = C \\ SA & \text{otherwise} \end{cases}$$

then

$$(P/S) \cong (P'/S')$$

Proof Note that hypothesis (iii) ensures that $S'A \subseteq (P'A)^2$ for all A in P'.

Let $\sim_A$ denote the quotient equivalence relation E_{SA} induced by S on P and $\sim'_A$ the quotient equivalence relation $E_{S'A}$ induced by S' on P' for each $A \in G$.

We define operators e (extension) and c (contraction) between equivalence relations on P and equivalence relations on P' as follows:
Given equivalence relations E_A on PA and $E'A$ on $P'A$ for each $A \in G$,

$$E_A^c = E_A \setminus \{(a, b) \mid a \text{ or } b = v\}$$

$$E'^e_A = E'_A \cup \{(v, a), (a, v) \mid (a, u) \in E'_A\}$$

Note that E_A^c and E'^e_A are clearly themselves equivalence relations, and that e and c both preserve inclusions.
Note that $E_A^{ce} = E_A$ and $E'^{ec}_A = E'_A$, hence in particular

$$E_{SA}^{ce} = E_{SA}$$
$$E_{S'A}^{ec} = E_{S'A}$$

Now $S'A \subseteq SA$ so $S'A \subseteq E_{SA}^c$.

Also, for all $g: A \to B$ in G,

$$\begin{aligned}(a,b) \in E^c_{SA} &\Rightarrow (a,b) \in E_{SA}\\ &\Rightarrow a \sim b\\ &\Rightarrow Pg(a) \sim Pg(b)\\ &\Rightarrow P'g(a) \sim P'g(b)\\ &\Rightarrow (P'g(a), P'g(b)) \in E^c_{SB} \quad \text{by property (ii).}\end{aligned}$$

Thus E^c_{SA} is closed under P' for each $A \in G$, hence

$$E_{S'A} \subseteq E^c_{SA} \text{ and so } E^e_{S'A} \subseteq E^{ce}_{SA} = E_{SA}$$

We now establish the reverse inclusions:
Firstly, $SA \subseteq E^e_{S'A}$ since $S'A \subseteq E^e_{S'A}$ and e adds (u,v) to $E_{S'C}$, as $(u,u) \in E_{S'C}$.
Now consider $(a,b) \in E^e_{S'A}$ and $g: A \to B$ in G:

If $a,b \neq v$ then $(a,b) \in E_{S'A}$
so $(Pg(a), Pg(b)) \in E_{S'B} \subseteq E^e_{S'B}$.

If $a = v$ but $b \neq v$, then by definition of e, we have $b \sim' u$.
and,

$$\begin{aligned}b \sim' u &\Rightarrow P'g(b) \sim' P'g(u)\\ &\Rightarrow Pg(b) \sim' Pg(u) \quad \text{by definition of } P'g.\end{aligned}$$

Also, it follows from (i) and (ii) that $Pg(u) — Pg(v)$ is in S' and hence

$$\begin{aligned}Pg(u) \sim' Pg(v) = Pg(a) &\Rightarrow Pg(a) \sim' Pg(b)\\ &\Rightarrow (Pg(a), Pg(b)) \in E^e_{S'B}.\end{aligned}$$

If $a = b = v$ then trivially $(Pg(a), Pg(b)) \in E^e_{S'B}$.
Thus $E^e_{S'}$ is closed under P, which means for all $A \in G$,

$$E_{SA} \subseteq E^e_{S'A} \text{ and so } E^c_{SA} \subseteq E^{ec}_{S'A} = E_{S'A}$$

Hence

$$E_{SA} = E^e_{S'A} \text{ and } E_{S'A} = E^c_{SA}$$

This allows us to consider the well-defined isomorphisms

$$\phi_A: (P'/S')A \to (P/S)A: [x]' \longmapsto [x]$$

where [ ] and [ ]$'$ represent equivalence class under $\sim$ and $\sim'$ respectively.

Then ϕ is natural since for all $g: A_1 \longrightarrow A_2$ in G,

$$\begin{aligned}(P/S)g\phi_{A_1}[x]' &= (P/S)g[x] \\ &= [Pg(x)] \\ &= \phi_{A_2}[Pg(x)]' \\ &= \phi_{A_2}[P'g(x)]' \\ &= \phi_{A_2}(P'/S')g[x]'.\end{aligned}$$

□

LEMMA 4.3 Given a complete presentation P and coincidences S on the graph G, and producing a presentation P' and equations S' using the quotient procedure, it follows that

$$(P/S) \cong (P'/S').$$

Proof We adopt the notation of §3. Noting that the definitions of ° and T in the procedure ensure that v is removable via u in P_1 and S_1,

$$\begin{aligned}(P/S) &\cong (P_1/S_1) \quad \text{by LEMMA 4.2} \\ (P_1/S_1) &\cong (P_2/S_2) \quad \text{by LEMMA 4.3.} \\ \text{But} \quad (P_2/S_2) &= (P'/S').\end{aligned}$$

□

LEMMA 4.4 Given a presentation P and equations S on the graph G, if we produce a presentation P' and equations S' by applying the quotient procedure to P and S, and a presentation $\overline{P}'$ and equations $\mathcal{S}$ using the same procedure on the $\overline{P}$ and S, it follows that

$$(\overline{P'}/S') \cong (\overline{P}'/\mathcal{S})$$

Proof In **Step 1**, we produce P_1,S_1 and $(\overline{P})_1$, $\mathcal{S}_1$.
Clearly $\overline{P_1} = (\overline{P})_1$ and for each $A \in G_B$,

$$\mathcal{S}_1 A = S_1 A \cup \{(\overline{P}g(u), \overline{P}g(v)) \mid Pg(u) \text{ or } Pg(v) \text{ undefined, for } \operatorname{cod} g = A\}$$

The pairs in the last set are required to be equivalent to obtain a *closed* equivalence relation since $u — v$, thus

$$(\overline{P})_1/\mathcal{S}_1 \cong (\overline{P})_1/S_1$$

and so

$$(\overline{P})_1/\mathcal{S}_1 \cong \overline{P_1}/S_1.$$

Following **Step 2** we also have

$$\mathcal{S}_2 A = S_2 A \cup \{(\overline{P}g(u), \overline{P}g(v)) \mid Pg(u) \text{ or } Pg(v) \text{ undefined, for } \operatorname{cod} g = A\}$$

and further,

$$(\overline{P})_2 A = \overline{P_2} A \cup \{g_n \dots g_1 v \mid Pg_1(v) \text{ undefined, } \operatorname{cod} g_n = A\}$$

All the terms $g_n \dots g_1 v$ in the last set will, however be made equivalent to $\overline{P}g_n \dots \overline{P}g_1(u)$ by the expression given above for S_2. Hence

$$(\overline{P})_2/S_2 \cong \overline{P_2}/S_2.$$

from which the result follows. □

THEOREM 2 Given a presentation P and equations S on the graph G, and producing a presentation P' and equations S' using the quotient procedure, it follows that

$$\overline{(P/S)} \cong \overline{(P'/S')}.$$

Proof We use the notation of the previous lemma, and

$$\begin{aligned} \overline{(P/S)} &\cong (\overline{P}/S) && \text{by THEOREM 1} \\ &\cong (\overline{P}'/S) && \text{by LEMMA 4.3} \\ &\cong (\overline{P'}/S') && \text{by LEMMA 4.4} \\ &\cong \overline{(P'/S')} && \text{by THEOREM 1.} \end{aligned}$$

□

We conclude the analysis of the procedure by considering the question of termination.

THEOREM 3 The procedure terminates after a finite number of steps.

Proof Each time the two steps of the procedure are carried out, one of the PA decreases in size by one. Since $(PA)_{A \in G}$ is a finite family of finite sets, the procedure must terminate after a finite number of steps. □

§5 Left Kan Extensions

The procedure for computing left Kan extensions described in [1] in the style of the traditional Todd-Coxeter procedure for Coset Enumeration (filling out tables etc.) deals with the following problem: given a graph G_A which generates a category $\mathbf{A}$ and a graph G_B along with equations which determine a category $\mathbf{B}$, and functors $F\colon \mathbf{A} \to \mathbf{B}$ and $X\colon \mathbf{A} \to \mathbf{Set}$ given on generators (i.e. on the graphs), compute the values of the left Kan extension $\mathrm{Lan}_F X$ on the generators of $\mathbf{B}$.

The left Kan extension procedure begins with a presentation on G_B given by

$$PB = \sum_{\substack{A \in \mathbf{A} \text{ s.t.} \\ FA = B}} XA \qquad \text{for all } B \in G_B$$

and

$$Pg = \emptyset \qquad \text{for all } g \text{ in } G_B$$

For this presentation,

$$\overline{P}B = \sum_{A \in \mathbf{A}} \mathcal{F}G_B(FA, B) \times XA$$

Thus it is clear that if S consists of the equations derived from relations between generators of $\mathbf{B}$ and ones of the form

$$(1_{FA_2}, Xf(x)) — (Ff, x) \quad \text{for each } f: A_1 \to A_2$$

then the quotient $\overline{P}/S$ is simply the coend formula for left Kan extensions, so

$$\begin{aligned}\overline{P}/S &\cong \int^{A} \mathbf{B}(FA, \_) \cdot XA \\ &\cong \mathrm{Lan}_F X.\end{aligned}$$

The procedure for calculating $\mathrm{Lan}_F X$ is to calculate $\overline{P}/S$ by progressively forming quotients of $\overline{P}$ by subsets of S using the procedure described in this paper. If done in a sufficiently systematically manner, this procedure will in fact terminate to give $\mathrm{Lan}_F X$ on the generators of $\mathbf{B}$ in the case when the sets $\mathrm{Lan}_F XB$ are finite for each $B \in \mathbf{B}$. This has a simple connection to the traditional Todd-Coxeter procedure for Coset Enumeration. If we have groups $H \leq G$ and consider them as categories $\mathbf{H}$ and $\mathbf{G}$ with one object $*$ with inclusion functor I and have the trivial one-point action $T: \mathbf{H} \to \mathbf{Set}$ then $\mathrm{Lan}_I T(*) \cong G/H$. That is to say, the left Kan extension is simply the action of G on its (left) cosets given by premultiplication.

References

1. S. Carmody, *The Todd-Coxeter Procedure & Left Kan Extensions*, Honours Essay, Pure Mathematics Department, University of Sydney, 1990.
2. S. Carmody and R. F. C. Walters, The Todd-Coxeter Procedure & Left Kan Extensions, Research Report 90-19, Pure Mathematics Dept., University of Sydney (1990).
3. H. S. M. Coxeter and W. O. J. Moser, *Generators and Relations for Discrete Groups*, Springer-Verlag, 1972.
4. L. A. Dimino, A Graphical Approach to Coset Enumeration, *Sigsam Bulletin*, 19 (1971) 8–43.
5. D. L. Johnson, *Presentation of Groups*, Cambridge University Press, 1976.
6. J. Leech, Coset Enumeration from *Computational Group Theory*, Lon. Math. Soc. Symposium on Comp. Group theory, (ed. M. D. Atkinson) Academic Press, 1984.
7. S. MacLane, *Categories for the Working Mathematician*, Springer Verlag, 1971.
8. N. S. Mendelsohn, An Algorithmic Solution for a Word Problem in Group Theory, *Canad. J. Mathematics*, 16 (1964) 509–516.

This paper is in final form and will not be published elsewhere.

A LONG EXACT SEQUENCE IN NON-ABELIAN COHOMOLOGY

Antonio M. Cegarra & Antonio R. Garzón
Department of Algebra, University of Granada
Granada 18071, Spain

The context , $\mathfrak{B}$, in which this paper is developed is a variety of Ω-groups , [19]. This includes the most usual algebraic categories (Groups, Associative algebras,...) in which a comprehensible and well motived non-abelian cohomology theory has been developed in dimensions ≤ 2 using crossed modules as coefficients. All of these theories have a natural interpretation using the homotopy theory of the corresponding simplicial category, [28]: crossed modules are equivalent to internal groupoids and then, by taking nerves, to simplicial objects whose Moore complex is trivial in dimensions >1, and there are natural bijections between $\mathbb{H}^2(X,\Phi)$ and $[F_\bullet, G(\Phi)]$, the set of homotopy classes of simplicial morphisms from a free simplicial resolution of an object X into the simplicial object associated to the crossed module Φ, [14],[6],[7].

In the general context $\mathfrak{B}$,the concept of crossed module has been explicitely given by Ellis, [16], who shows the equivalence between the category of crossed modules in $\mathfrak{B}$, **XM**($\mathfrak{B}$), the category of internal groupoids in $\mathfrak{B}$, **GPD**($\mathfrak{B}$), and the category of simplicial objects in $\mathfrak{B}$ with trivial Moore complex at dimensions >1, which we will denote by **1-HYPGD**($\mathfrak{B}$) and will call the category of 1-hypergroupoids in $\mathfrak{B}$, because their objects are just the internal 1-hypergroupoids in the sense of Duskin-Glenn, [17]. Using the fact that **Simpl**($\mathfrak{B}$), the category of simplicial objects in $\mathfrak{B}$, is a closed simplicial model category, [28], we study the sets $[\mathbb{G}_\bullet(X), B_\bullet]$ of homotopy classes of simplicial morphisms from the standard cotriple resolution of an object X into 1-hypergroupoid $B_\bullet$, and then apply our results to the classical non-abelian cohomology theories.

In fact, we will establish some facts about the sets $[\mathbb{G}_\bullet(X), B_\bullet]$, where $B_\bullet$ is an n-hypergroupoid in $\mathfrak{B}$, [17], which is just a simplicial object in $\mathfrak{B}$ whose Moore complex is trivial at dimensions >n, for an arbitrary $n \geq 1$. The main reason to do it is that for certain n-hypergroupoids $B_\bullet$ these sets will be used to provide an appropriate definition of non abelian $\mathbb{H}^n$ with coefficients in crossed modules in $\mathfrak{B}$. For $n \geq 1$, we will denote **n-HYPGD**($\mathfrak{B}$) the full subcategory of **Simpl**($\mathfrak{B}$) whose objects are the n-hypergroupoids..

Let us note too that the use of n-hypergroupoids in non-abelian cohomology appears in some recents papers [6], [7],[14],[15] ;and also in abelian cohomology since Duskin, [13], showed that the usual monadic cohomology of an object X in $\mathfrak{B}$ with coefficients in an internal X-module A, $H^n(X,A)$, is isomorphic to $[\mathbb{G}_\bullet(X), K(A \rtimes X, n)]_{I_X}$,the subset of $[\mathbb{G}_\bullet(X), K(A \rtimes X, n)]$ whose elements are those homotopy clases of simplicial morphisms $f_\bullet : \mathbb{G}_\bullet(X) \longrightarrow K(A \rtimes X, n)$ such that $\Pi_0(f_\bullet) = I_X$, where $K(A \rtimes X, n)$ is the n-hypergroupoid defined as the Eilenberg-Mac Lane complex associated to the X-module A (or to the internal abelian group in the comma category $\mathfrak{B}/X$

given by the semidirect product $A\rtimes X \xrightarrow{p_r} X)$,

$$K(A\rtimes X,n)=\operatorname{cosk}^{n+1}(\ A^{n+1}\rtimes X \rightrightarrows A\rtimes X \rightrightarrows X \rightrightarrows \ldots X \rightrightarrows X).$$

Moreover, it turns out that abelian cohomology is strongly related to the structure of the sets $[\mathbb{G}.(X),B.]$. The first section of our paper is essentially dedicated to showing that for any n-hypergroupoid B., the abelian group in the comma category of Sets over the set $(X,\Pi_0(B.))=\mathrm{Hom}_{\mathfrak{B}}(X,\Pi_0(B.))$, $[\mathbb{G}.(X),K(\Pi_n(B.)\rtimes\Pi_0(B.),n)]\longrightarrow(X,\Pi_0(B.))$, acts in an "adequate form" on $[\mathbb{G}.(X),B.]\longrightarrow(X,\Pi_0(B.))$.

In the second section we associate to any n-hypergroupoid B. an (n+1)-hypergroupoid C.(B.), in such a way that it generalizes, in a natural sense, the transition from the space $K(A\rtimes B,n)$ to $K(A\rtimes B,n+1)$, and using the actions of the abelian cohomology groups we construct a long exact sequence of sets with distinguished elements (neutral and null)

$$[\mathbb{G}.(X),B.]\xrightarrow{q_*}[\mathbb{G}.(X),B"]\xrightarrow{\delta_n}[\mathbb{G}.(X),K(C.(q.))]\xrightarrow{i_*}[\mathbb{G}.(X),C.(B.)]\longrightarrow\ldots$$

$$\ldots[\mathbb{G}.(X),C.^{(i}(B.)]\xrightarrow{q_*}[\mathbb{G}.(X),C.^{(i}(B")]\xrightarrow{\delta_{n+i}}[\mathbb{G}.(X),K(C.^{(i+1}(q.))]\ \ldots$$

associated to any fibration of n-hypergroupoids $q.:B.\longrightarrow B"$ for which $\operatorname{Cosk}^n(q.):\operatorname{Cosk}^n(B.)\longrightarrow\operatorname{Cosk}^n(B")$ is a trivial fibration.

In the last section, we define the (n+1)-cohomology set, $n\geq1$, of an object X in $\mathfrak{B}$ with coefficients in a crossed module Φ in $\mathfrak{B}$ by

$$\mathbb{H}^{n+1}(X,\Phi)=[\mathbb{G}.(X),C.^{n-1}(\mathfrak{G}(\Phi))]$$

where $\mathfrak{G}(\Phi)$ is the 1-hypergroupoid nerve of the groupoid associated to Φ. This cohomology reduces to the usual monadic abelian cohomology when one considers the crossed module $M\longrightarrow0$ defined by a trivial X-module, and associated to a short exact sequence of crossed modules $\Phi'\longrightarrow\Phi\longrightarrow\Phi"$ we obtain a long exact sequence

$$\mathbb{H}^2_\Phi(X,\Phi')\longrightarrow\mathbb{H}^2(X,\Phi)\longrightarrow\mathbb{H}^2(X,\Phi")\xrightarrow{\delta_1}\mathbb{H}^3_\Phi(X,\Phi')\longrightarrow\ldots$$

$$\ldots\mathbb{H}^n(X,\Phi")\xrightarrow{\delta_{n-1}}\mathbb{H}^{n+1}_\Phi(X,\Phi')\longrightarrow\mathbb{H}^{n+1}(X,\Phi)\longrightarrow\mathbb{H}^{n+1}(X,\Phi")\longrightarrow\ldots$$

which extends the low dimensional exact sequences in non-abelian cohomology classically established, [10], [12], [22].

1. ON THE STRUCTURE OF THE SETS $[\mathbb{G}.(X),B.]$.

Given B. an n-hypergroupoid in $\mathfrak{B}$ and X an object of $\mathfrak{B}$, there is a canonical map $\chi:[\mathbb{G}.(X),B.]\longrightarrow(X,\Pi_0(B.))$ given by $\chi([f.])=\Pi_0(f.)$. This map gives a decomposition of the set $[\mathbb{G}.(X),B.]$ as the disjoint union of the fibers $\chi^{-1}(f)=[\mathbb{G}.(X),B.]_f$, $f\in(X,\Pi_0(B.))$. In this section we will establish the following results:

$\Pi_n(B.)$ *is a* $\Pi_0(B.)$*-module in the variety* $\mathfrak{B}$, *and for each* $f\in(X,\Pi_0(B.))$ *such that the fiber* $[\mathbb{G}.(X),B.]_f$ *is not empty, the abelian group* $H^n_f(X,\Pi_n(B.))$, *where* $\Pi_n(B.)$ *is* X*-module via* f, *acts on* $[\mathbb{G}.(X),B.]_f$. *If*

$B_\bullet \longrightarrow Cosk^{n-1}(B_\bullet)$ *is a retraction and the canonical morphism* $Cosk^{n}(B_\bullet) \longrightarrow Cosk^{n-1}(B_\bullet)$ *is a trivial fibration, the actions are principal.*

Considering $B_\bullet \xrightarrow{d} Cosk^{n}(B_\bullet)$, *and* $[f_\bullet],[g_\bullet] \in [\mathbb{G}_\bullet(X),B_\bullet]_f$, *we have that* **df.** *is homotopic to* **dg.** *if and only if there is an element* $\alpha \in H^n_f(X,\Pi_n(B_\bullet))$ *such that* ${}^{\alpha}[f_\bullet]=[g_\bullet]$.

1.1. The structure of the homotopy groups.

We start recalling some facts about the homotopy groups of a simplicial object in the variety of Ω-groups $\mathfrak{B}$ where $\Omega=\Omega_0\cup\Omega_1\cup\Omega_2$ is a set of operators (none of weight greater than 2) with only one operator of weight zero (denoted by 0), just two of weight two (denoted by + and *, respectively), and at least one of weight one (denoted by −). The (Moore) homotopy groups of a simplicial object

$$B_\bullet = \ldots B_n \underset{d_0}{\overset{d_n}{\rightrightarrows}} B_{n-1} \ldots B_2 \underset{d_0}{\overset{d_2}{\rightrightarrows}} B_1 \underset{d_0}{\overset{d_1}{\rightrightarrows}} B_0 \qquad (s_0,\ldots,s_{n-1}: B_{n-1}\to B_n;\ s_0,s_1: B_1\to B_2;\ s_0: B_0\to B_1)$$

in $\mathfrak{B}$ are defined, as in the case of simplicial groups, by

$$\Pi_n(B_\bullet) = \frac{\bigcap_{i=0}^{n} Ker(d_i : B_n \longrightarrow B_{n-1})}{d_{n+1}\left(\bigcap_{i=0}^{n} Ker(d_i : B_{n+1} \longrightarrow B_n)\right)} \quad , \ n\geq 0 \ ,$$

or equivalently the homology groups of the Moore complex

$$N(B_\bullet) = \ldots \longrightarrow N_{n+1}(B_\bullet) \longrightarrow N_n(B_\bullet) \xrightarrow{\delta} N_{n-1}(B_\bullet) \longrightarrow \ldots$$

where $N_n(B_\bullet) = \bigcap_{i=0}^{n-1} Ker(d_i : B_n \longrightarrow B_{n-1})$ and δ is induced by d_n.

$\Pi_n(B_\bullet)$ is an object in $\mathfrak{B}$, and for all $x \in B_n$ and $a \in N_n(B_\bullet)$, $n\geq 1$, we have:

$$[x+a-x] = [s_0^n d_0^n(x) + a - s_0^n d_0^n(x)] \ ,$$
$$[x*a] = [s_0^n d_0^n(x)*a] \quad \text{and}$$
$$[a*x] = [a*s_0^n d_0^n(x)],$$

where square brackets are used to denote equivalence classes in $\Pi_n(B_\bullet)$ and consequently $\Pi_n(B_\bullet)$ is a $\Pi_0(B_\bullet)$-module in $\mathfrak{B}$, for all $n\geq 1$, with actions:

$${}^{[y]}[x] = [s_0^n(y)+x-s_0^n(y)] \ ,$$
$$[y]*[x] = [s_0^n(y)*x] \quad \text{and} \quad [x]*[y] = [x*s_0^n(y)] \ \text{for}$$

$x \in N_n(B_\bullet)$ and $y \in N_0(B_\bullet) = B_0$ (see [5]).

1.2. The fibrations $\bar{\xi}_\bullet$.

Let us now recall that an n-truncated simplicial object consists of objects B_0, B_1, ... , B_n and the usual face and degeneracy between them. If we designate by $Tr^n\mathbf{Simpl}(\mathfrak{B})$ the category of n-truncated simplicial objects of $\mathfrak{B}$,the functor "truncation at level n"

$$tr^n:\mathbf{Simpl}(\mathfrak{B}) \longrightarrow Tr^n\mathbf{Simpl}(\mathfrak{B})$$

admits a right adjoint $cosk^n$, called the n-coskeleton functor. A construction of the n-coskeleton of a truncated complex can be done by using simplicial kernels, as follows:

Given a n-truncated simplicial object $B_{\bullet tr}$, $n\geq1$, the n+1-simplicial kernel of $B_{\bullet tr}$ is an object, denoted by $\Delta_{n+1}(B_{\bullet tr})$, together with morphisms $d_i:\Delta_{n+1}(B_{\bullet tr}) \longrightarrow B_n$, $0\leq i\leq n+1$, which is universal respect to the property: $d_i d_j = d_{j-1} d_i$ for all $0\leq i<j\leq n+1$. There are degeneracy morphisms $s_j:B_n \rightarrow \Delta_{n+1}(B_{\bullet tr})$, $0\leq j\leq n$, in such a way

that $$\Delta_{n+1}(B_{\bullet tr}) \underset{d_0}{\overset{d_{n+1}}{\rightrightarrows}} B_n \underset{d_0}{\overset{d_n}{\rightrightarrows}} B_{n-1} \quad \cdots$$

is an (n+1)-truncated complex. Then, $cosk^n:Tr^n\mathbf{Simpl}(\mathfrak{B}) \longrightarrow \mathbf{Simpl}(\mathfrak{B})$ is obtained by iterating simplicial kernel construction. The unity of the adjunction $\mathbf{d}_\bullet:B_\bullet \longrightarrow cosk^n tr^n(B_\bullet)$ consists of identities at dimensions $\leq n$, and the canonical morphism

$$\mathbf{d}_{n+1}:B_{n+1} \longrightarrow \Delta_{n+1}(B_\bullet) \; ; \; \mathbf{d}_{n+1}(x) = (d_0(x),d_1(x),\ldots,d_{n+1}(x))$$

at dimension n+1.

The endofunctor $Cosk^n = cosk^n tr^n:\mathbf{Simpl}(\mathfrak{B}) \longrightarrow \mathbf{Simpl}(\mathfrak{B})$, $n\geq1$,is also called n-coskeleton functor.

<u>PROPOSITION 1</u> .- Let $B_\bullet$ be a simplicial object in $\mathfrak{B}$ and $n\geq1$. There exists a natural surjective derivation $\xi:\Delta_{n+1}(B_\bullet) \longrightarrow \Pi_n(B_\bullet)$ such that in the sequence

$$B_{n+1} \xrightarrow{\mathbf{d}_{n+1}} \Delta_{n+1}(B_\bullet) \xrightarrow{\xi} \Pi_n(B_\bullet)$$

$\mathbf{d}_{n+1}(B_{n+1})=\xi^{-1}(0)$.

<u>Proof</u>.-

It is given in [5]; we only recall how ξ is defined. Let us denote $\Theta_i=(I-s_i d_i):\Delta_{n+1}(B_\bullet) \longrightarrow \Delta_{n+1}(B_\bullet)$ the map

$$\Theta_i(x_0,x_1,\ldots,x_{n+1}) = (x_0,x_1,\ldots,x_{n+1}) - s_i(x_i),\ 0\leq i\leq n.$$

Observe, using the simplicial identities, that the element

$$\Theta_n\Theta_{n-1}\cdots\Theta_0(x_0,x_1,\ldots,x_{n+1})$$

has all its components, up to the last one, equal to zero. Then we write

$$\Theta_n\Theta_{n-1}\cdots\Theta_0(x_0,x_1,\ldots,x_{n+1}) = (0,0,\ldots,0,\Theta(x_0,x_1,\ldots,x_{n+1}))$$

for $\Theta(x_0,x_1,\ldots,x_{n+1})$ an element in $\bigcap_{i=0}^{n} Ker(d_i:B_n \rightarrow B_{n-1})$ and ξ is defined by

$$\xi(x_0,x_1,\ldots,x_{n+1}) = [\Theta(x_0,x_1,\ldots,x_{n+1})],$$

where square bracket denotes equivalence class in $\Pi_n(B_\bullet)$.∎

PROPOSITION 2.- Let B. be a simplicial object in $\mathfrak{B}$, $n\geq 1$ and ξ the derivation introduced in proposition 1. Then ξ determines a fibration in **Simpl($\mathfrak{B}$)**, $\bar{\xi}.:\mathrm{Cosk}^n(B.) \longrightarrow K(\Pi_n(B.)\rtimes\Pi_0(B.),n+1)$ in which

$$\bar{\xi}_{n+1}:\Delta_{n+1}(B.) \longrightarrow \Pi_n(B.)\rtimes \Pi_0(B.)$$

is given by $\bar{\xi}_{n+1}(x_0,x_1,\dots,x_{n+1}) = (\xi(x_0,x_1,\dots,x_{n+1}),d_0^{n+1}(x_0))$.

Moreover, in the sequence

$$B. \xrightarrow{\ \mathbf{d}\ } \mathrm{Cosk}^n(B.) \xrightarrow{\ \bar{\xi}.\ } K(\Pi_n(B.)\rtimes\Pi_0(B.),n+1)$$

considering $K(\Pi_0(B.),0) \subseteq K(\Pi_n(B.)\rtimes\Pi_0(B.),n+1)$ we have

$$\mathbf{d}(B.) = \bar{\xi}.^{-1}(K(\Pi_0(B.),0)).$$

In particular, if B. is an n-hypergroupoid, $\mathbf{d}$ is injective and therefore $B. = \bar{\xi}.^{-1}(K(\Pi_0(B.),0))$.

Proof.-

$N(\bar{\xi}.):N(\mathrm{Cosk}^n(B.)) \longrightarrow N(K(\Pi_n(B.)\rtimes\Pi_0(B.),n+1))$ is surjective because it is the zero morphism in all dimensions $\neq 0,n+1$; it is the projection $B_0 \longrightarrow \Pi_0(B.)$ in dimension 0 and it is the projection from $\mathrm{Ker}(\delta_n:N_n(B.)\longrightarrow N_{n-1}(B.))$ onto $\Pi_n(B.)=\mathrm{Ker}\delta_n/\mathrm{Im}\delta_{n+1}$ in dimension n+1. Therefore $\bar{\xi}.$ is surjective and so $\bar{\xi}.$ is a fibration, (see [28]).

On the other hand, in dimensions $k\leq n$, is clear that $\mathbf{d}(B.)=B_k=\bar{\xi}.^{-1}(K(\Pi_0(B.),0))$. In dimension n+1 we have, according to proposition 1, $\bar{\xi}^{-1}_{n+1}(\Pi_0(B.))= \xi^{-1}(0)=\mathbf{d}(B_{n+1})$. Finally, since for any $m>n+1$, $N_m(\mathbf{d})(N_m(B.))=0 = N_m(\bar{\xi}.)^{-1}(N_m(K(\Pi_0(B.),0))$,we have that $\bar{\xi}.^{-1}(K(\Pi_0(B.),0)) = \mathbf{d}(B.)$.

If B. is an n-hypergroupoid $N_m(B.)=0$ for all $m>n$ so that $N_m(\mathbf{d})$ is injective for all $m\geq 0$ and then $\mathbf{d}$ is injective (see [28]).■

Proposition 2 has a certain converse,

PROPOSITION 3.- Let $B._{tr}$ be a n-truncated simplicial object in $\mathfrak{B}$, $\Pi_0(B.)=B$ and A a B-module. Suppose $\bar{\xi}.:\mathrm{cosk}^n(B._{tr}) \longrightarrow K(A\rtimes B,n+1)$ a fibration with $\bar{\xi}_0:B_0\longrightarrow B$ the canonical projection and such that $N_{n+1}(\bar{\xi}.)$ is an isomorphism. Then $\bar{\xi}.^{-1}(K(B,0))$ is a n-hypergroupoid B., $\Pi_n(B.)\cong A$ as B-modules and the fibration associated to B. according to the above proposition identifies itself with the given one.

Proof:

$N_{n+1}(\bar{\xi}.)^{-1}N_{n+1}(K(B,0)) = N_{n+1}(\bar{\xi}.)^{-1}(0) = 0$ and in dimensions m higher than n+1 is clear that $N_m(B.)=0$.■

1.3.- The action of $H^n_f(X,\Pi_n(B.))$ on $[\mathbb{G}.(X),B.]_f$.

We now establish the announced main resuts of this section.

Let B. be a n-hypergroupoid in $\mathfrak{B}$; in what follows we will denote $\Pi_n(B.)=A$ and $\Pi_0(B.)=B$ and recall that A is, in a canonical way, a B-module.

For any object X in $\mathfrak{B}$, the cotriple resolution of X, $\mathbb{G}.(X)$, is a cofibrant object in **Simpl($\mathfrak{B}$)**, consequently the fibration

$$\bar{\xi}.:\mathrm{Cosk}^n(B.)\longrightarrow K(A\rtimes B,n+1)$$

induces another one between the function complex simplicial sets

$$\bar{\xi}_{\bullet}:\mathrm{Cosk}^{n}(B_{\bullet})^{\mathbb{G}_{\bullet}(X)} \longrightarrow K(A\rtimes B,n+1)^{\mathbb{G}_{\bullet}(X)}$$

for which

$$B_{\bullet}^{\mathbb{G}_{\bullet}(X)} = \bar{\xi}_{\bullet}^{-1}\big((K(B,0)^{\mathbb{G}_{\bullet}(X)}\big) .$$

Observe that $K(B,0)^{\mathbb{G}_{\bullet}(X)}$ is a constant simplicial set whose vertices are the simplicial morphisms in $\mathfrak{B}$ from $\mathbb{G}_{\bullet}(X)$ to $K(B,0)$ and these can be identified, via Π_0, with the morphisms in $\mathfrak{B}$ from X to B ; so we will identify $K(B,0)^{\mathbb{G}_{\bullet}(X)}$ with $K((X,B),0)$, and $B_{\bullet}^{\mathbb{G}_{\bullet}(X)}$ with $\bar{\xi}_{\bullet}^{-1}K((X,B),0)$.

Now, let $f_{\bullet}$ be a simplicial morphism from $\mathbb{G}_{\bullet}(X)$ to $B_{\bullet}$ and $f=\Pi_0(f_{\bullet}):X\longrightarrow B$. Then the fibre $\bar{\xi}_{\bullet}^{-1}(f)$ is isomorphic to the simplicial subset of $B_{\bullet}^{\mathbb{G}_{\bullet}(X)}$ whose simplices have all their 0-faces the simplicial morphisms $g_{\bullet}:\mathbb{G}_{\bullet}(X)\longrightarrow B_{\bullet}$ for which $\Pi_0(g_{\bullet})=f$. We will denote this simplicial set by $(B_{\bullet}^{\mathbb{G}_{\bullet}(X)})_f$ and let $[\mathbb{G}_{\bullet}(X),B_{\bullet}]_f=\Pi_0((B_{\bullet}^{\mathbb{G}_{\bullet}(X)})_f)$ which is just the set of homotopy classes of simplicial morphisms $g_{\bullet}$ from $\mathbb{G}_{\bullet}(X)$ to $B_{\bullet}$ with $\Pi_0(g_{\bullet})=f$.

Of course we can apply these facts to the n-hypergroupoid $K(A\rtimes B,n)$, so the fibration $\bar{\xi}_{\bullet}:\mathrm{Cosk}^{n}(K(A\rtimes B,n))^{\mathbb{G}_{\bullet}(X)}\longrightarrow K(A\rtimes B,n+1)^{\mathbb{G}_{\bullet}(X)}$ allows us to identify $(K(A\rtimes B,n)^{\mathbb{G}_{\bullet}(X)})_f$ with $\bar{\xi}_{\bullet}^{-1}(f)$; but in this case the canonical projection $\mathrm{Cosk}^{n}(K(A\rtimes B,n))\longrightarrow K(B,0)$ is a trivial fibration and therefore $(\mathrm{Cosk}^{n}(K(A\rtimes B,n))^{\mathbb{G}_{\bullet}(X)})_f\longrightarrow (K(B,0)^{\mathbb{G}_{\bullet}(X)})_f$ is also a trivial fibration, consequently all the homotopy groups of $(\mathrm{Cosk}^{n}(K(A\rtimes B,n))^{\mathbb{G}_{\bullet}(X)})_f$ are trivial and there are canonical isomorphisms $\Pi_1((K(A\rtimes B,n+1))^{\mathbb{G}_{\bullet}(X)},f)\cong$ $\cong \Pi_0(((K(A\rtimes B,n))^{\mathbb{G}_{\bullet}(X)})_f) \cong [\mathbb{G}_{\bullet}(X),K(A\rtimes B,n)]_f \cong H^n_f(X,A)$ the n-th cohomology group of X with coefficients in the X-module, via f, A.

Using the above results, we have

<u>PROPOSITION</u> <u>4</u>.- Let $B_{\bullet}$ be a n-hypergroupoid in $\mathfrak{B}$, $n\geq 1$. Let $\Pi_0(B_{\bullet})=B$ and $\Pi_n(B_{\bullet})=A$. Associated to any morphism in $\mathfrak{B}$, $f:X\longrightarrow B$, and any simplicial morphism $f_{\bullet}:\mathbb{G}_{\bullet}(X) \longrightarrow B_{\bullet}$ with $\Pi_0(f_{\bullet})=f$, the fibration

$$\bar{\xi}_{\bullet}:\mathrm{Cosk}^{n}(B_{\bullet}) \longrightarrow K(A\rtimes B,n+1)$$

induces an exact sequence of groups and pointed sets

$$\Pi_1(\mathrm{Cosk}^{n}(B_{\bullet})^{\mathbb{G}_{\bullet}(X)},f_{\bullet})\xrightarrow{\bar{\xi}_*} H^n_f(X,A)\xrightarrow{\omega} [\mathbb{G}_{\bullet}(X),B_{\bullet}]_f\xrightarrow{\mathbf{d}_*} [\mathbb{G}_{\bullet}(X),\mathrm{Cosk}^{n}B_{\bullet}]$$

and an action of $H^n_f(X,A)$ on $[\mathbb{G}_{\bullet}(X),B_{\bullet}]_f$ such that:

a) $\mathbf{d}_*([f_{\bullet}])=\mathbf{d}_*([g_{\bullet}])$ if and only if ${}^{a}[f_{\bullet}]=[g_{\bullet}]$, for some $a\in H^n_f(X,A)$, and

b) $\omega(a)=\omega(a')$ if and only if $a'= \bar{\xi}_*(z)+a$, for some z.■

The following proposition gives sufficient conditions for the actions of the groups $H^n_f(X,A)$ on the sets $[\mathbb{G}_{\bullet}(X),B_{\bullet}]_f$ to be principal.

PROPOSITION 5.- Let $B_\bullet$ be an n-hypergroupoid in $\mathfrak{B}$, $n\geq 1$, such that the morphism $\mathbf{d}:B_\bullet \longrightarrow Cosk^{n-1}(B_\bullet)$ is a retraction and the morphism $\mathbf{d}:Cosk^n(B_\bullet)\longrightarrow Cosk^{n-1}(B_\bullet)$ is a trivial fibration. Then the homomorphism

$\Pi_1(Cosk^n(B_\bullet)^{\mathbb{G}_\bullet(X)},f_\bullet) \xrightarrow{\bar{\xi}_*} H^n_f(X,A)$ is trivial for any simplicial morphism $f_\bullet:\mathbb{G}_\bullet(X)\longrightarrow B_\bullet$, $\Pi_0(f_\bullet)=f$ and the action of $H^n_f(X,A)$ on $[\mathbb{G}_\bullet(X),B_\bullet]_f$ is principal.

Proof:

Let $s_\bullet:Cosk^{n-1}(B_\bullet)\longrightarrow B_\bullet$ be such that $\mathbf{d}s_\bullet=I$. Then $\bar{\xi}_\bullet s_\bullet:Cos^{n-1}(B_\bullet)\longrightarrow K(A\rtimes B,n+1)$ factors through $K(B,0)$ since $B_\bullet=\bar{\xi}_\bullet^{-1}K(B,0)$. On the other hand, considering the induced morphisms

$$Cosk^n(B_\bullet)^{\mathbb{G}_\bullet(X)} \xrightarrow{s_*\mathbf{d}_*} Cosk^n(B_\bullet)^{\mathbb{G}_\bullet(X)} \xrightarrow{\mathbf{d}_*} Cosk^{n-1}(B_\bullet)^{\mathbb{G}_\bullet(X)}$$

we have that $\mathbf{d}_*s_*\mathbf{d}_*=\mathbf{d}_*$, so that $s_*\mathbf{d}_*$ is homotopic to the identity map since $\mathbf{d}_*$ is a trivial fibration. Thus $\bar{\xi}_*$ is homotopic to $\bar{\xi}_*s_*\mathbf{d}_*$ which factors through $K(B,0)^{\mathbb{G}_\bullet(X)}=K((X,B),0)$ and consequently $\Pi_1(\bar{\xi}_*)=0$.■

2.A LONG EXACT SEQUENCE FOR THE SETS OF HOMOTOPY CLASSES.

Classically, exact sequences in non-abelian cohomology appear associated to exact sequences of crossed modules, or more generally to quotient maps of groupoids in the sense of Higgins, [18]. In our context $\mathfrak{B}$ of a variety of Ω-groups, internal groupoids are equivalent to 1-hypergroupoids and quotient maps of groupoids to morphisms of 1-hypergroupoids $B_\bullet\longrightarrow B''$ such that the induced morphism $Cosk^1(B_\bullet)\longrightarrow Cosk^1(B'')$ is a trivial fibration.

In this section we are going to show the existence of a long exact sequence

$$[\mathbb{G}_\bullet(X),B_\bullet]\xrightarrow{q_*}[\mathbb{G}_\bullet(X),B'']\xrightarrow{\delta_n}[\mathbb{G}_\bullet(X),K(C_\bullet(q_\bullet))]\xrightarrow{i_*}[\mathbb{G}_\bullet(X),C_\bullet(B_\bullet)]\longrightarrow\ldots$$

associated to any morphism of n-hypergroupoids $q:B_\bullet\longrightarrow B''$ such that the induced morphism $Cosk^n(B_\bullet)\longrightarrow Cosk^n(B'')$ is a trivial fibration. This sequence will permit us in the subsequent section to obtain a long exact sequence in non abelian cohomology with coefficients in crossed modules, which extend the known classical ones and the 9-term exact sequences shown by the authors in [7].

Abelian cohomology is included in homotopy theory via the Eilenberg-Mac Lane complexes; with this point of view the coefficients for H^n and H^{n+1} are related, for any X-module A, by the equalizer diagram

$$K(A\rtimes X,n) \longrightarrow Cosk^n(K(A\rtimes X,n)) \underset{t_\bullet}{\overset{\bar{\xi}_\bullet}{\rightrightarrows}} K(A\rtimes X,n+1)$$

where $t_\bullet$ is the unique morphism which factors through $K(X,0)$, corresponding to I_X; and $\Pi_i(K(A\rtimes X,n+1)) = \Pi_i(K(A\rtimes X,n))$ for $i\neq n,n+1$, $\Pi_n(K(A\rtimes X,n+1))=0$ and $\Pi_{n+1}(K(A\rtimes X,n+1))=\Pi_n(K(A\rtimes X,n))=A$.

In the same way, given any n-hypergroupoid $B_\bullet$ in $\mathfrak{B}$, we will show that there exists an (n+1)-hypergroupoid $C.(B.)$ in $\mathfrak{B}$ and an equalizer diagram

$$B. \longrightarrow Cosk^n(B.) \overset{r.}{\underset{t.}{\rightrightarrows}} C.(B.)$$

with $\Pi_i(C.(B.))=\Pi_i(B.)$ for $i\neq n,n+1$, $\Pi_n(C.(B.))=0$,and $\Pi_{n+1}(C.(B.))=\Pi_n(B.)$.

2.1.The hypergroupoid $C.(B.)$.

Given an n-hypergroupoid $B.$, to build the (n+1)-hypergroupoid $C.(B.)$ we use proposition 3: Let $C.(B.)_{tr}$ be the (n+1)-truncated simplicial object in $\mathfrak{B}$ which consists of the same n-truncation as $B.$ and in dimension n+1 it is the subobject of B_n^{n+3} given by

$C_{n+1}(B.)=\{(b'_0,b_0,\dots,b_{n+1}) \mid d_ib'_0=d_ib_0, i=0,\dots,n \text{ and } d_ib_j=d_{j-1}b_i, i<j\}$

with the face morphisms $d_i:C_{n+1}(B.)\longrightarrow B_n$ given by

$$d_i(b'_0,b_0,\dots,b_{n+1})=b_i,\ 0\le i\le n+1$$

and the degeneracies $s_j:B_n\longrightarrow C_{n+1}(B.)$ given by

$$s_j(b)=(s_{j-1}d_0b,s_{j-1}d_0b,\dots,s_{j-1}d_jb,b,b,s_jd_{j+1}b,\dots,s_jd_nb)\ ,\ 0\le j\le n\ .$$

Now we use the canonical derivation $\xi:\Delta_{n+1}(B.)\longrightarrow\Pi_n(B.)$ to build the fibration $\bar{\xi}.:cosk^{n+1}(C.(B.)_{tr})\longrightarrow K(\Pi_n(B.)\rtimes\Pi_0(B.),n+2)$, which in dimension n+2 is defined as follows: Given an element $z=(z_0,\dots,z_{n+2})\in\Delta_{n+2}(C.(B.)_{tr})$, $z_i=(b'_{i0},b_{i0},b_{i1},\dots,b_{in+1})\in C_{n+1}(B.)$,

$$\bar{\xi}_{n+2}(z)=\Big(\sum_{i=0}^{n+2}(-1)^i\big(\xi(b_{i0},b_{i1},\dots,b_{in+1})-\xi(b'_{i0},b_{i1},\dots,b_{in+1})\big),\ d_0^{n+1}b_{00}\Big)$$

It is plain to see that in the Moore complex, $N_{n+2}(\bar{\xi}.)$ is an isomorphism and therefore, applying proposition 3 ,

$$C.(B.)=\bar{\xi}.^{-1}(K(\Pi_0(B.),0)$$

is an (n+1)-hypergroupoid. Clearly $\Pi_i(C.(B.))=\Pi_i(B.)$ for $i\neq n,n+1$, the canonical morphism $C_{n+1}(B.)\longrightarrow\Delta_{n+1}C.(B.)$ is surjective so that $\Pi_n(C.(B.))=0$, and by construction $\Pi_{n+1}(C.(B.))=\Pi_n(B.)$.

We have the equalizer diagram

$$B. \longrightarrow Cosk^n(B.) \overset{r.}{\underset{t.}{\rightrightarrows}} C.(B.)$$

where r. and t. are respectively determined by

$$t_{n+1}(b_0,\dots,b_{n+1})=(b_0,b_0,b_1,\dots,b_{n+1})\ ,\text{ and}$$

$$r_{n+1}((b_0,\dots,b_{n+1})=(\xi(b_0,\dots,b_{n+1})+b_0,b_0,b_1,\dots,b_{n+1}).$$

The following lemma shows how the above construction $C.(-)$ works on the hypergroupoids $K(A\rtimes B,n)$.

<u>LEMMA</u> <u>6</u>.- Let B be an object of $\mathfrak{B}$ and A a B-module. Then there is a trivial fibration $\nu.:C.(K(A\rtimes B,n))\longrightarrow K(A\rtimes B,n+1)$.

<u>Proof</u>:

The fibration is determined by

$\nu_{n+1}((a'_0,b),(a_0,b),(a_1,b),\dots,(a_{n+1},b))=(a'_0,b)$. ∎

2.2. Neutral and null elements.

Given any n-hypergroupoid B. in $\mathcal{B}$, we observe the existence of a canonical inclusion $j.:B. \longrightarrow C.(B.)$ determined by $j_{n+1}:B_{n+1} \longrightarrow C_{n+1}(B.)$ given by $j_{n+1}(b)=(d_0(b),d_0(b),d_1(b),\dots,d_{n+1}(b))$ This simplicial morphism j. allow us to study the "neutral cocycles" with coefficients in C.(B.), that is, those simplicial morphisms from $\mathbb{G}.(X)$ which factor through the simplicial subobject of C.(B.) generated by the degenerate (n+1)-simplices.

<u>PROPOSITION 7</u>.-The simplicial morphism j. is an isomorphism between B. and the simplicial subobject of C.(B.) generated by the degenerate (n+1)-simplices.

<u>Proof:</u>

Given $b \in B_n$,

$s_i(b) = (s_{i-1}d_0(b), s_{i-1}d_0(b), \dots, s_{i-1}d_i(b), b, b, s_i d_{i+1}(b), \dots, s_i d_n(b)) = $ $=j_{n+1}(s_i(b))$, $i=0,1,\dots,n$, and so any degenerate (n+1)-simplex of $C_{n+1}(B.)$ belongs to the image of j_{n+1}. On the other hand, because B. is an n-hypergroupoid, $N_{n+1}(B.)=0$ and so $((1-s_n d_n)\dots(1-s_0 d_0))(y) = 0$ for any $y \in B_{n+1}$ and any element of B_{n+1} is a sum of degenerate (n+1)-simplices.■

<u>COROLLARY 8</u>.- The simplicial morphism j. induces a bijection between $(\mathbb{G}.(X),B.)$ and the set of neutral (n+1)-cocycles over X under C.(B.).■

In order to establish the general long exact sequence it is necessary to point out two larger subsets of $(\mathbb{G}.(X),C.(B.))$ than that consisting of neutral cocycles. For this we observe that the simplicial morphism j. has two natural factorizations through the simplicial object $Cosk^n(B.)$ determined by the simplicial morphisms r. and t. of **2.1.** :

$$\begin{array}{ccc} B. & \longrightarrow & C.(B.) \\ & \mathbf{d} \searrow \quad \nearrow r. \nearrow t. & \\ & Cosk^n(B.) & \end{array}$$

Define the "r-null" (n+1)-cocycles (respectively the "t-null" (n+1)-cocycles) as those which factor through r. (resp. t.). The respective classes of such cocycles in $[\mathbb{G}.(X),C.(B.)]$ will permit us to describe the exactness of the long exact sequence.

If we denote by $\mathbb{C}^n_{\mathbb{G}}(X,B.)$ the set of n-cochains (i.e., n-truncated simplicial morphisms from $Tr^n(\mathbb{G}.(X))$ to $Tr^n(B.)$) the universal adjunction property between the functors Tr^n and $cosk^n$, (see **1.2**), gives a bijection

$$\mathbb{C}^n_{\mathbb{G}}(X,B.) \cong (\mathbb{G}.(X),Cosk^n(B.))$$

and now, because r. and t. are injective, we have

<u>PROPOSITION 9</u>.- There is a canonical bijection between the set of r-null (respectively t-null) (n+1)-cocycles over X under C.(B.) and the set $\mathbb{C}^n_{\mathbb{G}}(X,B.)$.■

2.3. The hypergroupoid kernel.

Let $q.:B. \longrightarrow B''$ be a morphism of n-hypergroupoids in $\mathcal{B}$, $n \geq 1$, and consider the induced morphism of (n+1)-hypergroupoids

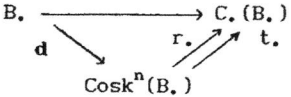

$C.(q.):C.(B.)\longrightarrow C.(B'')$. We define $K(C.(q.))$ the "(n+1)-hypergroupoid kernel" of $C.(q.)$ by the cartesian square

$$\begin{array}{ccc} K(C.(q.)) & \longrightarrow & Cosk^n(B'') \\ i.\big\downarrow & & \big\downarrow r. \\ C.(B.) & \xrightarrow{\ C.(q.)\ } & C.(B'') \end{array}$$

Note that the existence of the commutative square

$$\begin{array}{ccc} Cosk^n(B.) & \xrightarrow{\ Cosk^n(q.)\ } & Cosk^n(B'') \\ r.\big\downarrow & & \big\downarrow r. \\ C.(B.) & \xrightarrow{\ C.(q.)\ } & C.(B'') \end{array}$$

implies that r. factors through $K(C.(q.))$, i.e., there is a commutative diagram

$$\begin{array}{ccc} Cosk^n(B.) & \xrightarrow{\ \ r.\ \ } & C.(B.) \\ & \searrow \quad \nearrow i. & \\ & K(C.(q.)) & \end{array}$$

Thus, the neutral (n+1)-cocycles under $K(C.(q.))$ are the neutral ones under $C.(B.)$.

Note also that $\Pi_i(K(C.(q.)))=\Pi_i(B.)$ for $i\neq n,n+1$, $\Pi_n(K(C.(q.)))=0$ and $\Pi_{n+1}(K(C.(q.)))=Ker(\Pi_n(B.)\longrightarrow\Pi_n(B''))$.

2.4. The connecting map.

Let $q.:B.\longrightarrow B''$ be a morphism of n-hypergroupoids in $\mathfrak{B}$, $n\geq 1$, such that the induced morphism $Cosk^n(q.):Cosk^n(B.)\longrightarrow Cosk^n(B'')$ is a trivial fibration. There is map

$$\delta_n:[\mathbb{G}.(X),B'']\longrightarrow[\mathbb{G}.(X),K(C.(q.))]$$

defined as follows:

Let $f''\in(\mathbb{G}.(X),B'')$ and consider the following diagram of simplicial morphisms

$$\begin{array}{ccccccc} & & B. & \xrightarrow{\ \mathbf{d}\ } & Cosk^n(B.) & \xrightarrow{\ \ t.\ \ } & C.(B.) \\ & & q.\big\downarrow & & Cosk^n(q.)\big\downarrow \ \swarrow & K(C.(q.)) \nearrow & \big\downarrow C.(q.) \\ \mathbb{G}.(X) & \xrightarrow{\ f''.\ } & B'' & \xrightarrow{\ \mathbf{d}\ } & Cosk^n(B'') & \overset{r.}{\underset{t.}{\rightrightarrows}} & C.(B'') \end{array}$$

Now, since $Cosk^n(q.)$ is a trivial fibration and $\mathbb{G}.(X)$ is a cofibrant object in **Simpl**($\mathfrak{B}$), there exists a morphism $g.:\mathbb{G}.(X)\longrightarrow Cosk^n(B.)$ such that $Cosk^n(q.)g.=\mathbf{d}f''$. Moreover, the composition $t.g.:\mathbb{G}.(X)\longrightarrow C.(B.)$ factors through $K(C.(q.))$ because $C.(q.)t.g.=t.Cosk^n(q.)g.=t.\mathbf{d}f''=r.\mathbf{d}f''$ and so it defines a simplicial morphism to $K(C.(q.))$. We define $\delta_n[f'']=[t.g.]$.

That this definition is independent is an inmediate consequence of the "covering homotopy theorem", [28],applied to the cofibrant object $\mathbb{G}.(X)$ and the fibration $Cosk^n(q.)$.

Let us observe that

(a) $\Pi_0(B_\bullet)=\Pi_0(B'')=\Pi_0(C_\bullet(B_\bullet))=\Pi_0(C_\bullet(B''))=\Pi_0(K(C_\bullet(q_\bullet)))$,

(b) the sequence $0\longrightarrow\Pi_{n+1}(K(C_\bullet(q_\bullet)))\longrightarrow\Pi_n(B_\bullet)\longrightarrow\Pi_n(B'')\longrightarrow 0$ is exact, and

(c) for each $f\in(X,\Pi_0(B_\bullet))$ such that $[\mathbb{G}_\bullet(X),\Pi_0(B_\bullet)]_f$ is not empty, the morphism $\delta_n:[\mathbb{G}_\bullet(X),B'']_f \longrightarrow [\mathbb{G}_\bullet(X),K(C_\bullet(q_\bullet))]_f$ is equivariant with respect to the classical connecting morphism $\delta_n:H^n_f(X,\Pi_n(B''))\longrightarrow$ $\longrightarrow H^{n+1}_f(X,\Pi_{n+1}(K(C_\bullet(q_\bullet))))$ and the corresponding actions.

2.5. The exact sequence.

The proof of the exactness of the promised long exact sequence we stated at the first of this section will use the following general result about G-sets which is easy to prove.

LEMMA 10.- Let $G'\xrightarrow{f'} G \xrightarrow{f} G''$ be an exact sequence of groups and consider a commutative diagram in Sets

$$\begin{array}{ccccc} E' & \xrightarrow{q'} & E & \xrightarrow{q} & E'' \\ & & p\big\downarrow & & \big\downarrow p'' \\ & & X & \longleftarrow & X'' \end{array}$$

where E',E,E" are respectively G',G and G"-sets, such that (1) q' and q are equivariant maps (i.e. $q'({}^{g'}e')={}^{f'(g')}q'(e')$, $q({}^{g}e)={}^{f(g)}q(e)$), (2) the action of G" on E" is principal and (3) $p(e_1)=p(e_2)$ if and only if there exists $g\in G$ such that ${}^{g}e_1=e_2$. Then given any element $e\in E$, $q(e)=qq'(e')$ for some $e'\in E'$ if and only if there exists an element $a\in E'$ such that $q'(a)=e$. ∎

PROPOSITION 11.- Let $q_\bullet:B_\bullet\longrightarrow B''$ be a morphism of n-hypergroupoids in $\mathbb{S}$, $n\geq 1$, such that the induced morphism $Cosk^n(q_\bullet):Cosk^n(B_\bullet)\longrightarrow Cosk^n(B'')$ is a trivial fibration. Then the sequence

$$[\mathbb{G}_\bullet(X),B_\bullet]\xrightarrow{q_*}[\mathbb{G}_\bullet(X),B'']\xrightarrow{\delta_n}[\mathbb{G}_\bullet(X),K(C_\bullet(q_\bullet))]$$
$$\xrightarrow{i_*}[\mathbb{G}_\bullet(X),C_\bullet(B_\bullet)]\xrightarrow{q_*}[\mathbb{G}_\bullet(X),C_\bullet(B'')]$$

is exact in the following sense:

a)An element of $[\mathbb{G}_\bullet(X),B'']$ is in the image of q_* if and only if it is mapped by δ_n into a neutral class.

b)An element of $[\mathbb{G}_\bullet(X),K(C_\bullet(q_\bullet))]$ is in the image of δ_n if and only if its image by i_* is a t-null element.

c)An element of $[\mathbb{G}_\bullet(X),C_\bullet(B_\bullet)]$ is in the image of i_* if and only if it is applied by q_* into an r-null class.

Proof:

c) is an elementary application of the covering homotopy theorem, since $C(q_\bullet)$ is a fibration. To see a) and b), let us observe that for any $f:X\longrightarrow B=\Pi_0(B_\bullet)$, we have the exact sequence of groups

$$H^n_f(X,A)\longrightarrow H^n_f(X,A'')\longrightarrow H^{n+1}_f(X,A')\longrightarrow H^{n+1}_f(X,A)$$

where $A=\Pi_n(B.)$, $A''=\Pi_n(B'')$ and $A'=\Pi_{n+1}(K(C.(q.)))$, and the commutative diagram

$$\begin{array}{ccccccc}
[\mathbb{G}.(X),B.]_f & \xrightarrow{q_*} & [\mathbb{G}.(X),B'']_f & \xrightarrow{\delta_n} & [\mathbb{G}.(X),K(C.(q.))]_f & \xrightarrow{i_*} & [\mathbb{G}.(X),C.(B.)]_f \\
 & & \downarrow d_* & & \downarrow d_* & \swarrow d_* & \\
 & & [\mathbb{G}.(X),\mathrm{Cosk}^n(B'')] & \longleftarrow & [\mathbb{G}.(X),\mathrm{Cosk}^{n+1}C.(B'')] & &
\end{array}$$

where the morphisms $\mathbf{d}_*$ in the triangle are the compositions of the canonical $\mathbf{d}_*$ with the bijections

$$[\mathbb{G}.(X),\mathrm{Cosk}^{n+1}K(C.(q.))]\cong[\mathbb{G}.(X),\mathrm{Cosk}^{n+1}C.(B.)]\cong[\mathbb{G}.(X),\mathrm{Cosk}^{n+1}C.(B'')]$$

which are induced by the weak equivalences

$$\mathrm{Cosk}^{n+1}K(C.(q.))\longrightarrow\mathrm{Cosk}^{n+1}C.(B.)\longrightarrow\mathrm{Cosk}^{n+1}(C.(B''))$$

(see proposition below). The diagram satisfies the conditions of the lemma **10** with the above exact sequence of groups and the respective actions showed in **1.3** ; thus only remains to show that:

i) the image of $\delta_n q_*$ is the set of neutral elements ;

ii) an element in the image of i_* is t-null if and only if it is in the image of $i_*\delta_n$.

It is clear that $\delta_n q_*[g.]$ is a neutral element. On the other hand, if $[f.]\in[\mathbb{G}.(X),K(C.(q.))]$ is a neutral element $[f.]=[r.\mathbf{d}g.]$ for some $g.\in(\mathbb{G}.(X),B.)$ and therefore $\delta_n q_*[g.]=\delta_n[q.g.]=[t.\mathbf{d}g.]=[r.\mathbf{d}g.]=[f.]$, then any neutral element in $[\mathbb{G}.(X),K(C.(q.))]$ is in the image of $\delta_n q_*$.

Now image of $i_*\delta_n$ is clearly included in the set of t-null elements. To complete the proof we first observe the existence of the morphism $\nu.:C.(B'')\longrightarrow K(A''\rtimes B,n+1)$ which in dimension n+1 is given by

$$\nu_{n+1}(b'_0,b_0,\ldots,b_{n+1})=(\xi(b'_0,b_1\ldots,b_{n+1})-\xi(b_0,\ldots,b_{n+1}),d^{n+1}_0(b_0))$$

and has the property that $\nu.r.=\bar{\xi}.:\mathrm{Cosk}^n(B'')\longrightarrow K(A''\rtimes B,n+1)$ and $\nu.t.$ factors through $K(B,0)$. Then, if $g'':\mathbb{G}.(X)\longrightarrow\mathrm{Cosk}^n(B'')$ has the property that $r.g''$ is homotopic to $t.g''$, $\bar{\xi}.g''$ is homotopic to $\nu.t.g''$ which factors through $K(B,0)$ and so, by the covering homotopy theorem, there exists an $f'':\mathbb{G}.(X)\longrightarrow B''$ such that $\mathbf{d}f''$ is homotopic to g''. Now, let us suppose that $i_*[f.]$ is a t-null element, that is $i_*[f.]=[t.g.]$ for some $g.:\mathbb{G}.(X)\longrightarrow\mathrm{Cosk}^n(B.)$; then $C.(q.)t.g.$ represents a t-null class and also a r-null class,and therefore if $g''=\mathrm{Cosk}^n(q.)g.$, $r.g''$ is homotopic to $t.g''$ so that there exists $f'':\mathbb{G}.(X)\longrightarrow B''$ such that $\mathbf{d}f''$ is homotopic to $\mathrm{Cosk}^n(q.)g.$. It now follows that $i_*[f.]==i_*\delta_n[f'']$. ∎

The above proposition shows the exactness of the five first terms of the announced sequence. Now, in order to extend it to an infinite one we prove

<u>PROPOSITION</u> <u>12</u>.- Let $q.:B.\longrightarrow B''$ be a morphism of n-hypergroupoids in $\mathfrak{B}$, $n\geq1$, such that the induced morphism $\mathrm{Cosk}^n(q.):\mathrm{Cosk}^n(B.)\longrightarrow\mathrm{Cosk}^n(B'')$ is a trivial fibration. Then $q.$ is a fibration and the induced morphism $\mathrm{Cosk}^{n+1}(C.(q.)):\mathrm{Cosk}^{n+1}(C.(B.))\longrightarrow\mathrm{Cosk}^{n+1}(C.(B''))$ is a trivial fibration.

<u>Proof</u>:

Since $\mathrm{Cosk}^n(q.)$ is a trivial fibration it is surjective and so the induced morphism $N(\mathrm{Cosk}^n(q.))$ between the normal (Moore) complexes will be surjective too, but then $N(q.)$ is surjective and consequently $q.$ is a fibration.

On the other hand, if we consider the Moore complex of the morphism $\mathrm{Cosk}^{n+1}(C.(q.))$we have the morphism of complexes

$$\begin{array}{ccccccccccc}
\ldots 0 & \longrightarrow & \Pi_n(B.) & \longrightarrow & \Pi_n(B.)^2 & \longrightarrow & N_n(B.) & \longrightarrow \ldots \longrightarrow & N_1(B.) & \longrightarrow & B_0 \\
 & & \downarrow \Pi_n(q.) & & \downarrow \Pi_n(q.)^2 & & \downarrow N_n(q.) & & \downarrow N_1(q.) & & \downarrow q_0 \\
\ldots 0 & \longrightarrow & \Pi_n(B") & \longrightarrow & \Pi_n(B")^2 & \longrightarrow & N_n(B") & \longrightarrow \ldots \longrightarrow & N_1(B") & \longrightarrow & B"_0
\end{array}$$

which is clearly surjective and so $\mathrm{Cosk}^{n+1}(C.(q.))$is a fibration. Moreover, it is a trivial fibration because in both complexes the homology groups of the three non trivial first terms are themselves trivial and in the other terms they are isomorphic, since $\mathrm{Cosk}^n(q.)$ is a trivial fibration. ∎

Thus, according to the above propositions , we have finally

<u>THEOREM</u> <u>13</u>.- Let $q.:B. \longrightarrow B"$ be a morphism of n-hypergroupoids in $\mathfrak{B}$, $n\geq 1$, such that the induced morphism $\mathrm{Cosk}^n(q.):\mathrm{Cosk}^n(B.)\longrightarrow \mathrm{Cosk}^n(B")$ is a trivial fibration.Then there exists a long sequence

$$[\mathbb{G}.(X),B.]\xrightarrow{q_*}[\mathbb{G}.(X),B"]\xrightarrow{\delta_n}[\mathbb{G}.(X),K(C.(q.))]\xrightarrow{i_*}[\mathbb{G}.(X),C.(B.)]\longrightarrow\ldots$$

$$\ldots[\mathbb{G}.(X),C.^{(i}(B.)]\xrightarrow{q_*}[\mathbb{G}.(X),C.^{(i}(B")]\xrightarrow{\delta_{n+i}}[\mathbb{G}.(X),K(C.^{(i+1}(q.))]\ \ldots$$

in which $C.^{(i}(B.)=C.(C.^{(i-1}(B.))$, and which is exact in the sense that an element is in the image of q_* if and only if its image by the corresponding δ is neutral; an element is in the image of δ if and only if its image by i_* is a t-null element; an element is in the image of i_* if and only if its image by q_* is an r-null element. ∎

3. NON-ABELIAN $\mathbb{H}^n$ WITH COEFFICIENTS IN CROSSED MODULES: A LONG EXACT SEQUENCE.

Let Φ be a crossed module in $\mathfrak{B}$ and consider the associated groupoid $\mathfrak{G}(\Phi)$ and, via the functor Ner(-), the associated 1-hypergroupoid which we also denote $\mathfrak{G}(\Phi)$. Now, applying the construction $C.(-)$, we obtain a 2-hypergroupoid $C.(\mathfrak{G}(\Phi))$ (this is isomorphic to the 2-hypergroupoid $\mathfrak{G}^2(\Phi)$ used in [7]), and by iterating this process we have an embeding

$$\mathrm{XM}(\mathfrak{B}) \overset{\mathfrak{G}}{\simeq} \mathrm{GPD}(\mathfrak{B}) \xrightarrow{C.^{n-1}} n\text{-}\mathbf{HYPGD}(\mathfrak{B})$$

of the category of crossed modules into the category of n-hypergroupoids in $\mathfrak{B}$.

Note that, up to homotopical equivalence, this embeding, restricted to the abelian category of singular objects in $\mathfrak{B}$ (i.e. the internal abelian groups), is the same that the given by the Eilenberg-Mac Lane construction $K(-,n):\mathrm{Ab}(\mathfrak{B}) \longrightarrow n\text{-}\mathbf{HYPGD}(\mathfrak{B})$, since $\mathfrak{G}(M\longrightarrow 0)=K(M,1)$ and

$$C.^{n-1}(\mathfrak{G}(M\longrightarrow 0))\simeq C.^{n-2}(K(M,2)\simeq\ldots\simeq C.(K(M,n-1))\simeq K(M,n)$$

according to lemma 6.

Now given $X\in\mathfrak{B}$, we define the (n+1)-cohomology set of X with coefficients in Φ by

$$\mathbb{H}^{n+1}(X,\Phi)=[\mathbb{G}_\bullet(X),C_\bullet^{n-1}(\mathfrak{G}(\Phi))] \quad , \quad n\geq 1$$

Suppose now that $\Phi'=(B'\xrightarrow{\delta'}A',\mu')$, $\Phi=(B\xrightarrow{\delta}A,\mu)$, $\Phi''=(B''\xrightarrow{\delta''}A'',\mu'')$ are crossed modules in $\mathfrak{B}$ and recall that a sequence of morphisms of crossed modules $\Phi'\xrightarrow{(\text{in},1)}\Phi\xrightarrow{p=(p_1,p_0)}\Phi''$ is called a short exact sequence if $0\longrightarrow B'\longrightarrow B\xrightarrow{p_1}B''\longrightarrow 0$ is an exact sequence in $\mathfrak{B}$, p_0 is surjective and $\mathrm{Ker}(p_0)=\mathrm{Im}(\delta')$. This last condition (which is sometimes avoided in the definition of short exact sequences of crossed modules) links this concept of short exact sequence to the concept of short exact sequence of groupoids in the sense of Higgins, [18], via the equivalence of categories $\mathfrak{G}(-)$. We state this correspondence in the following proposition, which also allow us to apply the results of section 2 to obtain the announced long exact non-abelian cohomology sequence associated to a short exact sequence of crossed modules (theorem 15 below)

<u>PROPOSITION</u> <u>14</u>.- Let $p=(p_1,p_0):\Phi\longrightarrow\Phi''$ be a morphism of crossed modules in $\mathfrak{B}$, and $\mathfrak{G}(p):\mathfrak{G}(\Phi)\longrightarrow\mathfrak{G}(\Phi'')$ the induced morphism. Then the following conditions are equivalent:
i) p is a quotient map of crossed modules, i.e., $\mathrm{Ker}(p)\longrightarrow\Phi\longrightarrow\Phi''$ is a short exact sequence of crossed modules.
ii) $\mathfrak{G}(p)$ is a quotient map of groupoids in the sense of Higgins.
iii) p_0 is surjective, $\Pi_0(p)$ is an isomorphism and $\Pi_1(p)$ is surjective.
iv) $\mathrm{Cosk}^1(\mathfrak{G}(p)):\mathrm{Cosk}^1(\mathfrak{G}(\Phi))\longrightarrow\mathrm{Cosk}^1(\mathfrak{G}(\Phi''))$ is a trivial fibration.
<u>Proof</u>:
i) ⇔ ii) ⇔ iii) is essentially proved in [7] and iii) ⇔ iv) is straightforward.∎

Given $\Phi'\xrightarrow{(\text{in},1)}\Phi\xrightarrow{p=(p_1,p_0)}\Phi''$ a short exact sequence of crossed modules we can consider the induced morphism of n-hypergroupoids $C_\bullet^{n-1}(\mathfrak{G}(p)):C_\bullet^{n-1}(\mathfrak{G}(\Phi))\longrightarrow C_\bullet^{n-1}(\mathfrak{G}(\Phi''))$ and its n-hypergroupoid kernel $K(C_\bullet^{n-1}(\mathfrak{G}(p)))$ (see 2.3). Then, for any $X\in\mathfrak{B}$, we define the "relative (n+1)-cohomology set" as

$$\mathbb{H}_\Phi^{n+1}(X,\Phi') = [\mathbb{G}_\bullet(X),K(C_\bullet^{n-1}(\mathfrak{G}(p)))]$$

Let us note that if $\Phi'=(M'\longrightarrow 0)\longrightarrow\Phi=(M\longrightarrow 0)\xrightarrow{p}\Phi''=(M''\longrightarrow 0)$ is the short exact sequence of crossed modules associated to a short exact sequence of trivial X-modules, it is easy to see that the homotopical equivalence $C_\bullet^{n-1}(\mathfrak{G}(M\longrightarrow 0))\simeq K(M,n)$ restricts to an equivalence between $K(M',n)$ and $K(C_\bullet^{n-1}(\mathfrak{G}(p)))$, so that $\mathbb{H}_\Phi^{n+1}(X,\Phi')$ is the usual abelian cohomology with coefficients in M'.

Finally we have

<u>THEOREM</u> <u>15</u>.- Let $\Phi'\xrightarrow{(\text{in},1)}\Phi\xrightarrow{p=(p_1,p_0)}\Phi''$ be a short exact sequence of crossed modules in $\mathfrak{B}$ and X an object of $\mathfrak{B}$. There exists an exact sequence of sets with distinguished elements

$$\mathbb{H}_\Phi^2(X,\Phi')\longrightarrow\mathbb{H}^2(X,\Phi)\longrightarrow\mathbb{H}^2(X,\Phi'')\xrightarrow{\delta_1}\mathbb{H}_\Phi^3(X,\Phi')\longrightarrow\ldots$$

$$\ldots\mathbb{H}^n(X,\Phi'')\xrightarrow{\delta_{n-1}}\mathbb{H}_\Phi^{n+1}(X,\Phi')\longrightarrow\mathbb{H}^{n+1}(X,\Phi)\longrightarrow\mathbb{H}^{n+1}(X,\Phi'')\longrightarrow\ldots$$

Proof:

$\mathrm{Cosk}^1(\mathfrak{G}(p))$ is a trivial fibration by proposition 14 and then, by proposition 12, $\mathrm{Cosk}^n(C_\bullet^{n-1}(\mathfrak{G}(p))$ is a trivial fibration for all $n\geq 1$. Theorem 13 gives now the exactness of the sequence in the corresponding sense.■

REFERENCES

1. M.Barr, J.Beck. *Homology and standard constructions*, Lecture Notes in Math. *80* , Springer-Verlag (1969), 245-335.
2. R.Brown. *Some non abelian methods in homotopy theory and homological algebra* , in *Categorical Topology Proc. Conf. Toledo, Ohio, 1983,* ed. H.L.Bentley, et.al., Heldermann- Verlag, Berlin (1984), 108-146.
3. R.Brown. *Fibrations of Groupoids* , Journal of Algebra *15* (1970), 103-132.
4. R.Brown, N.D.Gilbert. *Algebraic models of 3-types and automorphisms structure for crossed modules*, Proc. London Math. Soc. *(3) 59* (1989), 51-73.
5. A.M.Cegarra, M.Bullejos. *Cohomology and higher dimensional Baer invariants*, Journal of Algebra 132 (2), (1990), 321-339.
6. A.M.Cegarra, M.Bullejos, A.R.Garzón, *Higher dimensional obstruction theory in algebraic categories*, Journal of Pure and Applied Algebra *49* (1987), 43-102.
7. A.M. Cegarra, A.R. Garzon, *Non-abelian cohomology of associative Algebras. The 9-term exact sequence*, Cahiers Top. et Geom. Diff. Catégoriques, vol XXX-4, (1989), 295-338.
8. E.B. Curtis, *Simplicial homotopy theory*, Advances in Mathe. *6* (1971) , 107-209.
9. P.Dedecker, *Sur la n-cohomologie non abelienne*, C.R. Acad.Sc. Paris, t. *260*, 1 (1965), 4137-4139.
10. P.Dedecker, *Cohomologie non-abelienne*, Mimeographie, Fac. Sc. Lille, 1965.
11. P.Dedecker, *Three dimensional non abelian cohomology for groups*, Lecture Notes in Math. *92*, Springer-Verlag (1969), 32-64.
12. P. Dedecker, A.S.T. Lue, *A non-abelian 2-dimensional cohomology for associative algebras*, Bull. A.M.S. 72, (1966), 1044-1050.
13. J.Duskin, *Simplicial methods and the interpretation of triple cohomology*. Memoir A.M.S. vol *3*, issue *2, 163 (1975)*.
14. J.Duskin, *Non-abelian monadic cohomology and low dimensional obstruction theory*. Math. Forschung Ins. Oberwolfach Tagungsbericht 33, (1976).
15. J.Duskin, *Group cohomology and torsors*, Mimeographed notes . (1984).
16. G.J. Ellis, *Crossed modules and their higher dimensional analogues*, Ph.D. thesis. Univ. of Wales (1984).
17. P. Glenn, *Realization of cohomology classes in arbitrary exact categories*, Journal of Pure and Applied Algebra 1 , 33-107, (1982).
18. P.J.Higgins, *Categories and groupoids*, Van Nostrand Reinhold Math. Estudes *32* (1971)
19. P.J.Higgins, *Groups with multiple operators*. Proc. London Math. Soc. 6, (1956), 366-416.
20. D. Kan, *On homotopy theory and c.s.s. groups*, Annals of Math. *68, 1* (1958), 38-53.
21. F.Keune, *Homotopical algebra and algebraic K-theory*, Thesis Univ. of Amsterdam (1972).
22. R. Lavendhomme, J.R. Roisin, *Cohomologie non abelienne de structures algebriques* . J. of Algebra *67* (1980), 385-414.

23. J.-L. Loday, *Spaces with finitely many non trivial homotopy groups*, J. of Pure and Applied Algebra *24* (1982), 179-202.

24. A.S.T. Lue, *Non-abelian cohomology of associative algebras*, J. Math. Oxford (2) 19, (1968), 159-180.

25. A. S.-T.Lue, *Cohomology of groups relative to a variety*, J. of Algebra *69* (1981), 155-174.

26. J.P. May, *Simplicial objects in Algebraic Topology*, Van Nostrand, (1976).

27. T. Porter, *Extensions, crossed modules and internal categories in categories of groups with operations*. Edinburgh Math. Soc. 30, (1987), 415-429.

28. D.Quillen, *Homotopical Algebra* Lecture Notes in Math. *43* , Springer-Verlag (1967).

29. J.P. Van Deuren, *Etude de l'obstruction au relevement d'un cocycle non abelien. Essai d'une definition d'une 3-cohomologie non abelienne*, Seminaire de mathematique pure, Univ. Catholique de Louvaine (1978).

30. J.H.C.Whitehead, *Combinatorial homotopy* ,II, Bull. A.M.S. *55*, (1949), 496-543.

31. G.W. Whitehead, *Elements of homotopy theory*, Springer-Verlag (1978).

ALGEBRAICALLY COMPLETE CATEGORIES

by
PETER FREYD
University of Pennsylvania

The title I used when I gave this talk was *Computer Science Contradicts Mathematics.* Many modern programming languages are inconsistent with standard mathematical foundations. The task of finding sound interpretations for what it is that computer scientists do strikes this writer as, perhaps, the highest type of applied mathematics. It is akin to the process that has been going on throughout the 20th Century with respect to physics. The interaction between the mathematicians and the practitioners in each case has resulted in the growth of both subjects. I wish here to report on just one side of this process: the effect of computer science on category theory, at least on my category theory. The effect has been pervasive and I can report here just one part of it, but a part for which I owe a lot to Italy and the Italians. I must particularly thank Pino Rosolini for the many conversations I had during a very productive visit to Parma.

In many programming languages one may create new "data types" as "minimal fixed points" of "data constructors". In the broad sense of the word "complete" this is a completeness condition on the family of things that are considered to be data types. A computer scientist typically state this in terms of "domain equations", that is, given a "type constructor", T, one seeks a minimal solution to the equation $D = TD$. Let me relax the conditions a bit. If we consider the category of "data types" and specialize to the case that the "data constructor" is a covariant functor and weaken the equality to an isomorphism we arrive at the following condition on a category:

IOP: For every covariant endo-functor, T, there is a ***T-INVARIANT OBJECT*** *that is, an object D and an isomorphism* $TD \to D$.

At the end of this lecture Adamek will tell me that in his recent book it is shown that this condition contradicts the usual categorical notion of completeness. To be precise, a category satisfying IOP and that has arbitrary limits must be a pre-ordered set, that is, there is never more than one map from one object to another. Here is a proof, albeit not the proof in Adamek's book: suppose that the hom-set (A,B) has more than one element; let T be defined by $TX = \Pi_{S(X)}B$, that is, the $S(X)$-fold cartesian power of B where S is the set-valued contravariant functor that sends X to the power set of the set of maps (A,X). If D were an invariant object for T then the number of maps (A,D) would be the same as the number of maps $(A,TD) \cong \Pi_{S(D)}(A,B)$ which is a clear contradiction in the classical foundations. (The classical Tarski theorem, it should be recalled, says that any complete lattice satisfies IOP.)

IOP does not catch the notion of "minimallity". We could strengthen it to the condition that the category of invariant objects has an initial object. But better, and closer to what is actually assumed in various programming languages, we consider the larger category of ***T*-ALGEBRAS**: to wit, the category whose objects are maps of the form $a{:}TA \to A$. and whose maps look like:

The author was supported by the Office of Naval Research when this research was performed.

$$\begin{array}{ccc} TA & \xrightarrow{Tx} & TB \\ a \downarrow & & \downarrow b \\ A & \xrightarrow[x]{} & B \end{array}$$

If $f{:}TF \to F$ is initial in this category it is called the **INITIAL**, or free, T-algebra. (The phrase "inductive type" is often used by computer scientists to describe initial algebras.) Lambek was the first to show that initiallity implies invariance, that is, f is necessarily an isomorphism. We are thus led to:

A category is called ***ALGEBRAICALLY COMPLETE*** *if for every covariant endofunctor, T, there is an initial T-algebra.*

The most famous initial algebra is for the functor $TX = 1 + X$. A T-algebra structure on an object A may be viewed as a "point" $1 \to A$ and an endomorphism $A \to A$. The Lawvere definition of **NATURAL NUMBERS** object may be restated as the initial algebra for this functor. (This works for any category with coproducts.)

The dual property is of almost equal interest in computer science. A **T-COALGEBRA** is a map of the form $a{:}A \to TA$. The co-Lambek lemma says that if $f'{:}F' \to TF'$ is a final T-coalgebra then f' is an isomorphism. In the category of sets a co-algebra structure on S for the functor $TX = 1 + X$, may be viewed as a partial endomorphism on S. The final coalgebra is a universal partial endomorphism. We may construct it as the "extended natural numbers" that is, the natural numbers with a point "at infinity" adjoined. The universal partial endomorphism is the predecessor function, understood to be undefined at zero and with infinity as a fixed point.

Objects that arise in programming are rarely good only for input or only for output. I have called this the **PRINCIPAL OF VERSALITY**: every universal mapping definition is equivalent to a dual definition. This principal is quite untrue in mathematics and is not claimed to be universally true in computer science. But it leads to:

A category is said to be ***ALGEBRAICALLY COMPACT*** *if every covariant endofunctor has an initial algebra and a final coalgebra and they are canonically isomorphic.*

The **canonical isomorphism** is defined as follows: for a covariant endofunctor T let $f{:}TF \to F$ be an initial T-algebra and $f'{:}F' \to TF'$ be a final T-coalgebra. Let $g{:}F \to F'$ be the unique map such that:

$$\begin{array}{ccc} F & \xrightarrow{g} & F' \\ f^{1} \downarrow & & \downarrow f' \\ TF & \xrightarrow[Tg]{} & TF' \end{array}$$

If g is an isomorphism it is called the canonical isomorphism. (We used the fact that f is an isomorphism and f' is final. Alternatively we could have used the fact that f' is an isomorphism and f is initial. The same map $g{:}F \to F'$ would be defined.) When the initial T-algebra and final T-coalgebra are canonically isomorphic we will, in practice, take the initial T-algebra structure $f{:}TF \to F$ as primitive and use $f^{1}{:}\,F \to TF$ as the final T-coalgebra.

The category of T-invariant objects appears as a full subcategory of both the categories of T-algebras and T-coalgebras. Algebraic compactness thus says, in particular, that the category of T-invariant objects is a **PUNCTUATED** category, that is, a category in which the terminator and coterminator coincide. As in any such category, the (bi)terminator appears as a proper retract of every non-terminator and is thus distinguished by the fact that it has only one idempotent. For an algebraically compact category, therefore, we have that each covariant endofunctor has a distinguished invariant object D which is a proper retract—as an invariant object—of every undistinguished invariant object. It came as a complete surprise to me that this remains true (as we will see) even when the word covariant is replaced with contravariant.

Note in the case $TX = 1 + X$ on the category of sets that F and F' are isomorphic (F' is a countably infinite set) but that g is not an isomorphism: that is, the principal of versality fails.

In our LICS90 paper on *Extensional PERS* Mulry, Rosolini, D.Scott and I show that the realizable topos has a small full reflective subcategory that is algebraically compact in the relevant sense, that is, the condition holds for every endofunctor that is definable as a functor in the topos. This example is a rich one in terms of other structure. If one wants an impoverished example of an algebraically compact category take the full subcategory of the category of groups of all those groups that have at most two elements (or better, chose a skeletal subcategory with just two objects and five maps). Note that the only posets that as categories are algebraically compact are one-element posets.

Let me go back to algebraic completeness. I will need some theorems true for any category. First, a standard lemma implicit in the proof of the Lambek lemma:

LEMMA:

If $g: TF \to F$ is an initial T-algebra then so is $Tg: T^2F \to TF$.

Let h be the unique map such that:

$$\begin{array}{ccc} TF & \xrightarrow{Th} & T^2F \\ g\downarrow & & \downarrow Tg \\ F & \xrightarrow[h]{} & TF \end{array}$$

We also have the diagram:

$$\begin{array}{ccc} T^2F & \xrightarrow{Tg} & TF \\ Tg\downarrow & & \downarrow g \\ TF & \xrightarrow[g]{} & F \end{array}$$

The uniqueness condition says that $h;g$ is the identity map on F (the semicolon is the standard notation in computer science for the composition of operations in execution order, that is, $h;g$ means first do h then do g). If we already know that g is an isomorphism then, of course, h is forced to be its inverse. But we may as well take the occasion to prove the Lambek lemma. Since $h;g$ is the identity on F, $Th;Tg$ is the identity map on TF. In the commutative diagram that defines h it says that $g;h = Th;Tg$, hence $g;h$ is the identity map and g is seen to have a two-sided inverse. ■

A new theorem of surprisingly many uses:

THE ITERATED SQUARE THEOREM:

For any category $\mathcal{A}$ and endofunctor $T: \mathcal{A} \to \mathcal{A}$ suppose that T^2 has an initial algebra. Then so does T.

Let $g: T^2F \to F$ be an initial T^2-algebra. Let h be the unique map such that:

$$\begin{array}{ccc} T^2F & \xrightarrow{Th} & T^3F \\ g\downarrow & & \downarrow Tg \\ F & \xrightarrow[h]{} & TF \end{array}$$

We may apply T to this diagram and paste to obtain:

$$\begin{array}{ccccc} T^2F & \xrightarrow{Th} & T^3F & \xrightarrow{T^2h} & T^4F \\ g\downarrow & & \downarrow & & \downarrow T^2g \\ F & \xrightarrow[h]{} & TF & \xrightarrow[Th]{} & T^2F \end{array}$$

The rightmost T^2-algebra is T^2 applied to the initial T^2-algebra. The lemma above says therefore that it is itself an initial T^2-algebra and the T^2-algebra-map $h;Th$ must be an isomorphism. Thus h has a right inverse and Th has a left inverse. But any functor preserves the existence of right inverses, hence Th has a right inverse. In any category any map with both inverses is, in fact, an isomorphism. Since both $h;Th$ and Th are isomorphisms it follows that h is an isomorphism.

Note that it follows from the argument above that $Tg:T^3F \rightarrow TF$ is also an initial T^2-algebra.

Define $f:TF \rightarrow F$ to be the inverse of h. We wish to show it is an initial T-algebra. Note that from the construction of h that $Tf;f = g$. Given an arbitrary T-algebra $a:TA \rightarrow A$ let $x:F \rightarrow A$ be the unique map such that:

$$\begin{array}{ccc} & T^2x & \\ T^2F & \rightarrow & T^2A \\ Tf\ \downarrow & & \downarrow\ Ta \\ TF & \rightarrow & TA \\ f\ \downarrow & & \downarrow\ a \\ F & \longrightarrow & A \\ & x & \end{array}$$

The middle horizontal map must exist since f is an isomorphism We temporarily denote it as $y:TF \rightarrow TA$. If we can show that $y = Tx$ we will be done. (The uniqueness condition is easy). If we apply T to the lower rectangle, remove it, and then paste it on top we obtain:

$$\begin{array}{ccc} & T^2y & \\ T^3F & \rightarrow & T^3A \\ T^2f\ \downarrow & & \downarrow\ T^2a \\ T^2F & \rightarrow & T^2A \\ Tf\ \downarrow & & \downarrow\ Ta \\ TF & \longrightarrow & TA \\ & y & \end{array}$$

Since $T^2f;Tf = Tg$ we know that the left T^2-algebra is initial and y is unique. By applying T to the previous diagram we obtain:

$$\begin{array}{ccc} & T^3x & \\ T^3F & \rightarrow & T^3A \\ T^2f\ \downarrow & & \downarrow\ T^2a \\ T^2F & \rightarrow & T^2A \\ Tf\ \downarrow & & \downarrow\ Ta \\ TF & \longrightarrow & TA \\ & Tx & \end{array}$$

The uniqueness of y thus forces $y = Tx$. ∎

There is more in this proof than was recorded in the statement of the theorem. We will need:

LEMMA:

If $f:TF \rightarrow F$ is initial and if there is an initial T^2-algebra then $(Tf;f):T^2F \rightarrow F$ is initial.

We showed that $h;Th$ is an isomorphism from the initial T^2-algebra to T^2 applied to the initial T^2-algebra. As in the last lemma, its inverse must be g. ∎

(Several comments about the theorem. First, the proof may be easily generalized to give *THE ITERATED POWERS THEOREM: if T^n has an initial algebra then so does T.* The converses are not true. Given T let N be the set of positive integers *{ n / there is an initial T-algebra }*. The theorem says that N is closed under divisors. For any such set of positive integers, N, we may construct a discrete category, $\mathcal{A}$ and a permutation T as follows: let $\mathcal{A}$ be the disjoint union of a fixed-point for T together with an m-cycle for T for each m <u>not</u> in N. ((A permutation on a discrete category has an initial algebra iff it has a unique fixed point.)

(If $\mathcal{A}$ has products then the converses do hold. For the case $n = 2$ argue as follows: given an initial T-algebra $f:TF \rightarrow F$ show that $(Tf;f):T^2F \rightarrow F$ is initial by taking any $a:T^2A \rightarrow A$ and considering the T-algebra $\langle Tr;a, Tl\rangle:T(A \times TA) \rightarrow A \times TA$ ((where l and r are the left and right projection maps from the product)); let $\langle x,y\rangle: F \rightarrow A \times TA$ be the unique map of T-algebras. It is routine to verify that necessarily $y = f^{-1};Tx$ and that x is the unique map of T^2-algebras.)

Back to algebraically complete categories:

THE PRODUCT THEOREM FOR ALGEBRAICALLY COMPLETE CATEGORIES:

If $\mathcal{A}'$ and $\mathcal{A}$ are algebraically complete then so is $\mathcal{A}' \times \mathcal{A}$

Any endofunctor on the product may be resolved into its coordinate functors. Hence we suppose that we are given

$T':\mathcal{A}'\times\mathcal{A}\to\mathcal{A}'$ and $T':\mathcal{A}'\times\mathcal{A}\to\mathcal{A}$. We wish to construct an initial algebra for the functor that sends an object $\langle A',A\rangle$ to $\langle T'A'A, TA'A\rangle$.

First define a functor $F':\mathcal{A}\to\mathcal{A}'$ as follows: given A in $\mathcal{A}$ let $f'_A:T'(F'A)A\to F'A$ be an initial algebra for the endofunctor $T'(-)A$ on $\mathcal{A}'$, that is, the functor that carries X' to $T'X'A$. (In the extensional PERs case a choice-free construction of initial algebras was given, thus obviating the use of the axiom of choice here.) Given a map $x:A\to B$, define $F'x$ to be the unique map of $T'(-)A$-algebras from $f'_A:T'(F'A)A\to F'A$ to $(T'1x)f'_B:T'(F'B)A\to F'B$. By a little re-diagramming, $F'x$ may be characterized as the unique map that fits in:

$$\begin{array}{ccc} T'(F'A)A & \xrightarrow{T'(F'x)x} & T'(F'B)B \\ f'_A\downarrow & & \downarrow f'_B \\ F'A & \xrightarrow[F'x]{} & F'B \end{array}$$

Similarly, we define $F:\mathcal{A}'\to\mathcal{A}$, so that $f_A:TA'(FA')\to FA'$ is initial for the endofunctor that sends X in $\mathcal{A}$ to $TA'X$.

The iterated square theorem says that the endofunctor on $\mathcal{A}'\times\mathcal{A}$ that sends $\langle A', A\rangle$ to $\langle F'A, FA'\rangle$ must have an initial algebra since its square is just a cartesian product of endofunctors (the pair of initial algebras will work fine). Accordingly let $\langle q',q\rangle:\langle F'D, FD'\rangle\to\langle D', D\rangle$ be its initial algebra. We will show that $\langle D',D\rangle$ has an initial algebra structure for the original functor. We need a pair of maps, $d':T'D'D\to D'$ and $d:TD'D\to D$. We define d' to be the unique map such that:

$$\begin{array}{ccc} T'(F'D)D & \xrightarrow{T'q'1} & T'D'D \\ f'_D\downarrow & & \downarrow d' \\ F'D & \xrightarrow[q']{} & D' \end{array}$$

(We are using only that $T'q'1$ is an isomorphism.) d is defined similarly.

Given $\langle a',a\rangle:\langle T'A'A, TA'A\rangle\to\langle A', A\rangle$ we first define $\langle y',y\rangle: \langle F'A,FA'\rangle\to\langle A',A\rangle$ by taking y' to be the unique map such that:

$$\begin{array}{ccc} T'(F'A)A & \xrightarrow{T'y'1} & T'A'A \\ f'_A\downarrow & & \downarrow a' \\ F'A & \xrightarrow[y']{} & A' \end{array}$$

using the initiallity of $F'A$. y is similarly defined. Define $\langle x',x\rangle$ to be the unique pair such that:

$$\begin{array}{ccc} \langle F'D, FD'\rangle & \xrightarrow{\langle F'x, Fx'\rangle} & \langle F'A, FA'\rangle \\ \langle q',q\rangle\downarrow & & \downarrow\langle y',y\rangle \\ \langle D',D\rangle & \xrightarrow[\langle x',x\rangle]{} & \langle A',A\rangle \end{array}$$

To show that $\langle x',x\rangle$ is a map of algebras for the original endofunctor on $\mathcal{A}'\times\mathcal{A}$ consider the pair of diagrams, only one of which need be shown:

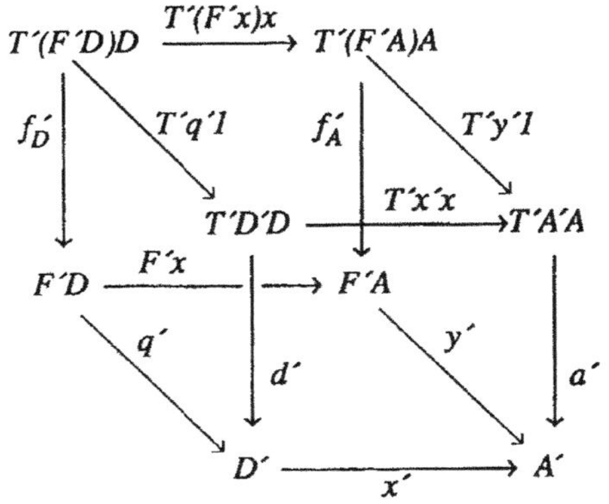

We wish to show that the front face commutes. The bottom face commutes by definition of $x';x$; the left face by definition of d'; the back face by definition of $F'x$; the right face by definition of y'; the top face because $T'(F'x;y')x = T'(q';x')x$ (the bottom face already says $F'x;y' = q';x'$). Since $T'q'1$ is an isomorphism we may conclude that the front face commutes. The uniqueness of $x';x$ is routinely verified from these same diagrams. ■

As just one corollary we obtain the "dinaturality" of the two-functor that delivers initial algebras. That is, given a functor $F:\mathcal{A} \to \mathfrak{B}$ the induced diagram of categories commutes in the relevant sense:

$$\begin{array}{ccccc} & & (\mathcal{A},\mathcal{A}) & \longrightarrow & \mathcal{A} \\ & \nearrow & & & \\ (\mathfrak{B},\mathcal{A}) & & & & \downarrow \\ & \searrow & & & \\ & & (\mathfrak{B},\mathfrak{B}) & \longrightarrow & \mathfrak{B} \end{array}$$

The commutativity at the object level says:

COROLLARY:

Given $F:\mathcal{A} \to \mathfrak{B}$ and $G:\mathfrak{B} \to \mathcal{A}$, F applied to the initial (F;G)-algebra is the initial (G;F)-algebra.

To prove it apply the last theorem to the endofunctor on $\mathcal{A} \times \mathfrak{B}$ that sends $<A,B>$ to $<GB,FA>$. ■

Returning to algebraically compact categories it is just a matter of proof inspection to obtain:

THE PRODUCT THEOREM FOR ALGEBRAICALLY COMPACT CATEGORIES:

If $\mathcal{A}'$ and $\mathcal{A}$ are algebraically compact then so is $\mathcal{A}' \times \mathcal{A}$

The proof inspection consists of showing that if each of the *isomorphisms* that arise in the the product theorem for algebraically complete categories is inverted the result is the construction for the final coalgebra. ■

Algebraic compactness is a self-dual property, that is, if $\mathcal{A}$ is compact then so is $\mathcal{A}^o$. Hence:

COROLLARY:

If $\mathcal{A}$ is algebraically compact then so is $\mathcal{A}^o \times \mathcal{A}$ ■

Now suppose that T is a *contravariant* endofunctor on an algebraically compact category $\mathcal{A}$ Consider the covariant endofunctor on $\mathcal{A}^o \times \mathcal{A}$ that sends $<A', A>$ to $<TA, TA'>$. If $<f',f>:<T'D,TD'> \to <D',D>$ is its initial algebra then we know that it is characterized by the fact that there is only one pair of endomorphisms $<x', x>$ such that:

$$\begin{array}{ccc} & Tx & \\ T'D & \to & T'D \\ f' \downarrow & & \downarrow f' \\ D' & \longrightarrow & D', \\ & x' & \end{array} \qquad \begin{array}{ccc} & Tx' & \\ TD' & \to & TD' \\ f \downarrow & & \downarrow f \\ D & \longrightarrow & D. \\ & x & \end{array}$$

But, of course, the same can be said about the pair $<D, D'>$ hence $<D', D> \cong <D, D'>$. All of which says that the contravariant functor itself has an invariant object and, in the relevant sense, an invariant object characterized by the fact that it has only one invariant idempotent.

We may extend the argument to functors of mixed variance and, as usual, it suffices to consider the case of a bifunctor, T, contravariant on the first variable, covariant on the second. By considering the covariant endofunctor that sends $<A', A>$ to $<TAA', TA'A>$ we obtain:

THEOREM:

For any bifunctor, T, on an algebraically compact category there is a minimal invariant object, $f:TDD \cong D$, characterized by the fact that there is only one idempotent (indeed, only one endomorphism) such that:

$$\begin{array}{ccc} & Txx & \\ TDD & \to & TDD \\ f \downarrow & & \downarrow f \\ D & \longrightarrow & D. \\ & x & \end{array}$$

■

We look at an easy lemma:

LEMMA:

For any category the identity functor has an initial algebra iff the category has a coterminator.

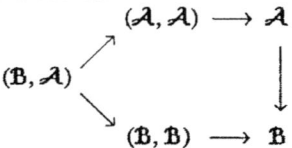

Any functor that preserves the coterminator is easily seen to have the coterminator as an initial algebra. Conversely, suppose that $f: F \to F$ is an initial I-algebra. The uniqueness condition says that it has only one endomorphism as an algebra, hence f must be the identity map (because f commutes with f). Thus F is a coterminator in the full subcategory of those I-algebras for which the structure map is the identity map. But that category is, of course, isomorphic to the given category. ■

As an immediate corollary:

LEMMA:

Any algebraically compact category is punctuated. ■

The tradition in mathematics is to denote the biterminator in a punctuated category as 0 and in computer science (and group theory) as 1. The unique map from 1 to an object A is denoted $\perp: 1 \to A$, pronounced "bottom" and thought of as the "undefined" or "never defined" state. Bottom-preserving maps are called **STRICT**. The purpose of the undefined state is often understood to allow partial maps to be encoded as maps.

Total maps may be viewed as special cases of partial maps, to wit, those maps that not only preserve the bottom but reflect it, that is those maps for which

$$\begin{array}{ccc} 1 & \to & 1 \\ \perp\downarrow & & \downarrow\perp \\ A & \to & B \end{array}$$

is a pullback. In my paper on the well-ordering of choice objects I said that such maps "strictly preserve the base point" and referred to them as "strict maps." Because of the widespread use of the word "strict" to mean bottom-preservation I am now forced to escalate the language. I will say that such maps are "restrictive" and the category of such maps to be the category of **RESTRICT MAPS.**

In many of the categories that arise when one tries to find mathematical semantics for programming languages it is the case that the lluf subcategory of restrict maps is coreflective. The coreflection of an object A is often denoted $A_\perp$ and pronounced "A lifted." In the extensional PERs example the category of restrict maps fails to be coreflective, indeed does not even have a terminator. There is, however, a maximal subterminator and it has been of interest for all sorts of reasons for many years and is traditionally denoted Σ. If we cut down to the category of restrict maps between objects that do have a restrict map to Σ we do obtain a coreflective subcategory. Indeed, the construction of the coreflection is not difficult: given A let ΣA be the maximal subobject of $\Sigma \times A$ such that $\Sigma A \to \Sigma \times A \to \Sigma$ is a restrict map.

As for any coreflective lluf subcategory the coreflection is an embedding functor and we may view the situation differently. Letting $\mathcal{A}$ denote the algebraically compact category and $\mathfrak{R}$ the coreflective subcategory of restrict maps we may view the embedding $\mathcal{A} \to \mathfrak{R}$ as an <u>enlargement</u> of $\mathcal{A}$. Computer scientists call the objects of the ambient category "pre-domains." The objects that are isomorphic to objects coming from $\mathcal{A}$ are then called "domains" and their full subcategory, $\mathfrak{D}$, is called the "category of domains." In this context, $\mathcal{A}$ is viewed as the lluf subcategory of strict maps.

I have taken $\mathcal{A}$ as primitive with $\mathfrak{R}$ and $\mathfrak{D}$ as constructs. If the truth be told, that is not the way it usually happens. Typically $\mathfrak{D}$ is first defined then the notion of strict maps and then the notion of pre-domains (by "bottom deletion"). When viewed this way $A_\perp$, or as I will prefer, ΣA is the <u>reflection of</u> a domain in the subcategory of strict maps.

In a punctuated category $1 + X$ is naturally equivalent to the identity functor and the natural numbers object is just the biterminator. We consider some variations. We use the topologists notation for the coproduct in a punctuated category, the "wedge" $A \vee B$ (called the "coalesced sum" in computer science). The variations on $1 \vee X$ to be considered are the three functors $\Sigma \vee X$, $1 \vee \Sigma X$, $\Sigma \vee \Sigma X$, where I write Σ for $\Sigma 1$.

In the traditional models, the objects are "CPO's", that is, posets with suprema for directed sets, and the maps are maps that preserve such suprema. The variations on the natural numbers have all been studied and given names. If I may quote from myself:

A $\Sigma \vee X$ algebra is an object A together with strict maps $\Sigma \circ\!\!\to A$ and $A \circ\!\!\to A$. (The decoration on the horizontal arrows is the standard way of indicating strict maps.) Because Σ is the reflection of 1 into the subcategory of strict maps, we may rephrase: an object A together with an "initial point" $1 \to A$ and a strict "successor map" $A \circ\!\!\to A$. The initial algebra is called the **FLAT** or **HORIZONTAL NATURAL NUMBERS** and is usually pictured as

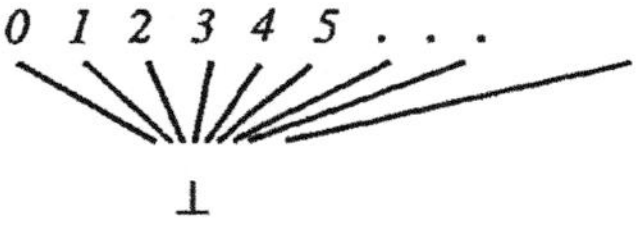

The successor map shifts the horizontal row to the right, keeping the bottom fixed. The left-most point on the horizontal row is the initial point.

A $1 \vee \Sigma X$ algebra is an object A together with strict maps $1 \circ\!\!\to A$ and $\Sigma A \circ\!\!\to A$. Of course, there is only one strict map of the form $1 \circ\!\!\to A$ (indeed, the functor in question is isomorphic to ΣX). Because ΣA is the reflection of A into the subcategory of strict maps we may rephrase: a $1 \vee \Sigma X$ algebra is an object A together with a (not necessarily strict) successor map $A \to A$. The initial algebra is called the **ORDERED** or **VERTICAL NATURAL NUMBERS** and is usually pictured as an ascending chain climbing forever to an invisible point at infinity:

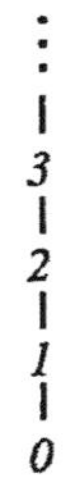

The successor map is the up-shift, keeping the point at infinity fixed.

A $\Sigma \vee \Sigma X$ algebra—following similar arguments—may be rephrased to be an object A together with an initial point $1 \to A$ and a successor map $A \to A$, neither of which need be strict. The initial algebra is called the **LAZY** or **OBLIQUE NATURAL NUMBERS** and is usually pictured as:

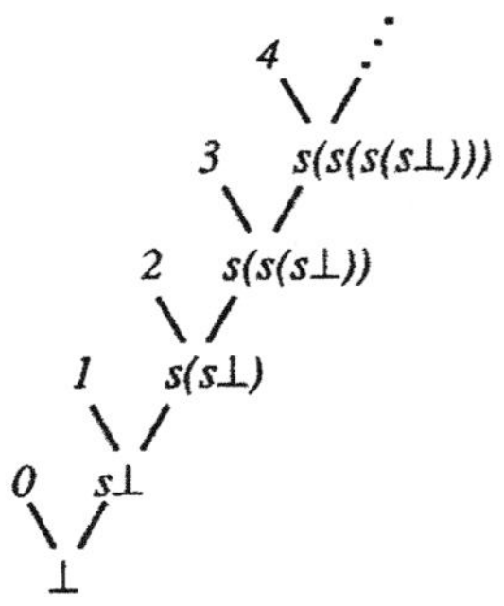

where it must be understood that there's a point at infinity above all the s-iterates of $\bot$. The points marked by numbers are each maximal elements. 0 is the initial point. The successor map is the north-east shift, again with the point at infinity fixed.

Every endofunctor that arises on either of the ambient categories seems always to have the subcategory of strict maps as an invariant subcategory. Indeed, in the extensional PERs case, it is provably the case that all functors definable in the topos have this property. All of which gives interest to the easily verified:

THE REFLECTIVE SUBCATEGORY LEMMA:
If $\mathcal{A}$ is a reflective subcategory of $\mathfrak{B}$ and if T is an endofunctor on $\mathfrak{B}$ which has $\mathcal{A}$ as an invariant subcategory then a final T-coalgebra for T viewed as an endofunctor on $\mathcal{A}$ remains a final T-coalgebra for T viewed as an endofunctor on $\mathfrak{B}$. ■

Computer scientists have been using fixed points for endomorphisms longer than invariant objects for functors. To close this paper I wish to show that algebraic compactness entails much of the standard material on fixed points.

Viewing Σ as a endofunctor on domains we note that the subcategory of strict maps is surely invariant and we will apply the reflective subcategory lemma to the vertical natural numbers. We define $s{:}N \to N$ as $u;f$ where we are using $u{:}N \to \Sigma N$ to denote the reflector map. It is the universal endomorphism in the category of domains, that is for every $a{:}A \to A$ there is a unique <u>strict</u> map $Ja : N \to A$ such that:

$$\begin{array}{ccc} & Ja & \\ N & \circ\!\!\to & A \\ s\downarrow & & \downarrow a \\ N & \circ\!\!\to & A \\ & Ja & \end{array}$$

Recall that the decoration on the horizontal indicate strict maps. It is important that the vertical maps needn't be strict and that the horizontal map must be. Note that if a is strict then Ja is the map constantly equal to the bottom of A.

Consider the Σ-coalgebra $u{:}1 \to \Sigma 1$. Using the reflective subcategory lemma we know that there exists a unique map $y{:}1 \to N$ such that:

$$\begin{array}{ccc} & y & \\ 1 & \to & N \\ u\downarrow & & \downarrow f^{-1} \\ \Sigma 1 & \to & \Sigma N. \\ & \Sigma y & \end{array}$$

Note that we also have:

$$\begin{array}{ccc} & y & \\ 1 & \to & N \\ u\downarrow & & \downarrow u \\ \Sigma 1 & \to & \Sigma N. \\ & \Sigma y & \end{array}$$

From $y;f^{-1} = y;u$ we obtain $y = y;f^{-1};f = y;u;f = y;s$. That is, y is a fixed point of the universal endomorphism s. It is the unique fixed point: for any y the last square still commutes, from which one may infer that the previous square commutes iff $y;f^{-1} = y;u$ hence iff $y = y;f^{-1};f = y;u;f = y;s$. We will use the traditional name for the unique fixed point of the universal endomorphism $\infty{:}1 \to N$.

Since the universal endomorphism has a fixed point it is the case that every endomorphism between domains has a fixed point. By positing good invariant objects for endofunctors we have achieved fixed points for endomorphisms. (Note that it is a complete triviality that strict endomorphisms have fixed points. The computational usefulness of fixed points requires bottom-raising maps. From this point of view it is more useful to think of the bottom not as "never defined" but as "not yet defined.")

We have more than the sheer existence of fixed points, we have a family of operators, traditionally denoted Y, one for each domain, which yield fixed points for endo-morphisms: given $a{:}A \to A$ we define $Ya{:}1 \to A$ to be $y;Ja$. **PLOTKIN'S AXIOM** for Y is now immediate, that is, if

$$\begin{array}{ccc} & x & \\ B & \circ\!\!\to & A \\ b\downarrow & & \downarrow a \\ B & \circ\!\!\to & A \\ & x & \end{array}$$

then $Yb;x = Ya$. And since the universal endomorphism has a unique fixed point, the Plotkin axiom characterizes Y. (Which, albeit in a different context, was why Plotkin stated the axiom.)

LEMMA:

$Y(a^2) = Ya$ *for any endomorphism* $a{:}A \to A$.

First we prove it for the special case of $s{:}N \to N$. Not surprisingly we use the iterated square theorem. We will show that s^2 has only one fixed point, the one we already know, $\infty{:}1 \to N$. Consider the functor Σ^2. The initial Σ^2-algebra may be taken to be $\Sigma f;f{:}\Sigma^2 N \to N$. The reflective subcategory theorem says that there is at most one map $y{:}1 \to N$ such that :

$$\begin{array}{ccc} 1 & \xrightarrow{y} & N \\ {\scriptstyle u;\Sigma u}\downarrow & & \downarrow{\scriptstyle f^{-1};\Sigma f^{-1}} \\ \Sigma^2 1 & \xrightarrow[\Sigma^2 y]{} & \Sigma^2 N \end{array} .$$

But for any y we have:

$$\begin{array}{ccc} 1 & \xrightarrow{y} & N \\ {\scriptstyle u;\Sigma u}\downarrow & & \downarrow{\scriptstyle u;\Sigma u} \\ \Sigma^2 1 & \xrightarrow[\Sigma^2 y]{} & \Sigma^2 N \end{array} .$$

Hence the first square commutes iff $y;f^{-1};\Sigma f^{-1} = y;u;\Sigma u$. We may follow each side with the isomorphism $\Sigma f;f$ to obtain: $y = y;u;\Sigma u;\Sigma f;f = y;u;\Sigma(u;f);f = y;u;\Sigma s;f = y;s;u;f = y;s^2$, that is, the first square commutes iff y is a fixed point for s^2. (We used the equation $u;\Sigma s = s;u$, the neutrality of u.) The uniqueness condition for final coalgebras thus forces $y = \infty$.

Given $a{:}A \to A$ consider:

$$\begin{array}{ccc} N & \overset{Ja}{\hookrightarrow} & A \\ {\scriptstyle s}\downarrow & & \downarrow{\scriptstyle a} \\ N & \hookrightarrow & A \\ {\scriptstyle s}\downarrow & & \downarrow{\scriptstyle a} \\ N & \underset{Ja}{\hookrightarrow} & A \end{array}$$

The Plotkin axiom says that Ja carries $Y(s^2)$ to $Y(a^2)$. But we have just seen that $Y(s^2) = \infty$. But by definition Ja carries ∞ to Ya.

Assuming that our algebraically compact category has products we can now easily prove the dinaturality of Y (just as for the initial-algebra operator), that is for any $f{:}A \to B$:

$$\begin{array}{ccccc} & & (A,A) & \xrightarrow{Y} & (1,A) \\ & \nearrow & & & \downarrow \\ (B,A) & & & & \\ & \searrow & & & \\ & & (B,B) & \xrightarrow[Y]{} & (1,B). \end{array}$$

The dinaturality of Y implies many things. It already implies that it is a fixed point operator: consider just the case that $A = B$ and chase the diagram starting with the identity map in the left-most hom-set. It implies that isomorphisms are automatically strict.

At a later time I will explore the many consequences of the condition that the subcategory of strict maps be multi-coreflective in the sense of Diers. It doesn't take too much else to get a representation theorem of algebraically compact categories into categories of CPO's.

This paper is in final form and will not be published elsewhere.

Order-Enriched Sketches for typed lambda calculi.

John W. Gray[1]
University of Illinois at Urbana Champaign
Urbana, Il, U.S.A.

Abstract: Order-enriched sketches are constructed for which the initial algebras are respectively, the simple typed λ-calculus, the PCF version of the λ-calculus, and the polymorphic typed λ-calculus.

1 Introduction

1.1 The context.

There are three sources for this present work. The first is a general concern with the relationships between logical type theory and algebraic type theory discussed in [7]. As was shown there, if algebraic type theory is identified with the category of finite limit sketches, then it is in a sense a model of logical type theory; in particular, it admits constructions that resemble dependent types. (See Barr and Wells [1], or Gray [7] for basic information on sketches.) One purpose of the present paper is to show that there is a relation in the other direction. If logical type theory is identified with the typed λ-calculus, then what is shown here is that the typed λ-calculus is an example of an algebraic data type; i.e., an initial algebra for a finite limit sketch. A second source is an attempt I made to implement the particular typed λ-calculus, PCF, in the rewrite rule language *Mathematica*. I found this difficult to do, mainly because there was no clear theoretical basis for producing such an implementation. The third source is a remark in [7] to the effect that instead of viewing the collections Terms(σ) of terms of type σ in a typed λ-calculus as a family of sets indexed by the set Types of types, it might be fruitful to consider the collection of all terms of all types, Terms = $\cup_{\sigma \in \text{Types}}$ Terms(σ) as a set *over* Types via a function type : Terms → Types. This turned out to be the key idea in constructing a sketch whose initial algebra is the typed λ-calculus. It also solved the second problem since, in [9] I had already shown how to implement the initial algebra for a sketch in *Mathematica*.

Note there there are no special theoretical tools to determine when a given mathematical structure is the initial algebra for a finite limit sketch. There are tools for deciding if a given category is the category of all models of such a sketch, but one of the problems here is that it is not clear what should be taken as the category of all models of a typed λ-calculus. The answer to this becomes clear once one has the appropriate sketch; namely, it is the category of all models of the sketch. A great deal is known about such categories. See, e.g., [1] or [13].

In this paper, explicit sketches are described in three cases: the pure typed λ-calculus, PCF, and the polymorphic typed λ-calculus F_2. What is constructed in each case is an order-enriched sketch whose initial algebra in the category of preordered sets consists of the raw terms of the appropriate λ-calculus together with preorder relations for terms of each sort corresponding to the rewrite rules for terms of that sort. The initial algebra in the category of sets consists of the

[1]This research was partially supported by the National Science Foundation.

normal forms of such terms. Each case presents its own difficulties. For the pure typed λ-calculus, the difficulty is to give a correct account of substitution. The most annoying problem is to make the choice of a new, or fresh, variable in the beta rule sketchable. This explains why we cannot use Stouten's treatment of substitution in [17]. PCF adds a mass of detail to this construction. Finally, the polymorphic typed λ-calculus adds four more basic constructions. In the last section of this paper there is an algorithm for converting such a sketch into a program implementing the initial algebra for the sketch. This algorithm is used in implementing PCF and F_2 in *Mathematica*. See [10] and [11]. The relation of these sketches to the well-known sketch for cartesian closed categories has not yet been clarified.

1.2 Typed lambda calculi

A typed l-calculus L consists of types and terms of each type.

i) The set Type of types is given recursively as follows: There is a finite or countable set B of basic types and if σ and τ are types then $[\sigma \to \tau]$ is a type.
ii) For each type τ, there are a countable set $Var(\tau)$ of variables of type τ and a finite or countable set $Const(\tau)$ of constants of type τ. We set $Atom(\tau) = Var(\tau) + Const(\tau)$.
iii) Write $f : \tau$ for "f is a term of type τ". The set $Terms(\tau)$ of terms of type τ is described recursively as follows:
    a) $Terms(\tau) \supseteq Atom(\tau)$
    b) If $f : [\sigma \to \tau]$ and $g : \sigma$, then $(f\ g) : \tau$;
    c) If $g : \tau$ and $x \in Var(\sigma)$, then $(\lambda x : \sigma . g) : [\sigma \to \tau]$;

1.3 Operational semantics of typed lambda calculi.

The operational semantics of a typed lambda calculus consists of some of the following rewrite rules. We use "⇒" here to mean the left hand side rewrites as the right hand side. Type information is omitted, but it is assumed that all terms are well-typed.

i) (α - conversion) $(\lambda x . f) \Rightarrow (\lambda y .[y / x] f)$ providing $y \notin FV(f)$.
ii) (β - conversion) $((\lambda x . f)\ g) \Rightarrow [g / x]f$. (See below.)
iii) (η - conversion) $((\lambda x . f)\ x) \Rightarrow f$ (or the reverse rule)
iv) (rewrite schemes)
    a) $\dfrac{h \Rightarrow k}{(h\ g) \Rightarrow (k\ g)}$ b) $\dfrac{h \Rightarrow k}{(f\ h) \Rightarrow (f\ k)}$ c) $\dfrac{h \Rightarrow k}{(\lambda x . h) \Rightarrow (\lambda x . k)}$.
v) (δ-conversion) Special rewrite rules for constants.

Here, the operation $[g / x]\ f$ of substituting g for x (where the type of g must be the same as the type of x) in f is defined recursively by the rules:

$[g / x]\ c = c$ if c is a constant
$[g / x]\ x = g$
$[g / x]\ y = y$ if $y \neq x$ and y is a variable
$[g / x]\ (h\ k) = (([g / x]\ h)\ ([g / x]\ k))$
$[g / x]\ (\lambda x . f) = (\lambda x . f)$ (i.e., don't substitute for bound variables.)
$[g / x]\ (\lambda y. f) = (\lambda y . [g / x]\ f)$ if $x \neq y$ and $y \notin FV(g)$
$[g / x]\ (\lambda y. f) = (\lambda z . [g / x]\ [z / y]\ f)$ if $x \neq y$ and $y \in FV(g)$, where $z \neq x$, $z \neq y$, $z \notin FV[f] \cup FV[g]$ and type(z) = type(y). (Prevent the capture of free variables.)

The set FV(g) of free variables of a term g is defined by the rules:

$FV(x) = \{x\}$
$FV(c) = \{\}$ if c is a constant
$FV((h\ k)) = FV(h)$ » $FV(k)$
$FV(\lambda x . f) = FV(f) - \{x\}$

The following notation will be used throughout this work.

Var $= \cup\,\{\mathrm{Var}(\tau) \mid \tau \in \mathrm{Type}\}$, Const $= \cup\,\{\mathrm{Const}(\tau) \mid \tau \in \mathrm{Type}\}$.

Terms $= \cup\,\{\mathrm{Terms}(\tau) \mid \tau \in \mathrm{Type}\}$.

type : Terms $\rightarrow$ Types denotes the function taking terms of type τ to τ.

In particular, type$(x : \tau) = \tau$.

2. THE PURE TYPED LAMBDA CALCULUS.

A pure typed lambda calculus is determined by a fixed set B of basic types, but no constant terms. We take it to have α-conversion, β-conversion and all three rewrite schemes in 1.3, iv).

2.1 The basic sketch for the pure typed lambda calculus.

A sketch is a graph whose nodes represent the objects of a date type and whose arrows represent the operations, together with extra structure. Sketches were introduced in a somewhat different form by Ehresmann [2] as presentations of theories. We follow the usage in Barr and Wells [1] and Gray [7]. The basic graph for the pure typed λ–calculus looks as follows:

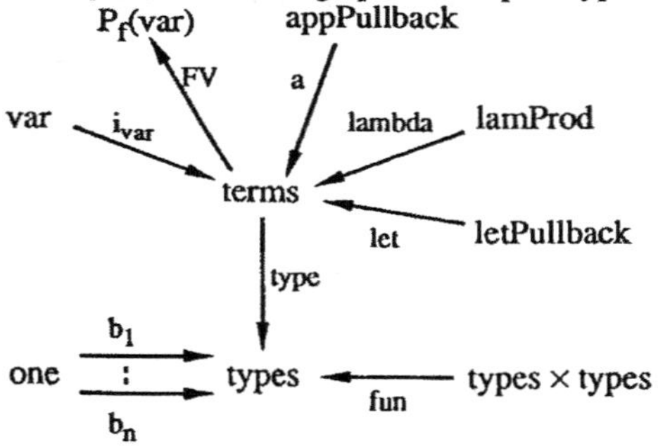

There an arrow from "one" to "types" for each element b_i of the fixed set B. The names of the nodes are chosen to correspond to the intended interpretation of the values of the initial model at each node. In a sketch, certain nodes are specified as *formal* limits of one kind or another. Here lamProd is specified as a formal product and appPullback and letPullback are formal pullbacks, to be completely described below. In a sketch there are also certain diagrams which are specified as *formal* commutative diagrams. A *model* of a sketch in a category **C** assigns an object of **C** to each node and a morphism of **C** to each arrow in such a way that actual limits are assigned to formal limits and formal commutative diagrams are taken to actual commutative diagrams. Homomorphisms of models are families of morphisms, one for each node, which commute with the operations. The category of models of a sketch S in a category C is denoted by $\mathrm{Mod}_C(S)$. If C satisfies reasonable properties (e.g., is accessible) then $\mathrm{Mod}_C(S)$ has an initial object. If C is the category SET of sets, then the initial object is the data type presented by the sketch.

2.2 Formal 2-cells.

A sketch contains more information about a data type than just the objects and operations. As described above, equations between composed operations in the graph are usually represented by diagrams of arrows in the graph which are declared to be formally commutative. However, in the present case, we are interested in representing the rewrite rules in the operational semantics of the pure typed λ–calculus, which have a direction. Thus, instead of formal commutative diagrams, we introduce the idea of a formal 2-cell in a diagram. This looks like:

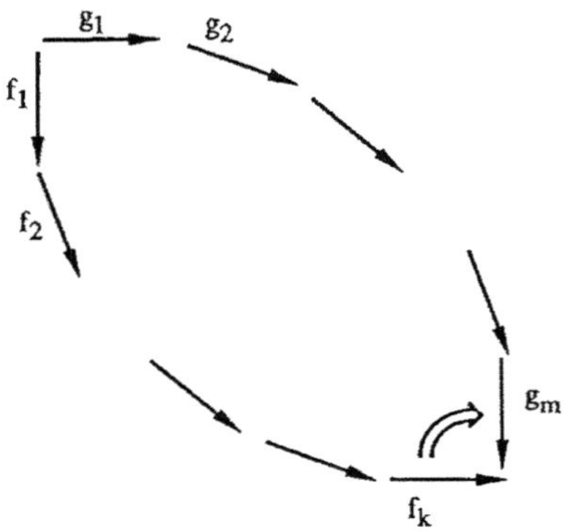

Such a 2-cell represents a formal order relation; i.e., here it represents the formal order relation $f_1 \bullet f_2 \bullet \ldots \bullet f_k \leq g_1 \bullet g_2 \bullet \ldots \bullet g_m$. More generally, a 2-cell can represent a "morphism" from its *domain*, $f_1 \bullet f_2 \bullet \ldots \bullet f_k$ to its *codomain*, $g_1 \bullet g_2 \bullet \ldots \bullet g_m$. Note that identically equal paths are considered to have an identity 2-cell between them. Essentially, there is a formal 2-cell for each rewrite rule in the operational semantics, except that there are also extra formal 2-cells for auxiliary arrows. If we consider the rewrite rules to be named, then we use the more general notion of a "morphism", whereas if we are just concerned with the existence of a rewrite rule, then we consider the 2-cells as order relations.

2.3 Formal quasi limits

There is still a further aspect of data types that is represented in a sketch. As described above, an object can be declared to be a formal product of other objects or, in general, a formal finite limit of a diagram of other objects. For instance, there is a node "types × types", which means that this node is declared to be a formal product of "types" with itself, and fun is a arrow from "types × types" to "types".

In category theory, limits of diagrams are described in terms unique morphisms making certain triangles commutative. Here we have rejected commutative diagrams in favor of 2-cells, so we are forced to make a corresponding change in the notion of limit. Recall that a 2-category, or a *CAT-enriched category*, **C** is an ordinary category enriched in the category of small categories; i.e., the hom sets of **C** are the sets of objects of hom categories and composition is extended to be a functor. (See Gray [6] or Kelly and Street [12].) Morphisms in such hom categories are called 2-cells in **C**. If these hom categories are all preordered categories (at most one morphism between any two objects), then **C** is called an *order-enriched* category.

A *quasi-limit cone* in a 2-category **C** over a diagram $\{A(i), a(i, j) | i \in I\}$ in **C** is an object X in **C** together with arrows $p(i) : X \to A(i)$ and 2-cells $a(i, j) \bullet p(i) \Rightarrow p(j)$ for all i and j, which are compatible with composition. Such a cone is called a *quasi-limit* of the diagram if given any other quasi-limit cone over the diagram with vertex Y, paths of arrows $q(i) : Y \to A(i)$, and 2-cells $a(i, j) \bullet q(i) \Rightarrow q(j)$ which are compatible with composition, there is a unique arrow $q : Y \to X$ and 2-cells $p(i) \bullet q \Rightarrow q(i)$ for all i, satisfying a composition condition which is clear from the diagram, and such that given any other arrow $f : Y \to X$ and 2-cells $p(i) \bullet r \Rightarrow q(i)$ satisfying the same condition, then there is a unique 2-cell $r \Rightarrow q$; i.e., q is the terminal such arrow.

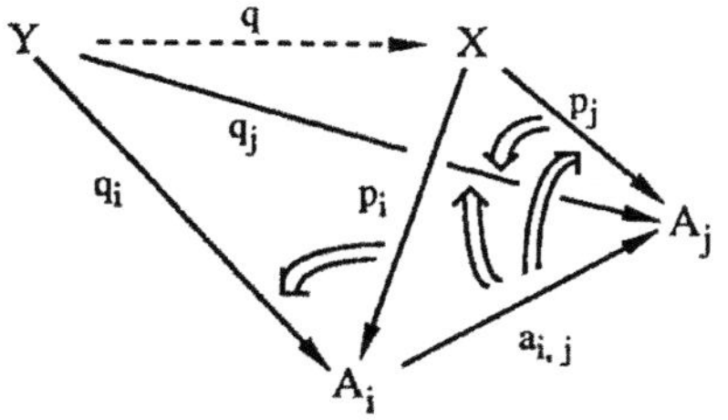

In the special case of an order-enriched category, all of the compatibility and commutativity conditions are automatically satisfied. Frequently q is written as <q(i) : i ∈ I> and is called the canonical arrow determined by {q(i) : i ∈ I}. Quasi-limits were introduced in Gray [6]; a special case, called cartesian quasi-limits in [6], corresponds to what are called lax limits in Kelly and Street [12]. See 5.1 for an example.

One can define a *2-graph*, or *CAT-enriched graph*, as an ordinary graph in which the set of arrows between any two fixed vertices is the set of objects of a category. Morphisms in these categories are called 2-cells in the graph. If these categories are all preordered categories, then the graph is called *order-enriched*. However, this does not capture the situation here. As indicated above, we want to consider 2-cells between paths in the graph. There is a somewhat complicated interaction between the free category generated by the graph and the addition of such 2-cells which is described in Section 5. Here we shall call this a *graph with added 2-cells*. We can then describe a *quasi-limit cone* in a graph G with added 2-cells over a diagram {A(i), a(i, j)| i ∈ I} in G exactly as in the 2-category case, except that there are no compatibility conditions. Such a quasi-limit cone can be declared to be a formal quasi-limit in G. The intention is that a model in a 2-category of this situation is required to take a formal quasi-limit in a sketch to an actual quasi limit. Given any other vertex Y, paths of arrows q(i) : Y → A(i), and 2-cells a(i, j) • q(i) ⇒ q(j), then an arrow q : Y → X is called the *formal canonical arrow* determined by this data if models are required to take it to an actual canonical arrow. Thus, we arrive at the following definition:

2.3.1 Definition. A *CAT-enriched, finite quasi-limit sketch* **A** consists of

i) A (directed) graph |A|, called the underlying graph of the sketch.
ii) A set of formal 2-cells between (directed) paths in |A|.
iii) A (possibly empty) set of finite formal quasi-cones in |A|.
iv) A (possibly empty) set of formal canonical arrows in |A|.

2.3.2 Definition. A (2-category) model of a CAT-enriched, finite quasi-limit sketch A in a 2-category **C** is a function M taking

i) vertices and arrows of |A| to objects and arrows in **C**,
ii) a formal 2-cell in **A** to an actual 2-cell in **C** from the composition of the values of M on the domain of the formal 2-cell to the composition of the values of M on the codomain.
iii) the chosen finite formal quasi-cones in **A** to quasi-limit cones in **C**, and
iv) the formal canonical arrows in **A** to actual canonical arrows in **C**.

In this paper, we are only concerned with the order-enriched case. The definition of an order-enriched sketch is exactly the same as the definition of a CAT-enriched sketch. The only difference is that we regard the 2-cells as order relations rather than as "morphisms". However, the intention is to restrict attention to models in an order-enriched category rather than a general 2-category. Such models are called *order* models of A. Note that the category **SET** of sets can be regarded as a discrete order-enriched category; i.e., all hom categories are discrete categories (no non-identity 2-cells). In this case, quasi-limits coincide with ordinary limits, since the only 2-cells are identites. Thus, a model of an order-enriched sketch in **SET** is the same as a model of the ordinary sketch determined by replacing all formal 2-cells by formal commutative diagrams, which turns quasi-limits into ordinary limits.

Standard techniques show that a CAT-enriched, finite quasi-limit sketch A generates a 2-category with finite quasi-limits, which we call the *2-theory of* A. As usual, there is a model of A in its 2-theory through which every other model factors by a quasi-limit preserving 2-functor. Similarly, A generates on order-enriched category with finite quasi-limits, which we call the *ordered-theory* of A. There is an order model of A in its ordered theory through which every other order model factors by a quasi-limit preserving order enriched functor. See 5.2 for an outline of the construction.

2.4 The formal quasi-limits for the pure typed lambda calculus.

There are three important formal quasi limits in the sketch for the pure typed lambda calculus.

2.4.1 Application, which is written (f g) in the operational semantics, is realized by an operation "a" and written in the form a[f, g]. "a" is an arrow whose codomain (i.e., target) is the object "terms" but whose domain (i.e., source) is a certain formal quasi limit called appPullback. The picture for this formal quasi limit is:

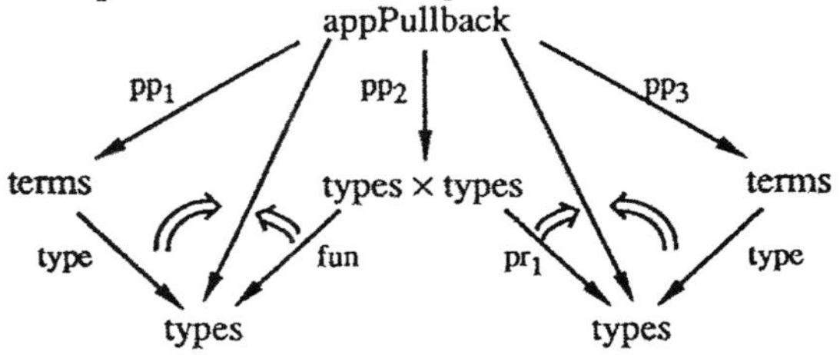

Canonical arrows into appPullback will always be specified as triples of paths of arrows, <qq_1, qq_2, qq_3>, corresponding to the three arrows pp_1, pp_2, and pp_3. The other two arrows into types will be taken to be fun and pr_1 respectively following qq_2, and the two inner 2-cells will be taken as identity 2-cells. A canonical arrow of this form exists exactly when there are 2-cells

$$\text{type} \bullet qq_1 \Rightarrow \text{fun} \bullet qq_2 \text{ and } \text{type} \bullet qq_3 \Rightarrow pr_1 \bullet qq_2.$$

This describes appPullback, and a : appPullback $\rightarrow$ terms. In the picture, some objects and arrows are repeated for clarity. AppPullback can be thought of as a "triple quasi pullback". Read in terms of its intended interpretation by rewrite rules, this says that an element of appPullback is a triple (f, (σ, τ), g), where f and g are terms and σ and τ are types, such that the type of f rewrite to the same type that fun(σ, τ) rewrites to, and the type of g and σ rewrite to the same type. This is the circumstance in which f can be applied to g.

2.4.2 Similarly, lambda abstraction, which is written λx . f in the operational semantics, is realized by an operation "lambda" and written in the form lambda[x, f]. "Lambda" is an arrow whose codomain is also the object "terms" but whose domain is the quasi product of the object of variables and the object of terms, called lamProd here. The picture for this formal quasi product is just:

$$\text{var} \xleftarrow{pr_1} \text{lamProd} \xrightarrow{pr_2} \text{terms}$$

so we can also write lamProd = var × terms when necessary. Canonical arrows into lamProd are of the form $<q_1, q_2>$ with no conditions. Note that there are then 2-cells $pr_1 \bullet <q_1, q_2> \Rightarrow q_1$ and $pr_2 \bullet <q_1, q_2> \Rightarrow q_2$. Again, these arrows are omitted from the basic graph.

2.4.3 Finally, β-reduction is described in terms of a substitution function called "let". "Let" is an arrow whose codomain is also the object "terms" but whose domain is a formal quasi limit realizing the requirement that in a term let[x, $expr_1$, $expr_2$], the type of x has to match the type of $expr_2$. Thus let : letPullback → terms where letPullback is the following quasi-limit (where 2-cells are omitted for simplicity):

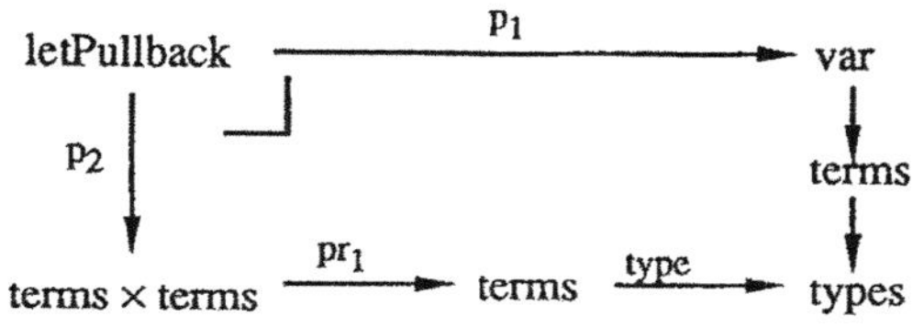

2.5 Rewrite rules for type.

Each of the three constructions requires a rewrite rule to specify its type. These are given by the following three 2-cells.

2.5.1 The rule: if f : [σ → τ] and g : σ, then (f g) : τ, says in the present terminology that type[a[f, g]] rewrites to "type(codomain(f))". Removing variables, this becomes the 2-cell:

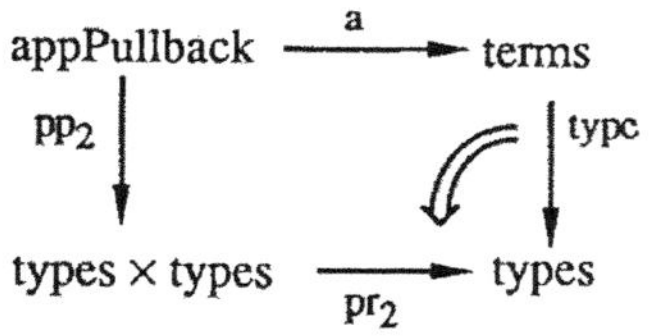

2.5.2 The rule: if g : τ and x ∈ Var(σ), then (λx : σ . g) : [σ → τ], says in the present terminology that type[λx : σ . g] rewrites to [type(x) → type(g)]. Removing variables, this becomes the 2-cell:

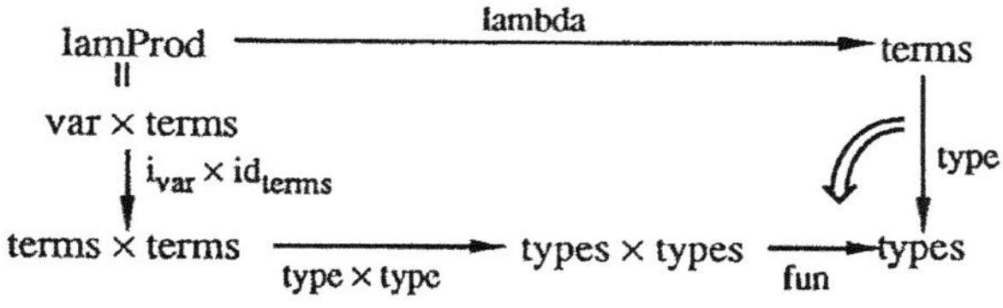

2.5.3 Since we have introduced a "let" construction into the language, we must also specify its type, which is clearly the type of the last component of the let expression. This becomes the 2-cell:

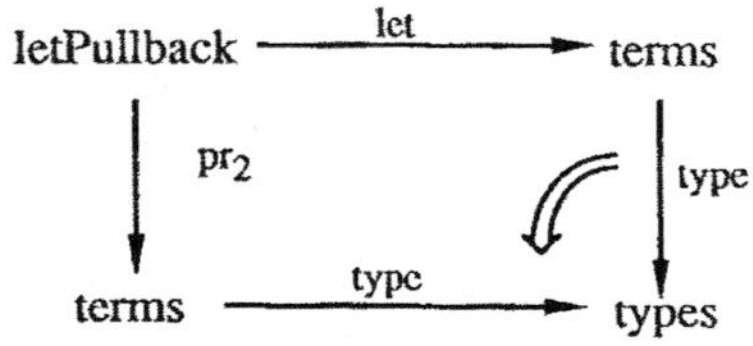

2.6 Rewrite rules for terms: β-reduction.

There is only one specific rewrite rule for the general lambda calculus here: the rule for β-reduction. However, this makes use of the let operation which has several rewrite rules of its own which are rather intricate to represent in sketch form. Recall that β - reduction is the rule: (λx . f) g ⇒ [g / x] f. Here this is written as a[lambda[x, f], g] ⇒ let[x, g, f]. To represent this as a 2-cell, consider the following confusing diagram:

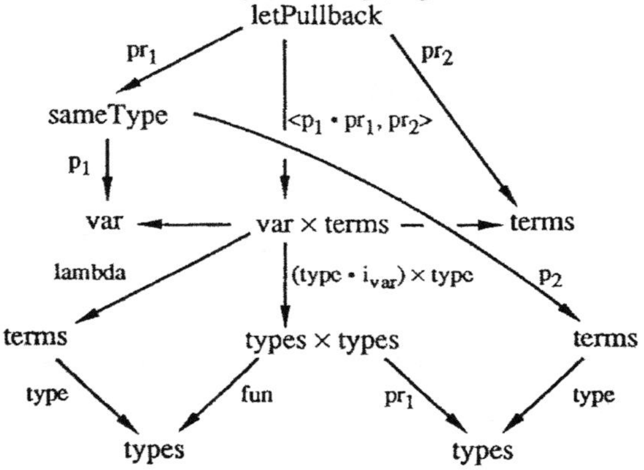

This illustrates a quasi cone over the diagram defining appPullback with vertex letPullback:

$<\text{lambda} \cdot <p_1 \cdot pr_1, pr_2>,$
$((\text{type} \cdot \text{id(var)}) \times \text{type}) \cdot <p_1 \cdot pr_1, pr_2>,$
$p_2 \cdot pr_1>.$

Let n : letPullback → appPullback denote the corresponding canonical arrow. Then the 2-cell representing β-reduction is a 2-cell a • n ⇒ let.

2.7 Rules for let.

The translation of the rules for let into the notation used here is:

let[x, g, x] ⇒ g, if x is a variable
let[x, g, y] ⇒ y, if y is a variable and y ≠ x
let[x, g, c] ⇒ c, if c is a const
let[x, g, a[h, k]] ⇒ a[let[x, g, h], let[x, g, k]]
let[x, g, lambda[x, f]] ⇒ lambda[x, f]
let[x, g, lambda[y, f]] ⇒ lambda[y, let[x, g, f]] if y ≠ x and y ∉ FV(g)
let[x, g, lambda[y, f]] ⇒ let[x, g, lambda[z, let[y, z, f]]] if y ≠ x and
  y ∈ FV(g), where z is not equal to x or y, and z ∉ FV[f] ∪ FV[g].

In all of these terms, x and y are variables and the type of g has to be the same as the type of x.

From the definition of letPullback as a formal quasi pullback, canonical arrows into letPullback are of the form $<<s_1, s_2>, s_3>$ where s_3 is an arrow into terms and $<s_1, s_2>$ is a pair of arrows with s_1 into var and s_2 into terms having the same type. Also, for any object, there is a unique canonical arrow to one, denoted polymorphically by $! : X \to$ one.

2.7.1 Equality of variables.

Several of the rules for let involve the inequality $y \neq x$. This requires some extra structure on the graph. First of all, we add a natural numbers object nat , given as a sketch by a zero arrow, 0, from one to nat and a successor arrow, suc, from nat to nat. We assume a formal quasi product cone making var the formal product of nat and type, so variables rewrite to a natural number and a type. We assume further a 2-cell rewriting type • i(var) as the second projection, so the type of $\langle n, t \rangle$ is rewritten to t. The purpose of this is to make the equality of variables decidable.

To describe equality, we need a boolean object bool in the graph. There are two arrows from one to bool called true and false and well as operations "or", "and", and "not" satisfying the usual rules for a boolean algebra. In order to describe equality for variables we first have to describe equalities for nat and type. They are arrows eq(nat) : nat $\times$ nat $\to$ bool and eq(type) : type $\times$ type $\to$ bool respectively satisfying the following rules:

eq(nat)[n, n] $\Rightarrow$ true, for all n
eq(nat)[suc(n), suc(m)] $\Rightarrow$ eq(nat)[n, m], for all n and m
eq(nat)[0, suc(n)] $\Rightarrow$ false, for all n
eq(nat)[suc(n), 0] $\Rightarrow$ false, for all n.

eq(type)[(τ, τ] $\Rightarrow$ true, for all τ
eq(type)[fun(σ, τ), fun(σ', τ')]
$\Rightarrow$ eq(type)[σ, σ'] and eq(type)[τ, τ'], for all σ, σ', τ, and τ'.
eq(type)[int, bool] $\Rightarrow$ false
eq(type)[bool, int] $\Rightarrow$ false
eq(type)[int, fun(σ, τ)] $\Rightarrow$ false, for all σ, τ
eq(type)[bool, fun(σ, τ)] $\Rightarrow$ false, for all σ, τ
eq(type)[fun(σ, τ), int] $\Rightarrow$ false, for all σ, τ
eq(type)[fun(σ, τ), bool] $\Rightarrow$ false, for all σ, τ.

It is easy to see how to represent these rules by 2-cells. Finally, equality for var is just the conjunction of equality for nat and for bool; i.e., eq(var) : var $\times$ var $\to$ bool is an arrow satisfying the rule:

eq(var)[$\langle n, \tau \rangle$, $\langle m, \sigma \rangle$] $\Rightarrow$ eq(nat)[n, m] and eq(type)[τ, σ].

This rule is represented by the 2-cell

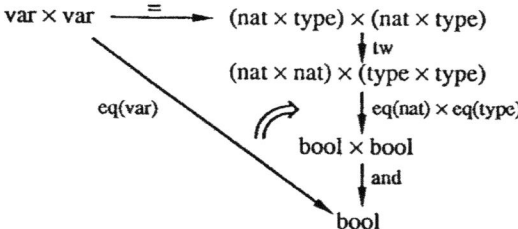

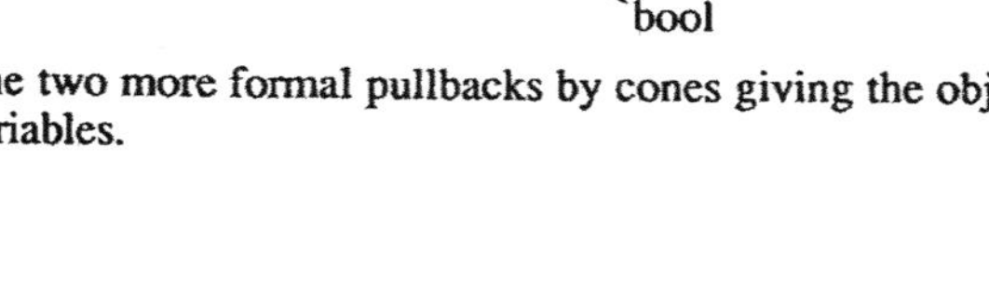

Now, define two more formal pullbacks by cones giving the objects of equal variables and of unequal variables.

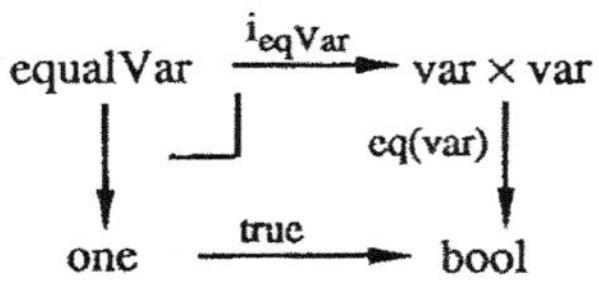

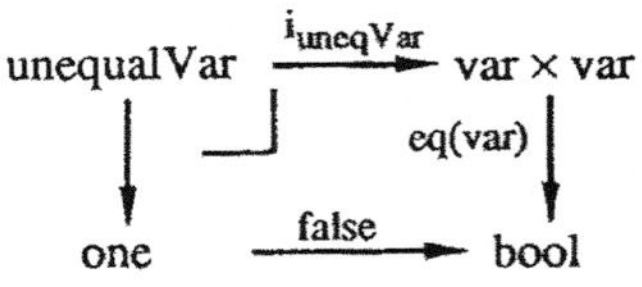

2.7.2 First rule for let

To describe the 2-cell for the first rule, let[x, g, x] ⇒ g, consider the quasi pullback:

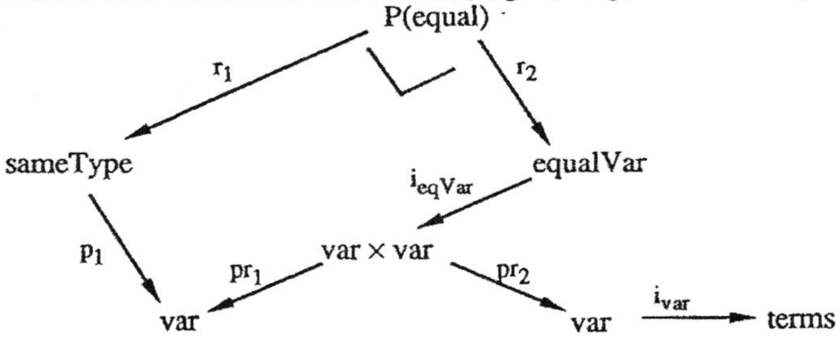

Here, sameType is the quasi pullback describing terms and variables of the same type. Using the notation in the diagram, this gives a canonical arrow

$\mu(eq) = i(var) \bullet pr_2 \bullet i(eqVar) \bullet r_2 : P(equal) \to terms$

and hence a canonical arrow $<r_1, \mu(eq)> : P(equal) \to letPullback$. The desired 2-cell is the following:

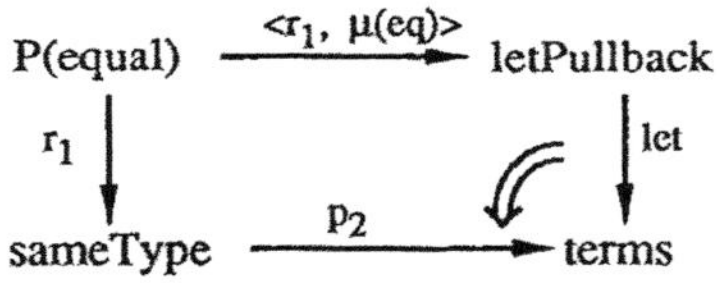

2.7.3 Second rule for let

To describe the 2-cell for the second rule, let[x, g, y] ⇒ y if y ≠ x, construct P(unequal) and

$\mu(uneq) = i(var) \bullet pr_2 \bullet i(uneqVar) \bullet r_2$

similarly, giving a canonical arrow $<r_1, \mu(uneq)> : P(unequal) \to letPullback$. The desired 2-cell is the following:

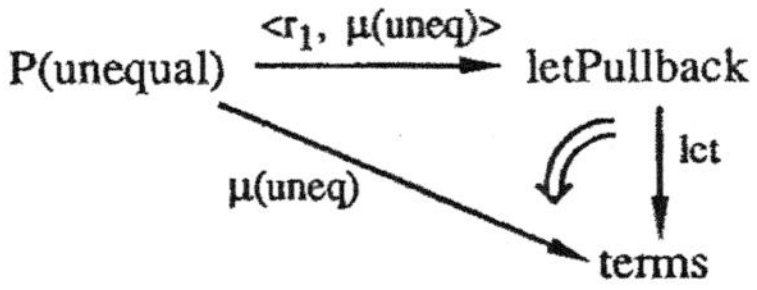

2.7.4 Third rule for let

The 2-cell for the third rule, let[x, g, c] ⇒ c, where c is a constant, is given by the following diagram:

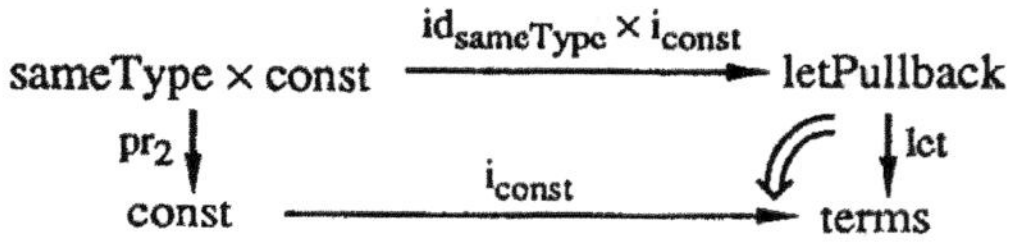

2.7.5 Fourth rule for let

To describe the 2-cell for the fourth rule,

let[x, g, a[h, k]] ⇒ a[let[x, g, h], let[x, g, k]],

let pp1 : appPullback → terms be the first projection, so

id(sameType) × pp_1 : sameType × appPullback → letPullback

and let • id(sameType) × pp_1 : sameType × appPullback → terms.

Using pp_3 : appPullback → terms in the same way gives a canonical arrow

v = <id(sameType) × pp_1, pp_2 • pr_2, id(sameType) × pp_3> :

sameType × appPullback → appPullback.

The desired 2-cell is then given by the diagram:

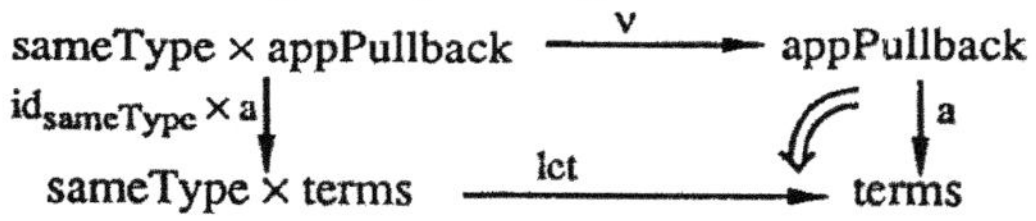

2.7.6 Fifth rule for let

We now come to the rules involving substitution in a lambda expression. These rules lie at the heart of the lambda calculus. The first rule involving lambda,

let[x, g, lambda[x, f]] ⇒ lambda[x, f]

says "do not substitute for bound variables". The point here is that the x in the let is the same as the x in the lambda, but the x in the lambda is not really there so substitution should not take place. Define Psame to be the indicated pullback:

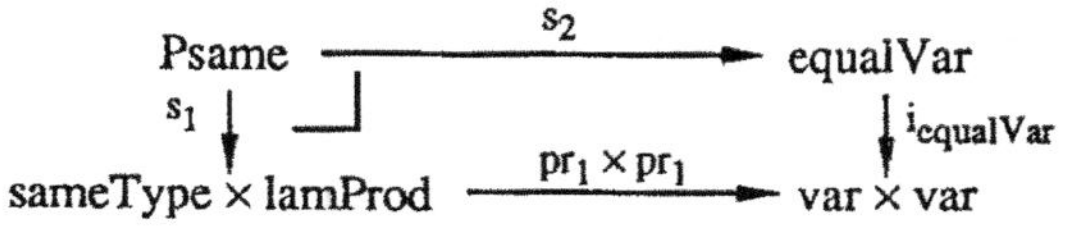

Then the following diagram describes the 2-cell corresponding to this rule

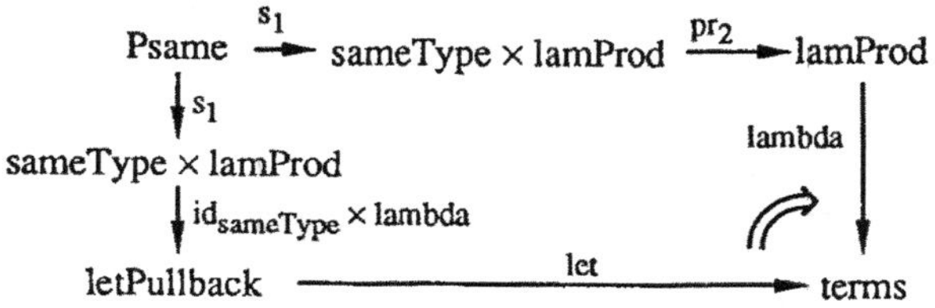

2.7.7 The sixth rule for let

The second rule involving lambda,

let[x, g, lambda[y, f]] ⇒ lambda[y, let[x, g, f]] if y ≠ x and y ∉ FV(g)

says "substitution passes inside a lambda expression if no variable capture takes place". This will involve us in a digression because we haven't described FV yet. FV is an arrow from terms to Pf(var) where Pf(var) is intended to describe the set of finite subsets of var. More structure is required in the graph to implement this. Thus, we assume that our graph contains arrows and a formal product object as illustrated:

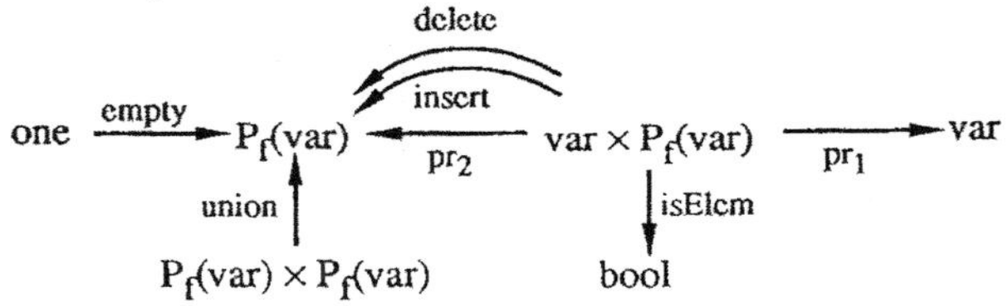

The arrow insert is subject to the two rewrite rules:

insert[n(τ), insert[m[σ], p]] ⇒ insert[m[σ], insert[n[τ], p]]
 if lessThan(var)[m(σ), n(τ)] ⇒ true
insert[n(τ), insert[n(τ), p]] ⇒ insert[n(τ), p]

Here, lessThan(var) is defined by the rewrite rule:

lessThan(var)[m(σ), n(τ)] ⇒
 lessThan(nat)[n, m] or (eqnat[n, m] and lessThan(type)[σ, τ])

where, in turn, lessThan(nat) is defined by the rewrite rules:

lessThan(nat)[0, 0] ⇒ false
lessThan(nat)[0, suc(n)] ⇒ true
lessThan(nat)[suc(n), 0] ⇒ false
lessThan(nat)[suc(n), suc(m)] ⇒ lessThan(Nat)[n, m]

and lessThan(type) is defined by the rewrite rules:

lessThan(type)[int, int] ⇒ false,
lessThan(type)[bool, bool] ⇒ false,
lessThan(type)[bool, int] ⇒ false,
lessThan(type)[int, bool] ⇒ true,
lessThan[int, fun[σ, τ]] ⇒ true,
lessThan(type)[bool, fun[σ, τ]] ⇒ true,
lessThan[fun[σ, τ], int] ⇒ false,
lessThan(type)[fun[σ, τ], bool] ⇒ false,
lessThan(type)[fun[σ, τ], fun[σ', τ']
 ⇒ lessThan(type)[σ, σ'] or
 (eqType[σ, σ'] and lessThan(type)[τ, τ'])

The arrow isElem is subject to the two rewrite rules:

isElem[n(τ), empty] ⇒ false
isElem[n(τ), insert(m(σ), p)] ⇒ eq(var)[n(τ), m(σ)] or isElem[n(τ), p].

The arrow union is subject to the two rewrite rules:

union[empty, p] ⇒ p
union[insert[n(τ), p], p'] ⇒ insert[n(τ), union[p, p']].

The arrow delete is subject to the rewrite rules:

delete[n(τ), empty] ⇒ empty
delete[n(τ), insert[(m(σ), p]] ⇒ p if eq(var)[n(τ), m(σ)] = true

delete[n(τ), insert[m(σ), p]] ⇒ insert[m(σ), delete[n(τ), p]]
if if eq(var)[n(τ), m(σ)] = false.

All of these rules are easily realized as 2-cells.

The arrow FV : terms → Pf(var) then is described by the rewrite rules:

FV[x] ⇒ insert[x, empty]
FV[c] ⇒ empty
FV[a[h, k]] ⇒ union[FV[h], FV[k]]
FV[lambda[x, f]] ⇒ delete[x, FV[f]]

These rules are also easily realized as 2-cells. We omit the preceding 24 pictures.
We are now ready for the sixth rule for let. Let $Q_{\notin}$ be the following formal quasi limit:

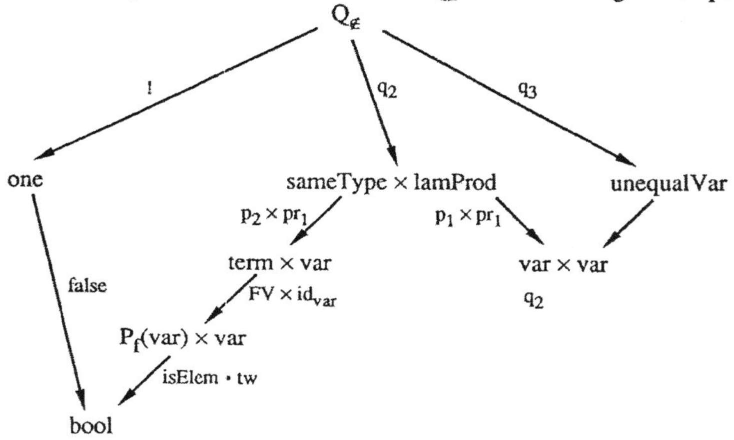

Then the following 2-cell represents the rule
let[x, g, lambda[y, f]] ⇒ lambda[y, let[x, g, f]] if y ≠ x and y ∉ FV(g).

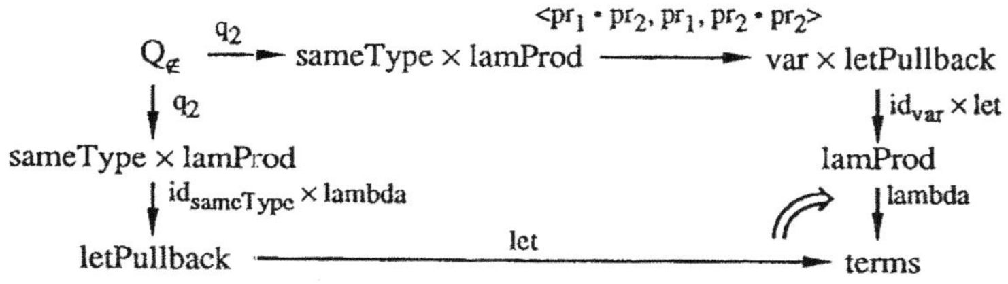

2.7.8 The seventh rule for let

The third rule for let involving lambda,

et[x, g, lambda[y, f]] ⇒ let[x, g, lambda[z, let[y, z, f]]]
if y ≠ x and y ∈ FV(g),
where z is not equal to x or y, and z ∉ FV[f] » FV[g]

says that if variable capture would take place, then the bound variable in the lambda expression has to be changed. In order to describe this, we have to add an arrow newvar : sameType × lamProd → var whose value will be the required variable z. This requires still more structure in the sketch.

First of all, construct Pf(nat) in the same way that Pf(var) was built. Then, describe an arrow $Pf(pr_1)$: Pf(var) → Pf(Nat) corresponding to the projection arrow pr_1 : var → Nat by the rewrite rules:

Pf(pr_1)[empty(var)] ⇒ empty(Nat)
Pf(pr1)[insert(var)[n(t), p] ⇒ insert(nat)[n, Pf(pr_1)[p]].

Here we have added parameters to empty and insert to keep things straight.

Addition for nat is given by an arrow plus : nat × nat → nat with the rewrite rules:

plus[n, 0] ⇒ n
plus[n, suc[m]] ⇒ suc[plus[n, m]].

Using plus, we describe an arrow sum : Pf(nat) → nat by the rewrite rules:

sum[empty(nat)] ⇒ 0
sum[insert(nat)[n, p]] ⇒ plus[n, sum[p]].

We omit the pictures for the six 2-cells representing these rewrite rules.

We need an abbreviation for a path of arrows:

singleton = insert(var) • (id(var) × empty(var)) • <id(var), !>.

The intended meaning in a model is that singleton[x] = {x}.

Now we are ready to construct the arrow newvar. The following picture illustrates two composed arrows from sameType × lamProd to Pf(var) that we call α and β:

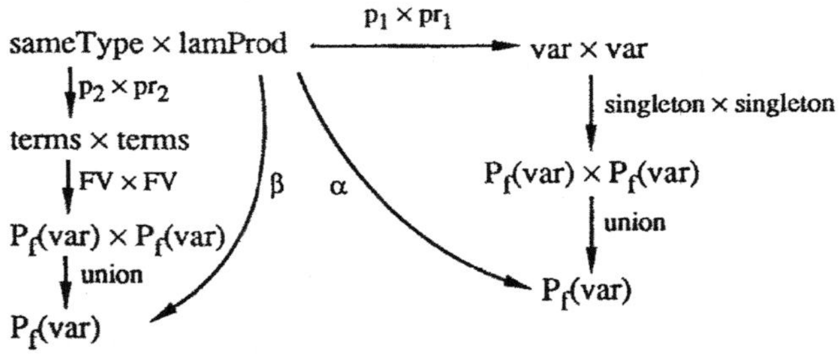

Using them, we define

γ = union • <α, β> : sameType × lamProd → Pf(var).

Then newvar is defined to be the indicated composed arrow:

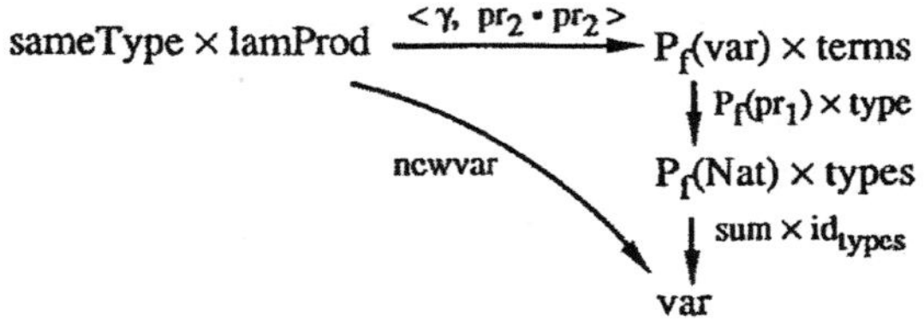

Here, we use the definition, var = nat × type. Now, construct Q(∈) in the same way that Q(∉) was constructed, replacing false by true. The the last rule for let is represented by the following 2-cell:

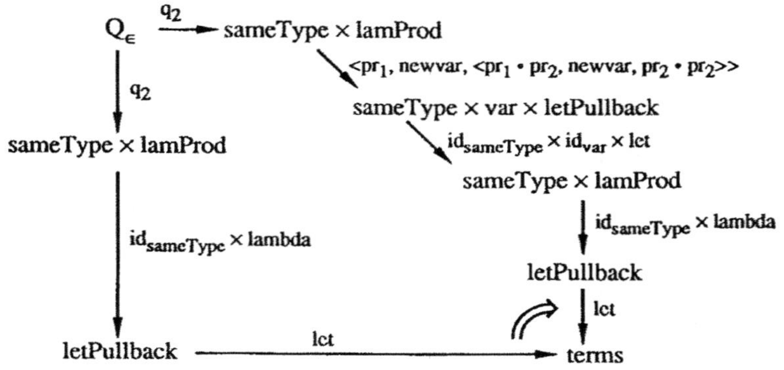

3 THE TYPED LAMBDA CALCULUS PCF.

3.1 The language PCF. An important example for theoretical work is the lambda calculus PCF (= Programming Computable Functions) which is part of the language LCF (= Logic of Computable Functions) created by D. Scott , and described in Plotkin [15]. This lambda calculus has the following description.

3.1.1 Definition of PCF.

The set of basic types is B = {bool, int}.

There are the following constant terms:

tt, ff	: bool
n	: int , for n an integer bigger than or equal to 0.
succ	: [int → int]
pred	: [int → int].
iszero	: [int → bool].
if(bool)	: bool → bool → bool → bool
if(int)	: bool → int → int → int.
fix(t)	: $[\tau \to \tau] \to \tau$ for each $\tau \in$ Type.

Note: We use the standard convention that → associates to the right; i.e.,

$$t_1 \to t_2 \to \ldots \to t_n = t_1 \to [t_2 \to [\ldots \to t_n] \ldots]],$$

as well as the standard convention that application associates to the left; i.e.,

$$f\ g\ h \ldots k = (((f\ g)\ h) \ldots k).$$

3.1.2 The operational semantics for PCF

The rewrite rules for PCF consist of

ii) (β - conversion) $((\lambda x . f)\ g) \Rightarrow [g / x]f$.

iv) a) $\dfrac{h \Rightarrow k}{(h\ g) \Rightarrow (k\ g)}$

b) $\dfrac{h \Rightarrow k}{(f\ h) \Rightarrow (f\ k)}$ only for f = succ, pred, or iszero,

v) the following special rewrite rules for the constants:

(succ n)	$\Rightarrow$ n + 1	for n bigger than or equal to 0
(pred n + 1)	$\Rightarrow$ n	for n bigger than or equal to 0
(iszero 0)	$\Rightarrow$ tt	
(iszero n + 1)	$\Rightarrow$ ff	for n bigger than or equal to 0
if(σ) tt f g	$\Rightarrow$ f	where σ = int or bool
if(σ) ff f g	$\Rightarrow$ g	where σ = int or bool
(fix(τ) f)	$\Rightarrow$ f (fix(τ) f)	for every type τ.

3.2 The basic sketch for PCF.

The basic sketch for the typed lambda calculus is augmented by the addition of several new nodes and arrows. There are just two basic types, called int and bool. (The boolean basic type in spelled with an "e" to distinguish it from the node "bool" which is not included in the basic graph.) There are many constants, and in particular, the natural number object has an inclusion arrow into const.

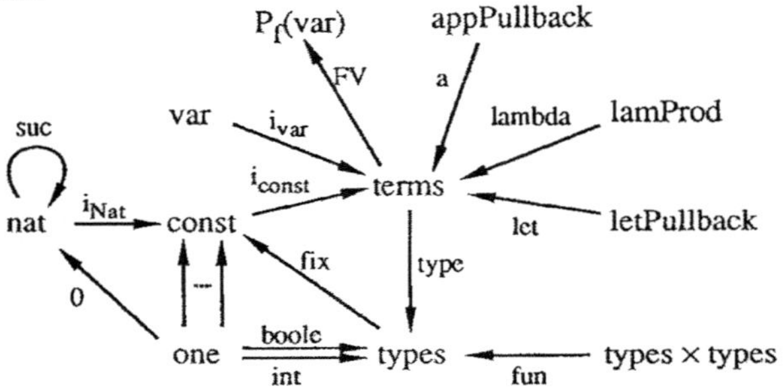

3.3 Rewrite rules for type in PCF.

Next we determine the rewrite rules for the constants in PCF. They require many 2-cells, but are all quite simple.

3.3.1 The constants tt and ff have a type which rewrites to boole.

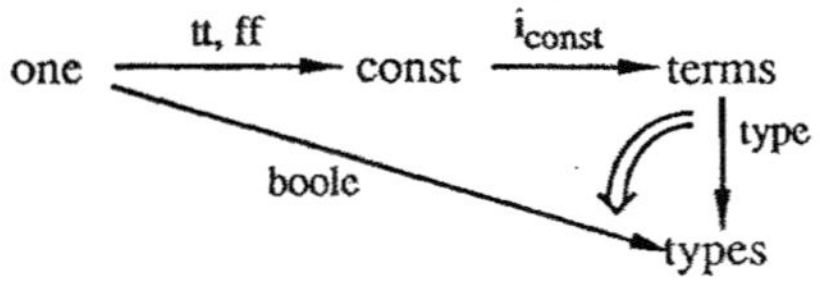

3.3.2 The constants 0 and suc(n) have a type which rewrites to int.

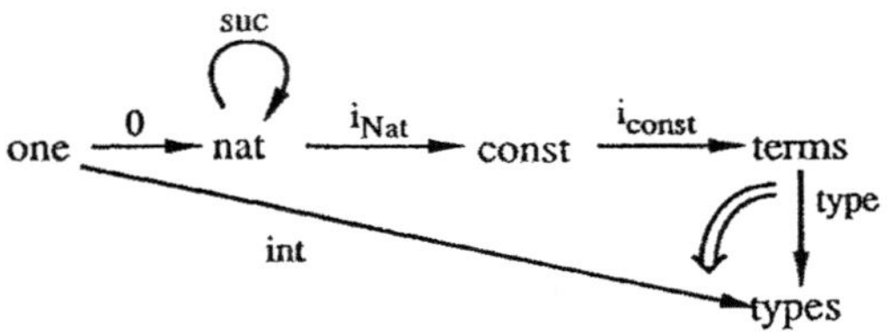

3.3.3 Succ and pred have a type which rewrites to fun[int, int].

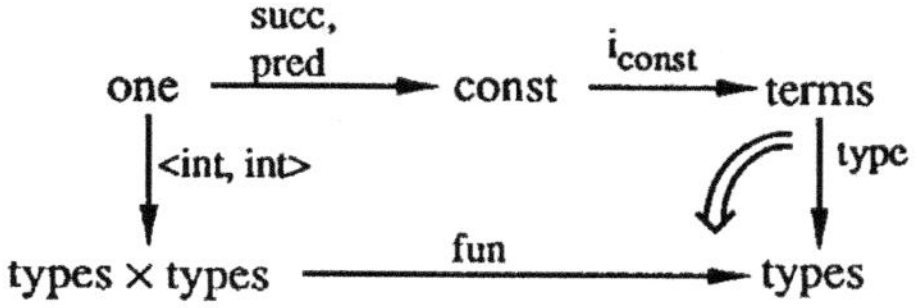

3.3.4 Iszero is in a similar diagram, except that <int, int> is replaced by <int, bool> since its type rewrites to fun[int, bool].

3.3.5 The type of if[boole] rewrites to fun[boole, fun[boole, fun[boole, boole]]], so there is a 2-cell:

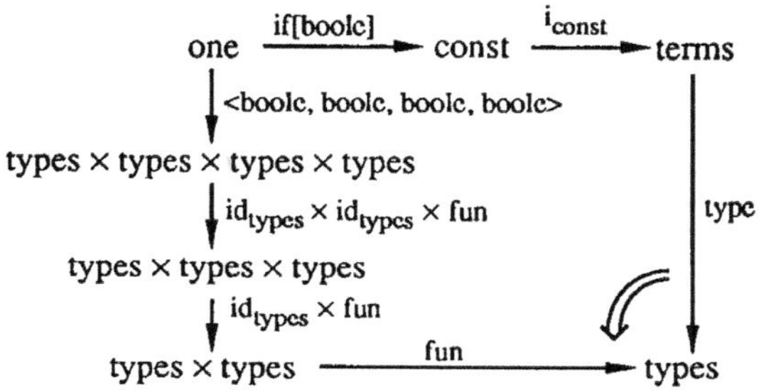

If[int] is in a similar diagram, except that <boole, boole, boole, boole> is replaced by <boole, int, int, int>.

3.3.6 Finally the type of fix[t] rewrites to fun[fun[t, t], t] so there is a 2-cell:

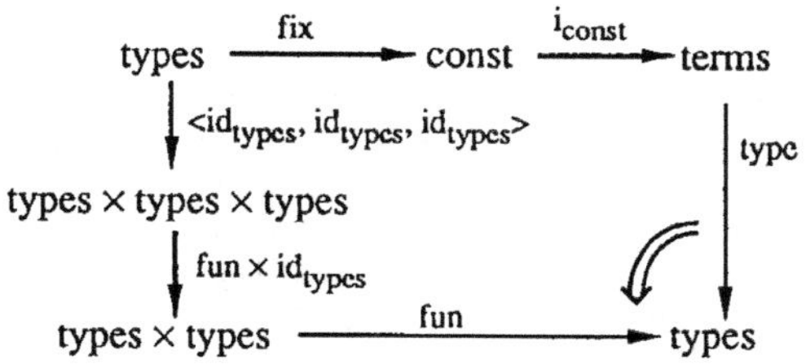

3.4 Rewrite rules for terms in PCF.

In this version of the lambda calculus, there are many rewrite rules for reducing terms to normal form.

3.4.1 Consider the first special rewrite rule for PCF:

(succ n) ⇒ n + 1 for n an integer bigger than or equal to 0, which in our present notation is a[succ, n] ⇒ n + 1. This is represented here as follows: let n^ denote to path of arrows given by i(const) • suc^n • 0 from one to terms and consider the following quasi cone on the diagram whose formal quasi limit is appPullback.

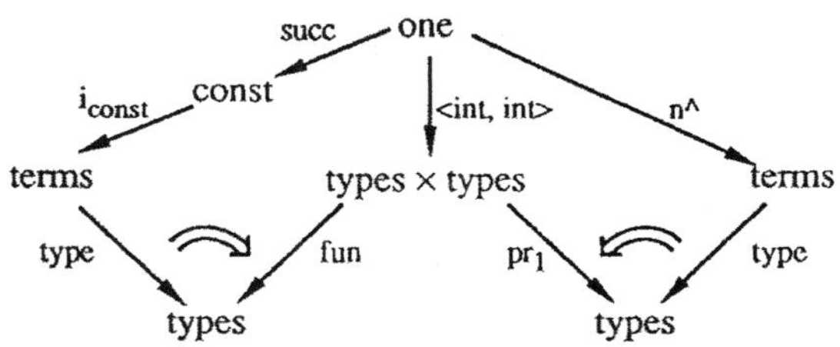

It corresponds to a canonical arrow we call $\alpha(n)$ from one to appPullback. The desired rewrite rule is represented by a 2-cell as illustrated:

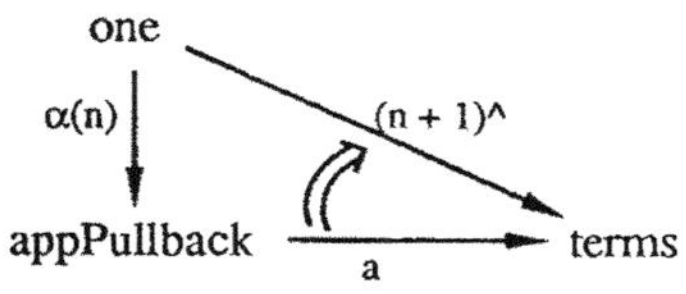

The rewrite rules

(pred n + 1) ⇒ n   for n bigger than or equal to 0
(iszero 0) ⇒ tt
(iszero n + 1) ⇒ ff   for n bigger than or equal to 0

are represented similarly.

3.4.2 Next, consider the rule

if(boole) tt f g ⇒ f

which here is written a[a[a[if[boole], tt], f], g] ⇒ f. The 2-cell for this is constructed in three steps because of the three applications. In the first, as with the diagram for a[succ, n], succ is replaced by if[boole], <int, int> is replaced by the path of arrows given by

(if(types) × fun) • (id(types) × id(types) × fun) • <boole, boole, boole, boole>

and n^ is replaced by tt. This yields an arrow from one to appPullback whose composition with "a" is a path from one to terms. This serves as the left hand path in another quasi cone over the diagram for appPullback whose middle path of arrows is

<boole, fun • <boole, boole>>

and whose right hand arrow is f. This yields another arrow from one to appPullback whose composition with "a" is a path from one to terms which serves as the left hand path in yet another quasi cone over the diagram for appPullback whose middle arrow is <boole, boole> and whose right hand arrow is g. This yields the third arrow from one to appPullback whose composition with "a" is finally the domain of a new 2-cell from it to f. The other "if" rules are treated similarly.

3.4.3 The final special rule for PCF is

$(\mathrm{fix}(\tau)\ f) \Rightarrow f\ (\mathrm{fix}(\tau)\ f)$ for every type τ.

Here this is written: $a[\mathrm{fix}[\tau], f] \Rightarrow a[f, a[\mathrm{fix}[\tau], f]]$. Note that f has to have the type $\mathrm{fun}[\tau, \tau]$ for this to make sense. Both the left and right sides of this rule have to be constructed as suitable arrows into terms. The domain of these arrows is the object P which is the formal quasi pullback in the following diagram.

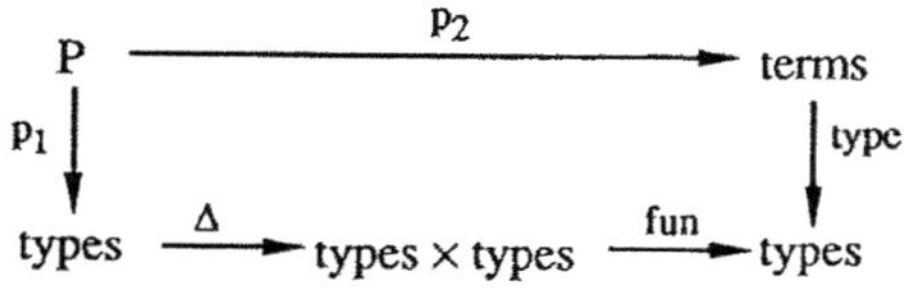

P represents terms whose type rewrites to the form fun[τ, τ] for some τ. As before, consider the quasi cone over the diagram whose quasi limit is appPullback in which the left hand path of arrows is fix • p_1, the middle path of arrows is

$$P \xrightarrow{p_1} \text{types} \xrightarrow{\Delta_3} \text{types} \times \text{types} \times \text{types} \xrightarrow{\text{fun} \times \text{id}_{\text{types}}} \text{types} \times \text{types}$$

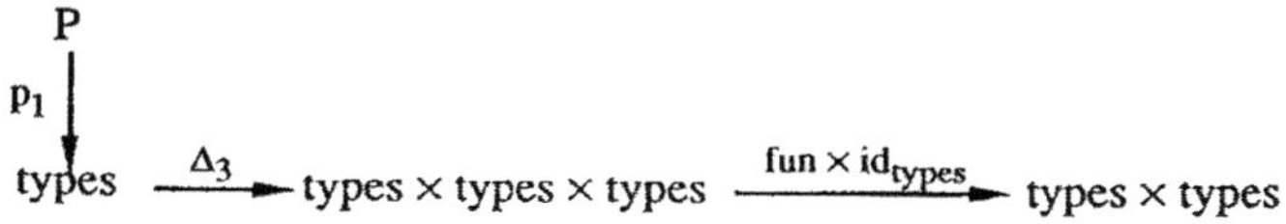

and the right hand path is p_2. Denote the canonically determined arrow into appPullback by γ = <fix • p_1, (fun × id$_{\text{types}}$) • Δ_3 • p_1, p_2>. Using the same triple notation, there is another canonically determined arrow from P into appPullback given by δ = <p_2, Δ • p_1, a • γ>. The 2-cell representing the rewrite rule for fix is then a 2-cell a • $\gamma \Rightarrow$ a • δ. This completes the treatment of the special rewrite rules for PCF.

3.5. The actual order-enriched sketch

The final order-enriched sketch is made up of all of the parts discussed above; i.e., all objects, all arrows, all formal quasi limits and all 2-cells that are introduce either in the pictures or implied by the rewrite rules that were not specifically illustrated. We think of the 2-cells as primitive, generating, order relations between paths of arrows.

4   THE POLYMORPHIC TYPED LAMBDA CALCULUS.

4.1   The language of the polymorphic lambda calculus.

The version or the polymorphic lambda calculus implemented here is called F_2. See Girard [4] and Reynolds [16]. In it, ordinary variables include their type as part of the expression. Since there are no constants, the only place that types occur explicitly in expressions is as the types of variables. The key point in using types is in instantiating a LAMBDA expression at some type. This must change the type variable in the body of the expression to the new type.

4.1.1 The definition of the polymorphic lambda calculus.

The set of basic types in the polymorphic lambda calculus consists of a set "typevar" of type variables. The rules for the formation of types in 1.2 i) are extended by the rule: If t is a type variable and σ is a types then $\forall$t . σ is a type, called a *universally quantified* type. A grammar for type is given as follows: $\sigma ::= t \mid [\sigma \rightarrow \sigma'] \mid \forall t . \sigma$.

The set Term(σ) of terms of type σ described in 1.2 iii) is augmented by two extra rules:
d)   If t $\in$ Tv and g : σ has the property that for all x : $\tau \in$ FV(g), t $\notin$ fv(τ), then

$\Lambda t \,.\, g : \forall t \,.\, \sigma$. Here, $\Lambda t \,.\, g$ is called a *type abstraction* of a term.

e) If $f : \forall t \,.\, \sigma$ and $\tau \in T$, then $f[\,\tau\,] : [\,\tau / t\,]\,\sigma$. Here, $f[\,\tau\,]$ is called a *type instantiation.*

FV : Terms $\rightarrow P_f(\text{var})$ assigns to any term its set of free (ordinary) variables as in 1.3, and fv : Types $\rightarrow P_f(\text{TypeVar})$ assigns to any type its set of free type variables. ($P_f(X)$ means the set of finite subsets of X.) Finally, $[\,\tau / t\,]\,\sigma$ means the result of substituting the type τ for all occurrences of the type variable t in σ.

4.1.2 The operational semantics for the polymorphic lambda calculus.

There is one additional rewrite rule for terms in the polymorphic lambda calculus involving the instantiation of a type abstraction.

$$\Lambda t \,.\, g[\tau] \Rightarrow [\,\tau / t\,]\, g$$

where $[\,\tau / t\,]\, g$ means substitute τ for all occurrences of t in g. Actually, substitution of a type for a type variable has to be defined both for types and for terms.

i) The definition of $[\,\tau / t\,]\,\sigma$, where σ is a type, is given by the rules:
   a) $[\,\tau/t\,]\,t \Rightarrow \tau$
   b) $[\,\tau/t\,]\,s \Rightarrow s$, if $s \neq t$.
   c) $[\,\tau/t\,]\,[\sigma \rightarrow \sigma'] \Rightarrow [\,[\,\tau/t\,]\,\sigma \rightarrow [\,\tau/t\,]\,\sigma'\,]$
   d) $[\,\tau/t\,]\,\forall t \,.\, \sigma \Rightarrow \forall t \,.\, \sigma$
   ε) $[\,\tau/t\,]\,\forall s \,.\, \sigma \Rightarrow \forall s \,.\, (\,[\,\tau/t\,]\,\sigma\,)$ if $s \neq t$ and $s \notin fv[\tau]$.
   f) $[\,\tau/t\,]\,\forall s \,.\, \sigma \Rightarrow \forall u \,.\, (\,[\,\tau/t\,]\,[\,u/s\,]\,\sigma\,)$ if $s \neq t$ and $s \in fv[\tau]$,
      where $u \notin fv[\sigma] \cup fv[\tau]$, $u \neq t$, and $u \neq s$.

ii) The definition of $[\,\tau / t\,]\, g$, where g is a term, is given by the rules:
   a) $[\,\tau/t\,]\,(x : \sigma) \Rightarrow x : [\,\tau/t\,]\,\sigma$, if x is a variable.
   b) $[\,\tau/t\,]\,(\,f\,g\,) \Rightarrow ((\,[\,\tau/t\,]\,f\,)\,(\,[\,\tau/t\,]\,g\,))$
   c) $[\,\tau/t\,]\,(\lambda\, x : \sigma \,.\, g) \Rightarrow \lambda\, x : [\,\tau/t\,]\,\sigma \,.\, [\,\tau/t\,]\,g$
   d) $[\,\tau/t\,]\,f[\,t\,] \Rightarrow f[\,\tau\,]$
   e) $[\,\tau/t\,]\,f[\,\sigma\,] \Rightarrow (\,[\,\tau/t\,]\,f)[\,[\,\tau/t\,]\,\sigma\,]$
   f) $[\,\tau/t\,]\,(\Lambda t \,.\, g) \Rightarrow \Lambda t \,.\, g$
   g) $[\,\tau/t\,]\,(\Lambda s \,.\, g) \Rightarrow \Lambda s \,.\, (\,[\,\tau/t\,]\,g)$, provided $s \neq t$ and $s \notin fv(\tau)$
   h) $[\,\tau/t\,]\,(\Lambda s \,.\, g) \Rightarrow [\,\tau/t\,]\,(\Lambda u \,.\, [\,u/t\,]\,g)$, provided $s \neq t$ and $s \in fv(\tau)$.

iii) There are also two additional rules for $[g / x]\, f$ that have to be added to the rules in 1.3.
   a) $[g/x]\,(\Lambda t \,.\, f) \Rightarrow \Lambda t \,.\, [\,g/x\,]\,f$
   b) $[g/x]\,f[\,\tau\,] \Rightarrow ([g/x]\,f)[\,\tau\,]$.

See [5] or [14] for examples of types and terms.

4.2 The basic sketch for the polymorphic lambda calculus.

The basic sketch for the polymorphic lambda calculus looks as follows. There are no basic types and no constant terms. However, there are type variables, another type constructor, forall (corresponding to $\forall$), and several more term constructors; namely, LAMBDA, LET, typelet, and inst. The term constructors correspond to the operations Λ, $[\,\tau / t\,]g$, $[\,\tau / t\,]\sigma$, and $f[\,\sigma\,]$ respectively. It is also necessary to consider finite sets of type variables and two free type variable operations, fv from types and FtV from terms.

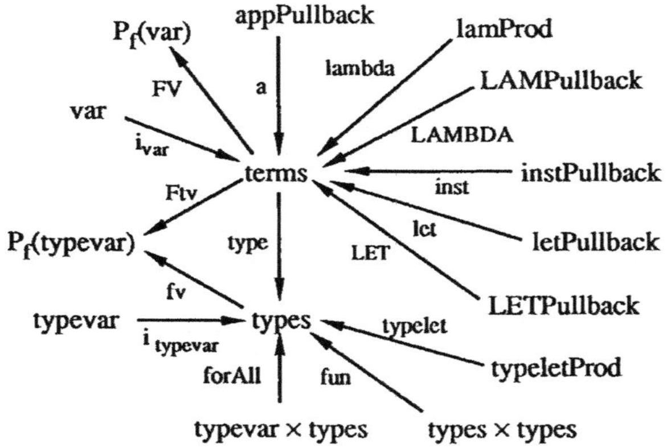

The actual sketch is much larger, of course. It contains all of the auxiliary structure of the sketch for the ordinary typed lambda calculus as well as special auxiliary structure for the polymorphic lambda calculus that will be described below.

4.3 The formal quasi limits for the polymorphic lambda calculus.

There are four new quasi limits in this sketch: LAMPullback, LETPullback, instProd, and typeletProd.

4.3.1 LAMPullback denotes the domain of the arrow LAMBDA,which corresponds to the rule:

If $t \in$ Tv and $g : \sigma$ has the property that for all $x : \tau \in FV(g)$, $t \notin fv(\tau)$ then $\Lambda t \,.\, g : \forall t \,.\, \sigma$.

The condition on g can be described by an arrow FtV : terms $\rightarrow$ P_f(typevar) whose meaning assigns to each term g the set

$$\{t \in \text{typevar} \mid \text{there is an } x : \tau \in FV(g) \text{ with } t \in fv(\tau)\}.$$

The rules for FtV are given below in 4.5.1. The construction of the diagram for LAMPullback requires an auxiliary structure for P_f(typevar) just like that for P_f(var) in 2.7.7. In particular, typevar has to have a decidable equality relation, so we assume that there is an initial typevar t_0 and a successor operation for typevar making typevar into a copy of the natural numbers. LAMPullback is the following formal quasi pullback:

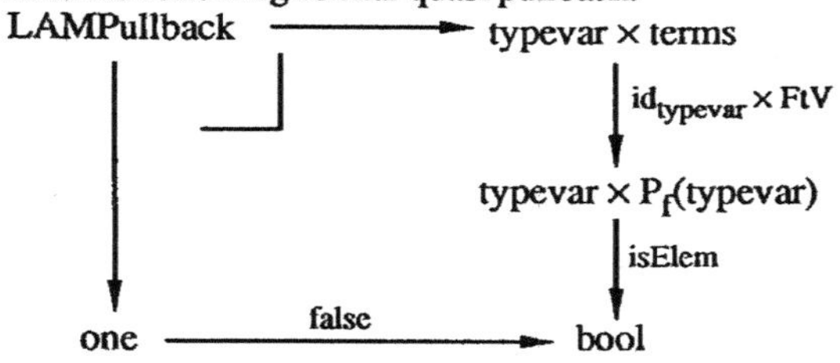

4.3.2 LETPullback is easily described as a further formal quasi pullback from LAMPullback.

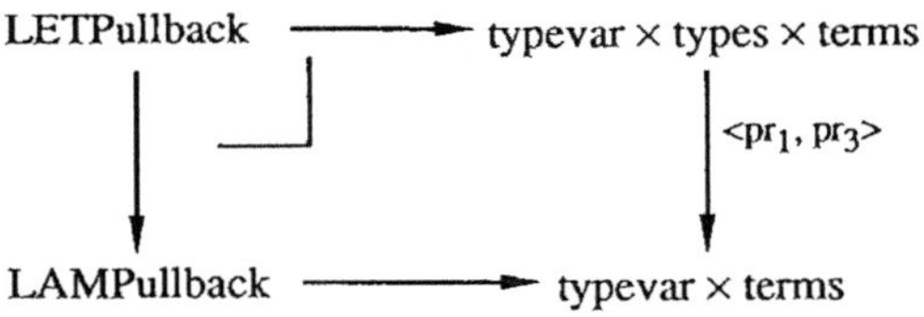

4.3.3 InstPullback is the object on which the instantiation of a LAMBDA terms at a type is defined. Such a term has a universally quantified type, so instPullback is the following formal quasi pullback:

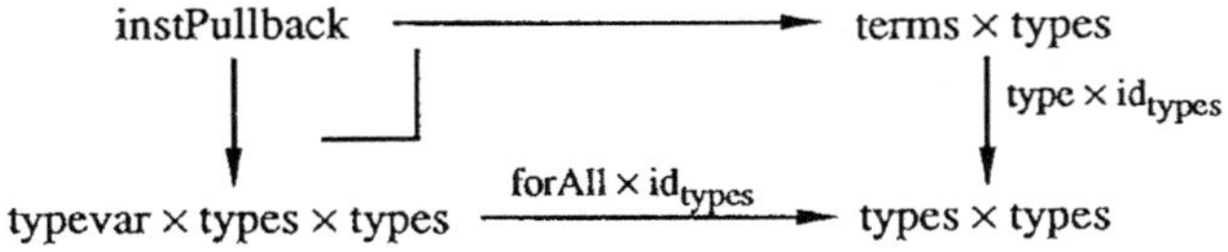

4.3.4 TypeletProd is just the formal quasi product typevar × types × types.

4.4 Rewrite rules for type in the polymorphic lambda calculus.

The types of a, lambda, and let are as in the pure typed lambda calculus. There are three new arrows whose composition with type has to be specified: LAMBDA, LET, and inst. Here there are the following rules:

type[LAMBDA[t_, expr_]] ⇒ forAll[t, type[expr]]
type[LET[t_, ss_, expr_]] ⇒ typelet[t, ss, type[expr]]
type[i[expr_, t_]] ⇒ typelet[type[expr][[1]], t, type[expr][[2]]]

Note that in the last case, expr has to be a universally quantified expression. These are represented by 2-cells as follows:

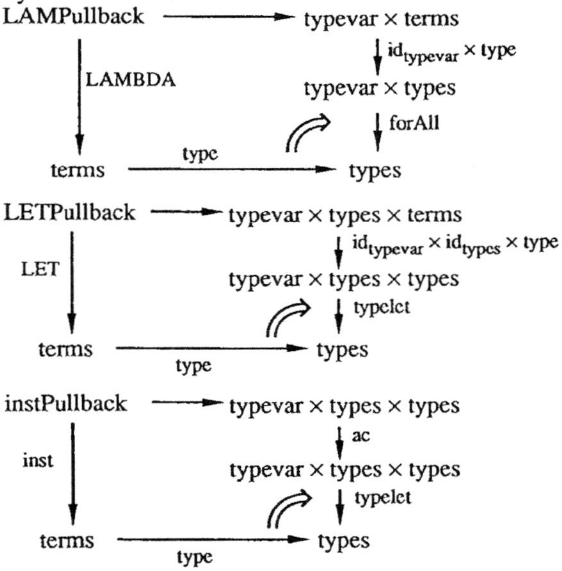

In the last diagram, ac is the associativity-commutativity arrow interchanging the last two factors.

Also, typelet is given by recursive rules:

typelet[t, τ, t] ⇒ τ
typelet[t, τ, s] ⇒ s, provided s ≠ t
typelet[t, τ, fun[σ, σ']] ⇒ fun[typelet[t, τ, σ], typelet[t, τ, σ']]
typelet[t, τ, forAll[t, σ]] ⇒ forAll[t, σ]
typelet[t, τ, forAll[s, σ]] ⇒ forAll[s, typelet[t, τ, σ]] provided s ≠ t and s ∉ fv(τ)
typelet[t, τ, forAll[s, σ]] ⇒ forAll[u, typelet[t, τ, typelet[s, u, σ]]
provided s ≠ t and s ∈ fv(τ) where u ∉ fv[σ] ∪ fv[τ], u ≠ t, and u ≠ s.

It is evident that these can all be represented by appropriate 2-cells.

4.5 Rewrite rules for terms in the polymorphic lambda calculus.

4.5.1 In order to describe the rules for evaluation, the three free variables operations have to be described. Fv is as before, with two additional cases which pose no problems. For fv, everything is evident. For FtV, we need a special auxiliary arrow in the graph, fvSet : Pf(types) → Pf(types)and the following two 2-cells for it:

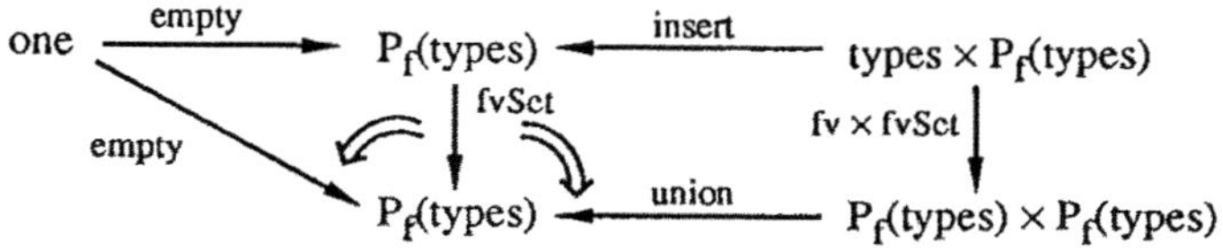

Then FtV : Terms → Pf(types) comes equipped with the following 2-cell:

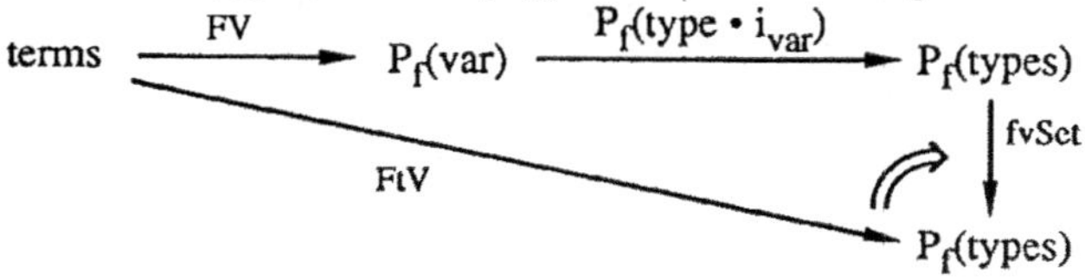

4.5.2 There is ordinary beta reduction plus the rule for instantiating a LAMBDA expression. i[LAMBDA[t, expr], s] ⇒ LET[t, s, expr]. The 2-cell for beta reduction is as in the pure typed lambda calculus. The 2-cell for instantiating a LAMBDA expression is as follows:

The left hand side is constructed by taking the product of the 2-cell for type • LAMBDA with the object types. This is a pseudo cone on the diagram whose pseudo limit is instPullback, so there is a canonical arrow γ : LAMPullback × types → instPullback. The left hand side is then inst • γ. Similarly, there is an obvious canonical arrow

δ : LAMPullback × types → LETPullback

whose composition with LET represents the right hand side. Thus the desired 2-cell is

inst • γ ⇒ LET • δ.

The rules for let have to be expanded to include the new terms and there are the many similar rules for LET. The new rules for let are the following:

let[x[t], expr1, LAMBDA[s, expr2]] ⇒ LAMBDA[s, let[x[t], expr1, expr2]];
let[x[t], expr1, i[expr2, s_]] ⇒ i[let[x[t], expr1, expr2], s].

These just say that let goes inside a LAMBDA expression and inside an instantiation. They obviously correspond to 2-cells.

There is a problem hidden here; the rules for let for the ordinary typed lambda calculus require that var has a decidable equality. This was arranged by declaring var to be the product of nat and type and observing that both of these had decidable equalities. However, type is now considerably more complicated. In particular, it is no longer a free signature, but has non trivial two cells involving typelet. We assert without proof here that in the initial algebra for types, typelet does not occur in any normal form, and that these normal forms are the same as the initial algebra for the sketch for types without typelet. This is a free signature, based on the collection of type variables which is isomorphic to the natural numbers, and hence has a decidable equality.

5 The initial model of an order enriched sketch.

A model of a sketch assigns a set (or, more generally, an object in a category) to each node of the graph and a function (or, more generally, a morphism in a category) to each arrow in such a way that formal quasi limits are taken to actual quasi limits and formal commutative diagrams are taken to actual commutative diagrams. Actual quasi limits make sense in an order-enriched category. An ordinary category can be regarded as order-enriched by making each homset into a discrete poset. In this case, quasi limits coincide with ordinary limits.

5.1 Example. As an example, consider the category **POSET** of posets and order-preserving functions. The quasi pullback of a pair of morphisms $f_1 : A_1 \to B$, and $f_2 : A_2 \to B$ in **POSET** consists of the set $\{<a_1, b, a_2> : f_i(a_i) \leq b \text{ for } i = 1, 2\}$ with the pointwise ordering on such triples. Given a poset X and order-preserving functions $p_i : X \to A_i$, $i = 1, 2$, and $q : X \to B$, with order relations $f_i\, p_i \leq q$, for $i = 1, 2$, then there is a unique order-preserving function r from X to the quasi pullback given by $r(x) = (p_1(x), q(x), p_2(x))$. Note that r need not be the only order-preserving function such that $pr_{A_i}\, r \leq p_i$ and $pr_B\, r \leq q$. However, it is the largest such function. The fact that r actually commutes with the projections shows that this quasi pullback is also a lax pullback. Other quasi limits in **POSET** are constructed similarly. Note that **POSET** also has all ordinary limits

5.2 Construction of the initial model. The initial model, or term model, assigns to each node the set of directed paths of arrows in the graph from one to that node. The operation of an arrow is by juxtaposition of arrows. The 2-cells generate a preorder relation on such paths of arrows. A path of arrows is evaluated by finding a maximal element in this preorder bigger than it.

In more detail, the initial model is constructed from the order-enriched theory generated by the sketch. Namely, start with any order-enriched sketch **A**; i.e., a graph |**A**| with specified formal quasi limits and 2-cells between specified paths of arrows. First construct the free ordinary category **F**(|**A**|) generated by the graph |**A**|. Its objects are the objects of |**A**|, its morphisms are (directed) paths of arrows in |**A**| and its composition is juxtaposition of paths of arrows. Make **F**(|**A**|) into an order enriched category as follows: If

$$p_1 \bullet \ldots \bullet p_n \Rightarrow q_1 \bullet \ldots \bullet q_m$$

is a 2-cell in **A**, then, for all paths h, k such that the following juxtapositions are defined, there is an order relation

$$h \bullet p_1 \bullet \ldots \bullet p_n \bullet k \leq h \bullet q_1 \bullet \ldots \bullet q_m \bullet k.$$

The reflexive, transitive closure of these relations on each homset **F**(|A|)(X, Y) makes **F**(|A|) into a preorder enriched category.

Next, adjoin all finite quasi limits to **F**(|A|) with the required induced canonical morphisms and order relations together with isomorphisms between the objects of **A** that have been declared to be formal quasi limits and the appropriate ones of these new adjoined quasi limits. The formal canonical induced morphisms are identified with the new adjoined canonical morphisms. This is a countable completion process; i.e., it has to be repeated countably often. (Note: these quasi limits are not necessarily assumed to be enriched since that can lead to automatic divergence.) The resulting preorder enriched category with all finite quasi limits is called the preorder enriched theory , **T**(A), of the sketch **A**. The initial model of **A** is then given by the formula:

$$I_A(X) = \mathbf{T}(A)(\text{one}, X);$$

i.e., $I_A(X)$ consists of the paths of arrows in **T**(A) from one to X, regarded as a preordered set. The set of maximal elements of $I_A(X)$ is the set of normal forms of closed expressions of sort X. Nothing in general is guaranteed about these sets of normal forms. It may or may not happen that every expression is less than or equal to a maximal element. There may be loops (divergent terms) and Church-Rosser may or may not hold. This procedure, restricted to the basic sketch for PCF, yields the language for PCF and the order relations realize the operational semantics of PCF. Hence, in this case, each expression is the start of a unique chain terminating in a maximal element.

5.3 Theorem. If I is the initial model for the pure typed λ-calculus, PCF, or the polymorphic typed λ-calculus in the category **POSET**, then I(terms) is the poset whose underlying set is the set of closed raw terms of the corresponding λ-calculus and whose order relation corresponds to the rewrite ordering. Also, I(types) is the poset of all type expressions of the corresponding λ-calculus with the rewrite ordering. Similarly, the initial model in **SET** has value at terms the set of normal forms of terms in the corresponding λ-calculus and value at types the set of all types of the corresponding λ-calculus.

6 Implementation of the initial algebra of a sketch in a conditional rewrite-rule programming language.

We describe here an automatic compilation of a sketch into the rewrite-rule language Mathematica. The same principles would presumably work in other languages as well. An order-enriched sketch consists of nodes, arrows, formal finite quasi limits, and formal 2-cells. We describe in turn how each constituent is represented in Mathematica.

6.1 Nodes. For each node in the graph there is a predicate `nodeQ` on the universe of all Mathematica expressions.

6.2 Arrows. For each arrow of the form: arrow : dom → cod, in the graph, there is a rewrite rule

```
codQ[arrow[x_]]  :=  True  /;  domQ[x]  ==  True
```

6.3 Finite quasi limits. For each finite quasi limit in the sketch, there is a rewrite rule

```
limitnodeQ[x1_,  .  .  .  ,  xn_]  :=  True  /;
        node1Q[x1]  ==  True        and
        - - - - - - - - -           and
        nodenQ[xn]  ==  True        and
```

```
comp[1,  1][x1]  ==  x1   and
comp[1,  2][x1]  ==  x2   and
- - - - - - - - - -       and
comp[i,  j][xi]  ==  xj   and
- - - - - - - - - -       and
comp[n,  n][xn]  ==  xn
```

6.4 **2-cells.** For each formal 2-cell $f_1 \bullet \ldots \bullet f_m \Rightarrow g_1 \bullet \ldots \bullet g_n$, where the two paths of arrows start at the node dom and end at the node cod, there is a rewrite rule

```
f1[...[fm[x_]...]  :=  g1[...[gn[x]...]/;  domQ[x]  ==  True
```

These constructions are applied to the sketches for PCF and the polymorphic lambda calculus in the papers [10] and [11].

7 References.

[1] M. Barr and C. Wells, *Categories for Computer Scientists*, Prentice Hall, 1990.

[2] C. Ehresmann, Esquisses et types de structures algébriques, Bull. Instit. Polit., Iasi XIV (1968), 1-14.

[3] P. Gabriel and F. Ulmer, *Lokal Präsentierbare Kategorien*, Lecture Notes in Math., 221, Springer-Verlag, New York, 1971.

[4] J.-Y. Girard, Une extension de l'interpretation de Gödel a l'analyse, et son application a l'elimination des coupures dans l'analyse et la theorie des types. In J.E.Fenstad, editor, *Proceedings of the Second Scandinavian Logic Symposium*, North-Holland Pub. Co., Amsterdam, London, 1971, 63-92.

[5] J.-Y. Girard, *Proofs and Types*, Translated and with appendices by P. Taylor and Y. Lafont, Cambridge University Press, Cambridge, 1989.

[6] J. W. Gray, *Formal Category Theory: Adjointness for 2-Categories*, Lecture Notes in Mathematics 391, Springer-Verlag, New York, 1974

[7] J. W. Gray, The Category of Sketches as a model for Algebraic Semantics, in *Categories in Computer Science and Logic*, J. W. Gray and A. Scedrov, Eds., Contemporary Mathematics 92, Amer. Math. Soc., Providence, R. I., 1989, 109 - 135.

[8] J. W. Gray, The Integration of Logical and Algebraic Types, in Categorical Methods in Computer Science with Aspects from Topology, H. Ehrig et al, Eds., Lecture Notes in Computer Science 393, Springer-Verlag, New York, 1989, 16 - 35.

[9] J. W. Gray, Executable specifications for data type constructors, in preparation.

[10] J. W. Gray, PCF in Mathematica, in preparation.

[11] J. W. Gray, Poly Lambda in Mathematica, in preparation.

[12] G. M. Kelly and R. Street, *Sydney Category Theory Seminar*, Lecture Notes in Mathematics 420, Springer-Verlag, New York, 1974.

[13] M. Makkai and R. Pare, *Accessible Categories: The foundations of Categorical Model Theory*, Contemporary Mathematics 104, Amer. Math. Soc., Providence RI, 1989.

[14] B. Pierce, S. Dietzen, S. Michaylov, Programming in higher-order typed lambda-calculi, Technical Report CMU-CS-89-111, Carnegie Mellon University.

[15] G. D. Plotkin, LCF considered as a programming language, Theoretical Computer Science 5 (1977), 223-255.

[16] J. C. Reynolds, Towards a theory of type structures, in *Colloque sur la Programmation*, Lecture Notes in Computer Science 19, Springer-Verlag 1974,408-425.

[17] A. Stouten, Substitution Revisited, Theoretical Computer Science 59 (1988), 317 - 325.

This paper is in final form and will not be published elsewhere.

First Steps in Synthetic Domain Theory

J. M. E. Hyland

Department of Pure Mathematics and Mathematical Statistics
16 Mill Lane, Cambridge CB2 1SB (GB)

1 Introduction

1.1 Aims of synthetic domain theory

Domain theory is the study of various concrete categories **C** typically of (directed) complete partial orders in which constructions fundamental to the analysis of computing can be performed.

- One can take fixed points of endofunctions $X \to X$, so as to give meaning to functions defined by general recursion.
- One can find fixed points for various operations $\mathbf{C} \to \mathbf{C}$ in order to provide interpretations for recursively defined types.

Domain theory is a well developed body of mathematics, in which the stress is on limits of increasing sequences. Continuity serves in the theory as a substitute for effectivity.

This paper is concerned with an approach to programming semantics of a different flavour. The motivating slogan is "domains are sets". (More exactly but less memorably, "domains are certain kinds of constructive sets".) An investigation of the general kind presented here was proposed by Dana Scott in a talk at a meeting of the Peripatetic Seminar on Sheaves and Logic in Sussex in 1980. He had in mind the example of Synthetic Differential Geometry where generalized manifolds are treated as (special kinds of) sets with the result that the development of the basic theory becomes highly intuitive: and he asked for a treatment of domain theory in a similar spirit. Initial progress was slow, but it now appears that the major conceptual advance was made by Scott's student Rosolini in his thesis [10]. There he took the recursively enumerable subobject classifier Σ as the lynch-pin of the theory. Rosolini also described axioms in the internal logic which unified features of the effective and recursive toposes, and took the first steps towards formulating a synthetic theory of computation. (Mention should also be made in this context of Lawvere's student Mulry who had identified the recursively enumerable subobject classifier, in the course of his study of the recursive topos, see [6].) In a theory of computation, effectivity must be analyzed directly. The category theoretic approach is to consider intrinsic structure of objects within a category (see the discussion in [7]); and as envisaged by Scott and Rosolini, this exploits the analogy with continuity. The result is the development of a kind of coding-free recursion theory, and it is this that is stressed in this paper.

1.2 Scope of the paper

The study of Synthetic Differential Geometry has two complementary aspects: a synthetic (or axiomatic) aspect and a semantic aspect (involving categorical models), and so it is also with Synthetic Domain Theory. In this paper the emphasis is on the first of these two aspects. The approach taken is to attempt to axiomatize effectivity in a categorical framework by means of properties of a classifier Σ of 'semi-decidable' subsets. (One might regard Σ as playing the same central role in Synthetic Domain Theory as the synthetic line R does in Synthetic Differential Geometry.)

The reader should perhaps be aware of limitations on the scope of the enquiry. We only reach the first aspect of domain theory: the existence of fixed points of endofunctions. Furthermore the word domain is used only in the general sense associated with Scott (though *not* as synonomous with what are usually called Scott domains) as distinct from the sense associated with Berry. This has a number of consequences. Computations may run in parallel and so the 'semi-decidable subsets' are closed under unions (Axiom 7). Also, for good objects the intrinsic order on function spaces is pointwise. Properties of Scott's topological notion of domain are reflected in all of the paper from section 3 onwards, though some aspects of the later sections are more widely applicable. Perhaps this is a drawback of the presentation; what we have is far from being an axiomatization of domains in the general sense. It remains a challenging problem to produce a synthetic theory of (the various flavours of) stable domains.

1.3 Background assumptions

We shall work within a *non-trivial* category of sets S, and need to state what we assume about this category. Of course one can assume that it is a topos, but the full structure of a topos will not play any part in our axiomatization. Hence it seems important to give some more precise indication of the kind of situation which one should have in mind. We assume that we work in a category with properties enjoyed by (amongst others) the category of modest sets within the effective topos.

In addition to the tacit assumption of non-triviality, the assumptions which we have in mind are as follows:

1. S is a locally cartesian closed subcategory of a topos $\mathcal{E}$ (in the internal sense);
2. S has a natural number object N;
3. S is the category of separated objects for a topology j on a subpretopos of the topos;
4. S is a small category contained within and complete relative to the category of j-separated objects of the topos.

A discussion of (1) for the particular case of the modest sets in the effective topos can be found in Hyland [2]. It seems best in view of some applications not to assume in (2) that N is the natural number object in the ambient topos. But we have the usual structure

$$1 \xrightarrow{0} \mathsf{N} \xrightarrow{s} \mathsf{N}$$

satisfying Lawvere's 'initial algebra' universal property. And in view of (3) we may assume Freyd's formulation of a natural number object in terms of a coproduct $1+\mathsf{N} \cong \mathsf{N}$ and coequalizer $\mathsf{N} \rightrightarrows \mathsf{N} \to 1$. As regards (3), Carboni and Mantovani have characterized the categories which arise as j-separated objects for a topology j on a pretopos. Such a category

- is regular (in the logical sense - it has finite limits and stable (regular epi, mono) factorization);
- has stable (epi, regular mono) factorization;
- has finite stable coproducts;
- has all 'quotients' of equivalence relations and these are quasi-effective (that is, the natural map from an equivalence relation to the kernel pair of its quotient is an epimorphism).

There is some analysis of (4) in Hyland, Robinson, Rosolini [3] and a fuller discussion in Robinson [9]. The full force of this assumption does not seem necessary for the theory developed here; and it does not hold in some of our models. It is important however if we wish to model strongly polymorphic type systems.

We give some examples of toposes in which we can find at least some of our background assumptions satisfied, and which support models for much of the synthetic domain theory which we shall describe.

1. Suitable realizability toposes. It seems that these may not all model the full theory which will be described below, but many do. For example
   - The effective topos. (Historically this was the motivating example.)
   - Toposes based on domain theoretic models for the lambda calculus.
2. Some toposes based on other notions of functional interpretation (modified realizability, Dialectica Interpretation). I have only made "back of an envelope" calculations in these cases, but they should be a good source of counter examples.
3. Topological toposes. These include
   - Johnstone's topological topos see [4].
   - Scott toposes in the general sense indicated in [10].
4. Recursive versions of topological toposes. The recursive topos of Mulry. (This was the first topos in which an r.e. subobject classifier was identified.) (The natural example is analogous to the simplest "well-adapted" models for synthetic differential geometry.

1.4 Conventions and notation

We shall use the usual set-theoretic language appropriate within (pre-)toposes, and refer to subobjects as subsets and to regular epis as surjections. A pretopos also provides good notions of (finite) intersections and unions of subobjects; and we suppose enough completeness to give us small (internal) intersections and unions. Note however that the existence of two factorization systems already adds a nuance to the set theory which reflects the j-modality. The regular monomorphisms are j-closed subobjects, while the epimorphisms are j-dense maps.

We mention a few standard category theoretic conventions. If A is an object of a category (with terminal object), then we shall write $A: A \to 1$ for the unique map to the terminal object 1. The identity map on A is denoted by $1_A: A \to A$, and the subscript will be dropped wherever possible. Also we use some non-standard notation. If

$a: 1 \to A$ is a point or element of A, then I shall write $k_a: X \to A$ for the map of the form $X \xrightarrow{X} 1 \xrightarrow{a} A$. Finally if $u: A \to C$ and $v: B \to C$ are maps, we write $[u, v]: A + B \to C$ for the induced map from the coproduct.

1.5 Acknowledgements

Recently many people have been taking an interest in aspects of Scott's challenging proposals to use domains in toposes as the semantics of programming languages. The work reflected in this paper was done as part of the EC Esprit BRA project 'Categorical Logic in Computer Science', and I have benefitted from the critical interest of Samson Abramsky, Eugenio Moggi, Andy Pitts, Pino Rosolini and Paul Taylor. Independently I owe intellectual debts on the one hand to Dana Scott and on the other to my student Wesley Phoa. Others from whom I have learnt in the course of the development of the effective topos as a context for programming semantics include Peter Freyd, Phil Mulry, Gordon Plotkin, and Edmund Robinson. Finally I acknowledge the use of Paul Taylor's useful diagram macros in the production of this paper.

2 Basic theory of partial maps

2.1 Semi-decidable or Σ-subsets

We follow Rosolini in making the notion of a Σ-subobject of an object the cornerstone of the theory. These Σ-subobjects should be thought of as recursively enumerable or (better) semi-decidable subsets, and we shall refer to them as Σ-subsets. Our first assumption is that this notion is classified.

Assumption 1 *We assume that we have an object Σ equipped with a subobject $t: 1 \to \Sigma$. The pullbacks of $t: 1 \to \Sigma$ are called Σ-subsets, and we assume that $t: 1 \to \Sigma$ is a generic Σ-subset in the sense that any Σ-subset $A \to X$ appears in a pullback diagram*

$$\begin{array}{ccc} X & \xrightarrow{\;a\;} & \Sigma \\ \uparrow & & \uparrow{\scriptstyle t} \\ A & \longrightarrow & 1 \end{array}$$

for a unique map $a: X \to \Sigma$. We shall write $A \subseteq_\Sigma X$ for A is a Σ-subset of X.

This assumption has the obvious consequences.

Proposition 2.1.1 *The collection of Σ-subsets satisfies the basic closure properies:*

(i) *For any object X, $X \subseteq_\Sigma X$, that is the maximal subset $1: X \to X$ is a Σ-subset.*

(ii) *The pullback of a Σ-subset is a Σ-subset.*

Note that the object Σ^X internalises the notion of a Σ-subset of X; its global sections correspond to the Σ-subsets.

2.2 The lift functor

Now we can exploit the background assumption that S is locally cartesian closed to obtain from the classifying map $1 \to \Sigma$ a further piece of structure, namely a Σ-partial map classifier. In the usual way, we let $\perp(X) \to \Sigma$ be $\Pi_t(X \to 1)$ and obtain a pullback

$$\begin{array}{ccc} \perp(X) & \longrightarrow & \Sigma \\ {\scriptstyle \eta_X}\uparrow & & \uparrow{\scriptstyle t} \\ X & \longrightarrow & 1 \end{array}$$

The map $\eta = \eta_X : X \to \perp(X)$ classifies partial maps whose domain is a Σ-subset. We refer to these as *Σ-partial maps*; we let $X \rightharpoonup_\Sigma Y$ denote a Σ-partial map from X to Y. If $A \subseteq_\Sigma X$ and $u : A \to Y$ is a Σ-partial map $X \rightharpoonup_\Sigma Y$, then there is a unique map $\bar{u} : X \to \perp(Y)$ such that

$$\begin{array}{ccc} X & \xrightarrow{\bar{u}} & \perp(Y) \\ \uparrow & & \uparrow{\scriptstyle \eta_Y} \\ A & \longrightarrow & Y \end{array}$$

is a pullback. In particular, $\perp(1) \cong \Sigma$ and $\eta : 1 \to \perp(1)$ corresponds to $t : 1 \to \Sigma$. The object $\perp(X)$ is called *the lift* of X and is sometimes written $X_\perp$.

$\perp$ extends to a functor in a natural way. If $u : X \to Y$ then $\perp(u) = u_\perp$ classifies the Σ-partial map $X_\perp \rightharpoonup_\Sigma Y$ defined on $X \subseteq_\Sigma X_\perp$ as u. Furthermore it is easy to see that $\perp$ is an S-functor. Finally note that the lift functor has familiar preservation properties.

Proposition 2.2.1 *The lift functor $\perp$ preserves connected limits in both the external and internal senses.*

We can of course iterate the lift functor, and so in particular can obtain a sequence of objects Σ_n defined by

$$\Sigma_0 = 1; \quad \Sigma_{n+1} = \perp(\Sigma_n).$$

Clearly the sequences of maps

$$1 \xrightarrow{\eta} \Sigma \xrightarrow{\eta} \Sigma_2 \ldots \Sigma_{n-1} \xrightarrow{\eta} \Sigma_n$$

classify sequences of subobjects

$$A_0 \subseteq_\Sigma A_1 \subseteq_\Sigma A_2 \ldots A_{n-1} \subseteq_\Sigma X$$

of an object X where each A_i is a Σ-subset of A_{i+1}.

2.3 The lift monad

Our first axiom states the closure of (representatives for) Σ-subsets under composition.

Axiom 2 *If A is a Σ-subset of B and B is a Σ-subset of X, then A is a Σ-subset of X.*

There is an immediate consequence.

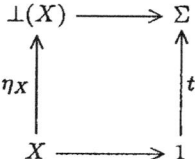

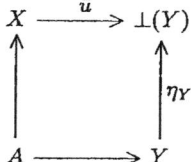

Proposition 2.3.1 *The collection of Σ-subsets of an object is closed under (finite) intersection.*

Corollary 2.3.2 *If $A \subseteq_\Sigma B$, then Σ^A is a retract of Σ^B.*

In the language of Rosolini [10], Axiom 2 says that the Σ-subsets form a dominance classified by $t\colon 1 \to \Sigma$. Axiom 2 has a useful alternative formulation in terms of the lift.

Proposition 2.3.3 *Given the classifying map $t\colon 1 \to \Sigma$ for Σ-subsets, the following are equivalent*

(i) *Σ-subsets form a dominance, that is Axiom 2 holds;*

(ii) *the composite $(1 \xrightarrow{t} \Sigma \xrightarrow{\eta} \Sigma_\perp)$ is a Σ-subset;*

(iii) *there is a (necessarily unique) natural transformation $\mu\colon (\perp)^2 \to \perp$ such that $(\perp, \eta, \mu)$ is a monad.*

Externally we have the usual 'subset ordering' or *inclusion ordering* on the Σ-subsets of X. Suppose that $a\colon X \to \Sigma$ and $b\colon X \to \Sigma$ classify $A \subseteq_\Sigma X$ and $B \subseteq_\Sigma X$ respectively. The inclusion relation induces a relation on classifying maps: if $A \subseteq B$, then we write $a \subseteq_\Sigma b$.

It is a consequence of Axiom 2 that we can identify in **S** a monic representing the subset order on Σ. We denote such a subobject of Σ^2 (which automatically exists in an ambient topos) by

$$\supseteq \rightarrow \Sigma^2 \quad \text{or} \quad \subseteq \rightarrow \Sigma^2$$

as appropriate. (It is convenient notationally to distinguish between the subobjects $\subseteq$ and $\supseteq$ of Σ^2 which are isomorphic via the twist map.) We have two maps $\perp(\Sigma)\colon \Sigma_\perp \to \Sigma$ and $\mu_1\colon \Sigma_\perp \to \Sigma$ and these induce a map $(\mu_1, \perp(\Sigma))\colon \Sigma_\perp \to \Sigma \times \Sigma$. It is a further consequence of Axom 2 that intersection is also represented in **S**. We let $\sqcap\colon \Sigma \times \Sigma \to \Sigma$ be the classifier of the Σ-subset $(t,t)\colon 1 \to \Sigma \times \Sigma$.

Proposition 2.3.4 (i) *A map $(a,b)\colon X \to \Sigma^2$ factors through $(\mu_1, \perp(\Sigma))$ if and only if $a \subseteq_\Sigma b$;*

(ii) *$(\mu_1, \perp(\Sigma))\colon \Sigma_\perp \to \Sigma^2$ represents the 'subset order' $\subseteq$ on Σ;*

(iii) *$(\mu_1, \perp(\Sigma))\colon \Sigma_\perp \to \Sigma^2$ is the equalizer of the maps* fst, $\sqcap\colon \Sigma^2 \to \Sigma$

(iv) *$(\Sigma, \top, \subseteq, \sqcap)$ forms a (meet) semilattice.*

Finally it is easy to see that the monad $(\perp, \eta, \mu)$ is strong. Hence the second part of 2.3.5 follows from the first.

Proposition 2.3.5 (i) *$(\Sigma_\perp \xrightarrow{\mu} \Sigma)$ is the free $\perp$-algebra on 1.*

(ii) *Each Σ^X has the structure of a $\perp$-algebra.*

2.4 The intrinsic order

The subset ordering on $\Sigma^{(\Sigma^X)}$ gives rise to a preorder on any object X of **S**.

Definition 1 The *intrinsic pre-order* on an object X is the relation $\leq$ which appears in the pullback

$$\begin{array}{ccc} X^2 & \longrightarrow & (\Sigma^2)^{(\Sigma^X)} \\ \uparrow & & \uparrow \\ \leq & \longrightarrow & \subseteq^{\Sigma^X} \end{array}$$

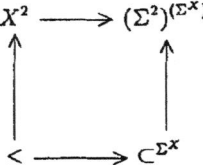

Using the internal logic, the definition says that

$$\mathsf{S} \models (x \leq y) \Longleftrightarrow (\forall R \in \Sigma^X . x \in R \Rightarrow y \in R).$$

Generally the pointwise preorder on function spaces need not coincide with the intrinsic preorder (see the discussion in [8]). We define the *pointwise preorder* $\preceq$ on a function space B^A in the internal logic by stipulating that

$$\mathsf{S} \models (f \preceq g) \Longleftrightarrow (\forall a \in A. f(a) \leq g(a)).$$

The intrinsic preorder on an object is analogous to the topological specialization order used in algebraic geometry. (Indeed the latter is a special case of the former.) Properties of the intrinsic preorder are most rapidly established using the internal logic. We now state a number of properties which we shall need and whose proofs are quite trivial in these terms. (I cannot resist remarking however that the intrinsic order seems to be a distraction. I would rather avoid reference to it where at all possible, and am distressed at my failure to do so more successfully.)

We shall eventually have an axiom ensuring that the intrinsic preorder on an object of the form Σ^A coincides with the inclusion ordering. One entailment is however both automatic and easy.

Proposition 2.4.1 *$R \leq S$ entails $R \subseteq S$ in Σ^A.*

Proposition 2.4.2 *Internally any function preserves the intrinsic preorder:*

$$\mathsf{S} \models \forall f \in B^A \, \forall a, c \in A (a \leq c \Longrightarrow f(a) \leq f(c)).$$

(And hence externally functions preserve the intrinsic order.)

Proof Essentially clear as f induces a map $\subseteq^{\Sigma^X} \to \subseteq^{\Sigma^Y}$. □

Corollary 2.4.3 *The intrinsic preorder on a product is the product of the intrinsic preorders.*

Note that the pointwise preorder $\preceq$ has a simple alternative characterization.

Proposition 2.4.4 *We have the following in the internal logic:*

$$\mathsf{S} \models \forall f, g \in B^A (f \preceq g \Longleftrightarrow (\forall S \in \Sigma^B) f^{-1}(S) \subseteq g^{-1}(S)).$$

Preorders are rather a bore. It is convenient to work with objects for which the intrinsic preorder is in fact an order.

Definition 2 For any object X there is a natural map $X \to \Sigma^{\Sigma^X}$. An object X is a *Σ-space* just when this map is a monic. X is *extensional* just when it is a regular monic.

Proposition 2.4.5 *If X is a Σ-space then the intrinsic preorder is an order.*

In what follows we shall often tacitly assume that we are working with Σ-spaces.

2.5 Objects with bottom

There are clearly a number of distinct notions of an object with a least or bottom element in the intrinsic order. In conformity with the 'algebraic spirit' of the categorical axiomatization, it seems best to work with a rather strong notion of an object equipped with a bottom element. As a conceit I introduce the definition before Axioms 3 and 6 which ensure that there really are bottom elements in these objects.

Definition 3 An object X *(equipped) with bottom* is an algebra $\bot(X) \to X$ for the lift monad. A *strict map* between such objects is a map of algebras.

Note that as we have presented the definition, having a bottom element is an additional piece of structure. However if we restrict attention to Σ-spaces, there is at most one such stucture on any X, so we may regard having a bottom element as a property. Since we may tacitly assume that the objects with which we are dealing are Σ-spaces, we shall not make any fuss over this.

There is considerable interest in the free algebras for the lift monad, that is, in objects which are themselves lifts of some object. We say that such an object $A_\bot$ is *a lift*. We already know that maps into such an object classify Σ-partial maps. But as $\bot^2(A) \xrightarrow{\mu} \bot(A)$ is a free $\bot$-algebra, we also know about strict maps from such an object into objects with bottom. In principle we want to regard these maps as maps preserving $\bot$, but clearly one has to say this in a positive way as one cannot yet identify the bottom elements! One particular case of interest is that of maps to an object of the form Σ^X.

Lemma 2.5.1 *Suppose that* $W \subseteq_\Sigma X \times \bot(A)$. *Then the diagram*

$$\begin{array}{ccc} X \times \bot^2(A) & \xrightarrow{1 \times \mu_A} & X \times \bot(A) \\ \uparrow & & \uparrow \\ W & = & W \end{array}$$

is a pullback if and only if $W \subseteq X \times A$.

Proposition 2.5.2 *Take a map* $u\colon \bot(A) \to \Sigma^X$ *and let* $U \subseteq_\Sigma X \times \bot(A)$ *be the subobject classified by the transpose* $\bar{u}\colon X \times \bot(A) \to \Sigma$ *of* u. *Then* u *is strict if and only if* $U \subseteq X \times A$.

2.6 The empty set

As things stand it is possible that $\Sigma = 1$ and that the collection of Σ-subsets is trivial. The next axiom ensures that this is not the case.

Axiom 3 *For any object* X, *the empty set is a* Σ*-subset of* X.

Proposition 2.6.1 *The following are equivalent:*

(i) *For any object* X, *the empty set is a* Σ*-subset of* X.

(ii) *The empty set is a* Σ*-subset of* 1.

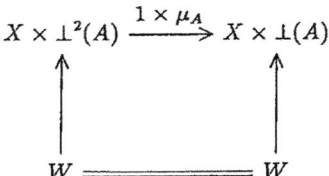

(iii) *There is a map $f\colon 1 \to \Sigma$ distinct from $t\colon 1 \to \Sigma$ in the sense that the diagram*

$$\begin{array}{ccc} 1 & \xrightarrow{t} & \Sigma \\ \uparrow & & \uparrow f \\ 0 & \longrightarrow & 1 \end{array}$$

is a pullback.

One simple consequence of Axiom 3 is that decidable subsets are semi-decidable. As usual a *decidable* subset of an object X is a subobject $A \subseteq X$ which is complemented: there is an isomorphism $A + B \cong X$ inducing $A \subseteq X$.

Proposition 2.6.2 *Decidable subsets are Σ-subsets.*

Proof Decidable subsets are classified by maps into $2 = 1+1$. As we have the pullback

$$\begin{array}{ccc} 2 & \xrightarrow{[t,f]} & \Sigma \\ \uparrow & & \uparrow \\ 1 & = & 1 \end{array}$$

they are also classified by maps into Σ. □

On the basis of Axiom 3 we start to get some recognizable structure. Note first that $\bot(0) = 1$ and the map $f\colon 1 \to \Sigma$ is equal to $\bot(0 \to 1)$. The diagram

$$\begin{array}{ccc} \Sigma & \xrightarrow{f_\bot} & \Sigma_\bot \\ t \uparrow & & \uparrow \eta \\ 1 & \longrightarrow & \Sigma \end{array}$$

is automatically a pullback. Thus we can identify three distinct points of $\Sigma_\bot$:

- the point $(\eta_\Sigma) \cdot t$ of $\Sigma_\bot$ which is the classifying map for $1 \subseteq_\Sigma 1 \subseteq_\Sigma 1$;
- the point $\eta \cdot f = (f_\bot) \cdot t$ of $\Sigma_\bot$ which is the classifying map for $0 \subseteq_\Sigma 1 \subseteq_\Sigma 1$;
- the point $\bot(0 \to \Sigma)$ of $\Sigma_\bot$ which is the classifying map for $0 \subseteq_\Sigma 0 \subseteq_\Sigma 1$.

The two maps $\bot(\Sigma)\colon \Sigma_\bot \to \Sigma$ and $\mu_1\colon \Sigma_\bot \to \Sigma$ are clearly distinct as they classify $\Sigma \subseteq_\Sigma \Sigma_\bot$ and $1 \subseteq_\Sigma \Sigma_\bot$ respectively. What's more, we can calculate the pullbacks of f along these maps. We get

$$\begin{array}{ccc} \Sigma_\bot & \xrightarrow{\bot(\Sigma)} & \Sigma \\ \uparrow & & \uparrow \\ 1 & = & 1 \end{array} \qquad \text{and} \qquad \begin{array}{ccc} \Sigma_\bot & \xrightarrow{\mu_1} & \Sigma \\ \uparrow & & \uparrow \\ \Sigma & \longrightarrow & 1 \end{array}$$

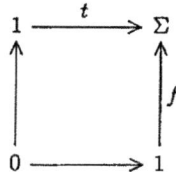

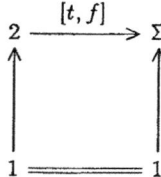

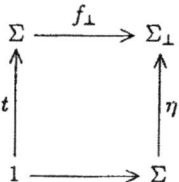

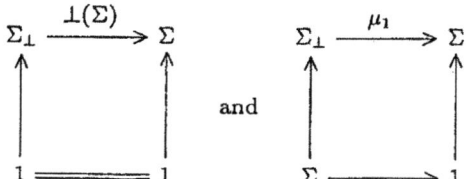

where $1 \to \Sigma_\perp$ is $\perp(0 \to \Sigma)$ and $\Sigma \to \Sigma_\perp$ is $\perp^2(0 \to 1)$. The left hand diagram is a pullback because $\perp$ preserves pullbacks. For the right hand one, note that $X \xrightarrow{u} \Sigma_\perp \xrightarrow{\mu} \Sigma$ factors through f if and only if u classifies subobjects $0 \subseteq_\Sigma U \subseteq_\Sigma X$; and that this happens if and only if $u = (\perp f) \cdot (\perp \Sigma) \cdot u$.

The reader will now readily see that the collection of objects $\Sigma_0, \Sigma_1, \ldots.$ and maps between them constructed from the maps t, f, μ by pullbacks and composition mirrors exactly the simplicial category of all nonempty finite ordinals and order-preserving maps.

3 Some dual structure

3.1 Co-Σ-subsets and the co-lift functor

Further structure of Σ can be conveniently described in terms of a notion of co-Σ-subset dual to that of Σ-subset. We assume that $f: 1 \to \Sigma$ classifies these subsets which we think of as the complements of semi-decidable subsets.

Assumption 4 *Call the pullbacks of $f: 1 \to \Sigma$ co-Σ-subsets. We assume that $f: 1 \to \Sigma$ is a generic co-Σ-subset in the sense that any co-Σ-subset A of X appears in a pullback diagram*

$$\begin{array}{ccc} X & \xrightarrow{\;a\;} & \Sigma \\ \uparrow & & \uparrow f \\ A & \longrightarrow & 1 \end{array}$$

for a unique $a: X \to \Sigma$. We shall write $A \subseteq_\Sigma X$ for A is a co-Σ-subset of X.

Note that now Σ^X also internalises the notion of co-Σ-subset. In fact more is true.

Proposition 3.1.1 *There is a bijective correspondence between Σ-subsets and co-Σ-subsets, where a co-Σ-subset of X corresponds to the Σ-subset with the same classifying map $X \to \Sigma$. This correspondence reverses the inclusion orders on co-Σ-subsets and Σ-subsets.*

Note that Axiom 3 enables us to 'see' the bottom element in a lift $A_\perp$: it is the map $k_\perp: 1 \to A_\perp$ lying in the unique pullback of form

$$\begin{array}{ccc} 1 & \longrightarrow & A_\perp \\ \uparrow & & \uparrow \\ 0 & \longrightarrow & A \end{array}$$

As a consequence of Proposition 3.1.1 one can deduce a more intuitive form of Proposition 2.5.2.

Proposition 3.1.2 *A map $u: \perp(A) \to \Sigma^X$ is strict if and only if $u \cdot k_\perp: 1 \to \Sigma^X$ names $\emptyset \subseteq_\Sigma X$. (That is, u is strict if and only if it preserves $\perp$.)*

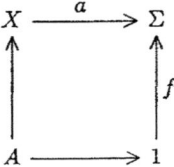

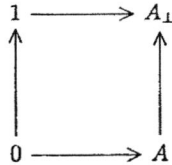

From the classifying map $f: 1 \to \Sigma$ we can obtain structure dual to that of 2.2. Thus there is a co-Σ-partial map classifier. This consists of a map $\zeta: X \to \top(X)$ such that if $A \subseteq_\Sigma X$ and $u: A \to Y$, then there is a unique map $\bar{u}: X \to \top(Y)$ such that

$$\begin{array}{ccc} X & \xrightarrow{\bar{u}} & \top(Y) \\ \uparrow & & \uparrow \zeta \\ A & \xrightarrow[u]{} & Y \end{array}$$

is a pullback. The object $\top(X)$ is called *the co-lift* of X and is sometimes written $X_\top$. The co-lift $\top$ extends to an S-functor; and $\top$ enjoys the standard preservation property.

Proposition 3.1.3 *The co-lift functor $\top$ preserves connected limits in both the external and internal senses.*

As things stand, there is some small interaction between the lift and co-lift. We have $\top(1) \cong \Sigma \cong \bot(1)$. However if we start iterating $\top$ there is no reason yet to suppose that we shall obtain a sequence of objects (and maps) isomorphic to the Σ_ns (and the face and degeneracy maps) of 2.6.

3.2 The co-lift monad

Next we introduce an axiom dual to Axiom 2.

Axiom 5 *If A is a co-Σ-subset of B and B is a co-Σ-subset of X, then A is a co-Σ-subset of X.*

As in the case of Axiom 2 there is an immediate consequence.

Proposition 3.2.1 *The collection of co-Σ-subsets of an object is closed under (finite) intersection.*

Corollary 3.2.2 *If $A \subseteq_\Sigma B$, then Σ^A is a retract of Σ^B.*

Furthermore as co-Σ-subsets form a dominance classified by $f: 1 \to \Sigma$, the duals of Propositions 2.3.3, 2.3.4 and 2.3.5 hold.

Proposition 3.2.3 *Given that $f: 1 \to \Sigma$ classifies co-Σ-subsets, the following are equivalent*

(i) *co-Σ-subsets form a dominance, that is, Axiom 5 holds;*

(ii) *the composite $(1 \xrightarrow{f} \Sigma \xrightarrow{\zeta} \Sigma_\top)$ is a co-Σ-subset;*

(iii) *there is a (necessarily unique) natural transformation $\nu: (\top)^2 \to \top$ such that $(\top, \zeta, \nu)$ is a monad.*

We have already seen in 3.1.1 that the bijective correspondence between Σ-subsets and co-Σ-subsets reverses the (external) inclusion orderings. Let $a: X \to \Sigma$ and $b: X \to \Sigma$ classify $A \subseteq_\Sigma X$ and $B \subseteq_\Sigma X$ respectively; then we have $A \supseteq B$ if and only if $a \subseteq_\Sigma b$. Thus there is essentially just one inclusion ordering on Σ. Using Axiom 5 we can find a representation of the subset order on Σ dual to that of Proposition 2.3.4. We have two maps $\top(\Sigma): \Sigma_\top \to \Sigma$ and $\nu_1: \Sigma_\top \to \Sigma$ and these induce a map $(\top(\Sigma), \nu_1): \Sigma_\top \to \Sigma \times \Sigma$. It is a further consequence of Axiom 5 that intersection of co-Σ-subsets is also represented in S. However this does not necessarily correspond to *union* of Σ-subsets. Hence we write $\vee: \Sigma \times \Sigma \to \Sigma$ for the classifier of the co-Σ-subset $(f, f): 1 \to \Sigma \times \Sigma$.

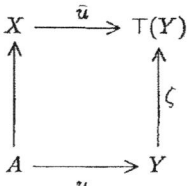

Proposition 3.2.4 (i) *A map* $(a, b): X \to \Sigma^2$ *factors through* $(\top(\Sigma), \nu_1)$ *if and only if* $a \subseteq_\Sigma b$;

(ii) $(\top(\Sigma), \nu_1): \Sigma_\top \to \Sigma^2$ *represents the 'subset order'* $\subseteq$ *on* Σ;

(iii) $(\top(\Sigma), \nu_1): \Sigma_\top \to \Sigma^2$ *is the equalizer of the maps* snd, $\vee: \Sigma^2 \to \Sigma$;

(iv) $(\Sigma, \bot, \subseteq, \vee)$ *forms a (join) semilattice.*

For completeness note that the monad $(\top, \zeta, \nu)$ is strong, and we have the dual of 2.3.5.

Proposition 3.2.5 (i) $(\Sigma_\top \xrightarrow{\nu} \Sigma)$ *is the free* $\top$*-algebra on* 1.

(ii) *Each* Σ^X *has the structure of a* $\top$*-algebra.*

As a consequence of 3.2.4 we have an isomorphism $\top(\Sigma) \cong \bot(\Sigma)$. Indeed now if we start with 1 and iterate $\top$ we do get a sequence of objects (and maps) isomorphic to those of 2.6. A complete collection of identifications between the presentations in terms of $\bot$ and $\top$ results. As an example note that

- $\top(t): \top(1) \to \top(\Sigma)$ corresponds to $\eta_\Sigma: \Sigma \to \bot(\Sigma)$.

Proposition 3.2.6 (i) *The co-lift functor* $\top$ *preserves* Σ*-subsets.*

(ii) *The lift functor* $\bot$ *preserves co-*Σ*-subsets.*

Proof Suppose that $u: X \to \Sigma$ classifies $U \subseteq_\Sigma X$. Consider the diagram

$$\begin{array}{ccccc}
\top(X) & \xrightarrow{\top(u)} & \top(\Sigma) \cong \bot(\Sigma) & \longrightarrow & \Sigma \\
\uparrow & & \top(t) \uparrow \eta_\Sigma & & \uparrow t \\
\top(U) & \longrightarrow & \top(1) \cong \Sigma & \longrightarrow & 1
\end{array}$$

where we use some of the identifications just mentioned. The left hand square is a pullback since $\top$ preserves pullbacks; and the right hand square is the classifying pullback diagram for $\Sigma \subseteq_\Sigma \bot(\Sigma)$. Hence the composite

$$\top(X) \xrightarrow{\top(u)} \top(\Sigma) \cong \Sigma_2 \longrightarrow \Sigma$$

classifies $\top(U) \subseteq_\Sigma \top(X)$. This proves (1), and (2) is just the dual. □

Finally note that we have the dual definitions of an object *equipped with a top element* (that is, the algebras for the co-lift monad), and of the *co-strict* maps between them. Objects of the form $A_\top$ (that is, the free algebras) are *co-lifts*. The duals of the result of 2.5 hold.

4 Finitary domain theory

4.1 A higher order axiom

After all the above axioms it still remains possible that t and f are the two distinct maps $1 \to 2$, so that the Σ-subsets, the co-Σ-subsets and the decidable subsets all coincide. This degeneracy is avoided by the main finitary axiom which genuinely exploits the higher order structure in our (locally) cartesian closed category S. By 2.6 there is a (monic) map $[t, f]: 2 \to \Sigma$.

Axiom 6 *The map*

$$1^{[t,f]}: \Sigma^\Sigma \to \Sigma^2$$

represents the inclusion order on Σ.

Let us spell out what this means in concrete terms. Write $e_\top$ and $e_\bot$ for the evaluation maps $1^t: \Sigma^\Sigma \to \Sigma$ and $1^f: \Sigma^\Sigma \to \Sigma$ respectively. Then $e_\bot \subseteq e_\top$ and there is a commutative diagram

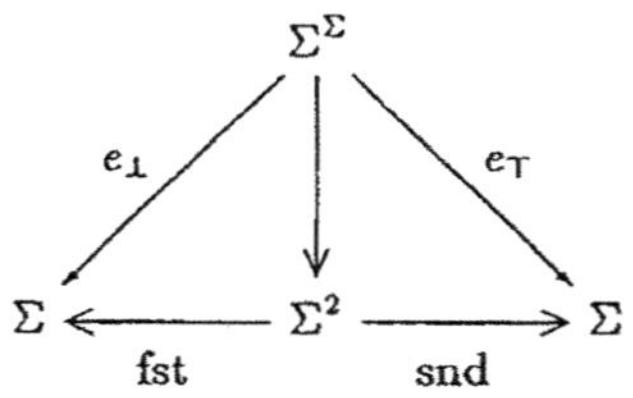

with the following universal property. Suppose that $a \subseteq b: X \to \Sigma$; then there is a unique map $(a; b): X \to \Sigma^\Sigma$ such that

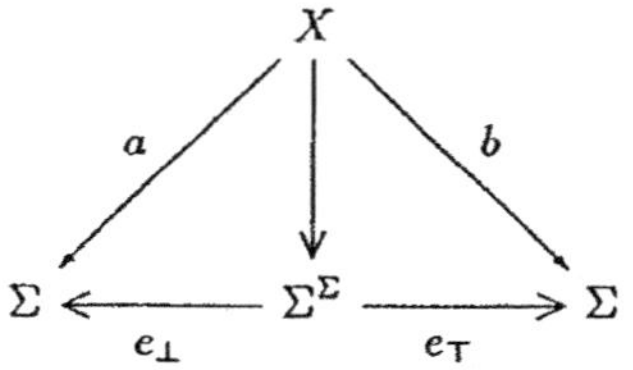

commutes.

Amongst many automatic pullbacks

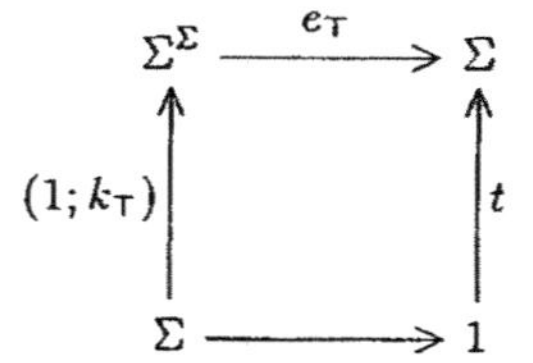

should be noted. This exhibits Σ as a Σ-subset of Σ^Σ.

As an application of the universal property, take $\mu \subseteq \bot(\Sigma): \Sigma_2 \to \Sigma$; thus there is a map $(\mu; \bot(\Sigma)): \Sigma_2 \to \Sigma^\Sigma$. It is easy to check that this map is inverse to the map $\Sigma^\Sigma \to \Sigma_\bot$ which classifies $1 \subseteq_\Sigma \Sigma \subseteq_\Sigma \Sigma^\Sigma$. Thus the content of Axiom 6 is given by isomorphisms

$$\Sigma^\Sigma \cong \Sigma_\bot \quad \text{and dually} \quad \Sigma^\Sigma \cong \Sigma_\top.$$

Modulo these isomorphism we can identify many maps. We give some examples:

- $\eta_\Sigma \circ f: \Sigma \to \Sigma_\bot$ corresponds to $(k_\bot; k_\top): \Sigma \to \Sigma^\Sigma$;
- $\eta_\Sigma: \Sigma \to \Sigma_\bot$ corresponds to $(1; k_\top): \Sigma \to \Sigma^\Sigma$;
- $\bot(f): \Sigma \to \Sigma_\bot$ corresponds to $(k_\bot; 1): \Sigma \to \Sigma^\Sigma$;
- $\mu: \Sigma_\bot \to \Sigma$ corresponds to $e_\bot: \Sigma^\Sigma \to \Sigma$;
- $\bot(\Sigma): \Sigma_\bot \to \Sigma$ corresponds to $e_\top: \Sigma^\Sigma \to \Sigma$.

4.2 Properties of the intrinsic order

One consequence of Axiom 6 is that it identifies the intrinsic order and the inclusion order.

Theorem 4.2.1 *The intrinsic order coincides with the inclusion order on all objects of the form Σ^X.*

Proof The diagram

$$\begin{array}{ccc} (\Sigma^A)^2 & \longrightarrow & (\Sigma^2)^{(\Sigma^{(\Sigma^A)})} \\ \uparrow & & \uparrow \\ (\Sigma^A)^\Sigma & \longrightarrow & \left(\Sigma^{(\Sigma^{(\Sigma^A)})}\right)^2 \end{array}$$

clearly commutes. But by Axiom 6 $(\Sigma^A)^\Sigma$ is the subset order on Σ^A. Hence it factors through the pullback of $(\Sigma^{(\Sigma^{(\Sigma^A)})})^2$ which is the intrinsic order. Thus the subset order entails the intrinsic order. But we already have the converse which was Proposition 2.4.1. □

Corollary 4.2.2 *$\bot$ is the least element in $A_\bot$.*

Proof In Σ, $\bot$ is the least element in the inclusion order and hence is least in the intrinsic order. Consider the map $\Sigma \times \bot(A) \to \bot(A)$ appearing in the unique pullback of form

$$\begin{array}{ccc} \Sigma \times \bot(A) & \longrightarrow & \bot(A) \\ \uparrow & & \uparrow \\ A & = & A \end{array}$$

We have $f \leq t$ in Σ and so $(f \times 1) \leq (t \times 1): \bot(A) \to \Sigma \times \bot(A)$. Hence by composition $k_\bot \leq 1: \bot(A) \to \bot(A)$. □

4.3 Σ-subsets of lifts and co-lifts

We return to analyze the general behaviour of maps $\bot(A) \to \Sigma^X$. In fact it is simpler to consider the case of a map $\top(A) \to \Sigma^X$. We know by the dual of 4.2.2 that $\top$ is the greatest element of $\top(A)$. But maps preserve the intrinsic order; hence if we let $s \in \Sigma^X$ be the image of $\top$ in Σ^X, s is the greatest element in the image in the intrinsic and hence in the subset order. It follows that if $s: X \to \Sigma$ classifies $S \subseteq_\Sigma X$, then $\top(A) \to \Sigma^X$ factors through the standard split monomorphism $\Sigma^S \to \Sigma^X$. Clearly the resulting map $\top(A) \to \Sigma^S$ is co-strict. Thus we have bijective correspondences:

$$\frac{\top(A) \to \Sigma^X}{S \subseteq_\Sigma X \quad \text{and a co-strict map} \quad \top(\mathrm{A}) \to \Sigma^{\mathrm{S}}}$$

$$\frac{}{S \subseteq_\Sigma X \quad \text{and} \quad U \subseteq_\Sigma S \times A.}$$

As a consequence we derive the following connections between the lift and co-lift.

Proposition 4.3.1 $\Sigma^{\top(A)} \cong \bot(\Sigma^A)$ *and dually* $\Sigma^{\bot(A)} \cong \top(\Sigma^A)$*, both these isomorphisms being natural in* A.

Proof The first isomorphism is a consequence of the bijective correspondences:

$$\frac{\frac{\frac{\frac{\frac{X \to \Sigma^{\top(A)}}{\top(A) \to \Sigma^X}}{S \subseteq_\Sigma X \quad \text{and} \quad U \subseteq_\Sigma S \times A}}{S \subseteq_\Sigma X \quad \text{and} \quad S \times A \to \Sigma}}{S \subseteq_\Sigma X \quad \text{and} \quad S \to \Sigma^A}}{X \to \bot(\Sigma^A)}$$

all natural in A. The second isomorphism is just the dual of the first. □

The attentive reader will have noticed that the argument from 4.2.2 onwards goes through simply under the assumption that the subset and intrinsic orders coincide. But Axiom 6 is a special case of Proposition 4.3.1; and so in fact it is equivalent to the coincidence of the two orders.

4.4 Rice's Theorem

One can regard Axiom 6 as a weak version of the undecidability of the halting problem. Hence we can now derive a little bit of non-trivial recursion theory.

Proposition 4.4.1 *For any objects* A*,* B*, and* C*, we have*

$$(B + C)^{\bot(A)} \cong B^{\bot(A)} + C^{\bot(A)}.$$

Proof Let $b: B + C \to \Sigma$ classify $B \subseteq_\Sigma B + C$. Consider the diagram

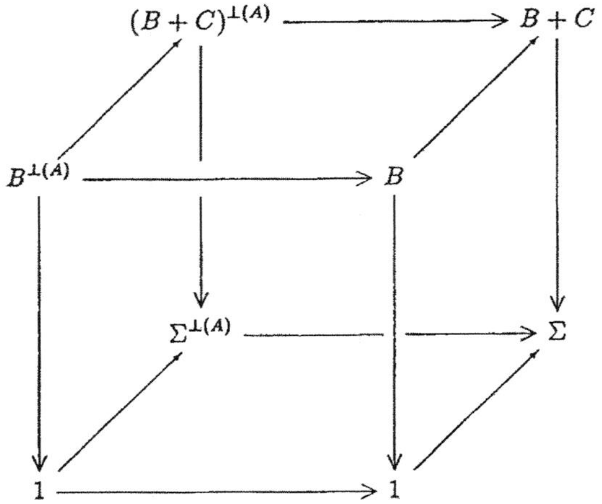

where the right hand face is the pullback exhibiting $B \subseteq_\Sigma B + C$, the left hand face is that pullback raised to the power $\perp(A)$ (and so is a pullback), and the horizontal maps are induced by $\perp: 1 \to \perp(A)$. By the naturality in 4.3.1, the map $\Sigma^{\perp(A)} \to \Sigma$ corresponds to $\top(\Sigma^A \to 1)$. Hence as $\top$ preserves pullbacks, the bottom face is a pullback. It follows at once that the top face is a pullback. The same argument works with C in place of B. So we have pullbacks

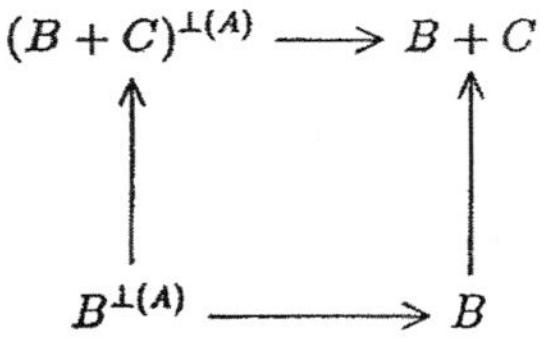

and

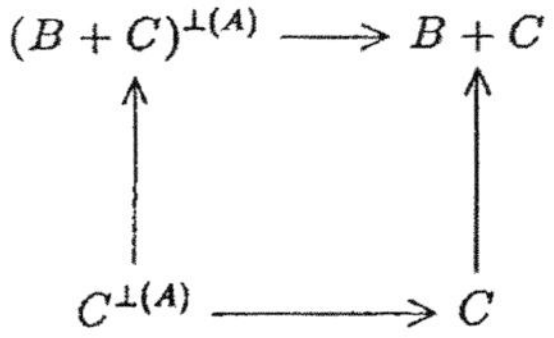

But coproducts are stable, so

$$(B + C)^{\perp(A)} \cong B^{\perp(A)} + C^{\perp(A)}.$$

□

Corollary 4.4.2 *If X is an object with bottom, then for any B and C*

$$(B + C)^X \cong B^X + C^X.$$

Proof Since X is a retract of $A = \perp(X)$, this is obvious. □

An immediate consequence is an abstract version of Rice's Theorem. Let us write $P = (\mathsf{N}_\perp)^{\mathsf{N}})$ for the object of Σ-partial functions from N to N.

Corollary 4.4.3 $2^P \cong 2$.

Proof P is a power of a lift and so is an object with bottom; and $2 = 1 + 1$. □

4.5 Further results

In this section we collect together some further consequences of Axiom 6, which we shall need.

From 4.3.1 we can derive an intuitively plausible characterization of the intrinsic order on a lift.

Proposition 4.5.1 *The intrinsic order on $\perp(B)$ is given in the internal logic by*

$$\mathbf{S} \models c \leq_{\perp(B)} d \Longleftrightarrow (c \in B \Rightarrow d \in B \wedge c \leq_B d).$$

Corollary 4.5.2 *The functor $\perp$ is externally order-preserving. If $f \preceq g\colon A \to B$ then $f_\perp \preceq g_\perp\colon A_\perp \to B_\perp$.*

The following proposition is related to the material of 5.3.

Proposition 4.5.3 *If A has a top, then Σ^A is a lift (and dually if A has a bottom, then Σ^A is a co-lift).*

Proof (Sketch.) Let $\top$ be the top element of A, and set $D = \{R \in \Sigma^A | \top \in R\} \subseteq_\Sigma \Sigma^A$. We obtain a map $\Sigma^A \to \perp(D)$ classifying the partial map $\Sigma^A \rightharpoonup_\Sigma D$ given by $1\colon D \to D$. Further we obtain a map $\perp(D) \to \Sigma^A$ as follows. Let $E \subseteq_\Sigma A \times D$ be classified by the obvious composite $A \times D \to A \times \Sigma^A \to \Sigma$. As $A \times D \subseteq_\Sigma A \times \Sigma^A$, we get $E \subseteq_\Sigma A \times \perp(D)$. Then we take the transpose of the corresponding classifying map $A \times \perp(D) \to \Sigma$. One checks that these two maps are inverses of one another. □

4.6 Finite unions

So far all our finitary domain theory is the consequence of just one axiom. But there is a further aspect of recursion theory which seems to require a further assumption. We note that while $\cap$ on Σ represents the usual notion of intersection of subobjects, we do not know that $\vee$ represents union. We make this our next axiom.

Axiom 7 *The class of all Σ-subsets is closed under finite unions.*

In recognition of this axiom we shall now write $\cup$ for $\vee$.

Axiom 7 can be expressed quite simply in diagrammatic terms, but we do not go into that here.

With Axiom 7 the misleading symmetry of our treatment is broken. We do not include its dual amongst our axioms, and indeed the dual is false in some of our standard models (in the effective topos for example). (This corresponds to the fact that while the collection of complements of recursively enumerable sets—the Π^0_1 sets—is closed under union, the logical equivalence on which this fact is based

$$(\forall x \phi(x) \vee \forall y \psi(y)) \iff (\forall x \forall y (\phi(x) \vee \psi(y)))$$

is not constructively valid.) Note that the failure of the dual axiom in the effective topos shows that Axiom 7 is independent of the earlier axioms.

The main consequence of Axiom 7 is the abstract version of the familiar fact that sets which are both semi-decidable and co-semi-decidable are decidable. First we exhibit 2 as a regular subobject of $\Sigma \times \Sigma$. It is natural to consider the two maps $(\cup, \cap), (k_\top, k_\perp)\colon \Sigma^2 \to \Sigma^2$.

Proposition 4.6.1 *The above two maps give rise to an equalizer diagram*

$$2 \longrightarrow \Sigma \times \Sigma \rightrightarrows \Sigma \times \Sigma$$

Corollary 4.6.2 *If both $A \subseteq_\Sigma X$ and $A \subseteq_{\bar{\Sigma}} X$, then $A \subseteq X$ is decidable.*

Proof If $a: X \to \Sigma$ classifies $A \subseteq_\Sigma X$ and $b: X \to \Sigma$ classifies $A \subseteq_{\bar{\Sigma}} X$, then it is easy to see that $(a, b): X \to \Sigma^2$ equalizes the maps $\Sigma^2 \rightrightarrows \Sigma^2$. Hence $A \subseteq X$ is decidable. □

5 Infinitary structure

5.1 A notion of ω-chain

Recall that we assumed that we have a natural number object N in our category S. Thus one can consider N-indexed sequences of Σ-subobjects of an object X: these are the maps $\mathsf{N} \to \Sigma^X$ or by transposition the Σ-subsets of $\mathsf{N} \times X$ or by further transposition the maps $X \to \Sigma^{\mathsf{N}}$. We are mainly interested in increasing (N-indexed) sequences of Σ-subobjects of an object X; these are maps $R: \mathsf{N} \to \Sigma^X$ such that $\forall n \in \mathsf{N}.R_n \subseteq R_{n+1}$ holds in S. Clearly we can represent the object of increasing sequences in Σ as the equalizer of maps $\Sigma^{\mathsf{N}} \rightrightarrows \Sigma^{\mathsf{N}}$, one being the identity, and the other the map which takes $\lambda n.\sigma_n \in \Sigma^{\mathsf{N}}$ to $\lambda n.\sigma_n \wedge \sigma_{n+1}$.

First let us show that there is an object ω which classifies increasing sequences in Σ in the sense that we can take Σ^ω to be the object of increasing sequences. (Morally ω corresponds to the object—usually called ω in domain theory—which consists of an infinite increasing sequence of points.) In essence ω is the colimit of the internal diagram

$$0 \to 1 \xrightarrow{f} \Sigma \xrightarrow{\bot(f)} \Sigma_2 \ldots \Sigma_n \xrightarrow{\bot^n(f)} \Sigma_{n+1} \ldots$$

and ω will be the initial algebra for the lift *functor*. We sketch a construction in category theoretic terms (avoiding completeness assumptions on S).

First we need some notation for indexed families. When we consider I-indexed families $X \to I$, it is often the case that we have a natural notation for the implicit fibres $X(i)$ but no notation for the corresponding object X. In these circumstances we shall write $(X(i)) \to I$ or more simply $(X(i))$ for the corresponding object of S/I. We shall use some specific families indexed over N. Let us adopt Martin-Löf's notation $(\mathsf{N}(n))$ for the N-indexed family of finite sets. From it we obtain an N-indexed family $(\Sigma^n) \to \mathsf{N}$. As a subobject of this one can construct an N-indexed family $(\Sigma_n) \to \mathsf{N}$ which internalizes the sequence Σ_n. It seems best to take the subobject

$$\Sigma_n = \{(p_0, \ldots, p_{n-1}) | \forall i (p_i \geq p_{i+1})\} \subset \Sigma^n$$

which is given by a simple equalizer in S/N.

Proposition 5.1.1 *The indexed family $(\Sigma_n) \to \mathsf{N}$ satisfies the recursion equations:*

$$\Sigma_0 \cong 1 \quad (\Sigma_{n+1}) \cong (\bot(\Sigma_n))$$

To obtain an N-indexed diagram, we need to describe the (indexed family) of maps $\bot^n(f)$; but these are induced by the maps $(1, f): \Sigma^n \to \Sigma^n \times \Sigma \cong \Sigma^{n+1}$. Now the colimit ω of the internal diagram will appear in a coequalizer diagram $\amalg_n(\Sigma_n) \rightrightarrows \amalg_n(\Sigma_n) \to \omega$ which lies over the standard coequalizer $\mathsf{N} \rightrightarrows \mathsf{N} \to 1$. (Note that the object ω is both the initial algebra for the lift functor and for the lift monad; but of course the structures are different!)

Recall from 2.6 that Σ_n has $n+1$ distinct global sections. We can internalize these as a family of maps $(\mathsf{N}(n+1)) \to (\Sigma_n)$, and this induces in the colimit, the standard enumeration of $\mathsf{N} \to \omega$. Hence in particular the global sections of ω include analogues of the finite ordinals.

Proposition 5.1.2 *The object Σ^ω is the object of increasing sequences in Σ: the enumeration $\mathsf{N} \to \omega$ induces an equalizer diagram $\Sigma^\omega \to \Sigma^{\mathsf{N}} \rightrightarrows \Sigma^{\mathsf{N}}$ where the two maps $\Sigma^{\mathsf{N}} \to \Sigma^{\mathsf{N}}$ are as described above.*

Proof (Sketch) The covariant functor $\Sigma^{()}$ transforms internal colimits into internal limits; the resulting equalizer diagram is not quite what we want, but we easily derive from it the fact that $\Sigma^\omega \to \Sigma^{\mathsf{N}} \rightrightarrows \Sigma^{\mathsf{N}}$ is an equalizer. □

Definition 4 For any object X of S the object X^ω is *the object of ω-chains in X.*

In general the ω-chains in X will not coincide with the sequences increasing in the intrinsic order. However it is an easy corollary of 5.1.2 that they do coincide in the case of retracts of powers of Σ. In fact they coincide for all linked Σ-spaces (see Phoa [8]).

5.2 Suprema of ω-chains

Suprema of ω-chains (in the traditional sense) play a major role in domain theory; so the next axiom is no surprise.

Axiom 8 *The collection of Σ-subsets of an object is closed under suprema of increasing (and N-indexed) sequences.*

Since ω classifies increasing sequences in Σ we can readily express our axiom in terms of Σ^ω.

Proposition 5.2.1 *The following are equivalent*

(i) *The class of Σ-subsets is closed under N-directed suprema.*

(ii) *There is a map $\bigvee : \Sigma^\omega \to \Sigma$ which is left adjoint to the constant map $\Sigma \to \Sigma^\omega$.*

We can identify ω with the subobject

$$\{(p_n)_{n\in\mathsf{N}} | \forall n(p_n \geq p_{n+1}) \wedge \exists n(p_n = \bot)\} \subseteq \Sigma^{\mathsf{N}}.$$

Perhaps it is worth stressing that this need not be a regular subobject! Clearly $\omega \subseteq \Sigma^{\mathsf{N}}$ is not closed under increasing sequences in Σ^{N}; we identify its closure as

$$\bar{\omega} = \{(p_n)_{n\in\mathsf{N}} | \forall n(p_n \geq p_{n+1})\} \subseteq \Sigma^{\mathsf{N}}.$$

For $\bar{\omega}$ is a retract of Σ^{N} via the map which sends $(p_n)_{n\in\mathsf{N}}$ to $(\bigwedge_{m\leq n}(p_m))_{n\in\mathsf{N}}$. Hence $\bar{\omega}$ is closed under limits of ω-chains in Σ^{N}: the supremum map $(\Sigma^{\mathsf{N}})^\omega \to \Sigma^{\mathsf{N}}$ restricts to a supremum map $\bar{\omega}^\omega \to \bar{\omega}$. Furthermore the pointwise intersection $(\Sigma^{\mathsf{N}}) \times \Sigma^{\mathsf{N}} \to \Sigma^{\mathsf{N}}$ restricts to a map $\omega \times \bar{\omega} \to \omega$; and its exponential transpose is a map $\bar{\omega} \to \omega^\omega$ which associates to every element of $\bar{\omega}$ a canonical increasing sequence in ω tending to it. In particular the composite $\bar{\omega} \to \omega^\omega \to \bar{\omega}$ is the identity. Other constructions of $\bar{\omega}$ are possible: for example we could set $\bar{\omega}$ to be the internal limit of the standard diagram of form

$$(1 \leftarrow \Sigma \leftarrow \bot(\Sigma)...).$$

This would be given externally as an equalizer.

5.3 The main infinitary axiom

We think of $\bar{\omega}$ as the object ω with a limit to the ω-chain added. So morally it should represent increasing sequences with limit point—at least in sufficiently good objects. In particular therefore we expect ω and $\bar{\omega}$ to have the same Σ-subsets, and we take this as our next axiom.

Axiom 9 *The map $\omega \to \bar{\omega}$ induces an isomorphism $\Sigma^{\bar{\omega}} \to \Sigma^{\omega}$.*

To understand the force of this axiom let us look closely at the object $\Sigma^{\bar{\omega}}$ of Σ-subsets of $\bar{\omega}$. Since $\bar{\omega}$ is a retract of Σ^{N}, it has a top; we write $\infty: 1 \to \bar{\omega}$ for this element. Write $e_\infty: \Sigma^{\bar{\omega}} \to \Sigma$ for the 'evaluation at top' map. From e_∞ we obtain the pullback

$$\begin{array}{ccc} \Sigma^{\bar{\omega}} & \xrightarrow{e_\infty} & \Sigma \\ \uparrow & & \uparrow t \\ E & \longrightarrow & 1 \end{array}$$

exhibiting $\Sigma^{\bar{\omega}} \cong \perp(E)$. (The reader should refer to 4.5.3.) Of course $\Sigma^{\bar{\omega}}$ is also an object with top (in fact a co-lift); indeed E has a top. Let us consider this situation in the abstract.

Proposition 5.3.1 *Suppose E has a top element $\uparrow$. Then the classifying map $\perp(E) \to \Sigma$ is left adjoint to the map $\perp(\uparrow): \Sigma \to \perp(E)$.*

Proof As $\uparrow$ is the top element of E the map $\uparrow: 1 \to E$ has $E \to 1$ as left adjoint. As $\perp$ is order-preserving in the sense of 4.5.2, it preserves adjunctions. Whence the result. □

Corollary 5.3.2 *The evaluation map $e_\infty: \Sigma^{\bar{\omega}} \to \Sigma$ is left adjoint to the constant map $\Sigma \to \Sigma^{\bar{\omega}}$.*

This shows, at least for objects of the form Σ^A and their retracts, that $\infty \in \bar{\omega}$ represents the suprema of ω-chains.

Proposition 5.3.3 *The supremum map $\bigvee: \Sigma^{\omega} \to \Sigma$ is equal to the composite $\Sigma^{\omega} \cong \Sigma^{\bar{\omega}} \xrightarrow{e_\infty} \Sigma$.*

Since sups of ω-chains are represented, they are automatically preserved (even in the internal sense) by maps $\Sigma^A \to \Sigma^B$. In fact the following are equivalent:

(i) $\omega \to \bar{\omega}$ induces an isomorphism $\Sigma^{\bar{\omega}} \to \Sigma^{\omega}$.

(ii) $\mathsf{S} \models (\forall f: \Sigma^A \to \Sigma^B)(\forall (R_n) \in (\Sigma^A)^{\omega})\, f(\bigvee(R_n)) = \bigvee(f(R_n))$.

We have just seen that (i) implies (ii), while (ii) implies (i) because $\bar{\omega}$ is the closure of ω under sups of ω-chains. We could also give an equivalent formulation of Axiom 9 in terms of a notion of *intrinsic limit* of ω-chains; but this makes more sense in connection with Axiom 10.

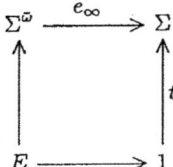

5.4 Directed unions

As we saw in connection with Axiom 7, there may be more to closure under unions than the existence of a suitable left adjoint. Hence there is a further infinitary axiom.

Axiom 10 *The collection of Σ-subsets of an object is closed under unions of increasing (and* N*-indexed) sequences.*

(As with Axiom 6 we could easily express this extra information in diagrammatic terms.) Obviously we can combine Axioms 7 and 10.

Proposition 5.4.1 *The class of Σ-subsets is closed under* N*-indexed unions.*

As a result we have a union map $\bigcup: \Sigma^{\mathsf{N}} \to \Sigma$. Using it we can exhibit N as a regular subobject of Σ^{N}. The essential map $\Sigma^{\mathsf{N}} \to \Sigma$ is that which takes $(p_n)_{n\in\mathsf{N}}$ to $\bigcup\{p_n \cap p_m | n \neq m\}$. If we pair this map with $\bigcup$ and $k_\perp$ with $k_\top$, we get two maps $\Sigma^{\mathsf{N}} \rightrightarrows \Sigma^2$.

Proposition 5.4.2 *There is an equalizer diagram* $\mathsf{N} \to \Sigma^{\mathsf{N}} \rightrightarrows \Sigma^2$*, where the maps* $\Sigma^{\mathsf{N}} \rightrightarrows \Sigma^2$ *are as just described.*

Corollary 5.4.3 N*-indexed partitions of an object X into Σ-subsets correspond to maps* $X \to \mathsf{N}$.

A further consequence of Axiom 10 is a form of the Rice-Shapiro Theorem.

Theorem 5.4.4 *(Rice-Shapiro) Σ-subsets of an object are Scott open with respect to ω-chains:*

$$\forall U \in \Sigma^A . \forall (a_n) \in A^\omega . \bigvee (A_n) \in U \Rightarrow \exists n . a_n \in U.$$

A full development of material in this area is contained in Rosolini's Thesis [10].

6 A category of predomains

6.1 The category of Σ-replete objects

A number of people have considered candidates for a good category of predomains within a topos (usually the effective topos). Phoa [8] considered the category of complete Σ-spaces, which consists of those objects which are complete with respect (N-indexed) sequences increasing with respect to the intrinsic order. Freyd et al [1] considered (in effect) the subcategory consisting of those complete Σ-spaces which are regular subobjects of a power of Σ. Here we consider a notion which is in some sense canonical. There are two perspectives which one can take. Since the object Σ^A of all Σ-subsets of A is to play the crucial conceptual role in the theory, we should consider those objects which are (in some sense) determined by their Σ-subsets. Alternatively we could argue that we need the object Σ and good completeness properties, but should not take more objects than we are forced to take. These two perspectives are equivalent; they lead to the same good category, the category of replete objects.

We can give an account of the category of replete objects by means of some internal category theory applied to the internal category S. The definitions which follow should be understood in that sense. (I hope that my giving translations into the internal logic of sets will help rather than confuse.)

Definition 5 A map $g: Q \to \Sigma$ is *Σ-equable* if the induced map $\Sigma^Q \to \Sigma^P$ is an isomorphism. (That is, in the internal logic of sets, any map $P \to \Sigma$ extends uniquely to a map $P \to \Sigma$.) We write **Equ** for the class of Σ-equable maps. A map $f: A \to B$ is *Σ-replete* if for any Σ-equable map $g: P \to Q$ the diagram

$$\begin{array}{ccc} A^Q & \longrightarrow & A^P \\ \downarrow & & \downarrow \\ B^Q & \longrightarrow & B^P \end{array}$$

is a pullback. (That is, in the internal logic of sets, given any commutative square as indicated below, there is a unique fill-in $Q \longrightarrow A$

$$\begin{array}{ccc} P & \longrightarrow & Q \\ \downarrow & \swarrow & \downarrow \\ A & \longrightarrow & B \end{array}$$

making the two triangles commute.) We write **Rep** for the class of all Σ-replete maps.

For the most part it is sufficient to restrict attention to the (internal) full subcategory of all objects A such that $A \to 1$ is a replete map. We write **R** for this category, the category of *replete objects*. (Of course maps between replete objects are automatically replete maps.)

Clearly (**Equ**, **Rep**) forms a prefactorization on S, and with enough (internal) completeness on that category we could show that it is a factorization system. However using rather little of the internal completeness of the ambient topos, we get what we chiefly need.

Theorem 6.1.1 **R** *is a reflective subcategory of* S.

Proof The unit of the reflection $A \to r(A)$ appears in $A \to r(A) \to \Sigma^{(\Sigma^A)}$ as the largest subobject of $\Sigma^{(\Sigma^A)}$ such that $A \to r(A)$ is Σ-equable. The image (and indeed the regular image) of A is necessarily such a subobject. □

Unfortunately it does not seem easy to give a simple concrete description of the replete objects in (for example) the effective topos. Indeed though Freyd was (of course) aware of the possibility of the above definition, he prefered for this reason to work with the more concrete category in [1]. Paul Taylor independently came to regard the replete objects as important and has carefully analyzed different equivalent descriptions of $r(A)$.

The significance of **R** is indicated by the following standard characterization.

Theorem 6.1.2 **R** *is the least internally full reflective subcategory of* S *which contains* Σ.

Note that the categories considered by Phoa [8] and Freyd et al [1] within the effective topos are complete and contain Σ. Hence they contain the category **R** of replete objects. However one can show that **R** is strictly contained in them. Rosolini and Scott considered a more topologically motivated category of *σ-spaces* (see [10] and [11]). In a Scott topos, Johnstone's example (see [5]) shows that the σ-spaces are strictly contained in the category of replete objects. Whether or not this holds in the effective topos remains an open question.

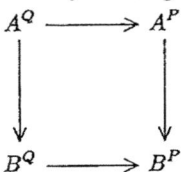

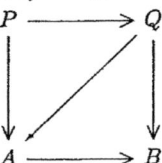

6.2 Simple closure properties of R

As R is reflective in S it inherits some properties.

Theorem 6.2.1 *The category* R *is as complete and cocomplete as is our initial category* S. *In particular under the assumptions of 1.3* R *is cartesian closed and closed under products indexed over separated objects.*

Warning: Local cartesian closedness is not a completeness property. It does not follow that R is locally cartesian closed. The problem was first made explicit by Thomas Streicher, see [12] where there is a counterexample.

Corollary 6.2.2 (i) *The objects* 0 *and* 2 *are in* R, *and* R *is closed under finite coproducts.*

(ii) *The object* N *is in* R *and* R *is closed under coproducts (internally) indexed by* N.

Proof R is closed under finite limits. Hence 0 is in yR by axiom 3, while 2 is in R by Proposition 4.6.1. More generally $A + B$ lies in an equalizer diagram of form

$$(A+B) \longrightarrow (A_\perp \times B_\perp) \rightrightarrows \Sigma \times \Sigma,$$

and so is in R. This deals with the first part. A similar argument involving Proposition 5.4.2 establishes the second part. □

The construction of R ensures that epimorphisms in R are easy to detect.

Proposition 6.2.3 *A map* $e: A \to B$ *in* R *is an epimorphism (in* R*) if and only if the induced map* $\Sigma^e: \Sigma^B \to \Sigma^A$ *is a monomorphism.*

Proof One direction is familiar enough. Take the pushout

$$\begin{array}{ccc} A & \xrightarrow{e} & B \\ {\scriptstyle e}\downarrow & & \downarrow{\scriptstyle f_0} \\ B & \xrightarrow[f_1]{} & C \end{array}$$

in R so that $C = (B +_A B)$. If e is epi then $f = f_0 = f_1$ are (equal) isos. Hence in the corresponding pullback

$$\begin{array}{ccc} \Sigma^C & \longrightarrow & \Sigma^B \\ \downarrow & & \downarrow \\ \Sigma^B & \longrightarrow & \Sigma^A \end{array}$$

the map $\Sigma^f: \Sigma^C \to \Sigma^B$ is an iso. Thus $\Sigma^e: \Sigma^B \to \Sigma^A$ is a mono. Conversely, if Σ^e is mono then Σ^f is an iso. But $f: B \to C$ is a map between replete objects. So if Σ^f is an iso, then f must be an iso. □

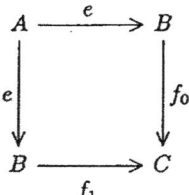

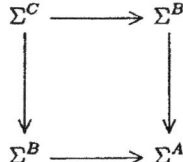

6.3 Lifts of replete objects

Next we mention a theorem which is needed for applications, and whose proof is pleasingly algebraic in character.

Theorem 6.3.1 *If A is replete, then so are $\perp(A)$ and $\top(A)$.*

Proof We give a proof for the case of $\perp(A)$ in the style of the internal category theory which we have been using. First we note the following trivial lemma. Suppose that $f: C \to D$ is Σ-equable and that $D' \subseteq_\Sigma D$ is a Σ-subset. Consider the pullback

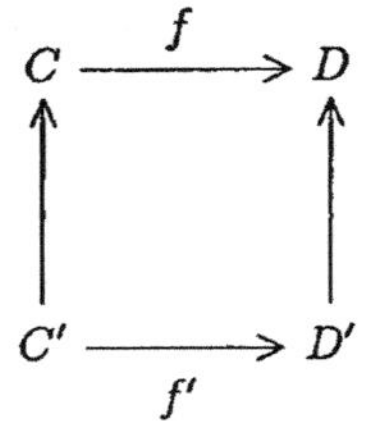

of $D' \to D$ along f. Then the map $f': C' \to D'$ is Σ-equable. Suppose now that $C \to D$ is Σ-equable and that we are given a map $C \to \perp(A)$. Clearly there is a unique map $D \to \Sigma$ making the diagram

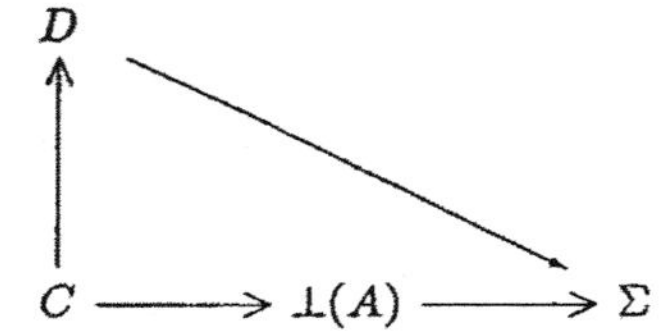

commute. If we pull this diagram back along $t: 1 \to \Sigma$, we obtain

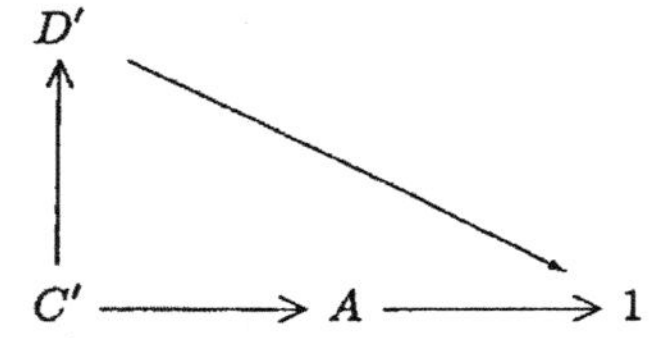

By the lemma $C' \to D'$ is Σ-equable and so there is a unique map $D' \to A$ such that

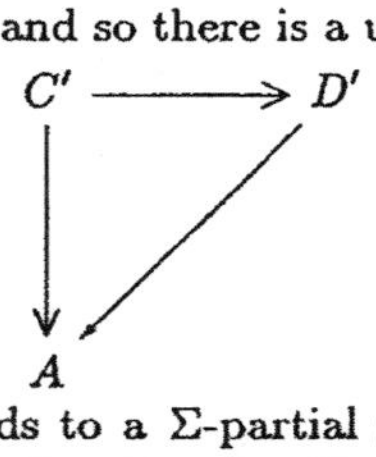

commutes. But this map corresponds to a Σ-partial map $D \rightharpoonup A$ and so induces a unique map $D \to \perp A$ fitting into the diagram of pullbacks:

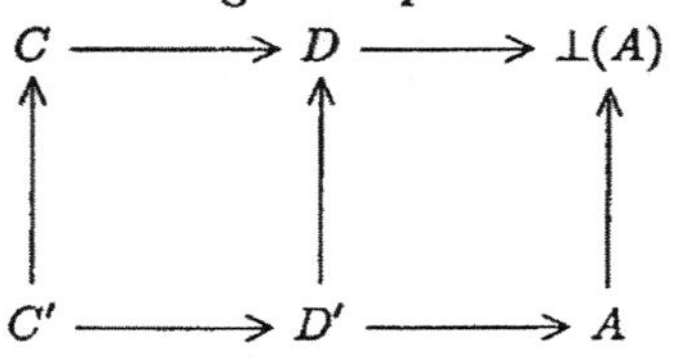

By the uniqueness property of $\bot(A)$, the map $D \to \bot(A)$ is the unique extension of the map $C \to \bot(A)$. □

6.4 Fixed points

Repleteness of an object is of course a kind of completeness condition. Indeed Axiom 9, which says in effect that the map $\omega \to \bar{\omega}$ is equable, has the immediate effect that replete objects have limits of ω-chains. This allows us to develop a traditional theory of fixed points. (Another somewhat fuller treatment has been given by Paul Taylor.)

Definition 6 An object A in a cartesian closed category $\mathbf{C}$ is in *the fixed point category* just when for any $I \in \mathbf{C}$, the power A^I is a *fixed point object*: any map $A^I \to A^I$ has a fixed point.

It does not matter for the purposes of this definition whether we take 'fixed point object' in the internal or external sense.

Proposition 6.4.1 *Suppose that A is replete and has a bottom element. Then every map $A \to A$ has a* least *fixed point. Hence as powers of replete objects with bottom are again replete objects with bottom, such an A is automatically in the fixed point category.*

Proof The usual proof works! □

Thus if we take as category of domains the objects of $\mathbf{R}$ with bottom element, we can interpret within it all definitions of functions by recursion. We have taken the first steps in a synthetic domain theory.

References

[1] P. Freyd, P. Mulry, G. Rosolini and D. S. Scott. 'Extensional PERs.' In: *Proceedings of the 5th Annual Symposium on Logic in Computer Science* (1990) 346-354.

[2] J. M. E. Hyland. 'A small complete category.' *Annals of Pure and Applied Logic,* 40 (1988) 135-165.

[3] J. M. E. Hyland, E. P. Robinson and G. Rosolini. 'The discrete objects in the effective topos.' *Proc. Lond. Math. Soc.* (3) 60 (1990) 1-36.

[4] P. T. Johnstone. 'On a topological topos.'*Proc. Lond. Math. Soc.* (3) 38 (1979) 237-271.

[5] P. T. Johnstone. 'Scott is not always sober.' In: *Lecture Notes in Mathematics* 871, Springer-Verlag, 1981, 282-283.

[6] P. Mulry. 'Generalized Banach-Mazur functionals in the topos of recursive sets.' *J.Pure and Appl. Alg.* 26 (1981) 71-83.

[7] W. Phoa. 'Effective domains and intrinsic structure.' In: *Proceedings of the 5th Annual Symposium on Logic in Computer Science*, (1990) 366-377.

[8] W. Phoa. *Domain Theory in Realizability Toposes.* Ph. D. Thesis, Cambridge, 1990.

[9] E. P. Robinson. 'How complete is PER?' In: *Proceedings of the 4th Annual Symposium on Logic in Computer Science*, (1989).

[10] G. Rosolini. *Continuity and effectiveness in topoi.* D.Phil. Thesis, Oxford, 1986.

[11] G. Rosolini. 'Categories and effective computations.' In: *Lecture Notes in Computer Science* 283, Springer-Verlag, 1987, 1-11.

[12] T. Streicher. 'Independence results for calculi of dependent types.' In: *Lecture Notes in Computer Science* 389, Springer-Verlag, 1989, 141-154.

This paper is in final form and will not be published elsewhere.

PRECATEGORIES AND GALOIS THEORY

GEORGE JANELIDZE

MATHEMATICAL INSTITUTE OF THE GEORGIAN
ACADEMY OF SCIENCES

Z. Rukhadze str. 1, 380093, Tbilisi, Georgia

Abstract

We give a new version of Galois theory in categories in which normal extensions are replaced by arbitrary extensions for which the "pullback functor" is monadic, and their Galois groupoids are replaced by internal pregroupoids; we obtain the "fundamental theorem of Galois theory" using just simple remarks on internal precategories and change of universe for internal functors.

I would like to thank Aurelio Carboni at Milan University for inviting me for extended visit in Summer 1990 during which the work on this paper was completed.

Introduction

The fundamental theorem of the classical Galois theory can be reformulated as follows:

THEOREM 0.1. *Let K/k be a Galois field extension with the Galois group G. Then the category* Spl(K/k) *of k-algebras split over K is antiequivalent to the category* (Fin. Sets)^G *of finite G-sets.* □

Observe that there are two equivalent definitions of "split by K":

DEFINITION 0.2. *A k-algebra S (commutative, with* 1*) is split over K iff*

$$K \otimes_k S \approx K \times \cdots \times K,$$

i.e. S is finite dimensional as a vector space and $K \otimes_k S$ is generated by idempotents over K. □

DEFINITION 0.3. *A k-algebra S is split over K iff*

$$S \approx K_1 \times \cdots \times K_n,$$

where $K_1, \ldots, K_n$ are subextensions, i.e. $k \subset K_i \subset K$ for each $i = 1, \ldots, n$ (and so $S = 0$ for $n = 0$). □

The reformulation above is well known — it is just a simple particular case of Grothendieck's Galois theory [2]. There are also many other generalizations and analogues of 0.1, in particular in the case of commutative rings the most general version of 0.1 is the Magid theorem (IV.31 in [13]).

Studying Magid's Galois theory I discovered that all results of such sort can be proved by relatively simple, purely-categorical arguments [6] (the main result was announced with a sketch of proof in 1984 [4]). The fundamental theorem of Galois theory in categories [6] is the following:

THEOREM 0.4. *Let $I : \underline{C} \to \underline{X}$ be a functor, and $\underline{C}$ an I-normal object in $\underline{C}$ with the Galois groupoid G. Then the category $\mathrm{Spl}_I(C)$ of objects in $\underline{C}$ split over C with respect to I is equivalent to the category $\underline{X}^G$ of internal functors from the internal groupoid G to the "universe" $\underline{X}$.* □

For the definitions and the connection with Magid theorem see [6] (which also uses [3] and Pierce theory [14]); we only recall here the definition of "I-normal" and "split":

DEFINITION 0.5. *Let $\underline{C}$, $\underline{X}$ be categories with finite limits and $I : \underline{C} \to \underline{X}$ a functor. An object C is I-normal iff the following conditions hold:*

(a) *the induced functor*

$$I^C : (\underline{C} \downarrow C) \to (\underline{X} \downarrow I(C))$$

has the right adjoint H^C and the counit

$$I^C H^C \to 1_{(X \downarrow I(C))}$$

is an isomorphism;

(b) *the right adjoint P^C of the forgetful functor*

$$U^C : (\underline{C} \downarrow C) \to \underline{C}$$

is monadic;

(c) *the image of the composition $U^C H^C$ is in $\mathrm{Spl}_I(C)$ which is the full subcategory of $\underline{C}$ with objects all $A \in \underline{C}$ such that the canonical morphism*

$$P^C(A) \to H^C I^C P^C(A)$$

is an isomorphism. □

REMARK 0.6. *As it was proved in* [9], 0.5(c) *is equivalent to the condition $C \in \mathrm{Spl}_I(C)$, and it is nice that in the case of* 0.1 *the conditions* 0.5(a) *and* 0.5(b) *are trivial, and* 0.5(c) *is exactly the Grothendieck definition of Galois extension.*

Since the Galois theory of [6] is very general it must have new examples. An important one is the theory of central extensions of groups (see § 3 of [6]) in which Galois groupoid is an internal goupoid in the category of abelian groups and is very far from the usual Galois groups (i.e. from automorphism group). However the results of [6] give a description of central extensions only for groups with perfect commutant, and, to do it for arbitrary groups, the Galois theory of [6] was generalized in [7], where, instead of an adjunction between $\underline{C}$ and $\underline{X}$, we consider so called "Galois structures".

In this paper we consider a new generalization: we work without the most important condition 0.5(c) using a new method — so called "change of universe" (see § 2 and § 3), and Galois groupoids are now internal precategories (actually pregroupoids) which are just suitable fragments of simplicial objects. Observe that we use internal functors from an internal precategory to the corresponding universe — the idea of the definition and change of universe actually comes from the theory of internal (discrete) fibrations, and we could write and we could write Fibr(P) instead of $\underline{A}^P$ in 1.1 (see § 1).

The results of this paper give improvements in all examples of Galois theories given in [4]–[10]; in particular for commutative rings we obtain:

THEOREM 0.7. *Let R be a commutative ring and E a commutative R-algebra (both with "the same" 1) which a faithfully flat R-module with the Galois pregroupoid G. Then the category* Spl(E/R) *of R-algebras split over E is antiequivalent to the category of internal functors from G to the universe (the universe in this case is the category of profinite, i.e. Boolean topological spaces).* □

The simplest definition of "split" in this case is the following:

DEFINITION 0.8. *An R-algebra S (commutative, with 1) is split over E iff $E \otimes_R S$ is freely generated by idempotents over E.* □

REMARK 0.9. *If $\underline{X}$ is a category like* Sets *(as in 0.7), then internal pregroupoids in $\underline{X}$ can be replaced by groupoids. Moreover a pregroupoid can be considered as a presentation of a groupoid by generators and identities. This in particular can be used to give a category-theoretical explanation of J. Kennison's definition of fundamental groups* [12].

It would be also interesting to compare the results of this paper with the results of the paper [1] *of M. Bunge.*

1. Internal precategories and functors

We start with the category τ generated by the graph

$$2 \;\begin{array}{c} \xrightarrow{p_1} \\ \xrightarrow{m} \\ \xrightarrow{p_2} \end{array}\; 1 \;\begin{array}{c} \xrightarrow{r} \\ \xrightarrow{d} \\ \xleftarrow{e} \end{array}\; 0 \tag{1.1}$$

and the identities

$$de = 1_0 = re, \quad dp_1 = rp_2, \quad dm = dp_2, \quad rm = rp_1. \tag{1.2}$$

DEFINITION 1.1. *Let $\underline{A}$ be an arbitrary category.*
Then:

(a) *an internal precategory P in $\underline{A}$ is a functor $P : \tau \to \underline{A}$;*

(b) *let P be an internal precategory in $\underline{A}$; an internal prefunctor $P \to \underline{A}$ is an object of the comma category $(\underline{A}^\tau \downarrow P)$, i.e. a pair (F, φ), where F is a functor from τ to $\underline{A}$ and $\varphi : F \to P$ is a natural transformation;*

(c) *an internal functor* $(F,\varphi) : P \to \underline{A}$ *is an internal prefunctor* $(F,\varphi) : P \to \underline{A}$ *such that the diagrams*

$$\begin{array}{ccc} F(2) & \xrightarrow{F(p_2)} & F(1) \\ {\scriptstyle\varphi_2}\downarrow & & \downarrow{\scriptstyle\varphi_1} \\ P(2) & \xrightarrow[P(p_2)]{} & P(1) \end{array} , \qquad \begin{array}{ccc} F(1) & \xrightarrow{F(d)} & F(0) \\ {\scriptstyle\varphi_1}\downarrow & & \downarrow{\scriptstyle\varphi_0} \\ P(1) & \xrightarrow[P(d)]{} & P(0) \end{array} \tag{1.3}$$

are pullbacks; the category of internal functors $P \to \underline{A}$ *will be denoted by* $\underline{A}^P$. □

Consider the main example. Let $\underline{A} =$ Sets be the category of sets, C a small category and $\Phi : C \to \underline{A}$ a functor. Then one can construct an internal precategory P in $\underline{A}$ and an internal functor $(F,\varphi) : P \to \underline{A}$ as follows:

$P(0)$ is the set of objects in C; $P(1)$ is the set of morphisms in C; $P(2)$ is the set of composable pairs of morphisms in C, i.e.

$$P(2) = \{(\beta,\alpha) \mid \beta\alpha \quad \text{is defined in} \quad C\};$$

and for $a \xrightarrow{\alpha} b \xrightarrow{\beta} c$ in C one has:

$$(\beta,\alpha) \begin{array}{l} \xmapsto{P(p_1)} \beta, \\ \xmapsto{P(m)} \beta\alpha, \\ \xmapsto{P(p_2)} \alpha, \end{array} \qquad \begin{array}{rl} \alpha & \xmapsto{P(r)} b, \\ & \xmapsto{P(d)} a, \\ 1_a & \xleftarrow{P(e)} a; \end{array}$$

$$\begin{aligned} F(0) &= \{(a,f) \mid a \in P(0), \quad f \in \Phi(a)\}; \\ F(1) &= \Big\{(\alpha,f) \mid \alpha \in P(1), \quad f \in \Phi\big(P(d)(\alpha)\big)\Big\}; \\ F(2) &= \Big\{(\beta,\alpha,f) \mid (\beta,\alpha) \in P(2), \quad f \in \Phi\big(P(d)(\alpha)\big)\Big\}; \end{aligned}$$

and for $a \xrightarrow{\alpha} b \xrightarrow{\beta} c$ as above and $f \in \Phi(a)$ one has:

$$(\beta,\alpha,f) \begin{array}{l} \xmapsto{F(p_1)} (\beta,\ \Phi(\alpha)(f)), \\ \xmapsto{F(m)} (\beta\alpha,\ f), \\ \xmapsto{F(p_2)} (\alpha,\ f) \end{array} \qquad \begin{array}{rl} (\alpha,\ f) & \xmapsto{F(r)} (b,\ \Phi(\alpha)(t)), \\ & \xmapsto{F(d)} (a,\ f), \\ (1_a,\ f) & \xleftarrow{F(e)} (a,\ f); \end{array}$$

$$\begin{aligned} &\varphi_0 : (a,\ f) \mapsto a; \\ &\varphi_1 : (\alpha,\ f) \mapsto \alpha; \\ &\varphi_2 : (\beta,\ \alpha,\ f) \mapsto (\beta,\ \alpha). \end{aligned}$$

This actually shows, that in this case $\underline{A}^P$ is equivalent to the category of usual functors $C \to$ Sets. And after that clearly we have:

PROPOSITION 1.2. *Let $\underline{A}$ be a category with pullbacks and $C = (C_0, C_1, \delta, \rho, \eta, \mu)$ an internal category in $\underline{A}$.*
Then the diagram

$$C_1 \times_{C_0} C_1 \;\begin{array}{c}\xrightarrow{\text{proj}_1}\\ \xrightarrow{\mu}\\ \xrightarrow{\text{proj}_2}\end{array}\; C_1 \;\begin{array}{c}\xrightarrow{\rho}\\ \xrightarrow{\delta}\\ \xleftarrow{\eta}\end{array}\; C_0$$

can be considered as an internal precategory in $\underline{A}$ in the sense of 1.1(a) and the category $\underline{A}^C$, constructed as in 1.1(c), is equivalent to the usual internal functor category $\underline{A}^C$. □

2. Change of universe for internal prefunctors

Consider an adjunction

$$(I,\, H,\, \eta,\, \varepsilon) : \underline{C} \to \underline{X} \tag{2.1}$$

between categories $\underline{C}$ and $\underline{X}$ with pullbacks.

The full subcategory of $\underline{C}$ with objects all $C \in \underline{C}$ such that $\eta_C : C \to HI(C)$ is an isomorphism, will be denoted by $\mathrm{Spl}_I(\underline{C})$ and the full subcategory of $\underline{X}$ with objects all $X \in \underline{X}$ such that $\varepsilon_X : IH(X) \to X$ is an isomorphism will be denoted by $\mathrm{Cospl}_I(X)$; we have the induced equivalence

$$\mathrm{Spl}_I(\underline{C}) \sim \mathrm{Cospl}_I(\underline{X}). \tag{2.2}$$

We have also the following induced adjunctions:

1°. The adjunction between the comma categories $(\underline{C} \downarrow C)$ and $(\underline{X} \downarrow I(C))$ for each $C \in \underline{C}$ — it will be written as

$$(I^C,\, H^C,\, \eta^C,\, \varepsilon^C) : (\underline{C} \downarrow C) \to (\underline{X} \downarrow I(C)); \tag{2.3}$$

by the definition,

$$I^C(A,\, \alpha) = (I(A),\, I(\alpha)),$$
$$H^C(X,\, \xi) = (C \times_{HI(C)} H(X),\, \mathrm{proj}_1)$$

with the picture

$$\begin{array}{ccc} C \times_{HI(C)} H(X) & \xrightarrow{\mathrm{proj}_2} & H(X) \\ {\scriptstyle \mathrm{proj}_1}\downarrow & & \downarrow{\scriptstyle H(\xi)} \\ C & \xrightarrow[\eta_C]{} & HI(C) \end{array} \quad ,$$

$$\eta^C_{(A,\alpha)} = \langle \alpha, \eta_A \rangle : A \to C \times_{HI(C)} HI(A),$$
$$\varepsilon^C_{(X,\xi)} = \varepsilon_X I(\mathrm{proj}_2) : I(C \times_{HI(C)} H(X)) \to X,$$

in an obvious notation.

2°. The adjunction between the functor categories $\underline{C}^{\tau}$ and $\underline{X}^{\tau}$ - it will be written as

$$(I^{\tau}, H^{\tau}, \eta^{\tau}, \varepsilon^{\tau}) : \underline{C}^{\tau} \to \underline{X}^{\tau}; \tag{2.4}$$

by the definition,

$$I^{\tau}(P) = IP, \quad H^{\tau}(Q) = HQ, \quad \eta^{\tau}_P = (\eta_{P(0)}, \eta_{P(1)}, \eta_{P(2)}),$$
$$\varepsilon^{\tau}_Q = (\varepsilon_{Q(0)}, \varepsilon_{Q(1)}, \varepsilon_{Q(2)})$$

for each $P \in \underline{C}^{\tau}$, $Q \in \underline{X}^{\tau}$.

3°. The adjunction between the internal prefunctor categories $(\underline{C}^{\tau} \downarrow P)$ and $(\underline{X}^{\tau} \downarrow IP)$, where P is a fixed internal precategory in $\underline{C}$ — using the notation above it should be written as

$$((I^{\tau})^P, (H^{\tau})^P, (\eta^{\tau})^P, (\varepsilon^{\tau})^P) : (\underline{C}^{\tau} \downarrow P) \to (\underline{X}^{\tau} \downarrow I^{\tau}(P)),$$

but we will write

$$(I^P, H^P, \eta^P, \varepsilon^P) : (\underline{C}^{\tau} \downarrow P) \to (\underline{X}^{\tau} \downarrow IP); \tag{2.5}$$

by the definition,

$$I^P(F, \varphi) = (IF, I\varphi),$$
$$H^P(G, \psi) = (P \times_{HIP} HG, \text{proj}_1),$$
$$\eta^P_{(F,\varphi)} = \langle \varphi, \eta^{\tau}_F \rangle = (\langle \varphi_0, \eta_{F(0)} \rangle, \langle \varphi_1, \eta_{F(1)} \rangle, \langle \varphi_2, \eta_{F(2)} \rangle)$$
$$= \left(\eta^{P(0)}_{(F(0),\varphi_0)}, \eta^{P(1)}_{(F(1),\varphi_1)}, \eta^{P(2)}_{(F(2),\varphi_2)} \right),$$

$$\varepsilon^P_{(G,\psi)} = \varepsilon^{\tau}_G I^{\tau}(\text{proj}_2) = \left(\varepsilon_{G(0)} I(\text{proj}_2^{(0)}), \varepsilon_{G(1)} I(\text{proj}_2^{(1)}), \varepsilon_{G(2)} I(\text{proj}_2^{(2)}) \right)$$
$$= \left(\varepsilon^{P(0)}_{(G(0),\psi_0)}, \varepsilon^{P(1)}_{(G(1),\psi_1)}, \varepsilon^{P(2)}_{(G(2),\psi_2)} \right),$$

where $\text{proj}_2^{(i)}$ $(i = 0, 1, 2)$ denotes the second projection

$$P(i) \times_{HIP(i)} HG(i) \to HG(i).$$

We will use the equivalence

$$\text{Spl}_{IP}(\underline{C}^{\tau} \downarrow P) \sim \text{Cospl}_{IP}(\underline{X}^{\tau} \downarrow IP), \tag{2.6}$$

obtained from (2.5) as (2.2) was obtained from (2.1); by the definition it is the restriction of (2.5) and we have:

PROPOSITION 2.1.
(a) $(F,\varphi)\in \mathrm{Spl}_{IP}(\underline{C}^{\tau}\downarrow P)$ *iff the morphisms* $\eta^{P(i)}_{(F(i),\varphi_i)}$ $(i=0,1,2)$ *are isomorphisms, i.e. the diagrams*

$$\begin{array}{ccc} F(i) & \xrightarrow{\ \eta_{F(i)}\ } & HIF(i) \\ {\scriptstyle \varphi_i}\downarrow & & \downarrow{\scriptstyle HI(\varphi_i)} \\ P(i) & \xrightarrow[\ \eta_{P(i)}\ ]{} & HIP(i) \end{array}$$

$(i=0,1,2)$ *are pullbacks;*
(b) $(G,\psi)\in \mathrm{Cospl}_{IP}(\underline{X}^{\tau}\downarrow IP)$ *iff the morphisms* $\varepsilon^{P(i)}_{(G(i),\psi_i)}$ $(i=0,1,2)$ *are isomorphisms, i.e. the compositions*

$$I(P(i)\times_{HIP(i)} HG(i)) \xrightarrow{I(\mathrm{proj}_2^{(i)})} IHG(i) \xrightarrow{\varepsilon_{G(i)}} G(i)$$

$(i=0,1,2)$ *are isomorphisms.* □

3. Change of universe for internal functors

Consider the adjunction (2.5).

Since H preserves pullbacks, the functor H^P induces a functor from $\underline{X}^{IP}$ to $\underline{C}^P$ and so (2.6) induces a category equivalence

$$\mathrm{Spl}_{IP}(\underline{C}^{\tau}\downarrow P)\cap \underline{C}^P\cap (I^P)^{-1}(\underline{X}^{IP})\sim \mathrm{Cospl}_{IP}(\underline{X}^{\tau}\downarrow IP)\cap \underline{X}^{IP},$$

which we will write as

$$\mathrm{Spl}_I(P,\underline{C})\sim \mathrm{Cospl}_I(IP,\underline{X}); \tag{3.1}$$

by the definition the left side consists of all $(F,\varphi): P\to \underline{C}$ such that:

$\eta^P_{(F,\varphi)}$ is an isomorphism,

(F,φ) is an internal functor,

and $I^P(F,\varphi)$ is an internal functor;

and the right side consists of all $(G,\psi): IP\to \underline{X}$ such that:

$\varepsilon^P_{(G,\psi)}$ is an isomorphism,

and (G,ψ) is an internal functor.

PROPOSITION 3.1. *Let* $(F,\varphi): P\to \underline{C}$ *be an internal functor. The following conditions are equivalent:*
(a) $(F,\varphi)\in \mathrm{Spl}_I(P,\underline{C})$;
(b) *the morphisms*

$$\eta^{P(0)}_{(F(0),\varphi_0)},\quad \varepsilon^{P(1)}_{(IP(1)\times_{IP(0)}IF(0),\,\mathrm{proj}_1)},\quad \varepsilon^{P(2)}_{(IP(2)\times_{IP(0)}IF(0),\,\mathrm{proj}_1)}$$

are isomorphisms.

PROOF. By the definition 1.1(c) we have:

Claim 1. $I^P(F, \varphi)$ is an internal functor iff I preserves the following pullbacks:

$$\begin{array}{ccc} P(2)\times_{P(1)} F(1) & \longrightarrow & F(1) \\ \downarrow & & \downarrow{\scriptstyle \varphi_1} \\ P(2) & \xrightarrow[P(p_2)]{} & P(1) \end{array}\quad , \qquad \begin{array}{ccc} P(1)\times_{P(0)} F(0) & \longrightarrow & F(0) \\ \downarrow & & \downarrow{\scriptstyle \varphi_0} \\ P(1) & \xrightarrow[P(d)]{} & P(0) \end{array}.$$

Then we have:

Claim 2. $I^P(F, \varphi)$ is an internal functor iff I preserves the following pullbacks:

$$\begin{array}{ccc} P(2)\times_{P(0)} F(0) & \longrightarrow & F(0) \\ \downarrow & & \downarrow{\scriptstyle \varphi_0} \\ P(2) & \xrightarrow[P(dp_2)]{} & P(0) \end{array}\quad , \qquad \begin{array}{ccc} P(1)\times_{P(0)} F(0) & \longrightarrow & F(0) \\ \downarrow & & \downarrow{\scriptstyle \varphi_0} \\ P(1) & \xrightarrow[P(d)]{} & P(0) \end{array}.$$

Now recall the following (the proof is easy and is given in [11], so we omit it):

Claim 3. (the lemma 3.5 in [11]). Let $\gamma_j : C_i \to C$ $(i = 1, 2)$ be morphisms in $\underline{C}$. If $\eta^{C}_{(C_2,\gamma_2)}$ is an isomorphism, then I preserves the pullback

$$\begin{array}{ccc} C_1\times_C C_2 & \longrightarrow & C_2 \\ \downarrow & & \downarrow{\scriptstyle \gamma_2} \\ C_1 & \xrightarrow[\gamma_1]{} & C \end{array}$$

iff the morphism $\varepsilon^{C_1}_{(I(C_1)\times_{(C)}I(C_2),\, \mathrm{proj}_1)}$ is an isomorphism.

After that the claim 2 gives:

Claim 4. If $\eta^{P(0)}_{(F(0),\varphi_0)}$ is an isomorphism, then $I^P(F, \varphi)$ is an internal functor iff the morphisms

$$\varepsilon^{P(1)}_{(IP(1)\times_{IP(0)}IF(0),\mathrm{proj}_1)}, \quad \varepsilon^{P(2)}_{(IP(2)\times_{IP(0)}IF(0),\mathrm{proj}_1)}$$

are isomorphisms.

So, it is sufficient to prove the following:

Claim 5. If (F, φ) and $I^P(F, \varphi)$ are internal functors and $\eta^{P(0)}_{(F(0),\varphi(0))}$ is an isomorphism, then $\eta^{P(1)}_{(F(1),\varphi_1)}$ and $\eta^{P(2)}_{(F(2),\varphi_2)}$ are isomorphisms too.

And to prove this is sufficient to observe that, in the diagram

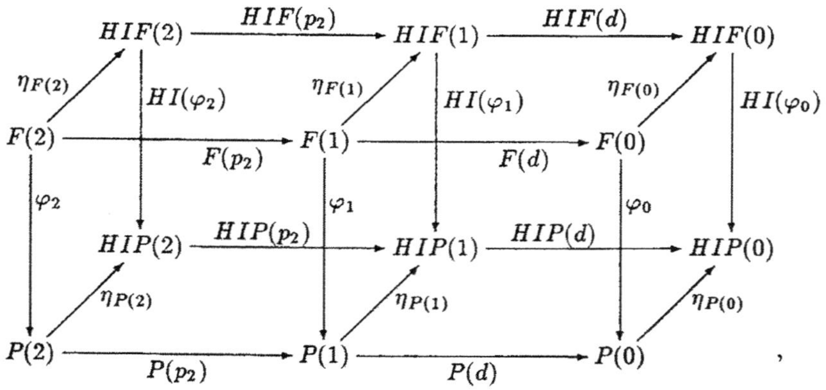

if the front, the rear and the right squares are pullbacks, then the middle and the left squares (we consider only vertical squares) are pullbacks too. □

PROPOSITION 3.2. *Let $(G, \psi) : IP \to \underline{X}$ be an internal functor. The following conditions are equivalent*

(a) $(G, \psi) \in \mathrm{Cospl}_I(IP, \underline{X})$;

(b) *the morphisms* $\varepsilon^{P(0)}_{(G(0),\psi_0)}$, $\varepsilon^{P(1)}_{(IP(1)\times_{IP(0)}G(0),\,\mathrm{proj}_1)}$, $\varepsilon^{P(2)}_{(IP(2)\times_{IP(0)}G(0),\,\mathrm{proj}_1)}$ *are isomorphisms.*

PROOF. It is sufficient to observe that

$$\varepsilon^{P(1)}_{(IP(1)\times_{IP(0)}G(0),\,\mathrm{proj}_1)} \quad \text{and} \quad \varepsilon^{P(2)}_{(IP(2)\times_{IP(0)}G(0),\,\mathrm{proj}_1)}$$

are the same (up to isomorphisms) as

$$\varepsilon^{P(1)}_{(G(1),\psi_1)} \quad \text{and} \quad \varepsilon^{P(2)}_{(G(2),\psi_2)} \,.\ \square$$

REMARK 3.3. *The results of* § 3 *of* [11] *follows from* 3.1 *and* 3.2*; the proofs of* 3.1 *and* 3.2 *are much easier than the proofs in* [11] *because now we use the "change of universe for prefunctors".* □

4. Internal antidiscrete groupoid

Let S be a set and S^{ad} the corresponding antidiscrete groupoid, i.e. the groupoid whose set of objects is S, and such that for each s, $s' \in S$ there exists the unique morphism $s \to s'$. For $S \neq \varnothing$ we have the following equivalent definition: S^{ad} is the category whose sets of objects is S, and which is equivalent to the "terminal category", i.e. the category with one object and one (identity) morphism.

Now, if $\underline{A}$ is a category with finite products and S is an object in $\underline{A}$, we still can define S^{ad} – as the unique internal category with the underlying graph

$$S \times S \underset{\text{proj}_1}{\overset{\text{proj}_2}{\rightrightarrows}} S,$$

and it will be equivalent to the "internal terminal category", i.e. the internal category whose objects of objects and morphisms are terminal objects in $\underline{A}$, iff there exists a morphism from the terminal object to S (which in the case $\underline{A}$ = Sets means that $S \neq \Phi$). In that case we have also a category equivalence

$$\underline{A} \sim \underline{A}^{S^{\text{ad}}}, \tag{4.1}$$

and we can ask about a necessary and sufficient condition for (4.1) in the general case.

Consider the (monadic) forgetful functor

$$\underline{A}^{S^{\text{ad}}} \to (\underline{A} \downarrow S) \tag{4.2}$$

— an easy computation shows that the corresponding monad is the same as the monad corresponding to the adjoint functors

$$\underline{A} \underset{\text{forgetful}}{\overset{A \mapsto (S \times A,\ \text{proj}_1)}{\rightleftarrows}} (\underline{A} \downarrow S), \tag{4.3}$$

and so we obtain the comparison functor

$$K^S : \underline{A} \to \underline{A}^{S^{\text{ad}}}, \tag{4.4}$$

and the following proposition (which can be considered as an answer on the question above):

PROPOSITION 4.1. *The following conditions are equivalent:*

(a) *the functor $K^S : \underline{A} \to \underline{A}^{S^{\text{ad}}}$ is a category equivalence;*

(b) *the functor $\underline{A} \to (\underline{A} \downarrow S)$ defined by $A \mapsto (S \times A,\ \text{proj}_1)$ is monadic.* □

We need the relative version: let $\underline{C}$ be a category with pullbacks and $f : E \to B$ a morphism in $\underline{C}$ — then we can put $\underline{A} = (\underline{C} \downarrow B)$ and $S = (E, f)$, which gives

$$(\underline{C} \downarrow B) \underset{f_!}{\overset{f^*}{\rightleftarrows}} (\underline{C} \downarrow E) \tag{4.5}$$

instead of (4.3) *(where f^* is the pullback functor and $f_!$ the "composition functor" defined by $(C, \gamma) \mapsto (C, f\gamma)$ and the following proposition instead of* 4.1*:*

PROPOSITION 4.2. *The following conditions are equivalent:*

(a) *the functor* $K^{(E,f)} : (\underline{C} \downarrow B) \to \underline{C}^{(E,f)^{\mathrm{ad}}}$ *is a category equivalence;*

(b) *the pullback functor* $f^* : (\underline{C} \downarrow B) \to (\underline{C} \downarrow E)$ *is monadic.* □

Observe that the underlying graph of $(E, f)^{\mathrm{ad}}$ *is*

$$E \times_B E \underset{\mathrm{proj}_1}{\overset{\mathrm{proj}_2}{\rightrightarrows}} E,$$

i.e. $(E, f)^{\mathrm{ad}}$ *is just the equivalence relation corresponding to* f *(which is considered now as an internal category in* $\underline{C}$ *— not in* $\underline{A} = (\underline{C} \downarrow B)$*).*

REMARK 4.3. *This section repeates a small part of § 1 of [7], which is written without enough explanation: the lemma 1.2 of [7] is easy but not well known.* □

5. Galois theory

In this section we consider again a fixed adjunction (2.1) between categories $\underline{C}$ and $\underline{X}$ with pullbacks, and we will use the notation of the previous sections.

We also fix a morphism $f : E \to B$ and by P_f denote the internal groupoid $(E, f)^{\mathrm{ad}}$ considered as an internal precategory in $\underline{C}$.

PROPOSITION 5.1. *For an object* (A, α) *of the comma category* $(\underline{C} \downarrow C)$ *the following conditions are equivalent:*

(a) $K^{(E,f)}(A; \alpha) \in \mathrm{Spl}_I(P_f, \underline{C})$;

(b) *the morphisms*

$$\eta^{E}_{(E\times_B A,\, \mathrm{proj}_1)},$$

$$\varepsilon^{E\times_B E}_{\left(I(E\times_B E)\times_{I(E)} I(E\times_B A),\, \mathrm{proj}_1\right)},$$

$$\varepsilon^{E\times_B E\times_B E}_{\left(I(E\times_B E\times_B E)\times_{I(E)} I(E\times_B A),\, \mathrm{proj}_1\right)}$$

are isomorphisms.

PROOF. This follows from 3.1 and the following two observations:

1. IP_f is

$$I(E \times_B E \times_B E) \;\substack{\longrightarrow\\ \longrightarrow\\ \longrightarrow}\; I(E \times_B E) \;\substack{\longrightarrow\\ \longrightarrow\\ \longleftarrow}\; I(E).$$

2. Denote $K^{(E,f)}(A, \alpha) = (F, \varphi)$ — then

$$(F(0), \varphi_0) = (E \times_B A, \mathrm{proj}_1).$$

The first of them is obvious, and the second one follows from the commutativity of the diagram

$$\begin{array}{ccc}
(\underline{C} \downarrow B) & \xrightarrow{K^{(E,f)}} & \underline{C}^{P_f} \\
{\scriptstyle f^*}\big\downarrow & & \big\downarrow{\scriptstyle \text{forgetful}} \\
(\underline{C} \downarrow E) & = & (\underline{C} \downarrow E),
\end{array}$$

where $f^*(A, \alpha) = (E \times_B A, \text{proj}_1)$ and the forgetful functor (the right vertical arrow) is defined by $(F, \varphi) \mapsto \big(F(0), \varphi_0\big)$. □

DEFINITION 5.2.

(a) *An object* (A, α) *of the comma category* $(\underline{C} \downarrow B)$ *is split over* (E, f) *with respect to* I *iff it satisfies the equivalent conditions of* 5.1. *The full subcategory of* $(\underline{C} \downarrow B)$ *with objects all such* (A, α) *denote by* $\text{Spl}_I(E, f)$*;*

(b) *we say that* (E, f) *is a monadic extension of* B *iff the functor* f^* *is monadic.* □

THEOREM 5.3. *If* (E, f) *is a monadic extension of* B*, then there exists a category equivalence*

$$\text{Spl}_I(E, f) \sim \text{Cospl}_I(IP_f, \underline{X}),$$

namely the composition of the equivalence

$$\text{Spl}_I(E, f) \sim \text{Spl}_I(P_f, \underline{C}),$$

induced by $K^{(E,f)}$*, and the equivalence* (3.1). □

COROLLARY 5.4. *Suppose that* $\varepsilon^C : I^C H^C \to 1_{(\underline{X} \downarrow I(C))}$ *is an isomorphism for each* $C \in \underline{C}$*. Then:*

(a) (A, α) *is split over* (E, f) *with respect to* I *iff* $\eta^E_{(E \times_B A, \text{proj}_1)}$ *is an isomorphism, which itself is equivalent to the existence of* $(X, \xi) \in (\underline{X} \downarrow I(E))$ *with an isomorphism*

$$\begin{array}{ccc}
E \times_B A & \approx & E \times_{HI(E)} H(X) \\
{\scriptstyle \text{proj}_1}\big\downarrow & & \big\downarrow{\scriptstyle \text{proj}_1} \\
E & = & E
\end{array}$$

in $(\underline{C} \downarrow E)$*;*

(b) $\text{Spl}_I(E, f) \sim \underline{X}^{IP_f}$. □

REMARK 5.5.

(a) *There are more general situations, when* f^* *is not necessary monadic, but the category equivalences considered in* 5.3 *and* 5.4 *still exist. For example it is obviously sufficient to have a full subcategories* $\underline{S} \subset (\underline{C} \downarrow B)$ *and* $\underline{S}' \subset (\underline{C} \downarrow E)$ *such that:*

(i) $\text{Spl}_I(E, f) \subset \underline{S}$*;*

(ii) *f^* induces a functor $\underline{S} \to \underline{S}'$, which is monadic;*

(iii) *$f_!$ induces a functor $\underline{S}' \to \underline{S}$.*

In this case it is natural to say that (E, f) is a $(\underline{S}, \underline{S}')$-monadic extension.

(b) *One can define the notion of an internal pregroupoid as follows. Let τ' be the extension of τ having the same objects and generated by τ and new morphisms given in the picture*

$$1 \xrightarrow{\ i\ } 1 \xrightarrow{\ j\ } 2,$$

and the identities

$$di = r, \quad ri = d, \quad p_1 j = 1_1, \quad p_2 j = i, \quad mj = er, \quad i^2 = 1_1.$$

Now an internal pregroupoid in a category $\underline{A}$ is a functor from τ' to $\underline{A}$. This notion corresponds to the usual notion of an internal groupoid: if C is an internal category in $\underline{A}$ and $P : \tau \to \underline{A}$ the corresponding internal precategory, then the following conditions are obviously equivalent:

(i) *C is an internal groupoid;*

(ii) *P is an internal pregroupoid, i.e. P can be extended to τ';*

(iii) *P is an internal groupoid in a unique way, i.e. there exists the unique extension of P to τ'.*

In particular P_f is an internal pregroupoid in $\underline{C}$ and so IP_f is an internal pregroupoid in $\underline{X}$. It is natural to call it the Galois pregroupoid of (E, f) and denote by $\mathrm{Gal}_I(E, f)$.

(c) *There are two questions coming from* (a) *and* (b) *respectively:*

(i) *under what conditions can we put $\underline{S} = \mathrm{Spl}_I(E, f)$ and $\underline{S}'$ = the full subcategory of $(\underline{C} \downarrow E)$ with objects all $(C, \gamma) \in (\underline{C} \downarrow E)$ such that $\eta^E_{(C,\gamma)}$ is an isomorphism?;*

(ii) *under what conditions is the Galois pregroupoid of (E, f) an (internal) groupoid?*

We have the following "reasonably complete" answers:

(i) *Let U be the set of all pairs $(\underline{S}, \underline{S}')$ as in* (a) *and $(\underline{S}_0, \underline{S}'_0)$ the pair considered in* (i)*; suppose that $U \neq \phi$. Our question is: is $(\underline{S}_0, \underline{S}'_0)$ an element of U? The condition* (i) *of* (a) *is trivial now and* (ii) *of* (a) *follows easily from $U \neq \phi$. Hence $(\underline{S}_0, \underline{S}'_0) \in U$ iff $f_!$ induces a functor $\underline{S}'_0 \to \underline{S}'$. In this case we will say that (E, f) is I-normal extension of C, which corresponds to the definition of normality given in* [7].

(ii) *As it follows from observations given in* [7] *(and the same in* [11]*), the canonical morphisms*

$$I((E \times_B E) \times_E (E \times_B E)) \to I(E \times_B E) \times_{I(E)} I(E \times_B E),$$

$$I((E \times_B E) \times_E (E \times_B E) \times_E (E \times_B E)) \to I(E \times_B E) \times_{I(E)} I(E \times_B E) \times_{I(E)} I(E \times_B E)$$

should be isomorphisms — in [7] *and* [11] *we say that (E, f) is Cartesian-I-normal. Moreover from the results of* [7] *(or* [11]*) it follows that "I-normal" implies "Cartesian-I-normal" (it can be easily proved directly in the style of the proof of* 3.1*).*

(d) *The results above give a "theory of torsors". For simplicity consider the situation of* 5.4. *Let G be a group in $(\underline{X} \downarrow I(B))$ and $\mathrm{Tors}^{(E,f)}(G)$ the set of isomorphic classes of $H^B(G)$-torsors in $(\underline{C} \downarrow E)$ split by a monadic extension (E,f) of B. The usual cohomology theory of categories has an obvious generalization for internal precategories in dimensions* 0 *and* 1, *and* 5.4 *gives*

$$\mathrm{Tors}^{(E,f)}(G) \approx H^1(\mathrm{Gal}(E, f), G)$$

— an isomorphism of pointed sets, which will be a group isomorphism for abelian G.

6. Indexed and Galois structures

The collection of comma categories $(\underline{C} \downarrow C)$ and functors of the form $f^* : (\underline{C} \downarrow B) \to (\underline{C} \downarrow E)$ form a $\underline{C}$-indexed category, and by an indexed structure on $\underline{C}$ we mean a subcategory of $\underline{C}$ which generates a $\underline{C}$-indexed subcategory of it; more precisely, we introduce the following

DEFINITION 6.1. *An indexed structure on $\underline{C}$ is a nonempty subcategory $\underline{E}$ of $\underline{C}$ such that, if*

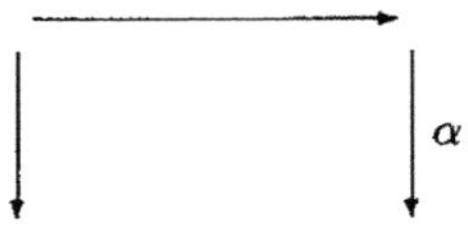

is a pullback in $\underline{C}$ with $\alpha \in \underline{E}$, then $\pi \in E$. □

The full subcategory of $(\underline{C} \downarrow C)$ with objects all (A, α) with $\alpha \in \underline{E}$ will be denoted by $\underline{E}(C)$. For a morphism $f : E \to B$ in $\underline{C}$, the functor f^* induces a functor

$$f^{*\underline{E}} : \underline{E}(B) \to \underline{E}(C), \tag{6.1}$$

and, if $f \in \underline{E}$ then also induces a functor

$$f_{!\underline{E}} : \underline{E}(E) \to \underline{E}(B), \tag{6.2}$$

which is the left adjoint of $f^{*\underline{E}}$.

EXAMPLE 6.2. *Let $\underline{E} = \mathrm{Iso}(\underline{C})$, i.e. $\underline{E}$ is the subcategory of $\underline{C}$ with morphisms all isomorphisms. This is the smallest indexed structure on $\underline{C}$.* □

EXAMPLE 6.3. *$\underline{E} = \underline{C}$ gives the largest one.* □

EXAMPLE 6.4. *Recall that an epimorphism α is called universal, if for each pullback*

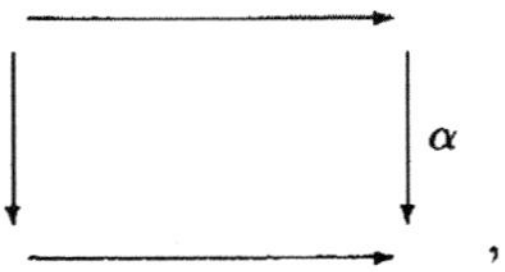

π is an epimorphism (and therefore a universal epimorphism). Clearly compositions of universal epimorphisms are universal, and all isomorphisms are universal epimorphisms. Hence we can take $\underline{E}$ to be the subcategory of $\underline{C}$ with morphisms all universal epimorphisms. □

EXAMPLE 6.5. *Let $\underline{C} = Top$ be the category of topological spaces and $\underline{E}$ its subcategory with morphisms all local homeomorphisms. The fact that it is an indexed structure follows from well known properties of local homeomorphisms.* □

DEFINITION 6.6. *A Galois structure is an adjunction* (2.1) *above with indexed structures $\underline{E}$ and $\underline{Z}$ on $\underline{C}$ and $\underline{X}$ respectively, such that $I(\underline{E}) \subset \underline{Z}$ and $H(\underline{Z}) \subset E$.* □

We will consider a fixed Galois structure Γ and so assume that an adjunction (2.1), $\underline{E}$ and $\underline{Z}$ are fixed.

For an object $C \in \underline{C}$ the adjunction (2.3) induces an adjunction

$$(I^{C,\Gamma}, H^{C,\Gamma}, \eta^{C,\Gamma}, \varepsilon^{C,\Gamma}) : \underline{E}(C) \to \underline{Z}(I(C)), \tag{6.3}$$

and we can reformulate the "Galois theory" above replacing (2.3) by (6.3).

DEFINITION 6.7. *Let $f : E \to B$ a morphism in $\underline{E}$; we say that (E, f) is a monadic extension of B iff the functor $f^{*\underline{E}}$ is monadic.* □

We fix also a monadic extension (E, f) of B, and we will use the notation given by the picture

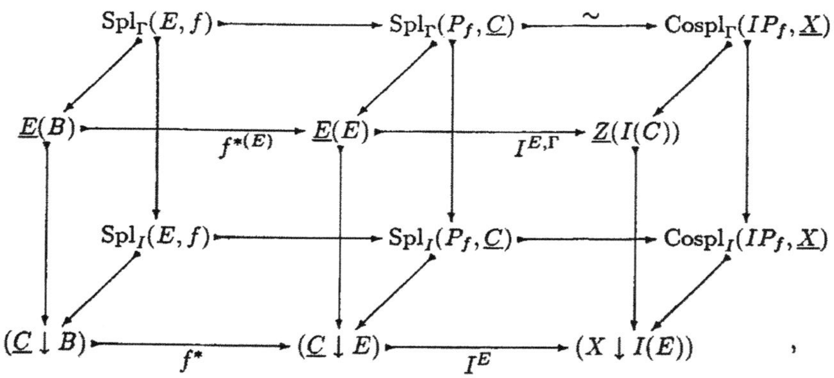

,

where the "three parallel diamonds" are pullbacks.

Instead of the theorem 5.3 and the corollary 5.4 we have:

THEOREM 6.8. *The composition*

$$\mathrm{Spl}_I(E, f) \to \mathrm{Spl}_I(P_f, \underline{C}) \sim \mathrm{Cospl}_I(IP_f, \underline{X})$$

induces a category equivalence

$$\mathrm{Spl}_\Gamma(E, f) \sim \mathrm{Cospl}_\Gamma(IP_f, \underline{X}). \ \square$$

COROLLARY 6.9. *Suppose that* $\varepsilon^{C,\Gamma} : I^{C,\Gamma}H^{C,\Gamma} \to 1_{\underline{Z}(I(C))}$ *is an isomorphism for each* $C \in \underline{C}$.
Then:

(a) *an object* $(A, \alpha) \in \underline{E}(B)$ *is in* $\mathrm{Spl}_\Gamma(E, f)$ *iff* $\eta^{E,\Gamma}_{(E \times_B A, \mathrm{proj}_1)}$ *is an isomorphism, i.e. iff there exists* $(X, \xi) \in \underline{Z}(I(E))$ *with an isomorphism*

$$\begin{array}{ccc} E \times_B A & \approx & E \times_{HI(E)} H(X) \\ \mathrm{proj}_1 \downarrow & & \downarrow \mathrm{proj}_1 \\ E & = & E \end{array}$$

in $\underline{E}(E)$;

(b) $\mathrm{Spl}_\Gamma(E, f) \sim \underline{X}^{IP_f} \cap \underline{Z}(I(E))$, *where the right side is defined by the pullback*

$$\begin{array}{ccc} \underline{X}^{IP_f} \cap \underline{Z}(I(E)) & \longrightarrow & Z(I(E)) \\ \downarrow & & \downarrow \\ \underline{X}^{IP_f} & \xrightarrow[\text{forgetful}]{} & (\underline{X} \downarrow I(E)) . \end{array}$$

REMARK 6.10.

(a) *One can obviously reformulate the remark* 5.5 *for this (more general) case.*

(b) *We still have the equivalence* $\mathrm{Spl}_\Gamma(E, f) \sim \mathrm{Cospl}_\Gamma(IP_f, \underline{X})$ *even in the case when* $\underline{C}$ *and* $\underline{X}$ *have only those pullbacks which we need in* 6.1. □

REFERENCES

[1] M. BUNGE, *Open covers and the fundamental localic groupoid of a topos*, McGill Univ. Math. Preprint 89–33 (1989).

[2] A. GROTHENDIECK, *Revêtements étales et groupe fondamental*, SGA 1, Springer Lecture Notes in Math. 269 (1972).

[3] G. JANELIDZE, *Galois extensions of commutative rings by profinite families of groups*, Transactions of Razmadze Math. Inst. of the Georgian Acad. Sci. 74 (1983), 39–51 (in Russian).

[4] G. JANELIDZE, *Magid theorem in categories*, Bull. Georgian Acad. Sci. 114(3) (1984), 497–500 (in Russian).

[5] G. JANELIDZE, *A generalization of the theory of covering spaces*, IV International Conf. in Topology and its Applications, Abstract of Reports, Dubrovnic (1985).

[6] G. JANELIDZE, (G.Z. DZHANELIDZE), The fundamental theorem of Galois theory, Math. USSR Sbornik 64(2) (1989), 359– 374.

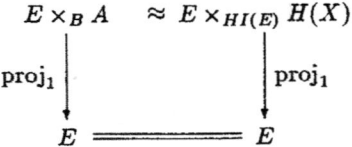

[7] G. Janelidze, *Pure Galois theory in categories*, U.C.N.W. Pure Math. Preprint 87.20 (Bangor 1987) (and to appear in Journal of Algebra).

[8] G. Janelidze, *Galois theory in categories: the new example of differential fields*, Categorical Topology (Proc. Conf. in Prague), World Scientific (1988), 369–380.

[9] G. Janelidze, *What is a double central extension — the question was given by Ronnie Brown*, to appear.

[10] G. Janelidze, *A note on Barr-Diaconescu covering theory*, to appear.

[11] G. Janelidze, *Change of universe in the category of internal functors*, McGill Univ. Math. Preprint 90–11 (1990).

[12] J. Kennison, What is the fundamental group?, J. Pure Appl. Algebra 59 (1989), 187–200.

[13] A.R. Magid, *The separable Galois theory of commutative rings*, Marsel Dekker (1974).

[14] R.S. Pierce, *Modules over commutative regular rings*, Mem. AMS 70 (1967).

This paper is in final form and will not be published elsewhere.

HOW ALGEBRAIC IS THE CHANGE-OF-BASE FUNCTOR?

George JANELIDZE
Mathematics Institute of the Georgian Academy of Sciences, Tbilisi 380093, U.S.S.R.

Walter THOLEN
Department of Mathematics and Statistics, York University, Toronto, Canada M3J 1P3

Dedicated to Max Kelly on the occasion of his sixtieth birthday.

Abstract

The continuous maps $p : E \to B$ for which $p^* : \mathbf{Top}/B \to \mathbf{Top}/E$ reflects isomorphisms are shown to coincide with the universal quotient maps as characterized by Day and Kelly. Monadicity of p^* turns out to be a local property. This is used to prove the main result of the paper, namely that p^* is monadic for every locally sectionable map $p : E \to B$. There are therefore important classes of maps p for which spaces over B are equivalently described as spaces over E which come equipped with a simple algebraic structure: local homeomorphisms, locally trivial quotient maps, surjective covering maps, etc. Finally, the monadic decomposition of p^* is examined for arbitrary maps p.

AMS Subject Classification: 18A30, 18C20, 18F15, 54C10, 55R10

Introduction

The change-of-base functor $p^* : \mathbf{Top}/B \to \mathbf{Top}/E$ induced by a continuous map $p : E \to B$ assigns to a space $(A, \alpha : A \to B)$ over B (i.e., a bundle) the fibre product (i.e., the pullback, or the Whitney sum) $E \times_B A$, considered as a space over E.

Partial financial support by NSERC (Canada) and by CNR (Italy) for both authors while working on this paper is gratefully acknowledged.

$$\begin{array}{ccc} E \times_B A & \xrightarrow{\pi_2} & A \\ {\scriptstyle \pi_1}\downarrow & & \downarrow{\scriptstyle \alpha} \\ E & \xrightarrow{p} & B \end{array} \qquad (1)$$

There are various ways of interpreting the question formulated in the title of the paper. First one may ask for particular algebraic features of the functor p^*, such as the *reflection of isomorphisms.* This means that one tries to characterize those maps p for which the following property holds: a map $f : A \to A'$ over B is a homeomorphism if

$$1_E \times_B f \;:\; E \times_B A \to E \times_B A'$$

is a homeomorphism. Such a characterization is given in Section 1, both in categorical and in topological terms, utilizing the Day-Kelly[5] result on universal quotient maps, i.e. the toplogical characterization of those quotient maps which are stable under pullback.

When interpreting algebraic as *monadic* (cf. [12]), the question of the title becomes more difficult. The main result of the paper says that if $p : E \to B$ is locally sectionable (i.e., if for every $b \in B$ there is a neighbourhood U and a section $s : U \to p^{-1}U$ of the restriction of p ; cf.[6]), then p^* is monadic. Hence, in this case, a space (A, α) over B can be equivalently described as a space (C, γ) over E which comes equipped with an action $\xi : E \times_B C \to C$ such that (in multiplicative notation)

$$\gamma(c) \cdot c = c\,, \quad \gamma(e \cdot c) = e\,, \quad e \cdot (e' \cdot c) = e \cdot c$$

holds for all $c \in C$ and for all $e, e' \in E$ which belong to the same p-fibre as $\gamma(c)$. In other words: spaces over B are spaces over E with an easily described algebraic structure.

Important classes of maps p are locally sectionable and give therefore monadic functors p^* : local homeomorphisms; locally sliceable surjective maps (i.e., for every $x \in E$ there is a neighbourhood U of $b = p(x)$ and a section $s : U \to p^{-1}U$ of the restriction of p with $s(b) = x$; cf.[7]), in particular: locally trivial surjections; covering maps; projections of completely regular G-spaces, with G a compact Lie group, onto their orbit spaces (cf.[6]). However, we do not know an example of a map p with p^* monadic, but p not locally sectionable. Monadicity of p^* does not imply openess of p since there are locally sectionable maps which are not open. However, we do not know whether p^* is monadic for p an open quotient map. (In the light of the Joyal-Tierney result [10] that open surjections of toposes are effective descent morphisms, one would anticipate such a result.)

There is finally a quantitative way of interpreting the question of the title: for every map p , investiagate the *monadic decomposition* (cf. [11], [1], [2]) of the functor p^* . In particular: how long can the decomposition get? For which p is p^* essentially monadic (cf.[11],[1]), i.e. the composite of a (possibly infinite) chain of monadic functors? These questions turn out to be very easy and answers are provided at the end of the paper.

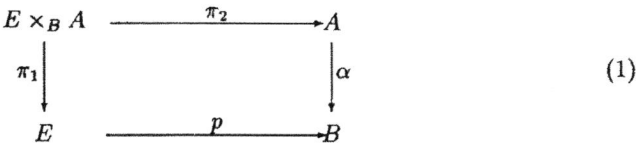

We deliberately use only elementary categorical methods in this paper in order to keep it accessible to those interested only in the topological aspects of the problem. The theme of the paper can be discussed in (at least) two more advanced settings, by presenting the category of Eilenberg-Moore algebras w.r.t. p^* as a category of *internal groupoid actions* (cf. [8]), or as the *category of descent data* of the (categorical) fibration $X \mapsto \mathcal{C}/X$ over the morphism p in a category $\mathcal{C}$ (Bénabou). In particular the first of these two settings allows for a very short proof of Proposition 2.4 below which is known to specialists.

A comprehensive presentation of these different viewpoints will be given in [9], with various generalizations, in particular with respect to subfibrations of $X \mapsto \mathcal{C}/X$. For instance, locally sectionable maps are shown to be exactly the maps for which a descent theorem with respect to topological fibrations can be proved, even when we consider spaces with structure, i.e. vector fibrations, etc. We especially discuss the case when p is the map $p : \amalg U_i \to B$ for an open cover (U_i) of B.

We would like to thank Michael Barr, Jean Bénabou and Myles Tierney for directing us to work previously done in this area.

1. Reflection of Isomorphisms

1.1 Let $\mathcal{E}$ be a class of maps. A map $p : E \to B$ in $\mathcal{E}$ is called a *universal $\mathcal{E}$-map* if the pullback of p along any other map belongs to $\mathcal{E}$, that is: in every pullback diagram (1), $\pi_2 \in \mathcal{E}$. With this terminology one has the following theorem, the difficult part of which is due to Day and Kelly[5]:

THEOREM *The following conditions are equivalent for a map $p : E \to B$ in* **Top**:

(i) *p^* reflects isomorphisms,*

(ii) *p is a universal quotient map,*

(iii) *p^* reflects quotient maps,*

(iv) *for every point $b \in B$ and for every family $(D_i)_{i \in I}$ of open sets in E which covers the fibre $p^{-1}b$, there are finitely many $i_1, \ldots, i_n \in I$ with $b \in \operatorname{int} p(D_{i_1} \cup \ldots \cup D_{i_n})$.*

PROOF (i) $\Rightarrow$ (ii) Given the pullback (1), let A' be the set $\pi_2(E \times_B A)$ provided with the quotient topology with respect to π_2. The injection $j : A' \to A$ is a map over B, and its pullback along π_2 is easily seen to be a homeomorphism. Hence j must be homeomorphism, that is: π_2 is a quotient map.

(ii) $\Rightarrow$ (iii) Let $f : A \to A'$ be a map over B such that $1_E \times_B f$ is a quotient map. In the diagram

$$\begin{array}{ccc} E \times_B A & \xrightarrow{\quad \pi_2 \quad} & A \\ {\scriptstyle 1_E \times_B f}\downarrow & & \downarrow{\scriptstyle f} \\ E \times_B A' & \xrightarrow{\quad \pi_2' \quad} & B \end{array} \qquad (2)$$

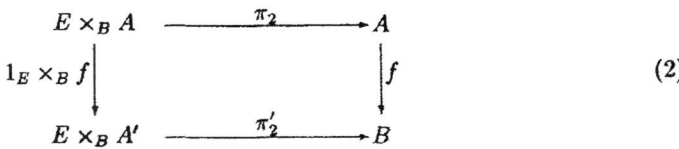

the map $\pi_2{}'$ is a quotient map by (ii), hence $\pi_2{}'(1_E \times_B f) = f\pi_2$ is a quotient map. Therefore, f must be quotient map.

(iii) $\Rightarrow$ (i) is trivial, and (ii) $\Longleftrightarrow$ (iv) was shown by Day and Kelly[5].

1.2 REMARKS Condition (iv) has been considered first by Michael[13] for Hausdorff spaces. If p is a perfect map (so that each fibre $p^{-1}b$ is compact), it reduces to the condition that, whenever $p^{-1}b \subseteq D$ with D open in E , then $b \in \text{int } p(D)$. Such maps are usually called ***pseudo-open*** and are fairly easily characterized as ***hereditary quotient maps*** (i.e., for each pullback diagram (1) with α a subspace embedding, π_2 is a quotient map), cf.[3]. In general, pseudo-open maps do not satisfy condition (iv). For example, let E be the sum of countably many copies of the Sierpinski dyad, i.e.

$$E = \coprod_{n \geq 1} S_n$$

with $S_n = \{(0,n),(1,n)\}$ and $\{(1,n)\}$ open but not closed in S_n. B is the set $\{0,1,2,\ldots\}$ which is topologized such that $B - \{0\}$ is discrete, but 0 does not belong to any open set except B. It is easy to see that $p : E \to B$ with $p(0,n) = 0$, $p(1,n) = n$ proves the claim.

We also wish to point out that maps satisfying condition (iv) need not be open: consider the restriction of the first projection $\mathbf{R} \times \mathbf{R} \to \mathbf{R}$ to the subspace

$$\{(x,1)|x \leq 0\} \cup \{(0,y)|0 \leq y \leq 1\} \cup \{(x,0)|x \geq 0\}$$

Therefore, the class of maps p for which p^* reflects isomorphisms represents a well-characterized class $\mathcal{Q}$ of maps with

$$\{\text{ open surjective maps }\} \subsetneqq \mathcal{Q} \subsetneqq \{\text{ pseudo-open surjective maps }\}$$

By the Theorem, it is easy to see that $\mathcal{Q}$ enjoys a number of pleasant general properties:

(a) $p \in \mathcal{Q}$, $q \in \mathcal{Q} \Longrightarrow qp \in \mathcal{Q}$,

(b) $qp \in \mathcal{Q} \Longrightarrow q \in \mathcal{Q}$,

(c) every map in $\mathcal{Q}$ is a universal $\mathcal{Q}$ – map ,

(d) $p : E \to B$ in $\mathcal{Q}$, $p' : E' \to B'$ in $\mathcal{Q} \Longrightarrow p \times p'$ in $\mathcal{Q}$. □

1.3 Recall that an epimorphism p in a category $\mathcal{C}$ is called *extremal* if $p = mf$ with m monic holds only if m is an isomorphism. (Obviously the latter property implies that p must be epic if $\mathcal{C}$ has equalizers.) The extremal epimorphisms in **Top** are precisely the quotient maps. Now we can generalize the easy part of Theorem 1.1:

PROPOSITION *Statements* (i), (ii), (iii) *of the Theorem remain equivalent for any morphism* p *in a finitely complete category* $\mathcal{C}$ *if "quotient map" is read as "extremal epimorphism"*

Proof We first observe that $p^* : \mathcal{C}/B \to \mathcal{C}/E$ has a left adjoint $p_!$ which assigns to an object $(C, \gamma : C \to E)$ the object $(C, p\gamma)$. For $(A, \alpha : A \to B) \in \mathcal{C}/B$, the co-unit

$$\varepsilon_{(A,\alpha)} : p_! p^*(A, \alpha) \to (A, \alpha)$$

is simply the map π_2 of (1), considered as a morphism over B. Now, *for every category $\mathcal{A}$ with equalizers and every functor $U : \mathcal{A} \to \mathcal{X}$ with a left adjoint, the following statements are equivalent* (cf.[4]):

(i) U *reflects isomorphisms,*

(ii) *the co-units $\varepsilon_A (A \in \mathcal{A})$ are extremal epimorphisms,*

(iii) U *reflects extremal epimorphisms.*

It is easy to check that a morphism in $\mathcal{C}/B$ is an extremal epimorphism (or an isomorphism) if and only if it has the same property in $\mathcal{C}$. Therefore the Proposition is an immediate consequence of the quoted more general fact applied to $U = p^*$. □

2. Premonadicity and Monadicity

2.1 We briefly recall the general framework of monadicity. Let $F \dashv U : \mathcal{A} \to \mathcal{X}$ be a pair of adjoint functors with unit $\eta : 1_{\mathcal{X}} \to UF$ and co-unit $\varepsilon : FU \to 1_{\mathcal{A}}$. For the induced monad $\tau = (T, \eta, \mu) = (UF, \eta, \mu)$ with $\mu = U\varepsilon F : T^2 \to T$ one forms the *Eilenberg-Moore category* $\mathcal{X}^\tau$ as follows: objects are τ-algebras (X, ξ) with an $\mathcal{X}$-morphism $\xi : TX \to X$ such that $\xi \cdot \eta_X = 1_X$ and $\xi \cdot \mu_X = \xi \cdot T\xi$; a morphism $h : (X, \xi) \to (Z, \zeta)$ in $\mathcal{X}^\tau$ is an $\mathcal{X}$-morphism $h : X \to Z$ with $h.\xi = \zeta \cdot Th$, and composition is as in $\mathcal{X}$.

There is the so-called comparison functor $K : \mathcal{A} \to \mathcal{X}^\tau$, $A \mapsto (UA, U\varepsilon_A)$. U is called *premonadic* if K is full and faithful, and *monadic* if K is an equivalence of categories.

If $\mathcal{A}$ has (sufficiently many) coequalizers, K has a left adjoint L: for a τ-algebra (X, ξ), the $\mathcal{A}$-object $Q = L(X, \xi)$ is given by the coequalizer

$$FUFX \underset{\varepsilon_{FX}}{\overset{F\xi}{\rightrightarrows}} FX \xrightarrow{q} Q . \tag{3}$$

The underlying $\mathcal{X}$-morphism of the unit $\kappa_{(X,\xi)} : (X, \xi) \to KL(X, \xi)$ is the composite

$$X \xrightarrow{\eta_X} UFX \xrightarrow{Uq} UQ .$$

The definition of q easily yields

$$U^\tau \kappa_{(X,\xi)} \cdot \xi = Uq \cdot \eta_X \cdot \xi = Uq ;$$

therefore, and since ξ is the (split) coequalizer of the pair $(T\xi, \mu_X) = (UF\xi, U\varepsilon_{FX})$, it is clear that $\kappa_{(X,\xi)}$ is an isomorphism if and only if U preserves the coequalizer diagram (3).

2.2 PROPOSITION *For a map p in* **Top**, *p^* is premonadic if and only if p is a universal quotient map. This holds true if* **Top** *is replaced by any finitely complete category and if "quotient map" is read as "regular epimorphism".*

PROOF   Regular epimorphism simply means coequalizer of any pair of morphisms. Very similarly to the proof of Prop. 1.3, the statement follows from the more general fact that *a functor $U : \mathcal{A} \to \mathcal{X}$ with left adjoint is premonadic iff the co-units $\varepsilon_A (A \in \mathcal{A})$ are regular epimorphisms*, applied to $U = p^*$. The statement for **Top** follows since regular epimorphism means quotient map here.   □

2.3   In order to examine monadicity of p^* , we continue to work in a finitely complete category $\mathcal{C}$. We described the left adjoint $p_!$ and the co-unit ε in 1.3. The underlying $\mathcal{C}$-morphism of the unit

$$\eta_{(C,\gamma)} : (C,\gamma) \to p^* p_!(C,\gamma)$$

is the morphism $e =< \gamma, 1_C >$ rendering the diagram

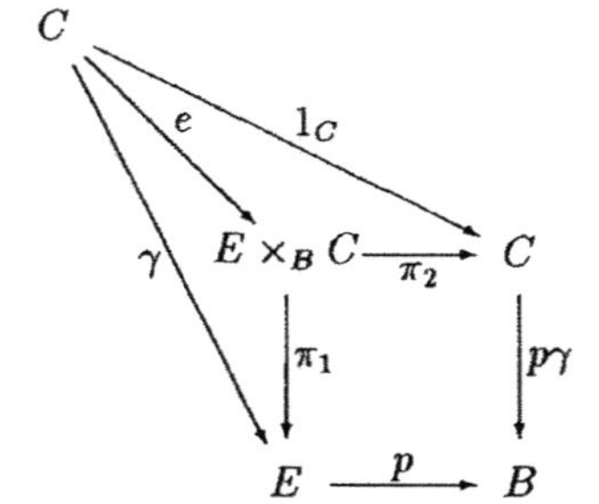

$$(4)$$

commutative.

Let τ be the monad induced by $p_! \dashv p^*$. A τ-algebra structure for (C, γ) is a map $\xi : E \times_B C \to C$ over E (hence $\gamma\xi = \pi_1$) such that $\xi < \gamma, 1_C >= 1_C$ and

$$
\begin{array}{ccc}
E \times_B (E \times_B C) & \xrightarrow{\;\; 1_E \times_B \xi \;\;} & E \times_B C \\
{\scriptstyle 1_E \times_B \pi_2} \downarrow & & \downarrow {\scriptstyle \xi} \\
E \times_B C & \xrightarrow{\;\;\;\; \xi \;\;\;\;} & C
\end{array}
\qquad (5)
$$

commutes. (This is the diagrammatic description of the equations given in the Introduction.)

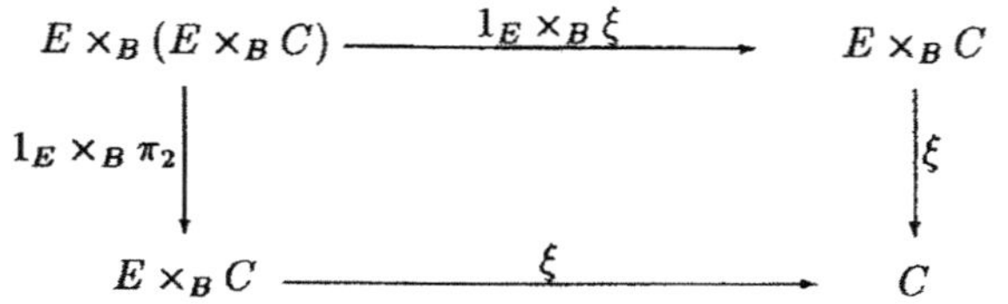

2.4 Proposition *For a split-epimorphism (= retraction) p in a finitely complete category, p^* is monadic.*

Proof Let s be a section of p, hence $ps = 1_B$. We give an ad-hoc construction of the left adjoint L of the comparison functor $K : \mathcal{C}/B \to (\mathcal{C}/E)^\tau$, with τ the monad induced by $p_! \dashv p^*$, letting L be the composite

$$(\mathcal{C}/E)^\tau \overset{U^\tau}{\to} \mathcal{C}/E \overset{s^*}{\to} \mathcal{C}/B$$

with U^τ the forgetful functor. Hence $L(C,\gamma;\xi)$ is given by the middle vertical arrow in

$$\begin{array}{ccccc} E\times_B(B\times_E C) & \longrightarrow & B\times_E C & \longrightarrow & C \\ \Big\downarrow & & \Big\downarrow & & \Big\downarrow{\scriptstyle\gamma} \\ E & \xrightarrow{\quad p\quad} & B & \xrightarrow{\quad s\quad} & E \end{array} \qquad (6)$$

Obviously, $LK \cong 1$. Hence we need only to show that $KL \cong 1$. This is done by giving an alternative construction for the pullback (6). It is easy to check that

$$\begin{array}{ccccc} C & \xrightarrow{<\gamma, 1_C>} & E\times_B C & \xrightarrow{(sp)\times_B 1_C} & E\times_B C \\ \Big\downarrow{\scriptstyle\gamma} & ① & \Big\downarrow{\scriptstyle 1_E\times_B\gamma} & ② & \Big\downarrow{\scriptstyle 1_E\times_B\gamma} \\ E & \xrightarrow{<1_E, 1_E>} & E\times_B E & \xrightarrow{(sp)\times_B 1_E} & E\times_B E \end{array}$$

$$\begin{array}{ccccc} E\times_B C & \xrightarrow{<\gamma\pi_2, \xi>} & E\times_B C & \xrightarrow{\quad\pi_2\quad} & C \\ \Big\downarrow{\scriptstyle 1_E\times_B\gamma} & ③ & \Big\downarrow{\scriptstyle 1_E\times_B\gamma} & ④ & \Big\downarrow{\scriptstyle\gamma} \\ E\times_B E & \xrightarrow{<\pi_2, \pi_1>} & E\times_B E & \xrightarrow{\quad\pi_2\quad} & E \end{array} \qquad (7)$$

commutes and that each ①, ② and ④ is a pullback square. In order to show that also ③ is a pullback square, it suffices to show that $<\gamma\pi_2, \xi>$ is an isomorphism since $<\pi_2, \pi_1>$ is one. But

$$\pi_1 <\gamma\pi_2, \xi><\gamma\pi_2, \xi> = \gamma\pi_2 <\gamma\pi_2, \xi> = \gamma\xi = \pi_1 ,$$

and with (5) one has

$$\begin{aligned} \pi_2 <\gamma\pi_2, \xi><\gamma\pi_2, \xi> &= \xi(1_E \times_B \xi) <\gamma\pi_2 , 1_{E\times_B C}> \\ &= \xi(1_E \times_B \pi_2) <\gamma\pi_2, 1_{E\times_B C}> \\ &= \xi <\gamma, 1_C> \pi_2 = \pi_2 ; \end{aligned}$$

therefore, $<\gamma\pi_2\,,\ \xi><\gamma\pi_2, \xi>=1$. Since the composite of the lower horizontal arrows in (7) is, as in (6), sp, the pullback diagrams (6) and (7) must coincide up to isomorphism. If we denote the middle vertical arrow in (6) by $\delta : Q \to B$, this means that there is an isomorphism $\varphi = \varphi_{(C,\gamma;\xi)} : C \to E \times_B Q$ in $\mathcal{C}$ such that $\pi_1\varphi = \gamma$ and $\tilde{\pi_2}\pi_2\varphi$ (with $\pi_2 : E \times_B Q \to Q$ and $\tilde{\pi_2} : Q = B \times_E C \to C$) is the composite of the upper horizontal arrows of (7). Two facts remain to be shown in order to complete the proof:

1. φ is a morphism of τ-algebras, i.e. $\varphi : (C,\gamma;\xi) \to KL(C,\gamma;\xi)$. This means that we must show that

$$\begin{array}{ccc} E \times_B C & \xrightarrow{\quad 1_E \times_B \varphi \quad} & E \times_B (E \times_B Q) \\ {\scriptstyle \xi}\downarrow & & \downarrow{\scriptstyle 1_E \times_B \pi_2} \\ C & \xrightarrow{\quad \varphi \quad} & E \times_B Q \end{array} \tag{8}$$

is commutative; this is done below.

2. φ is a natural transformation, i.e. for a τ-algebra homomorphism $h : (C,\gamma;\xi) \to (C',\gamma';\xi')$

$$\begin{array}{ccc} C & \xrightarrow{\quad h \quad} & C' \\ {\scriptstyle \varphi}\downarrow & & \downarrow{\scriptstyle \varphi'} \\ E \times_B Q & \xrightarrow{\quad 1_E \times_B h \quad} & E \times_B Q' \end{array} \tag{9}$$

is commutative. This amounts to showing that diagram (7) is "natural in $(C,\gamma;\xi)$" which is clear for parts ①, ②, ④. In case of ③, one must show that

$$\begin{array}{ccc} E \times_B C & \xrightarrow{\quad <\gamma\pi_2,\xi> \quad} & E \times_B C \\ {\scriptstyle 1_E \times_B h}\downarrow & & \downarrow{\scriptstyle 1_E \times_B h} \\ E \times_B C' & \xrightarrow{\quad <\gamma'\pi_2',\xi'> \quad} & E \times_B C' \end{array} \tag{10}$$

commutes; but this follows easily from the commutativity of

$$\begin{array}{ccc} E\times_B C & \xrightarrow{\quad\xi\quad} & C \\ {\scriptstyle 1_E\times_B h}\downarrow & & \downarrow{\scriptstyle h} \\ E\times_B C' & \xrightarrow{\quad\xi'\quad} & C' \end{array} \tag{11}$$

Hence we are left with showing that (8) commutes. Trivially,

$$\pi_1(1_E \times_B \pi_2)(1_E \times_B \varphi) = \pi_1 = \gamma\xi = \pi_1\varphi\xi\,.$$

Verifying that also $\tilde{\pi}_2\pi_2(1_E \times_B \pi_2)(1_E \times_B \varphi) = \tilde{\pi}_2\pi_2\varphi\xi$ holds amounts to showing that the composite c of the top horizontal arrows of (8) satisfies $c\pi_2 = c\xi$. In fact, since $c = \xi < sp\gamma, 1_C >$, one has with (5)

$$\begin{aligned} c\pi_2 &= \xi < sp\gamma\pi_2\,,\ \pi_2 >= \xi < sp\pi_1, \pi_2 > \\ &= \xi(1_E \times_B \pi_2) < sp\pi_1, 1_{E\times_B C} >= \xi(1_E \times_B \xi) < sp\pi_1, 1_{E\times_B C} > \\ &= \xi < sp\pi_1, \xi >= \xi < sp\gamma\xi, \xi >= c\,\xi\,. \end{aligned}$$

This completes the proof. □

2.5 We describe the left adjoint L of the comparison functor $K : \mathcal{C}/B \to (\mathcal{C}/E)^{\tau}$ (with τ the monad induced by $p_! \dashv p^*$) in case of an arbitrary morphism $p : E \to B$ in $\mathcal{C}$. According to 2.1, for a τ-algebra $(C, \gamma; \xi)$, one forms the coequalizer q of (ξ, π_2) in $\mathcal{C}$ and has an induced arrow δ such that

$$\begin{array}{ccccc} E\times_B C & \overset{\xi}{\underset{\pi_2}{\rightrightarrows}} & C & \xrightarrow{\quad q\quad} & Q \\ & & {\scriptstyle\gamma}\downarrow & & \downarrow{\scriptstyle\delta} \\ & & E & \xrightarrow{\quad p\quad} & B \end{array} \tag{12}$$

commutes; now $L(C, \gamma; \xi) = (Q, \delta)$. By 2.1 and (4), the unit

$$\kappa = \kappa_{(C,\gamma;\xi)} : (C, \gamma; \xi) \to K(Q, \delta)$$

is given by $\kappa = (1_E \times_B q) < \gamma, 1_C >=< \gamma, q >: C \to E \times_B Q$, considered as a morphism in $(\mathcal{C}/E)^{\tau}$.

2.6 We call a map $p : E \to B$ in **Top** *locally monadic* if for every $b \in B$ there is an open neighbourhood U in B such that p_U^* is monadic, with $p_U : p^{-1}U \to U$ the restriction of p.

PROPOSITION *p^* is monadic if and only if p is locally monadic.*

Proof Trivially, p is locally monadic if p^* is monadic. Vice versa, from the local characterization (iv) in Theorem 1.1 it follows with 2.2 that p^* is premonadic if p is locally monadic. In terms of the adjunction $L \dashv K$ as given in 2.5, this means that the co-units are isomorphisms; it remains to be shown that the units are isomorphisms as well.

Certainly, as a **Set**-map, $\kappa = \kappa_{(C,\gamma;\xi)}$ is bijective since $p^* : \mathbf{Set}/B \to \mathbf{Set}/E$ preserves coequalizers and is therefore monadic (cf. 2.1). Hence it suffices to show that the continuous bijection

$$\kappa = <\gamma, q>: C \to E \times_B Q$$

is a local homeomorphism, i.e. for $x \in C$ there is an open neighbourhood V of $\kappa(x)$ such that the restriction $\kappa^{-1}(V) \to V$ is a homeomorphism.

For $b = p\gamma(x) \in B$ one can choose U as in 2.6 and then consider

$$V = p^{-1}U \times_U \delta^{-1}U = p^{-1}U \times_B \delta^{-1}U \subseteq E \times_B Q\,.$$

Since p is assumed to be locally monadic, it suffices to check that diagram (12) can be restricted as follows:

$$\begin{array}{ccccc} p^{-1}U \times_U \kappa^{-1}V & \overset{\xi'}{\underset{\pi_2'}{\rightrightarrows}} & \kappa^{-1}V & \xrightarrow{\quad q' \quad} & \delta^{-1}U \\ & & \downarrow{\scriptstyle \gamma'} & & \downarrow{\scriptstyle \delta_U} \\ & & p^{-1}U & \xrightarrow{\quad p_U \quad} & U \end{array} \qquad (13)$$

with the top row a coequalizer diagram, and with $(\kappa^{-1}V, \gamma'; \xi')$ an algebra with respect to the monad induced by p_U^*. One easily verifies that the top row gives a coequalizer diagram in **Set**. Furthermore, q' is the restriction of q to open subspaces, with $\kappa^{-1}V \subseteq q^{-1}(\delta^{-1}U)$. Hence q' is a quotient map, and the top row of (13) yields a coequalizer diagram in **Top**. Checking that $(\kappa^{-1}V, \gamma'; \xi')$ is an algebra is easy: the diagrams describing it as such are just restrictions of the diagrams describing the algebra $(C, \gamma; \xi)$. □

2.7 A map $p : E \to B$ in **Top** is a *local split-epimorphism or locally sectionable* if for every $b \in B$ there is a neighbourhood U in B such that the restriction $p_U : p^{-1}U \to U$ is a split-epimorphism. Now Propositions 2.4 and 2.6 yield the key-result of the paper:

Theorem *For a locally sectionable map p in* **Top**, *p^* is monadic.* □

Corollary *p^* is monadic if p is a surjective map with one of the following properties: p is a local homeomorphism; p is locally trivial; p is a covering map; p is the projection of a completely regular G-space onto its orbit space, with G a compact Lie group* [6].

PROOF Each such map is surjective and locally sliceable [7], hence locally sectionable. □

REMARKS (1) A map p in **Top** with p^* monadic does not need to be open: the projection of $(-1,1]$ onto $\mathbf{R}/\mathbf{Z}$ is easily seen to be locally sectionable but not open (cf. [7]). Other examples with such properties are given by surjective polynomial maps $p:\mathbf{R}\to\mathbf{R}$.

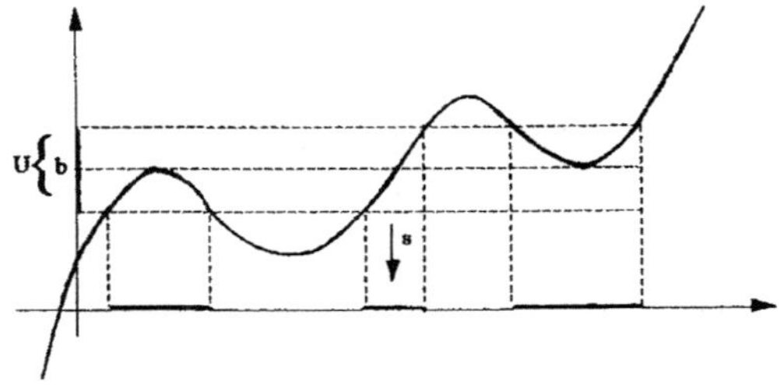

(14)

(2) Since monadicity implies premonadicity, locally sectionable maps are universal quotient maps. The example of a non-open universal quotient map given in 1.2 also shows that such maps need not be locally sectionable. Hence at least one of the implications

$$p \text{ locally sectionable} \Rightarrow p^* \text{ monadic} \Rightarrow p \text{ universal quotient map}$$

must be proper. We don't know which one(s).

3. The Monadic Decomposition of p^*

3.1 (Cf.[11],[1],[2]) If $\mathcal{A}$ is (sufficiently) cocomplete and n is an ordinal or ∞, a pair of adjoint functors $F \dashv U : \mathcal{A} \to \mathcal{X}$ can be factored as

$$\mathcal{A} \xrightarrow{U_n} \mathcal{X}_n \to \ldots \mathcal{X}_2 \xrightarrow{V_{2,1}} \mathcal{X}_1 \xrightarrow{V_{1,0}} \mathcal{X}_0 = \mathcal{X}$$

where $\mathcal{X}_{i+1} = \mathcal{X}_i^{\tau_i}$ with τ_i the monad induced by the comparison functor $U_i : \mathcal{A} \to \mathcal{X}_i$ $(U_0 = U)$, and $\mathcal{X}_i = \lim_{j<i} \mathcal{X}_j$ in case i is a limit ordinal. U *has a monadic decomposition* and n is the *length* of the decomposition provided n is the least ordinal or ∞ for which $\tau_n = (T_n, \eta_n, \mu_n)$ is idempotent, i.e. $\eta_n T_n = T_n \eta_n$; in this case the forgetful functor $V_{n+1,n} : \mathcal{X}_n^{\tau_n} \to \mathcal{X}_n$ is a full reflective embedding, and $U_{n+1} : \mathcal{A} \to \mathcal{X}_{n+1}$ induces a trivial monad on $\mathcal{X}_n$ (isomorphic to $1_{\mathcal{X}_n}$) hence $\mathcal{X}_{n+1} \cong \mathcal{X}_{n+2} \cong \ldots\ldots$ U is *essentially monadic* (with length n) if U_{n+1} is an equivalence of categories. Hence, for a categerory $\mathcal{A}$ with coequalizers, U *premonadic means* U *essentially monadic with length* 0 *or* 1 .

In [2] it was proved that *if* U *maps regular epimorphisms to epimorphisms, then the monadic decomposition of* U *has length at most* 1 . This result can be applied particularly in the case $U = p^*$, for $p : E \to B$ a map in **Top**. This way one obtains:

Proposition *For every map p in* **Top**, *p^* can be factored as $p^* = VW$, with V a premonadic functor and W a right adjoint functor which induces an idempotent monad.* □

3.2 The maps p for which the monadic decomposition has length 0 (i.e., when V in the Proposition above can be chosen as the identity functor) can be easily characterized:

Proposition *For a map p in a finitely complete category $\mathcal{C}$, the functor p^* induces an idempotent monad if and only if p is a monomorphism.*

Proof p is monic iff $\pi_2 = \varepsilon_{(A,\alpha)}$ is monic for all $(A,\alpha) \in \mathcal{C}/B$ (see 1.3). There is a pullback diagram

$$\begin{array}{ccc} E \times_B (E \times_B A) & \longrightarrow & E \times_B A \\ \downarrow & & \downarrow {\scriptstyle \pi_2 = \varepsilon_{(A,\alpha)}} \\ E \times_B A & \xrightarrow{\pi_2} & A \end{array} \qquad (15)$$

the projections of which are $p_! p^* \varepsilon_{(A,\alpha)}$ and $\varepsilon_{p_! p^*(A,\alpha)}$; hence they coincide iff π_2 is monic. Since p^* induces an idempotent monad iff $p_! p^* \varepsilon = \varepsilon p_! p^*$, this completes the proof. □

3.3 An arbitrary map p in **Top** can be factored as $p = iq$, with q a quotient map and i injective, hence $p^* \cong q^* i^*$. Since i^* induces an idempotent monad, for our purposes we may assume that $p = q$ is a quotient map. Since not every quotient map is universal, in general it is not possible to choose W as an identity functor in the factorization $p^* = VW$ of 3.1. In general this factorization cannot be naturally induced by a factorization of p, as $p = ju$, with u a universal quotient map and j an injective map. More precisely: the class $\mathcal{Q}$ of 1.2 does not belong to factorization system of **Top** since it is not stable under multiple pushout (=co-intersection). In fact, the hereditary but not universal quotient map described in 1.2 can easily be presented as co-intersection of countably many universal quotient maps.

References

1. J. Adámek, H. Herrlich and J. Rosický, Essentially equational categories, *Cahiers Topologie Géom. Différentielle Catégoriques* **29**(1988) 175-192.

2. J. Adámek, H. Herrlich and W. Tholen, Monadic Decompositions, *J. Pure Appl. Algebra* **59**(1989) 111-123.

3. A. Arhangel'skii, Bicompact sets and the topology of spaces, *Dokl. Akad. Nauk SSSR* **153**(1963) 743-746, English translation: *Soviet Math. Dokl.* **4**(1963) 1726-1729.

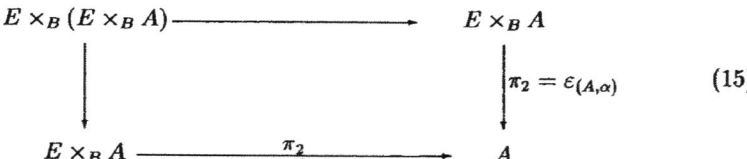

4. R. Börger and W. Tholen, Strong, regular, and dense generators, preprint (York University, Toronto 1989).
5. B.J. Day and G.M. Kelly, On topological quotient maps preserved by pullbacks or products, *Proc. Cambridge Phil. Soc.* **67**(1970) 553-558.
6. T. tom Dieck, *Transformation Groups*, de Gruyter (Berlin, New York 1987).
7. I.M. James, *Fibrewise Topology*, Cambridge University Press (Cambridge, New York 1989).
8. G. Janelidze, Precategories and Galois theory, preprint (Milan 1990).
9. G. Janelidze and W. Tholen, Pull-back functors, in preparation.
10. A. Joyal and M. Tierney, An extension of the Galois Theory of Grothendieck, *Memoirs Amer. Math. Soc.* **309**(1984).
11. J.L. MacDonald and A. Stone, The tower and regular decomposition, *Cahiers Topologie Géom. Différentielle Catégoriques* **23**(1982) 197-213.
12. S. MacLane, *Categories for the Working Mathematician*, Springer-Verlag (Berlin, New York 1971).
13. E. Michael, Bi-quotient maps and cartesian products of quotient maps, *Ann. Inst. Fourier* **18**(1968) 287-302.

Note added in proof (May 1991)

(1) M. Sobral informed the authors that she proved monadicity of p^* for an open quotient map p of topological spaces, and I. Moerdijk pointed out that this follows essentially from the corresponding Joyal-Tierney result [10] for locales – see the remark at the end of §1 in his paper on "Descent theory of toposes" (Bull. Soc. Math. Belgique 41 (1989) 373-391).

(2) J. Reiterman has constructed a universal quotient map p for which p^* is not monadic. This example, together with a topological characterization of "effective descent morphisms" p (i.e. of those maps p with p^* monadic), will be described in a forthcoming paper by J. Reiterman and W. Tholen.

This paper is in final form and will not be published elsewhere.

Fixpoint and loop constructions as colimits

C. Barry Jay *
Department of Computer Science, University of Edinburgh
The King's Buildings, Mayfield Road, Edinburgh, U.K. EH9 3JZ
e-mail: cbj@lfcs.ed.ac.uk

Abstract

The constructions of fixpoints and while-loops in a category of domains can be derived from the colimit, loop(f) of a diagram which consists of a single endomorphism $f : D \to D$. If f is increasing then the colimiting map is the least-fixpoint function Y and

$$\text{loop}(f) = fix(f)$$

the subobject of fixpoints. If $f = \text{cond}(b, g, 1)$ is the conditional of a while-program then

$$\text{loop}(f) = (D_{\neg b} + \text{loop}_\infty(g))_\perp$$

the lifted sum of the terminating values (where b is false) and the infinite loops.

1 Fixpoints

Let **Pos**(ω) the category of ω-complete partial orders and continuous functions (bottomless c.p.o.'s). A bottom element ($\perp$) can be adjoined to such a poset D by *lifting* it to $D_\perp$. Also, if $g : D \to D'_\perp$ is continuous then $g* : D_\perp \to D'_\perp$ is the strict extension of g i.e. agrees with g on D and preserves $\perp$ [4].

Let $f : D \to D$ be a continuous map in **Pos**(ω). Its fixpoints are defined by $\text{fix}(f) = \{d \in D \mid fd = d\}$. Categorically, this definition asserts that the inclusion $i : \text{fix}(f) \to D$ is the limit of the loop:

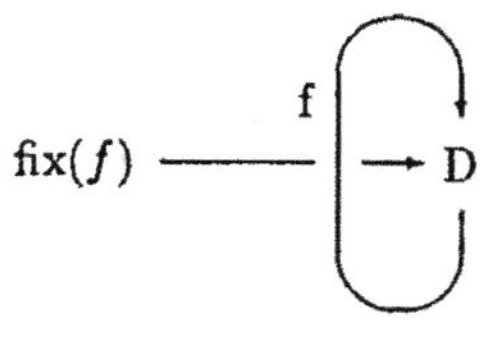

Equivalently, it is the equaliser of the pair $f, 1_D : D \to D$. (but not of the diagram $D \xrightarrow{f} D$ whose limit is just 1_D). Of course, this definition does not show how to construct any

*Supported by grants GR/E 78487 and GR/F 07866 from SERC, and OGPIN 016 from NSERC.

fixpoints. Of more use is the diagram's *colimit*, Lf

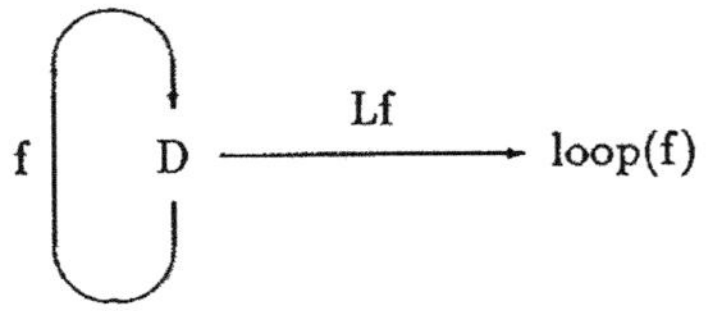

which will identify all of the elements of the sequence $\{f^n d\}$. The colimit exists since **Pos**(ω) is cocomplete [3] (though the colimit constructions are non-trivial).

Theorem 1.1 *Let $f \geq 1 : D \to D$ be increasing, i.e. $fd \geq d$ for all $d \in D$. Then the colimit of the loop is fix(f) and $Lf = Yf : D \to fix(f)$ is the least fixpoint function.*
Proof Yf is defined as usual [6]) by $Yf(d) = \bigvee f^n d$ and is denoted by Y when there is no risk of ambiguity. Recall [2] that it is the colimit if whenever $g : D \to D'$ is a continuous function such that $gf = g$ then there is a unique continuous function h : fix(f)$\to D'$ satisfying $h.Y = g$.

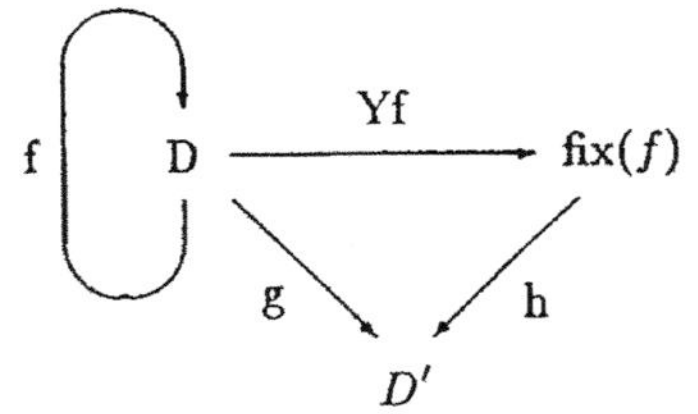

The value of h (and hence its uniqueness) is determined by $h = hYi = gi$. It remains to prove that $giY = g$. Now given $d \in D$ we have $giY(d) = g \bigvee f^n d = \bigvee gf^n d = \bigvee gd = gd$. //

More generally, we may consider an endofunctor $F : C \to C$ equipped with a natural transformation $\alpha : 1 \Rightarrow F$. Define fix(α) be its inverter [7] i.e. the full subcategory of objects for which α is an isomorphism, and $L(\alpha) : C \to$loop(α) to be the coinverter, i.e. the universal functor whose composite with α is invertible.

Theorem 1.2 *If C has, and F preserves, colimits of countable chains then we can take loop(α) = fix(α) with $L(\alpha)C = colim\{F^n\alpha_C : F^nC \to F^{n+1}C\}$.*
Proof Just generalise the previous proof in the style of [5]. //

A similar result holds in the category of complete metric spaces and distance-decreasing functions: if $f : X \to X$ is Lipschitz, i.e. there is a constant $\lambda < 1$ such that for all $x, y \in X$

$$d(fx, fy) \leq \lambda d(x, y)$$

then loop(f) = fix(f). Lawvere [1] observed that metric spaces can be thought of as *enriched* categories with distance-decreasing functions as functors. Also, in any category, if $f^2 = f : D \to D$ then loop(f) = fix(f) whenever they exist. Is there a common generalisation of these results?

Of course, we may want to compute fixpoints of a continuous function f even when it is not always increasing. One solution is to cut the domain down to those points where it is. The *postfix* points of f (pace *prefix points* [5]) are defined by

$$D_0 = \text{postfix}(f) = \{d \in D \mid d \leq fd\}$$

(Taking f as a functor, this is the *inserter* [7] from 1_D to f.) Now the postfix points are closed under the application of f and include the fixpoints. Thus the restriction of f to D_0 is an increasing endomorphism f_0 which has the same fixpoints as f, whence $Yf_0(d)$ is the least fixpoint over d.

For example, the factorial function fact : $N_\perp \rightarrow N_\perp$ on the lifted natural numbers is the least fixpoint of $h : (N_\perp \rightarrow N_\perp) \longrightarrow (N_\perp \rightarrow N_\perp)$ given by

$$\begin{aligned} h(g)(0) &= 1 \\ h(g)(n+1) &= (n+1) * g(n) \end{aligned}$$

Here postfix(h) consists of fact and its initial segments fact_n where

$$\begin{aligned} \text{fact}_n(k) &= k! \quad \text{if } 0 \leq k \leq n \\ &= \perp \quad \text{otherwise} \end{aligned}$$

Further $h_0(\text{fact}_n) = \text{fact}_{n+1}$ and so we may identify postfix(h) with $\omega+1$ (the free ω-chain $0 \leq 1 \leq \ldots \leq \omega$) and h_0 with the successor function. Clearly then Yh_0 is the unique map $\omega + 1 \rightarrow \{\omega\}$ where ω corresponds to $fact$.

Although this method recovers the fixpoints, as a method of assigning fixpoints it is quite limited. For example, if some tail of the sequence $\{f^n d\}$ is a chain then its supremum is a fixpoint naturally associated to d, even if $d \not\leq fd$.

In more detail, consider $f : N \times N \rightarrow N \times N$ defined by

$$\begin{aligned} f(0,n) &= (0,n) \\ f(m+1,n) &= (m,(m+1)n) \end{aligned}$$

Clearly fix(f) = postfix(f) = $\{(0,n) \mid n \in N\}$. The interesting question is how to map the point (m,n) to the 'limit' of the sequence $\{f^k(m,n)\}$ which becomes stationary at $(0,m!n)$.

Lemma 1.3 *$Lf = fact^* : N \times N \rightarrow N$ where $fact^*(m,n) = m!n$.*
Proof A simple induction shows that if $gf = g : N \times N \rightarrow D$ then $g(m,n) = g(0,m!n)$. Define $h : N \rightarrow D$ by $h(n) = g(0,n)$. //

Now $fact$ is

$$N \xrightarrow{<1_N, \Delta(1)>} N \times N \xrightarrow{fact^*} N$$

where $\Delta(k)$ is the function which is constantly k. Note that, unlike the usual fixpoint construction of fact, this doesn't require function spaces or $\perp$.

2 While-loops

The last example above has the pleasant property that every sequence $\{f^n d\}$ becomes stationary at a desired value (i.e. a pair (m, n) where $m = 0$) after finitely many steps. In general when evaluating a loop there will be other sequences which are either not stationary, or become so at a value where some undesirable condition holds. When defining the semantics of a while-loop, however, those undesirable sequences represent infinite loops which should perhaps be discarded (mapped to $\perp$). This proves to be easy as the loop object is a lifted sum of the finite and infinite loops.

Taylor [8] has also tried to give a semantics for while loops using colimits, specifically, via coequalisers in a topos.

First we need some notation. For any full sub-object D_1 of D (i.e. sub-c.p.o. whose order is the restriction of that of D) define the full sub-objects

$$
\begin{aligned}
f^{\forall}(D_1) &= \{d \in D \mid (\forall n \geq 0) f^n d \in D_1\} \\
f^{\exists}(D_1) &= \{d \in D \mid (\exists n \geq 0) f^n d \in D_1\}
\end{aligned}
$$

The set $f^{\forall}(D_1)$ is the greatest fixpoint of $f^{-1} : sub(D) \rightarrow sub(D)$ contained in D_1 while $f^{\exists}(D_1)$ is the least fixpoint containing it if $D_1 \subseteq \mathrm{fix}(f)$.

Let $B = \{tt, ff\}$ be the c.p.o. of booleans and consider a continuous function $b : D \rightarrow B_\perp$ which is to be thought of as a condition on D. Define

$$
\begin{aligned}
D_b &= b^{-1}\{tt\} \\
D_{?b} &= b^{-1}\{\perp\}
\end{aligned}
$$

which are open and closed, respectively, in D. Note that $D_{\neg b}$ is defined using $\neg : B \rightarrow B$ the negation.

For continuous functions $g, g' : D \rightarrow D'$ we can represent conditionals by $\mathrm{cond}(b, g, g') : D \rightarrow D'_\perp$ where

$$
\mathrm{cond}(b, g, g')(d) = \begin{cases} gd & \text{if } d \in D_b \\ g'd & \text{if } d \in D_{\neg b} \\ \perp & \text{if } d \in D_{?b} \end{cases}
$$

The loop 'while b do g' should be interpreted by a function $\mathrm{while}(b, g) : D \rightarrow D_\perp$ which satisfies

$$
\mathrm{while}(b, g) = \mathrm{cond}(b, \mathrm{while}(b, g)^* g, 1_D)
$$

Its finite and infinite loops are those of $f = \mathrm{cond}(b, g, 1_D)^*$.

Theorem 2.1 *Let*

$$
\begin{aligned}
D_1 &= f^{\exists}(D_{?b})_\perp \\
D_2 &= f^{\exists}(b^{-1}\{ff, \perp\}) \\
D_3 &= f^{\forall}(b^{-1}\{tt, \perp\})
\end{aligned}
$$

Then the following diagram of inclusions is a pushout

$$\begin{array}{ccc} D_1 & \longrightarrow & D_2 \\ \Big\downarrow & & \Big\downarrow \\ D_3 & \longrightarrow & D_\perp \end{array}$$

Further, f restricts to endomorphisms $f_i : D_i \to D_i$ for $i = 1, 2, 3$. Thus

$$\begin{array}{ccc} loop(f_1) & \longrightarrow & loop(f_2) \\ \Big\downarrow & & \Big\downarrow \\ loop(f_3) & \longrightarrow & loop(f) \end{array}$$

is also a pushout which implies (writing $loop_\infty(f)$ for $loop(f_3)$)

$$loop(f) = (D_{\neg b} + loop_\infty(f))_\perp$$

and

$$(Lf)d = \begin{cases} g^n d & \text{if } g^n d \in D_{\neg b} \\ L(f_3)d & \text{if } d \in D_3 \\ \perp & \text{if } d \in D_1 \end{cases}$$

Proof Let $h_i : D_i \to D'$ for $i = 2, 3$ be continuous functions whose restrictions to D_1 are equal. It suffices to show that they determine a unique continuous function $h : D \to D'$ whose restriction to D_i is h_i for $i = 2, 3$. Clearly there is a unique such function. Its continuity follows since the set-theoretic differences $D_2 - D_1$ and $D_3 - D_1$ are open subsets.

Now the colimit of a pushout is just the pushout of the colimits. Thus loop(f) is the pushout of the loop(f_i). //

Thus we can interpret the while-loop by

$$D \xrightarrow{\ j\ } D_\perp \xrightarrow{\ Lf\ } \text{loop}(f) \xrightarrow{\ p\ } (D_{\neg b})_\perp \xrightarrow{\ k_\perp\ } D_\perp$$

where j and k are the obvious inclusions, and p is the projection from the lifted sum (that maps all elements of $\text{loop}_\infty(f)$ to $\perp$).

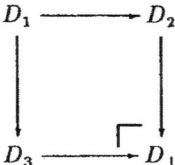

References

[1] F.W. Lawvere, Metric spaces, generalised logic, and closed categories, Rend. Sem. Mat. Fis. Milano 43 (1973) 135–166.

[2] S. Mac Lane, *Categories for the Working Mathematician* (Springer-Verlag, 1971).

[3] J. Meseguer, Completions, factorisations and colimits for ω-posets, in *Mathematical Logic in Computer Science*, Colloquia Mathematica Janos Bolyai, XXVI (Hungary, 1978) 509–545.

[4] E. Moggi, Computational Lambda-Calculus and Monads, Proc. Fourth Annual Symp. on Logic in Computer Science (1989) 14–23.

[5] M.B. Smyth & G.D. Plotkin, The category-theoretic solution of recursive domain equations, SIAM J. of Comp. 11 (1982) 761–783.

[6] D.S. Scott, Data types as lattices, SIAM J. of Computing 5 (1976) 522–587.

[7] R.H. Street, Limits indexed by category-valued 2-functors, J. Pure Appl. Alg. 8 (1976) 149–181.

[8] Paul Taylor, An exact interpretation of while (I,II), manuscript.

This work is in final form; no similar paper has been or is being submitted elsewhere.

PREFRAME PRESENTATIONS PRESENT

Peter Johnstone,
Department of Pure Mathematics and Mathematical Statistics,
University of Cambridge,
16 Mill Lane, Cambridge, CB2 1SB.

Steven Vickers,
Department of Computing, Imperial College,
180 Queen's Gate, London, SW7 2BZ.
email: sjv@doc.ic.ac.uk

Abstract

Preframes (directed complete posets with finite meets that distribute over the directed joins) are the algebras for an infinitary essentially algebraic theory, and can be presented by generators and relations. This result is combined with a general argument concerning categories of commutative monoids to give a very short proof of the localic Tychonoff Theorem.

It is also shown how frames can be presented as preframes, a result analogous to Johnstone's construction of frames from sites, and an application is given.

1. Introduction

It is well known (Joyal and Tierney [84]) that a coproduct of frames is a tensor product in a rather conventional sense, for it is universal for functions with a certain bilinear property:

> Let A and B be frames, let C be a sup-lattice, and let ϕ: A×B → C preserve all joins in each argument. Then there is a unique sup-lattice homomorphism ϕ': A⊗B → C such that $\phi'(a\otimes b) = \phi(a, b)$ for all $a \in A$, $b \in B$.

(A *sup-lattice* is a complete join semilattice, and a function between sup-lattices is a homomorphism if it preserves all joins.)

To put it another way, suppose for sup-lattices A and B the tensor product A⊗B is presented as

$$\mathrm{SupLat}\langle\, a\otimes b\ (a \in A, b \in B) \mid$$
$$\bigvee S \otimes b = \bigvee\{a\otimes b: a \in S\} \qquad (S \subseteq A)$$
$$a \otimes \bigvee T = \bigvee\{a\otimes b: b \in T\} \qquad (T \subseteq B) \quad \rangle$$

If A and B are actually frames, then this tensor product serves as a frame coproduct. (This is directly analogous to the case of commutative rings, where coproducts are constructed as tensor products of Abelian groups.)

The uniqueness part of this universal property is obvious. Every element of $A \otimes B$ is a join of elements $a \otimes b$, so a sup-lattice homomorphism is determined by its values on these elements.

There is another representation of elements of $A \otimes B$. Let us write $a ⅋ b$ (a "par" b) for the element $a \otimes \mathbf{true} \vee \mathbf{true} \otimes b$. It is easily verified that

- ⅋ preserves finite meets and non-empty joins (in particular, directed joins) in each argument.
- ⅋ preserves all joins jointly in its two arguments.
- $a \otimes b = a ⅋ \mathbf{false} \wedge \mathbf{false} ⅋ b$

Now a *finite* join of elements $a \otimes b$ is a finite join of finite meets of elements $a ⅋ b$, and hence, by finite distributivity, a finite meet of finite joins of elements $a ⅋ b$, hence a finite meet of elements $a ⅋ b$. Hence every element of $A \otimes B$ is a directed join of finite meets of elements $a ⅋ b$. We are going to investigate the possibility of defining functions from $A \otimes B$ that preserve finite meets and directed joins, by defining the values at elements $a ⅋ b$. A major purpose of this paper is to present the corresponding existence result, Theorem 4.3, and to use it to give what is probably the simplest proof yet discovered of the Localic Tychonoff Theorem.

What is happening here is that instead of the sup-lattice structure (joins) of frames, we are considering the *preframe* structure: following Banaschewski [88], we call a poset A a *preframe* if it has all finite meets and directed joins, with binary meets distributing over directed joins; and we call a function between preframes a *preframe homomorphism* if it preserves finite meets and directed joins. (Note – Gierz et al. [80] call preframes "meet continuous semilattices".)

The original result for the sup-lattice structure is quite cheap. The sup-lattice tensor product $A \otimes B$ can be presented as above, and it is then easy to prove that it is in fact a frame, the frame coproduct of A and B. The essential fact, that the presentation does indeed present a sup-lattice, comes from standard universal algebra, with only minor hitches from the infinitary joins in the theory of sup-lattices.

Our corresponding result for preframes follows a somewhat parallel route (put in a quite general setting in Section 4), but it is much harder to show that the same method of presentations works. The mathematical core of this paper lies in our result that for preframes also, a presentation by generators and relations does indeed present a preframe (Section 3). The underlying idea is that the theory of preframes is essentially algebraic. For *small* essentially algebraic theories, presentations by generators and relations do present, so for large theories (such as preframes) one can expect the same, provided that cardinality problems can be circumvented. For convenience, we summarize results about essentially algebraic theories in Section 2.

Finally, we prove a preframe analogue of the coverage theorem for frames (Johnstone [82]), which is best seen in a sup-lattice context, and use it in some applications.

2. Essentially algebraic theories

The infinitary analogue of finitary algebraic (or equational) theories has been fairly extensively studied by several people from Słominski [59] and Linton [66] onward; but much less attention has been paid to the infinitary analogue of what are variously called essentially algebraic theories (Freyd [72]), lim-theories (Coste [79]) or left exact theories (McLarty [86]): essentially the only works in this area are the paper of Isbell [72] and the monograph of Gabriel and Ulmer [71], neither of which takes an explicitly syntactic point of view (and the work of Gabriel and Ulmer applies only to small theories, which are insufficient for our purposes). However, in order to understand the theory of preframes, which is our main concern in this paper, it is necessary to regard it in this general context; so we devote this section to developing as much of the general theory as we shall need.

We shall follow the approach of Freyd [72] in presenting the syntax via partial operations whose domains are specified by equations, rather than that of Coste using primitive predicates and a "provably unique" existential quantifier. It is not hard to show that the expressive powers of the two approaches are identical, provided one allows many-sorted theories (partially ordered sets are a single-sorted theory in Coste's sense, but in our approach we must take two primitive sorts, one for the underlying set and one for the set of instances of the order relation).

An *essentially algebraic theory*, then, is specified by the following data:

(i) A set S of *sorts*.

(ii) A class O of *operation-symbols* ω, each of which is equipped with an *arity* f_ω which is a function from some index set I to S, and a *type* t_ω which is a single sort. (The intended interpretation is that, if the sorts have been interpreted by sets A_s, $s \in S$, then $\omega \in O$ will be interpreted as a partial function ω_A from $\prod_{i\in I} A_{f_\omega(i)}$ to A_{t_ω}.)

(iii) A well-founded partial ordering $<$ on O, with the property that the $<$-predecessors of any given operation-symbol form a set.

(iv) For each $\omega \in O$, a *domain declaration* which is a (possibly infinite) conjunction of equations $(t = t')$, where t and t' are terms of the same type, constructed in the usual way from variables and operation-symbols that preced ω in the ordering $<$. (The intended interpretation is that ω_A is defined exactly at those I-tuples where, for each equation $(t = t')$, the terms t_A and t'_A are defined and equal; if the domain declaration of ω is the empty conjunction, then the domain of ω_A is to be the whole of $\prod_{i\in I} A_{f_\omega(i)}$.)

(v) A class A of *axioms*, which are equations ($t = t'$) between pairs of terms of the same type. (The intended interpretation is that a structure A for the language, as so far defined, will be a *model* for the theory provided, for each axiom ($t = t'$), the interpretations t_A and t'_A are equal at all tuples of elements of A where they are both defined.)

Of course, we say the theory is *single-sorted* if the set S is a singleton; in what follows we shall generally deal with the single-sorted theories in order to simplify notation. We say that the theory is *small* if the class O is a set (in which case A will also be a set, since we have only a set of possible terms).

The reader may have been surprised to see that our axioms all have the form of equations rather than Horn sequents; however, the presence of domain declarations allows us to achieve the effect of Horn sequents, by introducing extra operations that are restrictions of projections. For example, if we wish to assert that an equation ($t = t'$) holds conditionally upon an equation ($s = s'$), we introduce a new operation α (whose arity includes the sorts of all the variables appearing in s or s', and whose type is that of one of the variables appearing in t or t'), with domain declaration ($s = s'$); we then add a new axiom ($\alpha(x_i)_{i\in I} = x_{i_0}$) and substitute $\alpha(x_i)_{i\in I}$ for x_{i_0} somewhere in t or t'.

The same device may also be used to turn all the axioms into directed equalities ($t \succcurlyeq t'$) in the sense of Freyd and Scedrov [90] (intended interpretation: "if t is defined, then t' is also defined and equal to it") – though we cannot, in general, reduce them to assertions of the form "one side is defined iff the other is, and then they are equal".

Example 2.1 As previously mentioned, our primary interest in this paper is in the notion of *preframe* introduced by Banaschewski [88]: a preframe is a partially ordered set having finite meets (including a top element 1) and directed joins, such that binary meets distribute over directed joins. We give here a presentation of the theory of preframes as a (single-sorted) essentially algebraic theory. At the lowest level of the ordering on operations, we have a constant 1 and a binary operation $\wedge$, both with empty domain declarations (and we shall have axioms that say that ($\wedge$, 1) defines a semilattice structure on our underlying set A). To handle directed joins, we next introduce an operation $\bigvee^{\uparrow}_P$ of arity $|P|$ (the underlying set of P) for each directed poset P; the domain declaration for $\bigvee^{\uparrow}_P$ will be the conjunction, over all pairs (p, q) in P with $p \leq q$, of the equations ($x_p \wedge x_q = x_p$), so that $\bigvee^{\uparrow}_P(f)$ is defined, for a function $f\colon P \to A$, iff f is order preserving. To ensure that $\bigvee^{\uparrow}_P(f)$ is the least upper bound of the image of f, it suffices to write down the axioms

$$\bigvee\nolimits^{\uparrow}_{\{p\}} (x_p) = x_p$$

for a singleton poset $\{p\}$, and

$$\bigvee\nolimits^{\uparrow}_P (x_p)_{p\in P} \wedge \bigvee\nolimits^{\uparrow}_Q (x_{h(q)})_{q\in Q} = \bigvee\nolimits^{\uparrow}_Q (x_{h(q)})_{q\in Q}$$

whenever $h\colon Q \to P$ is an order-preserving map between directed posets. Finally, we write down the distributive law as a scheme of axioms, one for each directed poset P:

$$x \wedge \bigvee\nolimits^{\uparrow}_P (y_p)_{p\in P} = \bigvee\nolimits^{\uparrow}_P (x \wedge y_p)_{p\in P}$$

For a small essentially algebraic theory $\mathbb{T}$, the forgetful functor from $\mathbb{T}$-models to **Set** (or to $\mathbf{Set}^n$ if $\mathbb{T}$ is many-sorted) has a left adjoint, just as in the algebraic case: the free $\mathbb{T}$-model on a set X is constructed in the usual way as the set of words (i.e. terms) in the elements of X, modulo $\mathbb{T}$-provable equality. The adjunction will not be monadic unless $\mathbb{T}$ is algebraic (i.e. has a presentation on which all the domain declarations are empty), but it will be possible to factor it as a tower of monadic adjunctions in the style of MacDonald and Stone [82], by expressing $\mathbb{T}$ as a union (indexed by some ordinal) of subtheories $\mathbb{T}_\alpha$, corresponding to the levels in the ordering on primitive operations, and successively constructing free functors from $\mathbb{T}_\alpha$-**Mod** to $\mathbb{T}_{\alpha+1}$-**Mod**. The tower may be of arbitrary height, as is shown by the following example:

Example 2.2 Let α be an ordinal, and let $\mathbb{T}_\alpha$ denote the theory having one unary operation ω_β for each $\beta < \alpha$ (ordered in the obvious way), the domain declaration of ω_β being the conjunction over all $\gamma < \beta$ of $(\omega_\gamma(x) = x)$, and no axioms. It is clear that, after β ordinal steps starting from **Set**, we cannot get any further than $\mathbb{T}_\beta$-**Mod**; in the free $\mathbb{T}_\alpha$-model on a $\mathbb{T}_\beta$-model, the operation $\omega_{\beta+1}$ has no fixed points, and so the later operations are never defined.

For large theories, even when the forgetful functor to **Set** has a left adjoint (and even when the arities of the generating operations, and the lengths of chains in the ordering on these operations, are bounded), it may not be possible to decompose the adjunction as a tower of monadic ones, as is shown by the following simple modification of the previous example: take a single unary operation ω_0 with empty domain declaration, and then a proper class of unary operations ω_α, all with domain declaration $(\omega_0(x) = x)$. The forgetful functor from $\mathbb{T}$-**Mod** to **Set** has a left adjoint, which sends a set X to the free $\mathbb{T}_1$-model on X; but the comparison functor from $\mathbb{T}$-**Mod** to $\mathbb{T}_1$-**Mod** has no left adjoint. Note, in particular, that the existence of free $\mathbb{T}$-models on sets is no guarantee of the existence of colimits (in particular, coequalizers) in $\mathbb{T}$-**Mod**.

Since the theory in which we are interested, that of preframes, is a large one, we shall have to investigate whether its category of models possesses this sort of structure. As far as colimits are concerned, Banaschewski [88] gave an explicit construction of coproducts; and we shall be able to infer the existence of coequalizers (though not to describe them all that explicitly) from our presentation theorem in the next section. For monadicity, the questions are more easily answered.

Lemma 2.3 The forgetful functor from **PreFrm** to **Set** has a left adjoint.

Moreover, the monadic length of the adjunction is 2.

Proof

The forgetful functor may be factored as

PreFrm $\rightarrow$ **SLat** $\rightarrow$ **Set**

where **SLat** is the category of meet-semilattices. Now **SLat** $\rightarrow$ **Set** has a left adjoint that sends a set X to the set $\wp_{\text{fin}}(X)$ of finite subsets of X, ordered by reverse inclusion, and

PreFrm → **SLat** has a left adjoint sending a semilattice P to the set Idl(P) of ideals of P, ordered by inclusion (see Vickers [89], Theorem 9.1.5). Now **SLat** → **Set** is monadic, since semilattices are an algebraic theory; but in a free semilattice $\wp_{\mathrm{fin}}(X)$ the upward closure of each element is finite, from which it follows easily that every ideal is principal, and so the monad on **Set** induced by the composite adjunction is (isomorphic to) the free semilattice monad. In other words, **SLat** is exactly the algebraic part of the essentially algebraic theory of preframes.

To complete the proof, we must show that the forgetful functor **PreFrm** → **SLat** is monadic; but we may do this directly, as follows. Let P be a semilattice, and suppose it has an algebra structure α: Idl(P) → P for the monad on **SLat** induced by the adjunction. Since α sends principal ideals to their generators and is order-preserving, we see that it must send each ideal of P to its join in P; so P has directed joins. Moreover, since α preserves binary meets, we see that binary meets distribute over directed joins in P, so it is a preframe; and its Idl-algebra structure is uniquely determined by its preframe structure.]

We note in passing that, once we have established the existence of coequalizers in **PreFrm**, Lemma 2.3 combined with Linton's theorem (Linton [69]) will enable us to "lift" more general colimits from **SLat** to **PreFrm**, without making use of Banaschewski's construction of coproducts.

In the case of a small theory $\mathbb{T}$, the decomposition of $\mathbb{T}$-**Mod** → **Set** into a tower of monadic functors, plus the fact that these functors have rank (i.e preserve α-filtered colimits for some cardinal α – just take a regular cardinal greater than the arities of all the generating operations of $\mathbb{T}$), ensure that the category $\mathbb{T}$-**Mod** is locally presentable (cf. Gabriel-Ulmer [71], Satz 10.3). Indeed, there is a converse: any locally α-presentable category is equivalent to the category of α-continuous set-valued functors on a small α-complete category C, and the theory of such functors may readily be presented as a (many-sorted) essentially algebraic theory (cf. Coste [79], Theorem 2.3.2, for the case $\alpha = \aleph_0$). Thus the categories of models of small essentially algebraic theories are, up to equivalence, exactly the locally presentable categories. In particular, all such categories are cocomplete.

Given a morphism f: A → B of models of an essentially algebraic theory $\mathbb{T}$, the set-theoretic image I of f is not in general a $\mathbb{T}$-model: the identifications made in passing from A to I may create "new" tuples of elements satisfying the domain declaration of some operation of $\mathbb{T}$, none of whose pre-images in A do so. However, $\mathbb{T}$-**Mod** does have image factorizations (even if $\mathbb{T}$ is large): that is, every morphism factors as a strong epimorphism followed by a monomorphism. To obtain the image of f: A → B in $\mathbb{T}$-**Mod**, we simply take the sub-$\mathbb{T}$-model I of B generated by the set-theoretic image I, i.e. the intersection of all submodels that contain I. This factorization is not, in general, stable under pullback (and so the category $\mathbb{T}$-**Mod** is not in general regular), but it is at least functorial: that is, a commutative square

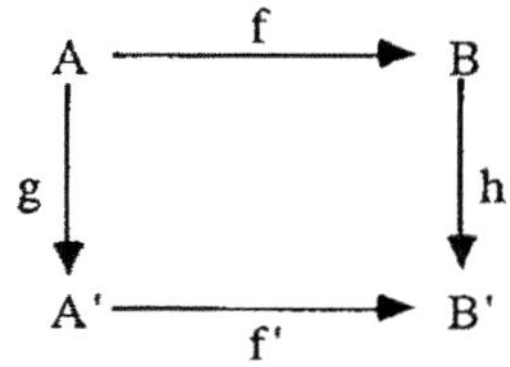

gives rise to a (unique) morphism from the $\mathbb{T}$-model image of f to that of f', since the inverse image of the latter under h is a sub-$\mathbb{T}$-model of B containing the set-theoretic image of f. (We shall need this observation, in the case of preframes, in our proof of the presentation theorem.)

Another point that should be noted is that strong epimorphisms in **$\mathbb{T}$-Mod** (that is, morphisms $f: A \to B$ that do not factor through any proper sub-$\mathbb{T}$-model of their codomain) need not be regular epimorphisms (that is, coequalizers): the coequalizer of the kernel-pair of f, if it exists, will be the $\mathbb{T}$-model freely generated by the set-theoretic image I, modulo the preservation of as much of the $\mathbb{T}$-model structure as exists in I, and this may not map injectively to B. This is another indication that, for large theories $\mathbb{T}$, the construction of coequalizers in **$\mathbb{T}$-Mod** can be expected to be a delicate matter.

Leaving this question aside for the moment, we conclude this section by briefly considering the notion of commutativity for essentially algebraic theories. For algebraic theories, the notion is well understood, but it is perhaps less widely appreciated that it makes perfectly good sense in our more general context. We say that a single-sorted essentially algebraic theory $\mathbb{T}$ is *commutative* if any two of its operations commute with each other, or, equivalently, if each operation is a homomorphism of $\mathbb{T}$-models. (To make sense of this second formulation, note that the commutativity of the subtheory $\mathbb{T}_1$ generated by the operations that precede a given operation ω ensures that the domain of ω_A is a sub-$\mathbb{T}_1$-model of the power of A corresponding to its arity.)

Proposition 2.4 Let $\mathbb{T}$ be a commutative essentially algebraic theory.

(i) The category **$\mathbb{T}$-Mod** has a symmetric closed structure, in which the internal hom [A, B] is the set of $\mathbb{T}$-model homomorphisms from A to B with operations defined pointwise.

(ii) If **$\mathbb{T}$-Mod** has a free functor and coequalizers (for instance, if $\mathbb{T}$ is small), then it also has a symmetric monoidal structure $(\otimes, I)$, where I is the free $\mathbb{T}$-model on one generator and $(-)\otimes A$ is left adjoint to $[A, -]$.

Proof

(i) Commutativity of $\mathbb{T}$ (plus an induction over the ordering on operations) implies that if ω is an operation of $\mathbb{T}$ (of arity J, say) and $(f_j)_{j\in J}$ is a family of homomorphisms from A to B that (pointwise) satisfy the domain declaration of ω, then the function

$$a \mapsto \omega_B(f_j(a))_{j\in J}$$

is again a homomorphism from A to B. So [A, B] has the structure required for a $\mathbb{T}$-model, and it satisfies the axioms since B does. Moreover, the assignment $(A, B) \mapsto [A, B]$ is easily seen to be a bifunctor, contravariant in the first argument and covariant in the second. For the symmetry, we observe that homomorphisms from A to [B, C] correspond to functions from $A \times B$ to C that are *bihomomorphisms*, i.e. are homomorphic in each variable provided the other is held constant, and these in turn correspond to homomorphisms from B to [A, C], yielding a natural isomorphism

$$[A, [B, C]] \cong [B, [A, C]]$$

(ii) To obtain the monoidal structure, we need to construct a universal bihomomorphism from $A \times B$ to $A \otimes B$, i.e. one through which every bihomomorphism from $A \times B$ to C factors by a unique homomorphism from $A \otimes B$ to C. Under the extra hypotheses on $\mathbb{T}$, we may do this by first forming the free $\mathbb{T}$-model F on $A \times B$, and then forming the coequalizer $F \to A \otimes B$ of $R \rightrightarrows F$, where R is the smallest congruence on F such that the composite $A \times B \to F \to A \otimes B$ is bihomomorphic. The remaining details are straightforward.]

Example 2.5 The theory of preframes, as defined earlier in this section, is commutative. It is well known that the theory of meet-semilattices is commutative, and the directed join operations commute with each other (and with themselves); so we need only verify that directed joins commute with finite meets. Now the distributive law tells us that, if $(x_i)_{i \in I}$ and $(y_i)_{i \in I}$ are monotone nets indexed by the same directed set I, we have

$$\bigvee^{\uparrow}{}_I (x_i)_{i \in I} \wedge \bigvee^{\uparrow}{}_I (y_i)_{i \in I} = \bigvee^{\uparrow}{}_{I \times I} (x_i \wedge y_j)_{(i,j) \in I \times I}$$

Then directedness of I tells us that the $(x_i \wedge y_i)$, $i \in I$, are cofinal among the $(x_i \wedge y_j)$ and so have the same join.

Although the commutativity of the theory of preframes does not seem to have been explicitly observed before, it is not exactly a new idea; in particular, it underlies the "Lawson duality" of continuous semilattices (see Johnstone [82], VII 2.11). A continuous semilattice is just a preframe which satisfies the additional requirement of continuity (expressed in the usual way in terms of the way-below relation); the *dual* of a continuous semilattice A is the poset $\hat{A}$ of Scott open filters of A (i.e. subsets that are upper closed, closed under finite meets and inaccessible by directed joins), ordered by inclusion. But a subset of A is a Scott open filter iff its characteristic function is a preframe homomorphism from A to $\mathbf{2} = \{0, 1\}$, so we may identify $\hat{A}$ with $[A, \mathbf{2}]$ as defined in Proposition 2.4. Lawson duality is then the assertion that, for a continuous semilattice A, $[A, \mathbf{2}]$ is also continuous, and the canonical mapping from A to $[[A, \mathbf{2}], \mathbf{2}]$ is an isomorphism. (We note in passing that this duality cannot be extended to any larger class of preframes, since a preframe is continuous iff its Scott-open sets separate its points.)

The theory of frames, unlike that of preframes, is not commutative: although the distributive law implies that finite meets commute with directed joins, finite meets and finite joins can never commute in a nontrivial lattice. (If $a = d \leq b = c$, then $(a \wedge b) \vee (c \wedge d) =$

a, but $(a \vee c) \wedge (b \vee d) = b$. Also, the presence of two distinct constants 0 and 1 is incompatible with commutativity.) In fact, the theory of preframes is a maximal commutative subtheory of frames. It is not unique with this property. Another such is the theory of complete join-semilattices (*sup-lattices*), obtained by retaining all the join operations (including 0) and discarding the finite meets. We shall see that in many ways the relationship between frames and preframes is similar to that between frames and sup-lattices. The latter was extensively investigated by Joyal and Tierney [84], and their work has served as a model for a large part of ours.

3. Preframe presentations present

First, recall Banaschewski's [88] theory of *prenuclei* on a frame: if A is a frame and ν_0 is a function from A to itself that is monotone and inflationary ($\nu_0(a) \geq a$), then ν_0 is a *prenucleus* if

$$\forall a, b \in A.\ \nu_0(a) \wedge b \leq \nu_0(a \wedge b)$$

If, in addition, ν_0 is idempotent, then it is a nucleus in the usual sense (see Johnstone [82]).

Banaschewski proves that for each prenucleus ν_0, there is a nucleus ν characterized by the property that each $\nu(a)$ is the least fixpoint of ν_0 greater than a. ν_0 and ν have the same fixpoints, which – by the standard theory for the nucleus ν – form a frame; it has the universal property of

$$\mathrm{Frm} \langle A \text{ (qua Frm)} \mid \nu_0(a) \leq a \ (a \in A) \rangle$$

Next, let us summarize some free constructions. Note for all of them that the concrete constructions show that the injections of generators are all 1-1.

Proposition 3.1

(i) The free preframe over a meet semilattice S is the ideal completion Idl(S).

(ii) The free frame over a meet semilattice S is the set $\mathrm{Alex}(S^{op})$ of lower closed sets of S.

(iii) The free frame over a preframe A is the set of Scott closed subsets of A.

Proof (i) This has already been mentioned; it is Theorem 9.1.5 in Vickers [89].
(ii) This comes immediately from the coverage theorem (Johnstone [82] Proposition II.2.11).
(iii) Banaschewski [88], Proposition 1.]

Proposition 3.2 Let S be a meet semilattice, and let R be a set each of whose elements has the form (X, a) where $X = (x_i)_{i \in P}$ is a monotone net in S and a is an upper bound in S for $\{x_i : i \in P\}$.

Then $\mathrm{PreFrm} \langle S \text{ (qua meet semilattice)} \mid \bigvee^{\uparrow} X = a \ ((X, a) \in R) \rangle$ exists.

Proof

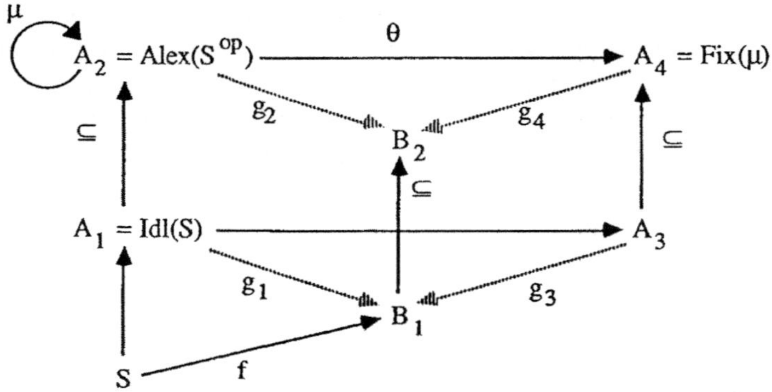

Let A_1 and A_2 be the free preframe and frame over the meet semilattice S. We define $\mu: A_2 \to A_2$ by

$$\mu(U) = U \cup \{a \wedge b: b \in S, \exists X. (X, a) \in R, \forall i.\ x_i \wedge b \in U\}$$

Then μ is monotone and inflationary. It's also a prenucleus, for suppose $a \wedge b \in \mu(U) \cap V$ with (X, a) in R and all $x_i \wedge b$ in U. Then each $x_i \wedge b \leq a \wedge b \in V$, so $a \wedge b \in \mu(U \cap V)$.

Let A_4 be the set of fixpoints of μ, the corresponding sublocale, let $\theta: A_2 \to A_4$ be the natural frame homomorphism, and let A_3 be the subpreframe of A_4 generated by the image of A_1 under θ (the preframe image of A_1 under θ, in the sense of Section 2). A_3 is the preframe we are trying to present. $A_1 \to A_3$ is a preframe epimorphism, and this will prove uniqueness in the universal property.

For existence, suppose B_1 is a preframe, and $f: S \to B_1$ is a meet semilattice homomorphism such that if $(X, a) \in R$, then $f(a) \leq \bigvee^{\uparrow} (f \circ X)$. f factors via a preframe homomorphism $g_1: A_1 \to B_1$. Let B_2 be the free frame over the preframe B_1; note that the concrete construction of Proposition 3.1 tells us that the injection $B_1 \to B_2$ is 1-1. g_1 lifts to a frame homomorphism $g_2: A_2 \to B_2$, and $g_2 \circ \mu = g_2$, so that g_2 factors as $g_2 = g_4 \circ \theta$, where g_4 is a frame homomorphism. The inverse image of B_1 under g_4 is a subpreframe of B_4 containing the image of A_1, and hence containing A_3; so g_4 restricts to a preframe homomorphism $g_3: A_3 \to B_1$.]

Theorem 3.3 PreFrm has coequalizers.
Proof Let f, g: A → B be two preframe homomorphisms. Let h: B → S be the meet semilattice coequalizer, and let R be generated by the set $\{(h(X), h(a)): X \subseteq B, X \text{ directed}, \bigvee^{\uparrow} X = a\}$. Then apply the previous Proposition.]

It follows that one can present preframes by generators and relations.

4. Frames as monoids in **PreFrm**

We have seen that the theory of preframes is commutative; so, thanks to the presentation theorem of the last section, we can now assert that the category **PreFrm** has a symmetric monoidal structure, left adjoint to its closed structure.

To construct the preframe tensor product A⊗B of A and B, we first construct their meet-semilattice tensor product (P, say), and then equip it with the coverage R generated by all pairs (X⊗b, a⊗b) and (a⊗Y, a⊗b) where X and Y are monotone nets in A and B, with joins a and b, X⊗b denotes the monotone net (x⊗b | x ∈ X) and a⊗b denotes (for the moment) the image of (a, b) under the universal bihomomorphism from A×B to P. (Actually, as in the proof that the theory of preframes is commutative, we could simplify the presentation by considering only covers of the form $((x_i \otimes y_i \mid i \in I), a \otimes b)$ where $(x_i \mid i \in I)$ and $(y_i \mid i \in I)$ are monotone nets in A and B with the same index set I and with joins a and b. But we shall not have any use for this simplification.) Then the preframe A⊗B is that presented as ⟨P | R⟩. For the rest of this section, a⊗b will denote the image of (a, b) under the universal preframe bihomomorphism from A×B to A⊗B.

The unit of the monoidal structure is the free preframe on one generator; this is simply the two-element frame **2** = {0, 1}, the generator being the bottom element 0. The fact that it is a frame and not just a preframe is no accident, as we shall see in a moment.

First, we need a general result on symmetric monoidal categories. This is surely well-known, although we have not been able to find an explicit reference to it.

Lemma 4.1 Let *C* be a symmetric monoidal category. Then the category **CMon**(*C*) of commutative monoids (with respect to the tensor product ⊗) in *C* has finite coproducts, which are given by ⊗ and the unit I.

Proof

First, we observe that **CMon**(*C*) is closed under tensor products, in an obvious sense: if M and N are commutative monoids in *C* (with multiplications * and units e), then M⊗N has multiplication and unit

$$(M\otimes N)\otimes(M\otimes N) \xrightarrow{\cong} (M\otimes M)\otimes(M\otimes M) \xrightarrow{*\otimes *} M\otimes N$$
$$I \xrightarrow{\cong} I\otimes I \xrightarrow{e\otimes e} M\otimes N$$

Similarly, I carries a unique commutative monoid structure.

Moreover, there are canonical maps

$$M \xrightarrow{\cong} M\otimes I \xrightarrow{Id\otimes e} M\otimes N$$
$$N \xrightarrow{\cong} I\otimes N \xrightarrow{e\otimes Id} M\otimes N$$

and a codiagonal map from M⊗M to M given by * itself; it is easy to verify that these are monoid homomorphisms, and that they define the unit and counit of an adjuction between ⊗ (as a bifunctor from **CMon**(*C*) × **CMon**(*C*) to **CMon**(*C*)) and the diagonal functor.

Hence M⊗N is the coproduct of M and N in **CMon**(C); and a similar argument establishes that I is initial in this category.]

The best known application of this Lemma occurs when C is the category of Abelian groups. It then becomes the assertion that the coproduct of two commutative rings is (founded on) the tensor product of their additive groups. Interpreted when C is the category of sets (with its cartesian monoidal structure), it becomes the assertion that finite products and coproducts coincide in the category of commutative monoids (or in any full subcategory thereof that is closed under products, such as abelian groups or semilattices).

Another application of Lemma 4.1 was exploited by Joyal and Tierney [84]. There they considered frames as commutative monoids in the monoidal category **SupLat** of complete join semilattices (their *suplattices*), the monoid structure being given by finite meets. Of course, the general commutative monoid is a commutative quantale, not necessarily a frame; but, as Joyal and Tierney observed, frames may be characterized as those commutative quantales for which the multiplication is idempotent and the unit is the top element. These conditions cannot be expressed by commutative diagrams in **SupLat** – since the monoidal structure is not Cartesian, there is no diagonal map from A to A⊗A, which would be needed to express the idea of idempotency. However, they do suffice to recover the result (already known, of course, before the work of Joyal and Tierney – cf. Wigner [79]) that the coproduct of two frames coincides with their tensor product in **SupLat**.

Exactly the same arguments apply when we consider frames in relation to preframes. In fact, we have –

Lemma 4.2

(i) Let A be a frame. Then the binary join map $\vee$: $A\times A \to A$ is a preframe bihomomorphism, and the induced map from A⊗A to A gives A the structure of a commutative monoid in **PreFrm**.

(ii) A commutative monoid (A, *, e) in **PreFrm** is (derived in this way from) a frame iff $e(0) \leq a$ and $*(a\otimes a) = a$ for all $a \in A$.

(iii) The category **Frm**, regarded via (i) as a full subcategory of **CMon(PreFrm)**, is closed under finite coproducts.

Proof

(i) That $\vee$ is a bihomomorphism for finite meets is just the (finite) distributive law; that it is a bihomomorphism for directed joins is a consequence of the fact that joins are idempotent and commute with other joins. The rest follows from the fact that join is associative and commutative, and has the bottom element of A as a unit.

(ii) This is easy, and is exactly like Proposition 1 on p. 21 of Joyal and Tierney [84].

(iii) Let A and B be frames. To verify that A⊗B, equipped with its commutative monoid structure, is a frame, we note first that its unit is $0_A\otimes 0_B$, which is clearly its least element (let us write it as 0). Thus we have $u*u \geq u*0 = u$ for all $u \in A\otimes B$, and it suffices to verify that $u*u \leq u$. But the set

$\{u \in A\otimes B \mid u^*u \leq u\}$

is a sub-preframe of $A\otimes B$ that contains all the generators $a\otimes b$.]

Theorem 4.3 The (underlying preframe of the) coproduct of two frames is the tensor product of their underlying preframes.]

We note in particular that if A and B are frames, then their tensor product in **PreFrm** is order isomorphic to their tensor product in **SupLat**, because both are isomorphic to the frame coproduct. However, the generators of the two tensor products are different. If we think of A and B as the open-set lattices of locales X and Y, and of a and b as corresponding to the open sublocales U and V, then the **SupLat** generator $a\otimes b$ is the "open rectangle" $U\times V \subseteq X\times Y$. However, the **PreFrm** generators correspond to the complements of closed rectangles: if we (temporarily) write $a \parr b$ for the **PreFrm** generator corresponding to (a,b), to distinguish it from the **SupLat** generator $a\otimes b$, then we have

$$a \parr b = (a\otimes 1_B) \vee (1_A\otimes b)$$

In other words, $a \parr b$ corresponds to the complement of the closed rectangle $(X\backslash U) \times (Y\backslash V)$. Such complements of closed rectangles have been seen before, notably in the description of the Vietoris locale (Johnstone [85]). (They also make a brief appearance in the construction of the weak product of locales in Johnstone and Sun [88], although this seems to be largely coincidental.)

A further useful result, again analogous to one observed by Joyal and Tierney, is

Theorem 4.4 The forgetful functor from **Frm** to **PreFrm** creates filtered colimits.

Proof

Consider a diagram of frames (A_i), indexed by some filtered category I, with colimit A in **PreFrm**. Since the monoidal structure on **PreFrm** is closed, $A\otimes(-)$ preserves colimits, whence we deduce that $A\otimes A$ is the $I\times I$-indexed colimit of the $A_i\otimes A_j$, and hence (by filteredness of I) that it is the I-indexed colimit of the $A_i\otimes A_i$. Thus the $\otimes$-monoid structures on the A_i induce a $\otimes$-monoid structure on A, and arguments similar to those of Lemma 4.2 (ii) ensure that it is a frame structure. The fact that A, with this structure, is the colimit of the A_i in **CMon(PreFrm)** and hence in **Frm** is easily verified.]

As an application of the last two results, we give what is probably the simplest proof yet discovered of the Tychonoff theorem for locales (cf. Ehresmann [57], Papert [67], Dowker and Strauss [76], Johnstone [81], Kříž [85], Banaschewski [88], Vermeulen [90], Coquand [90]). We note, following Banaschewski, that compactness is easily defined for preframes: a preframe A is *compact* iff its top element 1 is inaccessible by directed joins, or, equivalently, iff $\{1\}$ is a Scott open filter in A. But a subset $U \subseteq A$ is a Scott open filter iff its characteristic function is a preframe homomorphism from A to **2**, from which we deduce –

Lemma 4.5

(i) A tensor product of two compact preframes is compact.

(ii) A colimit of a diagram of compact preframes and injective preframe homomorphisms is compact.

Proof

(i) Let A and B be compact preframes. The characteristic functions of $\{1_A\}$ and $\{1_B\}$ induce a preframe homomorphism from $A \otimes B$ to $\mathbf{2} \otimes \mathbf{2}$; but $\mathbf{2} \otimes \mathbf{2} \cong \mathbf{2}$, the isomorphisms being induced by the binary join map from $\mathbf{2} \times \mathbf{2}$ to $\mathbf{2}$. Thus we have a homomorphism h from $A \otimes B$ to $\mathbf{2}$ such that $h(a \otimes b) = 1$ iff either $a = 1$ or $b = 1$, i.e. iff $a \otimes b$ is the top element of $A \otimes B$. The fact that, for an arbitrary $u \in A \otimes B$, we have $h(u) = 1$ iff $u = 1$, now follows from the fact that any such u may be reached from the generators $a \otimes b$ by taking finite meets and directed joins, and h preserves directed joins.

(ii) Let the vertices of the diagram be $(A_i \mid i \in I)$. Since the transition maps $A_i \to A_j$ in the diagram are all injective, the characteristic functions $h_i\colon A_i \to \mathbf{2}$ of the top elements of the A_is form a cone under the diagram, and so induce a homomorphism from its colimit to $\mathbf{2}$. the fact that this homomorphism is the characteristic function of the top element is proved as in (i).]

For the particular case of coproducts, the result of Lemma 4.5 (ii) is in Banaschewski's paper [88], although his proof is different – it involves an explicit construction of coproducts in **PreFrm**.

Theorem 4.6 *(The Localic Tychonoff Theorem)*

A coproduct of compact frames is compact.

Proof

For finite coproducts, this is simply a special case of Lemma 4.5 (i), using Lemma 4.2 (iii). Just as in the case of rings (see, e.g., Bourbaki [70]), we may extend the result to infinite coproducts by regarding an infinite coproduct as a filtered colimit of finite coproducts (the transition maps being injective unless one of the factors in the product is degenerate – in which case the coproduct is degenerate, and so certainly compact), and using Theorem 4.4 and Lemma 4.5 (ii).]

Let us finish this section with an extension of Tychonoff that covers arbitrary Scott open filters of A and B, not just the case (compactness) when $\{1\}$ is Scott open.

Proposition 4.7 (cf. Vickers [89], Lemma 6.4.3.)

Let A and B be frames. Then there is an order isomorphism between Scott open filters of $A \otimes B$ and subsets $U \subseteq A \times B$ satisfying –

- U is upper closed
- if $(a, b) \in U$ for all $a \in S$, where $S \subseteq_{fin} A$, then $(\bigwedge S, b) \in U$
- if $(\bigvee^{\uparrow} X, b) \in U$, where $X \subseteq A$ is directed, then $(a, b) \in U$ for some $a \in X$
- if $(a, b) \in U$ for all $b \in T$, where $T \subseteq_{fin} B$, then $(a, \bigwedge T) \in U$
- if $(a, \bigvee^{\uparrow} Y) \in U$, where $Y \subseteq B$ is directed, then $(a, b) \in U$ for some $b \in Y$

Proof Scott open filters of A⊗B are just preframe homomorphisms from A⊗B to the two-element frame **2**. The subsets U described are just the preframe bihomomorphisms from A×B to **2**.]

Corollary 4.8 Let A and B be frames, and let F and G be Scott open filters in A and B. Then there is a Scott open filter H in A⊗B such that $a⅋b \in H$ iff $a \in F$ or $b \in G$.
Proof Apply Proposition 4.7 to the set $U = \{(a, b): a \in F \text{ or } b \in G\}$.]

This is actually another localic version of Tychonoff's Theorem, for the following reason. Let D and E be locales, and let F and G be Scott open filters in ΩD and ΩE. By the Hofmann-Mislove Theorem ([81] – the restriction to the spatial case is unnecessary; or see Vickers [89], Theorem 8.2.5), these correspond to compact saturated sets C_F and C_G of points of D and E, and H corresponds to a set C_H of points of D×E.

$$
\begin{aligned}
(x, y) \in C_H &\Leftrightarrow \forall u \in H.\ (x, y) \models u \\
&\Leftrightarrow \forall a \in A, b \in B.\ (a \in F \text{ or } b \in G \Rightarrow (x, y) \models a⅋b) \\
&\Leftrightarrow \forall a \in A.\ (a \in F \Rightarrow (x, y) \models a⅋\mathbf{false}) \text{ and } \forall b \in B.\ (b \in G \Rightarrow (x, y) \models \mathbf{false}⅋b) \\
&\Leftrightarrow x \in C_F \text{ and } y \in C_G
\end{aligned}
$$

Hence $C_H = C_F \times C_G$. In other words, the product of compact saturated sets of points of D and E is still compact.

Although apparently more general than the finite case of Theorem 4.6, Corollary 4.8 could alternatively have been deduced from it using Lemma 3.4 of Johnstone [85] which enables one to reduce to compact sublocales of A and B.

5. The Preframe Version of the Coverage Theorem

In [82], Proposition II.2.11, Johnstone shows how to construct a frame as the set C-Idl(P) of "C-ideals" in a meet semilattice P, where C, a "coverage", is a set of relations of the form "X covers u" where $X \subseteq P$ and $u \in P$. C must also satisfy certain "meet stability" properties. It is also shown that C-Idl(P) has the universal properties of

$$\text{Frm}\langle\, P \text{ (qua meet semilattice)} \mid u \le \bigvee X \qquad \text{(whenever X covers u in C)} \,\rangle$$

("qua meet semilattice" means that all meet semilattice relations holding in P are to hold also in the frame being presented.)

Abramsky and Vickers [90] show how this construction has a specific technical meaning in the contex of sup-lattices. For regardless of whether P is a meet semilattice or C has the meet stability properties, the same definition of C-ideals leads to

$$\text{C-Idl(P)} \cong \text{SupLat}\langle\, P \text{ (qua poset)} \mid u \le \bigvee X \qquad \text{(whenever X covers u in C)} \,\rangle$$

Hence the content of the result can be seen as being that provided P is a meet semilattice and C is meet stable, then

Frm ⟨ P (qua meet semilattice) | $u \le \bigvee X$ (whenever X covers u in C) ⟩
$\cong$ SupLat ⟨ P (qua poset) | $u \le \bigvee X$ (whenever X covers u in C) ⟩

This facilitates the definition of *sup-lattice* homomorphisms out of frames, and is particularly useful in the work of Abramsky and Vickers, where functions are defined between frames and quantales. In practice, it is easy to work any frame presentation into the required form by putting in all finite meets of generators and any extra relations needed to give meet stability.

This understanding is related to a well-known result from the theory of rings. If R is a ring and I is a subgroup (of R as an additive group), then, *provided that I is an ideal,* we have

Ring ⟨ R (qua ring) | r = 0 (whenever $r \in I$) ⟩
$\cong$ Abelian Group ⟨ R (qua Abelian Group) | r = 0 (whenever $r \in I$) ⟩

The purpose of this section is to give an analogous result enabling one to present frames as preframes, instead of sup-lattices.

Theorem 5.1 *The Preframe Version of the Coverage Theorem.*
Let P be a poset, and let C be a set of preframe relations of the form

$$\bigwedge S \le \bigvee\nolimits^{\uparrow}_i \bigwedge S_i$$

where the sets S, S_i are all finite subsets of P. Let

$$A_1 = \text{PreFrm} \langle P \text{ (qua poset)} \mid C \rangle$$

(Every preframe presentation can be reduced to this form.)

Suppose in addition that

- P is a join semilattice,
- C is *join stable,* i.e. if $\bigwedge S \le \bigvee^{\uparrow}_i \bigwedge S_i$ is a relation in C, and $x \in P$, then the relation

$$\bigwedge\{x \vee y: y \in S\} \le \bigvee\nolimits^{\uparrow}_i \bigwedge\{x \vee y: y \in S_i\}$$

  is also in C.

Then A_1 is isomorphic to A_2 = Frm ⟨ P (qua $\vee$-semilattice) | C ⟩, the generators corresponding under the isomorphism in the obvious way.

Proof

First, we show that A_1 is a frame.

$0 \in P$ is bottom in A_1, for $\{a \in A_1: a \ge 0\}$ is a subpreframe containing the generators. (Hence there is a preframe homomorphism from A_1 to this set which, when composed with the inclusion, gives the identity map on A_1: so the inclusion is onto.)

Next, if $x \in P$, we can define a preframe endomorphism θ_x of A_1 by $y \mapsto x \vee y$. $x \mapsto \theta_x$ is monotone and respects the relations in C, and so defines a preframe homomorphism ϕ from A_1 to $[A_1, A_1]$. Writing $a \oplus b$ for $\phi(a)(b)$, $\oplus$ is a preframe bihomomorphism and $x \oplus y = (x \vee y)$ for $x, y \in P$. Now,

- $b \le a \oplus b$, i.e. $Id \le \phi(a)$, i.e. $y \le \phi(a)(y)$ for all $y \in P$: for the set of such a is a subpreframe containing the generators.
- $a \oplus b = b \oplus a$. For first, if $x \in P$ then $\{b: x \oplus b = b \oplus x\}$ is a subpreframe containing the generators; then, fixing b, $\{a: a \oplus b = b \oplus a\}$ is a subpreframe containing the generators.
- $a \oplus a = a$, $\{a: a \oplus a = a\}$ being a subpreframe containing the generators. For $1 \oplus 1 \ge 1$; if $a \oplus a = a$ and $b \oplus b = b$, then

  $$(a \wedge b) \oplus (a \wedge b) = a \oplus a \wedge b \oplus b \wedge a \oplus b \wedge b \oplus a = a \wedge b \wedge a \oplus b = a \wedge b$$

  and if S is a directed set of such elements, then

  $$\bigvee^{\uparrow} S \oplus \bigvee^{\uparrow} S = \bigvee^{\uparrow}\{a \oplus b: a, b \in S\} = \bigvee^{\uparrow}\{c \oplus c: c \in S\} = \bigvee^{\uparrow} S$$

We have now shown that $\oplus$ is a binary join in A_1. We might as well write it as $\vee$.

$\vee$ in A_1 distributes over $\wedge$, so A_1 is a distributive lattice, and hence a frame; also, the injection of generators preserves finite joins.

We now know that we can define a frame homomorphism $\alpha: A_2 \to A_1$ and a preframe homomorphism $\beta: A_1 \to A_2$, both mapping generators to generators in the obvious way. α is a preframe homomorphism, so $\beta;\alpha$ is the identity on A_1. It remains to show that β is a frame homomorphism, after which we know that $\alpha;\beta$ is also the identity. Fixing $y \in P$, the set $\{a \in A_1: \beta(a \vee y) = \beta(a) \vee y\}$ is a subpreframe containing the generators; and then fixing a, the set $\{b \in A_1: \beta(a \vee b) = \beta(a) \vee \beta(b)\}$ is a subpreframe containing the generators.]

This result could in fact have been used in the proof of Theorems 4.3 and 4.4, though we preferred to put those results in a more general context. We give here another sample application concerning the upper and lower power locales. These, decomposing the Vietoris construction into two parts, were first studied as topologies in Michael [51]. They are also well-known in computer science following the work of Smyth [78, 83]; see, for instance, Vickers [89].

Suppose D is a locale. Following Vickers [89], we write ΩD for the corresponding frame "of opens", and if $f: D \to E$ is a continuous map between locales, we write $\Omega f: \Omega E \to \Omega D$ for its inverse image map.

The *upper power locale* $P_U D$ is defined by $\Omega P_U D = \mathrm{Frm}\langle \Omega D \text{ (qua preframe)} \rangle$.

The *lower power locale* $P_L D$ is defined by $\Omega P_L D = \mathrm{Frm}\langle \Omega D \text{ (qua sup-lattice)} \rangle$.

If $a \in \Omega D$, then we write $\Box a$ and $\Diamond a$ for the corresponding generators of $\Omega P_U D$ and $\Omega P_L D$; so $\Box$ preserves finite meets and directed joins, while $\Diamond$ preserves all joins.

Proposition 5.2 *The upper and lower power locale functors commute.*

Let D be a locale. Then $P_U P_L D \cong P_L P_U D$.

Proof We define mutually inverse frame homomorphisms

$$\Omega\theta: \Omega P_L P_U D \to \Omega P_U P_L D \qquad \Omega\phi: \Omega P_U P_L D \to \Omega P_L P_U D$$

for which $\Omega\theta(\Diamond \Box a) = \Box \Diamond a$ and $\Omega\phi(\Box \Diamond a) = \Diamond \Box a$.

$\Omega\theta$: This must be equivalent to a sup-lattice homomorphism from $\Omega P_U D$ to $\Omega P_U P_L D$ taking $\Box a$ to $\Box \Diamond a$. Johnstone's coverage theorem for frames, i.e. the "sup-lattice version", tells us that

$$\Omega P_U D \cong \text{Frm}\langle \Box a \ (a \in \Omega D \text{ qua } \wedge\text{-semilattice}) \mid \Box \text{ preserves } \bigvee^{\uparrow} \rangle$$
$$\cong \text{SupLat}\langle \Box a \ (a \in \Omega D \text{ qua poset}) \mid \Box \text{ preserves } \bigvee^{\uparrow} \rangle$$

and $\Box a \mapsto \Box \Diamond a$ does indeed preserve directed joins.

$\Omega\phi$: This must be equivalent to a preframe homomorphism from $\Omega P_L D$ to $\Omega P_L P_U D$. The preframe version of the coverage theorem tells us that

$$\Omega P_L D \cong \text{Frm}\langle \Diamond a \ (a \in \Omega D \text{ qua } \vee\text{-semilattice}) \mid \Diamond \text{ preserves } \bigvee^{\uparrow} \rangle$$
$$\cong \text{PreFrm}\langle \Diamond a \ (a \in \Omega D \text{ qua poset}) \mid \Diamond \text{ preserves } \bigvee^{\uparrow} \rangle$$

and then, as before, $\Diamond a \mapsto \Diamond \Box a$ preserves directed joins.]

6. Bibliography

S. ABRAMSKY and S.J. VICKERS
[90] *Quantales, Observational Logic and Process Semantics,* Department of Computing Report DOC 90/1, Imperial College of Science, Technology and Medicine, London, 1990.

B. BANASCHEWSKI
[88] "Another look at the Localic Tychonoff Theorem", pp. 647-656 in *Commentationes Mathematicae Universitatis Carolinae* **29(4)**, Charles University, Prague, 1988.

N. BOURBAKI
[70] *Éléments de Mathématique: Algèbre,* Hermann, Paris, 1970.

T. COQUAND
[90] "An intuitionistic proof of Tychonoff's theorem", preprint (1990).

M. COSTE
[79] "Localization, spectra and sheaf representation", pp. 212-238 in M.P.Fourman, C.J.Mulvey and D.S.Scott (eds) *Applications of Sheaves,* Springer Lecture Notes in Mathematics **753**, 1979.

C.H. DOWKER and D. STRAUSS
[76] "Sums in the category of frames", pp. 17-32 in Houston J. Math. **3** (1976).

C. EHRESMANN
[57] "Gattungen von lokalen Strukturen", pp. 59-77 in Jber. Deutsch. Math.-Verein. **60** (1957).

P. FREYD
[72] "Aspects of topoi", pp. 1-76 in Bull. Austral. Math. Soc. **7** (1972).

P. FREYD and A. SCEDROV
[90] *Categories, Allegories,* North-Holland 1990.

P. GABRIEL and F. ULMER
[71] "Lokal präsentierbare Kategorien", Springer Lecture Notes in Mathematics **221**, 1971.

G. GIERZ, K.H. HOFMANN, K. KEIMEL, J.D. LAWSON, M. MISLOVE and D.S. SCOTT
[80] *A Compendium of Continuous Lattices*, Springer-Verlag, Berlin,1980.

KARL H. HOFMANN and MICHAEL W. MISLOVE
[81] "Local compactness and continuous lattices", pp. 209-248 in B. Banaschewski and R.-E. Hoffmann (eds) *Continuous Lattices: Proceedings, Bremen 1979*, Lecture Notes in Mathematics **871**, Springer-Verlag, Berlin, 1981.

J.R. ISBELL
[72] "General functorial semantics I", pp. 535-596 in Amer. J. Math **94** (1972).

P.T. JOHNSTONE
[81] "Tychonoff's theorem without the axiom of choice", pp. 21-35 in Fund. Math. **113** (1981).
[82] *Stone Spaces*, Cambridge University Press, Cambridge, 1982.
[85] "Vietoris locales and localic semilattices", pp. 155-180 in R.-E. Hoffmann and K.H. Hofmann (eds) *Continuous Lattices and their Applications*, Marcel Dekker 1985.

P.T. JOHNSTONE and S.H. SUN
[88] "Weak products and Hausdorff locales", pp. 173-193 in F. Borceux (ed.) *Categorical Algebra and its Applications*, Springer Lecture Notes in Mathematics **1348**,1988.

ANDRÉ JOYAL and MYLES TIERNEY
[84] *An Extension of the Galois Theory of Grothendieck*, Memoirs of the American Mathematical Society **309**, 1984.

I. KŘÍŽ
[85] "A constructive proof of the Tychonoff theorem for locales", pp. 619-630 in *Commentationes Mathematicae Universitatis Carolinae* **26**, Charles University, Prague, 1985.

F.E.J. LINTON
[66] "Some aspects of equational categories", pp. 84-94 in S.Eilenberg et al. (eds) *Proceedings of the Conference on Categorical Algebra*, Springer-Verlag1966.
[69] "Coequalizers in categories of algebras", pp. 75-90 in *Seminar on Triples and Categorical Homology Theory*, Springer Lecture Notes in Mathematics **80**,1969.

J.L. MacDONALD and A. STONE
[82] "The tower and regular decomposition", pp. 197-213 in Cahiers Top. Géom. Diff. **23** (1982).

C. McLARTY
[86] "Left exact logic", pp. 63-66 in J. Pure Appl. Alg. **41** (1986).

ERNEST MICHAEL
[51] "Topologies on spaces of subsets", pp. 152-182 in Trans. Amer. Math. Soc. **71** (1951).

S. PAPERT
[64] "An abstract theory of topological spaces", pp. 197-203 in Proc. Camb. Phil. Soc. **60** (1964).

J. SLOMINSKI
[59] "The theory of abstract algebras with infinitary operations", Rozprawy Mat.**18** (1959).

M. SMYTH
[78] "Power domains", in JCSS **16**, 1978.
[83] "Powerdomains and predicate transformers: a topological view", pp. 662-675 in J.Diaz (ed.) *Automata, Languages and Programming,* Lecture Notes in Computer Science **154**, Springer-Verlag, Berlin, 1983.

J.J.C VERMEULEN
[90] "Tychonoff's theorem in a topos", preprint (1990.

STEVEN VICKERS
[89] *Topology via Logic,* Cambridge University Press, Cambridge, 1989.

D. WIGNER
[79] "Two notes on frames", pp. 257-268 in J.Austral.Math.Soc. A**28** (1979).

This paper is in final form and will not be published elsewhere.

Strong Stacks and Classifying Spaces

André Joyal* and Myles Tierney**

* Département de mathématiques, UQAM, C.P. 8888, Succ 'A', Montréal, Canada, H3C 3P8

** Department of Mathematics, Rutgers University, New Brunswick, NJ 08903

Introduction

Let $\mathcal{E}$ be a Grothendieck topos, and denote by Cat($\mathcal{E}$), respectively Gpd($\mathcal{E}$), the category of categories, respectively groupoids in $\mathcal{E}$. In this paper we show there are Quillen homotopy structures on Cat($\mathcal{E}$) and Gpd($\mathcal{E}$), in which the weak equivalences are the (internal) categorical equivalences, the cofibrations are the functors injective on objects, and the fibrations have the right lifting property with respect to the cofibration weak equivalences. The fibrant objects for these structures are called *strong stacks*, since they represent a strengthening of the notion of *stack* introduced by Grothendieck and Giraud [Giraud 1971]. When the topos $\mathcal{E}$ is the category of simplicial sheaves for a Grothendieck topology, strong stacks have an intimate connection with the theory of classifying spaces for sheaves of simplicial groups, or groupoids.

The paper begins in section 1 with a general discussion of torsors and stacks for groupoids in $\mathcal{E}$.

In section 2 we introduce the concept of a *strong stack* in Gpd($\mathcal{E}$), and treat the problem of strong stack completions by establishing the above-mentioned Quillen homotopy structure on Gpd($\mathcal{E}$). We finish the section with a number of examples.

Section 3 is concerned with strong stacks in Cat($\mathcal{E}$), and the development is parallel to that of section 2. Some of the equivalences of Theorem 3 overlap with results of [Bunge 1979], though the treatment given here is quite different.

Section 4 contains some applications of strong stacks to the problem of the existence of classifying spaces for sheaves of simplicial groupoids. For a full account of this topic, the reader should consult [Joyal-Tierney (to appear)].

1. Torsors and Stacks

Let G be a group in a topos $\mathcal{E}$. A (right) G-*torsor* in $\mathcal{E}$ is a non-empty object E (meaning $E \longrightarrow 1$ is surjective) equipped with a free (right) G-action a: $E\times G \longrightarrow E$, which is transitive. (Free and transitive is expressed by requiring that the map $(x, g) \longmapsto (xg, x)$ from $E\times G$ to $E\times E$ be an isomorphism.) A *mapping* f: $E' \longrightarrow E$ of G-torsors is a function f: $E' \longrightarrow E$ compatible with the G-actions. It is always an isomorphism.

Torsors solve the problem of finding all objects T locally isomorphic to a given object S of $\mathcal{E}$. Recall that T is locally isomorphic to S iff there exists a covering K (meaning K is non-empty) and an isomorphism $K\times T \longrightarrow K\times S$ over K. Equivalently, T is locally isomorphic to S iff Iso(S, T) is non-empty. In fact, letting G = Aut(S), if T is locally isomorphic to S, then

E = Iso(S, T) is a right G-torsor, and T can be recovered from E since the evaluation mapping defines an isomorphism $E\otimes_G S \longrightarrow T$. Moreover, the correspondence $T \longmapsto$ Iso(S, T) defines a bijection between the set of isomorphism classes of objects T locally isomorphic to S and the set $H^1(G)$ of isomorphism classes of G-torsors in $\mathcal{E}$. The proofs are clear by localization, but can also be found in [Giraud 1971], which is a general reference for this section.

For a general object X of $\mathcal{E}$, a (right) G-*torsor over* X is a G-torsor in $\mathcal{E}/X$. That is, it is an object $E \longrightarrow X$ over X provided with a free (right) G-action $a: E\times G \longrightarrow E$ over X (free meaning the map $(x, g) \longmapsto (xg, x)$ from $E\times G$ to $E\times E$ is injective), such that $E/G \longrightarrow X$ is an isomorphism. When $\mathcal{E}$ is the category of simplicial sets, a G-torsor over X is a principal G-bundle over X [May 1967]. Let $H^1(X, G)$ be the set of isomorphism classes of G-torsors over X. When S is a fixed object of $\mathcal{E}$, and G = Aut(S), the above correspondence (interpreted in $\mathcal{E}/X$), yields a bijection between $H^1(X, G)$ and the set of isomorphism classes of objects $T \longrightarrow X$ locally isomorphic (in $\mathcal{E}/X$) to the projection $X\times S \longrightarrow X$. (Such an object in simplicial sets is just a fibre bundle with fibre S.)

In what follows, we shall have to consider not only groups, but also groupoids and their torsors. We recall some fundamental definitions. A *groupoid* $\mathbb{G}$ in $\mathcal{E}$ is a reflexive graph

$$G_0 \xrightarrow{u} G_1 \overset{s}{\underset{t}{\rightrightarrows}} G_0$$

in $\mathcal{E}$, i.e. a diagram such that su = tu = id, provided with an associative composition

$$c: G_1\times_{G_0}G_1 \longrightarrow G_1$$

for which the elements of G_0 are units (via u), and each element of G_1 is invertible. (These statements, of course, are interpreted in the internal language of $\mathcal{E}$.) If $\mathbb{G}$ and $\mathbb{H}$ are groupoids, a functor $f: \mathbb{G} \longrightarrow \mathbb{H}$ is a morphism of reflexive graphs, which respects composition. f is a *categorical equivalence* if it is full, faithful, and representative in the sense that each object of $\mathbb{H}$ is isomorphic to the image of an an object of $\mathbb{G}$ under f. f is a *strong categorical equivalence* if there is a functor $g: \mathbb{H} \longrightarrow \mathbb{G}$ such that $gf \simeq id_{\mathbb{G}}$ and $fg \simeq id_{\mathbb{H}}$. (See [Bunge-Paré 1979] for a discussion of various different notions of equivalence for internal categories.)

Let $\mathbb{G}$ be a groupoid in $\mathcal{E}$. A (right) $\mathbb{G}$-*torsor* is a non-empty object E over G_0 equipped with a free (contravariant) action

$$a: E\times_{G_0}G_1 \longrightarrow E$$

which is transitive. A $\mathbb{G}$-torsor $E \longrightarrow X$ over X is a $\mathbb{G}$-torsor in $\mathcal{E}/X$. The set of isomorphism classes of $\mathbb{G}$-torsors over X is denoted by $H^1(X, \mathbb{G})$ (we write just $H^1(\mathbb{G})$ when X is 1). $H^1(X, \mathbb{G})$ is contravariant in X and covariant in $\mathbb{G}$. In fact, if $g: Y \longrightarrow X$ is a map, and $E \longrightarrow X$ is a $\mathbb{G}$-torsor over X, then $g^*(E)$ is a $\mathbb{G}$-torsor over Y, and if $f: \mathbb{G} \longrightarrow \mathbb{H}$ is a functor then $E\otimes_{\mathbb{G}}\mathbb{H}$ is an $\mathbb{H}$-torsor over X. Note that $H^1(X, \mathbb{G})$ is invariant under categorical equivalence of groupoids.

Torsors for a groupoid solve the problem of finding all objects T locally isomorphic to a member of a given family $S \longrightarrow I$, by which we mean there is a cover K and a mapping $k: K \longrightarrow I$ such that $K \times T \simeq k^*(S)$ in $\mathcal{E}/K$. In fact, let $\mathbb{G}$ be the groupoid $\mathrm{Iso}(S, S) \longrightarrow I \times I$ of isomorphisms of the fibres of $S \longrightarrow I$. Recall that $\mathrm{Iso}(S, S) \longrightarrow I \times I$ is the object $\mathrm{Iso}(S \times I, I \times S)$ in $\mathcal{E}/I \times I$. Since pullback preserves the construction of Iso, $\mathbb{G}$ has the property that given two mappings $f,g: J \longrightarrow I$ in $\mathcal{E}$, there is a mapping $h: J \longrightarrow \mathrm{Iso}(S, S)$ such that

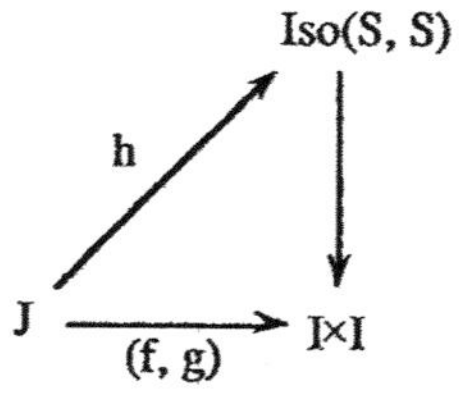

commutes iff $f^*(S) \simeq g^*(S)$ in $\mathcal{E}/J$. Now if T is locally isomorphic to a member of $S \longrightarrow I$ as above, then $E = \mathrm{Iso}(S, T) \longrightarrow I$ is a $\mathbb{G}$-torsor, and evaluation yields an isomorphism $E \otimes_{\mathbb{G}} S \longrightarrow T$. In this way we obtain a bijection between the set of isomorphism classes of objects T locally isomorphic to a member of $S \longrightarrow I$ and $H^1(\mathbb{G})$. When X is an arbitrary object of $\mathcal{E}$, we get a bijection between the set of isomorphism classes of objects T in $\mathcal{E}/X$ locally isomorphic to a member of $S \longrightarrow I$ and $H^1(X, \mathbb{G})$.

Let $\mathbb{G}$ be a groupoid in $\mathcal{E}$. If X is an object of $\mathcal{E}$, we denote by $\hom(X, \mathbb{G})$ the groupoid (in Sets) whose objects are maps $f: X \longrightarrow G_0$. A morphism between f and g is a map $h: X \longrightarrow G_1$ such that

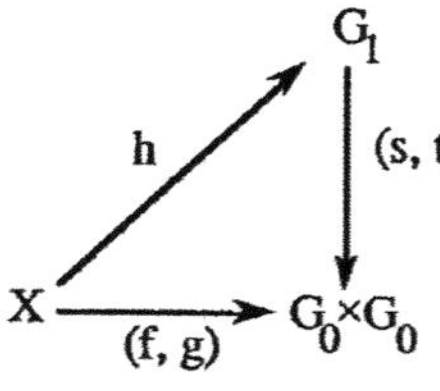

commutes. Let $H(X, \mathbb{G})$ denote the category of $\mathbb{G}$-torsors over X. ($H^1(X, \mathbb{G})$ is the set of connected components of $H(X, \mathbb{G})$.) $t: G_1 \longrightarrow G_0$ is a right $\mathbb{G}$-torsor over G_0, so there is a functor $\hom(X, \mathbb{G}) \longrightarrow H(X, \mathbb{G})$ defined on objects by $f: X \longrightarrow G_0 \longmapsto f^*(G_1)$. It is easy to see that $(s, t): G_1 \longrightarrow G_0 \times G_0$ is canonically isomorphic to $\mathrm{Iso}_{\mathbb{G}}(G_1, G_1) \longrightarrow G_0 \times G_0$, so this function extends in an obvious way to a full and faithful functor . $\mathbb{G}$ is said to be a *stack* if this functor is an equivalence of categories. Clearly, $\mathbb{G}$ is a stack iff the functor is representative. That is, $\mathbb{G}$ is a stack iff for each $\mathbb{G}$-torsor $E \longrightarrow X$ there is a map $f: X \longrightarrow G_0$ such that

$E \simeq f^*(G_1)$. In the example above, $\mathbb{G} = \mathrm{Iso}(S, S) \longrightarrow I \times I$ is a stack iff the family $S \longrightarrow I$ is *complete*, i.e. for any $T \longrightarrow X$ locally isomorphic to a member of $S \longrightarrow I$, there is a map $f: X \longrightarrow I$ such that $T \simeq f^*(S)$.

Theorem 1 The following are equivalent for a groupoid $\mathbb{G}$ in $\mathcal{E}$.
(i) $\mathbb{G}$ is a stack.
(ii) For each X in $\mathcal{E}$, every $\mathbb{G}$-torsor $E \longrightarrow X$ has a section $X \longrightarrow E$.
(iii) Every diagram of groupoids

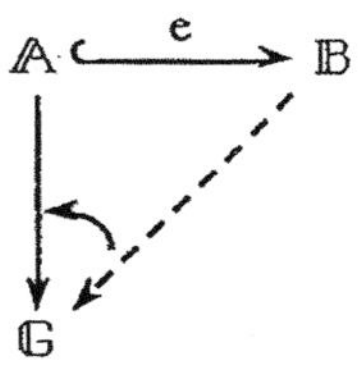

with e a categorical equivalence injective on objects has a dotted filler making the resulting triangle commute up to isomorphism.
(iv) Every diagram of groupoids

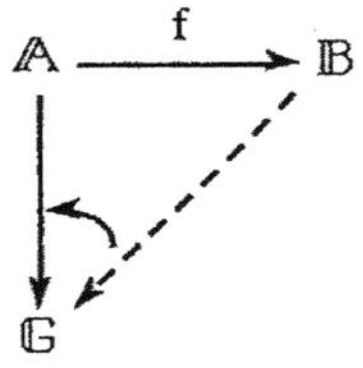

with f a categorical equivalence has a dotted filler making the resulting triangle commute up to isomorphism.

Proof: (i) $\Leftrightarrow$ (ii): $t: G_1 \longrightarrow G_0$ has the section $u: G_0 \longrightarrow G_1$, so if every torsor is a pullback of $t: G_1 \longrightarrow G_0$, every torsor has a section. On the other hand, let $E \longrightarrow X$ be a $\mathbb{G}$-torsor over X with a section $s: X \longrightarrow E$. Then if f denotes the composite of s with the structure map $E \longrightarrow G_0$ of E, $f^*(G_1) \simeq E$.
(ii) $\Rightarrow$ (iii): Categorical equivalences injective on objects are stable under pushout, so to prove (iii) it suffices to show that if (ii) holds, then any categorical equivalence $e: \mathbb{G} \hookrightarrow \mathbb{H}$ injective on objects has a retract up to isomorphism. For this, let

$$\begin{array}{ccc} E & \longrightarrow & H_1 \\ \downarrow & & \downarrow \\ G_0 \times H_0 & \xrightarrow[e\times 1]{} & H_0 \times H_0 \end{array}$$

be a pullback. We claim that $E \longrightarrow H_0$ is a $\mathbb{G}$-torsor over H_0. We argue with elements, which can be justified as in [Joyal-Tierney 1984], and will be clearer for the reader. First,

$$E = \{(g, h, e(g) \longrightarrow h) \mid g \in G_0, h \in H_0, \text{ and } e(g) \longrightarrow h \text{ is an arrow in } H_1\}$$

$E \longrightarrow H_0$ is the function $(g, h, e(g) \longrightarrow h) \longmapsto h$, and is surjective, since e: $\mathbb{G} \hookrightarrow \mathbb{H}$ is representative. The action a: $E \times_{G_0} G_1 \longrightarrow E$ of $\mathbb{G}$ on $E \longrightarrow G_0$ is given by composition:

$$a((g, h, e(g) \longrightarrow h), g' \longrightarrow g) = (g', h, e(g') \longrightarrow e(g) \longrightarrow h)$$

It is free by the faithfulness of e. Finally, if $(g', h, e(g') \longrightarrow h)$ and $(g, h, e(g) \longrightarrow h)$ in E both project to $h \in H_0$, then since e is full and faithful, there is a unique $g' \longrightarrow g$ such that $e(g') \longrightarrow h$ is the composite $e(g') \longrightarrow e(g') \longrightarrow h$. Hence, H_0 is the quotient of E by the action of $\mathbb{G}$. By (ii), $E \longrightarrow H_0$ has a section s: $H_0 \longrightarrow E$. Such an s is a choice, for each $h \in H_0$, of an element $s(h) = (g, h, e(g) \longrightarrow h)$ in E. Let r: $H_0 \longrightarrow G_0$ be the composite $H_0 \longrightarrow E \longrightarrow G_0$. Thus, if $h \in H_0$ and $s(h) = (g, h, e(g) \longrightarrow h)$ then $r(h) = g$. Clearly, there is an isomorphism $re \simeq id_{G_0}$. Finally, suppose $h' \longrightarrow h$ is an arrow of H_1. Since e: $\mathbb{G} \hookrightarrow \mathbb{H}$ is full and faithful, there is a unique arrow $rh' \longrightarrow rh$ in G_1 such that

$$\begin{array}{ccc} h' & \longrightarrow & h \\ {\scriptstyle s(h')}\uparrow & & \uparrow{\scriptstyle s(h)} \\ e(rh') & \longrightarrow & e(rh) \end{array}$$

commutes. Setting $r(h' \longrightarrow h) = rh' \longrightarrow rh$ completes the definition of the retraction r.

(iii) $\Rightarrow$ (iv): Let f: $\mathbb{A} \longrightarrow \mathbb{B}$ be a functor between groupoids in $\mathcal{E}$. Denote by $\mathbb{I}$ the groupoid in $\mathcal{E}$ with two objects 0 and 1, and one isomorphism between them. The *cylinder* on f is defined to be the pushout

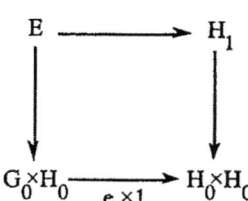

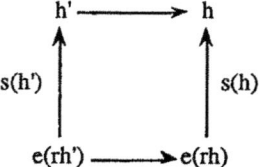

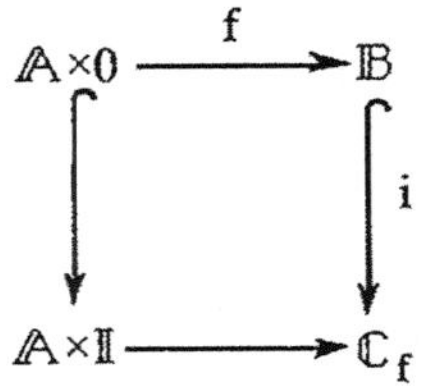

There is a unique functor r: $\mathbb{C}_f \longrightarrow \mathbb{B}$ given by the identity on $\mathbb{B}$, and the composite $f\pi$ on $\mathbb{A}\times\mathbb{I}$, where π: $\mathbb{A}\times\mathbb{I} \longrightarrow \mathbb{A}$ is the projection. Let e: $\mathbb{A} \hookrightarrow \mathbb{C}_f$ be the composite $\mathbb{A}\times 1 \hookrightarrow \mathbb{A}\times\mathbb{I} \longrightarrow \mathbb{C}_f$. Clearly, f = re, and e is injective on objects. r is a strong categorical equivalence since ri = id and ir $\simeq$ id. It follows that e is a categorical equivalence if f is. Thus, a functor $\mathbb{A} \longrightarrow \mathbb{G}$ can be extended up to isomorphism over e: $\mathbb{A} \hookrightarrow \mathbb{C}_f$ iff it can be extended up to isomorphism over f: $\mathbb{A} \longrightarrow \mathbb{B}$.

(iv) $\Rightarrow$ (i): Let $E \longrightarrow X$ be a $\mathbb{G}$-torsor over X, with action a: $E\times_{G_0}G_1 \longrightarrow E$ and structural map $E \longrightarrow G_0$. The pair

$$(a, \pi_1): E\times_{G_0}G_1 \longrightarrow E\times E$$

is a groupoid $\mathbb{E}$ with objects E (the *category of elements* of the action). It is as well the equivalence relation determined by the surjection $E \longrightarrow X$. The diagram

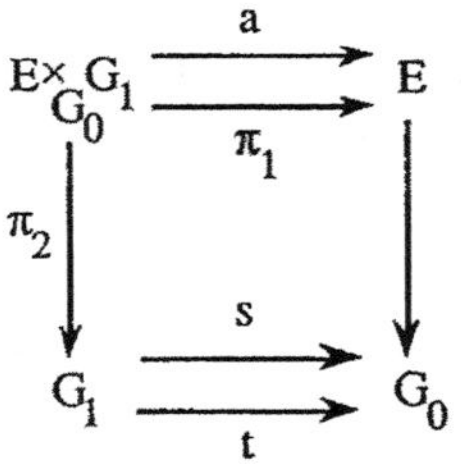

is a functor $\mathbb{E} \longrightarrow \mathbb{G}$. Letting disX denote the discrete groupoid in $\mathcal{E}$ determined by the object X, we obtain a categorical equivalence $\mathbb{E} \longrightarrow$disX, and a diagram

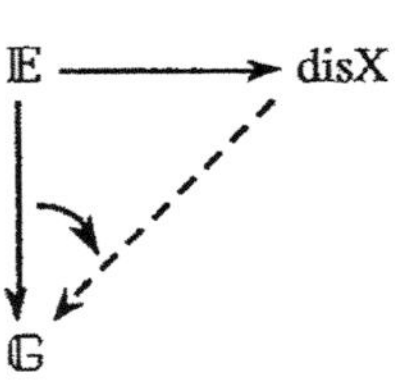

which has a dotted filler up to isomorphism by (iv). The filler disX $\longrightarrow \mathbb{G}$ is a map f: $X \longrightarrow G_0$, and the isomorphism, on objects, is a map φ: $E \longrightarrow G_1$ such that the diagrams

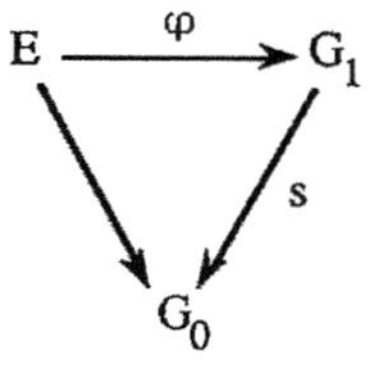

and

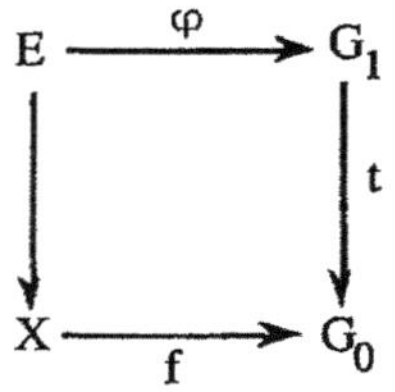

commute. The statement that φ is a natural transformation is equivalent to the statement that φ is compatible with the right $\mathbb{G}$-actions. It follows that the second diagram above is a pullback, i.e. $E \simeq f^*(G_1)$, and $\mathbb{G}$ is a stack. ■

Let $\mathbb{G}$ be a groupoid in $\mathcal{E}$. A *stack completion* of $\mathbb{G}$ is a categorical equivalence $\mathbb{G} \longrightarrow \mathbb{G}^*$ with $\mathbb{G}^*$ a stack. Stack completions are defined up to strong equivalence of groupoids. We postpone the discussion of their existence until the next section, where a stronger existence result will be proved. Notice, however, that if $\mathbb{G} \longrightarrow \mathbb{G}^*$ is a stack completion of $\mathbb{G}$, then since $H(X, \mathbb{G})$ is equivalent to $H(X, \mathbb{G}^*)$, it follows that $H(X, \mathbb{G})$ is equivalent to $\hom(X, \mathbb{G}^*)$, i.e. $\mathbb{G}^*$ "represents" $H(\ , \mathbb{G})$.

2. Strong Stacks

Definition 1 A groupoid $\mathbb{G}$ in $\mathcal{E}$ is a *strong stack* if condition (iii) of Theorem 1 holds on the nose. That is, if each diagram

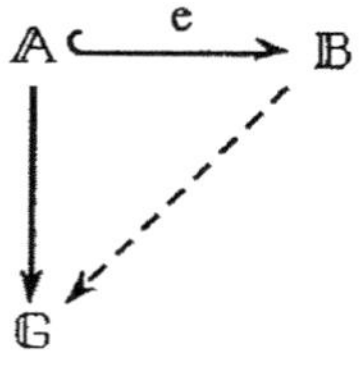

with e a categorical equivalence injective on objects has a dotted filler making the resulting triangle commute.

Definition 2 Let $\mathbb{G}$ be a groupoid in $\mathcal{E}$. A *strong stack completion* of $\mathbb{G}$ is a categorical equivalence injective on objects $\mathbb{G} \hookrightarrow \mathbb{G}^*$ such that $\mathbb{G}^*$ is a strong stack.

When $\mathcal{E}$ is a Grothendieck topos, every groupoid $\mathbb{G}$ in $\mathcal{E}$ has a strong stack completion. In fact, we prove more: strong stacks are the fibrant objects for a Quillen homotopy structure [Quillen 1969] on the category Gpd($\mathcal{E}$) of groupoids in $\mathcal{E}$.

Recall that a Quillen homotopy structure on a category $\mathcal{C}$ with finite limits and colimits consists of three classes of maps: weak equivalences, cofibrations and fibrations. These are required to satisfy the following axioms.

Q1: (*Composition*) Given two morphisms f: A$\longrightarrow$B and g: B$\longrightarrow$C in $\mathcal{C}$, if any two of f, g or gf, are weak equivalences, so is the third.

Q2 : (*Retracts*) Weak equivalences, cofibrations and fibrations are closed under retracts. More precisely, if

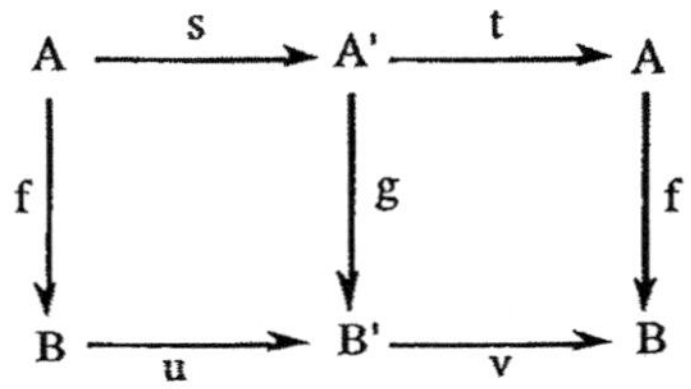

is a commutative diagram in $\mathcal{C}$ such that ts = id_A and vu = id_B, then if g is a weak equivalence, cofibration or fibration, so is f.

Q3 : (*Lifting*) A commutative square

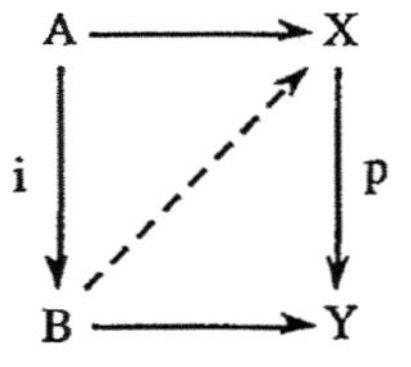

in $\mathcal{C}$ such that i is a cofibration and p is a fibration has a dotted filler making both triangles commute if either i or p is a weak equivalence.

Q4 : (*Factorization*) Any morphism f: X$\longrightarrow$Y in $\mathcal{C}$ can be factored in two ways as

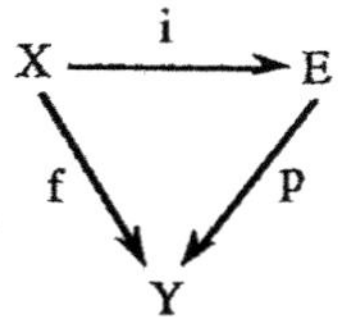

where i is a cofibration and p is a fibration. One in which i is a weak equivalence, and one in which p is a weak equivalence.

Theorem 2 There is a Quillen homotopy structure on Gpd($\mathcal{E}$), in which the weak equivalences are the categorical equivalences, the cofibrations are the functors injective on objects, and the fibrations have the right lifting property with respect to the cofibration weak equivalences.

Before proving the theorem, we prove two lemmas. First, call p: $\mathbb{E} \longrightarrow \mathbb{B}$ in Gpd($\mathcal{E}$) a *trivial fibration* if p has the right lifting property with respect to all cofibrations.

Lemma 1 If p: $\mathbb{E} \longrightarrow \mathbb{B}$ is a trivial fibration, then $E_0 \longrightarrow B_0$ is an injective object in $\mathcal{E}/B_0$, and p is a strong categorical equivalence.

Proof: Let p: $\mathbb{E} \longrightarrow \mathbb{B}$ be a trivial fibration. A diagram

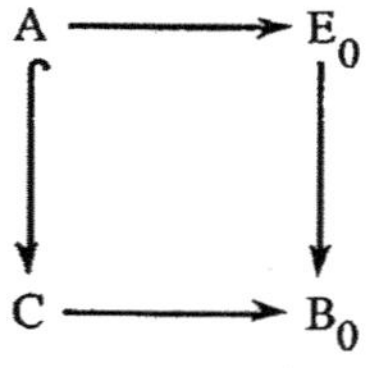

in $\mathcal{E}$ is equivalent to a diagram

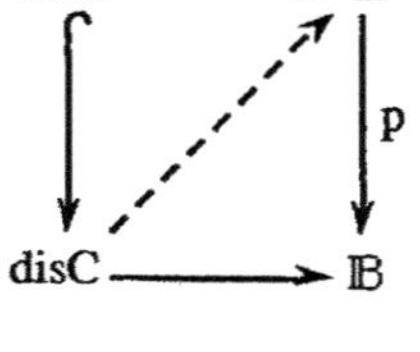

in Gpd($\mathcal{E}$), which has a dotted filler since disA$\hookrightarrow$disC is a cofibration. Thus the first square has a dotted filler, so $E_0 \longrightarrow B_0$ is an injective object in $\mathcal{E}/B_0$. Let $\mathbb{O}$ denote the empty groupoid. $\mathbb{O} \longrightarrow \mathbb{B}$ is a cofibration, so there is a dotted filler for the square

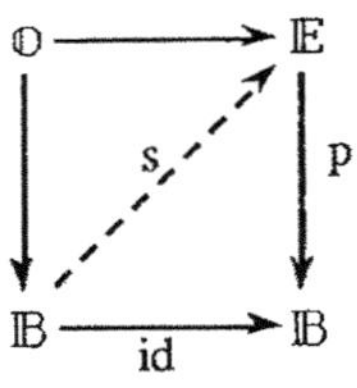

In the commutative square

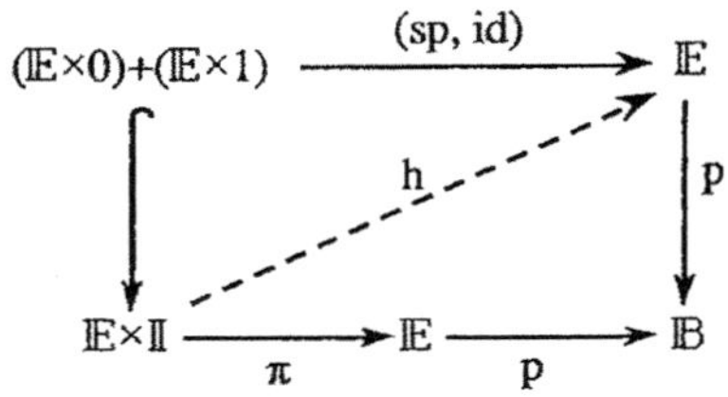

where π is the first projection, the left-hand vertical mapping is a cofibration, so the square has a dotted filler h, which provides an isomorphism $sp \simeq id$. ■

Lemma 2 If p: $\mathbb{E} \longrightarrow \mathbb{B}$ is such that p is full and faithful, and $E_0 \longrightarrow B_0$ is injective in $\mathcal{E}/B_0$, then p is a trivial fibration.

Proof: Let $\mathbb{A} \hookrightarrow \mathbb{C}$ be a cofibration in Gpd($\mathcal{E}$), and suppose

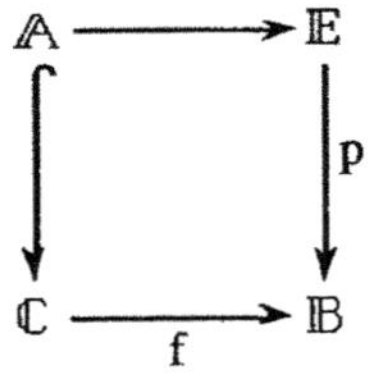

is a commutative square. The square

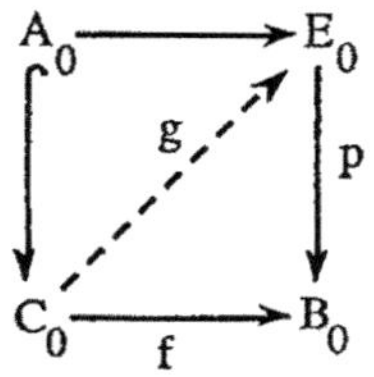

has a dotted filler g since $E_0 \longrightarrow B_0$ is injective. Again, we indicate with elements how to extend g to the arrows of $\mathbb{C}$. Namely, suppose $c' \longrightarrow c$ is an arrow in C_1. $f(c' \longrightarrow c) = f(c') \longrightarrow f(c) = pg(c') \longrightarrow pg(c)$. But p is full and faithful, so there is a unique arrow $g(c') \longrightarrow g(c)$ in E_1 such that $p(g(c') \longrightarrow g(c)) = f(c') \longrightarrow f(c)$. We set $g(c' \longrightarrow c) = g(c') \longrightarrow g(c)$. ∎

Proof of Theorem 2: The finite limits and colimits in Gpd($\mathcal{E}$) are clear, as are Q1 and Q2, whose verification we leave to the reader. Let f: $\mathbb{G} \longrightarrow \mathbb{H}$ be an arbitrary functor in Gpd($\mathcal{E}$). Embed $G_0 \longrightarrow H_0$ into an injective $E_0 \longrightarrow H_0$ over H_0.

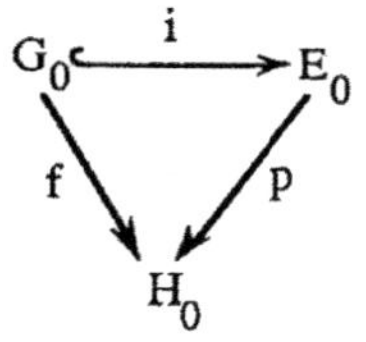

For example, we could take i to be the singleton mapping $f \hookrightarrow \Omega^f$ in $\mathcal{E}/H_0$. Now pull back the arrows of $\mathbb{H}$ over E_0 to obtain a commutative triangle of groupoids

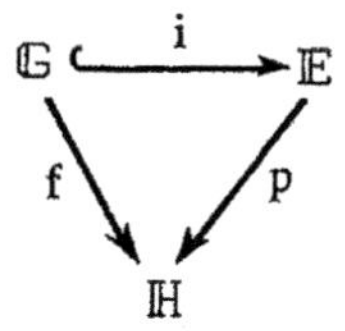

i is a cofibration, and p is a trivial fibration by Lemma 2. By Lemma 1, a trivial fibration is a fibration and a categorical equivalence, so we have established half of Q4. Moreover, suppose f: $\mathbb{G} \longrightarrow \mathbb{H}$ is a fibration and a categorical equivalence. Factor f as above into a cofibration i followed by a trivial fibration p. By Q1, i is a cofibration weak equivalence, so the square

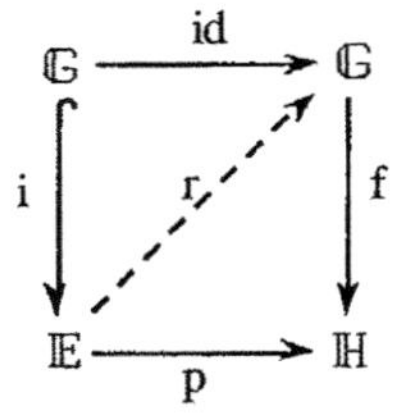

has a dotted filler r, making $f\colon \mathbb{G} \longrightarrow \mathbb{H}$ a retract of $p\colon \mathbb{E} \longrightarrow \mathbb{B}$. Since trivial fibrations are closed under retracts, f is a trivial fibration. But this establishes Q3, for the first part holds by definition, and the second is just the statement that a fibration which is a weak equivalence, is a trivial fibration.

It remains to show that an arbitrary $f\colon \mathbb{G} \longrightarrow \mathbb{H}$ can be factored as a cofibration weak equivalence followed by a fibration. However, cofibration weak equivalences are stable under pushout, and as in [Joyal (to appear)] they have a small set of generators, so we can use "the small object argument" [Quillen 1967] to obtain the desired factorization by repeated pushouts and ordinal colimits. ■

We consider some examples of fibrations, strong stacks and strong stack completions. First, recall that a functor $f\colon \mathbb{E} \longrightarrow \mathbb{B}$ in Gpd($\mathcal{E}$) is called a *discrete fibration* if the square

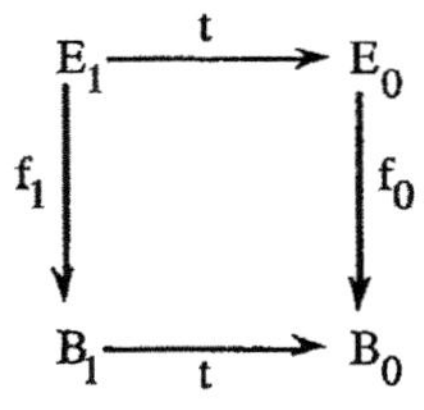

is a pullback. When this is the case, $\mathbb{E}$ is always the category of elements (as defined in the proof of Theorem 1) for a (contravariant) action of $\mathbb{B}$ on $E_0 \longrightarrow B_0$ [Johnstone 1977]. We claim that a discrete fibration is a fibration for the Quillen homotopy structure on Gpd($\mathcal{E}$). In fact, we prove more.

Proposition 1 A discrete fibration $f\colon \mathbb{E} \longrightarrow \mathbb{B}$ has the unique right lifting property with respect to the cofibration weak equivalences in Gpd($\mathcal{E}$).

Proof: Let $f\colon \mathbb{E} \longrightarrow \mathbb{B}$ be a discrete fibration. We want to show that if $\mathbb{A} \hookrightarrow \mathbb{C}$ is a cofibration weak equivalence, then a commutative square

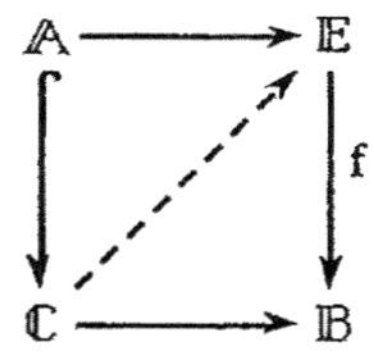

has a unique dotted filler. Since discrete fibrations are stable under pullback, it suffices to show that if $i\colon \mathbb{A} \hookrightarrow \mathbb{B}$ is a categorical equivalence injective on objects, and if $i^*(\mathbb{E})$ has a section s over $\mathbb{A}$, then there is a unique section t of f over $\mathbb{B}$, which extends s. But the category of discrete fibrations over $\mathbb{B}$ coincides with the category $\mathcal{E}^{\mathbb{B}^{op}}$ of (contravariant) $\mathbb{B}$-actions, and $i^*\colon \mathcal{E}^{\mathbb{B}^{op}} \longrightarrow \mathcal{E}^{\mathbb{A}^{op}}$ is an (ordinary, external) equivalence of categories. In particular, it is full and faithful. A section s of $i^*(\mathbb{E})$ is a map $id_{\mathbb{A}} \longrightarrow i^*(\mathbb{E})$ in $\mathcal{E}^{\mathbb{A}^{op}}$. Since $id_{\mathbb{A}} = i^*(id_{\mathbb{B}})$, there is a unique $t\colon id_{\mathbb{B}} \longrightarrow \mathbb{E}$ in $\mathcal{E}^{\mathbb{B}^{op}}$ such that $i^*(t) = s$. ■

Proposition 1 provides a number of examples of strong stacks. In fact, we know that a groupoid $\mathbb{G}$ is a strong stack iff $\mathbb{G}$ is fibrant in the Quillen structure on Gpd($\mathcal{E}$), i.e.iff $\mathbb{G} \longrightarrow \mathbb{1}$ is a fibration, where $\mathbb{1}$ is the terminal groupoid. Fibrations are stable under composition, so if $\mathbb{E} \longrightarrow \mathbb{G}$ is a discrete fibration and $\mathbb{G}$ is a strong stack, it follows that $\mathbb{E}$ is a strong stack. Thus, the category of elements of any $\mathbb{G}$-action is a strong stack. In particular, the category of elements of a $\mathbb{G}$-torsor is a strong stack. Also, for any object X of $\mathcal{E}$ $\text{dis}X \longrightarrow \mathbb{1}$ is a discrete fibration, so disX is a strong stack with unique lifting. Another class of examples can be constructed as follows.

Proposition 2 Let $p\colon E \longrightarrow B$ be a surjective mapping in $\mathcal{E}$, and denote by $\mathbb{E}$ the equivalence relation $E\times_B E \hookrightarrow E\times E$ on E determined by p, considered as a groupoid in $\mathcal{E}$. Then $\mathbb{E}$ is a strong stack iff $p\colon E \longrightarrow B$ is injective in $\mathcal{E}/B$.

Proof: The categorical equivalence $\mathbb{E} \longrightarrow \text{dis}B$ induced by p is a trivial fibration in Gpd($\mathcal{E}$) iff $p\colon E \longrightarrow B$ is injective in $\mathcal{E}/B$ by Lemmas 1 and 2. disB is always a strong stack, so if $p\colon E \longrightarrow B$ is injective in $\mathcal{E}/B$, $\mathbb{E}$ is a strong stack. On the other hand, suppose $\mathbb{E}$ is a strong stack, $\mathbb{A} \hookrightarrow \mathbb{C}$ is a cofibration weak equivalence, and we are given a commutative square

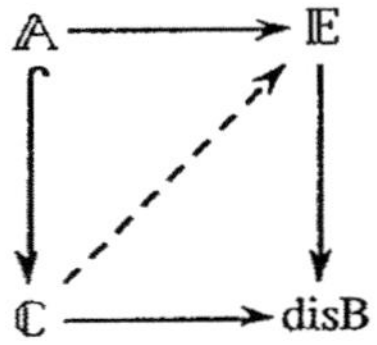

The square has a dotted filler making the upper triangle commute since $\mathbb{E}$ is a strong stack. The two functors $\mathbb{C} \longrightarrow$disB agree on $\mathbb{A}$ so they agree on $\mathbb{C}$ by the unique lifting property of disB. It follows that $\mathbb{E} \longrightarrow$disB is a fibration in Gpd($\mathcal{E}$). But it is also a weak equivalence, so it is a trivial fibration and p: E$\longrightarrow$B is injective in $\mathcal{E}$/B. ■

As an example of Proposition 2, let X be a non-empty object of $\mathcal{E}$, i.e. X$\longrightarrow$1 is surjective. Then the full equivalence relation id: X×X$\longrightarrow$X×X, which we call $\mathbb{X}$ considered as a groupoid of $\mathcal{E}$, is a strong stack iff X is an injective object of $\mathcal{E}$. In general, the strong stack completion of $\mathbb{X}$ is obtained by embedding X into an injective object Y, and taking the groupoid $\mathbb{Y}$. If $\mathbb{G}$ is any groupoid in $\mathcal{E}$, a functor $\mathbb{G} \longrightarrow \mathbb{X}$ is just a mapping $G_0 \longrightarrow$X in $\mathcal{E}$, and any two such are isomorphic. Thus, it follows from, say, condition (iv) of Theorem 1 that if X has an element 1$\longrightarrow$X, then $\mathbb{X}$ is a stack, which, by the above, is not strong in general.

Proposition 3 A groupoid $\mathbb{G}$ in $\mathcal{E}$ is a strong stack iff for any X in $\mathcal{E}$, each $\mathbb{G}$-torsor E$\longrightarrow$X is injective in $\mathcal{E}$/X, i.e. for any monomorphism A$\hookrightarrow$B in $\mathcal{E}$, each commutative square

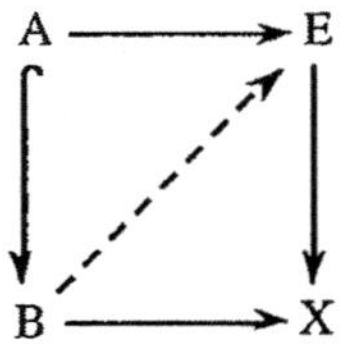

has a dotted filler.

Proof: If $\mathbb{G}$ is a strong stack, and E$\longrightarrow$X is a $\mathbb{G}$-torsor over X, then the category of elements of the $\mathbb{G}$-action on E$\longrightarrow G_0$ is a strong stack by Proposition 1. But the category of elements is also the equivalence relation on E determined by the surjection E$\longrightarrow$X, so E$\longrightarrow$X is injective in $\mathcal{E}$/X by Proposition 2.

In the other direction, suppose each $\mathbb{G}$-torsor E$\longrightarrow$X is injective in $\mathcal{E}$/X. As before, since categorical weak equivalences injective on objects are stable under pushout, to show that

$\mathbb{G}$ is a strong stack, it suffices to show that each categorical weak equivalence injective on objects e: $\mathbb{G} \hookrightarrow \mathbb{H}$ has a retract. As in Theorem 1, let

$$\begin{array}{ccc} E & \longrightarrow & H_1 \\ \downarrow & & \downarrow \\ G_0 \times H_0 & \xrightarrow[e \times 1]{} & H_0 \times H_0 \end{array}$$

be a pullback. We have seen that $E \longrightarrow H_0$ is a $\mathbb{G}$-torsor over H_0. Furthermore, since e is full and faithful, $e^*(E) \simeq G_1$, i.e. we have a pullback diagram

$$\begin{array}{ccc} G_1 & \hookrightarrow & E \\ {\scriptstyle t}\downarrow & & \downarrow \\ G_0 & \xhookrightarrow[e]{} & H_0 \end{array}$$

Now t has the section u: $G_0 \longrightarrow G_1$, and since $E \longrightarrow H_0$ is injective over H_0 it has a section s: $H_0 \longrightarrow E$ which extends u. As a result, the retraction r: $\mathbb{G} \longrightarrow \mathbb{H}$ constructed in Theorem 1 satisfies re = id, and we are done. ■

To finish this section, we give an example of a strong stack completion. Namely, let A be an abelian group in $\mathcal{E}$ considered as a groupoid in $\mathcal{E}$ with one object. Embed A in an injective abelian group I and denote the quotient I/A by B. Write p: $I \longrightarrow B$ for the quotient mapping. Let s,t: $B \times I \longrightarrow B$ denote, respectively, the projection, and the map $(x, y) \longmapsto x+p(y)$.

Proposition 4 The groupoid (s, t): $B \times I \longrightarrow B \times B$, with composition given by addition, is a strong stack completion of A.

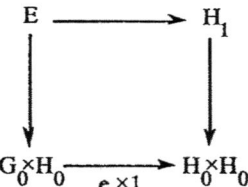

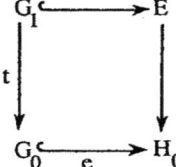

Proof: There is a functor from A to (s, t): $B \times I \longrightarrow B \times B$ given by sending the single object of A to the element 0 in B, and an arrow a in A to the arrow (0, a) in $B \times I$. This is clearly a categorical equivalence injective on objects, so it remains to show that (s, t): $B \times I \longrightarrow B \times B$ is a strong stack. To do this, recall that, for any abelian group, say C, in $\mathcal{E}$, there is an isomorphism $\mathrm{Ext}^1(\mathbb{Z}X, C) \simeq H^1(X, C)$, where $\mathbb{Z}X$ is the free abelian group on the object X of $\mathcal{E}$. In fact, the correspondence is given as follows: if $0 \longrightarrow C \longrightarrow E \longrightarrow \mathbb{Z}X \longrightarrow 0$ is an extension of $\mathbb{Z}X$ by C, then the pullback $T \longrightarrow X$ of $E \longrightarrow \mathbb{Z}X$ along the inclusion $X \hookrightarrow \mathbb{Z}X$ of the generators is an C-torsor over X. We want to show first that the injective abelian group I is a strong stack considered as a groupoid with one object. (It is a stack by Theorem 1 since $\mathrm{Ext}^1(\mathbb{Z}X, I) = 0$.) We show that I-torsors satisfy the condition of Proposition 3. So, let $T \longrightarrow X$ be an I-torsor over X, which has a splitting s over Y, where $Y \hookrightarrow X$. Let $0 \longrightarrow I \longrightarrow E \longrightarrow \mathbb{Z}X \longrightarrow 0$, be the extension corresponding to $T \longrightarrow X$. The extension corresponding to the restriction of T to Y is the pullback $E' \longrightarrow \mathbb{Z}Y$ of $E \longrightarrow \mathbb{Z}X$ along $\mathbb{Z}Y \hookrightarrow \mathbb{Z}X$. In the diagram

$$\begin{array}{ccccccccc} 0 & \longrightarrow & I & \underset{\overset{s}{\dashleftarrow}}{\longrightarrow} & E' & \longrightarrow & \mathbb{Z}Y & \longrightarrow & 0 \\ & & {\scriptstyle id}\downarrow & & \downarrow & & \downarrow & & \\ 0 & \longrightarrow & I & \longrightarrow & E & \longrightarrow & \mathbb{Z}X & \longrightarrow & 0 \end{array}$$

the splitting $s: E' \longrightarrow I$ can be extended to a splitting $t: E \longrightarrow I$ since I is injective as an abelian group. It follows that $T \longrightarrow X$ is an injective object in $\mathcal{E}/X$, so I is a strong stack by Proposition 3. Now, notice that I acts on B via $p: I \longrightarrow B$. In fact, the action is just the map $t: B \times I \longrightarrow B$, i.e. $(x, y) \longmapsto x+p(y)$. The category of elements of the action is the groupoid $(s, t): B \times I \longrightarrow B \times B$, which is a strong stack by Proposition 1. ■

3. Strong Stacks in Categories

Until this point, we have only considered stacks and strong stacks for groupoids, since this is where all our applications lie. The theory has a perfectly good extension to arbitrary categories, however, which we now proceed to sketch. We begin with the classical definition of stack, rather than than the one adopted in section 1, so that the reader can more easily compare our treatment with that of [Giraud 1971] or [Bunge 1979].

Thus, let $\mathcal{E}$ be a Grothendieck topos, and suppose $p: \mathcal{F} \longrightarrow \mathcal{E}$ is a categorical fibration over $\mathcal{E}$. Then $\mathcal{F}$ is a *stack* (for the canonical topology on $\mathcal{E}$) if

(1) For each set I, and each I-indexed family $\{ X_i \mid i \in I \}$ of objects of $\mathcal{E}$, the canonical functor

$$\mathcal{F}(\sum_{i \in I} X_i) \rightarrow \prod_{i \in I} \mathcal{F}(X_i)$$

is an equivalence of categories, and

(2) For each surjection $q: X \twoheadrightarrow Y$ of $\mathcal{E}$, the canonical functor $\mathcal{F}(Y) \longrightarrow \mathrm{des}\mathcal{F}(X)$ is an equivalence of categories, where $\mathrm{des}\mathcal{F}(X)$ is the category of objects of $\mathcal{F}(X)$ provided with descent data relative to the kernel pair of q.

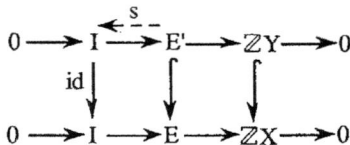

A category $\mathbb{C}$ in $\mathcal{E}$ is called a *stack* if its externalization $\mathcal{F}(X) = \hom(X, \mathbb{C})$ is a stack. Note that condition (1) is automatic in this case. To see more explicitly what condition (2) means, let $q\colon X \twoheadrightarrow Y$ be a surjection of $\mathcal{E}$, and denote by $\mathbb{X}$ the equivalence relation $X \times_Y X \hookrightarrow X \times X$ considered as a category in $\mathcal{E}$. q determines a categorical equivalence $\mathbb{X} \longrightarrow \mathrm{dis}Y$, and descent data on X with respect to q and $\mathbb{C}$ is simply a functor $\mathbb{X} \longrightarrow \mathbb{C}$. Thus, $\mathbb{C}$ is a stack in the above sense iff for any q, each diagram

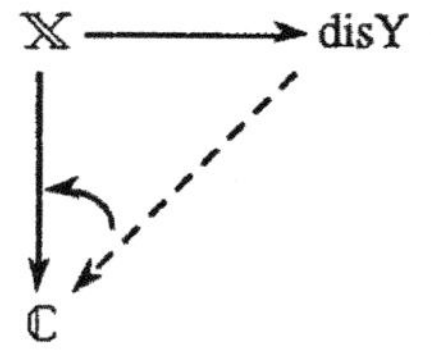

has a dotted filler making the resulting triangle commute up to isomorphism.

If $\mathbb{C}$ is a category in $\mathcal{E}$, let Iso($\mathbb{C}$) denote the groupoid in $\mathcal{E}$ whose objects are those of $\mathbb{C}$ and whose morphisms are the isomorphisms of $\mathbb{C}$. Let Cat($\mathcal{E}$) denote the category of categories in $\mathcal{E}$. For any surjection $q\colon X \twoheadrightarrow Y$, both $\mathbb{X}$ and disY are groupoids, so the following proposition is immediate.

Proposition 5 $\mathbb{C}$ is a stack in Cat($\mathcal{E}$), iff Iso($\mathbb{C}$) is a stack (in the present sense) in Gpd($\mathcal{E}$).■

For any category $\mathbb{A}$ in $\mathcal{E}$, let $T(\mathbb{A}) \hookrightarrow A_0$ be defined by

$$T(\mathbb{A}) = \{t \in A_0 \mid \forall a \in A_0\, \exists! a \longrightarrow t \text{ in } A_1\}$$

$T(\mathbb{A})$ is the *collection of terminal objects of* $\mathbb{A}$.

Passing from groupoids to categories in $\mathcal{E}$, torsors are replaced by locally representable functors. In fact, let $\mathbb{C}$ be a category in $\mathcal{E}$, and F an internal, $\mathcal{E}$-valued, contravariant functor on $\mathbb{C}$, with with structural map $F \longrightarrow C_0$, and action $a\colon F \times_{C_0} C_1 \longrightarrow F$. As before, the pair

$$(a, \pi_1)\colon F \times_{C_0} C_1 \longrightarrow F \times F$$

is a category $\mathbb{F}$ in $\mathcal{E}$ with objects F - the *category of elements* of the action. F is said to be *locally representable* if $T(\mathbb{F})$ is non-empty, i.e. $T(\mathbb{F}) \longrightarrow 1$ is a surjection. Note that when $\mathbb{C}$ is a groupoid and the action is free and transitive, then $T(\mathbb{F}) = F$. Thus, if F is a torsor it is locally representable. On the other hand, again when $\mathbb{C}$ is a groupoid, if F is locally representable then the action is free and transitive, so F is a torsor. For a general $\mathbb{C}$, F is locally representable iff $T(\mathbb{F})$ is an Iso($\mathbb{C}$)-torsor.

An action of $\mathbb{C}$ in Cat($\mathcal{E}$) on a family $F \longrightarrow X$ indexed by X is an action of $X \times \mathbb{C}$, considered as a category in $\mathcal{E}/X$, on $F \longrightarrow X$. Equivalently, this is an action of $\mathbb{C}$ on F, for which the map $F \longrightarrow X$ is constant. Given such an action, we say $F \longrightarrow X$ is a *locally representable family indexed by* X if it is locally representable in $\mathcal{E}/X$. That is, we regard the collection $T(\mathbb{F})_X$ of terminal objects of $\mathbb{F}$ computed fibrewise over X, and we ask that $T(\mathbb{F})_X \longrightarrow X$ be a surjection.
As above, this occurs iff $T(\mathbb{F})_X \longrightarrow X$ is an Iso($\mathbb{C}$)-torsor over X. When $T(\mathbb{F})_X \longrightarrow X$ is a surjection, it follows that $X \simeq \pi_0(\mathbb{F})$, so that F is a locally representable family iff there exists a terminal object (in the internal sense) in each component of the category of elements of F. F is said to be an X-*indexed family of representable functors* if $T(\mathbb{F})_X \longrightarrow X$ splits.

When $\mathbb{C}$ is a category in $\mathcal{E}$, and X is an object of $\mathcal{E}$, the category hom(X, $\mathbb{C}$) is defined as in section 1. Let H(X, $\mathbb{C}$) denote the category of locally representable families indexed by X. $t: C_1 \longrightarrow C_0$ is a representable family indexed by C_0, so $(f: X \longrightarrow C_0) \longmapsto f^*(C_1)$ induces a full and faithful functor hom(X, $\mathbb{C}$)$\longrightarrow$H(X, $\mathbb{C}$) as before.

Theorem 3 The following are equivalent for a category $\mathbb{C}$ in $\mathcal{E}$.
(i) $\mathbb{C}$ is a stack.
(ii) Every locally representable family of functors is representable.
(iii) For each X in $\mathcal{E}$, hom(X, $\mathbb{C}$)$\longrightarrow$H(X, $\mathbb{C}$) is an equivalence of categories.
(iv) Every diagram of categories

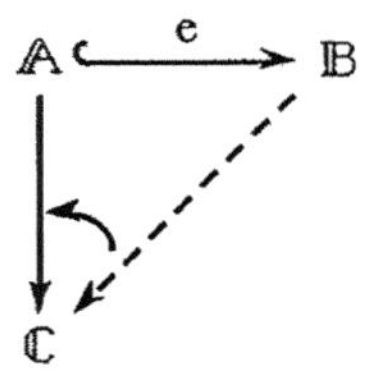

with e a categorical equivalence injective on objects has a dotted filler making the resulting triangle commute up to isomorphism.
(v) Every diagram of categories

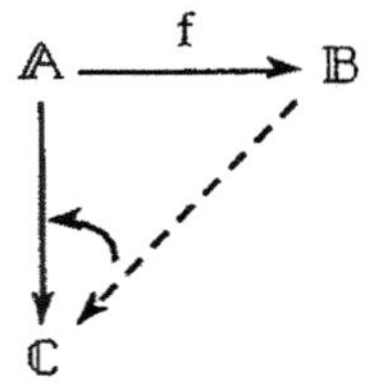

with f a categorical equivalence has a dotted filler making the resulting triangle commute up to isomorphism.

Proof: (i) ⇒ (ii): Let $F \longrightarrow X$ be a locally representable family indexed by X. As above, $T(\mathbb{F})_X \longrightarrow X$ is an Iso($\mathbb{C}$)-torsor over X. If $\mathbb{C}$ is a stack, so is Iso($\mathbb{C}$) - in the sense defined here. But this shows that any Iso($\mathbb{C}$)-torsor splits (see (iv) ⇒ (i) in the proof of theorem 1). Thus, $T(\mathbb{F})_X \longrightarrow X$ splits, and $F \longrightarrow X$ is a representable family.
(ii) ⇔ (iii) is similar to (i) ⇔ (ii) of Theorem 1.
(iii) ⇒ (iv) follows the same pattern as (ii) ⇒ (iii) of Theorem 1, except that the torsor $E \longrightarrow H_0$ therein defined is now a locally representable family. The fact that this family is then representable allows us to define the retraction in the same way.
(iv) ⇒ (v) is the same as (iii) ⇒ (iv) of Theorem 1 - this argument does not depend on the fact that the categories involved are groupoids.
(v) ⇒ (i) is obvious. ■

Let $\mathbb{C}$ be a category in $\mathcal{E}$. A *stack completion* of $\mathbb{C}$ is a categorical equivalence $\mathbb{C} \longrightarrow \mathbb{C}^*$, such that $\mathbb{C}^*$ is a stack. As before, stack completions are defined up to strong equivalence of categories. To obtain the stack completion of a category $\mathbb{C}$, it is enough to have the stack completion of its groupoid of isomorphisms. In fact, suppose $\text{Iso}(\mathbb{C}) \longrightarrow \text{Iso}(\mathbb{C})^*$ is a stack completion of the groupoid Iso($\mathbb{C}$). Taking the pushout

$$\begin{array}{ccc} \text{Iso}(\mathbb{C}) & \longrightarrow & \text{Iso}(\mathbb{C})^* \\ \downarrow & & \downarrow \\ \mathbb{C} & \longrightarrow & \mathbb{C}^* \end{array}$$

provides a stack completion $\mathbb{C} \longrightarrow \mathbb{C}^*$ of $\mathbb{C}$. The proof is immediate by Proposition 5.

Definition 3 A category $\mathbb{C}$ in $\mathcal{E}$ is a *strong stack* if condition (iv) of Theorem 3 holds on the nose. That is, if each diagram of categories

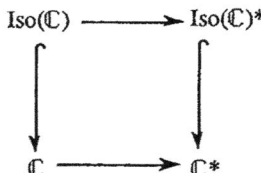

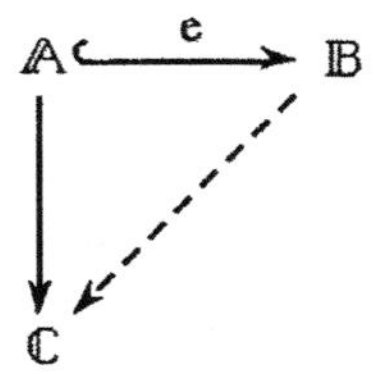

with e a categorical equivalence injective on objects has a dotted filler making the resulting triangle commute.

Proposition 6 $\mathbb{C}$ is a strong stack in Cat($\mathcal{E}$) iff Iso($\mathbb{C}$) is a strong stack in Gpd($\mathcal{E}$).

Proof: Clearly, if $\mathbb{C}$ is a strong stack in Cat($\mathcal{E}$), then Iso($\mathbb{C}$) is a strong stack in Gpd($\mathcal{E}$). The other direction follows immediately from the fact that if e: $\mathbb{A} \hookrightarrow \mathbb{B}$ is a categorical equivalence injective on objects, then

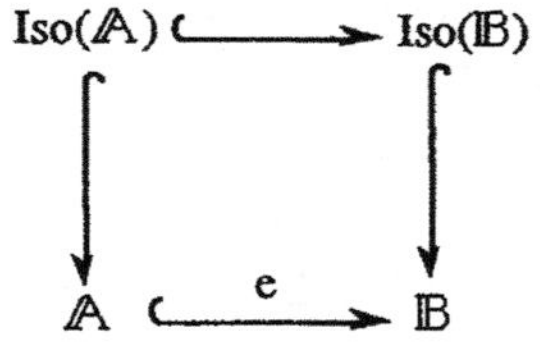

is a pushout of categories. ■

Definition 4 A *strong stack completion* of a category $\mathbb{C}$ in $\mathcal{E}$ is a categorical equivalence $\mathbb{C} \hookrightarrow \mathbb{C}^*$ injective on objects, such that $\mathbb{C}^*$ is a strong stack.

In view of Proposition 6, the strong stack completion of a category $\mathbb{C}$ can be obtained from the strong stack completion of its groupoid of isomorphisms Iso($\mathbb{C}$) as above.

As is the case for groupoids, strong stacks in Cat($\mathcal{E}$) are the fibrant objects for a Quillen homotopy structure on Cat($\mathcal{E}$). Namely, we have

Theorem 4 There is a Quillen homotopy structure on Cat($\mathcal{E}$), in which the weak equivalences are the categorical equivalences, the cofibrations are the functors injective on objects, and the fibrations have the right lifting property with respect to the cofibration weak equivalences.

Proof: There are two ways to prove Theorem 4: either repeat verbatim the proof of Theorem 2, which does not depend on groupoids, or repeat the first part, up to the point of showing that an arbitrary functor $f: \mathbb{C} \longrightarrow \mathbb{D}$ in $Cat(\mathcal{E})$ can be factored as a cofibration weak equivalence followed by a fibration. At this point, factor Iso(f) in $Gpd(\mathcal{E})$ as

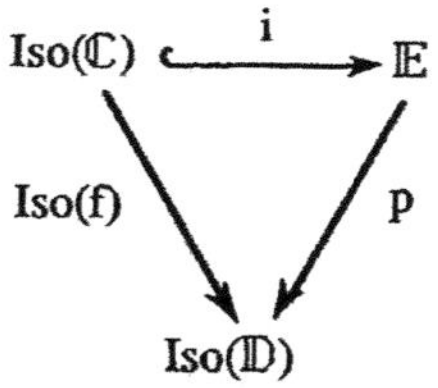

where i is a cofibration weak equivalence in $Gpd(\mathcal{E})$ and p is a fibration. Now take the pushout in $Cat(\mathcal{E})$

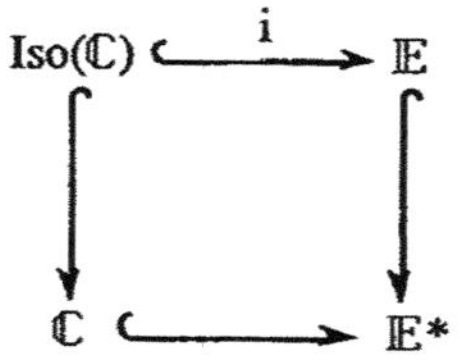

and use the obvious generalization of Proposition 6, which states that f is a fibration in $Cat(\mathcal{E})$ iff Iso(f) is a fibration in $Gpd(\mathcal{E})$. ■

4. Strong Stacks and Classifying Spaces

Let $\mathcal{E}$ be a Grothendieck topos, and let $S(\mathcal{E})$ denote the topos of simplicial objects in $\mathcal{E}$. Recall [Joyal (to appear)] that there is a Quillen homotopy structure on $S(\mathcal{E})$, in which the weak equivalences are maps $f: X \longrightarrow Y$ inducing isomorphisms on the homotopy sheaves, the cofibrations are the monomorphisms, and the fibrations are maps having the right lifting property with respect to the cofibration weak equivalences, which we call *anodyne extensions*. The *homotopy category* of $S(\mathcal{E})$ is obtained by formally inverting the weak equivalences (or just the anodyne extensions) of $S(\mathcal{E})$. If X and Y are objects of $S(\mathcal{E})$, and Y is fibrant, then the set of maps from X to Y in the homotopy category is in 1-1 correspondence with the set [X,Y] of *homotopy classes* of maps from X to Y. (A homotopy is a mapping $X \times I \longrightarrow Y$ with I the constant simplicial sheaf on the 1-simplex $\Delta[1]$ of S.)

We often use the principle of *boolean localization* to transfer results from the homotopy theory of simplicial sets to simplicial sheaves. Namely, for any Grothendieck topos $\mathcal{E}$ there exists a surjective geometric morphism p: $\mathbb{B} \longrightarrow \mathcal{E}$ such that $\mathbb{B}$ is boolean and satisfies the axiom of choice [Barr 1974]. The inverse image functor p* provides a faithful embedding of $\mathcal{E}$ into $\mathbb{B}$, whose logic is classical (i.e. $\mathbb{B}$ is a boolean valued model of ZF set theory). Not all constructions are preserved by p*, but those using only colimits and finite limits are (the so-called *geometric constructions*). For example, p* preserves the construction of the homotopy groups of a simplicial sheaf. Thus, a mapping f: $X \longrightarrow Y$ of $S(\mathcal{E})$ is a weak equivalence iff p*(f) is. As a result, geometric constructions yielding weak equivalences in simplicial sets, yield weak equivalences in $S(\mathcal{E})$.

The *nerve* of a groupoid $\mathbb{G}$ in $\mathcal{E}$ is a simplicial sheaf $N\mathbb{G}$ described as follows. $(N\mathbb{G})_n$ is the object G_n of composable strings of length n, $\sigma = x_n \longrightarrow x_{n-1} \longrightarrow \ldots x_1 \longrightarrow x_0$, of arrows of $\mathbb{G}$. Faces and degeneracies are given by $d^0\sigma = x_n \longrightarrow x_{n-1} \longrightarrow \ldots \longrightarrow x_1$, $d^n\sigma = x_{n-1} \longrightarrow \ldots \longrightarrow x_0$, and $d^i\sigma = x_n \longrightarrow \ldots x_{i+1} \longrightarrow x_{i-1} \ldots \longrightarrow x_0$ for $0 < i < n$, where $x_{i+1} \longrightarrow x_{i-1}$ is the composite of the pair $x_{i+1} \longrightarrow x_i \longrightarrow x_{i-1}$. $s^i\sigma = x_n \longrightarrow \ldots \longrightarrow x_i \longrightarrow x_i \longrightarrow \ldots \longrightarrow x_0$, where $x_i \longrightarrow x_i$ is the identity on x_i. In particular, $(N\mathbb{G})_0 = G_0$, the objects of $\mathbb{G}$, $(N\mathbb{G})_1 = G_1$, the morphisms of $\mathbb{G}$, and $d^0(\alpha: x_1 \longrightarrow x_0) = x_1 = s\alpha$, $d^1(\alpha: x_1 \longrightarrow x_0) = x_0 = t\alpha$, and $s^0x = id: x \longrightarrow x = ux$.

When $\mathbb{G}$ is a groupoid in $S(\mathcal{E})$, $N\mathbb{G}$ is a double simplicial object of $\mathcal{E}$, whose n-th column is the simplicial object G_n. Letting d: $S^2(\mathcal{E}) \longrightarrow S(\mathcal{E})$ denote the *diagonal complex* defined by $d(X)_n = X_{n\,n}$, we write $B\mathbb{G}$ for $d(N\mathbb{G})$. In arguments involving $B\mathbb{G}$, we often use the following, fundamental property of the diagonal complex. Namely, let f: $X \longrightarrow Y$ be a mapping in $S^2(\mathcal{E})$. We call f a *vertical weak equivalence* if $f_{n\,*}: X_{n\,*} \longrightarrow Y_{n\,*}$ is a weak equivalence for each $n \geq 0$, and a *horizontal weak equivalence* if $f_{*\,m}: X_{*\,m} \longrightarrow Y_{*\,m}$ is a weak equivalence for each $m \geq 0$. Then it follows by boolean localization, and the corresponding fact about double simplicial sets, that if f: $X \longrightarrow Y$ is either a vertical or horizontal weak equivalence, df: $dX \longrightarrow dY$ is a weak equivalence.

Proposition 7 Let $\mathbb{G}$ be a groupoid in $S(\mathcal{E})$. If $\mathbb{G} \hookrightarrow \mathbb{G}^*$ is a strong stack completion of $\mathbb{G}$, then G^*_0 is weakly equivalent (in $S(\mathcal{E})$) to $B\mathbb{G}$.

Proof: Proposition 3 says that each $\mathbb{G}^*$ torsor is a trivial fibration (in $S(\mathcal{E})$) over its base. In particular, t: $G^*_1 \longrightarrow G^*_0$ (and therefore also s) is a trivial fibration. Thus, the inclusion u: $G^*_0 \longrightarrow G^*_1$ of the units of $\mathbb{G}^*$ is a weak equivalence. In fact, u defines a functor $disG^*_0 \longrightarrow \mathbb{G}^*$, such that $N(disG^*_0) \longrightarrow N(\mathbb{G}^*)$ is a vertical weak equivalence. Taking the diagonal complex yields a weak equivalence $G^*_0 \cong B(disG^*_0) \longrightarrow B\mathbb{G}^*$. But since $\mathbb{G} \hookrightarrow \mathbb{G}^*$ is a categorical equivalence, $N\mathbb{G} \hookrightarrow N\mathbb{G}^*$ is a horizontal weak equivalence by boolean localization, so $B\mathbb{G} \longrightarrow B\mathbb{G}^*$ is also a weak equivalence, giving the result. ■

Definition 5 A groupoid $\mathbb{G}$ in $S(\mathcal{E})$ is said to be *amenable* if the functor $H^1(, \mathbb{G})$ inverts anodyne extensions, i.e. passes to the homotopy category.

Note that amenability is invariant under categorical equivalence of groupoids.

Proposition 8 If $\mathbb{G}$ is an amenable strong stack in $S(\mathcal{E})$, then G_0 is fibrant in the Quillen structure on $S(\mathcal{E})$.

Proof: Let $\mathbb{G}$ be an amenable strong stack in $S(\mathcal{E})$, $i: A \hookrightarrow B$ an anodyne extension, and $f: A \longrightarrow G_0$ a map. The pullback of $t: G_1 \longrightarrow G_0$ along f is a $\mathbb{G}$-torsor $T \longrightarrow A$ over A. Since $\mathbb{G}$ is amenable, there is a $\mathbb{G}$-torsor $S \longrightarrow B$ whose restriction to A is isomorphic to T. Since $\mathbb{G}$ is a strong stack, there is a map $g: B \longrightarrow G_0$ such that $g^*(G_1) \simeq S$. Thus, we obtain a mapping $h: A \longrightarrow G_1$ such that $sh = gi$ and $th = f$. As above, $s: G_1 \longrightarrow G_0$ is a trivial fibration in $S(\mathcal{E})$, so the commutative square

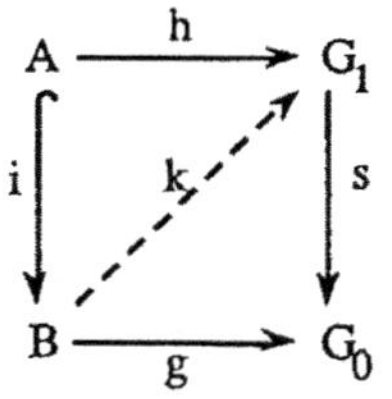

has a dotted filler k. The map tk is the desired extension of f to B. ■

Theorem 5 Let $\mathbb{G}$ be an amenable groupoid in $S(\mathcal{E})$, and $\mathbb{G} \hookrightarrow \mathbb{G}^*$ a strong stack completion of $\mathbb{G}$. Then for any X in $S(\mathcal{E})$, $H^1(X, \mathbb{G}) \cong [X, G^*_0]$. That is, G^*_0 is a classifying space for $\mathbb{G}$-torsors.

Proof: Since $\mathbb{G} \hookrightarrow \mathbb{G}^*$ is a categorical equivalence, $H^1(X, \mathbb{G}) \simeq H^1(X, \mathbb{G}^*)$, $\mathbb{G}^*$ is amenable, and G^*_0 is fibrant by Proposition 8. Since $\mathbb{G}^*$ is a strong stack, $H^1(X, \mathbb{G}^*) \simeq \pi_0(\hom(X,\mathbb{G}^*))$. Thus, we are left with showing that $\pi_0(\hom(X, \mathbb{G}^*)) \simeq [X, G^*_0]$. More precisely, we want to show that for two maps $f,g: X \longrightarrow G^*_0$, the map $(f, g): X \longrightarrow G^*_0 \times G^*_0$ lifts to G^*_1, i.e. the pullback of $t: G^*_1 \longrightarrow G^*_0$ along f is isomorphic to its pullback along g, iff f is homotopic to g. For this, notice that the projection $X \times I \longrightarrow X$ is a weak equivalence. Hence, any $\mathbb{G}^*$-torsor over $X \times I$ is constant in I, by the amenability of $\mathbb{G}^*$. It follows that if f is homotopic to g, then the pullback of $t: G^*_1 \longrightarrow G^*_0$ along f is isomorphic to its pullback along g. In the other direction, consider the diagram

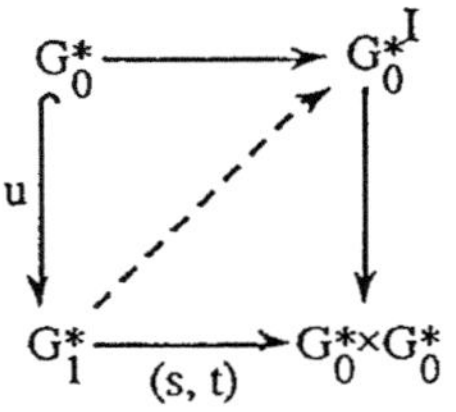

The right-hand vertical map is a fibration since G^*_0 is fibrant. u is a cofibration weak equivalence as above, so the diagram has a dotted filler. As a result, if (f, g): $X \longrightarrow G^*_0 \times G^*_0$ lifts to G^*_1, then f is homotopic to g, and the theorem is proved. ■

We remark that in the category of simplicial sets, a locally transitive groupoid $\mathbb{G}$ is amenable, where locally transitive means that (s,t): $G_1 \longrightarrow G_0 \times G_0$ is a Kan fibration. In fact, for simplicial sets the two concepts are equivalent. Thus, Theorem 5 provides a classifying space for any locally transitive simplicial groupoid, e.g. any simplicial group. See [Joyal-Tierney (to appear)] for a full discussion of amenability, local transitivity and classifying spaces.

References

[1] M. Barr, Toposes without points, J. Pure App. Alg. 5, 1974, pp 265-280.

[2] M. Bunge, Stack completions and Morita equivalence for categories in a topos, Cahiers Top. Geom. Diff. xx-4, 1979, pp 401-435.

[3] M. Bunge and R. Pare, Stacks and equivalence of indexed categories, Cahiers Top. Geom. Diff. xx-4, 1979, pp 373-399.

[4] J. Giraud, *Cohomologie non abélienne*, Grundlehren #179, Springer Verlag, 1971.

[5] P. T. Johnstone, *Topos Theory*, L.M.S. Monographs #10, Academic Press, London, 1977.

[6] A. Joyal, Homotopy theory of simplicial sheaves, to appear.

[7] A. Joyal and M. Tierney, An extension of the galois theory of Grothendieck, Memoirs of the AMS vol. 51 No. 309, 1984

[8] ________, Classifying spaces for sheaves of simplicial groupoids, JPAA, to appear.

[9] J. P. May, *Simplicial Objects in Algebraic Topology*, Van Nostrand Math.Studies #11, 1967.

[10] D. G. Quillen, *Homotopical Algebra*, Springer Lecture Notes in Mathematics #43, 1967.

[11] ________, Rational homotopy theory, Annals of Math.90, 1969, pp205-295.

This paper is in final form and will not be published elsewhere.

Trees in Distributive Categories

S. Kasangian* S. Vigna*
Dipartimento di Matematica, Università di Milano,
Via Saldini 50, I-20133 Milano MI, Italy
email: kasan@imiucca.unimi.it

Dedicated to Max Kelly in the occasion of his 60th birthday.

1 Introduction

This note presents a brief account of one aspect of the talk given at the conference, and its aim is to relate work done by the authors on a labelled version of Bénabou's ***motors*** with ***distributive categories***, viewed as a conceptual framework (as well as a practical tool) for the foundations of Computer Science (see [Law90, Coc90, Wal89, Wal]).

Motors, introduced by Jean Bénabou in his talk at this conference (and also in lectures held in Rome and Milan [Bén90]) have been shown to provide a deep algebraic insight on the inductive nature of *forests* and also of *labelled trees* [KV91]. Hence they represent a highly valuable tool for manifold applications, specially in Computer Science. We would like to mention, for instance, the algebraic characterization of Milner's ***observational equivalence*** shown in [DKV].

The relevance and ubiquity of *distributive categories* in many fields of Mathematics has been stressed by Lawvere and Schanuel (see e.g. [Law90]). They promise to become the conceptual, unifying framework for several aspects of Computer Science, as shown, e.g., in [Wal] and [Wal89]. In particular the Theory of Concurrency seems to benefit from the contrast between the "petit" and the "gros" aspects in distributive categories.

In analogy with the classical formula for the free monoid [Mac71] we give a characterization of the initial X-motor in a distributive category $\mathcal{D}$ (where X is an object of $\mathcal{D}$) as a power series in X with Catalan numbers as coefficients. Further, we extend Bénabou's ***unique decomposition theorem*** [Bén90, KV91] to distributive categories.

We would like to thank Jean Bénabou, who taught us motors and gave us advice and support, and Bob Walters, since the idea behind this work arose in conversations with him, as witnessed (at least) by the flavour of his "apparently illegitimate calculations" [Wal89]. We would like also to thank the referee for his useful suggestions.

*Work partially supported by the italian M.U.R.S.T. and C.N.R., in particular by Progetto Finalizzato Informatica e Calcolo Parallelo, obiettivo LAMBRUSCO. This paper is in final form and will not appear elsewhere.

2 X-Motors

Before introducing the general setting in distributive categories, we want to recall briefly the definition of X-motor as given in terms of set with operators.

Definition 2.1 *Given a set X (to be thought of as an alphabet[1]), an X-motor is a tuple $\langle M, \oplus, 0, f\rangle$, where $\langle M, \oplus, 0\rangle$ is a monoid, and*

$$f : X \times M \longrightarrow M$$

is a map.

The definition above generalizes Bénabou's notion of *motor*, which we recover when $X = \{*\}$, so that f becomes an endoarrow of M. If $\alpha, \beta, \gamma, \ldots$ are elements of X, and t is an element of M, we will write $\alpha(x)$ (or even αx) for $f(\alpha, x)$, according to the tradition of Ω-group and R-module theory.

By cartesian closedness, an alternative definition could be given using an X-indexed family of endoarrows of M. The resulting structure (which is basically a one-sorted algebra given by a monoid and by $|X|$ unary operators) is soundly established in Computer Science as the basic tool for the study of *synchronization trees* (see, for instance, [Mil80, Mil89]). Indeed, this line of thought leads to a powerful calculus of trees which is basically shaped as the classical one, but is far more general (see [KV91]). Here, however, we are concerned with distributive categories, so we prefer Definition 2.1.

As for the morphisms between X-motors, let us give the following

Definition 2.2 *A morphism between two X-motors $\langle M, \oplus, 0, f\rangle$ and $\langle M', \oplus', 0', f'\rangle$, is an arrow $\phi : M \longrightarrow M'$ which is a monoid morphism and satisfies $\phi(\sigma t) = \sigma\phi(t)$, for all $t \in M$ and $\sigma \in X$. Denote by $\mathbf{M}_X$ the initial object of X-***Mot***, the category of X-motors.*

In terms of the alternative definition (see above), the condition $\phi(\sigma t) = \sigma\phi(t)$ simply means that ϕ is *equivariant* with respect to all the endofunctions. This viewpoint is certainly more intuitive, but Definition 2.2 generalizes to distributive categories (see Definition 3.2). Further, the axioms for an X-motor as given in Definition 2.1 are, after all, a weakening of the axioms for a left R-module.

New we want to sketch some of the reasons justifying the use of X-motors to manage trees. Indeed, at first glance one could think that $\mathbf{M}_X$ is "too simple": after all, the initial object of **Mon** (the category of monoids) is the trivial monoid. But recall that $0 \in \mathbf{M}_X$, and that necessarily $\alpha(0) \neq 0$ for all $\alpha \in X$, by the initiality of $\mathbf{M}_X$. Further, still by initiality, we have that $\alpha(0) \oplus \beta(0)$ is a "new" element. Repeating these considerations one sees that $\mathbf{M}_X$ has actually many elements, but perhaps one does not have an intuitive insight of its structure. So we start with some heuristic remarks, which are essentially an adaptation of Bénabou's account of unlabelled forests.

Let us consider an "object" $\mathbf{T}_X$ whose members are to be thought of as trees ordered and labelled on their arcs. This object could have been defined in many different ways, and should also have been equipped with some operations. However, all we need to know

[1] Elsewhere we used Σ to denote the alphabet, according to the tradition of Computer Science. Here, however, Σ will be widely used to denote sums.

is that we can generate a new tree either by *joining the roots* of a pair of trees (preserving the order), or by adding a new labelled arc at the top of a tree. This second operation (left-prefix) can of course be performed with each label. The following picture illustrates these basic operations:

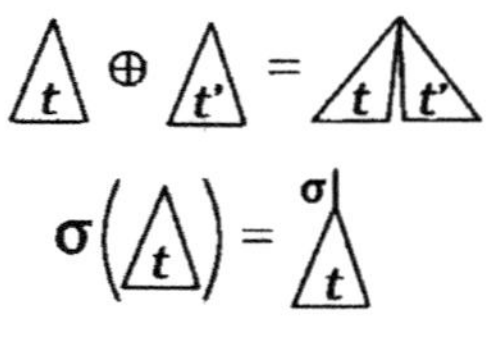

Of course, the "join" operation should be associative, and the "root-only" tree should play the rôle of the identity. So we would end up with a monoid. On the other hand, the "arc-creating" operation would be an X-indexed set of endofunctions, i.e., a map from $X \times M$ to M, so we would get a X-motor. Thus, we would have a unique arrow in X-**Mot**

$$\mathbf{M}_X \xrightarrow{\text{Pict}} \mathbf{T}_X$$

since $\mathbf{M}_X$ is initial.

Now, let us show some of the basic properties of labelled trees. For instance, we might wish to count how many *arcs* of a tree $t \in \mathbf{T}_X$ are labelled by a given label α. This means a map $n_\alpha : \mathbf{T}_X \longrightarrow \mathbb{N}$. But if $t = t_1 \oplus t_2$ then $n_\alpha(t) = n_\alpha(t_1) + n_\alpha(t_2)$, and when one creates a new tree by adding a labelled arc at the top of a tree, $n_\alpha(\alpha(t)) = n_\alpha(t) + 1$, while $n_\alpha(\beta(t)) = n_\alpha(t)$ for every $\beta \neq \alpha$. This means that

$$n : \mathbf{T}_X \longrightarrow < \mathbb{N}, +, 0, s_\alpha >,$$

where $s_\alpha(\alpha, n) = n + 1$ and $s_\alpha(\beta, n) = n$ for all $\beta \neq \alpha$, is an X-motor map. But then, the *unique* map

$$\nu_\alpha : \mathbf{M}_X \longrightarrow < \mathbb{N}, +, 0, s_\alpha >$$

must factor through Pict, since by the initiality of $\mathbf{M}_X$ the diagram

$$
\begin{array}{ccc}
\mathbf{M}_X & \xrightarrow{\text{Pict}} & \mathbf{T}_X \\
& \searrow{\nu_\alpha} & \downarrow{n_\alpha} \\
& & \mathbb{N}
\end{array}
$$

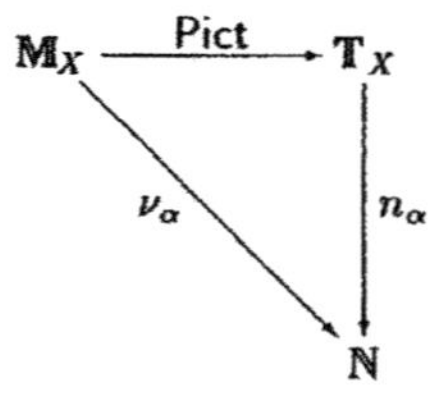

necessarily commutes. Hence the slogan: "count the labels in $\mathbf{M}_X$ rather than in $\mathbf{T}_X$".

This process can be carried on again with entirely different X-motors: many other counting maps can be created this way (see [KV91]).

It is now obvious in which sense $\mathbf{M}_X$ is, indeed, the object of X-labelled trees. The operation f creates new labelled arcs, $\oplus$ joins the roots and 0 is the root-only tree. Thus, $\alpha(0) \oplus \beta(0)$ and $\gamma(\alpha(0) \oplus \beta(0))$ represent just:

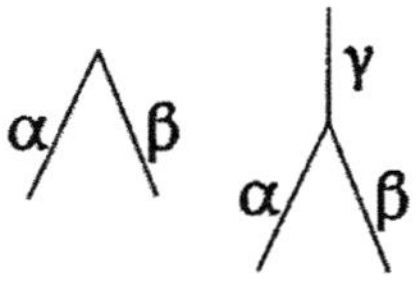

It is worth observing that we disregard the real nature of $\mathbb{T}_X$. We suggest that most (if not all) of the interesting properties of trees can be studied in $\mathbb{M}_X$, since most (if not all) of them should factor through Pict.

Many interesting theorems can be proved about $\mathbb{M}_X$. The fundamental one is that every element $t \in \mathbb{M}_X$ has a unique decomposition $\sigma_1(x_1) \oplus \sigma_2(x_2) \oplus \cdots \oplus \sigma_n(x_n)$: this property is the link between the completely invariant definition and a more descriptive calculus which can be easily developed (see [KV91]).

3 Distributive categories

There is no established terminology about distributive categories. We give here a few definitions to encompass the cases which are of interest for us.

Definition 3.1 *A category $\mathcal{D}$ is called* distributive *(in the sense of Walters) iff it has finite products and coproducts, and the finite products distribute with respect to the finite coproducts, i.e., the arrow*

$$A \times B + A \times C \longrightarrow A \times (B + C),$$

arising from the two arrows $1_A \times \mu_B$ and $1_A \times \mu_C$, is a natural isomorphism (μ_B and μ_C denote the injections).

Since we want to be able to write power series, we will consider *countably* (resp. *infinitely*) distributive categories, i.e., distributive categories which have countable (resp. infinite) coproducts, and finite products which distribute over them. Further, a (countably, infinitely) distributive category will be said to be *monoidal* if it satisfies Definition 3.1 when the product is weakened to a tensor product.

Extremely important are also distributive categories in the sense of Lawvere-Schanuel. Recall that in their definition every slice category is required to be distributive: the relevance to Computer Science is due to the fact that in order to model labelled transition systems, one needs a category distributive in this sense.

From now on we will often omit the product symbol. Integer indices without bounds are assumed to range over $\mathbb{N}$. When only natural isomorphisms are involved, we will denote them by equality.

3.1 X-Motors in a distributive category

Let us fix a countably distributive category $\mathcal{D}$ (actually, our results hold even in a *monoidal* countably distributive category). We are interested in the most proper lifting of Definition 2.1 to this setting. It turns out that the appropriate definition is the following one:

Definition 3.2 *A motor on the object (data type) X, or simply an X-motor, in a distributive category $\mathcal{D}$ is a tuple $\langle M, \mu, \eta, f\rangle$, where $\langle M, \mu, \eta\rangle$ is a monoid and*

$$f : XM \longrightarrow M$$

is an arrow of $\mathcal{D}$. A morphism between two X-motors $\langle M, \mu, \eta, f\rangle$ and $\langle M', \mu', \eta', f'\rangle$ is an arrow $\psi : M \longrightarrow M'$ in $\mathcal{D}$ such that the following diagrams commute:

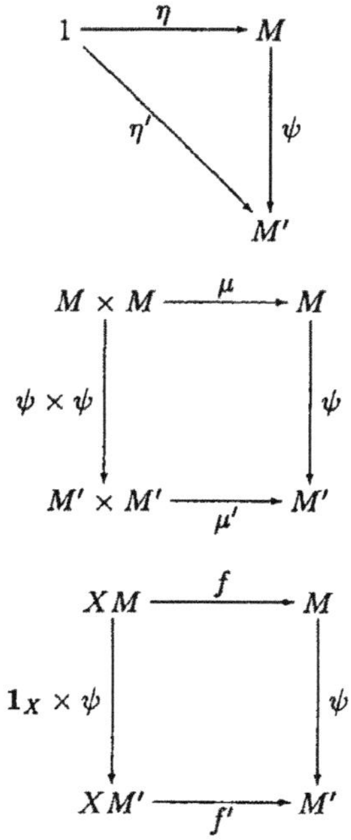

The reason for using in this context the term *data type* can be found in [Wal89]: an immediate naive motivation, however, is that in practical applications trees are "labelled" exactly in the sense that they carry data on each arc.

This definition is completely internal, and allows us to speak of X-motors in $\mathcal{D}$: the main result we will prove is the *existence of the initial X-motor:*

Theorem 3.1 *Given an object X of a distributive category $\mathcal{D}$, the initial X-motor in $\mathcal{D}$ is*

$$M_X = 1 + X + 2X^2 + 5X^3 + 14X^4 + \cdots = \sum_{n\geq 0} C_n X^n$$

together with suitable μ, η and f; C_n is here the n-th Catalan number, i.e.,

$$C_n = \binom{2n}{n}\frac{1}{n+1}$$

Before proving the theorem, we would like to give an insight. Catalan numbers are well-known in combinatorics, because C_n is the number of binary trees with n nodes. However, we began with a heuristic discussion of trees, which are different from binary trees, even if the two concepts are known to be equivalent: indeed, in practical programming trees are always managed as binary trees[2]. This correspondence is important, because (as we will see in the proof) the (apparently) complicated monoid operation of M_X can be really understood only keeping binary trees in mind.

To define μ, η and f we need a simple property of Catalan numbers, namely that if $n > 0$ then

$$C_n = \sum_{0 \le k \le n-1} C_k C_{n-1-k} = C_0 C_{n-1} + C_1 C_{n-2} + \cdots + C_{n-1} C_0.$$

We start by defining $f : XM \longrightarrow M$. Take the nth term of XM after a distribution, XC_nX^n, and consider the $(n+1)$th term of M:

$$C_{n+1}X^{n+1} = C_0 C_n X^{n+1} + \cdots + C_n C_0 X^{n+1}$$

Since $C_0 = 1$, we just inject XC_nX^n in the last summand, and then in M_X:

$$XC_nX^n = C_nX^{n+1} \xrightarrow{i_n} C_0C_nX^{n+1} + \cdots + C_nC_0X^{n+1} = C_{n+1}X^{n+1} \xrightarrow{i_{n+1}} M_X.$$

The morphism $\mu : M_X \times M_X \longrightarrow M_X$ is defined recursively: the distribution of the product $M_X \times M_X$ gives

$$M_X \times M_X = 1 \times 1 + 1 \times X + X \times 1 + X \times X + 1 \times 2X^2 + \cdots = \sum_{n,m \ge 0} C_nX^n \times C_mX^m$$

and we will denote with $\mu_{n,m}$ the (n,m)th component of μ. Since we want to use induction, we note that $\mu_{n,m}$ will land in $C_{n+m}X^{n+m}$ and eventually in M_X via the suitable injection, that will be understood with a slight abuse of notation.

The map $\mu_{0,0}$ is just the canonical isomorphism $1 \times 1 = 1$. Now, let us define $\mu_{k,n-k}$ for each integer $0 \le k \le n$, and a fixed n.

If $k = 0$, let $\mu_{0,n} : C_0X^0 \times C_nX^n \longrightarrow C_nX^n$ be the canonical isomorphism. If $k > 1$, given the term $C_kX^k \times C_{n-k}X^{n-k}$, we know that

$$\begin{aligned} &C_kX^k \times C_{n-k}X^{n-k} = \\ &\qquad C_0C_{k-1}X^k \times C_{n-k}X^{n-k} + C_1C_{k-2}X^k \times C_{n-k}X^{n-k} + \cdots + C_{k-1}C_0X^k \times C_{n-k}X^{n-k} \end{aligned}$$

so we will break again $\mu_{k,n-k}$ in k maps $\mu^p_{k,n-k}$, $0 \le p \le k-1$, from the last sum, and since

$$C_nX^n = \sum_{0 \le r \le n-1} C_rC_{n-r-1}X^n$$

we can simply define $\mu^p_{k,n-k}$ as

$$\begin{aligned} C_pC_{k-p-1}X^k \times C_{n-k}X^{n-k} &= C_pX^{p+1} \times C_{k-p-1}X^{k-p-1} \times C_{n-k}X^{n-k} \\ &\xrightarrow{C_p1_X^{p+1} \times \mu_{k-p-1,n-k}} C_pX^{p+1} \times C_{n-p-1}X^{n-p-1} \\ &= C_pC_{n-p-1}X^n \end{aligned}$$

[2]The correspondence between trees and binary trees is described in [Knu73], and has an algebraic formalization which can be found in [KV91].

and being $C_pC_{n-p-1}X^n$ just the pth summand of C_nX^n, we can inject in it.

The intuition behind this definition arises from the identification of trees with labelled arcs with binary trees with labelled nodes, and the corresponding identification of the sum of trees with the ***bottom-right*** product of binary trees [KV91]. The bottom-right product of binary trees means appending the second tree to the extreme right node of the first, as illustrated in the following picture, where it is denoted by $\oslash$:

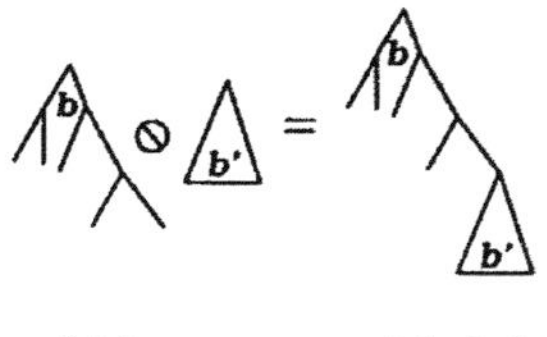

Now think of C_kX^k as the set of binary trees with k labelled nodes. Our description of μ is a description of the bottom-right product. The "labels" in X^k are "divided" in the first one (the top node), the following p ones (the nodes of the left subtree) and the remaining ones (the nodes of the right subtree). We use identities on the top node and on the left subtree (which, in a sense, remain "unchanged"), and then we take recursively the product of the right subtree with $C_{n-k}X^{n-k}$. Note that binary trees equipped with the bottom-right product and with the map illustrated in the following picture, where $\square$ denotes the empty binary tree:

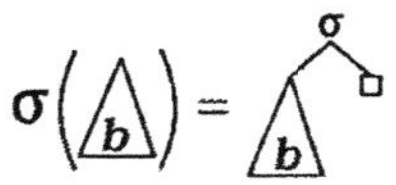

form an X-motor isomorphic to $\mathbf{M}_X$, as has been shown in [KV91]. This is the real reason why Catalan numbers work. Further, these pictures give an insight on the definition of μ and f.

A very special situation appears if we take $p = k - 1$. The general definition then yields

$$\begin{array}{rcl} C_{k-1}C_0X^k \times C_{n-k}X^{n-k} & = & C_{k-1}X^k \times C_0X^0 \times C_{n-k}X^{n-k} \\ & \xrightarrow{C_{k-1}1^k_X \times \mu_{0,n-k}} & C_{k-1}X^k \times C_{n-k}X^{n-k} \\ & = & C_{k-1}C_{n-k}X^n \end{array}$$

and this is just an isomorphism. Now, $\mu_{k,n-k}$ is the map defined by the $\mu^p_{k,n-k}$'s. Taking all the $\mu_{k,n-k}$'s together (composed with the injections i_n in M_X) we get the desired map $\mu : M_X \times M_X \longrightarrow M_X$. Of course, $\eta : 1 \longrightarrow M_X$, the identity, is simply the injection i_0. It is straightforward to show that μ and η satisfy the monoid axioms.

Proof of Theorem 3.1. All we need to prove is the initiality of $\langle M_X, \mu, \eta, f\rangle$, i.e., that given any X-motor $\langle M', \mu', \eta', f'\rangle$ in the category, there is a unique X-motor morphism $\psi : M_X \longrightarrow M'$. Suppose such a map exists: we will write ψ_k, $k \geq 0$, for the kth component of ψ, and we start by observing that the first condition

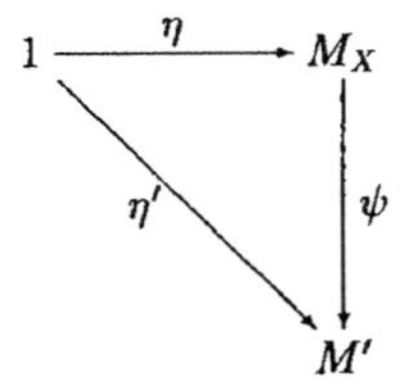

defines $\psi_0 : 1 \longrightarrow M'$. We will use f and the monoid structure to extend the map to M_X. We can rewrite the equivariance condition using only maps from the first component of M_X:

$$
\begin{array}{ccc}
X \times 1 = X & \xrightarrow{\ f_0\ } & M_X \\
\downarrow{\scriptstyle 1_X \times \eta'} & & \downarrow{\scriptstyle \psi} \\
X M' & \xrightarrow{\ f'\ } & M'
\end{array}
$$

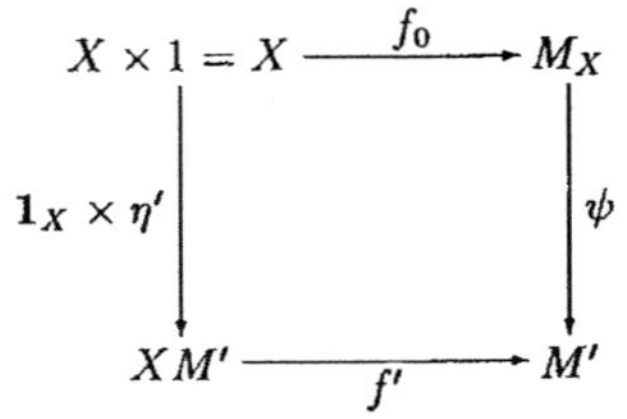

and since f_0 is just the injection in the coproduct, this means $\psi_1 = f' \circ (1_X \times \eta)$.

Consider now the general ψ_n. We decompose again the map in the parts corresponding to the decomposition of $C_n X^n$, and we write ψ_n^k, with $0 \leq k \leq n-1$. Suppose all ψ_m, $m < n$, have been shown to be derived from the first two maps ψ_0, ψ_1: then the equivariance condition, read on the $(n-1)$th term is

$$
\begin{array}{ccc}
X \times C_{n-1} X^{n-1} & \xrightarrow{\ f_{n-1}\ } & M_X \\
\downarrow{\scriptstyle 1_X \times \psi_{n-1}} & & \downarrow{\scriptstyle \psi} \\
X M' & \xrightarrow{\ f'\ } & M'
\end{array}
$$

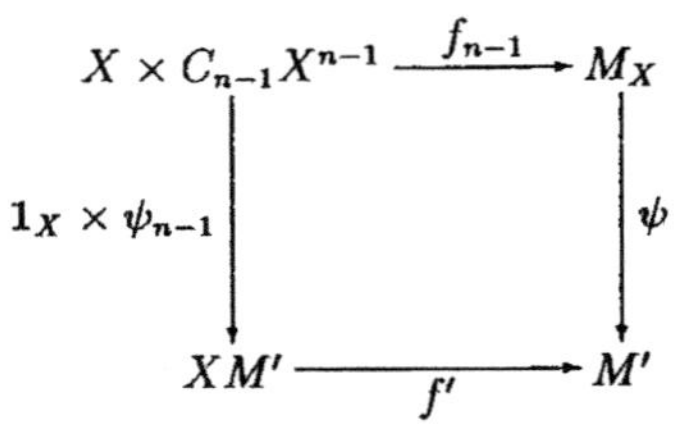

and since f_{n-1} is the injection in the coproduct in the last term of the decomposition of C_n the diagram can be rewritten as

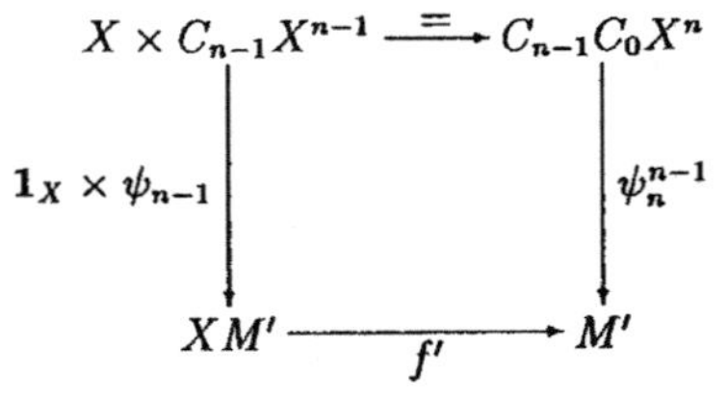

$$
\begin{array}{ccc}
X \times C_{n-1} X^{n-1} & \xrightarrow{\ =\ } & C_{n-1} C_0 X^n \\
\downarrow{\scriptstyle 1_X \times \psi_{n-1}} & & \downarrow{\scriptstyle \psi_n^{n-1}} \\
X M' & \xrightarrow{\ f'\ } & M'
\end{array}
$$

so $\psi_n^{n-1} = f' \circ (1_X \times \psi_{n-1})$.

On the other hand, take a ψ_n^k, with $0 \leq k \leq n-2$. The preservation of the monoid product diagram, when restricted to the pair $C_{k+1} X^{k+1} \times C_{n-k-1} X^{n-k-1}$ becomes

$$\begin{array}{ccc} C_{k+1}X^{k+1} \times C_{n-k-1}X^{n-k-1} & \xrightarrow{\mu_{k+1,n-k-1}} & M_X \\ \Big\downarrow{\scriptstyle \psi_{k+1} \times \psi_{n-k-1}} & & \Big\downarrow{\scriptstyle \psi} \\ M' \times M' & \xrightarrow[\mu']{} & M' \end{array}$$

but we observed that if we decompose C_{k+1} and take just the last component, the map μ becomes an isomorphism (plus an injection), so

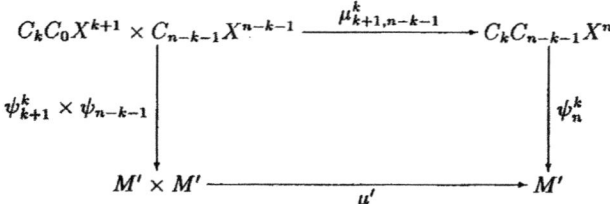

$$\begin{array}{ccc} C_kC_0X^{k+1} \times C_{n-k-1}X^{n-k-1} & \xrightarrow{\mu^k_{k+1,n-k-1}} & C_kC_{n-k-1}X^n \\ \Big\downarrow{\scriptstyle \psi^k_{k+1} \times \psi_{n-k-1}} & & \Big\downarrow{\scriptstyle \psi^k_n} \\ M' \times M' & \xrightarrow[\mu']{} & M' \end{array}$$

implies that ψ^k_n is defined by ψ^k_{k+1} and ψ_{n-k-1}. Thus, if ψ exists it is unique.

Now, as usual, we define ψ as we did above, using equivariance and preservation of the product, and we show it is an X-motor map. This will be proved by decomposing ψ on the sum.

We start observing that there is a class of diagrams which commute trivially, namely the diagrams with which we defined ψ. The class includes all equivariance diagrams and the product diagrams

$$\begin{array}{ccc} C_pC_{k-p-1}X^k \times C_{n-k}X^{n-k} & \xrightarrow{\mu^p_{k,n-k}} & M_X \\ \Big\downarrow{\scriptstyle \psi^p_k \times \psi_{n-k}} & & \Big\downarrow{\scriptstyle \psi} \\ M' \times M' & \xrightarrow[\mu']{} & M' \end{array}$$

with $p = k-1$. All we have to prove is that these diagrams commute for $0 \leq p \leq k-2$. We will use induction on the sum of the exponents of X in the top left corner of the diagram above, so that the base step is trivial, recalling that $\psi_0 = \eta'$.

Looking at the definition of μ, we see that the previous diagram is actually

$$\begin{array}{ccc} C_pC_{k-p-1}X^k \times C_{n-k}X^{n-k} & \xrightarrow{\mu^p_{k,n-k}} & C_pC_{n-p-1}X^n \\ \Big\downarrow{\scriptstyle \psi^p_k \times \psi_{n-k}} & & \Big\downarrow{\scriptstyle \psi^p_n} \\ M' \times M' & \xrightarrow[\mu']{} & M' \end{array}$$

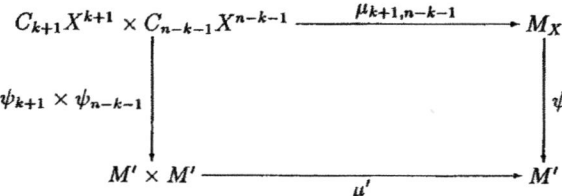

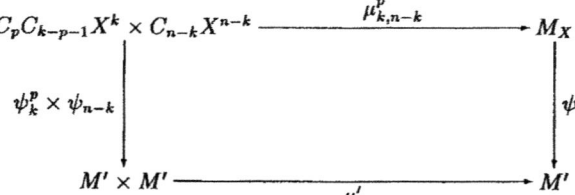

and using again the definitions of the maps involved in the diagram we get

$$
\begin{array}{ccccc}
(C_pX^{p+1} \times C_{k-p-1}X^{k-p-1}) \times C_{n-k}X^{n-k} & & \xrightarrow{\equiv} & & C_pX^{p+1} \times (C_{k-p-1}X^{k-p-1} \times C_{n-k}X^{n-k}) \\
\Big\downarrow {\scriptstyle (\psi^p_{p+1} \times \psi_{k-p-1}) \times \psi_{n-k}} & & & & \Big\downarrow {\scriptstyle C_p\mathbf{1}_X^{p+1} \times \mu_{k-p-1,n-k}} \\
(M' \times M') \times M' & & & & C_pX^{p+1} \times C_{n-p-1}X^{n-p-1} \\
\Big\downarrow {\scriptstyle \mu' \times \mathbf{1}_M} & & & & \Big\downarrow {\scriptstyle \psi^p_{p+1} \times \psi_{n-p-1}} \\
M' \times M' & \xrightarrow[\mu']{} & M' & \xleftarrow[\mu']{} & M' \times M'
\end{array}
$$

Now, by induction the following diagram commutes:

$$
\begin{array}{ccc}
C_{k-p-1}X^{k-p-1} \times C_{n-k}X^{n-k} & \xrightarrow{\mu_{k-p-1,n-k}} & C_{n-p-1}X^{n-p-1} \\
\Big\downarrow {\scriptstyle \psi_{k-p-1} \times \psi_{n-k}} & & \Big\downarrow {\scriptstyle \psi_{n-p-1}} \\
M' \times M' & \xrightarrow[\mu']{} & M'
\end{array}
$$

so we end up with

$$
\begin{array}{ccccc}
(C_pX^{p+1} \times C_{k-p-1}X^{k-p-1}) \times C_{n-k}X^{n-k} & & \xrightarrow{\equiv} & & C_pX^{p+1} \times (C_{k-p-1}X^{k-p-1} \times C_{n-k}X^{n-k}) \\
\Big\downarrow {\scriptstyle (\psi^p_{p+1} \times \psi_{k-p-1}) \times \psi_{n-k}} & & & & \Big\downarrow {\scriptstyle \psi^p_{p+1} \times (\psi_{k-p} \times \psi_{n-k})} \\
(M' \times M') \times M' & & & & M' \times (M' \times M') \\
\Big\downarrow {\scriptstyle \mu' \times \mathbf{1}_{M'}} & & & & \Big\downarrow {\scriptstyle \mathbf{1}_{M'} \times \mu'} \\
M' \times M' & \xrightarrow[\mu']{} & M' & \xleftarrow[\mu']{} & M' \times M'
\end{array}
$$

which obviously commutes. □

4 The unique decomposition property

As we pointed out in Section 2, the main property of $\mathbf{M}_X$ was a unique decomposition theorem. This theorem can be also proved in the present setting, in a more general form.

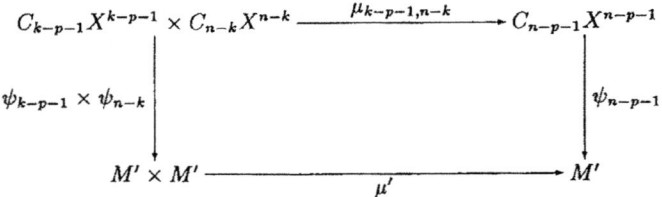

In our hypotheses, there is a left adjoint $(-)^*$ to the forgetful functor $U : \mathbf{Mon}(\mathcal{D}) \longrightarrow \mathcal{D}$, and if ψ is an arrow of $\mathcal{D}$ ending at $U(M)$ for some $M \in \mathbf{Obj}(\mathbf{Mon}(\mathcal{D}))$, we will write $h(\psi)$ for the map in $\mathbf{Mon}(\mathcal{D})$ corresponding to ψ in the adjunction. X-$\mathbf{Mot}(\mathcal{D})$ will denote the category of X-motors in $\mathcal{D}$.

Theorem 4.1 *Let* $\mathbf{M}_X = \langle M_X, \mu, \eta, f\rangle$ *be the initial* X*-motor, i.e., the initial object of* X-$\mathbf{Mot}(\mathcal{D})$. *Consider the functor* $\Lambda_X : X\text{-}\mathbf{Mot}(\mathcal{D}) \longrightarrow X\text{-}\mathbf{Mot}(\mathcal{D})$, *which to each* X*-motor* $\langle M, \mu, \eta, f\rangle$ *associates the* X*-motor given by the monoid* $(XM)^* = \sum_n (XM)^n$ *endowed with the arrow* $f' = j \circ (\mathbf{1}_X \times U(h(f)))$, *where* j *is the injection of generators*

$$XM \xrightarrow{j} (XM)^*,$$

and to each arrow ψ *in* X-$\mathbf{Mot}(\mathcal{D})$ *associates* $(\mathbf{1}_X \times U(\psi))^* = \sum_n (\mathbf{1}_X \times U(\psi))^n$. *Then,*

$$\mathbf{M}_X \cong \Lambda_X(\mathbf{M}_X).$$

Proof. First of all, we have to prove that Λ_X takes effectively morphisms of X-motors to morphisms of X-motors. Given an X-motor morphism

$$\psi : \langle M, \mu, \eta, f\rangle \longrightarrow \langle M', \mu', \eta', g\rangle,$$

$\Lambda_X(\psi)$ is trivially a monoid morphism. Moreover (dropping the U's with an abuse of notation),

$$\begin{aligned}
\Lambda_X(\psi) \circ f' &= \Lambda_X(\psi) \circ j \circ (\mathbf{1}_X \times h(f)) \\
&= j \circ (\mathbf{1}_X \times \psi) \circ (\mathbf{1}_X \times h(f)) \\
&= j \circ (\mathbf{1}_X \times (\psi \circ h(f))) \\
&= j \circ (\mathbf{1}_X \times (h(\psi \circ f))) \\
&= j \circ (\mathbf{1}_X \times (h(g \circ (\mathbf{1}_X \times \psi)))) \\
&= j \circ (\mathbf{1}_X \times (h(g) \circ (\mathbf{1}_X \times \psi)^*)) \\
&= j \circ (\mathbf{1}_X \times h(g)) \circ (\mathbf{1}_X \times \Lambda_X(\psi)) \\
&= g' \circ (\mathbf{1}_X \times \Lambda_X(\psi))
\end{aligned}$$

So that $\Lambda_X(\psi)$ is also an X-motor map. We have of course a unique morphism

$$\mathbf{M}_X \xrightarrow{!} \Lambda_X(\mathbf{M}_X)$$

To build an inverse, take the map $(XM_X)^* \longrightarrow M_X$ which corresponds to f in the adjunction $(-)^* \dashv U$, i.e., $h(f)$. It is an X-motor map, because

$$h(f) \circ f' = h(f) \circ j \circ (\mathbf{1}_X \times h(f)) = f \circ (\mathbf{1}_X \times h(f)).$$

Now, $h(f) \circ ! = \mathbf{1}_{\mathbf{M}_X}$ trivially because $\mathbf{M}_X$ is initial, and we have

$$\begin{aligned}
! \circ h(f) &= h(! \circ f) \\
&= h(f' \circ (\mathbf{1}_X \times !)) \\
&= h(j \circ (\mathbf{1}_X \times h(f)) \circ (\mathbf{1}_X \times !)) \\
&= h(((\mathbf{1}_X \times h(f)) \circ (\mathbf{1}_X \times !))^* \circ j) \\
&= (\mathbf{1}_X \times (h(f) \circ !))^* \\
&= (\mathbf{1}_X \times \mathbf{1}_{\mathbf{M}_X})^* \\
&= \mathbf{1}_{\Lambda_X(\mathbf{M}_X)}
\end{aligned}$$

and this proves $\mathbf{M}_X \cong \Lambda_X(\mathbf{M}_X)$. □

As a final remark, we want to stress that Theorem 4.1 shows in which sense trees are to be thought of as a "recursive closure" of the "labelled list of" operator. $(XM)^*$ can be built in **Sets** as the monoid of lists $\langle x_1, m_1\rangle, \langle x_2, m_2\rangle, \ldots, \langle x_k, m_k\rangle$ where $x_j \in X$ and $m_j \in M$, and thus represents the obvious generalization to distributive categories of the construction "labelled list of". Hence, to say that $\mathbb{M}_X \cong \Lambda_X(\mathbb{M}_X)$ is exactly to say that every X-labelled tree is a finite list of X-labelled trees, each one with an associated label in X (see also [Wal89]).

References

[Bén90] J. Bénabou. Lectures held at the Mathematics and Computer Science Departments of the University of Milano, and talk at the Category Theory '90 Conference in Como. 1990.

[Coc90] J.R.B. Cockett. Introduction to distributive categories. Technical Report 90–0052C, Macquarie University, May 1990.

[DKV] P. Degano, S. Kasangian, and S. Vigna. Applications of the calculus of trees to process description languages. Submitted.

[Knu73] D.E. Knuth. *The Art of Computer Programming*. Addison-Wesley, 1973.

[KV91] S. Kasangian and S. Vigna. Introducing a calculus of trees. In *Proceedings of the International Joint Conference on Theory and Practice of Software Development (TAPSOFT/CAAP '91)*, number 493 in Lecture Notes in Computer Science, pages 215–240, 1991.

[Law90] F.W. Lawvere. Categories of space and of quantity. In *Proceedings on Structures and Mathematical Theories, S. Sebastian*, 1990.

[Mac71] S. Mac Lane. *Categories for the Working Mathematician*. Springer-Verlag, 1971.

[Mil80] R. Milner. *A Calculus of Communicating Systems*. Number 92 in Lecture Notes in Computer Science. Springer-Verlag, 1980.

[Mil89] R. Milner. *Communication and Concurrency*. International Series in Computer Science. Prentice Hall, 1989.

[Wal] R.F.C. Walters. An imperative language based on distributive categories. *Mathematical Structures in Computer Science*. To appear.

[Wal89] R.F.C. Walters. Data types in a distributive category. *Bull. Austr. Math. Soc.*, 40:79–82, 1989.

This paper is in final form and will not be published elsewhere.

A NOTE ON RELATIONS RELATIVE TO A FACTORIZATION SYSTEM

G.M. KELLY

Pure Mathematics Department F07, University of Sydney,
NSW 2006, Australia

A number of authors have observed that regularity of a category $\mathcal{A}$ is not necessary for the existence of a "calculus of relations" in $\mathcal{A}$, with an associative composition of relations giving a 2-category Rel $\mathcal{A}$; it suffices that the finitely-complete $\mathcal{A}$ have a proper factorization system $(\mathcal{E},\mathcal{M})$ whose class $\mathcal{E}$ is stable by pullbacks, the classical regular-category case being that where $\mathcal{M}$ consists of all the monomorphisms. We show that this generalization is in a sense illusory : if $\mathcal{B}$ is the category of "maps" in Rel $\mathcal{A}$, then $\mathcal{B}$ is a regular category, and Rel $\mathcal{A}$ is isomorphic to the classical Rel $\mathcal{B}$.

1. **Introduction**

There is a rich literature on the important topic of relations in a category $\mathcal{A}$, most of it devoted to the case where $\mathcal{A}$ is a *regular category* in the sense of Barr and of Grillet in [1]. Indeed, regularity of $\mathcal{A}$ is necessary if we are to have an associative composition of relations — so long as, by a relation P from A to B, we mean just a subject P of $A \times B$. Yet we are not obliged to mean this, and a number of authors — see Klein [11], Meisen [12], and Richter [13] — have observed that the theory admits of a considerable extension, as follows.

Suppose that $\mathcal{A}$ is a category with finite limits, and that $(\mathcal{E},\mathcal{M})$ is a *proper factorization system* on $\mathcal{A}$, in the sense of Freyd and Kelly [7]; if the $(\mathcal{E},\mathcal{M})$-factorization of $f : A \to B$ is

$$A \xrightarrow[p]{} I \xrightarrow[i]{} B, \qquad (1.1)$$

we shall call i or I the *image* im f of f. By a *relation* P from A to B let us mean a subobject P of $A \times B$ whose inclusion $\langle p_1, p_2\rangle : P \to A \times B$ *lies in* $\mathcal{M}$. We compose relations $P : A \to B$ and $Q : B \to C$ by forming the diagram

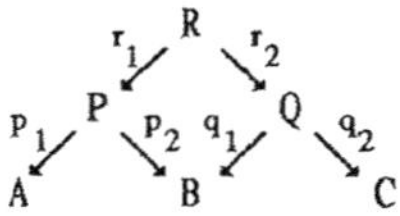

(1.2)

where the diamond is a pullback, and taking for QP the image of $\langle p_1r_1, q_2r_2\rangle : R \to A \times C$. This composition, as we shall see, is associative if (and only if — see [11, p.543]) the class $\mathcal{E}$ is stable

---

The author acknowledges with gratitude the support of the Australian Research Council.

under pulling back — whereupon we shall call $(\mathcal{E},\mathcal{M})$ a *stable* factorization system. Then the objects of $\mathcal{A}$ and the relations between these form a category Rel $\mathcal{A}$ — in fact a 2–category with the obvious definition of $P \le P'$ — into which $\mathcal{A}$ is embedded by identifying the morphism $f : A \longrightarrow B$ with the relation $\langle 1_A, f\rangle$ which is its graph (such a relation being aptly called a *function*).

To say that a finitely–complete $\mathcal{A}$ has a stable factorization system with $\mathcal{M}$ being the class of all monomorphisms is, as Joyal pointed out long ago — see, for instance, Carboni and Street [5], or Section 2 below — just to say that $\mathcal{A}$ is a regular category; this, then, is a special case of the above. But the categories of topological spaces and of hausdorff spaces, neither of which is regular, have each a stable factorization system with $\mathcal{M}$ the subspace–inclusions and $\mathcal{E}$ the surjections; in the former case, but not in the latter, $\mathcal{E}$ is the class of epimorphisms and $\mathcal{M}$ that of strong (see [10]) monomorphisms. Again, in the category of short exact sequences of abelian groups, a morphism

$$\begin{array}{ccccccccc} 0 & \longrightarrow & A' & \longrightarrow & A & \longrightarrow & A'' & \longrightarrow & 0 \\ & & {\scriptstyle f'}\downarrow & & \downarrow{\scriptstyle f} & & \downarrow{\scriptstyle f''} & & \\ 0 & \longrightarrow & B' & \longrightarrow & B & \longrightarrow & B'' & \longrightarrow & 0 \end{array} \qquad (1.3)$$

is a monomorphism when f (and consequently f') is monomorphic, and is a strong monomorphism when f'' too is monomorphic — with dual results for epimorphisms and strong epimorphisms; since both the epimorphisms and the strong epimorphisms are stable by pullbacks[*], this category *is* regular, but has a *second* stable factorization system in which $\mathcal{M}$ is the strong monomorphisms. In such cases Rel $\mathcal{A}$ is of course ambiguous : it is really short for Rel $(\mathcal{A};\mathcal{E},\mathcal{M})$. A more trivial example of this kind is obtained by taking for $\mathcal{A}$ an ordered set with finite infima : every morphism is monomorphic, while the strong monomorphisms are the identities, and dually for epimorphisms; $\mathcal{A}$ is regular since the strong epimorphisms are stable by pullbacks — but so too are the mere epimorphisms. Still further examples of this kind are the categories of separated objects for a topology on a pretopos or on an abelian category, studied by Carboni and Mantovani in [6]; we shall say a word about these in Section 5 below.

In fact the generalization is more limited than might be thought : although any respectable $\mathcal{A}$ admits proper factorization systems — if only the extreme ones (strong epimorphisms, monomorphisms) and (epimorphisms, strong monomorphisms), which may coincide — the existence of a *stable* factorization system is a severe restriction on $\mathcal{A}$; it implies (see [10] or Section 2 below) – at least when $\mathcal{A}$ admits coequalizers — that every strong epimorphism is regular, or equally that regular epimorphisms are closed under composition. So $\mathcal{A}$ cannot be, for example, the category of small categories.

---

[*] We can identify this category with that of the monomorphisms $A' \longrightarrow A$; the latter being a reflective full subcategory of the "arrow category" **[2, Ab]** of *all* $A' \longrightarrow A$ in **Ab** , limits in it are formed pointwise; so that in looking at pullbacks of the morphisms (1.3), we are looking at pullbacks of f and of f' in **Ab**.

The purpose of the present article is to point out that, even so far as it goes, the generalization is somewhat illusory : Rel $\mathcal{A}$ = Rel ($\mathcal{A}$;$\mathcal{E}$,$\mathcal{M}$) is nothing but Rel $\mathcal{B}$, in the classical sense, for a certain regular category $\mathcal{B}$ — to wit, the category Map Rel $\mathcal{A}$ that we now describe.

Recall that an arrow in a 2-category is often called a *map* if it has a right adjoint, the origin of this name being the observation that the maps in Rel $\mathcal{B}$ for a regular $\mathcal{B}$ are precisely the *functions*, so that Map Rel $\mathcal{B}$ = $\mathcal{B}$. Given now any finitely-complete $\mathcal{A}$ with a stable factorization system ($\mathcal{E}$,$\mathcal{M}$) as above, set $\mathcal{B}$ = Map Rel $\mathcal{A}$; an easy calculation below shows that $<p_1, p_2> : P \longrightarrow A \times B$ lies in $\mathcal{B}$ precisely when p_1 lies in the class

$$\Sigma = \mathcal{E} \cap \text{Mono} \tag{1.4}$$

consisting of the monomorphisms in $\mathcal{E}$ — which reduces to the isomorphisms, giving $\mathcal{B} = \mathcal{A}$, only in the classical case where $\mathcal{M}$ is the monomorphisms and consequently $\mathcal{A}$ is regular. In the general case, the inclusion $\mathcal{A} \longrightarrow \mathcal{B}$ turns out to be the universal $\mathcal{A} \longrightarrow \mathcal{A}[\Sigma^{-1}]$ inverting the class Σ — which not only admits a calculus of right fractions in the sense of Gabriel and Zisman [9], but is even what Bénabou [2] calls a *pullback congruence*. It is easily seen that $\mathcal{B}$ is a regular category (as a 2-category, it is locally discrete), and that $\mathcal{A} \longrightarrow \mathcal{B}$ (which, in a suitable 2-category, is a reflexion of $\mathcal{A}$ = ($\mathcal{A}$;$\mathcal{E}$,$\mathcal{M}$) into the regular categories) induces an isomorphism Rel $\mathcal{A} \cong$ Rel $\mathcal{B}$ of 2-categories.

Note that, in the examples above where $\mathcal{A}$ is topological or hausdorff spaces, Σ consists of the continuous bijections, and $\mathcal{B}$ is equivalent to the category of sets. In the first of these examples, $\mathcal{A} \longrightarrow \mathcal{B}$ has both adjoints; in the second, only a left adjoint. In the example of short exact sequences with $\mathcal{E}$ being the epimorphisms, Σ consists of those morphisms (1.3) with f invertible, and $\mathcal{B}$ is equivalent to the category of abelian groups; here $\mathcal{A} \longrightarrow \mathcal{B}$ again has both adjoints. When, however, $\mathcal{A}$ is an ordered set with finite infima, Σ consists of all morphisms, $\mathcal{B}$ is equivalent to 1, and $\mathcal{A} \longrightarrow \mathcal{B}$ has a right adjoint, but not in general a left one. So neither adjoint to $\mathcal{A} \longrightarrow \mathcal{B}$ exists by necessity.

In each of our examples the full subcategory $\mathcal{C} = \Sigma^{\perp}$ of $\mathcal{A}$, given by those C in $\mathcal{A}$ for which $\mathcal{A}(\alpha, C)$ is invertible for each $\alpha \in \Sigma$, is in fact reflective in $\mathcal{A}$, and the composite $\mathcal{C} \longrightarrow \mathcal{A} \longrightarrow \mathcal{B}$ is an equivalence; we end with some words about this phenomenon, especially in the case studied by Carboni and Mantovani in [6].

We turn now to the details and the proofs.

2. On stable factorization systems

We place ourselves throughout in a finitely-complete category $\mathcal{A}$. Recall from [11] the notion of *the pullback of a parallel pair* x, y along a morphism g ; that is, the limit

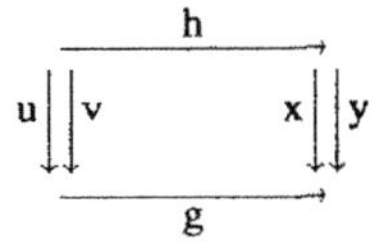

(2.1)

of the diagram given by x , y , and g . Since it is formed by taking the three pullbacks in

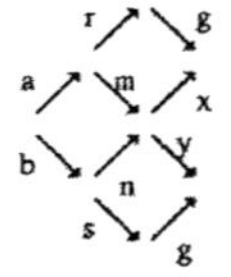

and setting $h = ma = nb$, $u = ra$, and $v = sb$, it follows that

h *is epimorphic if every pullback of* g *is epimorphic.* (2.2)

Lemma 2.1 *If* ig *has the same kernel–pair as* g , *and if every pullback of* g *is epimorphic, then* i *is monomorphic.*

Proof. It is easy to see that, if x , y is the kernel–pair of i , then u , v in the pullback (2.1) is the kernel–pair of ig . By hypothesis, this is the kernel–pair of g ; so that $xh = gu = gv = yh$, giving $x = y$ since h is epimorphic by (2.2).

Recall from [10] that, since $\mathcal{A}$ is finitely complete, p is a strong epimorphism precisely when it factorizes through no proper subobject of its codomain, and is a regular epimorphism precisely when it is the coequalizer of some parallel pair; every regular epimorphism is of course strong. To say that (strong epimorphisms, monomorphisms) is a factorization system is to say that, for every morphism f , there is a smallest subobject of its codomain through which it factorizes — we might call this its *absolute image*, in contrast to our use above of "image" relative to a general factorization system. Part (a) of the following is in [10]:

Proposition 2.2 *Suppose that pullbacks of strong epimorphisms are epimorphisms. Then strong epimorphisms are regular if* $\mathcal{A}$ *admits either* (a) *coequalizers or* (b) *absolute images.*

Proof (a) Let g be the coequalizer of the kernel–pair of the strong epimorphism p , so that $p = ig$; since the kernel–pair of g is that of p , it follows from Lemma 2.1 that i is monomorphic, and hence invertible. (b) Let $f : A \longrightarrow C$ coequalize such pairs as are coequalized by the strong epimorphism $p : A \longrightarrow B$; we are to show that f factorizes through p . Let the (strong epimorphism, monomorphism) factorization of $\langle p,f\rangle : A \longrightarrow B \times C$ be $\langle m,n\rangle g$. The kernel–pair of $mg = p$ coincides with that of $\langle p,f\rangle$ and hence with that of g , so that by Lemma 2.1 m is monomorphic and therefore (p being a strong epimorphism) invertible — giving $f = ng = nm^{-1}p$.

From (b) we get Joyal's characterization, mentioned above, of a regular category as a finitely–complete one with absolute images in which strong epimorphisms are stable by pullbacks. From (a) comes the confirmation of the assertion in the Introduction, that $\mathcal{A}$ cannot have a stable factorization system $(\mathcal{E},\mathcal{M})$ unless the strong epimorphisms coincide with the regular ones — for every strong epimorphism must be in $\mathcal{E}$. See [10] for the proof that this coincidence is, at least when coequalizers exist, equivalent to the closedness under composition of the regular epimorphisms, or again to the assertion that every f is ir where i is monomorphic and r is the coequalizer of the kernel–pair of f.

Suppose henceforth that $(\mathcal{E},\mathcal{M})$ is a stable factorization system on $\mathcal{A}$, noting that, when $\mathcal{M}$ is the class of all monomorphisms, $\mathcal{E}$ consists of the strong epimorphisms, so that $\mathcal{A}$ is a regular category. Consider the class Σ given by (1.4). Of course Σ contains the isomorphisms, and is closed under composition; moreover it is clearly stable by pullbacks. Recall that, if $fg \in \mathcal{E}$ we have $f \in \mathcal{E}$, while g is monomorphic if fg is monomorphic. Note that, since the $(\mathcal{E},\mathcal{M})$–factorization $f = ip$ of a monomorphism f has $p \in \Sigma$, the only case in which Σ consists of the isomorphisms alone is that in which every monomorphism lies in $\mathcal{M}$ — the classical regular–category case.

Proposition 2.3 (a) *If* $fg \in \mathcal{E}$ *and* f *is monomorphic, then* $g \in \mathcal{E}$; *in particular,* $g \in \Sigma$ *if* $fg \in \Sigma$ *and* $f \in \Sigma$. (*b*) *If* fg *is monomorphic and* $g \in \mathcal{E}$, *then* f *is monomorphic; in particular* $f \in \Sigma$ *if* $fg \in \Sigma$ *and* $g \in \Sigma$.

Proof For (a), observe that g is the pullback of fg along f; (b) is immediate from Lemma 2.1 since fg and g are monomorphic.

It follows that Σ not only admits a calculus of right fractions, but is a *pullback congruence* in the sense of Bénabou [2, §1]. Elegant though his theory of these is, we have no need to refer to it explicitly, since in our case there is a simple direct proof that $\mathcal{A}[\Sigma^{-1}] = \mathrm{Map\ Rel}\ \mathcal{A}$.

3. On relations

We continue to suppose that $(\mathcal{E},\mathcal{M})$ is a stable factorization system on the finitely–complete $\mathcal{A}$.

Lemma 3.1 $g : C \longrightarrow B$ *factorizes through the image of* $f : A \longrightarrow B$ *if only if we have* $ge = ft$ *for some* $e \in \mathcal{E}$ *and some* t.

Proof Let $f = ip$ be the $(\mathcal{E},\mathcal{M})$–factorization. If e and t as above exist, we have an s making commutative

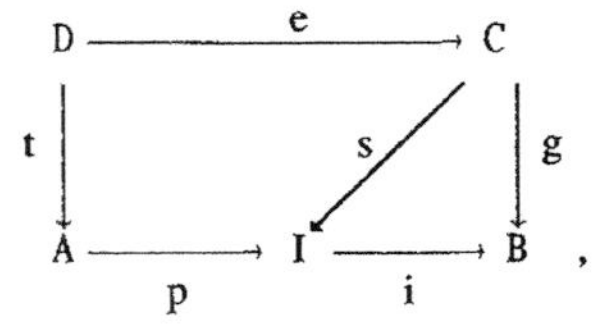

since $e \in \mathcal{E}$ and $i \in \mathcal{M}$; conversely, if $g = is$, we take for e the pullback of p along s.

Given a relation $P : A \to B$ as defined in the Introduction, we say that a span $\langle a,b\rangle : X \to A \times B$ *belongs to* P, written $b(P)a$, if it factorizes through the inclusion $\langle p_1,p_2\rangle : P \to A \times B$. Note that the graph $\langle 1_A,f\rangle$ of a morphism $f : A \to B$, being a coretraction and hence certainly in $\mathcal{M}$, is indeed a relation from A to B; as we said, we identify this relation with f, and call it a *function*; note that $b(f)a$ means $b = fa$. In particular we have the identity relation $1_A : A \to A$, tabulated by the diagonal $\langle 1,1\rangle : A \to A \times A$. Recall from the Introduction the definition of the composite $QP : A \to C$ of the relations $P : A \to B$ and $Q : B \to C$; Lemma 3.1 easily gives (*cf.* [13, p.456] and [5, p.277]) :

Lemma 3.2 *For a span* $\langle a,c\rangle$ *we have* $c(QP)a$ *if and only if, for some* $e \in \mathcal{E}$ *and some* b, *we have* $ce(Q)b$ *and* $b(P)ae$.

Associativity of this composition follows at once, so that the objects of $\mathcal{A}$ and the relations between these form a category Rel $(\mathcal{A}; \mathcal{E}, \mathcal{M})$, or Rel $\mathcal{A}$ for short, whose identities are the 1_A; it is in fact a 2–category, when we order the relations from A to B in the usual way as subobjects. Since every pullback of an $\mathcal{M}$ is an $\mathcal{M}$, this 2–category has local finite infima, $P \wedge P'$ being the usual intersection; of course the top element of $(\mathrm{Rel}\,\mathcal{A})(A,B)$, being the subobject $1 : A \times B \to A \times B$, lies in $\mathcal{M}$. This 2–category also has the anti-involution sending $P : A \to B$ to $P^o : B \to A$ given by $\langle p_2,p_1\rangle : P \to B \times A$. It contains a copy of $\mathcal{A}$ embedded as the functions : clearly the composition of relations extends that of functions. It follows very simply from Lemma 3.2 that, for $P : A \to B$, $Q : B \to C$, and $S : A \to C$ we have (*cf.* [8, p. 80]) Freyd's "modular law" $QP \wedge S \leq Q(P \wedge Q^oS)$.

When the relation P in (1.2) is a function $f : A \to B$, there is no need to pass to an image when forming the composite Qf, since $\langle r_1, q_2 r_2\rangle$ is already in $\mathcal{M}$, being the pullback of $\langle q_1,q_2\rangle$ along $f \times 1_C$. Thus for functions f and g the composite g^of is the relation R tabulated by r_1 and r_2 in the pullback

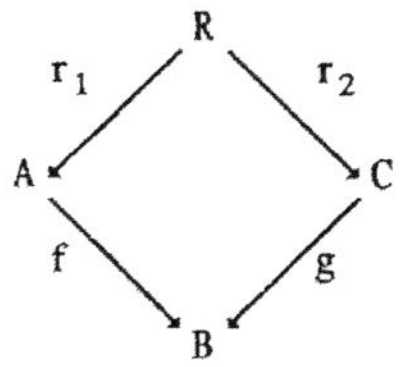

(3.1)

In particular, $f^\circ f$ is tabulated by the kernel–pair of f, whence

$$1_A \le f^\circ f \text{, } \textit{with equality iff } f \textit{ is monomorphic.} \tag{3.2}$$

When, however, P in (1.2) is arbitrary but Q is a function $k : B \longrightarrow C$, the pullback in (1.2) is trivial but we are obliged to take an image : the composite kP is the image of $\langle p_1, kp_2\rangle : P \longrightarrow A \times C$. Taking P here to be h° for a function $h : B \longrightarrow A$, we see that kh° is the image of $\langle h,k\rangle : B \longrightarrow A \times C$; equivalently, for any relation $S : A \longrightarrow C$ we have $k(S)h$ if and only if $kh^\circ \le S$. In particular, if S is tabulated by s_1 and s_2, we have

$$S = s_2 s_1^\circ . \tag{3.3}$$

Moreover, taking $k = h : B \longrightarrow A$ above, we see that hh° is the image of $\langle h,h\rangle : B \longrightarrow A \times A$, which is just $\langle i,i\rangle : I \longrightarrow A \times A$ where $i : I \longrightarrow A$ is $\operatorname{im} h$. Accordingly

$$hh^\circ \le 1_A \quad , \textit{ with equality iff } h \in \mathcal{E} . \tag{3.4}$$

It follows from (3.2) and (3.4) that each function f is a *map* in Rel $\mathcal{A}$, having f° as its right adjoint. It is well known (see [4, Proposition 5]) that, in the classical case of a regular $\mathcal{A}$ with $\mathcal{M}$ the monomorphisms, these are the only maps. In our present generality, that is false. We again use Σ in the sense of (1.4).

Theorem 3.3 *The relation* $P : A \longrightarrow B$ *tabulated by* p_1 *and* p_2 *is a map if and only if* $p_1 \in \Sigma$; *whereupon* P *has the right adjoint* P°.

Proof If $p_1 \in \Sigma$ we have $P \dashv P^\circ$ since then (3.2)–(3.4) give

$$PP^\circ = p_2 p_1^\circ p_1 p_2^\circ = p_2 p_2^\circ \le 1_A \qquad , \qquad P^\circ P = p_1 p_2^\circ p_2 p_1^\circ \ge p_1 p_1^\circ = 1_B . \tag{3.5}$$

Suppose conversely that P has a right adjoint $Q : B \longrightarrow A$. Since $1_A \le QP$, it follows, on taking $a = c = 1_A$ in Lemma 3.2, that for some $e \in \mathcal{E}$ and some b we have $b(P)e$; in other words, $e = p_1 t$ and $b = p_2 t$ for some t ; so p_1 like e lies in $\mathcal{E}$. It remains to show that p_1 is monomorphic. Let $x,y : K \longrightarrow P$ with $p_1 x = p_1 y$. Using $1_A \le QP$ again, but now taking $a = c = p_1 x$ in Lemma 3.2, we get some $e \in \mathcal{E}$ and some b with $p_1 xe(Q)b$; and since trivially we have $p_2 xe(P)p_1 xe$, Lemma 3.2 gives $p_2 xe(PQ)b$, whence $p_2 xe = b$ since $PQ \le 1_B$. Because $p_1 x = p_2 y$, we equally have $p_2 ye = b$. Since $e \in \mathcal{E}$ is epimorphic, we have $p_2 x = p_2 y$; and combining this with $p_1 x = p_1 y$ gives $x = y$, as desired.

Since an invertible arrow in a 2–category is certainly a map, it follows from (3.2)–(3.5) that:

Corollary 3.4 $P : A \longrightarrow B$, *tabulated* by p_1 *and* p_2, *is invertible in* Rel $\mathcal{A}$ *if and only if* $p_1 \in \Sigma$ *and* $p_2 \in \Sigma$; *whereupon the inverse of* $P = p_2 p_1{}^{o}$ *is* $P^{o} = p_1 p_2{}^{o}$. *In particular the function* $f : A \longrightarrow B$ *is invertible in* Rel $\mathcal{A}$ *if and only if* $f \in \Sigma$, *whereupon its inverse is* f^{o}.

4. The properties of Map Rel $\mathcal{A}$

Let us now write $\mathcal{B}$ for the category Map Rel $\mathcal{A}$ of maps in Rel $\mathcal{A}$, with $J : \mathcal{A} \longrightarrow \mathcal{B}$ for the inclusion. It will be convenient henceforth to use Greek letters for elements of Σ, retaining lower–case Roman letters for general morphisms of $\mathcal{A}$. A morphism $P : A \longrightarrow B$ in $\mathcal{B}$, being a relation with a tabulation $\langle\alpha,f\rangle : P \longrightarrow A \times B$ where $\alpha \in \Sigma$, has by (3.3) the form

$$P = f\alpha^{o} . \tag{4.1}$$

If $Q : A \longrightarrow B$ in $\mathcal{B}$ has the tabulation $\langle\beta,g\rangle$, to say that $Q \le P$ is to say that we have

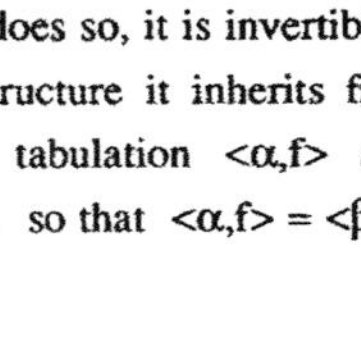

(4.2)

for some ε in $\mathcal{A}$; but since, by Proposition 2.3, ε like α and β lies in Σ and hence in $\mathcal{E}$, and since it also lies in $\mathcal{M}$ because $\langle\beta,g\rangle$ does so, it is invertible — so that $Q = P$. In other words, $\mathcal{B}$ is a mere category : the 2–category structure it inherits from Rel $\mathcal{A}$ is locally discrete. (It is convenient to speak as if the α in the tabulation $\langle\alpha,f\rangle$ above were always taken to be a *chosen* representative of an $\mathcal{M}$–subobject of A, so that $\langle\alpha,f\rangle = \langle\beta,g\rangle$ gives $\alpha = \beta$ and $f = g$.)

Of course *every* composite $g\beta^o$ in Rel $\mathcal{A}$ lies in $\mathcal{B}$, whether or not $\langle\beta,g\rangle$ lies in $\mathcal{M}$ and hence constitutes a relation. Then, as we saw in the last section, $g\beta^o$ is the relation $f\alpha^o$ tabulated by the *image* $\langle\alpha,f\rangle$ of $\langle\beta,g\rangle$; that is to say, we have a diagram (4.2) with $\langle\alpha,f\rangle \in \mathcal{M}$ and $\varepsilon \in \mathcal{E}$ — here too, ε is in fact in Σ by Proposition 2.3. More generally, consider an equation $f\alpha^o = g\beta^o$ where neither $\langle\alpha,f\rangle$ nor $\langle\beta,g\rangle$ need lie in $\mathcal{M}$. Let us systematically use the following notation, with bars, for pullbacks of the form

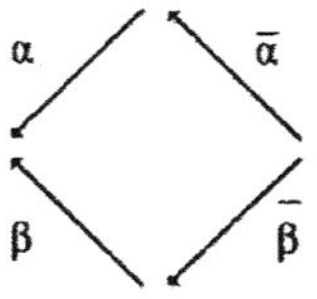

(4.3)

recalling that $\bar{\alpha}$ and $\bar{\beta}$ like α and β lie in Σ. By (3.1) and (3.3) we have $\beta^o\alpha = \bar{\beta}\,\bar{\alpha}^o$, so that (using Corollary 3.4 and the faithfulness of $\mathcal{A} \to \mathcal{B}$)

$$f\alpha^o = g\beta^o \ \text{in}\ \mathcal{B} \qquad \text{iff} \qquad f\bar{\alpha} = g\bar{\beta}\ \text{in}\ \mathcal{A}. \tag{4.4}$$

We saw in Corollary 3.4 that Σ consists precisely of the arrows inverted by $J : \mathcal{A} \to \mathcal{B}$. In fact:

Theorem 4.1 *$J : \mathcal{A} \to \mathcal{B}$ is the projection of $\mathcal{A}$ to its category of fractions $\mathcal{A}[\Sigma^{-1}]$.*

Proof We are to show that any $T : \mathcal{A} \to \mathcal{C}$ which inverts the elements of Σ is SJ for a unique $S : \mathcal{B} \to \mathcal{C}$. Uniqueness is clear: S must agree with T on objects, and by Corollary 3.4 it must be defined on the morphism $P = f\alpha^o$ of (4.1) with tabulation $\langle\alpha,f\rangle$ by $S(f\alpha^o) = Tf.(T\alpha)^{-1}$. It remains to show that S so defined is a functor. Clearly it preserves identities. As for composites, first consider $g\beta^o$ where $\langle\beta,g\rangle$ need not be in $\mathcal{M}$, its image being $\langle\alpha,f\rangle$ as in (4.2); we have $S(g\beta^o) = S(f\alpha^o) = Tf.(T\alpha)^{-1} = Tf.T\varepsilon.(T\varepsilon)^{-1}.(T\alpha)^{-1} = Tg.(T\beta)^{-1}$. It follows that, for a general composite as in (1.2), with p_1 and q_1 (and hence the pullback r_1 of q_1) in Σ, we have $S(QP) = T(q_2r_2).(T(p_1r_1))^{-1} = Tq_2.Tr_2.(Tr_1)^{-1}.(Tp_1)^{-1} = Tq_2.(Tq_1)^{-1}.Tp_2.(Tp_1)^{-1} = SQ.SP$.

The following is an instance of a general fact — see [9, Ch. I, §3] or [2, §1.7] — about calculi of right fractions, but here a direct proof is simple:

Theorem 4.2 *$\mathcal{B}$ has finite limits and J preserves these.*

Proof The terminal object of 1 of $\mathcal{A}$ is terminal also in $\mathcal{B}$; for if $g : A \longrightarrow 1$ is the unique morphism in $\mathcal{A}$ and $f\alpha^{\circ} : \mathcal{A} \longrightarrow 1$ is any morphism in $\mathcal{B}$, we have $f\alpha^{\circ} = g$ in $\mathcal{B}$ because $f = g\alpha$ in $\mathcal{A}$. It remains to show that a pullback (3.1) in $\mathcal{A}$ is also a pullback in $\mathcal{B}$; for then *arbitrary* morphisms $f\alpha^{\circ}$ and $g\beta^{\circ}$ of $\mathcal{B}$ admit a pullback, since α° and β° are invertible in $\mathcal{B}$. Suppose, then, that $h\gamma^{\circ} : D \longrightarrow A$ and $k\delta^{\circ} : D \longrightarrow B$ satisfy $fh\gamma^{\circ} = gk\delta^{\circ}$ in $\mathcal{B}$; in the notation of (4.3), (4.4) gives $fh\bar{\gamma} = gk\bar{\delta}$, whence $h\bar{\gamma} = r_1 m$ and $k\bar{\delta} = r_2 m$ for some m in $\mathcal{A}$; so that $h\gamma^{\circ} = r_1 P$ and $k\delta^{\circ} = r_2 P$ where $P = m\,\bar{\gamma}^{\circ}\gamma^{\circ} = m\,\bar{\delta}^{\circ}\delta^{\circ}$. As for uniqueness, suppose that $r_1 m\varepsilon^{\circ} = r_1 n\eta^{\circ}$ and $r_2 m\varepsilon^{\circ} = r_2 n\eta^{\circ}$; now (4.4) gives $r_1 m\bar{\varepsilon} = r_1 n\bar{\eta}$ and $r_2 m\bar{\varepsilon} = r_2 n\bar{\eta}$, so that $m\bar{\varepsilon} = n\bar{\eta}$, giving $m\varepsilon^{\circ} = n\eta^{\circ}$ by (4.4) again.

Theorem 4.3 *The morphism* $f\alpha^{\circ} : A \longrightarrow B$ *of* $\mathcal{B}$ *is a monomorphism in* $\mathcal{B}$ *if and only if* f *is monomorphic in* $\mathcal{A}$, *and is a strong epimorphism in* $\mathcal{B}$ *if and only if* f *lies in* $\mathcal{E}$. *The category* $\mathcal{B}$ *is regular, and the subobjects of* B *in* $\mathcal{B}$ *may be identified with the* $\mathcal{M}$*–subobjects of* B *in* $\mathcal{A}$.

Proof Since α° is invertible in $\mathcal{B}$, $f\alpha^{\circ}$ is monomorphic in $\mathcal{B}$ precisely when f is so. Let the pullback in $\mathcal{A}$ of f by itself be given, as in (3.1), by $r_1 : R \longrightarrow A$ and $r_2 : R \longrightarrow A$; by Theorem 4.2, this is also the pullback in $\mathcal{B}$. To say that f is monomorphic in $\mathcal{B}$ is to say that $r_1 = r_2$ in $\mathcal{B}$; which is equally to say that $r_1 = r_2$ in $\mathcal{A}$, or that f is monomorphic in $\mathcal{A}$. If the $(\mathcal{E},\mathcal{M})$–factorization of the monomorphism f is ip, we have $p \in \Sigma$, so that p is invertible in $\mathcal{B}$, and the arrow i of $\mathcal{M}$ represents the same subobject of B in $\mathcal{B}$ as f, or as $f\alpha^{\circ}$; of course different $\mathcal{M}$–subobjects of B in $\mathcal{A}$ give different subobjects of B in $\mathcal{B}$. The morphism $f\alpha^{\circ}$ is a strong epimorphism in $\mathcal{B}$ precisely when it — or equally f — factorizes in $\mathcal{B}$ through no proper subobject of B in $\mathcal{B}$; but if $f = ig\beta^{\circ}$ where $i \in \mathcal{M}$, so that $f\beta = ig$ in $\mathcal{A}$, then f factorizes through i in $\mathcal{A}$, since $\beta \in \mathcal{E}$ and we have a diagonal fill–in; thus $f\alpha^{\circ}$ is a strong epimorphism in $\mathcal{B}$ precisely when $f \in \mathcal{E}$. Clearly we get a (strong epimorphism, monomorphism) factorization of a general $f\alpha^{\circ}$ in $\mathcal{B}$ by taking the $(\mathcal{E},\mathcal{M})$–factorization of f in $\mathcal{A}$. That strong epimorphisms in $\mathcal{B}$ are stable by pullbacks, so that $\mathcal{B}$ is a regular category, is immediate from the above and Theorem 4.2, since $\mathcal{E}$ is stable by pullbacks in $\mathcal{A}$.

Write $\mathcal{K}$ for the 2–category of which an object is a finitely–complete $\mathcal{A}$ with a stable factorization system $(\mathcal{E},\mathcal{M})$, an arrow $T : \mathcal{A} \longrightarrow \mathcal{A}'$ is a left–exact functor with $T\mathcal{E} \subset \mathcal{E}'$ and $T\mathcal{M} \subset \mathcal{M}'$, and a 2–cell is a natural transformation. It has a full sub–2–category **Reg** given by the regular categories with $\mathcal{M}$ consisting of the monomorphisms; an arrow in $\mathcal{K}$ between regular categories is just a left–exact functor that preserves strong epimorphisms. By Theorems 4.2 and 4.3, the 2–functor J is an arrow in $\mathcal{K}$.

Proposition 4.4 $J : \mathcal{A} \longrightarrow \mathcal{B}$ *is the reflexion of the 2–category* $\mathcal{K}$ *into* **Reg**.

Proof Let $T : \mathcal{A} \longrightarrow \mathcal{C}$ in $\mathcal{K}$ where T is regular. Since T, being left exact, preserves monomorphisms, and since it takes an $\mathcal{E}$ to a strong epimorphism, it inverts each element of Σ. By Theorem 4.1, therefore, $T = JS$ for a unique $S : \mathcal{B} \longrightarrow \mathcal{C}$. By Theorems 4.2 and 4.3, S is an arrow in $\mathcal{K}$. That we have the universal property of J also for 2–cells is classical — see [9, Ch. I, Prop. 2.4] — for any category of fractions.

It is clear that any $T : \mathcal{A} \longrightarrow \mathcal{A}'$ in $\mathcal{K}$ induces a 2–functor $\mathrm{Rel}\, T : \mathrm{Rel}\, \mathcal{A} \longrightarrow \mathrm{Rel}\, \mathcal{A}'$; this sends A to TA, and sends $P : A \longrightarrow B$ to $TP : TA \longrightarrow TB$ tabulated by $\langle Tp_1, Tp_2 \rangle : TP \longrightarrow TA \times TB$; note that this lies in $\mathcal{M}'$, because it is in effect $T\langle p_1,p_2 \rangle : TP \longrightarrow T(A \times B)$; note further that $\mathrm{Rel}\, T$ is indeed a 2–functor, preserving composition by the properties of T, and preserving inequalities.

Theorem 4.5 *For* $\mathcal{B} = \mathrm{Map\, Rel}\, \mathcal{A}$, *the* $\mathrm{Rel}\, J : \mathrm{Rel}\, \mathcal{A} \longrightarrow \mathrm{Rel}\, \mathcal{B}$ *induced by* $J : \mathcal{A} \longrightarrow \mathcal{B}$ *is an isomorphism of 2–categories.*

Proof On objects, $\mathrm{Rel}\, J$ is bijective; and on each hom–category it is an isomorphism, since by Theorem 4.3 the subobjects of $A \times B$ in $\mathcal{B}$ can be taken to be the $\mathcal{M}$–subobjects of $A \times B$ in $\mathcal{A}$.

5. Comparison of $\mathcal{B}$ with $\Sigma^{\perp}$

Write $\mathcal{C}$ for the full subcategory $\Sigma^{\perp}$ of $\mathcal{A}$ given by those $C \in \mathcal{A}$ for which $\mathcal{A}(\alpha,C)$ is invertible for all $\alpha \in \Sigma$, with $I : \mathcal{C} \longrightarrow \mathcal{A}$ for the inclusion. Then $\mathcal{C}$ is closed under limits in $\mathcal{A}$, and so tends in favourable cases to be reflective in $\mathcal{A}$; it is so for each of our examples in the Introduction.

Proposition 5.1 $JI : \mathcal{C} \longrightarrow \mathcal{B}$ *is an equivalence of categories if and only if* $\mathcal{C}$ *is reflective in* $\mathcal{A}$ *and the units of this reflexion lie in* Σ.

Proof If JI is an equivalence, let $T : \mathcal{B} \longrightarrow \mathcal{C}$ be an equivalence–inverse. Then $JIT \cong 1$, so that for each A we have $A \cong TA$ in $\mathcal{B}$. By Corollary 3.4, this isomorphism has the form $\beta\alpha^{\circ}$ where $\alpha : B \longrightarrow A$ and $\beta : B \longrightarrow TA$. Since $TA \in \mathcal{C}$ and $\alpha \in \Sigma$, we have $\beta = \gamma\alpha$ for some $\gamma : A \longrightarrow TA$; and because $\gamma \in \Sigma$ by Proposition 2.3, it is clearly the reflexion of A in $\mathcal{C}$. Suppose now that $\mathcal{C}$ is reflective, I having the left adjoint R, with identity counit $RI = 1$ and the unit $\rho : 1 \longrightarrow IR$; and suppose that each $\rho_A : A \longrightarrow RA$ lies in Σ. Since R inverts the elements of Σ we have $R = TJ$ for a unique T. Now $TJI = RI = 1$, and $J\rho : J \longrightarrow JITJ$ is of the form σJ for some $\sigma : 1 \longrightarrow JIT$

(either by a simple direct argument, or by the two–dimensional part of the universal property of J). Since $\sigma_A = \rho_A$ is invertible in $\mathcal{B}$, we have $1 \cong JIT$, and T is an equivalence–inverse of JI.

It is easy to see that, whenever $\mathcal{C}$ is reflective, the units $\rho_A : A \to RA$ lie in $\mathcal{E}$. It suffices to see that $\mathcal{C}$ is closed under $\mathcal{M}$–subobjects in $\mathcal{A}$. However if $i : D \to C$ lies in $\mathcal{M}$ and $C \in \mathcal{C}$, then for any $\alpha : X \to Y$ in Σ and any $f : X \to D$, we have $if = g\alpha$ for some g; now because $i \in \mathcal{M}$ and $\alpha \in \mathcal{E}$ we have a diagonal fill–in giving $f = h\alpha$ (and h is unique because α is epimorphic); thus $D \in \mathcal{C}$. The author cannot see, however, why ρ_A should be monomorphic in general.

Carboni and Mantovani characterize in [6] those categories $\mathcal{A}$ that appear as the categories of separated objects for a topology on a category $\mathcal{P}$ that is either a pretopos or else abelian. Such an $\mathcal{A}$ is certainly regular, as they show, and has a second stable factorization system (epimorphisms, strong monomorphisms). When we take this as our $(\mathcal{E}, \mathcal{M})$, Σ is the class of monomorphisms that are also epimorphisms. Moreover it follows from [6, Lemma 1.5.2] that $\mathcal{C} = \Sigma^{\perp}$ is just the category of sheaves for the topology; and since $\mathcal{A}$ is reflective in $\mathcal{P}$, $\mathcal{C}$ is reflective in $\mathcal{A}$ if and only if it is reflective in $\mathcal{P}$. Even when this is the case, however, the author still fails to see why $\rho_A : A \to RA$ should be monomorphic for a separated A; for he knows of no proof, in this generality, that the separated objects are subobjects of sheaves, when the sheaves are reflective.

What can certainly be said is the following. Suppose that $\mathcal{C}$ is a localization of $\mathcal{P}$, and that we take that topology whose dense monomorphisms are precisely those inverted by the reflexion of $\mathcal{P}$ into $\mathcal{C}$. Now $\mathcal{C}$ is in fact, by [3, Proposition 2.3], the category of sheaves for this topology, and we can take $\mathcal{A}$ to be the category of separated objects. Here it is indeed the case that ρ_A is monomorphic for a separated A, since the equalizer of the kernel–pair x, y of ρ_A is inverted by R and thus, as a dense monomorphism in $\mathcal{A}$, is epimorphic in $\mathcal{A}$, giving $x = y$. In this case, therefore, $JI : \mathcal{C} \to \mathcal{B}$ is indeed an equivalence. A typical example, with $\mathcal{P}$ abelian, is that above where $\mathcal{A}$ is the category of short exact sequences of abelian groups.

REFERENCES

[1] M. Barr, P.A. Grillet, and D.H. van Osdol, *Exact Categories and Categories of Sheaves*, Lecture Notes in Math. 236 (Springer–Verlag, Berlin, Heidelberg, New York, 1971).

[2] J. Bénabou, Some remarks on 2–categorical algebra (Part I), *Bull. Soc. Math. Belgique* 41(1989), 127–194.

[3] F. Borceux and G.M. Kelly, On locales of localizations, *J. Pure Appl. Algebra* 46(1987), 1–34.

[4] A. Carboni, S. Kasangian, and R. Street, Bicategories of spans and relations, *J. Pure Appl. Algebra* 33(1984), 259–267.

[5] A. Carboni and R. Street, Order ideals in categories, *Pacific J. Math.* 124(1986), 275–288.

[6] A. Carboni and S. Mantovani, An elementary characterization of categories of separated objects, Quaderno no. 47/1990, Dip. di Matematica, Univ. di Milano.

[7] P.J. Freyd and G.M. Kelly, Categories of continuous functors I, *J. Pure Appl. Algebra* 2(1972), 169–191.

[8] P.J. Freyd and A. Scedrov, *Categories, Allegories,* (North-Holland, Amsterdam, New York, Oxford, Tokyo, 1990).

[9] P. Gabriel and M. Zisman, *Calculus of Fractions and Homotopy Theory* (Springer-Verlag, Berlin, Heidelberg, New York, 1967).

[10] G.M. Kelly, Monomorphisms, epimorphisms, and pull-backs, *J. Austral. Math. Soc.* 9(1969), 124–142.

[11] A. Klein, Relations in categories, *Illinois J. Math.* 14(1970), 536–550.

[12] J. Meisen. On bicategories of relations and pullback spans, *Communications in Algebra* 1(1974), 377–401.

[13] G. Richter, Mal'cev conditions for categories, in : *Categorical Topology, Proc. Conference Toledo, Ohio 1983* (Helderman Verlag, Berlin, 1984), 453–469.

This paper is in final form and will not be published elsewhere.

Algebras for the Partial Map Classifier Monad

Anders Kock
Matematisk Institut, Aarhus Universitet
DK 8000 Aarhus C, Denmark

Dedicated to the always helpful Max Kelly

Introduction

The category of algebras for the partial map classifier monad is shown to be the category of posets which have suprema for all subsets with at most one element, and which are *shallow*, by which we mean that for any pair of elements a, b,

$$a \leq b \text{ iff } a \text{ is the supremum of the subset } \{a\} \cap \{b\}.$$

On any elementary topos $\mathcal{E}$, one has the functor which to an object A associates the object $TA = \tilde{A}$ which classifies partial maps into A (cf e.g. [J] 1.2). This functor T carries a monad structure $\mathbf{T}= (T, \eta, \mu)$; it is a submonad of the power "set" monad $\mathbf{P}= (P, \eta, \mu)$, as described in, say, [AL],[Mi], or [J] 5.3. We shall analyze the category of algebras for $\mathbf{T}$, for the category of sets, but our arguments and constructions will be intuitionistically pure, so that everything carries over to an arbitrary elementary topos. In fact, when applied to the "usual" boolean category of sets, our notions are rather trivial, or even laughable; for instance, the category of $\mathbf{T}$-algebras is, in this case, just the category of pointed sets, and the proof of this fact is trivial.

From [Ma] , we know that the category of algebras for the power set monad $\mathbf{P}$ is the category of cocomplete posets, with the structure map being the formation of suprema. Since $TA \subseteq PA$ consists of those subsets of A which have at most one element, it has been conjectured that the category of algebras for T should be some category of posets with the (weak!) cocompleteness property that any subset with at most one element has a supremum, but the question was how to construct an order on A out of an algebra structure $\xi : TA \rightarrow A$; for the full power set monad such order is given by:

$$a \leq b \iff b = \xi(\{a, b\}),$$

(which it must be if ξ is to be supremum formation), but for $\mathbf{T}$, this does not work, since $\{a, b\}$ in general has more than one element. The theory of actions by the subobject classifier Ω, developed in Section 2, is the tool that is going to give us the desired order.

This theory of Ω-actions may have some independent interest. In fact, the main technical theorem is Theorem 2.3, where we derive a shallow ordering out of any "admissible" Ω-action; the result about the category of **T**- algebras is really a corollary, and appears as Theorem 5.3. A final paragraph deals with the question of products of families of weakly cocomplete posets. A preliminary version, containing Theorem 5.3, but Theorem 2.3 only in implicit form, appeared as [K] ; we should warn the readers of [K] that we use the words 'subsingleton' and 'shallow' in a slightly different sense here. I would like to thank Michel Thiebaud, Jørn Schmidt, and Wesley Phoa for stimulating discussions leading to the preliminary version; in particular, Wesley Phoa called my attention to actions of Ω on partial map classifiers, an issue which had been discussed in correspondence between him and Dana Scott in the context of effective domains. And I would like to thank Reinhard Börger and Bill Lawvere for interest and comments that were valuable for the present expanded version.

This article is in final form, and will not be published elsewhere.

1 Some order theoretic notions

A set U with at most one element is called a *subterminal* set, because having at most one element is equivalent to the unique map $U \to 1$ being monic. This property may also be described:

$$\forall x, y \in U : x = y.$$

A subset $U \subseteq A$ of a set A is called a *subsingleton* if there exists some $a \in A$ with $U \subseteq \{a\}$,

$$\exists a \in A \, \forall x \in U : \; x = a.$$

This clearly implies that U is subterminal. (Some sets have the "flabbyness" property that subterminal subsets of it are subsingletons ; but for instance the empty set does not.)

A subset U of a partially ordered set A is called *bounded* if there exists some upper bound for it. Clearly subsingleton subsets of an ordered set are bounded; subterminal subsets need not be (again look at the empty subset of the empty set). A partially ordered set $(A, \leq)$ will be called *subterminal cocomplete*, respectively *subsingleton cocomplete*, respectively *bounded-cocomplete*(=*conditionally cocomplete*) if every subterminal respectively subsingleton, respectively bounded subset $U \subseteq A$ has a supremum . Since

	$\Rightarrow$	subterminal
subsingleton		
	$\Rightarrow$	bounded

the corresponding cocompleteness notions for a poset $(A, \leq)$ relate the opposite way; so that subsingleton cocompletenes (s.s cocompleteness, for short) is the weakest among those considered. The most important for the present purpose is the subterminal-cocompleteness(=s.t. cocompleteness), for which we also use the word *weak* cocompleteness. An order preserving map between s.s. cocomplete posets will be called s.s. cocontinuous if it preserves suprema of s.s. subsets, and similarly for the other notions. Note that the empty set is s.s. cocomplete but not s.t. cocomplete.

We may remark without proof that s.s. cocompleteness for a poset A is equivalent to A being a *tensored* Ω-enriched category, cf. [Ke] for this notion.

For an s.s. cocomplete poset, the supremum in the following definition is apriori known to exist, but the definition makes sense for any poset:

Definition 1.1 *A poset* $(A, \leq)$ *is* shallow *if for any* $a, b \in A$

$$a \leq b \qquad \text{iff} \qquad a = \sup\{x \mid x = b \ \wedge a = b\}.$$

Note that the set over which we are forming supremum here is a subsingleton, and it can be written in various ways:

$$\{x \mid x = b \text{ and } a = b\}$$
$$= \{x \mid x = a \text{ and } a = b\}$$
$$= \{x \mid x = a \text{ and } x = b\} = \{a\} \cap \{b\}.$$

Leaving the intuitionistic purity aside for a moment, a weakly cocomplete poset is just a poset with a bottom element $\perp$, namely the supremum of the empty subset, which is the only subterminal subset which is not a singleton. Among the posets with bottom, the shallow ones are characterized by

$$a \leq b \text{ iff } (a = b \text{ or } a = \perp),$$

so they look like

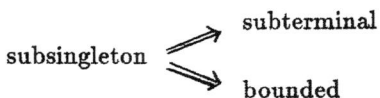

i.e. the bottom is not deep, whence the name "shallow". The word "flat domain" has also been used for such posets, in the boolean case.

2 Order derived from Ω−actions

We consider the set Ω of truth values (Ω = the subobject classifier). It is a monoid under conjunction $\wedge$, with the truth value 'true' as unit. Also, Ω carries a natural partial order, $a \leq b$ iff $a \wedge b = a$.

Definition 2.1 *An action by the monoid Ω on a set A is called* admissible *if for all $a, b \in A$,*

$$[a = b] \cdot a = [a = b] \cdot b, \tag{1}$$

where $[a = b]$ is the truth value of the statement $a = b$.

(Generally, we let the symbol ' [...]' mean 'truth value of'.)

Lemma 2.2 *Given an admissible Ω−action on a set A, and let $a, b \in A$. Then the following three conditions are equivalent*

$$a = [a = b] \cdot a \tag{2}$$

$$a = [a = b] \cdot b \tag{3}$$

$$a = \lambda \cdot b \text{ for some } \lambda \in \Omega. \tag{4}$$

Proof. (2) and (3) are equivalent, by the admissibility assumption. Clearly (3) implies (4). Assume (4). To prove (2) means to prove the first equality sign in (5)

$$\lambda \cdot b = [\lambda \cdot b = b] \cdot \lambda \cdot b = ([\lambda \cdot b = b] \wedge \lambda) \cdot b \tag{5}$$

(the second follows by associativity of the action). But we have $\lambda \leq [\lambda \cdot b = b]$; for, if λ, then $\lambda \cdot b = b$, since true $\cdot\, b = b$. So $[\lambda \cdot b = b] \wedge \lambda = \lambda$, and thus the right hand side in (5) equals $\lambda \cdot b$.

Theorem 2.3 *Given an admissible Ω-action on a set A. For $a, b \in A$, write $a \leq b$ if the equivalent conditions in Lemma 2.2 hold. Then*

1) The binary relation $\leq$ thus defined makes A into a shallow partial order;

2) the action $\Omega \times A \to A$ is order preserving in each variable separately;

3) for each $a \in A$, we have an adjointness

$$\Omega \rightleftharpoons A : \quad - \cdot a \dashv [a \leq -];$$

4) A is conditionally cocomplete, with

$$\sup U = [b \in U] \cdot b$$

for any bounded set U, and any bound b for it.

5) For each $a \in A$, the poset $\downarrow a$ of elements below a is a frame, isomorphic to a quotient frame of Ω.

(In 2)-5), the notions refer to the order $\leq$ given in 1).)

Proof. We use the criterion (4) in Lemma 2.2 for almost all the arguments. To prove 1): Since true $\cdot\, a = a$, $a \leq a$, proving reflexivity. To prove transitivity, assume $a \leq b$ and $b \leq c$, so $a = \lambda \cdot b$ and $b = \mu \cdot c$, for suitable $\lambda, \mu \in \Omega$. So

$$a = \lambda \cdot b = \lambda \cdot (\mu \cdot c) = (\lambda \wedge \mu) \cdot c,$$

so $a \leq c$. Finally, to prove antisymmetry, let $a \leq b$ and $b \leq a$, so $a = \lambda \cdot b$ and $b = \mu \cdot a$. Then

$$\lambda \cdot a = \lambda \cdot (\lambda \cdot b) = (\lambda \wedge \lambda) \cdot b = \lambda \cdot b = a \tag{6}$$

so

$$\begin{aligned} b &= \mu \cdot a = \mu \cdot \lambda \cdot a \\ &= (\mu \wedge \lambda) \cdot a = \lambda \cdot \mu \cdot a = \lambda \cdot b = a, \end{aligned}$$

using associativity of the action, and commutativity of $\wedge$. This proves 1), except for shallowness, which will be proved together with 4) below.

2) If $\lambda \leq \mu$, then $\lambda = \lambda \wedge \mu$, and so

$$\lambda \cdot a = (\lambda \wedge \mu) \cdot a = \lambda \cdot (\mu \cdot a) \leq \mu \cdot a.$$

If $a \leq b$, $a = \mu \cdot b$ for some μ, and so

$$\lambda \cdot a = \lambda \cdot \mu \cdot b = \mu \cdot (\lambda \cdot b) \leq \lambda \cdot b.$$

Before proving the adjointness assertion, we prove

Lemma 2.4 *The partial order $\leq$ on A has binary inf-formation given by*

$$a \wedge b := [a = b] \cdot a = [a = b] \cdot b, \tag{7}$$

(the last equality by the admissibility assumption).

Proof. Clearly $a \wedge b$ thus defined is $\leq a$ and $\leq b$ in virtue of the equations in (7), using (4). Conversely, if $c \leq a$ and $c \leq b$

$$c = [a = c] \cdot c = [b = c] \cdot c$$

by (2), whence

$$\begin{aligned} c &= ([a = c] \wedge [b = c]) \cdot c \\ &= ([a = b] \wedge [a = c] \wedge [b = c]) \cdot c \\ &\leq [a = b] \cdot c \leq [a = b] \cdot b = a \wedge b, \end{aligned}$$

using assertion 2) of the Theorem for the two inequality signs. This proves the Lemma.

3) Clearly $[a \leq -] : A \rightarrow \Omega$ is order preserving, and $- \cdot a : \Omega \rightarrow A$ is order preserving, by assertion 2) of the Theorem. We have, for $\lambda \in \Omega$,

$$\lambda \leq [x \leq \lambda \cdot x]; \tag{8}$$

for, if λ, $x = \lambda \cdot x$, since $x = \text{true} \cdot x$ (cf. the proof of Lemma 2.2). So if λ, $x \leq \lambda \cdot x$, proving (8). To get the other inequality for adjointness, we use existence of binary infs, as asserted by Lemma 2.4; we have

$$\begin{aligned} [x \leq y] \cdot x &= [(x \wedge y) = x] \cdot x \\ &= [(x \wedge y) = x] \cdot (x \wedge y) \text{ by admissibility} \\ &\leq [(x \wedge y) = x] \cdot y \leq y. \end{aligned}$$

4) Let U be bounded by b. To prove that $[b \in U] \cdot b$ is $\sup(U)$, we first prove that it is an upper bound for U; so let $a \in U$. Then $a = \lambda \cdot b$, since $a \leq b$ by assumption. We should prove $a \leq [b \in U] \cdot b$, or equivalently $\lambda \cdot b \leq [b \in U] \cdot b$. It suffices to prove $\lambda \leq [b \in U]$. But if λ, $b = \lambda \cdot b = a \in U$, so $b \in U$. Conversely, let c be an upper bound for U. To prove $[b \in U] \cdot b \leq c$, it suffices by the adjointness (assertion 4)) to prove $[b \in U] \leq [b \leq c]$. But if $b \in U$, $b \leq c$, since c was asssumed to be a bound for U.

Shallowness of the order is now easy: for any $a, b, \{a\} \cap \{b\}$ is bounded by a, so by 4),

$$\sup(\{a\} \cap \{b\}) = [a \in \{a\} \cap \{b\}] \cdot a = [a = b] \cdot a$$

which equals a iff $a \leq b$, by (2).

To prove the final assertion 5), we first prove that for fixed $a \in A$, the map $- \cdot a : \Omega \rightarrow A$ preserves $\wedge$. But

$$\begin{aligned} (\lambda \cdot a) \wedge (\mu \cdot a) &= [\lambda \cdot a = \mu \cdot a] \cdot \lambda \cdot a \\ &= [\lambda \cdot a = \mu \cdot a] \cdot \mu \cdot a \\ &= [\lambda \cdot a = \mu \cdot a] \cdot \lambda \cdot \mu \cdot a, \end{aligned} \tag{9}$$

the last equality since anything of the form $\rho \cdot \lambda \cdot a$ is fixed by multiplication by λ, by the same argument as in (6). From (9) follows

$$(\lambda \cdot a) \wedge (\mu \cdot a) \leq \lambda \cdot \mu \cdot a;$$

the other inequality is obvious, so

$$(\lambda \cdot a) \wedge (\mu \cdot a) = \lambda \cdot \mu \cdot a = (\lambda \wedge \mu) \cdot a$$

Since $\lambda \cdot a \leq a$, it follows that $- \cdot a : \Omega \rightarrow A$ factors through $\downarrow a \subseteq A$. As a map $\Omega \rightarrow \downarrow a$, it not only preserves $\wedge$ (by the above), but also top element. Also, from assertion 3) in the Theorem, it follows that it has a right adjoint, namely the restriction of $[a \leq -]$ to $\downarrow a$.. Hence it is a frame map; and it is surjective, by the criterion (4) for inequality. So it makes $\downarrow a$ a quotient frame of Ω. (The corresponding nucleus on Ω evidently is the composite of the two adjoints in assertion 3).)

This proves the Theorem.

Let us apply the Theorem to the case of Ω acting on itself. The derived order is then the natural one. From assertion 4), we then get the following, probably well known,

Corollary 2.5 *Let $U \subseteq \Omega$. Then* $\sup U$ *exists and equals the truth value* [true $\in U$].

3 Admissible Ω-actions derived from order

In the following, more than one partial order on the same set will be considered, prompting us to use the symbol $\sqsubseteq$ for one of these orders.

Theorem 3.1 *Let $A, \sqsubseteq$ be a poset which is subsingleton cocomplete. For $\lambda \in \Omega$ and $a \in A$, we put*

$$\lambda \cdot a := \sup\{x \mid x = a \text{ and } \lambda\}; \tag{10}$$

this is a unitary and associative action by the monoid $\Omega, \wedge$. It is admissible, and the partial order $\leq$ derived from the action satisfies

$$a \leq b \text{ implies } a \sqsubseteq b, \tag{11}$$

i.e. is weaker than the original order.

Proof. (All suprema here are with respect to the order $\sqsubseteq$.) We have

$$\text{true} \cdot a = \sup\{x \mid x = a \text{ and true}\} = \sup\{a\} = a,$$

proving that the action is unitary. To prove the associativity assertion,

$$\begin{aligned} \lambda \cdot (\mu \cdot a) &= \sup\{x \mid x = \mu \cdot a \text{ and } \lambda\} \\ &= \sup\{x \mid x = \sup\{y \mid y = a \text{ and } \mu\} \text{ and } \lambda\} \\ &= \sup\{y \mid y = a \text{ and } \mu \text{ and } \lambda\} \\ &= \sup\{y \mid y = a \text{ and } \lambda \wedge \mu\} \end{aligned}$$

by the standard rewriting of a sup of sups as a single sup. To prove the admissibility condition

$$[a = b] \cdot a = [a = b] \cdot b, \tag{12}$$

note that the two sides here are formed as suprema of the two sets

$$\{x \mid x = a \text{ and } a = b\}, \text{ resp. } \{x \mid x = b \text{ and } a = b\},$$

but these two sets are equal, by transitive law for equality.

Finally, to prove (11), recall that $a \leq b$ means that for some $\lambda \in \Omega$, $a = \lambda \cdot b$, so

$$a = \lambda \cdot b = \sup\{x \mid x = b \text{ and } \lambda\} \sqsubseteq \sup\{x \mid x = b\} = b,$$

the inequality because we are forming supremum over a larger set.

Recall from Section 2 that the order derived from an admissible Ω–action is always conditionally cocomplete. So whether or not $A, \sqsubseteq$ is conditionally cocomplete, it will be conditionally cocomplete in its new (weak) order $\leq$; furthermore

Proposition 3.2 *Let $A, \sqsubseteq$ and $A, \leq$ be as in Theorem 3.1. The identity map on A defines a map*

$$(A, \leq) \longrightarrow (A, \sqsubseteq) \tag{13}$$

(order preserving by (11)), which is conditionally cocontinuous. In particular, the Ω–actions derived from $\leq$ and $\sqsubseteq$ agree.

Proof. Let $U \subseteq A$ be bounded, by $b \in A$, say, w.r.to the order $\leq$. Then by Theorem 2.3, 4),

$$\begin{aligned} \sup_{\leq}(U) &= [b \in U] \cdot b \\ &= \sup_{\sqsubseteq}\{x \mid x = b \text{ and } b \in U\} \\ &\sqsubseteq \sup_{\sqsubseteq}\{x \mid x \in U\} \\ &= \sup_{\sqsubseteq}(U) \end{aligned}$$

The other inequality is clear, just because we are considering an order preserving map (12). The last assertion follows because the $\lambda \cdot a$'s for the two actions are defined by suprema, for the two orders, of $\{x \mid x = a \text{ and } \lambda\}$, but a set of this form is bounded (by a) in whatever order, so the suprema, for the two orders, of this set agree, by the first assertion of the Proposition.

Proposition 3.3 *Among all orders on A which are subsingleton cocomplete, and which induce a given Ω–action on A, there is a weakest one; and it is shallow, in fact the only shallow one among this class of orders.*

Proof. The weak order $\leq$ induced by the Ω–action is weaker than the others, by (11), and is shallow, by Theorem 2.3. If a shallow order $\sqsubseteq$ is subsingleton cocomplete, and induces the given action,

$$a \sqsubseteq b$$

iff

$$a = \sup\{x \mid x = b \text{ and } a = b\} = [a = b] \cdot b$$

iff

$$a \leq b.$$

4 Admissible Ω-actions derived from T-algebra structures

In this paragraph, $\mathbf{T}=(T,\eta,\mu)$ denotes the partial map classifier monad discussed in the introduction. We have

$$T1 = P1 = \Omega = \text{ set of truth values.}$$

The monoid structure $\wedge$ on this set may be understood in a more sophisticated manner: the monad $\mathbf{T}$ is a monoidal monad, in fact a sub-monoidal-monad of the power set monad, whose monoidal structure $\psi : PA \times PB \longrightarrow P(A \times B)$ is described by $\psi(X,Y) = X \times Y \subseteq A \times B$, as in [K1]; the same description yields $\psi : TA \times TB \longrightarrow T(A \times B)$. Since $\mathbf{T}$ is a monoidal monad, $T1$ acquires a monoid structure

$$\psi_{1,1} : T1 \times T1 \longrightarrow T(1 \times 1) \cong T1,$$

which is just the $\wedge$ on Ω. More generally, the monoid $\Omega = T1$ acts on any TA

$$\psi_{1,A} : \Omega \times TA = T1 \times TA \longrightarrow T(1 \times A) \cong TA. \tag{14}$$

The action of $\Omega = T1$ on any TA may be described set theoretically by

$$\lambda \cdot X = \{a \mid a \in X \text{ and } \lambda\}, \tag{15}$$

(for $X \subseteq A$ a subterminal subobject). If $f : A \longrightarrow B$, $Tf : TA \longrightarrow TB$ will preserve the Ω-action.

Now let $\alpha : TA \longrightarrow A$ be a $\mathbf{T}$-algebra structure on A. Then we provide A with an Ω–action by posing, for $\lambda \in \Omega$, $x \in A$

$$\lambda \cdot x := \alpha(\lambda \cdot \{x\}),$$

where $\lambda \cdot \{x\}$ is the action by Ω on TA given by (14) or (15). Let us for the moment call this the *structure induced* action on the algebra (A, α). The question arises whether the structure induced action on a (free) algebra (TA, μ_A) equals the action given by (14) or (15). The answer is yes: it amounts to proving that for any subterminal subset $X \subseteq A$,

$$\{x \mid x \in X \text{ and } \lambda\} = \bigcup\{Y \mid Y = X \text{ and } \lambda\}$$

(the right hand side being the one induced by the structure μ_A, since μ is just union formation). If x belongs to the left hand side, $X = \{x\}$ (since X is subterminal), and λ; so $\{x\}$ appears as a Y participating in the union on the right, hence x belongs to the union. Conversely, if x belongs to the union, we have for some subterminal Y that $x \in Y$ and $Y = X$ and λ. So $x \in X$ and λ, proving that x belongs to the left hand side.

It is easy to prove that any $\mathbf{T}$-algebra homomorphism $f : (A, \alpha) \to (B, \beta)$ preserves the structure-induced Ω-actions: for $x \in A$,

$$\begin{aligned} f(\lambda \cdot x) &= f(\alpha(\lambda \cdot \{x\})) = \beta(Tf(\lambda \cdot \{x\})) \\ &= \beta(\lambda \cdot Tf(\{x\})) \text{ since } Tf \text{ preserves action} \\ &= \beta(\lambda \cdot \{f(x)\}) \\ &= \lambda \cdot f(x). \end{aligned}$$

Since $\alpha : TA \to A$ is a $\mathbf{T}$-algebra homomorphism $(TA, \mu_A) \to (A, \alpha)$, it follows that it preserves Ω-action, and since α is a (split) surjection and the Ω-action on TA is an associative and unitary action by the monoid $(\Omega, \wedge)$, it follows that the Ω-action on A is likewise associative and unitary.

Thus we have proved the first two assertions in

Proposition 4.1 *For any $\mathbf{T}$-algebra (A, α), (14) or (15) defines an associative and unitary action by $(\Omega, \wedge)$ on A. Any $\mathbf{T}$-algebra homomorphism commutes with such actions. And such actions are admissible in the sense of Definition 2.1.*

Proof of the last assertion. Let $a, b \in A$. We have

$$\begin{aligned} [a = b] \cdot \{a\} &= \{x \mid x = a \text{ and } a = b\} \\ &= \{x \mid x = b \text{ and } a = b\} \\ &= [a = b] \cdot \{b\}. \end{aligned}$$

Applying α to this equation yields $[a = b] \cdot a = [a = b] \cdot b$, as desired.

With the Ω-action thus defined, the theory developed in Section2 furnishes the underlying set A of any $\mathbf{T}$-algebra (A, α) with a partial order $\leq$, which is conditionally cocomplete and shallow. Explicitly,

$$a \leq b \text{ iff } a = \alpha(\{x \mid x = b \text{ and } \lambda\}) \tag{16}$$

for some $\lambda \in \Omega$. But the order (16) is in fact also weakly cocomplete with α as sup-formation:

Theorem 4.2 *Let (A, α) be a $\mathbf{T}$-algebra. Let $X \subseteq A$ be a subterminal subset with $\alpha(X) = x$. Then $x = \sup X$ with respect to the order $\leq$ described by (16). In particular, $(A, \leq)$ is weakly cocomplete.*

Proof. We first prove that

$$[x \in X] \cdot x = x \tag{17}$$

by proving the following equality of subsets of A

$$[x \in X] \cdot \{x\} = X, \tag{18}$$

from which (17) follows by applying α (which commutes with action). To see (18), let $y \in [x \in X] \cdot \{x\}$. This means that $x \in X$ and $y \in \{x\}$, so $y = x \in X$, so $y \in X$. Conversely let $y \in X$. Then $X = \{y\} = \eta(y)$, (since X is subterminal), so $\alpha(X) = y$, so $x = y$. So $y \in \{x\}$ and $x \in X$. This means $y \in [x \in X] \cdot \{x\}$.

From (17), we can prove the Theorem. Let $y \in X$. Then $X = \{y\}$, whence (as above) $x = y$, whence $y \leq x$, so x is a bound for X. Conversely, suppose z is a bound for X. Then $[x \in X] \leq [x \leq z]$. Since the action is order preserving in the first variable (Theorem 2.3), $[x \in X] \cdot x \leq [x \leq z] \cdot x \leq z$,the last inequality by the back adjunction for the adjointness of Theorem 2.3, 3).

The Theorem is proved.

Proposition 4.3 *Let A be the underlying set of a* $\mathbf{T}$*-algebra. Then A is flabby, in the sense that every subterminal subset of it is a subsingleton.*

Proof. Let X be a subterminal subset; its supremum x exists by the theorem. In the proof of the theorem, we observed that if $y \in X$, then $y = x$; hence $X \subseteq \{x\}$, so X is a subsingleton.

Remark 4.4 There is a rather evident converse of the Theorem. If $(A, \sqsubseteq)$ is a weakly cocomplete poset, supremum formation for subterminals provides a map

$$\sup : TA \longrightarrow A;$$

this map is a $\mathbf{T}$-algebra structure on A. The order $\leq$ on A induced from the $\mathbf{T}$-algebra structure (via the Ω-action) is weaker than the original order, by Proposition 3.3, since it has the same supremum formation for subterminals, hence the same Ω-action, as the given order $\sqsubseteq$. Also, $\leq$ is shallow, by Theorem 2.3 (1). So if the original order $\sqsubseteq$ is shallow as well, it agrees with $\leq$, by Proposition 3.3.

This and related results will be formulated in functorial terms in the following section.

5 Functorality, and fullness of the functors

We first argue that the construction of Section 2 defines a functor (which preserves underlying sets)

$$\text{Admissible } \Omega\text{-actions} \longrightarrow \text{Subsingleton cocomplete posets;} \qquad (19)$$

the morphisms of the two categories here are, respectively, action preserving maps, and order preserving maps which preserve suprema of subsingleton subsets ('s.s. cocontinuous maps'). If $f : A \to B$ is action preserving, it is immediate from the characterization (4) of the constructed order relation $\leq$ that f is order preserving.

Also, if U is a subsingleton subset of A, say $U \subseteq \{a\}$, then U is bounded by a and $f(U)$ by $f(a)$, so by Theorem 2.3 (4),

$$\sup U = [a \in U] \cdot a \text{ and } \sup f(U) = [f(a) \in f(U)] \cdot f(a),$$

so that

$$f \sup U = f([a \in U] \cdot a) = [a \in U] \cdot f(a) \leq [f(a) \in f(U)] \cdot f(a),$$

using $[a \in U] \leq [f(a) \in f(U)]$, so that $f(\sup U) \leq \sup f(U)$. The other inequality is obvious. (The argument actually works for any bounded U.) So f is s.s.cocontinuous, and we have a functor (19).

Proposition 5.1 *The functor (19) is full and faithful.*

Proof. It is faithful because it preserves underlying sets. To see that it is full, let $f : (A, \leq) \to (B, \leq)$ be s.s.cocontinuous, where the orders $\leq$ on A and B are derived from an admissible Ω-action. For $\lambda \in \Omega, a \in A$, we have, by Theorem 2.3 4), applied to the subsingleton $U := \lambda \cdot \{a\} \subseteq \{a\}$:

$$\sup(\lambda \cdot \{a\}) = [a \in \lambda \cdot \{a\}] \cdot a = \lambda \cdot a, \tag{20}$$

so

$$f(\lambda \cdot a) = f(\sup(\lambda \cdot \{a\}) = \sup(f(\lambda \cdot \{a\})),$$

but since $f(\lambda \cdot \{a\}) = \lambda \cdot \{f(a)\}$,

$$\sup(f(\lambda \cdot \{a\}) = \sup(\lambda \cdot \{f(a)\}) = \lambda \cdot f(a),$$

the last equality in analogy with (20). This proves that f commutes with action.

The construction of Section 3 defines a functor, likewise preserving underlying sets, in the opposite direction of (19) (and which is actually left inverse of (19)). It is again trivially faithful, but evidently not full: in the category of boolean sets, a map may preserve bottom element without being order preserving.

Next we consider the functorality of the constructions of Section 4. We remarked here that $\mathbf{T}$ -algebra homomorphisms are also homomorphisms of Ω-actions, so that we have a functor (preserving underlying sets)

$$\mathbf{T} - \mathit{algebras} \longrightarrow \Omega\text{-actions}. \tag{21}$$

Proposition 5.2 *The functor (21) is full and faithful.*

Proof. 'Faithful' is again clear. To prove that it is full, let $f : A \to B$ preserve the Ω-action defined by $\mathbf{T}$-algebra structures α and β on A and B, cf. (14) or (15). Then by (19) f is s.s.cocontinuous for the order derived from the action. Since subterminals in A and B are subsingletons, by Proposition 3.3, it follows that f is weakly cocontinuous; but by Theorem 4.2, α and β agree with supremum formation for subterminal subsets, so f is an algebra homomorphism.

The results proved so far can be put together in a main theorem:

Theorem 5.3 *The category of* $\mathbf{T}$*-algebras is equivalent to the full subcategory of the category of weakly cocomplete posets consisting of those that are furthermore shallow. The equivalence preserves underlying sets.*

Proof. The full and faithful functors of (21) and (19) combine to give a full and faithful

$$\mathbf{T}-\text{algebras} \longrightarrow \text{Admissible } \Omega\text{-actions} \longrightarrow \text{S.s.cocomplete posets.}$$

But its values have underlying sets which are flabby, (Proposition 4.3), and for a flabby set the notions of s.s cocomplete/cocontinuous and weakly cocomplete/cocontinuous evidently agree. So we have a full and faithful functor from $\mathbf{T}$-algebras to weakly cocomplete posets. Its values are shallow posets, by Theorem 2.3 (1), since the order comes via an admissible Ω-action. On the other hand, by Remark 4.4, every shallow weakly cocomplete poset arsises this way. This proves the Theorem.

The codomain category in (21) is a topos, so (21) exhibits the category of $\mathbf{T}$-algebras as a full subcategory of a topos, raising the question about a pure Ω-action theoretic characterization of this full subcategory. We have intermediate full subcategories

$$\mathbf{T}\text{-algebras} \subseteq \text{Inhabited admissible } \Omega\text{-actions}$$

$$\subseteq \text{Connected } \Omega\text{-actions} \subseteq \Omega\text{-actions},$$

where 'connected' in this case means existence of a unique fixpoint for the action (it is clear that if $a \in A$ is an element of an Ω-action, then false$\cdot a$ is a fixpoint for the action; and if a and b are fixpoints for an *admissible* Ω-action, then $a \wedge b = \lambda \cdot a = a$, so $a \leq b$ and similarly $b \leq a$, so fixpoints are unique). These inclusions are all proper.We shall argue this only for the first of them, thereby in particular answer in the negative a question posed by Lawvere (private communication) about the relationship between $\mathbf{T}$-algebras and connected Ω-actions. We do this through some general observations.

For any set X, we have a map

$$\Omega \times X \longrightarrow TX \tag{22}$$

which to $\lambda \in \Omega$ and $a \in X$ associates $\{x \mid x = a \text{ and } \lambda\}$. Now TX consists of the subterminal subsets of X, and the image of (22) consists of the subsingletons. For, if $U \subseteq \{a\}$, then

$$U = \{x \mid x \in U\} = \{x \mid x \in U \text{ and } x = a\} = \lambda \cdot \{a\},$$

where $\lambda = [x \in U]$. So (22) is surjective precisely when X is flabby in the sense stated in Corollary 4.3. So if X is not flabby, the image $\Omega \cdot X$ of (22) is a proper sub-Ω-action of TX (containing X), and it will be admissible since TX is. But, being a proper subset of TX (containing X), $\Omega \cdot X$ cannot be a sub **T**-algebra, since TX is generated as a **T**-algebra by X.

If now X is inhabited, the image $\Omega \cdot X$ of (22) will be inhabited, and will thus be an inhabited, admissible, and hence connected Ω-action, but it will not be a **T**-algebra, unless X is flabby. So to argue that an inhabited admissible Ω-action need not be a **T**-algebra, one just has to exhibit a topos and an inhabited object X which is not flabby. Even in the Sierpinski topos Sets^2, it is easy to find such X, say $X = 1 + \frac{1}{2}$ (where $\frac{1}{2}$ is the subobject of 1 intermediate between 0 and 1).

6 On products of cocomplete posets

The following section provides an application of notions from the theory of weakly cocomplete posets to an understanding of products of cocomplete posets $\prod\{A_i \mid i \in I\}$ over an index set which is not necessarily decidable.

It is clear that for any well behaved cocompleteness notion (s.s., weak, conditional, or finitely cocomplete,say; or just cocomplete without qualification), a product $\prod A_i$ of cocomplete posets (with coordinatewise ordering) is again cocomplete, and the projections $\text{proj}_i : \prod A_i \to A_i$ preserve the relevant sup's . Hence they have, in the unqualified case, right adjoints; but they have also left adjoints $\text{in}_i \dashv \text{proj}_i$; these left adjoints exist just under the assumption that the A_i's are weakly (=subterminal-) cocomplete. For the case where I is decidable, these left adjoints are well known, and given by

$$\begin{aligned} \text{proj}_j \text{in}_i(a) &= a \text{ if } i = j \\ &= \bot_j \text{ (bottom element of } A_j) \text{ if not,} \end{aligned}$$

cf e.g. [JT] p.2 or p.3.

Theorem 6.1 *Let $\{A_i \mid i \in I\}$ be a family of weakly cocomplete posets. Then for each i, $proj_i : \prod A_i \to A_i$ has a left adjoint in_i (w.r.to the coordinatewise ordering of the product). If $proj_j \circ in_i : A_i \to A_j$ is denoted $\delta_{i,j}$, we have*

$$\delta_{i,i} = \textit{identity} \tag{23}$$

$$\delta_{j,i}(\delta_{i,j}(a)) = [i = j] \cdot a. \tag{24}$$

$$\delta_{i,j} \textit{ commutes with the action by } \Omega \tag{25}$$

(where the action by the monoid $\Omega = (\Omega, \wedge)$ is constructed as in Section 3).

Proof. The main thing is to construct the $\delta_{i,j}$s. To this end we reformulate the notion of weakly cocomplete into more diagrammatic terms: $(A, \leq)$ is weakly cocomplete iff for every *monic* $u : U \to V$, the order preserving map $\hom(V, A) \to \hom(U, A)$ has a left adjoint lan_u, (and such that furthermore a Beck-Chevalley condition holds). Now consider the family $\{A_i \mid i \in A\}$ of weakly cocomplete posets. For $i, j \in I$, form the equalizer $u : U \to 1$ of (the names of) i and j. For $a \in A_i$, we consider the diagram

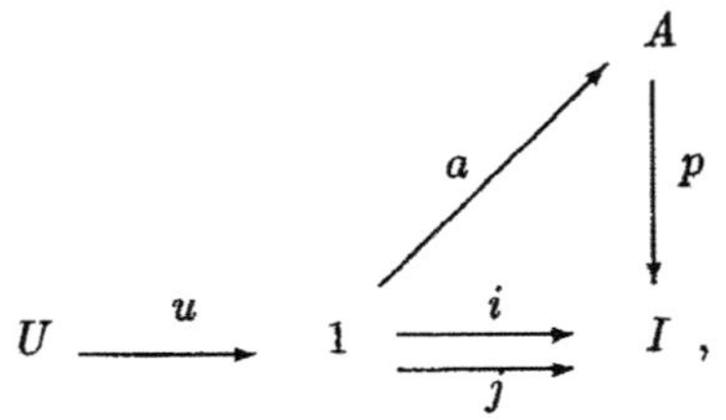

where A is the disjoint union of the A_i's (more precisely, to give an I-indexed family of sets A_i *by definition* amounts to give a map $A \to I$). We have $p \circ a = i$ since $a \in A_i$. But then

$$p \circ a \circ u = i \circ u = j \circ u$$

so that $a \circ u$ factors through the inclusion $A_j \to A$. We then define

$$\delta_{i,j}(a) := \mathrm{lan}_u(a \circ u).$$

If $i = j$, u is the identity map, so $\mathrm{lan}_u(a \circ u) = a$, proving (23). To prove (24), we prove two inequalities. To see

$$[i = j] \cdot a \leq \delta_{j,i}(\delta_{i,j}(a))$$

is by the adjointness of Theorem 2.3 equivalent to proving

$$[i = j] \leq [a \leq \delta_{j,i}(\delta_{i,j}(a))].$$

But if $i = j$, we have equality inside the second square bracket, by (23), in particular inequality.

To prove the other inequality, we first give a different description of $\delta_{i,j}(a)$, namely

$$\delta_{i,j}(a) = \bigwedge\{b \in A_j \mid i = j \text{ implies } a \leq b\}; \tag{26}$$

for, $\delta_{i,j}(a)$ itself appears as a b in the intersection, since if $i = j$, $a \leq \delta_{i,j}(a)$ (since $a = \delta_{i,i}(a)$). The other inequality $\leq$ in (26) follows because for each b satisfying $i = j \implies a \leq b$, we have $a \circ u \leq b \circ u$, and so, by the adjointness that defines lan_u, $\mathrm{lan}_u(a \circ u) \leq b$.

So to prove the inequality $\leq$ in (24), we note that by (26)

$$\delta_{j,i}(\delta_{i,j}(a)) = \bigwedge\{b \in A_i \mid j = i \text{ implies } \delta_{i,j}(a) \leq b\}.$$

But inside this infimum participates $b := [i = j] \cdot a$; for, if $j = i$, $\delta_{i,j}(a) = a = [i = j] \cdot a$.

Let now $\text{in}_i : A_i \to \prod A_i$ be given by $\text{proj}_j \circ \text{in}_i; = \delta_{i,j}$, then (23) provides the front adjunction inequality (an equality, actually), whereas for $\underline{a} = (a_i)_{i \in I}$ in $\prod A_i$, it suffices to prove the back adjunction inequality coordinatewise, i.e. to prove

$$\text{proj}_j(\text{in}_i(\text{proj}_i(\underline{a}))) \leq a_j \;\; \forall j \in I.$$

The left hand side here is by (26)

$$\delta_{i,j}(a_i) = \bigwedge\{b \in A_i \mid i = j \Rightarrow a_i \leq b\},$$

and a_j itself participates in this infimum. This proves the adjointness.

For the last assertion, $\delta_{i,j}$ commutes with Ω-action, since $\delta_{i,j} = \text{proj}_j \circ \text{in}_i$, and both proj_j and in_i preserves supremum formation over subterminals (in_i by being a left adjoint), and these suprema provide the Ω-action. This proves the Theorem.

Remark 6.2 Besides the coordinatewise order on $\prod A_i$, which is usually not shallow, there is also a shallow, weakly cocomplete order on $\prod A_i$, induced by the coordinatewise **T**-algebra structure. The proj_i preserve the **T**-induced order as well, since they are **T**-homomorphisms. It is not difficult to prove that the in_i, constructed as in the proof of the theorem, preserve order, even with the **T**-induced shallow order on the product. But the back adjunction inequality will not hold in general for the shallow order here.

REFERENCES

[AL] C.Anghel and P.Lecouturier,Generalisation d'un resultat sur le triple de la reunion, Ann.Fac.sci. de Kinshasa, Section Mat.Phys.1 (1975), 65-94

[J] P.T.Johnstone, Topos Theory, Academic Press 1977

[JT] A.Joyal and M.Tierney, An extension of the Galois Theory of Grothendieck, Mem.A.M.S. 309 (1984)

[Ke] M.Kelly, Basic Concepts of Enriched Category Theory, London Math. Soc. Lecture Notes Series 64, Cambridge Univ.Press 1982

[K1] A.Kock, Strong functors and monoidal monads, Archiv der Math.23 (1972), 113-120

[K] A.Kock, Algebras for the partial map classifier monad, Aarhus Preprint 89/90 No.6, Sept 1989

[Ma] E.Manes, Algebraic Theories, Springer GTM 26, 1976

[Mi] C.J.Mikkelsen, Lattice theoretic and logical aspect of elementary topoi, Aarhus Various Publ.Series 25 (1976)

Aarhus May 1990

This paper is in final form and will not be published elsewhere.

Intrinsic Co-Heyting Boundaries and the Leibniz Rule in Certain Toposes

F. William Lawvere
Department of Mathematics, S.U.N.Y. at Buffalo
Buffalo, NY 14214

Certain lattices, such as that of all closed subsets of a topological space or that of all subtoposes of a given topos, may be called "co-Heyting algebras" in that they enjoy a "subtraction" operator left adjoint to join, dually to the "implication" operator right adjoint to meet which Heyting algebras have. The cited examples illustrate that such lattices may occur in practice directly, not only as formal opposites of Heyting algebras. In particular, there is for each element A of a co-Heyting algebra a smallest element *non A* whose join with A is the top element, and the meet *A and non A* is a further element which deserves to be called the *boundary* of A.

For example, given a retract-closed locally finite category, the subtoposes of the topos of presheaves on it correspond to the retract-closed subcategories A of the locally finite category itself, and the boundary of A is the category of those objects which are simultaneously objects in A and retracts of objects not in A. The subtoposes classify the models of positive (or "geometric") extensions of the theory for which the full presheaf category is classifying. For example, any positive extension of the theory of distributive lattices (such as the theory of Boolean algebras or the theory of totally-ordered sets with end points) corresponds to a subcategory A of the category of finite posets; the boundary of each such theory is another theory which in principle can be calculated by the method just described. It would be interesting to compute the boundary classes in this sense of various known classes of abelian groups. However, in this note I concentrate on another issue and on a different kind of example, the algebra of subobjects of a given object in a special kind of topos.

In any co-Heyting algebra, the boundary operator satisfies a Leibniz product rule: the boundary of any meet of two elements is the join of two meets, each involving one of the elements and the boundary of the other. (This relationship is evident in the usual diagram showing a meet as the intersection of overlapping ovals, if one takes seriously the bounding curves of the ovals). There are several other universally-valid relationships between the co-Heyting operation "non" and the lattice operations, for example any element A is the join of its boundary with its "core" non non A.

In any presheaf topos (and more generally any essential subtopos of a presheaf topos), the lattice of all subobjects of any given object is another example of a co-Heyting algebra (as well as a Heyting algebra). The co-Heyting operations are in general not preserved by substitution (inverse image) along maps, unlike the Heyting "not" and the "possibility" operators provided by Grothendieck topologies. More like those "necessity" operators for which *de dicto/de re* is a genuine distinction, we have for non in such a topos in general only the "lax" preservation

$$\text{non}(Af) \subseteq (\text{non } A)f$$

for subobjects A of the codomain of a map f. Thus in particular, the non operator is not induced by an endomap of the truth-value object. Nonetheless, it is subject to some control because of the adjointness to join, which *is* preserved by substitution.

In a presheaf topos, the elements of kind C in non A (where A is a subobject of X) are easily seen to be just those elements x of kind C in X for which there exists some map u from C to another representable D and some z in X of kind D for which x = zu and z is not in A(D). For example, in the topos of directed graphs (where u can be either "source" or "target"), the boundary of a subgraph A consists of all its nodes which are either sources or targets of arrows (in the ambient graph) which are not in the subgraph A.

Proposition: In the topos of presheaves on a small category, the co-Heyting non is preserved by substitution along all maps iff the small category is a groupoid (so that the Heyting and co-Heyting structures collapse to a Boolean one).

Some toposes support the intuition that their objects are arbitrary "spaces" of a specified kind, and this suggests another Leibniz rule for the boundary of a *cartesian* product: the boundary of a cylinder (such as the tin enclosing a tin can) is the union of two cartesian products, each involving the boundary of one of the two factors. This turns out to be true in *some* presheaf toposes; in the others, *de dicto/de re* is a real distinction even for *projection* maps, starkly underlining the need for "declaration of variables " in order to have meaningful formulas. The connection between the two issues comes from the fact that a sub-cartesian product *is* a meet, namely of inverse images along projections.

Theorem: If a small category has the property that every map factors as a split mono followed by a split epi, then, in the topos of presheaves on it, the co-Heyting non is preserved by substitution along any product projection, and in particular the Leibniz product rule for the boundary of a cartesian product (within a larger cartesian product) holds.

Since one way to obtain such a factorization of a map is to use its "graph" together with the graph of any map in the reverse direction, we have

Corollary: If a small category has all hom-sets nonempty and has binary products (*or* coproducts), then in the topos of presheaves on it, the Leibniz product rule for the co-Heyting boundary of cartesian products is valid.

There is at least one important example in which the models have neither products nor coproducts but yet have the factorization property of the theorem:

Corollary: In the topos of simplicial sets, if $A \subseteq X$, $B \subseteq Y$ then

$$\partial(A \times B) = (\partial A) \times B \cup A \times \partial B$$

for the co-Heyting boundaries ∂ in $X \times Y$, X, Y respectively.

Reference: F. W. Lawvere, Introduction to *Categories in Continuum Physics*, Springer Lecture Notes in Mathematics **1174**, 1986.

This paper is in final form and will not be published elswhere.

CONCRETELY FUNCTORIAL PROGRAMMING

John L. MacDonald
Department of Mathematics, University of British Columbia
Vancouver, B. C., Canada V6T 1Y4

We present a theory of dynamic data types using 2-categorical methods. In the form given here the theory shows how programs written in assembly language are concretely functorial. Pertinent features are illustrated by a program written for an M88000 based machine, which carries out a simple calculation using RISC instructions[13].

The sequencing of program steps becomes simply the composition of concrete functors. Program loops are determined by the "shape" of a diagram of concrete functors. The "shape" is given by a control category. By concrete functors we mean here functors to the category of pointed sets. A morphism of concrete functors is a pair consisting of a functor and a natural transformation. It exists on two levels: change of data type and evaluation of partially defined function.

To perform a computation one needs a program and an input or initial configuration. The appropriate input here consists of *structured elements* of certain concrete functors. Programs and program steps can be ordered in various ways exploiting the ordering in the category of partial maps. For example, one can order by extent of definition on a given set of structured elements.

The first section gives the necessary background on comma categories in the 1- and 2- categorical setting emphasizing in particular the category $(\mathbf{C}\downarrow\underline{E})$ of objects over a given object $\underline{E}$. Ordered categories (cf. [7]) are introduced and their significance suggested for the case when the object $\underline{E}$ is itself an ordered category. The second section introduces program functors and control categories. In particular the abstract and concrete value level of each program step is emphasized. The third section shows what data should be recorded by a control category for an assembly language program by comparing the situation with that of a register machine (cf.[4]). This is illustrated by introducing a specific program written for the M88000 microprocessor using RISC instructions [13] and then presenting the associated control category, as well as indicating how to carry out this process in general.

The fourth section discusses the notion of input elements for a program and demonstrates that the appropriate notion here is that of structured element, rather than a global element(cf[3]). A structured element is a morphism whose domain is a certain lax terminal object of **Tot**$(\mathbf{C}\downarrow\underline{E})$, the full subcategory of total objects in $(\mathbf{C}\downarrow\underline{E})$ such total objects being defined with respect to total morphisms in $\underline{E}$. A result relating the ordering in $(\mathbf{C}\downarrow\underline{E})$ to that of $\underline{E}$ is presented(cf. [7]). We indicate how the set of structured elements determines (i.e. denotes) the partially defined function computed by the program. The final section shows how to assign a program functor to an assembly language program in general and gives a detailed presentation of structured elements and program steps for the 88000 program given in section 3.

This paper is dedicated to Max Kelly on his 60th birthday and in acknowledgement of his many influential ideas. I would also like to thank John Gray for his ideas in computer science and Art Stone for many helpful conversations and ideas during early versions of this work.

1. Background.

In [11] Mac Lane describes the *comma category* $(T\downarrow S)$ associated with functors T and S of common codomain. It has objects and morphisms

$$\langle e,d,f\rangle \quad \begin{array}{c} eT \\ f\big\downarrow \\ dS \end{array} \qquad\qquad \langle k,h\rangle \quad \begin{array}{ccc} eT & \xrightarrow{\;kT\;} & e'T \\ f\big\downarrow & & \big\downarrow f' \\ dS & \xrightarrow{\;hS\;} & d'S \end{array} \tag{1.1}$$

respectively with square commutative, and componentwise composition.

For each object a of a category C an instance of this construction is given by the category $(C\downarrow a)$ of objects over a. This is also called the *slice category* C/a (cf.[2]). It has objects f: c--->a and morphisms h: f --->f' with hf' =h. In this case $T = 1_C$, the identity functor on C, and S: **1** ---> C is the functor with value a on the category **1** which has one object and one morphism.

The comma category construction for 2-categories is described by Gray in [5], page 29. We will be working with an instance of this construction, namely, that of the 2-category $(\mathbf{C}\downarrow\underline{E})$ of objects over $\underline{E}$. Given an object $\underline{E}$ of a 2-category **C**, the category $(\mathbf{C}\downarrow\underline{E})$ has objects U: $\underline{C}$--->$\underline{E}$ and 1-cells (D,μ) with D: $\underline{C}$--->$\underline{D}$ and μ: U===>DV a 2-cell of **C**. The 2-cells ß of $(\mathbf{C}\downarrow\underline{E})$ are 2-cells of **C** with $\mu'..\beta;V = \mu$.

In the 1-category case there are projection functors P, R: $(C\downarrow a) \dashrightarrow C$ defined on objects f: c--->a by fP = c and fR = a and on morphisms h: f--->f' by hP = h: c--->c' and hR = 1: a--->a.

<u>Proposition 1.2</u>: Each functor T: D ---> $(C\downarrow a)$ determines a natural transformation π: TP ===> TR. This transformation π in turn may be regarded as the same thing as a functor π: D ---> $(TP\downarrow TR)$ with $\pi P' = \pi R' = 1$ for P', R' the projections of $(TP\downarrow TR)$.

<u>Proof</u>: A transformation ø: P ===> R is defined on objects f: c--->a of $(C\downarrow a)$ by f = fø: fP = c---> fR = a. Given h: f--->f' in $(C\downarrow a)$ we have hP.f'ø = h.f' = f = f.1a = fø.hR and naturality follows. Given T: D ---> $(C\downarrow a)$ the natural transformation π: TP ===> TR is given by π = Tø. The functor π: D ---> $(TP\downarrow TR)$ is given on objects d by dπ = <d,d,dT> where dT: dTP ----> dTR as in (1.1)(cf. Mac Lane[11], page 48).

Results from Proposition 1.2 can be generalized to $(\mathbf{C}\downarrow\underline{E})$ provided we replace the notion of natural transformation by *lax natural transformation*(cf. Gray [5]). Such a transformation Φ: F ===\ G between (strict) 2-functors F, G: **C** ---> **D** consists of the following data:

a. For each object $\underline{C}$ of **C** there is a 1-cell $\underline{C}\Phi$: $\underline{C}$F ---> $\underline{C}$G of **D**.

b. For each 1-cell f: $\underline{C}$ ---> $\underline{D}$ of **C** there is a 2-cell fΦ: $\underline{C}\Phi$;fG ===>fF;$\underline{D}\Phi$ in **D**

The following axioms hold:

1. The 2-cell fΦ is the identity if f = $1_{\underline{C}}$.

2. The 2-cell fgΦ is the horizontal composite of fΦ and gΦ, that is, fgΦ = fΦ;gG..fF;gΦ.

3. Given ß: f ===> f' in **C**, then fΦ.. ßF;$\underline{D}\Phi$ = $\underline{C}\Phi$;ßG..f'Φ.

As in (1.2) there are projection functors A, E: $(\mathbf{C}\downarrow\underline{E})$ ---> **C** defined on objects U: $\underline{C}$ ---> $\underline{E}$ by UA = $\underline{C}$, UE = $\underline{E}$, on 1-cells (D,μ) by (D,μ)A = D and (D,μ)E = $1_{\underline{E}}$ and on 2-cells ß: (D',μ') ===> (D,μ) by ßA = ß: D' ===> D and ßE = E^, the identity 2-cell on $1_{\underline{E}}$.

<u>Proposition 1.3</u>: Each strict 2-functor T: **K** ---> $(\mathbf{C}\downarrow\underline{E})$ determines a lax natural transformation π: TA ===\ E. This transformation is the composite

$$K \xrightarrow{\;T\;} (\mathbf{C}\downarrow\underline{E}) \underset{E}{\overset{A}{\rightrightarrows}} C \quad (\mathbf{u}\Downarrow) \tag{1.4}$$

<u>Proof</u>: The value of π will be π = T**u** for suitable **u**: A ===\ E. We define **u** on the object U: $\underline{C}$ ---> $\underline{E}$ of $(\mathbf{C}\downarrow\underline{E})$ by U**u** = U: $\underline{C}$ = UA ---> $\underline{E}$ = UE. The value of **u** on

the 1-cell (D,μ) of $(\mathbf{C}\downarrow\underline{E})$ must be a 2-cell

$$\begin{array}{ccc} UA = \underline{C} & \xrightarrow{D=(D,\mu)A} & VA = \underline{D} \\ {\scriptstyle U\mathbf{u} = U}\Big\downarrow & {\scriptstyle (D,\mu)\mathbf{u}}\;\nearrow\!\!\!\nearrow & \Big\downarrow{\scriptstyle V\mathbf{u} = V} \\ UE = \underline{E} & \xrightarrow[1_{\underline{E}}=(D,\mu)E]{} & VE = \underline{E} \end{array} \qquad (1.5)$$

from the definition. Let $(D,\mu)\mathbf{u} = \mu$. The lax naturality axioms for $\mathbf{u}$ hold as a consequence of the definition of 1-cell composition in $(\mathbf{C}\downarrow\underline{E})$ and the commutativity in $\mathbf{C}$ required of 2-cells in $(\mathbf{C}\downarrow\underline{E})$.

An *ordered category* is a 1-category O whose hom sets are ordered with order preserved by composition (cf.Jay[7]). It is thus a simple kind of 2-category. If $\underline{E}$ is an ordered category and $\mathbf{C}$ is **Cat** then $(\mathbf{C}\downarrow\underline{E})$ inherits an order.

2. Program functors.

Let $\underline{E} = \underline{Sets}^*$ be the category of pointed sets. Its objects are pairs $(X,*)$ where X is a set and $*$ is a fixed element of X. A morphism f: $(X,*) \dashrightarrow (Y,*)$ in E is a function f: X ---> Y with $f(*) = *$. Such a morphism corresponds to a partially defined function f_p from $X \setminus \{*\}$ to $Y \setminus \{*\}$ where $f_p(x)$ is defined if $f(x) \neq *$ and if defined $f_p(x) = f(x)$.

Suppose $\mathbf{C} = \mathbf{Cat}$. Then the objects of $(\mathbf{C}\downarrow\underline{E})$ are functors of codomain $\underline{E}$ and 1-cells (D,μ) are of the form

$$\begin{array}{ccc} \underline{C} & \xrightarrow{\quad D \quad} & \underline{D} \\ & {\scriptstyle U}\searrow \;\overset{\mu}{\Longrightarrow}\; \swarrow{\scriptstyle V} & \\ & \underline{E} & \end{array} \qquad (2.1)$$

where D is a functor and μ: U ===> DV a natural transformation. For each object c of $\underline{C}$ we have a morphism $c\mu$: cU ---> cDV from the underlying pointed set of c to the underlying pointed set of its image cD under D.

Let a *program* be a functor P: $\mathbf{K}$ ---> $(\mathbf{C}\downarrow\underline{E})$ where the domain is a 2-category called the *control* category. The 2-category $\mathbf{K}$ may have only identity 2-cells. The 2-cell structure might be used to model phenomena such as the passing of messages between threads running on separate processors. Note that the ordering in $\underline{E}$ allows us to compare programs according to the extent of their definition. By (1.4) there is a corresponding lax transformation Pu: PA===\ PE.

The *program steps* are the values of P on the 1-cells of **K**. Thus given j: $\underline{K}_i$ ---> $\underline{K}_j$ in **K** there is a program step jP: $\underline{K}_iP$ ---> $\underline{K}_jP$ in ($C\downarrow\underline{E}$). By (2.3) and (1.9) we write jP = (jPA, jPu) and picture it by

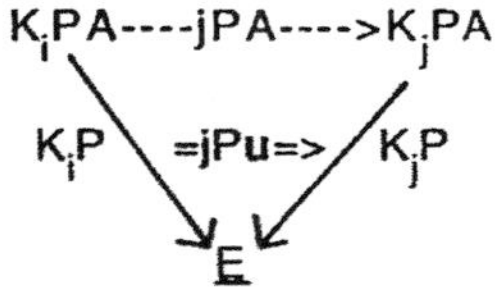

$$K_iPA\text{----}jPA\text{---->}K_jPA \qquad (2.2)$$

$$K_iP \searrow \overset{=jPu=>}{} \swarrow K_jP$$

$$\underset{E}{\searrow\swarrow}$$

A program step jP determines action on two levels. There is the *abstract level* A on which the functor jPA connects the categories $\underline{K}_iPA$ and $\underline{K}_jPA$ of abstract data structures and there is the underlying or concrete value level **u** on which the natural transformation jPu assigns a function c(jPu) connecting the underlying pointed set of each object c of the abstract data structure category $\underline{K}_iPA$ with the underlying set of its image object c(jPA) in $\underline{K}_jPA$. The control category **K** is used in exactly the same way that sketches are used. That is, certain cones are required to be taken to limit cones by the models (i.e. the programs). See the section 3 for an instance of this (cf. Barr-Wells[2], p228, and Guitart[6]).

3.   Control categories for assembly language programs.

The control category **K** determines the flow of control in the program. It is an obvious category theoretic translation of the old idea of a flowchart. By examining one of the mathematical idealizations of the computer, namely, the unlimited register machine (URM) of Cutland [4], it is easy to see theoretically what data should be recorded by a control category **K**. In Cutland's model there are an infinite number of registers R_1, ..., R_n, ... each containing a natural number. Contents of registers are altered by instructions and a finite list of instructions is called a program. There are four types of instructions, namely, zero (replace contents of R_n by 0), successor(replace contents r_n of R_n by r_n+1), transfer $T(m,n)$ (replace contents of R_n by r_m, the contents of R_m) and jump $J(m,n,q)$ (if $r_m=r_n$ proceed to qth instruction of P, if not, proceed to next instruction). We remark that URM computable is the same as Turing computable (cf.[4], page 57).

On theoretical grounds the control category **K** and program functor P: **K** --->

(C↓E) should thus deal with arithmetic operations, transfer (i.e. load and store) and jump (i.e. branch and loop) in the appropriate context. Implementation on actual computers appears more complex because of the variety of programming languages, microprocessors and machines available. For this reason we restrict our attention to programs written in assembly language and leave for later an analysis of "higher" imperative languages in the Algol family like C or Pascal.

Instructions for assembly language programs are "close to the machine" and as such can be seen as variations on or as closely related to the types of instructions used in the URM. The disadvantage is the large number of instructions used as well as the fact that the specific form of the instructions themselves vary, as for example between those used for the M68000 family of microprocessors and the RISC instructions for the M88000 which we examine next (cf. [12] and [13]).

For example, consider the following 14 line program in assembly language. This program computes a function, the Boolean AND of two bit vectors. It has one loop, calling for one label ("loop") and one branch point. The input is a linked list A,B,Tail where A and B hold arbitrary precision unsigned integers. For simplicity we assume A and B to have the same length. Each begins with a pointer to the next object on the list, so that the lengths of A and B are determined by these pointers. The output is a linked list C,Tail where C holds the logical AND of the arbitrary precision integers that were in A and B.

```
        ld      r1, sp, r0      ;r1=adr(B)=adr(B,Tail)              (3.1)
        ld      r2, r1, r0      ;r2 = adr(Tail)
        subu    r3, r2, r1      ;r3 =len(B), will be loop counter
        add     sp, r2, r0      ;sp=r2=adr(Tail)
loop:
        subu    r1, r1,#4       ;r1=adr(LSW(A)),A shrinks
        subu    sp, sp,#4       ;sp=adrLSW(B);Bshrinks,Cgrows
        ld      r4, r1, r0      ;load argument from A
        ld      r5, sp, r0      ;load argument from B
        and     r4, r4, r5      ;r4=A∧B, boolean operation
        st      r4, sp, r0      ;store A∧B in C
        subu    r3, r3,#4       ;decrement loop counter
        bcnd    ne0,r3,loop     ;branch on cond. loop counter ≠0
output:
        subu    sp, sp,#4       ;make space for pointer to Next
        st      r2, sp, r0      ;storepointer
```

Only six different machine instructions are used in this bit vector program, **ld, subu, add, and, st**, and **bcnd**. These are instructions for the MC88100, which is the processor chip in the current version of Motorola's M88000 family of processors(cf [13]). We call this simply the 88000. There are 52 machine instructions available on the 88000 if we count all the distinct mnemonic roots in the user's manual. Some of these instructions are just variations on each other. For example, included among the 52 are four variations on add, namely,

add, **add.ci**, **add.co** and **add.cio**

in which **.ci** stands for carry in , **.co** for carry out and **.cio** for carry in and out . For **add.ci**, the current value of the (one-bit) carry register (left there by some earlier computation) is added to the sum; for **add.co** the carry resulting from the sum is stored in the carry register; **add.cio** does both and the unaltered **add** does neither.

Most of the 52 instructions are sufficiently like the six in the program (3.1) so that our analysis go through unchanged except for details and hence for other programs on the 88000 as well. Other microprocessors, such as the MC68000, use instruction sets written in different form although these can be analysed in the same way by dividing them into those of arithmetic, transfer or branch type. We remark that the above program does contain instructions representing *arithmetic* (**subu**, **and**, **add**), *transfer* (**ld**, **st**) and *jump or branch* (**bcnd**) phenomena, hence representing the types of instructions found in the URM.

Here, are the machine instructions used in the program above.

ld **a, b, c**

Load register a with data pointed to by register **b**, offset by the address in register **c**. So if **c** above is **r0**, which always stores zero, then the instruction simply loads register **a** with the data stored at the location pointed to by register **b**.

add **a, b, c**

Add the numbers stored in registers **b** and **c** and put the result in register **a**.

subu **a, b, c**

Subtract the number stored in **c** from that in **b** and put the result in register **a**.

and **a, b, c**

Perform the bit-wise Boolean **and** of the numbers (now regarded as elements of the boolean algebra with 2^{32} elements) in registers **b** and **c**, and put result in **a**.

st **a, b, c**

Store the contents of register **a** at the location pointed to by register **b**, offset by the address in register **c**.

bcnd **neo, b, label**

Branch, if this condition is satisfied:

the number in register **b** is not equal to zero,

to the point in the code indicated by **label**.

We remark that there are restrictions in the 88000 instructions on the values that **a**, **b**, and **c** can take, particularly for the arithmetic operations. For example, for **add** we must have **b** and **c** take values in the processor registers.

For this 14 line example the control category **K** is the free category generated by the following diagram.

K (3.2)

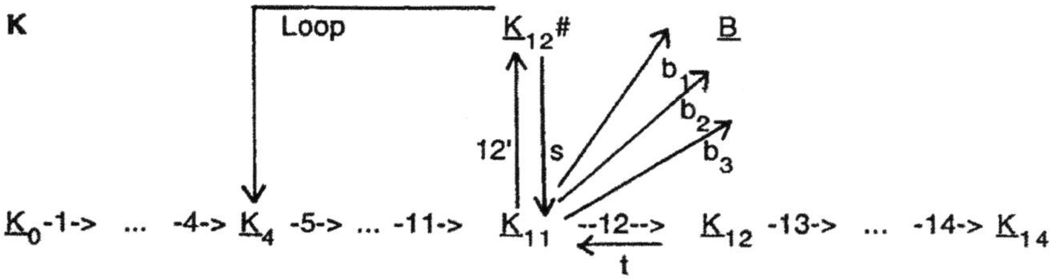

The 1-cells of the control category are named with the line numbers of the program. Nodes of the diagram, labelled $\underline{K}_i$, $0 \le i \le 14$, will correspond to states of the store "between" machine instructions. There is a 1-cell in **K** for each line and each label of the assembly language program, and a bit more for every branch point in the program. For a program functor P: **K** ---> $(\mathbf{C} \downarrow \underline{\mathbf{E}})$ we require, at a branch point, that sP be the equalizer of b_1P and b_2P and that tP be the equalizer of b_2P and b_3P and that $\underline{B}$P be the Boolean concrete category. This last requirement can be phrased differently by saying that we require P to be a model of **K**, regarded as a sketch. In this case **K**, as a sketch, has two cones involving s, b_1, b_2 and t, b_2, b_3, respectively, which must be taken by P to equalizer cones in $(\mathbf{C} \downarrow \underline{\mathbf{E}})$ (cf[2] pp 204 & 229). More details are given in section 5.

4. Computing values for structured elements.

In order to carry out a computation a program functor P: **K** ---> $(\mathbf{C} \downarrow \underline{\mathbf{E}})$ must be provided with an input or structured element , as defined below. A structured element is not the same as a global element in the categorical sense (cf.[3]) because a lax terminal object rather than a terminal object is involved.

The idea is similar to the one expressed by Cutland [4] for the Universal Register Machine as follows: "To perform a computation the URM must be provided with a program P and an initial configuration - i.e. a sequence a_1, a_2, a_3, ,,, of natural numbers in the registers R_1, R_2, R_3, ..." A structured element in our sense corresponds to an initial configuration. Expressed this in terms of programming languages we have the view of Allison [1], "...A denotational semantics of a programming language gives the mapping from programs in a language to the functions denoted..." Here the idea is the same. That is, we show here that there is a mapping from program functors P to the (partially defined) functions determined on the structured elements. In the next section we show how to assign a program functor P to an assembly language program.

Let **C** be a category with terminal object **1**. Then a well known definition [3] states that a *global element* b of an object $\underline{B}$ of a category **C** is a morphism b: $\underline{1}$ ---> $\underline{B}$. If **C** = **Cat**, then **1** is the category with 1 object and 1 morphism. A global element of **C** is then a functor **1** ---> **C**, i.e. a category.

Lemma 4.1: Let *: **1** ---> $\underline{E}$ be the functor with value {*} on the object of **1**. Then *: **1** ---> $\underline{E}$ is a terminal object of (**C**$\downarrow\underline{E}$).

Proof: Given an object U: $\underline{C}$ ---> $\underline{E}$ of (**C**$\downarrow\underline{E}$) we note that **1** is terminal in **Cat** so there is a unique functor $t_{\underline{C}}$: $\underline{C}$ ---> **1**. Given an object X of $\underline{C}$ we have $Xt_{\underline{C}}{}^*$ = {*} and there is a unique øU: XU ---> {*} since {*} is terminal in $\underline{E}$.

Lemma 4.2: Suppose (L,λ) is a global element of an object U of (**C**$\downarrow\underline{E}$) pictured by

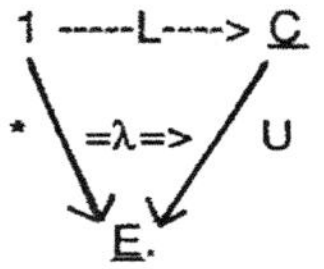

(4.3)

Then the value of λ at object 0 of **1** is the unique function 0* = {*} ---> OLU in $\underline{E}$.

Proof: {*} is a zero object in $\underline{E}$. Given an object Y of $\underline{E}$, then {*} ---> Y is unique since it must preserve *.

Thus global elements of U: $\underline{C}$ ---> $\underline{E}$ provide access to the elements L of $\underline{C}$ but not to elements of the underlying pointed set of OL (except for the point of OLU). For the right property we look at the structure of $\underline{E}$ as a 2-category.

Proposition 4.4: $\underline{E}$ is an ordered category.

Proof: We show that the hom sets are ordered with order preserved by composition. This shows that $\underline{E}$ is ordered in the sense of [7]. Given the hom set $\underline{E}$(X,Y) and f,g: X ---> Y belonging to it we say $f \leq g$ iff $xf \neq *$ implies xf = xg. Then by a standard argument for f', g' : Y ---> Z with $f' \leq g'$ we have $xff' \neq *$ implies xff' = xfg' = xgg'.

A *lax terminal object* of an ordered category O is an object s such that for each object r of O there is a morphism r ---> s which is largest in O (r,s).

Proposition 4.5: $(C\downarrow\underline{E})$ is an ordered category.

Proof: Given (D,μ), (D',μ'): U ---> V in $(C\downarrow\underline{E})$. Define $(D,\mu) \le (D',\mu')$ if $D = D'$ and $X\mu \le X\mu'$: XU ---> XDV in $\underline{E}$ for all objects X of $\underline{C}$. If in addition $(B,\beta) \le (B',\beta')$: V---> W, then we show $(D,\mu)(B,\beta) = (DB, \mu.D\beta) \le (D',\mu')(B',\beta') = (D'B', \mu'.D'\beta')$. But $D = D'$, $B = B'$ and we show $(DB, \mu.D\beta) \le (DB, \mu'.D\beta')$, i.e. show $X(\mu.D\beta) \le X(\mu'.D\beta')$. But $\underline{E}$ is ordered. Thus $X\mu \le X\mu'$ and $XD\beta \le XD\beta'$ implies $X\mu.XD\beta \le X\mu'.XD\beta'$.

Let s: **1** ---> $\underline{E}$ be the (global) element of $\underline{E}$ whose value is the set {s',*}, some $s' \neq *$. A morphism f: Y ---> Y' of $\underline{E}$ is *total* if $xf = *$ implies $x = *$. An object U: $\underline{C}$ ---> $\underline{E}$ of $(C\downarrow\underline{E})$ is total if for each morphism g of $\underline{C}$ we have gU total in $\underline{E}$. Let **Tot**$(C\downarrow\underline{E})$ be the full subcategory of total objects in $(C\downarrow\underline{E})$.

Proposition 4.6: s: **1** ---> $\underline{E}$ is a lax terminal object of **Tot**$(C\downarrow\underline{E})$.

Proof: Given U an object of **Tot**$(C\downarrow\underline{E})$ there is a unique functor $t_{\underline{C}}$: $\underline{C}$ ---> **1** since **1** is terminal in **Cat**. Then there is $X\tau$: XU ---> {*,s'} = $Xt_{\underline{C}}s$ which takes value s' on all y in XU with $y \neq *$. Furthermore t is natural precisely because fU is total in $\underline{E}$ for each f in $\underline{C}$. Clearly $(t_{\underline{C}}, \tau)$ is largest in its hom set.

The 1-cells (L, λ_s) of $(C\downarrow\underline{E})$ of the form

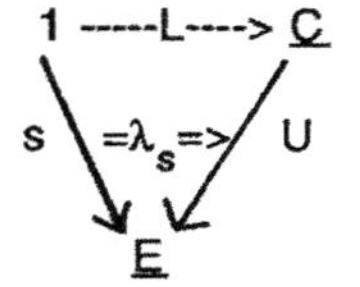

(4.7)

are called the structured elements of U. They pick out an object of $\underline{C}$ and an element of the underlying set of that object.

A program step is the value of P: **K** ---> $(C\downarrow\underline{E})$ on a morphism of **K**. As such it is a morphism of the form

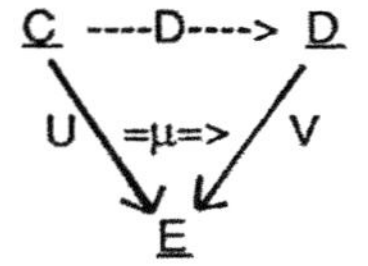

(4.8)

in $(C\downarrow\underline{E})$. A program step thus evaluates all the structured elements of U and in particular this means that each object X of $\underline{C}$ goes to XD of $\underline{D}$ (data structure changes) and each element of underlying pointed set of XU goes to an element of XDV (i.e. a function is calculated).

At each step of the program the program functor calculates a partially defined function. For arithmetic and move operations the description is straightforward since there is only one path to follow. For a branch instruction a structured element has its underlying element taken to a unique nonpointed element or to * in a manner described using equalizers. We give details for the program (3.1) in the next section and present the abstract version in later work.

5. Program functors for assembly language programs.

We will concentrate on the description of this functor for the M88000 program (3.1) and control category (5.2) and bring out some of the more theoretical aspects in later work.

We begin with M, the machine category, which contains (at least) finite products of certain basic objects M_i which all have isomorphic underlying pointed sets. Certain objects of M will be referred to as the *processor registers* Ri. We usually require a functor U: M ---> E which takes products in M to the appropriate lax product in E. In simplest terms $(M_1 \Pi\ M_2)U$ is the union of the set $(M_1U \setminus \{*\})$ X $(M_2U \setminus \{*\})$ with $\{(*,*)\}$ where X is the cartesian product in Sets. For example, if M_1 and M_2 are basic objects (spaces) in the machine we might think of M_1U and M_2U each as the set of 2^{32} possible integer values as well as a point *. Then $(M_1 \Pi\ M_2)U$ is the set of all (m_1,m_2) together with point (*,*) where m_1 and m_2 range over all non * values of M_1U and M_2U, respectively. This is a lax product in E (i.e. pointed sets) and is the appropriate notion for ordered categories (like E) (cf.[7], p6-7). We do not require this type of product preservation for all concrete functors which are values on objects of a control category but only as specified in the following analysis.

We now turn to the details of the example of an assembly language program for the 88000. We recall that the input to the program is a pair of bitvectors (x, y). The output is a bitvector $z = x \wedge y$. The bitvectors are each stored in n 32-bit words where n is not known at compile time. For simplicity we assume that x and y have the same length.

In effect the only first order algebras defined by the machine are objects that reside in the processor registers. That is, the instruction set available for the 88000 allows addition, subtraction etc. to occur only relative to a processor register. Objects stored in RAM or on disk(in their first order machine given structure) are only sets. Let X, Y, Z denote objects storing the bit vectors x, y, z. From the concrete functorial point of view the desired operation $\Lambda : X \prod Y \longrightarrow Z$ on $X \prod Y$ and its underlying set is given by applying a composite of concrete functors, that is, functors to $\underline{E}$ = pointed sets. The value on (x, y) is given by a component of one of the underlying natural transformations μ of this composite.

The program constructs that operation out of given operations available in registers, by manipulating objects X, Y, Z and others with functors (e.g. using load, store and register operations). Basically for this particular assembly language numbers are moved from storage to registers and operated on and the results put back in storage keeping track of everything with functors.

Going back to program description P: $\mathbf{K} \longrightarrow (\mathbf{C} \downarrow \underline{E})$ for $\mathbf{K}$ the control category we see that corresponding to the first morphism $\underline{K}_0$ -1-> $\underline{K}_1$ of $\mathbf{K}$ there is the first program step, as in (2.2), which is the morphism $\underline{K}_0 P$ -1P-> $\underline{K}_1 P$ of $(\mathbf{C} \downarrow \underline{E})$ pictured by

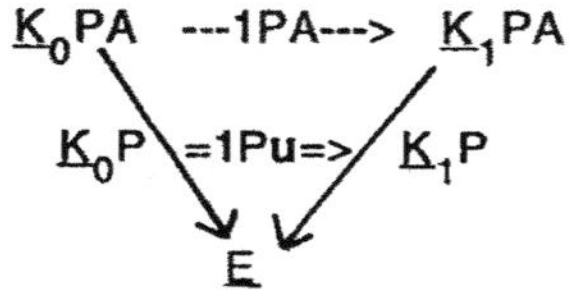

$$\begin{array}{ccc} \underline{K}_0 PA & \text{---1PA--->} & \underline{K}_1 PA \\ \underline{K}_0 P & \text{=1Pu=>} & \underline{K}_1 P \\ & \underline{E} & \end{array} \qquad (5.1)$$

as a diagram in $\mathbf{C} = \mathbf{Cat}$, where 1P = (1PA, 1Pu) and $(\mathbf{C} \downarrow \underline{E})$ --A--> $\mathbf{C}$ is the projection. We now arrive at the standard programmer's picture of the objects in the first addressed category $\underline{K}_0 PA$, namely,

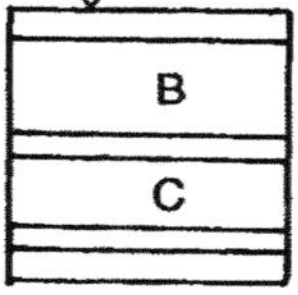

$$\boxed{\begin{array}{c} B \\ \hline C \end{array}} \qquad (5.2)$$

These have underlying sets with elements of the form <b,c,tail>. The objects of $\underline{K}_{14} PA$ are of the form

$$\boxed{\begin{array}{c} D \end{array}} \qquad (5.3)$$

These have structured elements of the form <d,tail>. We assume these categories $\underline{K}_i PA$ to be closed under projections.

In effect, for each object of $\underline{K}_0 PA$ of the form appearing in (5.2) we have a set of structured elements corresponding to elements of the underlying set $(B \prod C)U$ and we construct a function $(B \prod C)U \longrightarrow DU$ by applying the underlying natural transformations of the program steps 1Pu to 14Pu. The following pages give some more detailed information about the starting category, the first few program steps plus the crucial loop step.

We assume that when the program begins, a structured element e of $\underline{K}_0P$ is given. This picks out an object of the category $\underline{K}_0A$ and an element of the underlying set of that object. The objects of that category have the form $B\Pi C\Pi T$ where B itself is a product $B = Adr\Pi B_n \Pi B_{n-1} \dots \Pi B_1$ and likewise for C and T = Tail. The elements of the underlying set of $B\Pi C$ have the form <b,c,tail> where b and c are pairs b = <b. pntr, b. bitv> and c = <c. pntr, c. bitv>.

The structured element e = ($B\Pi C\Pi$Tail, <b,c,tail>) has the picture

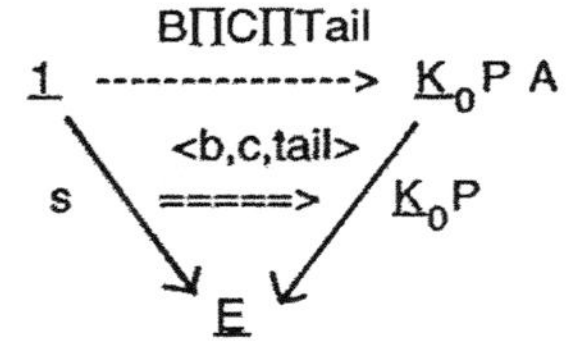

(5.4)

where $\underline{K}_0P$ is product preserving where product in $\underline{E}$ is the one for ordered categories referred to before [7]. Thus we may regard the program as acting initially upon the structured elements of the concrete functor $\underline{K}_0P$, which is itself the value of the program on the initial control object $\underline{K}_0$.

It should be noted that .pntr and .bitv are underlying functions for operators called ;Pntr and ;Bitv where ;Pntr takes value Adr on $B = Adr \Pi B_n \Pi \dots \Pi B_1$ and ;Bitv takes value $B_n \Pi \dots \Pi B_1$ on B (that is, they are projections).

<u>The lines of the program</u>. In each of the following we show the interesting parts of the abstract functor PA and below that the interesting part of the underlying natural transformation. Aside from the "interesting" part, everything is pretty much automatic, namely, we have identity mappings and mappings determined by preservation of products.

Given 1: $\underline{K}_0$ -----> $\underline{K}_1$ of **K** the program takes value

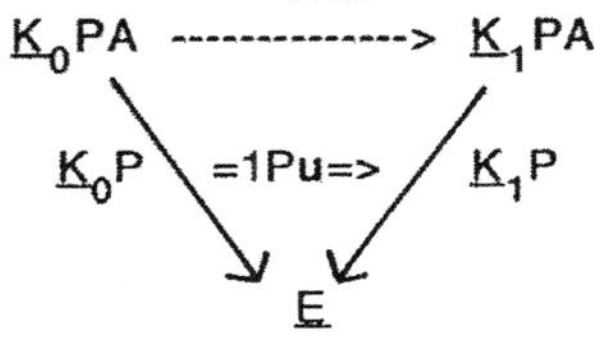

(5.5)

in (C$\downarrow\underline{E}$). The abstract functor 1PA: $\underline{K}_0$PA ---> $\underline{K}_1$PA takes B;Pntr to B;Pntr Π R1 (for a register object R1 of $\underline{K}_1$PA) and 1Pu takes value B;Pntr; $\underline{K}_1$P on object B;Pntr and maps b to <b,b>. Given 2: $\underline{K}_1$ ---> $\underline{K}_2$, C;Pntr goes to C;Pntr Π R2 under 2PA and C;Pntr;2Pu takes c to <c,c>. Given 3: $\underline{K}_2$ ---> $\underline{K}_3$, 3PA: R1 Π R2 --> R1 Π R2 Π R3 and R1 Π R2; 3Pu takes <b,c> to <b,c,b-c>. Given 4: $\underline{K}_3$ ---> $\underline{K}_4$, 4PA: R31 ---> Nil, 4PA: R2 ---> R2 Π R31 (where Nil has underlying set {*}) and b ~~~> <b,b> under R2;4Pu Given 5: $\underline{K}_4$ ---> $\underline{K}_5$, 5PA: R1 ---> R1 and R1;5Pu takes r~~~> r-4 and 5PA: B_n Π ... Π B_k ~~~> B_n Π ... Π B_{k+1} and the corresponding component of 5Pu is the projection.

Lines 6 through 11 follow a similar pattern as follows. 6: $\underline{K}_5$ ---> $\underline{K}_6$, 6PA: R2 ---> R2 and R2;6Pu takes r ~~~> r-4 (interpretation C shrinks, D grows. That is, also 6PA: C_n Π ... Π C_k ~~~> C_n Π ... Π C_{k+1} and 6PA: D_{k-1} Π ... Π D_1 ~~~> D_k Π ... Π D_1 with values 6Pu taking $(c_n, ..., c_k)$ ~~~> $(c_n,...,c_{k+1})$ and $(d_{k-1}, ..., d_1)$ ~~~> $(*, d_{k-1},...,d_1)$. 7: $\underline{K}_6$ ---> $\underline{K}_7$, R4 ---> Nil and B_n ---> B_n Π R4 with b ~~~> <b,b>. 8: $\underline{K}_7$ ---> $\underline{K}_8$, R5 ---> Nil, and B_n ---> B_n Π R5 with b~~~> <b,b>. 9: $\underline{K}_8$ ---> $\underline{K}_9$, R4 Π R5 ---> R4 Π R5 with <b,c> ~~~> b AND c 10: $\underline{K}_9$ ---> $\underline{K}_{10}$, D_n ---> Nil (D_n pointed to by R31), R5 ---> R5 Π D_n with b~~~> <b,b>. 11: $\underline{K}_{10}$ ---> $\underline{K}_{11}$, R3 ---> R3, b~~~> b-4 (decrement loop counter in R3).

The next program step12 uses two equalizers in (C$\downarrow\underline{E}$). More precisely program step 12 is **bcnd**, branch on condition. The machine branches if the value in R3 is not zero. The part of the control category **K** involved is

K (5.6)

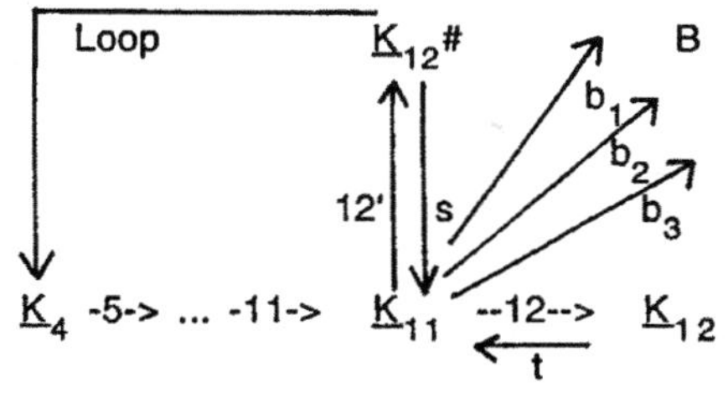

The value of P: **K** -----> (C$\downarrow\underline{E}$) on $\underline{B}$ is called the <u>boolean</u> concrete category and may be written as $\underline{B}$PA ---BP--> $\underline{E}$ where the abstract (or addressed) category $\underline{B}$PA has two objects B and Nil. The underlying pointed set functor maps B to the pointed set {0,1,*} and maps Nil to {*}. This B corresponds to a register the programmer doesn't see, something used by the processor itself, in its own computations in the background. $\underline{B}$PA is a pointed category with BΠNil = B.

Now we return to program step 12. The control 2-category **K** has objects

$\underline{K}_{12}\#$, $\underline{K}_{12}$ and $\underline{B}$ as above. The functor PA maps these to $\underline{K}_{12}\#PA$, $\underline{K}_{12}PA$ and $\underline{B}PA$ where $\underline{K}_{12}PA$ and $\underline{K}_{12}\#PA$ are copies of the abstract category $\underline{K}_{11}A$. As before we mention only the "interesting" parts of each abstract functor and of each underlying natural transformation. There are three 1-cells b_1, b_2, $b_3 : \underline{K}_{11}$ ---> $\underline{B}$ in **K**. The abstract 2-functor PA maps these to abstract 1-functors that are identical, mapping R3 ---> B and everything else that is not a product with R3 to Nil and an object R3 Π X gets mapped to B Π Nil = B.

(5.7)

b_iP:

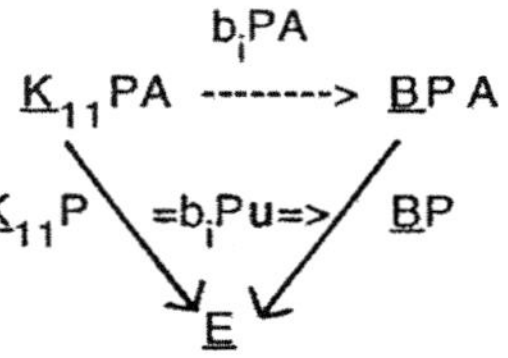

The concrete functor $\underline{P}$ puts different underlying natural transformations under these 1-cells. That is, $b_1PA = b_2PA = b_3PA$ but $b_iPu \neq b_jPu$ for $i \neq j$. The interesting components are defined on R3 which has for underlying set under $\underline{K}_{11}P$, a pointed set with, say 2^{32}(defined) elements and the 1 undefined element or point. The interesting components map

a. every defined element to 0 (b_1Pu).

b. all nonzero defined elements to 0 and 0 to 1 (b_2Pu).

c. every defined element to 1 (b_3Pu).

The first and third are "constant" pointed functions.

We require sP: $\underline{K}_{12}\#P$ ---> $\underline{K}_{11}P$ to be the equalizer of $b_1P = (b_1PA, b_1P\mathbf{u})$ and b_2P and tP: $\underline{K}_{12}P$ ---> $\underline{K}_{11}P$ to be the equalizer of b_2P and b_3P. That is, we require them to be equalizers in ($C\Downarrow\underline{E}$). In this sense we regard **K** as a sketch with cones based on s, b_1, b_2 and t, b_2, b_3 and require P (but not PA) to be a model of **K**. For the specific form taken by the equalizer in this case we first note that $b_iP = (b_iPA, B_iP\mathbf{u})$ has been described in (5.8) and now describe sP = (sPA, sP**u**) and tP = (tPA, tP**u**).

The abstract 2-functor PA takes the same category as value on control objects $\underline{K}_{11}$, $\underline{K}_{12}$ and $\underline{K}_{12}\#$ and the functors sPA: $\underline{K}_{12}\#PA$ ---> $\underline{K}_{11}PA$ and tPA: $\underline{K}_{12}PA$ ---> $\underline{K}_{11}PA$ are identity functors. However, the concrete 2-functor P does not take the same value on $\underline{K}_{11}$, $\underline{K}_{12}$, and $\underline{K}_{12}\#$ since the concrete categories $\underline{K}_{11}P$: $\underline{K}_{11}PA$ -----> E, $\underline{K}_{12}P$ and $\underline{K}_{12}\#P$ differ in their values on R3. The value of $\underline{K}_{11}P$ on R3 has already been described. In the concrete category $\underline{K}_{12}\#P$ the object R3 has an underlying set with 2^{32}-1 defined elements(nonzero elements) plus point while in $\underline{K}_{12}P$ the underlying set of R3 has one defined element plus point. Then sPu and tPu are the inclusion of nonzero elements and inclusion of zero, respectively, when applied to R3.

The horizontal composite in $(C \downarrow \underline{E})$ of the path from the beginning, starting with the initial input (that is, with the underlying set of the structured element) can be continued nontrivially through only one of the concrete categories $\underline{K}_{12}\#P$ and $\underline{K}_{12}P$. This is the case because the composites of the whole path with the parallel pairs (b_1Pu, b_2Pu) and (b_2Pu, b_3Pu) will be equal for exactly one of the parallel pairs. Once started, the program can continue only in one way once it reaches the branch point. Finally program step 13 is like step 3 and step 14 is like step 10.

References

[1] L. Allison, A Practical Introduction to Denotational Semantics, Cambridge Computer Science Texts 23, Cambridge University Press, 1986.

[2] M. Barr, C. Wells, Category Theory for Computing Science, Prentice Hall (1990).

[3] M. Barr, C. Wells, Toposes, Triples and Theories, Springer(1985).

[4] N. Cutland, Computability, Cambridge University Press, 1980.

[5] J. W. Gray, Formal Category Theory I: Adjointness for 2-Categories. Springer Lecture Notes in Mathematics 391 (1974).

[6] R. Guitart, *On the geometry of computations*, Cahiers Top. et Geom. Diff. XXVII-4(1986), 107-136.

[7] C. B. Jay, *Extending properties to categories of partial maps.* Laboratory for Foundations of Computer Science, Report 90-107, Edinburgh(1990).

[8] L.A. Leventhal, D. Hawkins, G. Kane, W.D. Cramer, 68000 Assembly Language Programming, 2nd Edition, Osborne McGraw-Hill (1986).

[9] J. MacDonald, A. Stone, *Soft adjunctions between 2-categories.* J. Pure Applied Algebra 60 (1989), 155-203.

[10] J. MacDonald, A. Stone, *A class of 2-adjunctions invariant under perturbation.* To appear.

[11] S. Mac Lane, Categories for the Working Mathematician, Springer(1971).

[12] MC 68000 16-Bit Microprocessor user's manual, 3rd Edition, Prentice Hall (1982).

[13] MC 88100 RISC Microprocessor user's manual, 2nd Edition, Prentice Hall (1990).

This paper is in final form and no similar paper has been or is being submitted elsewhere.

WEAK PRODUCTS OVER A LOCALLY HAUSDORFF LOCALE

Susan B. Niefield
Union College, Department of Mathematics
Schenectady, NY 12308, USA

Introduction

Let **Top** and **Loc** denote the categories of T_0 topological spaces and locales (in the sense [3]), respectively. For an introduction to locales, we refer the reader to [5] and [9].

It is well-known that if X and Y are topological spaces, then the locale product $\Omega(X) \times \Omega(Y)$ of their open set lattices may differ from $\Omega(X\ timesY)$, i.e. the induced locale morphism $f\colon \Omega(X \times Y) \to \Omega(X) \times \Omega(Y)$ need not be an isomorphism. In particular, Isbell [3] showed that f is not an isomorphism, when $X = Y = \mathbf{Q}$, the rationals with the subspace topology of the real line. In a positive direction, it is not difficult to show (and part of the folk-lore of the subject) that if X is locally compact, then f is an isomorphism for all spaces Y. For a partial converse to this result, the reader is referred to [4].

Instead of further investigating the relationship between the products of spaces and of locales, Peter Johnstone and Sun Shu-Hao [6] introduced a monoidal structure $\otimes$ on **Loc** called the weak product and showed that $\Omega(X) \otimes \Omega(Y) \cong \Omega(X \times Y)$, for all spaces X and Y. They also considered the largest coreflective subcategory **WSp** of **Loc** for which $\otimes$ is the categorical product.

Now, if X and Y are spaces over a base space T, then one can consider the induced locale morphism $f\colon \Omega(X \times_T Y) \to \Omega(X) \times_{\Omega(T)} \Omega(Y)$. It is well-known that if X is a locally closed subset of T (i.e. the intersection of an open and a closed), then f is an isomorphism for every space Y over T. Generalizing this result and the Folk Theorem mentioned above, Niefield [8] showed that if X is exponentiable as a space over T, then f is an isomorphism for every space Y over T. Of course, the exponentiable sober spaces are precisely the locally compact ones (see [1] and [2]).

In this paper, we consider a generalization of the weak product to **Loc/L**. After a brief summary of the relevant results from [6], in Section 2 we consider the question of preservation of equalizer by the functor $\Omega\colon \mathbf{Top} \to \mathbf{Loc}$. In Section 3, we investigate locally Hausdorff locales, providing examples of locales satisfyinga condition arising in the previous section. Next, a definition of $\otimes_L$ is introduced in Section 4 and shown to give the product in $\mathbf{WSp}/L$. We conclude by showing that $\otimes_L$ is an associative bifunctor on $\mathbf{Loc}/L$, but that L is not a unit.

1. Weak Products and Hausdorff Locales

In this section, we recall the properties of the weak product $\otimes$ which will be used throughout this article. For details, we refer the reader to [6].

Property 1.1 $\otimes$ is a symmetric monoidal structure on **Loc**.

Property 1.2 $\otimes$ is a subfunctor of $\times$.

Throughout this paper, we will write $A \otimes B \to A \times B$ to denote this subfunctor. Note that since we will not work directly with elements of $A \otimes B$, we will omit a description of the nucleus on $A \times B$ which defines it.

An inclusion $A' \hookrightarrow A$ of locales is called *locally closed* if A' is the intersection of an open and a closed sublocale of A, or equivalently, if it is the composition (in either order) of an open and a closed inclusion.

Property 1.3 If $A' \to A$ and $B' \to B$ are locally closed inclusions, then the following diagram is a pullback in **Loc**.

$$\begin{array}{ccc} A' \otimes B' & \longrightarrow & A \otimes B \\ \downarrow & & \downarrow \\ A' \times B' & \longrightarrow & A \times B \end{array}$$

A locale A is called *weakly spatial* if the diagonal morphism $\Delta: A \to A \times A$ factors through $A \otimes A$. The full subcategory of **Loc** consisting of the weakly spatial locales is denoted by **WSp**. For any locale A, we can consider the following pullback.

$$\begin{array}{ccc} \breve{A} & \longrightarrow & A \otimes A \\ \downarrow & & \downarrow \\ A & \longrightarrow & A \times A \end{array}$$

We will refer to the morphism $\breve{A} \to A \otimes A$ as the *weak diagonal* of A. Though $\breve{A}$ need not be weakly spatial, repeating this process (infinitely often, if necessary) gives rise to:

Property 1.4 **WSp** is a coreflective subcategory of **Loc** containing all spatial locales.

Property 1.5 $\otimes$ is the product in **WSp**.

Property 1.6 For any T_0 spaces X and Y, the induced map $\Omega(X \times Y) \to \Omega(X) \otimes \Omega(Y)$ is an isomorphism.

A locale A is called *Hausdorff* if whenever a and b are elements with $a \neq 1$ and $a \not\leq b$, there exist x and y in A such that $x \wedge y = 0$, $x \not\leq a$, and $y \not\leq b$.

Property 1.7 If A is Hausdorff, then the weak diagonal of A is locally closed.

Note that although the converse to Property 1.7 does not hold, an additional condition can be added to obtain a characterization of this Hausdorff property [6].

Property 1.8 For a T_0 space X, $\Omega(X)$ is Hausdorff if and only if X is Hausdorff.

Property 1.9 If A is Hausdorff, then $\check{A}$ is the spatial part of A. Moreover, a Hausdorff locale is spatial if and only if it is weakly spatial.

2. Equalizers

In this section, we consider which equalizers of **Top** (respectively, **WSp**) are preserved by Ω (respectively, the inclusion **WSp** $\to$ **Loc**). In either case, if the morphisms $X \rightrightarrows Y$ form the cokernel-pair of $E \hookrightarrow X$, then the equalizer $E \hookrightarrow X \rightrightarrows Y$ is preserved, since the functor in question preserves pushouts being a left adjoint. Also, every equalizer in **Top** (as well as **Loc**) factors through one of this form. However, not every equalizer if preserved, for consider the following example.

Example 2.1 Let $A \hookrightarrow \Omega(X)$ be any regular monomorphism in **Loc**, where A is not spatial and X is a topological space, for example, $\Omega(\mathbf{R})_{\neg\neg} \hookrightarrow \Omega(\mathbf{R})$. Let $f, g: \Omega(X) \to L$ be the cokernel-pair of $A \hookrightarrow \Omega(X)$ in **Loc**. Note that L is spatial since it is a quotient of a spatial locale. Write $L = \Omega(Y)$, where Y is a sober space and consider the equalizer

$$E \hookrightarrow X \underset{\hat{g}}{\overset{\hat{f}}{\rightrightarrows}} Y \qquad (\star)$$

where $\Omega(\hat{f}) = f$ and $\Omega(\hat{g}) = g$. Since $A \hookrightarrow \Omega(X)$ is the equalizer of f and g in **Loc**, it follows that Ω does not preserve the equalizer $(\star)$. If A is not weakly spatial, a similar argument gives rise to an equalizer in **WSp** which is not an equalizer in Loc. Note that $\Omega(\mathbf{R})_{\neg\neg}$ is not weakly spatial by 1.8 and 1.9.

Next, we will consider condition on Y such that every equalizer diagram with codomain Y is preserved. We begin with a lemma.

Lemma 2.2 A locally closed sublocale of a (weakly) spatial locale is (weakly) spatial.

Proof For the spatial case, we refer the reader to [5]. The weakly spatial case follows directly from Property 1.3.

Proposition 2.3 Suppose $E \hookrightarrow A \rightrightarrows L$ is an equalizer in **Loc**, where L is a locale whose weak diagonal is locally closed. Then E is a locally closed sublocale of L. Moreover, if A is (weakly) spatial, then so is E.

Proof Suppose A is (weakly) spatial and $f, g: A \to L$. Then $(f, g): A \to L \times L$ factors through $L \otimes L$, since the diagonal $A \to A \times A$ factors through $A \otimes A$. Since the equalizer of f and g can be obtained by pulling back (f, g) along the diagonal of L, we have the following diagram of pullbacks in **Loc**

$$\begin{array}{ccccc} E & \longrightarrow & \check{L} & \longrightarrow & L \\ \downarrow & & \downarrow & & \downarrow \\ A & \longrightarrow & L \otimes L & \longrightarrow & L \times L \end{array}$$

Since the weak diagonal of L is locally closed, it follows that E is a locally closed sublocale of A, and so E is (weakly) spatial by Lemma 2.2.

Corollary 2.4 If $E \hookrightarrow A \rightrightarrows L$ is an equalizer in **WSp** and L has a locally closed weak diagonal, then it is an equalizer in **Loc**.

Proof Since **WSp** is a coreflective subcategory of **Loc**, equalizers in **WSp** are obtained by applying the coreflection to the equalizer of the morphisms in **Loc**. Thus, the desired result follows from the Proposition.

In the next section, we will show that every spatial locale with a locally closed weak diagonal is of the form $\Omega(T)$, where T is locally Hausdorff. Thus, we have the following corollary.

Corollary 2.5 If T is locally Hausdorff and $E \hookrightarrow X \rightrightarrows T$ is an equalizer in **Top**, then $\Omega(E) \hookrightarrow \Omega(X) \rightrightarrows \Omega(T)$ is an equalizer in **Loc**.

3. Locally Hausdorff Locales

Next, we consider a class of examples of locales with locally closed weak diagonals, including the (weakly) spatial ones. A locale L is called *locally Hausdorff* if L is a join of Hausdorff open sublocales, i.e. $L = \bigvee L_\alpha$, where each L_α is a Hausdorff open sublocale of L.

Lemma 3.1 If $L = \bigvee L_\alpha$, where each L_α is an open sublocale of L, then $\check{L}_\alpha$ is open in $\check{L}$ and $\check{L} = \bigvee \check{L}_\alpha$.

Proof Consider the following commutative cube, where the front and back faces are pullbacks by definition of the weak diagonal and the right face is a pullback by Property 1.3.

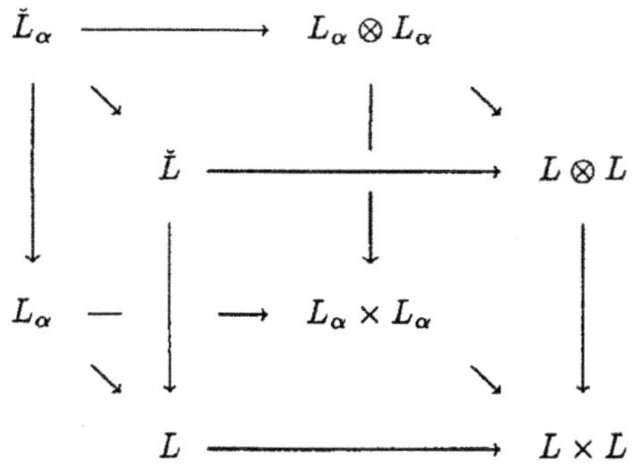

Thus, the left face is also a pullback and so $\check{L}_\alpha$ is an open sublocale of $\check{L}$ (since pullbacks preserve open sublocales) and $\check{L} = \bigvee \check{L}_\alpha$ (since pullbacks preserve sups of open sublocales).

Proposition 3.2 If L is locally Hausdorff, then the weak diagonal is locally closed.

Proof Write $L = \bigvee L_\alpha$, where each L_α is a Hausdorff open sublocale of L. Since L_α is Hausdorff, we know that the weak diagonal $\check{L}_\alpha \to L_\alpha \otimes L_\alpha$ is closed by Property 1.7, and so $\check{L}_\alpha \to L \otimes L$ is locally closed. Thus, each $\check{L}_\alpha$ is open in the closure of $\check{L}$ in $L \otimes L$. Since $\check{L} = \bigvee \check{L}_\alpha$ by Lemma 3.1, it follows that $\check{L}$ is open in its closure.

Proposition 3.3 The following are equivalent for a space T:

(1) Every point of T has a Hausdorff neighborhood

(2) T is a union of Hausdorff open subspaces

(3) The diagonal $T \to T \times T$ is locally closed

(4) The diagonal $T \to T \times T$ is exponentiable in $\mathbf{Top}/T \times T$

Proof The equivalence of (1) - (3) is left as an exercise and (3) $\leftrightarrow$ (4) appeared in [7].

A space satisfying the equivalent conditions of Proposition 3.3 is called *locally Hausdorff*. It is not difficult to show that such spaces are necessarily sober.

Corollary 3.4 For a T_0 space T, $\Omega(T)$ is locally Hausdorff if and only if T is locally Hausdorff.

Proof If T is locally Hausdorff, then (2) of Proposition 3.3 implies that $\Omega(T)$ is locally Hausdorff by Property 1.8. Conversely, if $\Omega(T)$ is locally Hausdorff the weak diagonal $\Omega(T) \to \Omega(T \times T)$ is locally closed, and so the diagonal $T \to T \times T$ is locally closed, as desired.

Proposition 3.5 A locally Hausdorff locale is spatial if and only if it is weakly spatial.

Proof Suppose $L = \bigvee L_\alpha$, where each L_α is a Hausdorff open sublocale of L. If L is weakly spatial, it follows that L_α is weakly spatial (Property 1.3), and so L_α is spatial (Property 1.9). Therefore, L is spatial.

4. Products in WSp/L

As in any slice category, products in $\mathbf{WSp}/L$ can be obtained via an equalizer

$$E \hookrightarrow A \otimes B \rightrightarrows L$$

in $\mathbf{WSp}$. If L has a locally closed weak diagonal, then we know that this is just the equalizer in $\mathbf{Loc}$ (Corollary 2.4).

Now, if A and B are any locales over L, then we define their ***weak product*** $A \otimes_L B$ by the following equalizer in **Loc**.

$$A \otimes_L B \hookrightarrow A \otimes B \rightrightarrows L$$

Proposition 4.1 The canonical diagram

$$\begin{array}{ccc} A \otimes_L B & \longrightarrow & A \otimes B \\ \downarrow & & \downarrow \\ A \times_L B & \longrightarrow & A \times B \end{array}$$

is a pullback.

Proof Consider the diagram

$$\begin{array}{ccccc} A \otimes_L B & \longrightarrow & A \otimes B & \searrow\!\!\searrow & \\ \Big\downarrow & & \Big\downarrow & & L \\ A \times_L B & \longrightarrow & A \times B & \nearrow\!\!\nearrow & \end{array}$$

Since the bottom row is an equalizer, the desired result follows.

Theorem 4.2 If X and Y are spaces over a locally Hausdorff space T, then the canonical morphism $f: \Omega(X \times_T Y) \to \Omega(X) \otimes_{\Omega(T)} \Omega(Y)$ is an isomorphism.

Proof Consider the diagram

$$\begin{array}{ccccc} \Omega(X \times_T Y) & \hookrightarrow & \Omega(X \times Y) & \searrow\!\!\searrow & \\ \Big\downarrow f & & \Big\downarrow g & & \Omega(T) \\ \Omega(X) \otimes_{\Omega(T)} \Omega(Y) & \hookrightarrow & \Omega(X) \otimes \Omega(Y) & \nearrow\!\!\nearrow & \end{array}$$

where the rows are equalizers by Corollary 2.5 and the definition of $\otimes_{\Omega(T)}$. Since g is an isomorphism by Property 1.6, it follows that f is an isomorphism.

Theorem 4.3 If L has a locally closed weak diagonal, then $\otimes_L$ is the product in **WSp**$/L$.

Proof Since **WSp** is a coreflective subcategory of **Loc**, equalizers in **WSp** are obtained by applying the coreflection to the equalizer formed in **Loc**. Thus, Corollary 2.4 gives the desired result.

5. Associativity of $\otimes_L$

We know that if L has a locally closed weak diagonal, then $\otimes_L$ gives a symmetric monoidal structure on $\mathbf{WSp}/L$, being the product. We conclude by considering $\otimes_L$ as a bifunctor on $\mathbf{Loc}/L$. Clearly, $\otimes_L$ is a symmetric bifunctor.

Theorem 5.1 If L has a locally closed weak diagonal, then $\otimes_L$ is associative.

Proof Consider the following diagram

$$\begin{array}{ccccc} A \otimes_L (B \otimes_L C) & \longrightarrow & A \otimes (B \otimes_L C) & \longrightarrow & A \otimes (B \otimes C) \\ \downarrow & & \downarrow & & \downarrow \\ A \times_L (B \otimes_L C) & \longrightarrow & A \times (B \otimes_L C) & \longrightarrow & A \times (B \otimes C) \\ \downarrow & & \downarrow & & \downarrow \\ A \times_L (B \times_L C) & \longrightarrow & A \times (B \times_L C) & \longrightarrow & A \times (B \times C) \end{array}$$

where the northwest square is the pullback defining $A \otimes_L (B \otimes_L C)$ using Proposition 4.1, the northeast square is a pullback by Property 1.3 and Proposition 2.3, the southwest square is a pullback since both horizontal arrows are equalizers of morphisms with codomain L, and the southeast square is a pullback since $A \times -$ preserves pullbacks. Thus, we obtain a pullback

$$\begin{array}{ccc} A \otimes_L (B \otimes_L C) & \longrightarrow & A \otimes (B \otimes C) \\ \downarrow & & \downarrow \\ A \times_L (B \times_L C) & \longrightarrow & A \times (B \times C) \end{array}$$

Similarly, $(A \otimes_L B) \otimes_L C$ is given by a pullback. Using the associativity of $\otimes$, $\times$, and $\times_L$, it follows that $\otimes_L$ is associative.

Finally, we note that L is not a unit for $\otimes_L$. In fact, $A \otimes_L L \cong A$ if and only if the graph $A \to A \times L$ of the structure morphism $A \to L$ factors through $A \otimes L$. In particular, L would have to be weakly spatial. But, even when L is spatial this condition presents a problem, for example, when $L = \Omega(\mathbf{R})$ and $A = \Omega(\mathbf{R})_{\neg\neg}$. Moreover, any definition of $\otimes_L$ which is compatible with "change of base" gives rise to a morphism $A \otimes_L B \to A \otimes B$, and hence the same difficulty.

References

[1] B.J. Day and G.M. Kelly, On topological quotients preserved by pullbacks or products, **Proc. Camb. Phil. Soc. 67** (1970), 553-558.

[2] K.H. Hofmann and J.D. Lawson, The spectral theory of distributive continuous lattices, **Trans. Amer. Math. Soc. 246** (1978), 285-310.

[3] J.R. Isbell, Atomless parts of spaces, **Math. Scand. 31** (1972), 5-32.

[4] J.R. Isbell, Product spaces in locales, **Proc. Amer. Math. Soc. 81** (1981), 116-8.

[5] P.T. Johnstone, **Stone Spaces**, Cambridge University Press, 1982.

[6] P.T. Johnstone and Sun Shu-Hao, Weak products and Hausdorff locales, in **Categorical Algebra and its Applications**, Lecture Notes in Mathematics 1348, Springer-Verlag, 1988.

[7] S.B. Niefield, Cartesianness: topological spaces, uniform spaces, and affine schemes, **J. Pure Appl. Alg. 23** (1982), 147-67.

[8] S.B. Niefield, Cartesian spaces over T and locales over $\Omega(T)$, **Cahiers de Top. et Géom. Diff. 23** (1982), 157-67.

[9] S. Vickers, **Topology via Logic**, Cambridge University Press, 1989.

This paper is in final form and will not be published elsewhere.

Categorical Interpolation: Descent and the Beck-Chevalley Condition without Direct Images

Duško Pavlović

Zevenwouden 223, 3524 CR Utrecht, The Netherlands

Fibred categories have been introduced by Grothendieck (1959, 1971), as the setting for his theory of descent. The present paper contains (in section 4) a characterisation of the effective descent morphisms under an arbitrary fibred category. This essentially geometric result complements a logical analysis of the Beck-Chevalley property (section 1) – which was crucial in the well-known theorem on sufficient conditions for the descent under bifibrations, due to Bénabou-Roubaud (1970) and Beck (unpublished). We describe the notion of *interpolants* (sections 2 and 3) as the common denominator of the concepts of descent and the Beck-Chevalley property.

(For the basic notions and facts about fibred categories, the reader can consult Gray 1966, or Bénabou 1985. A survey can also be found in Pavlović 1990.)

1. The Beck-Chevalley condition/property

11. Proposition. Let $E: \mathcal{E} \longrightarrow \mathcal{B}$ be a bifibration, $Q = (f,g,s,t)$ a square in $\mathcal{B}$ such that $f \circ g = s \circ t$, and $\Theta = (\varphi,\gamma,\sigma,\vartheta)$ a square in $\mathcal{E}$ such that $\varphi \circ \gamma = \sigma \circ \vartheta$, with $E\varphi = f$, $E\gamma = g$, $E\sigma = s$ and $E\vartheta = t$. The following conditions are equivalent:

a) if ϑ and φ are cartesian <u>and</u> if σ is cocartesian <u>then</u> γ must be cocartesian;
b) if σ and γ are cocartesian <u>and</u> if ϑ is cartesian <u>then</u> φ must be cartesian;
c) if ϑ is cartesian <u>and</u> if σ is cocartesian <u>then</u> φ is cartesian <u>iff</u> γ is cocartesian.

If some inverse image functors f^* and t^* and some direct image functors $g_!$ and $s_!$ are chosen, then every square Θ over Q satisfies conditions (a-c) iff there is a canonical natural isomorphism

d) $\quad f^*s_! \simeq g_!t^*$.

12. Definition. A square Q in the base of a bifibration $E:\mathcal{E} \longrightarrow \mathcal{B}$ satisfies the *Beck-Chevalley condition* if every square Θ over Q satisfies conditions (a-c). E is said to have the *Beck-Chevalley property* if all the pullback squares in $\mathcal{B}$ satisfy the Beck-Chevalley condition. – "Beck-Chevalley" will be abbreviated to "BC".

13. Proposition. A bifibration E has the BC-property iff the cocartesian arrows are stable under those pullbacks along cartesian arrows which E preserves.

14. Sources. The Beck-Chevalley condition has arisen in the theory of descent – as developed from Grothendieck 1959. Jon Beck and Claude Chevalley studied it independently from each another. The former expressed it in the form 11(d), the latter as in 11(a). It is conspicuous that neither of them ever published anything on it. Early references are: Bénabou-Roubaud 1970, Lawvere 1970.

The proofs of propositions 11 and 13 are elementary. They can be found in my thesis (1990).

15. Logical meaning of the BC-property. Consider a fibration $E:\mathcal{E} \longrightarrow \mathcal{B}$ as a "category of predicates": the base category $\mathcal{B}$ is to be thought of as a category of "sets" and "functions", while the objects and arrows of a fibre $\mathcal{E}_I$ represent "predicates" $\alpha(x^I)$ over the "set" I, and "proofs" between them. In this setting, the logical operation of substitution is interpreted by the inverse images. An inverse image functor over a "function" $t: I \longrightarrow J$ in $\mathcal{B}$ can be understood as mapping

$$t^*:\mathcal{E}_J \longrightarrow \mathcal{E}_I: \beta(y^J) \longmapsto \beta(t(x^I)).$$

Lawvere (1969) noticed that *the quantifiers are adjoint to the substitution*:

$$\alpha(x^I) \vdash \beta(t(x^I)) \quad \Leftrightarrow \quad \exists x^I\big(t(x^I)=y^J \wedge \alpha(x^I)\big) \vdash \beta(y^J),$$

$$\beta(t(x^I)) \vdash \alpha(x^I) \quad \Leftrightarrow \quad \beta(y^J) \vdash \forall x^I\big(t(x^I)=y^J \rightarrow \alpha(x^I)\big),$$

so that the logical picture of the direct image functors ($t_! \dashv t^* \dashv t_*$) becomes

$$t_! : \mathcal{E}_I \longrightarrow \mathcal{E}_J: \alpha(x^I) \longmapsto \exists x^I\big(t(x^I)=y^J \wedge \alpha(x^I)\big), \text{ and}$$

$$t_*: \mathcal{E}_I \longrightarrow \mathcal{E}_J: \alpha(x^I) \longmapsto \forall x^I\big(t(x^I)=y^J \rightarrow \alpha(x^I)\big).$$

What does the Beck-Chevalley property mean in this context? The simplest case is when the commutative square Q consists of projection arrows. A direct image functor along a projection $\pi: K \times M \longrightarrow K$ just quantifies a variable, while an inverse image functor adds a dummy:

$$\pi^*: \mathcal{E}_K \to \mathcal{E}_{K\times M}: \quad \chi(x^K) \mapsto \chi(x^K, \not{z}^M)$$
$$\pi_! : \mathcal{E}_{K\times M} \to \mathcal{E}_K : \psi(x^K, z^M) \mapsto \exists z^M.\psi(x^K, z^M),$$
$$\pi_*: \mathcal{E}_{K\times M} \to \mathcal{E}_K : \psi(x^K, z^M) \mapsto \forall z^M.\psi(x^K, z^M).$$

The picture of the BC-condition is:

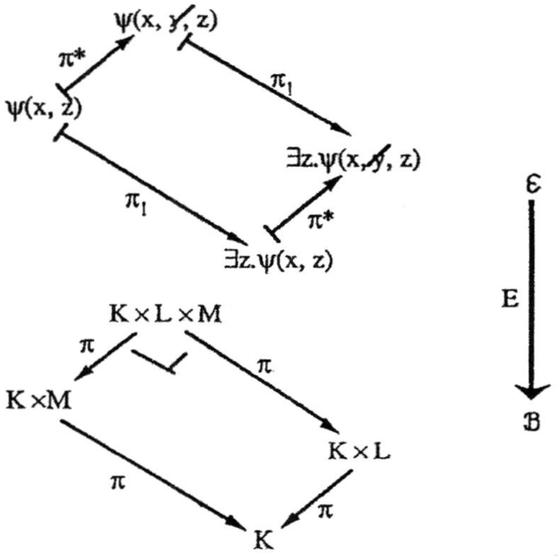

"The quantifier $\exists z$ and the variable y do not interfere" – says the BC-condition here. If we apply $\exists z$ on $\psi(x, \not{y}, z)$, we get the same result as when when we apply it on $\psi(x,z)$ and then add y.

Over a general pullback square Q, this picture becomes

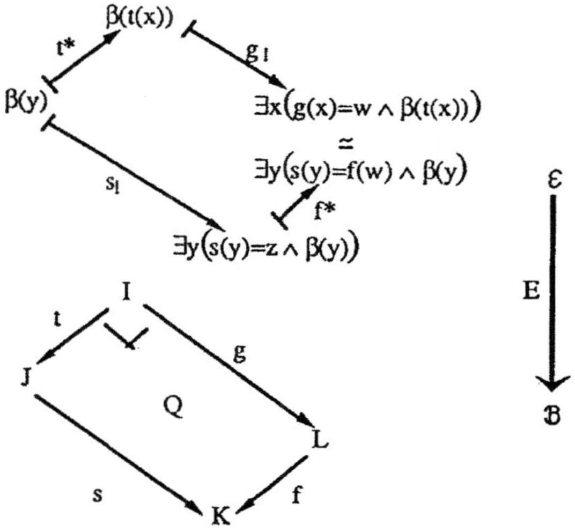

A proof $\exists x(g(x)=w \wedge \beta(t(x))) \vdash \exists y(s(y)=f(w) \wedge \beta(y))$ can be derived from a proof that Q is commutative:

$$\vdash f(g(x))=s(t(x)).$$

The converse proof $\exists y(s(y)=f(w) \wedge \beta(y)) \vdash \exists x(g(x)=w \wedge \beta(t(x)))$ – follows from

$$s(y)=f(w) \vdash \exists x(t(x)=y \wedge g(x)=w),$$

which tells that Q is a (weak) pullback. In this way, logic suggests the demand for an isomorphism $f^*s_!(\beta) \simeq g_!t^*(\beta)$ when Q is a pullback.

2. Interpolation condition

21. Motivation. The fact that variables do not interfere with each other can be expressed in a different way, without quantifiers:

$\alpha(x,y) \vdash \gamma(y,z) \Leftrightarrow$

there is an *interpolant* $\beta(y)$, such that $\alpha(x,y) \vdash \beta(y) \vdash \gamma(y,z)$.

At the first sight, this seems to be a *different idea* of the independence of variables. Surprisingly, it is not. We show in the sequel that the BC-property – whenever it can be expressed – is equivalent with the existence of a certain kind of interpolants.

22. Notation. For a given fibration $E:\mathcal{E} \longrightarrow \mathcal{B}$, an arrow $t \in \mathcal{B}(I,J)$ and an object $Y \in |\mathcal{E}_J|$ (i.e. $EY=J$), $\vartheta^t_Y: t^*Y \longrightarrow Y$ denotes an *arbitrary* cartesian lifting of t at Y; if E is a cofibration (i.e. if its dual $E^o:\mathcal{E}^o \longrightarrow \mathcal{B}^o$ is a fibration), then $\sigma^t_X: X \longrightarrow t_!X$ will be an *arbitrary* cocartesian lifting of t at an object X over I (i.e. a cartesian lifting with respect to E^o). In general, we do not choose the whole (co)cleavages, but we do use these generic symbols for an arbitrarily chosen cartesian or cocartesian arrow. Moreover, the unique vertical arrow by which σ^t_X factorizes through $\vartheta^t_{t_!X}$ will be $\eta: X \longrightarrow t^*t_!X$ – for the obvious reason that this arrow would be a component of the unit of the adjointness $t_! \dashv t^*$ *if* functors $t_!$ and t^* were chosen. Similarly, given a vertical arrow $q: t_!X \longrightarrow Y$, we denote by $q`: X \longrightarrow t^*Y$ the unique vertical arrow such that $\vartheta^t_Y \circ q` = q \circ \sigma^t_X$: this is the "right transpose" of q by $t_! \dashv t^*$. Given vertical $p: X \longrightarrow t^*Y$, its "left transpose" $`p: t_!X \longrightarrow Y$ is the unique vertical arrow such that $\vartheta^t_Y \circ p = `p \circ \sigma^t_X$.

The unique vertical isomorphisms between various inverse images of an object along an arrow will all be denoted by τ. E.g., if $f \circ g = s \circ t$, g^*f^*B and t^*s^*B are inverse images of B along the same arrow, and there is a unique vertical iso $\tau: t^*s^*B \longrightarrow f^*g^*B$.

Note, finally, that the thick points •...• enclose the (sketches of) proofs.

23. Definition. Let $E: \mathcal{E} \to \mathcal{B}$ be a fibration, and $Q = (f,g,s,t)$ a commutative square in $\mathcal{B}$. An (Q-)*interpolant* of an arrow $d \in \mathcal{E}_I(t^*A, g^*C)$ is a triple $\langle a,B,c \rangle$, where $B \in |\mathcal{E}_K|$, $a \in \mathcal{E}_J(A, s^*B)$, $c \in \mathcal{E}_L(f^*B, C)$, such that

$$d = g^*(c) \circ \tau \circ t^*(a).$$

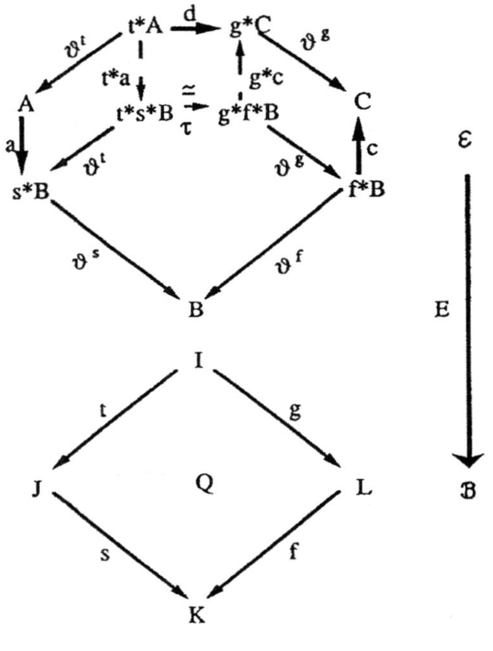

A square Q in the base of a fibration E satisfies the *interpolation condition* if there is a Q-interpolant for every arrow $d \in \mathcal{E}_I(t^*A, g^*C)$.

24. Proposition. Let $E: \mathcal{E} \to \mathcal{B}$ be a bifibration. A commutative square $Q = (f,g,s,t)$ in $\mathcal{B}$ satisfies the interpolation condition <u>iff</u> the vertical arrow $\rho = \rho_A : g_!t^*A \to f^*s_!A$ is a split mono for every $A \in |\mathcal{E}_J|$. (The arrow ρ is defined by the equation: $\vartheta^f_{s_!A} \circ \rho \circ \sigma^g_{t^*A} = \sigma^s_A \circ \vartheta^t_A$.)

• <u>If</u>: Suppose $e \circ \rho = id$. We claim that an interpolant $\langle a,B,c \rangle$ is given by:

$B := s_!A$,

$a := \eta : A \to s^*s_!A$,

$c := {}^{\backprime}d \circ e : f^*s_!A \to g_!t^*A \to C$.

On the diagram

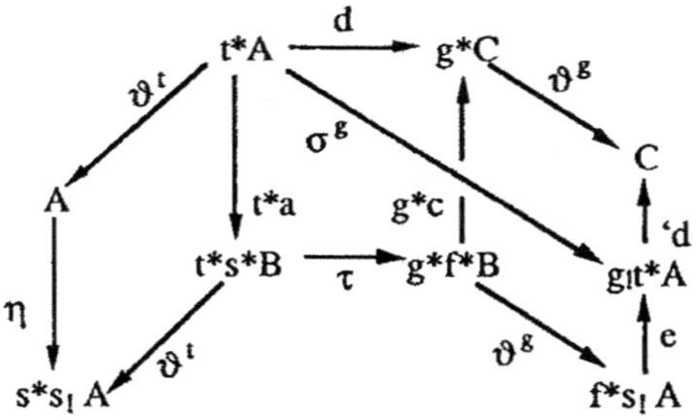

we see that

$$\sigma^g_{t^*A} = e\circ\vartheta^g_{f^*B}\circ\tau\circ t^*(a) \Rightarrow d = g^*(c)\circ\tau\circ t^*(a).$$

But the antecedens of this implication is a consequence of the fact that $e\circ\rho = \mathrm{id}$ and of lemma 25 below.

<u>Then</u>: Let $\langle a_\eta, B_\eta, c_\eta\rangle$ be an interpolant of the "unit"

$$\eta : t^*A \longrightarrow g^*g_!t^*A,$$

(defined as in 22) and let $`a_\eta : s_!A \longrightarrow B$ be the "left transpose" of a_η. By lemma 26, the arrow

$$e := c_\eta\circ f^*(`a_\eta) : f^*s_!A \longrightarrow f^*B \longrightarrow g_!t^*A$$

is then a left inverse of ρ.•

Lemmas. The following statements are true for any bifibration E.

25. $\rho\circ\sigma^g = \vartheta^g\circ\tau\circ t^*(\eta)$.

• By the definition of ρ, the left side is the unique factorisation over g of $\sigma^s_A\circ\vartheta^t_A$ through $\vartheta^f_{s_!A}$. But the right side is such a factorisation too, as the next diagram shows.

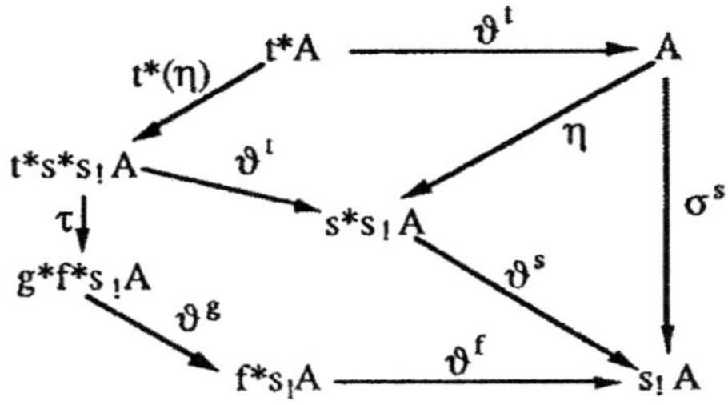

26. Each of two squares below commutes <u>iff</u> the other one does.

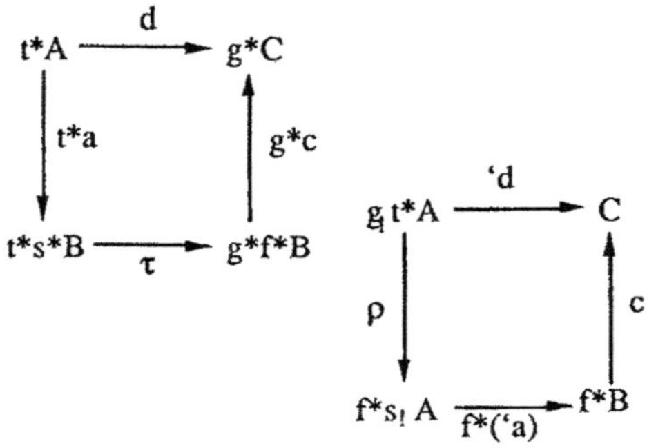

• In the following diagram, each of the triangles clearly commutes <u>iff</u> the other one does.

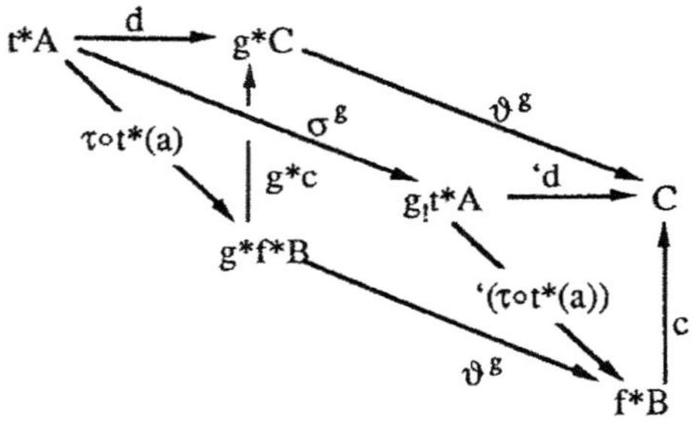

Thus we are done if we prove $f^*(‘a)\circ\rho = ‘(\tau\circ t^*(a))$.

As for this equality, compare the following two diagrams:

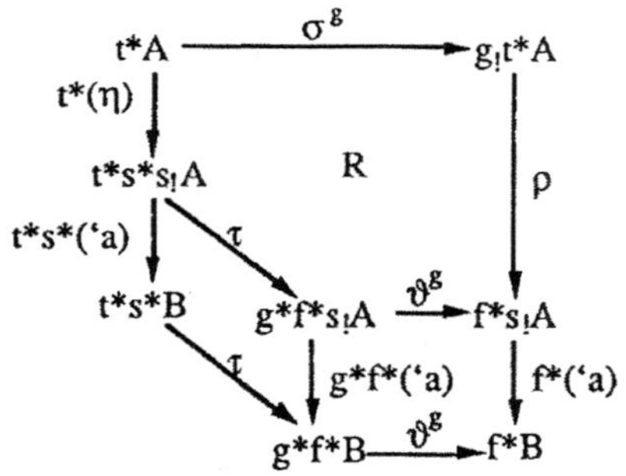

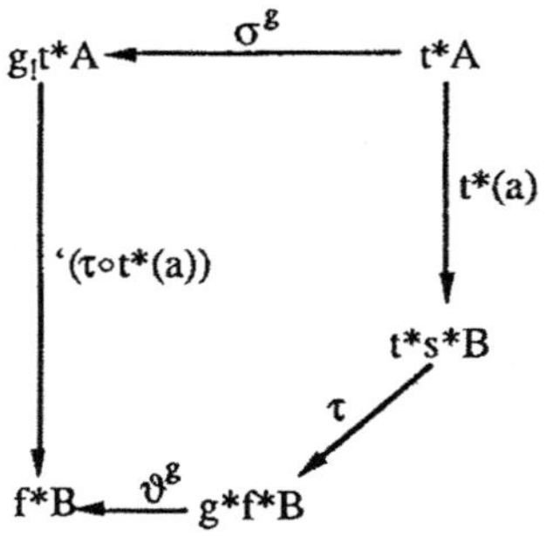

The pentangle R commutes by lemma 25, the rest by definitions. It is easy to see that $a = s^*(‘a)\circ\eta$; hence $t^*(a) = t^*s^*(‘a)\circ t^*(\eta)$. The arrows $‘(\tau\circ t^*(a))$ and $f^*(‘a)\circ\rho$ are thus the vertical factorizations of the same arrow $\vartheta^{g}\circ\tau\circ t^{*}(a) = \vartheta^{g}\circ\tau\circ t^{*}s^{*}(`a)\circ t^{*}(\eta)$ through σ^{g}. By the uniqueness, they must be equal.•

27. Corollary. For fibred preorders, the BC-condition is equivalent with the interpolation condition.

28. Remark. The connection of interpolation and the Beck-Chevalley condition in the category of Heyting algebras has been noticed by A.M. Pitts (1983a). He also studied the interpolation condition for a special sort of fibred Heyting algebras (1983b), showing how Craig's Interpolation Theorem and Beth's Definability Theorem can be presented in this setting.

3. Uniform interpolation

31. Definition. For functors $S:\mathcal{K} \to \mathcal{J}$ and $F:\mathcal{K} \to \mathcal{C}$, the *category of interpolants* $/S,F/$ consists of:

- the triples $\langle a,B,c\rangle$, where $B\in|\mathcal{K}|$, $a\in \mathcal{J}(A, SB)$, $c\in \mathcal{C}(FB,C)$;
- a morphism $\langle a,B,c\rangle \to \langle a',B',c'\rangle$ is a triple $\langle p,q,r\rangle$, such that the squares on the following diagram commute.

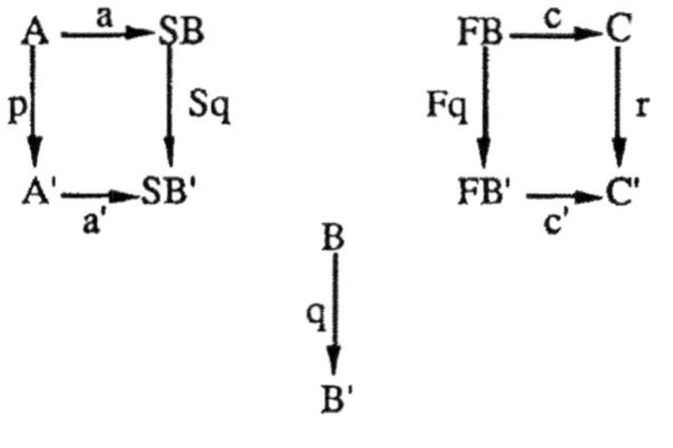

32. Comments. $/S,F/$ (with obvious projections) is a certain type of lax limit of the diagram $\mathcal{J}\overset{S}{\leftarrow}\mathcal{K}\overset{F}{\rightarrow}\mathcal{C}$ in the category of categories. It can also be obtained by strict pullbacks, as the next picture shows.

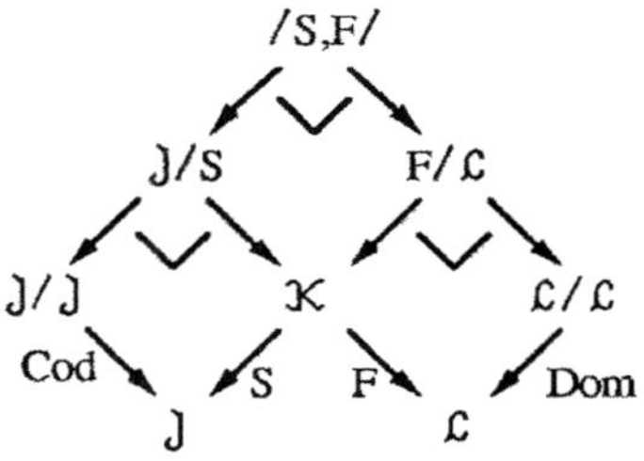

Given functors $T:\mathcal{J}\to\mathcal{I}$ and $G:\mathcal{L}\to\mathcal{I}$, every natural transformation $\varphi:TS\to GF$ induces a functor from the category of interpolants of S and F to the comma category of T and G:

$$R: /S,F/\to T/G: \langle a, B, c\rangle\mapsto\langle A, Gc\circ\varphi_B\circ Ta, C\rangle$$
$$\langle p, q, r\rangle\mapsto\langle p, r\rangle.$$

In the obvious sense, $\langle a,B,c\rangle$ is an interpolant of $R\langle a,B,c\rangle$, relative to φ.

33. Definition. Given a square of functors $Q=(F,G,S,T)$ as above, with a natural transformation $\varphi:TS\to GF$, we define *initial interpolants* to be those objects of /S,F/ which are initial among the interpolants of the same arrow. In other words, $\langle a, B, c\rangle$ is an initial interpolant <u>if</u> for any other interpolant $\langle a',B',c'\rangle$, such that $R\langle a',B',c'\rangle = R\langle a,B,c\rangle$, there is unique $q\in\mathcal{K}(B, B')$, with

$$a' = Sq\circ a \text{ and } c'\circ Fq = c.$$

We consider, thus, the initiality in R-fibres.

We say that the *interpolation in* Q *is uniform* if an initial interpolant can be recognized by the first component. In other words, with uniform interpolation, an interpolant $\langle a, B, c\rangle$ must be initial <u>whenever</u> for every $\langle a',B',c'\rangle$, such that $R\langle a',B',c'\rangle = R\langle a,B,c\rangle$, there is unique $q\in\mathcal{K}(B, B')$, with

$$a' = Sq\circ a.$$

34. Back to fibrations. For a fibration $E:\mathcal{E}\to\mathcal{B}$ and a commutative square $Q=(f,g,s,t)$ in $\mathcal{B}$, the categories $/s^*,f^*/$ obtained for various choices of the inverse image functors s^* and f^* are all isomorphic. Moreover, these categories are equivalent with the category of *all* the interpolants for all the possible inverse images along s and f. Similarly, the comma category t^*/g^* for some chosen t^* and g^* is equivalent with the category of triples $\langle A, d, C\rangle$, where $d: t^*A\to g^*C$ is a vertical arrow from an *arbitrary* inverse image of A along t to an *arbitrary* inverse image of C along g.

Note that Q satisfies the interpolation condition <u>if</u> the functor $R:/s^*,f^*/\to t^*/g^*$ induced by the canonical isomorphism $\tau: t^*s^*\to g^*f^*$ is a retraction, i.e. if there is a functor

$M: t^*/g^* \longrightarrow /s^*,f^*/$

such that RM=id. This functor M gives a choice of initial interpolants if it is left adjoint to R. On the other hand, the interpolation condition does not mention the arrows of t^*/g^*, so that it does not seem to imply the existence of M.

35. Terminology. We shall say that a commutative square Q in the base of a fibration satisfies the *uniform interpolation condition* if it satisfies the interpolation condition, and if the interpolation over it is uniform.

For the next proposition – which explains what is uniform about the uniform interpolation – we need the notion of *trifibration*. The notion of *bifibration* is standard: A functor $E:\mathcal{E} \longrightarrow \mathcal{B}$ is a bifibration if both E and its dual $E^o:\mathcal{E}^o \longrightarrow \mathcal{B}^o$ are fibrations. (The dual categories and functors are, of course, obtained by formally changing the directions of all morphisms.) We say that E is a *trifibration* if both $E:\mathcal{E} \longrightarrow \mathcal{B}$ and $E^{op}: \mathcal{E}^{op} \longrightarrow \mathcal{B}$ are bifibrations. The category $\mathcal{E}^{op}$, fibred over $\mathcal{B}$, is obtained by changing the direction of all the *vertical* arrows in $\mathcal{E}$. (The arrows of $\mathcal{E}^{op}$ are the equivalence classes of spans with a vertical arrow pointing at the source, and a cartesian arrow pointing at the target.)

A fibration $E:\mathcal{E} \longrightarrow \mathcal{B}$ is a bifibration <u>iff</u> every inverse image functor $t^*: \mathcal{E}_J \longrightarrow \mathcal{E}_I$ has a left adjoint $t_!: \mathcal{E}_I \longrightarrow \mathcal{E}_J$. It is a trifibration <u>iff</u> there is also a right adjoint $t_*: \mathcal{E}_I \longrightarrow \mathcal{E}_J$ of t^*.

A cocartesian lifting by E^{op} of t at X is generically denoted $\psi^t: X\backslash o(\longrightarrow,| \;)t_*X$. (The barred arrows $\not\longrightarrow$ always belong to $\mathcal{E}^{op}$.)

36. Proposition. Let $E: \mathcal{E} \longrightarrow \mathcal{B}$ be a trifibration, and Q = (f,g,s,t) a commutative square in $\mathcal{B}$, satisfying the interpolation condition. The interpolation is uniform <u>iff</u>

$c`o`a = \tilde{c}`o`\tilde{a}$

is true for any two interpolants $\langle a, B, c\rangle$ and $\langle \tilde{a}, \tilde{B}, \tilde{c}\rangle$ of the same arrow.

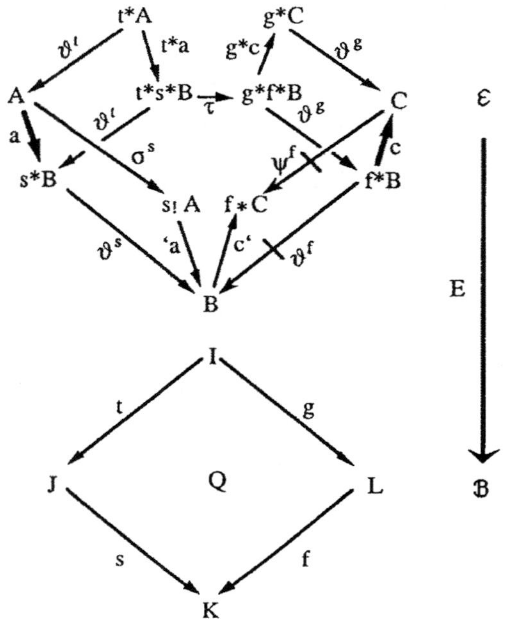

• Then: If $\langle a,B,c\rangle$ is an interpolant of d, the triple $\langle \eta, s_!A, c\circ f^*(`a)\rangle$ is another interpolant of d – since $a = s^*(`a)\circ\eta$, so that the next diagram commutes.

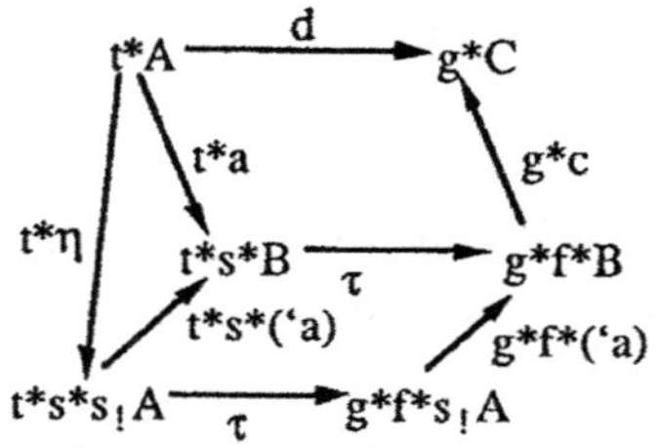

It follows from the uniformity that $\langle \eta, s_!A, c\circ f^*(`a)\rangle$ is an initial interpolant. Hence an arrow

$$\langle id_A, `\tilde{a}, id_C\rangle : \langle \eta, s_!A, c\circ f^*(`a)\rangle \longrightarrow \langle \tilde{a}, \tilde{B}, \tilde{c}\rangle$$

for each interpolant $\langle \tilde{a}, \tilde{B}, \tilde{c}\rangle$ of d.

On the other hand, for any two interpolants $\langle \hat{a}, \hat{B}, \hat{c}\rangle$, $\langle \tilde{a}, \tilde{B}, \tilde{c}\rangle$ of d, the existence of an arrow $\langle id_A, q, id_C\rangle : \langle \hat{a}, \hat{B}, \hat{c}\rangle \longrightarrow \langle \tilde{a}, \tilde{B}, \tilde{c}\rangle$ (in $/s^*,f^*/$), implies $\tilde{c}`\circ`\tilde{a} = \hat{c}`\circ`\hat{a}$.

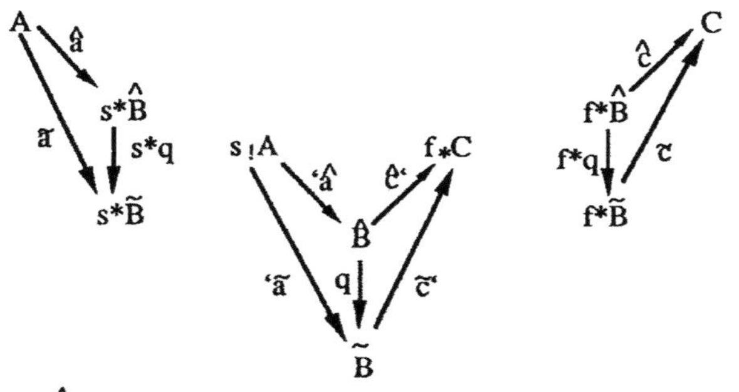

Putting $\hat{a} := \eta$ and $\hat{c} := c\circ f^*(\text{‘}a)$, we get

$$\tilde{c}\text{‘}\circ\text{‘}\tilde{a} = (c\circ f^*(\text{‘}a))\text{‘}\circ\text{‘}\eta = (c\circ f^*(\text{‘}a))\text{‘} = c\text{‘}\circ\text{‘}a.$$

<u>If</u>: Let $\langle a,B,c\rangle$ be an interpolant such that for every interpolant $\langle\tilde{a},\tilde{B},\tilde{c}\rangle$ of the same arrow there is unique $\tilde{q}\in\mathcal{E}_K(B,\tilde{B})$ with $\tilde{a}=s^*(\tilde{q})\circ a$. We must prove that $\tilde{c}\circ f^*(\tilde{q})=c$.

Since $\langle\eta, s_!A, c\circ f^*(\text{‘}a)\rangle$ is an interpolant of the same arrow, there is $q_\eta\in\mathcal{E}_K(B,s_!A)$, such that $\eta=s^*(q_\eta)\circ a$. It is easy to see that

1) $\quad q_\eta\circ\text{‘}a = id$ <u>and</u> $\text{‘}a\circ q_\eta = id$.

On the other hand, from $s^*(\tilde{q})\circ a=\tilde{a} : A\to s^*\tilde{B}$ follows $\tilde{q}\circ\text{‘}a=\text{‘}\tilde{a}: s_!A\to\tilde{B}$. Using (1), we get

2) $\quad \tilde{q} = \text{‘}\tilde{a}\circ q_\eta$.

Finally, the "left transpose" of $c\text{‘}\circ\text{‘}a$: $s_!A\to f_*C$ is $c\circ f^*(\text{‘}a)$: $f^*s_!A\to C$; the hypothesis $c\text{‘}\circ\text{‘}a=\tilde{c}\text{‘}\circ\text{‘}\tilde{a}$ <u>implies</u>

3) $\quad c\circ f^*(\text{‘}a) = \tilde{c}\circ f^*(\text{‘}\tilde{a})$.

Now we can derive:

$$\tilde{c}\circ f^*(\tilde{q}) \overset{(2)}{=} \tilde{c}\circ f^*(\text{‘}\tilde{a}\circ q_\eta) \overset{(3)}{=} c\circ f^*(\text{‘}a\circ q_\eta) \overset{(1)}{=} c.\bullet$$

37. Theorem. A commutative square $Q = (f,g,s,t)$ in the base of a bifibration $E:\mathcal{E}\to\mathcal{B}$ satisfies the uniform interpolation condition <u>iff</u> it satisfies the Beck-Chevalley condition.

• In section 2 we proved that the interpolation condition is satisfied <u>iff</u> $\rho: g_!t^*A\to f^*s_!A$ is a split mono. It is now sufficient to show that the interpolation is uniform <u>iff</u> $\rho: g_!t^*A\to f^*s_!A$ is an epi.

<u>If</u>: The <u>if</u>-part of the previous proof can be copied almost completely. The only difference is that equality (3) must be derived in a different way this time: namely, from lemma 26 and the fact that $\rho: g_!t^*A\to f^*s_!A$ is an epi.

<u>Then</u>: For arbitrary arrows c_1, c_2: $f^*s_!A\to C$, the equivalence

Then: For arbitrary arrows $c_1, c_2: f^*s_!A \to C$, the diagram

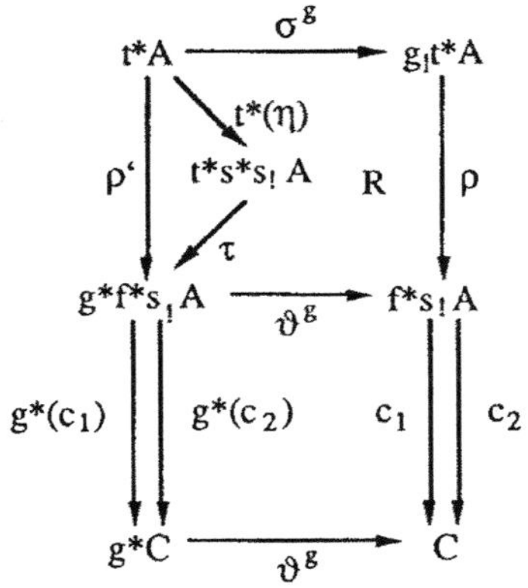

shows that

$$c_1 \circ \rho = c_2 \circ \rho \Leftrightarrow g^*(c_1) \circ \tau \circ t^*(\eta) = g^*(c_2) \circ \tau \circ t^*(\eta).$$

But this means that $c_1 \circ \rho = c_2 \circ \rho$ implies that $\langle \eta, s_!A, c_1 \rangle$ and $\langle \eta, s_!A, c_2 \rangle$ must be interpolants of the same arrow. The uniformity now implies that these interpolants must be initial; thus $c_1 = c_2$. So we derived that $c_1 \circ \rho = c_2 \circ \rho$ implies $c_1 = c_2$.•

4. Descent by interpolation

41. Geometric motivation. Let $\mathfrak{B}$ be a site (cf. Artin et al.), and $D: \mathfrak{D} \to \mathfrak{B}$ a discrete fibration. If $f: L \to K$ is a covering morphism in $\mathfrak{B}$, and $\partial_0, \partial_1: M \to L$ its *kernel pair* – obtained by pulling back f along itself – the sheaf condition (ibid.) on D tells that for every $A \in \mathfrak{D}_L$, such that $\partial_0{}^*A = \partial_1{}^*A$, there must exist a unique $B \in \mathfrak{D}_K$ with $A = f^*B$. In other words, every "vertical arrow" $\partial_0{}^*A \to \partial_1{}^*A$ (which is of course identity, since the fibration D is discrete) must have a unique interpolant over the kernel square $(f, \partial_0, f, \partial_1)$ of f. For arbitrary covering family $\{f_n: L_n \to K \mid n \in N\}$, the sheaf condition can be expressed by saying that every family of "vertical arrows" $\{{}^n_m\partial_0{}^*A_n \to {}^n_m\partial_1{}^*A_m \mid n, m \in N\}$ determines a unique common interpolant $B \in \mathfrak{D}_K$. The arrows ${}^n_m\partial_0: M_{nm} \to L_n$ and ${}^n_m\partial_1: M_{nm} \to L_m$ here are obtained in a pullback of f_n and f_m.

The notion of *descent* lifts the sheaf condition from the discrete fibrations to fibrations $E: \mathcal{E} \to \mathfrak{B}$ in general. For simplicity, we shall consider covering by one arrow; the passage on covering *families* only requires some more involved formulations. The question will be: When can one *descend* along a morphism $f: L \to K$, and represent $\mathcal{E}_K$ in terms of $\mathcal{E}_L$? If this is possible, f is said to be an *effective descent* morphism.

42. Notation, terminology. To fix the notation, consider the following cube of pullback squares.

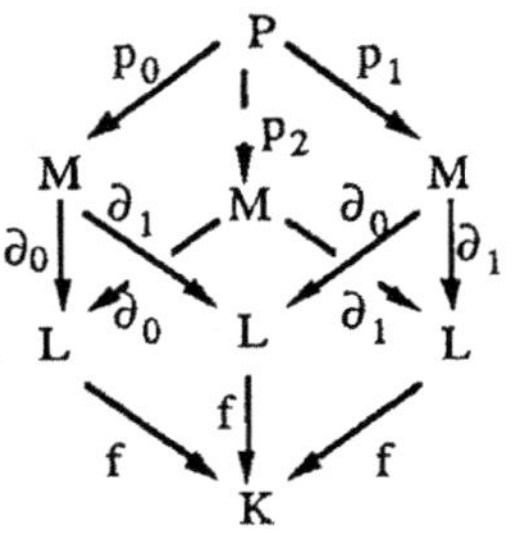

A *kernel square* of f is a pullback $(f) := (f, \partial_0, f, \partial_1)$. Note that the diagram above contains not only (f), but also (∂_0) and (∂_1). Three different pullback squares are obtained by pulling back (f) along f: (∂_0), (∂_1) and $S := (\partial_0, p_1, \partial_1, p_0)$.

By $\eta: L \to M$ we shall now denote the unique arrow such that $\partial_0 \circ \eta = \partial_1 \circ \eta = id_L$. For cartesian liftings of ∂_0 we shall use $\vartheta^0: \partial_0{}^*A \to A$ (instead of ϑ^{∂_0}). Let $\upsilon^0: A \to \partial_0{}^*A$ be the unique splitting of ϑ^0 over η (i.e., $\vartheta^0 \circ \upsilon^0 = id$ and $E\upsilon^0 = \eta$). It is straightforward to show that υ^0 must be cartesian. – Idem for the splitting υ^1 over η of the cartesian lifting ϑ^1 of ∂_1.

Given a fibration $E: \mathcal{E} \to \mathcal{B}$ and an (f)-interpolant $\langle a,B,c\rangle$ of $d \in \mathcal{E}_M(\partial_0{}^*A, \partial_1{}^*C)$, the triple

$$f^*\langle a,B,c\rangle := \langle \tau \circ \partial_0{}^*(a),\ f^*B,\ \partial_1{}^*(c) \circ \tau\rangle, \text{ where}$$

$$\tau \circ \partial_0{}^*(a): \partial_0{}^*A \to \partial_0{}^*f^*B \to \partial_1{}^*f^*B,$$

$$\partial_1{}^*(c) \circ \tau: \partial_0{}^*f^*B \to \partial_1{}^*f^*B \to \partial_1{}^*C,$$

is clearly an S-interpolant of

$$d^S := \tau \circ p_2{}^*(d) \circ \tau: p_0{}^*\partial_0{}^*A \to p_2{}^*\partial_0{}^*A \to p_2{}^*\partial_1{}^*C \to p_1{}^*\partial_1{}^*C.$$

We say that the interpolant $\langle a,B,c\rangle$ is f-*simple* if

$$\langle id, id, id\rangle: f^*\langle a,B,c\rangle \to f^*\langle a,B,c\rangle$$

is the only arrow in the category of S-interpolants. In other words, the object $f^*\langle a,B,c\rangle$ has just one endomorphism in its R-fibre.

(In general, every pullback square Q and arrow q with the same target span a cube of pullbacks as above; and every Q-interpolant induces an interpolant over the square opposite to Q in this cube. So we can speak of q-simple interpolants in this general situation. For Q:=(f) and q:=f, we should actually consider two more interpolants induced by $\langle a,B,c\rangle$: namely, those over (∂_0)

and (∂_1) – and include them in the definition of f-simplicity. This would, however, only add a couple of inessential sentences to the proofs below.)

43. Definition. Let a fibration $E\colon \mathcal{E} \to \mathcal{B}$ and an arrow $f \in \mathcal{B}(L,K)$ be given. An *f-descent data* (for E) is a pair $\langle A,d\rangle$, $A \in |\mathcal{E}_L|$, $d \in \mathcal{E}_M(\partial_0{}^*A, \partial_1{}^*A)$, such that

RE) $\vartheta^1 \circ d \circ \upsilon^0 = id$; i.e. $\eta^*(d)=id$:

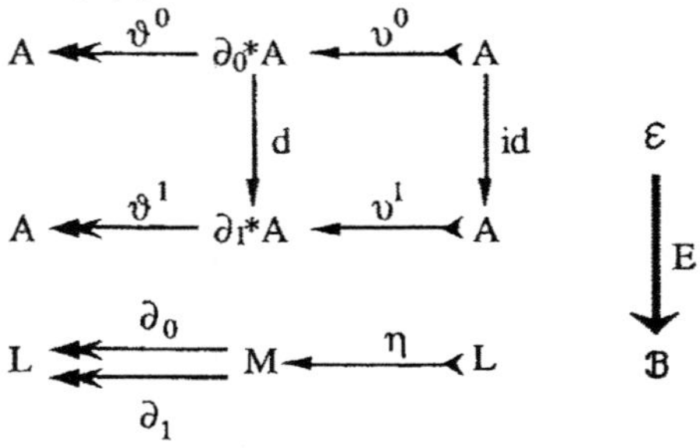

TR) $\langle d,A,d\rangle$ is an S-interpolant of $d' := \tau \circ p_2{}^*(d) \circ \tau\colon p_0{}^*\partial_0{}^*A \to p_1{}^*\partial_1{}^*A$; i.e., the following diagram commutes

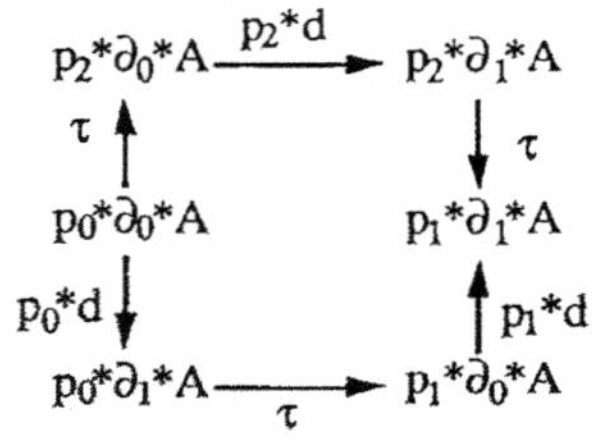

The descent data constitute a category, denoted $\underline{\mathrm{Des}}_E(f)$, or just $\underline{\mathrm{Des}}(f)$. A morphism $\langle A,d\rangle \to \langle \tilde{A},\tilde{d}\rangle$ in this category is an arrow $h \in \mathcal{E}_L(A,\tilde{A})$, such that $\partial_1{}^*(h) \circ d = \tilde{d} \circ \partial_0{}^*(h)$.

44. Remarks. The arrow d occuring as a descent data must be an isomorphism. • Since $\partial_0 = \partial_0 \circ \eta \circ \partial_0$, and the square $(\partial_0, p_0, \partial_0, p_2)$ is a pullback, there is a unique arrow $m_0\colon M \to P$ in $\mathcal{B}$, such that $p_0 \circ m_0 = id$ <u>and</u> $p_2 \circ m_0 = \eta \circ \partial_0$. It is routine to show that

$$\partial_0 \circ p_2 \circ m_0 = \partial_1 \circ p_2 \circ m_0 = \partial_0 \circ p_0 \circ m_0 = \partial_1 \circ p_1 \circ m_0 = \partial_0, \text{ and}$$
$$\partial_1 \circ p_0 \circ m_0 = \partial_0 \circ p_1 \circ m_0 = \partial_1.$$

If the inverse images of objects along m_0 are appropriately chosen, one obtains the following diagram.

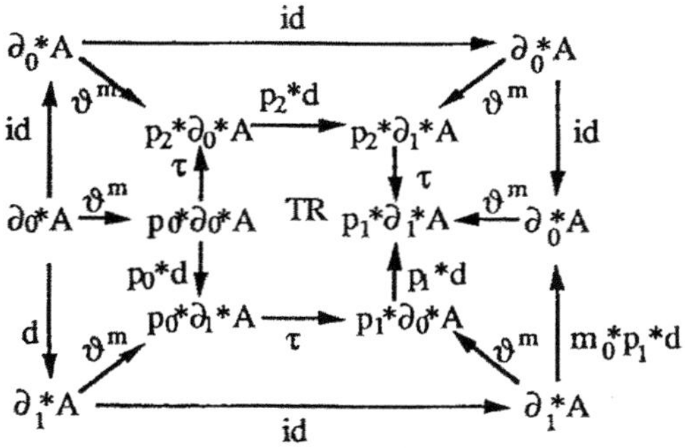

The commutativity of the part TR now <u>implies</u>

$$m_0{}^*p_1{}^*(d)\circ d = \mathrm{id}.$$

The arrow d is thus a split mono. An analogous argument with the arrow $m_1: M \longrightarrow P$, such that $p_1 \circ m_1 = \mathrm{id}$ <u>and</u> $p_2 \circ m_1 = \eta \circ \partial_1$, shows that d is split epi too.•

An f-descent data for E is, in a sense, an action in $\mathcal{E}$ of the equivalence relation (internal groupoid) $\partial_0, \partial_1: M \longrightarrow L$ induced by f. Namely, $\langle A,d\rangle$ can be viewed as an internal groupoid in $\mathcal{E}$, with $\vartheta^0, \vartheta^1 \circ d: \partial_0{}^*A \longrightarrow A$ as the domain and codomain arrows, and with $\upsilon^0: A \longrightarrow \partial_0{}^*A$ as the arrow of identities. The category <u>Des</u>(f) is equivalent with the category of internal groupoids in $\mathcal{E}$ which are built from cartesian liftings of ∂_0, ∂_1, η etc. (Cf. Pavlović 1990, III.1.2.)

Note that for every $B \in |\mathcal{E}_K|$, the pair $\langle f^*B, \tau_B\rangle$ is a descent data (where f^*B is any inverse image, while $\tau_B: \partial_0{}^*f^*B \longrightarrow \partial_1{}^*f^*B$ is the vertical isomorphism between two inverse images of B along $f\circ\partial_0 = f\circ\partial_1$). Every arrow $b \in \mathcal{E}_K(B,B')$, induces a unique morphism $f^*b: \langle f^*B,\tau_B\rangle \longrightarrow \langle f^*B',\tau_{B'}\rangle$ in <u>Des</u>(f). Every choice of inverse image functors f^*, $\partial_0{}^*$ and $\partial_1{}^*$ determines a functor

$$f^\# : \mathcal{E}_K \longrightarrow \underline{\mathrm{Des}}(f) : B \longmapsto \langle f^*B,\tau_B\rangle.$$

45. Definition. (Grothendieck 1959) An f-descent data $\langle A,d\rangle$ (for E) is *effective* if it is isomorphic (in $\underline{\mathrm{Des}}_E(f)$) with one in the form $\langle f^*B,\tau_B\rangle$. An arrow f is said to be *effective* if all the f-descent data are effective.

f is a *descent* morphism if for all B, $\tilde{B} \in |\mathcal{E}_K|$ each arrow $\langle f^*B,\tau_B\rangle \longrightarrow \langle f^*\tilde{B},\tau_{\tilde{B}}\rangle$ in $\underline{\mathrm{Des}}_E(f)$ is in the form f^*b for a unique $b \in \mathcal{E}_K(B,\tilde{B})$.

In other words, f is effective <u>iff</u> each $f^\#$ is essentially surjective; f is a descent morphism <u>iff</u> each $f^\#$ is full and faithful; f is an *effective descent* morphism <u>iff</u> each $f^\#$ is an equivalence of categories.

46. Terminology. An interpolant $\langle a,B,c\rangle$ of a descent data $d: \partial_0{}^*A \to \partial_1{}^*A$ is *natural* if $a: A \to f^*B$ and $c: f^*B \to A$ are morphisms of descent data, i.e. if they satisfy

$$\partial_1{}^*(a) \circ d = \tau_B \circ \partial_0{}^*(a) \ \underline{\text{and}}\ \partial_1{}^*(c) \circ \tau_B = d \circ \partial_0{}^*(c).$$

An arrow φ from a fibred category $\mathcal{E}$ is an E-*coequalizer* of a pair δ_0,δ_1 of parallel arrows if

- $\varphi \circ \delta_0 = \varphi \circ \delta_1$ <u>and</u>
- for every χ in $\mathcal{E}$, $\chi \circ \delta_0 = \chi \circ \delta_1$ <u>and</u> $E\chi = E\varphi$ <u>imply</u> that there is a unique vertical arrow b with $\chi = b \circ \varphi$.

47. Theorem. Let $E: \mathcal{E} \to \mathcal{B}$ be a fibration and f an arrow in its base. (Notation as above!)

i) f is an effective morphism <u>iff</u> each f-descent data has a natural, f-simple interpolant.

ii) f is a descent morphism <u>iff</u> every cartesian lifting $\vartheta^f: f^*B \to B$ is an E-coequalizer of its kernel pair.

• **i)** <u>If</u>: Let $\langle a,B,c\rangle$ be a natural, f-simple interpolant of f-descent data $\langle A,d\rangle$. So we have $d: \partial_0{}^*A \to \partial_1{}^*A$, $a: A \to f^*B$ and $c: f^*B \to A$ such that

1) $\quad d = \partial_1{}^*(c) \circ \tau \circ \partial_0{}^*(a)$,

2) $\quad \partial_1{}^*(a) \circ d = \tau \circ \partial_0{}^*(a)$, and

3) $\quad \partial_1{}^*(c) \circ \tau = d \circ \partial_0{}^*(c)$.

We shall prove $c \circ a = id_A$ <u>and</u> $a \circ c = id_{f^*B}$.

<u>$c \circ a = id$</u> : As before, choose the cartesian liftings υ^0 and υ^1 of η to be the splittings of ϑ^0 and ϑ^1 respectively.

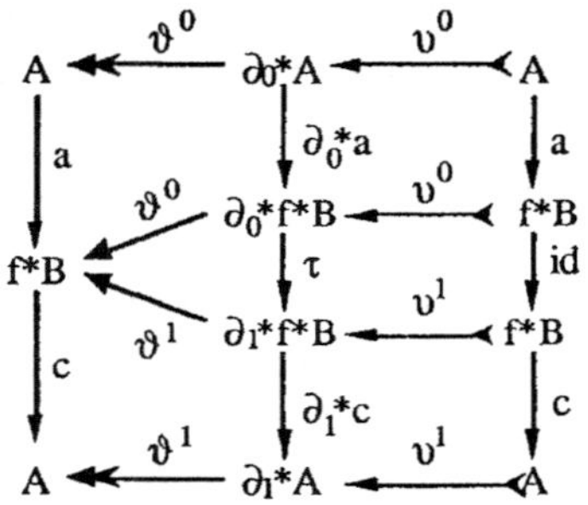

We see that

$$\vartheta^1_A \circ \upsilon^1_A \circ c \circ a = \vartheta^1_A \circ \partial_1{}^*(c) \circ \tau \circ \partial_0{}^*(a) \circ \upsilon^0_A = \vartheta^1_A \circ d \circ \upsilon^0_A.$$

But now $\vartheta^1_A \circ \upsilon^1_A = id_A$ holds by the definition, while $\vartheta^1_A \circ d \circ \upsilon^0_A = id_A$ is just condition (RE).

<u>$a \circ c = id$</u> : By condition (TR), $\langle d,A,d\rangle$ is an interpolant of

$d' := \tau \circ p_2{}^*(d) \circ \tau : p_0{}^*\partial_0{}^*A \longrightarrow p_1{}^*\partial_1{}^*A.$

Of course, $f^*\langle a,B,c\rangle = \langle \tau \circ \partial_0{}^*(a), f^*B, \partial_1{}^*(c) \circ \tau\rangle$ is another interpolant of the same arrow. The equalities (2) and (1) mean that

$\langle id, a, id\rangle : \langle d,A,d\rangle \longrightarrow f^*\langle a,B,c\rangle.$

is a morphism of interpolants. On the other hand, (1) and (3) tell the same for

$\langle id, c, id\rangle : f^*\langle a,B,c\rangle \longrightarrow \langle d,A,d\rangle$

Hence, $\langle id, a \circ c, id\rangle$ is an arrow $f^*\langle a,B,c\rangle \longrightarrow f^*\langle a,B,c\rangle$. The hypothesis that $\langle a,B,c\rangle$ is f-simple now <u>implies</u> that

$a \circ c = id.$

<u>Then</u>: Suppose that $\langle A,d\rangle \simeq \langle f^*B, \tau_B\rangle$ for some B; i.e., arrows $a : \langle A,d\rangle \longrightarrow \langle f^*B, \tau_B\rangle$ and $c : \langle f^*B, \tau_B\rangle \longrightarrow \langle A,d\rangle$ are given, such that $a \circ c = id$ <u>and</u> $c \circ a = id$. $\langle a, B, c\rangle$ is then an interpolant of d because

$d = \partial_1{}^*(c) \circ \partial_1{}^*(a) \circ d = \partial_1{}^*(c) \circ \tau \circ \partial_0{}^*(a).$

This interpolant is obviously natural. To show that it is f-simple, consider a morphism $h : f^*\langle a,B,c\rangle \longrightarrow f^*\langle a,B,c\rangle$. So we have an arrow $h \in \mathcal{E}_L(f^*B, f^*B)$, such that

$\tau \circ \partial_0{}^*(a) = \partial_1{}^*(h) \circ \tau \circ \partial_0{}^*(a)$ <u>and</u> $\partial_1{}^*(c) \circ \tau \circ \partial_0{}^*(h) = \partial_1{}^*(c) \circ \tau.$

Since c is an isomorphism, $\partial_1{}^*(c)$ is, and $\partial_0{}^*(h)$ must be an identity. Using this, we calculate:

$\vartheta_B^f \circ h \circ a \circ \vartheta_A^0 = \vartheta_B^f \circ \vartheta_{f^*B}^0 \circ \partial_0{}^*(h) \circ \partial_0{}^*(a) = \vartheta_B^f \circ \vartheta_{f^*B}^0 \circ \partial_0{}^*(a) = \vartheta_B^f \circ a \circ \vartheta_A^0,$

and conclude that $h \circ a = a$, because ϑ_A^0 is an epimorphism (split by υ^0) and ϑ_B^f is cartesian, while $h \circ a$ and a are vertical arrows. Since a is an isomorphism, $h = id$.

ii) It follows from lemma 48 that $\vartheta^0, \vartheta^1 \circ \tau : \partial_0{}^*f^*B \longrightarrow f^*B$ is a kernel pair of $\vartheta^f : f^*B \longrightarrow B$ (for every B).

Chasing the next diagram – where h is vertical, and $\chi = \tilde{\vartheta}^f \circ h$ – one easily proves that

$\partial_1{}^*(h) \circ \tau = \tilde{\tau} \circ \partial_0{}^*(h) \Leftrightarrow \chi \circ \vartheta^0 = \chi \circ \vartheta^1 \circ \tau.$

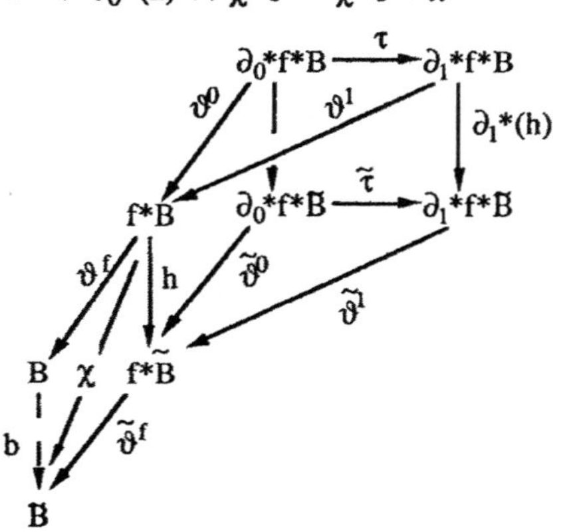

Thus, if h is a morphism of descent data, the assumption that ϑ^f is an E-coequalizer of its kernel pair gives a unique arrow $b \in \mathcal{E}_K(B, \tilde{B})$, such that $\chi = b \circ \vartheta^f$. In other words, $h = f^*b$. Conversely, if an arrow $\chi: f^*B \longrightarrow \tilde{B}$ over f satisfies $\chi \circ \vartheta^0 = \chi \circ \vartheta^1 \circ \tau$, its vertical part h must be a morphism of descent data $\langle f^*B, \tau_B \rangle \longrightarrow \langle f^*\tilde{B}, \tau_{\tilde{B}} \rangle$. The hypothesis that f is a descent morphism means that there is a unique arrow $b \in \mathcal{E}_K(B, \tilde{B})$, such that $h = f^*b$. Thus, $\chi = b \circ \vartheta^f$.

48. Lemma. Let Q = (f,g,s,t) be a pullback square in $\mathcal{B}$ and $\Theta = (\varphi, \gamma, \sigma, \vartheta)$ a commutative square in $\mathcal{E}$ over Q (i.e., $\varphi \circ \gamma = \sigma \circ \vartheta$, $E\varphi = f$, $E\gamma = g$, $E\sigma = s$, $E\vartheta = t$). If φ and ϑ are cartesian, then Θ is a pullback square.

49. Comment. This descent theorem is a far descendant of the method which Joyal and Tierney (1984) used to prove that open surjections of toposes are effective descent morphisms – *in absence* of the Beck-Chevalley property. More recently, Moerdijk (1989) observed that an appropriately saturated class $\mathcal{O}$ of arrows in $\mathcal{B}$ must consist of effective descent morphisms with respect to the fibration Cod: $\mathcal{B}/\mathcal{B} \longrightarrow \mathcal{B}$ if it satisfies the following axioms:

i) A coequalizer of every parallel pair of arrows from $\mathcal{O}$ exists, and it is stable under pullbacks;

ii) Each arrow belonging to $\mathcal{O}$ is a coequalizer of its kernel pair.

The two parts of theorem 47 clearly correspond to these axioms.

On the other hand, it is perhaps interesting to put theorems 37 and 47 together, aligning the Beck-Chevalley property and descent – on the common ground of interpolation. This way, one can analyze how Bénabou-Roubaud-Beck's theorem (cf. Hyland-Moerdijk 1990) provides some sufficient conditions for descent *in presence* of the Beck-Chevalley property.

References

Artin, M. Grothendieck, A., Verdier, J.L.

(1972) *Théorie des Topos* (Exposés I-IV from SGA 4, 1963/64), Lecture Notes in Mathematics 269 (Springer, Berlin)

Bénabou, J.

(1985) Fibered categories and the foundations of naive category theory, *J. Symbolic Logic* 50(1), 10-37

Bénabou, J., Roubaud, J.

(1970) Monades et descente, *C.R. Acad. Sc. Paris (Série A)*, t. 270, 96-98

Giraud, J.

(1964) Théorie de la descente, *Bull. Soc. math. France, Mémoire 2*

Gray, J.W.

(1966) Fibred and Cofibred Categories, *Proceedings of the Conference on Categorical Algebra, La Yolla 1965*, (Springer, Berlin) 21-84

Grothendieck, A.

(1959) Technique de descente et théorèmes d'existence en géométrie algébrique, I. Généralités. Descente par morphismes fidèlement plats, *Séminaire Bourbaki* 190

(1971) Exposés VI, VIII, IX (from SGA 1, 1960/61) in: *Revêtements Etales et Groupe Fondamental*, Lecture Notes in Mathematics 224 (Springer, Berlin), 145-260

Hyland, J.M.E., Moerdijk, I.

(1990) An application of Beck's Theorem, Manuscript

Joyal, A., Tierney, M.

(1984) *An extension of the Galois theory of Grothendieck*, Memoirs of the A.M.S., No. 309 (American Mathematical Society, Providence, RI)

Lawvere, F.W.

(1970) Equality in hyperdoctrines and comprehension schema as an adjoint functor, *Applications of Category Theory, Proceedings of A.M.S. Symposia on Pure Mathematics XVII* (American Mathematical Society, Providence RI), 1-14

Moerdijk, I.

(1989) Descent Theory for Toposes, *Bull. Soc. Math. de Belgique (Série A)*, XLI(2), 373-391

Pavlović, D.

(1990) *Predicates and Fibrations: From Type Theoretical to Category Theoretical Presentation of Constructive Logic*, Thesis, State University Utrecht

Pitts, A.M.

(1983a) Amalgation and Interpolation in the Category of Heyting Algebras, *J. Pure Appl. Algebra* 29, 155-165

(1983b) An Application of Open Maps to Categorical Logic, *J. Pure Appl. Algebra* 29, 313-326

This paper is in final form and will not be published elsewhere.

AN n-CATEGORICAL PASTING THEOREM

A.J.Power*
Department of Computer Science
University of Edinburgh
King's Buildings
Edinburgh EH9 3JZ
Scotland.

dedicated to Max Kelly on the occasion of his 60th birthday

Abstract

In order to facilitate the study of 2-categories with structure, we state and prove an n-categorical pasting theorem. This is based upon a new definition of n-pasting scheme that generalises Johnson's definition of a well-formed loop-free pasting scheme by weakening his no direct loops condition. We define n-pasting, prove the theorem, and show that for n=3, it incorporates all possible composites of n-cells. We show that that is not true for higher n. We define the horizontal n-category of an (n+1)-category to generalise that of a 2-category, we define horizontal and vertical composition for an (n+1)-category and we state and prove an interchange law. We also study further conditions on a pasting diagram and their impact upon how one may evaluate a composite, and we express Street's free n-categories in terms of left adjoints.

1 Introduction

"Pasting" has become a well known and valuable tool in the study of 2-categories. In [8], a "2-categorical pasting theorem" was proved in order to make precise the notion of pasting as outlined in [7]. Now, in various articles such as [2], [9], [10] and implicitly in

*This research was supported by the Australian Research Council and by ESPRIT Basic Research Action 3245: Logical Frameworks Design, Implementation and Experiment.

[11], 3-categories or mild generalisations of them are used to study 2-categories with extra structure. So we seek a 3-categorical pasting theorem in order to facilitate that. In fact, an n-categorical pasting theorem for each positive natural number n is only a little more difficult, and it seems a more natural level of generality; so that is the aim of this paper.

The idea of pasting is to unify the n different sorts of composite in the definition of n-category into one general notion of "pasting composite". For instance, for n=3, given 3-cells $\gamma : \alpha \to \alpha'$ and $\gamma' : \beta \to \beta'$, we should like to know directly from the definition of pasting composite that a diagram such as

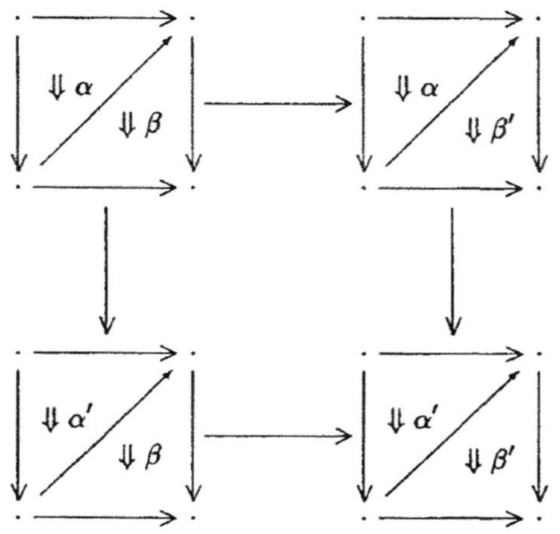

commutes, where the 3-cells are the evident ones generated by γ and γ'.

There has been and there continues to be much work on this problem. Michael Johnson produced an n-categorical pasting theorem in his thesis [3] and published the result in [4]. Johnson investigated two classes of schemes. The first class, loop-free schemes, are of interest because the free n-category on an n-dimensional such admits a particularly easy construction. We do not handle these: a counterexample appears in Section 2. The second class, well-formed loop-free schemes, are the pasting schemes of interest to us. These parametrize composable diagrams in n-categories. Johnson realised that his definition was not broad enough to include all possible composites of n-cells. In particular, it was not broad enough to allow all of the constructions he sought, specifically that of a certain tensor product of pasting schemes, as outlined in his lecture [5]. The difficulty was that his "no direct loops" condition was too strong. Here, we generalise Johnson's definition. Specifically, we drop most of the no direct loops condition, and what remains of it is true of all composites of n-cells. This allows us to capture all possible composites for n=3; we also capture a large class of composites for higher n. Alas, we do not capture all: an example appears in Section 3. That example is interesting as it is not merely a technical example but acts as a counterexample to a large class of conjectures. For instance, a variant of that example exhibits a composite of 4-cells that fails to satisfy the fourth (and most substantial other than no direct loops) requirement of Johnson's "loop-freeness" condition. It is not immediately evident from the two definitions that the definition here generalises that of Johnson, but it does follow from an analysis of the early results of Johnson's paper [4] as he has indicated in a private communication for general n and shown in detail for n=2 in a Sydney Category seminar in 1988.

It should be noted that greater generality for pasting schemes does not necessarily imply a more general pasting theorem. For instance, our Example 3.12 is not one of Johnson's well-formed loop-free pasting schemes, but diagrams of that form can be handled by Johnson's pasting theorem by parametrizing them by loop-free schemes.

Our approach is to give a topological foundation and to proceed by induction. Typically, one does not try to use 3-pasting schemes until one already has 2-pasting schemes to act as domains and codomains of the 3-cells; similarly for higher dimensions. So it seems reasonable to use the definition and results of n-pasting schemes in the definition of an (n+1)-pasting scheme. This is a convenient way to order the information so that it is less overwhelming in its detail. Just as a 2-cell can be seen as two parallel directed paths in a graph, together with a homotopy in 2-space from one to the other [8], we wish to regard a 3-cell as a homotopy in 3-space, and similarly for higher n. Using this paradigm, one may regard Ross Street's pictures [13] as n-pasting schemes together with a choice of the order in which some n-cells are composed.

Central to our approach is our "interchange law". Just as one speaks of horizontal and vertical composition for a 2-category, the various n+1 sorts of composite in the definition of an (n+1)-category may also be viewed naturally in terms of two sorts of composition: vertical composition, which is the top level composition of the (n+1)-category; and horizontal composition, which is the n-pasting composition of the (n+1)-category. These two sorts of composition satisfy an appropriately generalised interchange law given in Section 6. It is from this result that we deduce the unicity of a pasting composite.

There is other current work on giving a topological analysis of n-categories. Specifically, Kapranov and Voevodsky [6] are also working on pasting results that are topological in flavour, and Al-Agl and Steiner [1] have characterised ∞-categories in terms of simplexes. However, in both cases, the work is somewhat different in direction to ours. Of course, much of the work here was ultimately inspired by Street's fundamental paper [13], as explained in Section 8. It is possible to exhibit in terms of left adjoints the free n-categories described internally by Street [13]. A consequence of our result is that one can do so for a large class of them by means of pasting diagrams, hence in topological terms.

In Section 2, we commence the induction with the definitions for ordinary categories, i.e. in the case n=1, and we recall the definitions for 2-categories as in [8]. The definitions for 2-categories are not needed for the induction, but they illustrate the first nontrivial case. In Section 3, we give the inductive definitions needed to define pasting diagrams. In Section 4, we give the simple topological results that we require. In Section 5, we define horizontal pasting. In Section 6, we define horizontal and vertical composition in an (n+1)-category and give our "interchange law" and our "n-categorical pasting theorem". In Section 7, we show that we have captured all possible composites in a 3-category, we discuss our failure for higher n, and we illustrate a strengthening of the conditions that allow us greater flexibility in describing a composite. Finally, in Section 8, we explain limiting behaviour as n approaches infinity, and show how the work relates to that of Street [13], specifically to his free ω-categories.

Throughout, we shall assume that n-space is oriented for every n, and all embeddings into n-space are piecewise linear. This avoids irrelevant topological detail.

I should like to offer warm thanks to Michael Johnson and to Ross Street for valuable discussions on this work. This paper is in final form and no similar paper has been or is being submitted elsewhere.

2 Commencing the induction

A *0-pasting scheme* is the point of 0-space.

A *1-computad* is a finite directed graph; a *1-computad morphism* is a map of directed graphs.

A *1-pasting scheme* is a finite non-empty set G together with an embedding g of G into the oriented line. We identify G with its image, and we call the elements of G *0-cells*. G partitions the rest of the line into a finite number of disconnected open intervals and two rays. We call the open intervals *1-cells* and we define the *domain* and *codomain* of a 1-cell f to be its endpoints, written $dom f$ and $cod f$, with $dom f < cod f$ in the order determined by the orientation of the line. The *domain* of the pasting scheme, denoted $domG$, is the least element of G in the order determined by the orientation of the line; dually, the *codomain* of (G, g), denoted $codG$, is the greatest element. The 0-cells and 1-cells of (G, g) together with the domain and codomain functions form a directed graph, which we call the *underlying* 1-computad of (G, g), and which we denote by $C(G)$.
A *1-pasting diagram* in a 1-computad H consists of a 1-pasting scheme (G, g) together with a 1-computad morphism $h : C(G) \to H$. The *domain* of a 1-pasting diagram $((G, g), h)$ is $domG$ together with the restriction of h to $C(domG)$; dually for *codomain*. A 1-pasting diagram $((G, g), h)$ is called simply a *1-pasting scheme in H* if h is a monomorphism in the category of directed graphs. Every 1-pasting scheme with at least two 0-cells is called *spherical*. A 1-computad is *locally spherical* if each 1-cell together with its domain and codomain is a spherical 1-pasting scheme, i.e. if every 1-cell has its domain and codomain different 0-cells. A *1-subpasting scheme* of a 1-pasting scheme (G, g) is a subset J of G for which given $j < g < j'$ with j and j' in J and g in G, it follows that g lies in J.

A *labelling* of a 1-pasting scheme (G, g) in a category A is a 1-pasting diagram $((G, g), a)$ in A. A *1-pasting composite*, or equally a *strong composite*, of a labelled 1-pasting scheme in a category A is the composite in A of the sequence of arrows determined by the diagram. We call any composite in a category A a *vertical composite*.

We now recall the definitions and a few results for n=2, as appear in [8].

Definition 2.1 A *2-computad* consists of a directed graph G, together with a set G_2 and two functions $dom, cod : G_2 \to diags(G)$, where $diags(G)$ is the set of directed paths in G, such that $dom\,dom = dom\,cod$ and $cod\,dom = cod\,cod$. A ***2-computad morphism*** from (G_2, G, dom, cod) to (H_2, H, dom, cod) consists of a morphism of directed graphs $f : G \to H$ and a function $f_2 : G_2 \to H_2$ such that $dom\,f_2 = f\,dom$ and $cod\,f_2 = f\,cod$. The category 2-*Computad* is the category of 2-computads and 2-computad morphisms, with the evident composition.

Definition 2.2 A *2-pasting scheme* consists of a (non-empty) connected finite directed graph G together with an embedding g of G into oriented 2-space subject to the following conditions. The image of G divides the rest of 2-space into finitely many disconnected regions, called faces. The boundary of each interior face F must be of the form $\sigma(F)(\tau(F))^{-1}$ for directed paths $\sigma(F)$ and $\tau(F)$ in G of non-zero length, if moving clockwise around the boundary; and there must exist vertices $s(G)$ and $t(G)$ in the exterior face for which given any vertex v, there exist directed paths from $s(G)$ to v and from v to $t(G)$.

Remark 2.3 In [8], it was assumed in the definition of 2-pasting scheme that $s(G)$ and $t(G)$ are distinct, but that assumption was inessential to the argument. If they are the same, then the 2-pasting scheme must be trivial.

With the above notation, the interior faces F of (G, g) are called *2-cells*, and *domF* and *codF* are defined to be $\sigma(F)$ and $\tau(F)$ respectively. So, (G, g) has an underlying 2-computad given by the graph G and the 2-cells of (G, g); we denote this 2-computad by $C(G)$. It also follows that the exterior face E also has boundary of the form $\sigma(E)(\tau(E))^{-1}$ for directed paths $\sigma(E)$ and $\tau(E)$ of non-zero length, if moving around the boundary with E always on one's right side; these are called *codG* and *domG* respectively.

Proposition 2.4 If (G, g) is a 2-pasting scheme, then G has no directed cycle.

Proposition 2.5 If a 2-pasting scheme (G, g) has at least one interior face, then there exists an interior face F such that *domF* lies entirely on *domG*.

Definition 2.6 A *2-pasting diagram* in a 2-computad H consists of a 2-pasting scheme (G, g) together with a 2-computad morphism $h : C(G) \to H$. A *labelling* of a 2-pasting scheme (G, g) in a 2-category A is a 2-pasting diagram $((G, g), a)$ in A.

Theorem 2.7 (A 2-categorical pasting theorem) Every labelling of a 2-pasting scheme has a unique composite.

Proof: The existence of a composite follows from Proposition 2.5 and induction on the number of interior faces; unicity also follows directly by induction. □

Remark 2.8 For n=2, pasting composition describes all possible composites of a set of 2-cells. This is evident because a horizontal composite of labelled 2-pasting schemes is a labelled 2-pasting scheme, and a vertical composite of labelled 2-pasting schemes is a labelled 2-pasting scheme.

In the introduction, it was mentioned that our pasting schemes generalise the well-formed loop-free pasting schemes of Johnson, but not his loop-free schemes in general. For n=2, our pasting schemes are precisely his well-formed loop-free pasting schemes as he showed in a lecture in Sydney in 1988, but even then, we do not include all of his loop-free schemes, for instance the following.

Example 2.9 Consider the following diagram:

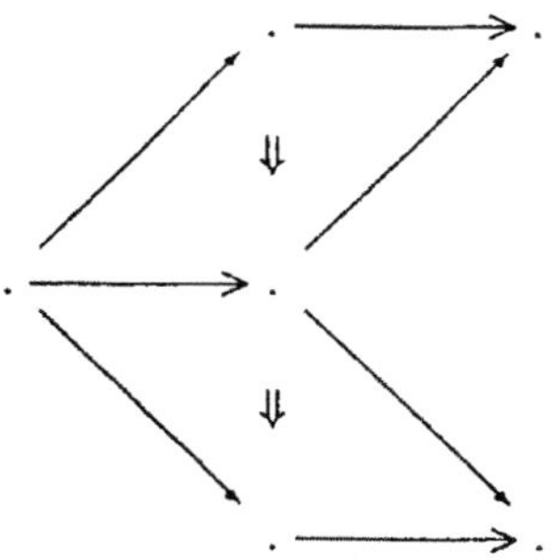

(2.1)

Observe that this is not one of our 2-pasting schemes since there is no candidate for $t(G)$. It is however one of Johnson's loop-free schemes. It is not well-formed since its domain is not even connected.

This generality allows Johnson to give an explicit description of the free 2-category on some data, but it evidently does not parametrize a composable diagram in general since the two 2-cells are not composable.

In order to study higher dimensions, we add the following definitions. The *domain* of a 2-pasting diagram $((G,g),h)$ in a 2-computad H is $domG$ together with the restriction of h to $C(domG)$; dually for *codomain*. A 2-pasting diagram $((G,g),h)$ is called simply a *2-pasting scheme in H* if h is a monomorphism in 2-*Computad*, i.e. if h is injective on k-cells for k=0 to 2. A 2-pasting scheme is called *spherical* if $domG \cup codG$ is homeomorphic to the 1-sphere. A 2-computad G is called *locally spherical* if each 2-cell in it is a spherical 2-pasting scheme containing just one 2-cell. A *2-pasting inclusion* into a 2-pasting scheme (G,g) is a directed graph inclusion j into G such that the composite gj is a 2-pasting scheme for which every 2-cell is a 2-cell of (G,g). A *2-subpasting scheme* is an equivalence class of 2-pasting inclusions under the evident equivalence.

3 Continuing the induction

This section is devoted to defining the terms n-computad and n-pasting scheme. In order to simplify the exposition, it is convenient to define several subsidiary terms. We have already commenced with n=1 and exhibited the ideas to follow by recalling the definitions for n=2.

Generalising the situation for n=2, an (n+1)-computad will be an (n+1)-dimensional cellular complex, together with an orientation assigned to each k-cell for k=0 to n+1, subject to some axioms. Each (n+1)-cell will have a domain and a codomain, each of which will be an n-pasting scheme, hence the induction. An (n+1)-pasting scheme will consist of an n-computad embedded in (n+1)-space: this is why we have a mutual induction on pasting schemes and computads. An (n+1)-pasting scheme (G, g) will be called spherical if its domain and codomain, which will be subcomputads of G, are spherical n-pasting schemes and if $domG \cup codG$ is homeomorphic to the n-sphere. An (n+1)-computad will be called locally spherical if for each (n+1)-cell, the subcomputad generated by taking its domain and codomain is a spherical (n+1)-pasting scheme (with only one (n+1)-cell in it). Given an (n+1)-pasting scheme (G, g), it will generate an (n+1)-computad by taking the embedding and the region enclosed by it; this (n+1)-computad will be denoted by $C(G)$. An (n+1)-pasting diagram in an (n+1)-computad H will be an (n+1)-pasting scheme (G, g) together with an (n+1)-computad morphism $h : C(G) \rightarrow H$.

Definition 3.1 An *(n+1)-computad* consists of an n-computad G_n, a set G_{n+1}, and two functions $dom, cod : G_{n+1} \rightarrow diags(G_n)$, where $diags(G_n)$ is the set of n-pasting diagrams (See Definition 3.13) in G_n, such that $dom\, dom = dom\, cod$ and $cod\, dom = cod\, cod$. An *(n+1)-computad morphism* from (G_{n+1}, G_n, dom, cod) to (H_{n+1}, H_n, dom, cod) consists of an n-computad morphism $f_n : G_n \rightarrow H_n$ and a function $f_{n+1} : G_{n+1} \rightarrow H_{n+1}$ such that $dom\, f_{n+1} = f_n\, dom$ and $cod\, f_{n+1} = f_n\, cod$. The category $(n+1)$-*Computad* is the category of (n+1)-computads and (n+1)-computad morphisms, with the evident composition.

Proposition 3.2 Let (G, g) be a locally spherical (See Definition 3.15) n-computad G together with an embedding g of G into oriented (n+1)-space. Then the orientation of (n+1)-space determines an orientation for each n-cell of G, giving it a "left hand side" and a "right hand side".

Proof: Since G is locally spherical, every n-cell in G is given by an inclusion of an n-pasting scheme with just one n-cell into G, and so composing this inclusion with g gives a piecewise linear embedding of an n-ball into oriented (n+1)-space. So the orientation of n-space specifies an orientation for the embedded n-cell. □

Notation 3.3 Given a finite locally spherical n-computad G together with an embedding g of G into (n+1)-space, the image of G divides the rest of (n+1)-space into finitely many disconnected regions. For each such region F, we write inF for the set of n-cells of G that are incident to F and with orientation into F, as given by Proposition 3.2. Dually, we write $outF$ for the set of n-cells of G that are incident to F and with orientation out of F. We shall call the interior regions *(n+1)-cells* of (G, g). Let the exterior region determined by such an embedding be denoted by E.

Definition 3.4 An *(n+1)-embedding* is a (non-empty) connected, finite, locally spherical n-computad G together with an embedding g of G into (n+1)-space subject to the following conditions:

1. if F is an interior region, then F is homeomorphic to the (n+1)-ball, and inF and $outF$, together with their domains and codomains, each forms a spherical n-pasting scheme, denoted $domF$ and $codF$ respectively, such that $domF \cup codF$ is the boundary of F; $dom(domF) = dom(codF)$; and $cod(domF) = cod(codF)$.

2. there exist n-pasting schemes $domG$ and $codG$ in G such that the n-cells of $domG$ are precisely the elements of $outE$; the n-cells of $codG$ are precisely the elements of inE; $dom(domG) = dom(codG)$; $cod(domG) = cod(codG)$; and $domG \cup codG$ is the boundary of E. Moreover, $domG$ extends to an n-sphere $S(domG)$ such that $G - domG$ lies strictly inside $S(domG)$.

Notation 3.5 Given an (n+1)-embedding (G, g), we shall denote $dom(domG)$ by S and $cod(domG)$ by T, and we shall denote the region enclosed by $S(domG)$ by $B(domG)$. Observe, more generally, that if X is any n-pasting scheme in G with domain S and codomain T, then X together with $S(domG) - domG$ forms an n-sphere. We shall denote this n-sphere by $S(X)$ and we shall denote the region it encloses by $B(X)$. Note that every n-cell of X is an element of $in(B(X))$. If F is an (n+1)-cell of (G, g), we shall denote by $d(F)$ the union over all k$\leq$n of the set of those k-cells of G that lie on the boundary of F but not on $codF$. Johnson called this the set of "beginnings" of F: it is equally the set of those k-cells in the interior of $domF$. Dually, we shall denote by $c(F)$ the union over all k$\leq$n of the set of those k-cells of G that lie on the boundary of F but not on $domF$. Johnson called this the set of "ends" of F. The n-computad G, the set of regions it encloses, and dom and cod as defined above, yield an (n+1)-computad, which we shall denote by $C(G)$.

Example 3.6 The following is an example of a 2-embedding that is not a 2-pasting scheme as defined in Section 2.

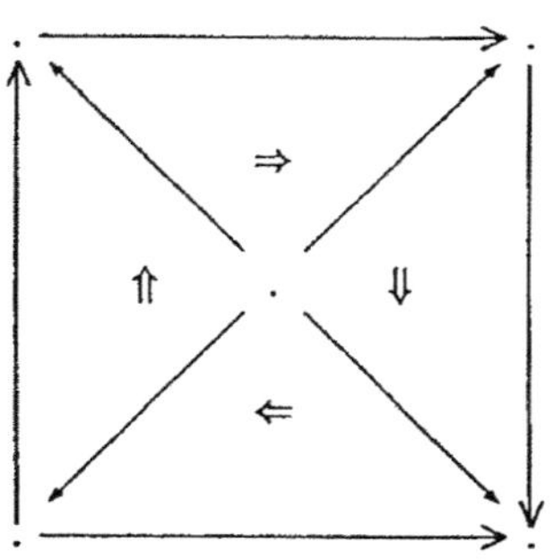

(3.1)

Observe that this 2-embedding has no candidate for $s(G)$, which by definition must lie in the exterior face.

Owing to this example, we evidently need an extra condition in order to obtain pasting schemes. Thus, we introduce the following.

Definition 3.7 An (n+1)-embedding (G, g) satisfies the *n-cell extension* condition if every n-cell x extends to an n-pasting scheme X in G with $domX = S$ and $codX = T$.

Remark 3.8 A 2-embedding that satisfies the 1-cell extension condition is exactly a 2-pasting scheme as defined in Section 2: that a 2-pasting scheme is a 2-embedding that satisfies 1-cell extension follows from Proposition 2.4; the converse is trivial. However, even for n=3, an n-embedding that satisfies (n-1)-extension does seem to suffice to determine what we require of an n-pasting scheme. The reason is that such a 3-embedding will have a "topmost" 3-cell F, and for any such, if $d(F)$ is removed from G, we want the domain of the new 3-embedding to be a 2-pasting scheme embedded in 3-space. That does not seem to be the case in general for n=3, although we do not have a counterexample. It is certainly false for higher n, as Example 3.11 shows. So we require a further definition. In fact, the following definition can be simplified substantially for n=3, as shown by Theorem 4.14.

Definition 3.9 An (n+1)-embedding (G, g) satisfies the *domain replacement* condition if for every n-pasting scheme X in G with domain S and codomain T,

1. if F is an (n+1)-cell with $domF$ a subpasting scheme of X, then replacing $domF$ by $codF$ in X gives another n-pasting scheme embedded in (n+1)-space, and also in X, replacing $domF$ by a single cell gives an n-pasting scheme

2. if F and F' are different (n+1)-cells such that $domF$ and $domF'$ are both subpasting schemes of X, then in X, replacing both $domF$ and $domF'$ by single cells gives an n-pasting scheme.

Note the difference in the two parts of condition 1 of domain replacement. By definition, X is a pair $(\overline{X}, j_X)$ consisting of an n-pasting scheme $\overline{X}$ and a monomorphism $j_X : C(X) \to G$. Since G is locally spherical, $domF$ is a pair $(\overline{domF}, j_{domF})$ consisting of a spherical n-pasting scheme $\overline{domF}$ and a monomorphism j_{domF}; similarly for $codF$. We assume that $\overline{domF}$ is a subpasting scheme of $\overline{X}$ and that j_X restricts to j_{domF}. If we replace $\overline{domF}$ by $\overline{codF}$ in $\overline{X}$, we would have an n-embedding $\overline{X'}$, and if we make the corresponding replacement of j_{domF}, we would have a new map $j_{X'}$ in *n-Computad*. The first part of condition 1 demands that $\overline{X'}$ is an n-pasting scheme and that $j_{X'}$ is a monomorphism, and hence, by composition with g, embeds $\overline{X'}$ into (n+1)-space. The second part of condition 1 only demands that the replacement of $\overline{domF}$ by a single cell is an n-pasting scheme.

Domain replacement is a strong condition. It is not true of all possible composites of (n+1)-cells, as illustrated by Example 3.11. So, one would dearly like to weaken it, but it is not possible to weaken it enough to capture all possible composites while still allowing the inductive argument outlined in Remark 3.8. Nevertheless, it is evidently true of all 2-pasting schemes, by Proposition 2.4, and we shall show by Theorem 4.14 and Proposition 7.1 that it is true of all 3-pasting schemes too. In Section 5, we shall re-express it in terms of horizontal pastings of (n+1)-cells.

The apparent asymmetry in the definition of domain replacement is, in the presence of the n-cell extension condition, illusory. One could say that that an (n+1)-embedding (G,g) satisfies the *codomain replacement* condition if (G,g) with the reverse orientation of (n+1)-space satisfies domain replacement. Corollary 4.8 shows that in the presence of the n-cell extension condition, the domain and codomain replacement conditions are equivalent. So we can now make a definition that is provably symmetric.

Definition 3.10 An *(n+1)-pasting scheme* is an (n+1)-embedding that satisfies the n-cell extension condition and the domain replacement condition.

Example 3.11 This example of a 3-pasting scheme will be used for several purposes. It will exhibit the need for and the independence of the various parts of the domain replacement condition. Also, it will be used to show that there exist composites of 4-cells that do not form 4-pasting schemes, and that there exist 4-pasting schemes that fail the fourth requirement of Johnson's "loop-freeness" condition. Consider the following diagram: it is to be understood to contain two 3-cells F and F'. The idea is to imagine F' as sitting in space below and to the right of F, and to imagine γ as directed vertically downwards. The domain of F is α, and the codomain is given by γ and $\overline{\alpha}$, which is not explicitly indicated. The domain of F' is given by α' and γ, and the codomain is given by $\overline{\alpha'}$, which is not depicted explicitly.

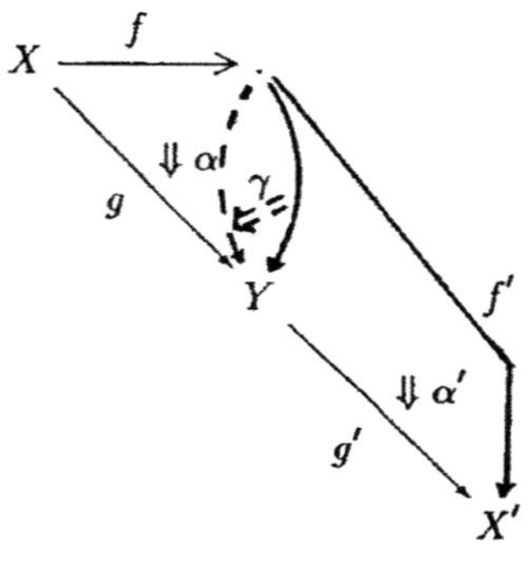

(3.2)

We shall denote this 3-pasting scheme by F_1. Moreover, we shall denote the somewhat trivial 3-pasting scheme given by removing γ from (3.2) by F_0.

Now consider the following diagram, which is similar to the first. It is to be understood to contain two 3-cells H and H'. The idea is that H sits to the left of and below

H', and that δ is directed vertically upwards. The domain of H is given by β and δ, and the codomain is given by $\overline{\beta}$, which is not depicted explicitly. The domain of H' is β', and the codomain is given by δ and $\overline{\beta'}$, which is not depicted explicitly.

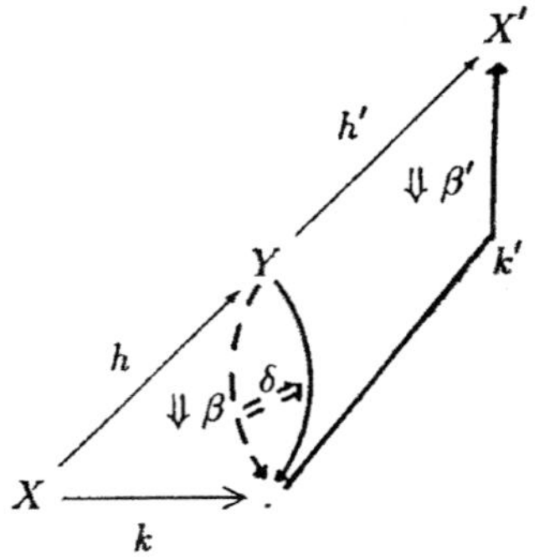

(3.3)

We shall denote this 3-pasting scheme by H_1; and we shall denote the 3-pasting scheme given by removing δ from (3.3) by H_0.

Now identify (3.2) and (3.3) at X, Y and X′ by gradually curving (in a piecewise linear way) the two diagrams. This gives the following picture.

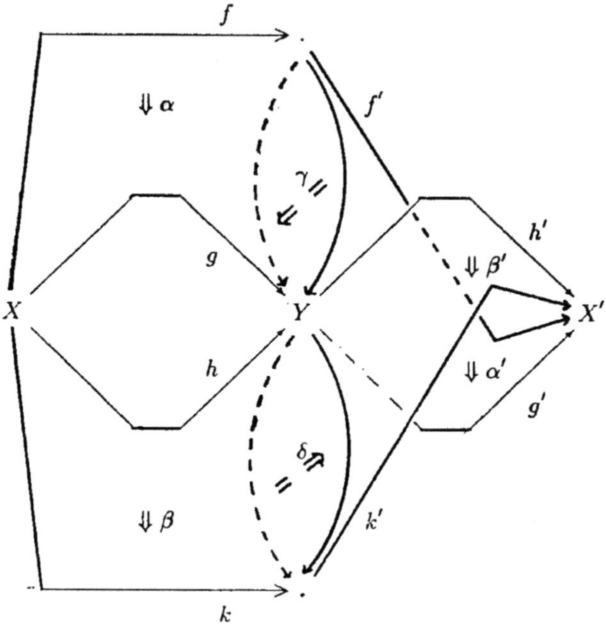

(3.4)

To the left side of the picture, we add two new 2-cells. The domain of one of them is given by g and $cod\,\delta$, and its codomain is k. The domain of the other is given by f and $cod\gamma$, and its codomain is h. Then, these two new 2-cells, together with $\overline{\alpha}$ and β enclose a region K with domain given by $\overline{\alpha}$ and one of the new 2-cells, and with codomain given by β and the other new 2-cell. Dually, on the right side of the picture, we can form another 3-cell K' whose domain includes $\overline{\beta'}$ and whose codomain includes α'.

Having made these constructions, we have a 3-pasting scheme, which we shall call G. The domain of G is given by α, β', and two of the new 2-cells. That it is a 3-embedding is easy; to check the 2-cell extension condition, observe that it is true for any 2-cell in the domain, then by removing a topmost 3-cell, say F, one can see that it is true for all 2-cells in $codF$, etcetera. For instance, the 2-pasting scheme given by starting with $domG$ and removing F and H' is depicted by

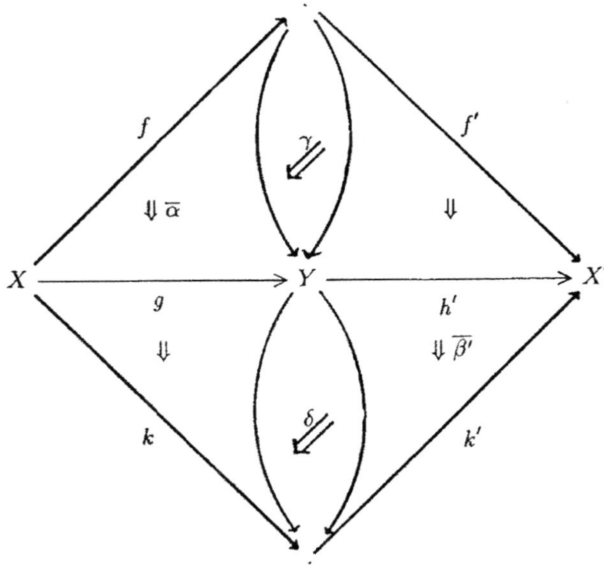

(3.5)

where the unlabelled 2-cells are two of the new 2-cells. The domain replacement condition can be most easily seen as a result of Theorem 4.14.

The 3-pasting scheme G has several 3-sequences, i.e. sequences of all the 3-cells in G such that the first has domain on $domG$, then removing it from G, one continues inductively. A formal definition appears in Corollary 4.6. Two important 3-sequences in G are H', K', F, F', K, H and F, K, H', H, K', F'. Observe that every 3-sequence must have both F and H' occurring before F' and H.

If one removes γ from G, then one still has a pasting scheme which we shall call G_1. A 3-sequence in G_1 is given by H', K', F_0, K, H, i.e. the replacement of the subsequence F, F' in our first 3-sequence above by the singleton F_0. Dually, if one removes δ from G, one also has a pasting scheme G_2, and G_2 has a 3-sequence given by F, K, H_0, K', F'. However, if one removes both γ and δ simultaneously from G, then one no longer has either a composite or a pasting scheme: any 3-sequence must have both F_0 before H_0 and H_0 before F_0, and the 2-cell extension condition fails. The latter is easiest seen as a consequence of Proposition 4.3.

Given all of these definitions, one may have a 4-embedding with domain G_1, and with two 4-cells Y and Z. The domain of Y is F_0 and its codomain is F_1. The domain of Z is H_1 and its codomain is H_0. So the codomain of the 4-embedding is G_2. The 4-composite given by composing first the evident extension of Y followed by that of Z is certainly possible in any 4-category, but the 4-embedding does not satisfy condition 1 of domain replacement since Z is a topmost 4-cell. Note that the 4-embedding does satisfy the 3-cell extension condition. So this shows both that condition 1 of domain replacement is not redundant, and that there exist composites of 4-cells that do not form 4-pasting schemes.

Also, one may have a 4-embedding with domain G, and with 4-cells whose domains are F_1 and H_1, and with codomains given by new copies of the interiors of F_1 and H_1. This would satisfy the 3-cell extension condition and condition 1 of domain replacement, but it would not satisfy condition 2.

Finally, supposing one reverses the direction of γ in G, one would still have a 3-pasting scheme, and it may appear as part of the domain of a 4-cell in a 4-embedding; but then, the 4-embedding would fail to satisfy the fourth (and most substantial other than no direct loops) requirement of Johnson's "loop-freeness" condition, as one may take H_1 to be the subpasting scheme to act as a counterexample. Again, the 4-embedding could still satisfy the 3-cell extension condition and could in fact form a 4-pasting scheme with just one 4-cell and hence even satisfy the no direct loops condition.

Example 3.12 We now give an example of a 3-pasting scheme with just one 3-cell that does not satisfy Johnson's "no direct loops" condition since its underlying graph has a cycle.

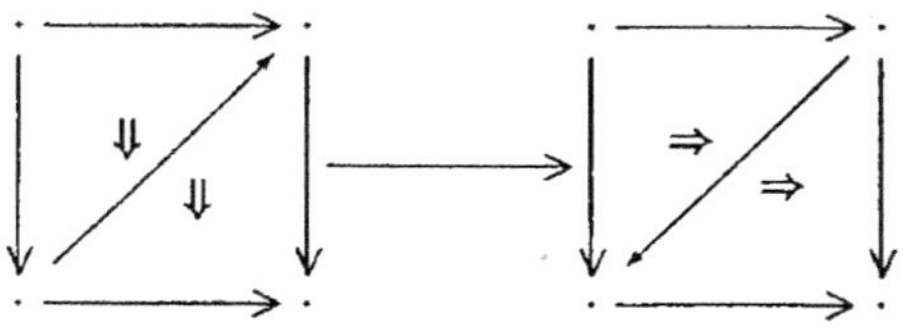

(3.6)

Such a scheme could arise in practice as follows. In [13], Street defines a "2-adjunction" in a 3-category. Some of the data is depicted by

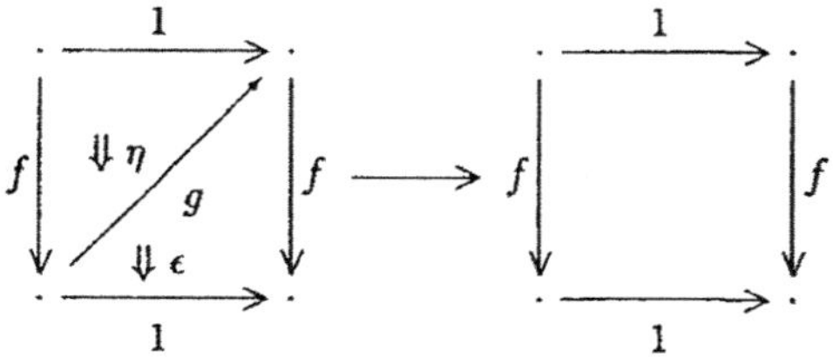

(3.7)

where the unmarked interior region represents an identity 2-cell. Then, if one appends an identity 3-cell, one obtains

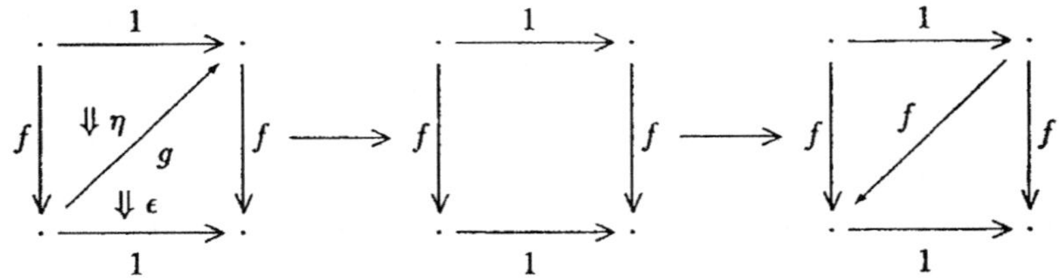

(3.8)

where the unmarked interior regions represent identity 2-cells. In fact, Vaughan Pratt observed some time ago that (3.7) itself fails the no direct loops condition, but his argument is somewhat more complex than ours.

Finally, to end this section, we need four more definitions, which have been used above for n, in order to allow us to continue the induction.

Definition 3.13 An *(n+1)-pasting diagram* $((G, g), h)$ in an (n+1)-computad H is an (n+1)-pasting scheme (G, g) together with an (n+1)-computad morphism $h : C(G) \to H$. The *domain* of an (n+1)-pasting diagram $((G, g), h)$ is given by $domG$ and the restriction of h to $domG$; dually for the *codomain*. We shall call an (n+1)-pasting diagram $((G, g), h)$ simply an *(n+1)-pasting scheme in H* if h is a monomorphism in the category $(n + 1)$-*Computad*, i.e. if h is injective on k-cells for all k=0 to n+1.

Definition 3.14 An (n+1)-embedding (G, g) is *spherical* if $domG$ and $codG$ are spherical n-pasting schemes and $domG \cup codG$ is homeomorphic to the n-sphere.

Definition 3.15 An (n+1)-computad G is *locally spherical* if each (n+1)-cell together with its domain and codomain is a spherical (n+1)-pasting scheme in G containing one (n+1)-cell.

Definition 3.16 An *(n+1)-pasting inclusion* into an (n+1)-pasting scheme (G, g) is an n-computad inclusion j into G such that the composite gj is an (n+1)-pasting scheme for which every (n+1)-cell is an (n+1)-cell of (G, g). An *(n+1)-subpasting scheme* is an equivalence class of (n+1)-pasting inclusions under the evident equivalence.

4 Simple topological results

The main result of this section is Theorem 4.5, which states that given any (n+1)-pasting scheme (G, g) with at least one interior region, there exists an interior region F such that $domF$ is a subpasting scheme of $domG$, and such that if one removes $d(F)$ from G, then one has a new (n+1)-pasting scheme containing one fewer interior regions. We now give a series of Propositions leading to that result.

Definition 4.1 An (n+1)-embedding (G,g) is *acyclic* if there exists no alternating sequence $F_0, x_o, F_1, \ldots, F_m$ with m>0, with each F_i an (n+1)-cell and each x_i an n-cell of (G,g), with $F_0 = F_m$, and x_i an element of $outF_i \cap inF_{i+1}$.

Acyclicity is the highest dimensional part of Johnson's "no direct loops" condition, and it is the only part of the condition that we keep, as we shall show that it is a consequence of the n-cell extension condition. Example 3.12 shows that we do not require lower dimensional parts of the no direct loops condition.

Proposition 4.2 If an (n+1)-embedding (G,g) is acyclic and has at least one interior region, then there exists an interior region F for which $domF$ is a subpasting scheme of $domG$.

Proof: Let F_0 be any interior region. If every n-cell of inF lies on $domG$, then since $domF$ is spherical, it follows that $domF$ is a subpasting scheme of $domG$. If not, then let x_0 be an n-cell of inF that is not on $domG$. It must therefore be an element of $outF_1$ for some interior region F_1. Continue inductively. Since (G,g) is finite and acyclic, there must exist some m for which $domF_m$ is a subpasting scheme of $domG$. □

We shall call any interior region F of (G,g) such that $domF$ is a subpasting scheme of $domG$ a *topmost* region of (G,g).

Proposition 4.3 If an (n+1)-embedding (G,g) satisfies the n-cell extension condition, then it is acyclic.

Proof: Suppose there is a cycle $F_0, x_0, \ldots, F_m$. Then, x_0 extends to an n-pasting scheme X from S to T in G, which in turn extends to $S(X)$ as in 3.5, enclosing the region $B(X)$. Recall that every n-cell of X is an element of $in(B(X))$, and $S(X)$ extends X through the exterior region of (G,g). So F_0 is outside $B(X)$ but F_1 is inside it. Since $F_0 = F_m$, it follows that there exists an i such that F_i is inside $B(X)$ but F_{i+1} is outside it, but this is a contradiction since every n-cell on X is oriented into $B(X)$, so cannot be an element of $outF_i$. □

In the presence of condition 1 of the domain replacement condition, Proposition 4.3 has a converse given by Proposition 4.9, but before proving that, we shall develop some of the consequences of Proposition 4.3.

Corollary 4.4 If an (n+1)-embedding (G,g) with at least one interior region satisfies the n-cell extension condition, then there exists an interior region F for which $domF$ is a subpasting scheme of $domG$.

Proof: Propositions 4.2 and 4.3. □

This leads us directly to the main theorem of this section. It allows us to use induction to show that any labelled (n+1)-pasting scheme has a composite.

Theorem 4.5 For any (n+1)-pasting scheme (G,g) with at least one interior region, there exists an interior region F such that $domF$ is a subpasting scheme of $domG$, and such that removing $d(F)$ from G gives a new (n+1)-pasting scheme containing one fewer interior regions.

Proof: This follows immediately from the definitions and Corollary 4.4 given the observation that any n-pasting scheme X from S to T in G that includes any n-cell in the domain of a topmost (n+1)-cell F necessarily includes every element of inF. □

Corollary 4.6 For any (n+1)-pasting scheme (G,g), the (n+1)-cells of (G,g) can be ordered $F_0 < F_1 \ldots < F_m$ such that there exists a sequence $G_0, G_1, \ldots, G_m$ of (n+1)-pasting schemes with $(G,g) = G_0$ and such that G_{i+1} is the result of removing $d(F_i)$ from G_i.

We shall call any such sequence an *(n+1)-sequence.*

Note that we have not yet used condition 2 of the domain replacement condition. It will be used later to show the unicity of the possible composites of a labelled pasting scheme.

Several other results follow from Proposition 4.3. One is that the apparent asymmetry in the definition of (n+1)-pasting scheme is illusory.

Proposition 4.7 Let (G,g) be an acyclic (n+1)-embedding. Then (G,g) satisfies domain replacement if and only if it satisfies codomain replacement.

Proof: Suppose (G,g) satisfies domain replacement. Let X be an n-pasting scheme from S to T in G and let F be an (n+1)-cell of G with $codF$ a subpasting scheme of X. We shall show that if one replaces $codF$ by $domF$ in X, one has an n-pasting scheme embedded in (n+1)-space. We proceed by induction on the number of (n+1)-cells of (G,g) that lie outside $B(X)$. If it is 1, then the (n+1)-cell must be F, so the replacement of $codF$ by $domF$ in X is $domG$, hence an n-pasting scheme embedded in (n+1)-space. If it is m+1, then there exists an (n+1)-cell F_1 different to F and outside $B(X)$. If F_1 is not a topmost (n+1)-cell, then there exists F_2 outside $B(X)$ such that $outF_2 \cap inF_1$ is nonempty. Note that F_2 is not F since $codF$ is a subpasting scheme of X but F_2 lies outside $B(X)$. So, by induction, since (G,g) is acyclic, there exists a topmost (n+1)-cell F' of (G,g) that lies outside $B(X)$ and is not F. Then, if we remove $d(F')$ from G, we are done by induction. Thus we have proved part of codomain replacement. The rest follows directly from this part together with domain replacement. The converse is dual. □

Corollary 4.8 Let (G,g) be an (n+1)-embedding that satisfies the n-cell extension condition. Then (G,g) satisfies domain replacement if and only if it satisfies codomain replacement.

Proof: Propositions 4.3 and 4.7. □

Now we give the promised converse to Proposition 4.3.

Proposition 4.9 If an (n+1)-embedding (G, g) satisfies condition 1 of the domain replacement condition and is acyclic, then it satisfies the n-cell extension condition.

Proof: This follows by induction on the number of (n+1)-cells in (G, g). If (G, g) has no (n+1)-cells, the result is trivial since every n-cell lies on *domG*. If (G, g) has m+1 (n+1)-cells, let F be a topmost (n+1)-cell: one such exists by Proposition 4.2. Then, given an n-cell x, either x lies on *domF*, therefore on *domG*, or x is in G_1, the acyclic (n+1)-embedding given by removing $d(F)$ from G. Trivially, G_1 satisfies condition 1 of domain replacement, so by induction we are done. □

Corollary 4.10 Let (G, g) be an (n+1)-embedding that satisfies condition 1 of domain replacement. Then (G, g) satisfies the n-cell extension condition if and only if it is acyclic.

Proof: Propositions 4.3 and 4.9. □

Corollary 4.11 Let (G, g) be an (n+1)-pasting scheme and let X be an n-pasting scheme in G from S to T. Then the collection of (n+1)-cells of (G, g) that lie in $B(X)$ form an (n+1)-subpasting scheme of (G, g) with domain given by X and codomain given by *codG*.

Proof: Domain replacement and the fact of being an (n+1)-embedding are immediate. By Corollary 4.10, (G, g) is acyclic, and so our collection of (n+1)-cells is also acyclic, and hence by Corollary 4.10 again, satisfies the n-cell extension condition. □

Now, as indicated in Example 3.11, we shall exhibit, for the case of 3-embeddings, a simpler condition than domain replacement, but which in the presence of the 2-cell extension condition, is equivalent to it. Consider the following condition on an (n+1)-embedding (G, g)

(*) if x is a k-cell of G with k<n, then x appears in $c(F)$ for at most one interior region F of (G, g); and if x appears on *domG*, then it appears in $c(F)$ for no interior region of (G, g).

Proposition 4.12 Every (n+1)-pasting scheme (G, g) satisfies (*).

Proof: We proceed by induction on the number of (n+1)-cells of (G, g). If (G, g) has no (n+1)-cells, then the result is trivial. If (G, g) has m+1 (n+1)-cells, then let F be a topmost (n+1)-cell of (G, g) and let G_1 be the (n+1)-pasting scheme given by removing $d(F)$ from G. Then, by induction, G_1 satisfies condition (*). By definition of (n+1)-embedding, since F is a topmost (n+1)-cell, no k-cell in $d(F)$ appears in $c(F')$ for any (n+1)-cell F'. The result follows directly. □

In general, there is no converse to Proposition 4.12. The first 4-embedding in Example 3.11 satisfies both the 3-cell extension condition and condition (*), but it does not satisfy condition 1 of domain replacement. However, for n=3, we do have a converse.

Lemma 4.13 Let (G,g) be an acyclic 3-embedding that satisfies condition (*). Then if x is a k-cell on *codG* for any k<2, then x either lies on *domG* or x lies in $c(F)$ for some interior region F of (G,g).

Proof: The proof is by induction on the number of 3-cells in (G,g). If (G,g) has no 3-cells, the result is trivial. If (G,g) has m+1 3-cells, then by Proposition 4.2, (G,g) has a topmost region, say F. Note that if one replaces any spherical subpasting scheme of a 2-pasting scheme by another 2-pasting scheme with the same domain and codomain, then the result is a 2-pasting scheme. Moreover, by condition (*), no 0-cell or 1-cell in $c(F)$ lies on *domG*. Also, no 2-cell can lie on both *outF* and *outE*. Hence, if one removes $d(F)$ from G, one obtains a new 3-embedding G_1. Trivially, G_1 is acyclic. Moreover, G_1 satisfies condition (*) and $codG_1 = codG$. So, by induction, we are done. □

Theorem 4.14 Let (G,g) be a 3-embedding that satisfies the 2-cell extension condition. Then (G,g) satisfies the domain replacement condition if and only if it satisfies condition (*).

Proof: First assume that (G,g) satisfies condition (*). If one replaces any spherical subpasting scheme of a 2-pasting scheme by another 2-pasting scheme with the same domain and codomain, then the result is a 2-pasting scheme. So it suffices to show that if X is a 2-pasting scheme from S to T in (G,g), and F is a 3-cell in (G,g) with *domF* lying on X, then no k-cell in $c(F)$ lies on X. The collection of 3-cells of (G,g) that lie outside $B(X)$ forms a 3-embedding with domain *domG* and codomain X. We shall call this 3-embedding X'. Since (G,g) satisfies the 2-cell extension condition, it is acyclic, so X' is also acyclic. Moreover, X' also satisfies condition (*). So, by Lemma 4.13, for each k<2, every k-cell on $X = codX'$ occurs either on *domG* or in $c(F')$ for some F' outside $B(X)$. So, since (G,g) satisfies (*), it follows that no k-cell on X lies in $c(F)$. For k=2, no k-cell can lie on both *outF* and *outX*. The converse holds by Proposition 4.12. □

Theorem 4.14 allows us to reformulate the definition of a 3-pasting scheme as follows.

Definition 4.15 A 3-pasting scheme is a 3-embedding (G,g) that satisfies the 2-cell extension condition and condition (*).

5 Horizontal pasting

In Section 4, we showed that for any (n+1)-pasting scheme, there is at least one (n+1)-sequence of (n+1)-cells from the domain to the codomain of the pasting scheme. That will be used to show that every labelled pasting scheme has at least one composite. Note that we never used condition 2 of domain replacement in order to deduce the existence of such a sequence. Now, we do use condition 2 to provide the topological ideas required to prove unicity of a composite.

Definition 5.1 An *horizontal (n+1)-pasting* is an (n+1)-embedding (G,g) such that every interior region F has domain a subpasting scheme of $domG$ and codomain a subpasting scheme of $codG$; and such that if one simultaneously replaces every interior region F by a single n-cell from $dom(domF)$ to $cod(domF)$, then one has an n-pasting scheme G_h. We shall call G_h the *horizontal n-pasting scheme* of (G,g).

Remark 5.2 Every horizontal (n+1)-pasting evidently satisfies the n-cell extension condition. In fact, for n=1 to 4, every horizontal n-pasting is an n-pasting scheme: because (as will be shown in Proposition 7.1 for the case of n=4) if (G_1,g_1) and (G_2,g_2) are (n-1)-pasting schemes with $domG_2 = codG_1$ but with (G_1,g_1) and (G_2,g_2) otherwise disjoint, then the union of (G_1,g_1) and (G_2,g_2) is an (n-1)-pasting scheme. The result follows by induction on the number of (n-1)-cells. However, for n=5, there exist horizontal n-pastings that are not n-pasting schemes: condition 1 of domain replacement fails. A counterexample exists with just two 5-cells and four 4-cells, given by an analysis of Example 3.11.

Definition 5.3 An *(n+1)-whisker* is an (n+1)-pasting scheme that contains precisely one (n+1)-cell.

Observe that an (n+1)-whisker is necessarily an horizontal (n+1)-pasting. So, we can now reformulate domain replacement as follows.

Proposition 5.4 An (n+1)-embedding (G,g) satisfies the domain replacement condition if and only if for every n-pasting scheme X from S to T in G,

1. if F is an (n+1)-cell with $domF$ a subpasting scheme of X, then X together with F is an (n+1)-whisker.

2. if F and F' are different (n+1)-cells with $domF$ and $domF'$ both subpasting schemes of X, then X together with F and F' is an horizontal (n+1)-pasting.

6 Composition and the Interchange Law

In this section, we shall define pasting composition and horizontal composition, state and prove the interchange law for horizontal and vertical (top-level) composition, and state and prove the main result of the paper, that every labelled pasting scheme has a unique pasting composite.

Definition 6.1 A *labelling* of an (n+1)-embedding (G,g) in an (n+1)-category A is an (n+1)-computad morphism from $C(G)$ to A.

Note that Definition 6.1 incorporates the idea of labelling for both (n+1)-pasting schemes and horizontal (n+1)-pastings.

Notation 6.2 Given an (n+1)-category A, we shall denote by A_h the n-category given by all of the elements of A and all of the structure of A except for the n-source, n-target, and n-composition structure. We shall call A_h the *horizontal n-category* of A.

Observe that A_h is not the ordinary underlying n-category of A. For instance, if A is a 2-category, then the non-trivial arrows of A_h are the non-trivial 1-cells and 2-cells of A: in this case, our use of the term *horizontal category* agrees with that in the literature for 2-categories. If A is an (n+1)-category, the non-trivial n-cells of A_h are the non-trivial n-cells and (n+1)-cells of A. Recall that in Definition 5.1, we also defined G_h where (G,g) is an horizontal (n+1)-pasting. Given a labelling a of (G,g) in an (n+1)-category A, we shall denote the restriction of a to the n-pasting scheme G_h by a_h.

Definition 6.3 An *horizontal composite* of a labelled horizontal (n+1)-pasting $((G,g),a)$ in an (n+1)-category A is the n-pasting composite in A_h of the labelled n-pasting scheme (G_h, a_h).

Note that, by induction, an horizontal composite has n-domain and n-codomain in A given by the n-pasting composites of *domG* and *codG* with the restricted labellings. Moreover, since every whisker is an horizontal (n+1)-pasting, every labelled whisker has a unique horizontal composite. We shall refer to this simply as the composite of the labelled whisker.

Definition 6.4 A *vertical composite* in an (n+1)-category A is precisely an n-composite, i.e. a top level composite, in A.

Observe that the definitions of horizontal and vertical composition for (n+1)-categories agree with those for 2-categories if n=1.

We now move towards the interchange law, and using that, state and prove our n-categorical pasting theorem.

Lemma 6.5 Let $((G_1,g_1),a_1)$ and $((G_2,g_2),a_2)$ be labelled whiskers containing (n+1)-cells F_1 and F_2 respectively, such that $domG_2 = codG_1$, and $domF_2 = codF_1$, and (G_1,g_1) and (G_2,g_2) are otherwise disjoint, and the labellings agree on $codG_1$. Then the vertical composite of the composites of the labelled whiskers equals the composite of the labelled whisker given by the vertical composite of F_1 and F_2.

Proof: Take any composite of $((G_1,g_1),a_1)$. By induction, it is expressible in terms of k-composition for k=0 to n-1 in A_h. Take such an expression. If one replaces $a_1(F_1)$ by $a_2(F_2)$ in this expression, then by the n-categorical pasting theorem, one obtains an expression for the composite of $((G_2,g_2),a_2)$. Since the k-sources and k-targets of $a_1(F_1)$ and $a_2(F_2)$ agree for k=0 to n-1, it follows from the interchange laws between n-composition and k-composition for each k<n that the vertical composite of these expressions equals

the composite determined by replacing $a_1(F_1)$ in the original expression by the vertical composite of $a_1(F_1)$ and $a_2(F_2)$, i.e. the composite of the labelled whisker as claimed. □

Theorem 6.6 (The interchange law)
Let $((G_1, g_1), a_1)$ and $((G_2, g_2), a_2)$ be labelled horizontal (n+1)-pastings such that $domG_2 = codG_1$, and the labellings agree on $codG_1$, and such that there is a bijection $\natural : (n+1)\text{-}cells((G_1, g_1)) \to (n+1)\text{-}cells((G_2, g_2))$ such that for each (n+1)-cell F_1 in (G_1, g_1), we have $codF_1 = dom\natural F_1$. Then the vertical composite of the horizontal composites of $((G_1, g_1), a_1)$ and $((G_2, g_2), a_2)$ equals the horizontal composite of the labelled horizontal (n+1)-pasting given by the vertical composite of each pair $(F_1, \natural F_1)$.

Proof: This follows by induction on the number of (n+1)-cells in (G_1, g_1) and by Lemma 6.5. If (G_1, g_1) has no (n+1)-cells, the result is trivial. If (G_1, g_1) has one (n+1)-cell, the result is given by Lemma 6.5. If (G_1, g_1) has m+1 (n+1)-cells, take any horizontal composite of $((G_1, g_1), a_1)$. By induction, it is given by an (n-1)-composite, whose arguments consist of the composite of a labelled whisker and an horizontal composite of an horizontal (n+1)-pasting containing m (n+1)-cells. So, by the interchange law for n-composition and (n-1)-composition together with induction, it follows that the vertical composite of horizontal composites as in the statement of the theorem is given by an (n-1)-composite, whose arguments consist of the vertical composite of composites of two labelled whiskers and an horizontal composite of an (n+1)-pasting containing m (n+1)-cells. Moreover, the two labelled whiskers match as in the statement of Lemma 6.5. So, by Lemma 6.5 and by induction on n applied to Theorem 6.16, we are done. □

Corollary 6.7 Let $((G, g), a)$ be a labelled (n+1)-pasting scheme that is also an horizontal (n+1)-pasting, and assume that (G, g) contains at least one (n+1)-cell. Then its horizontal composite is the vertical composite of labelled whiskers. Moreover, the whiskers may be taken in any order.

Proof: We proceed by induction on the number of (n+1)-cells in (G, g). If (G, g) has one (n+1)-cell, the result is trivial. If (G, g) has m+1 (n+1)-cells, choose any one of them, say F. Then consider the vertical composite of the evident labellings of (G_1, g_1) and (G_2, g_2), where (G_2, g_2) is given by removing $domF$ from G and replacing it by another copy of $codF$, with the labelling of the new (n+1)-cell to be given by the identity, and where (G_1, g_1) is given similarly, by replacing $codF'$ for every other (n+1)-cell F' appropriately. By the interchange law and since (G_2, g_2) has m (n+1)-cells, we have the result. The final claim follows from the fact that we have allowed any (n+1)-cell F to give a top whisker. □

Now we proceed to our n-categorical pasting theorem. For simplicity of exposition, we shall first prove a weak version of the result: this weak version allows us to use a simple but quite strong definition of composite of a labelled n-pasting scheme, which we shall call a strong composite. Afterwards, we shall discuss the more general natural definition of composite, which we shall call an n-pasting composite. At that point, we can state our n-categorical pasting theorem in its natural level of generality, and we shall use the weak result to prove it.

Definition 6.8 A *strong composite* of a labelled (n+1)-pasting scheme $((G,g),a)$ is the vertical composite of the composites of labelled whiskers determined by any (n+1)-sequence of (G,g) if (G,g) has at least one (n+1)-cell. If (G,g) has no (n+1)-cells, it is any n-pasting composite of $domG$.

Lemma 6.9 Let $((G,g),a)$ be a labelled (n+1)-pasting scheme containing exactly two (n+1)-cells F and F', each with domain on $domG$. Then both (n+1)-sequences of (G,g) determine the same (n+1)-cell in A, and this agrees with the horizontal composite of $((G,g),a)$.

Proof: By condition 2 of domain replacement, (G,g) is an horizontal (n+1)-pasting. So, by Corollary 6.7, we have the result. □

Theorem 6.10 (A weak n-categorical pasting theorem)
For any positive natural number n, every labelled n-pasting scheme in an n-category A has a unique strong composite.

Proof: If n=1, the result is trivial. So we may assume the result is true for n and prove it for n+1. Existence is immediate since every (n+1)-pasting scheme has an (n+1)-sequence by Corollary 4.6 and since every labelled whisker has a unique composite. For unicity, we proceed by induction. If (G,g) contains at most one (n+1)-cell, unicity holds by induction on n and by the definition of the composite of a labelled whisker. If (G,g) has m+1 (n+1)-cells, suppose we have two (n+1)-sequences in (G,g). Either they commence with the same (n+1)-cell, in which case we are done by induction, or they commence with different (n+1)-cells. In the latter case, the result follows by induction and Lemma 6.9. □

We now make precise the general notion of n-pasting composite. The idea is that in composing a labelled n-pasting scheme, we could conceivably not merely find a sequence of whiskers, then evaluate the vertical composite of them, but we could take an horizontal composite of each equivalence class of a partition of them, then the vertical composite of these horizontal composites. Moreover, this process could be refined many times. Of course, the interchange law essentially shows us that we still determine the same composite as any strong composite. However, we now make that precise.

Definition 6.11 An (n+1)-pasting scheme (G,g) is called *indecomposable* if for some $k \leq n$, (G,g) consists of one k-cell and its domain and codomain.

Notation 6.12 Given an (n+1)-pasting scheme (G,g), we will define dom^kG inductively by $dom^nG = domG$ and $dom^kG = dom^k(domG)$ if k<n. We define cod^kG dually.

Definition 6.13 A *simple decomposition* of an (n+1)-pasting scheme (G, g) consists of two (n+1)-subpasting schemes (G_1, g_1) and (G_2, g_2) such that

1. for some k≤n, $dom^k G_2 = cod^k G_1$, and (G_1, g_1) and (G_2, g_2) are otherwise disjoint
2. $(G_1, g_1) \cup (G_2, g_2) = (G, g)$ and
3. neither (G_1, g_1) nor (G_2, g_2) is a subpasting scheme of the other.

We define the notion of decomposition of an (n+1)-pasting scheme (G, g) by induction on the sum over all k≤n of the number of k-cells in (G, g).

Definition 6.14 A *decomposition* of an (n+1)-pasting scheme (G, g) consists of

1. (G, g), if (G, g) is indecomposable
2. a simple decomposition of (G, g) together with decompositions of (G_1, g_1) and (G_2, g_2) otherwise.

We now define the notion of an (n+1)-pasting composite of a labelled (n+1)-pasting scheme $((G, g), a)$ by induction on the sum over all k≤n of the number of k-cells in (G, g) and using the notion of a decomposition of (G, g).

Definition 6.15 An *(n+1)-pasting composite* of a labelled (n+1)-pasting scheme $((G, g), a)$ consists of the result of choosing a decomposition of (G, g) and evaluating

1. the labelling of the highest dimensional non-trivial cell in (G, g), if (G, g) is indecomposable
2. the k-composite in A of the (n+1)-pasting composites of (G_1, g_1) and (G_2, g_2) labelled by the restriction of a, otherwise.

Theorem 6.16 (An n-categorical pasting theorem)
For every positive natural number n, every labelled n-pasting scheme in an n-category A has a unique n-pasting composite.

Proof: For n=1, the result is given by associativity of composition in the category A. So we may assume the result true for n and prove it for n+1. First observe that every strong composite of a labelled (n+1)-pasting scheme is an (n+1)-pasting composite: if (G, g) has at most one (n+1)-cell, this holds by the definition of strong composite; if (G, g) has more than one (n+1)-cell, then a strong composite is a vertical composite of whiskers, each containing fewer cells than (G, g), so by induction we are done. So, by the weak (n+1)-categorical pasting theorem, we have existence. For unicity, it suffices by the weak (n+1)-categorical pasting theorem to show that every (n+1)-pasting composite is a strong composite. So suppose we have an (n+1)-pasting composite of $((G, g), a)$. If (G, g) is indecomposable or contains no (n+1)-cells, the result is trivial. If our pasting composite is a vertical composite of the (n+1)-pasting composites of $((G_1, g_1), a_1)$ and $((G_2, g_2), a_2)$, then by induction, it is the vertical composite of strong composites, so taking the (n+1)-sequence of (G, g) given by an (n+1)-sequence of (G_1, g_1) followed by an (n+1)-sequence of (G_2, g_2) gives the result. So suppose our composite is a k-composite of (n+1)-pasting composites for some k<n. At least one of (G_1, g_1) and (G_2, g_2) contains an (n+1)-cell F. So by induction on the number of (n+1)-cells in (G, g) and by the interchange law for k-composition and n-composition in A, our k-composite equals the vertical composite of the extensions of whiskers of (G_1, g_1) and (G_2, g_2) along $domG_2$ and $codG_1$ respectively. □

Observe what the n-categorical pasting theorem does not say: it does not say that given a labelled n-pasting scheme $((G, g), a)$, one can take an n-ball in it, evaluate the n-ball in A, then replace the n-ball by a single n-cell in (G, g) labelled by its composite in A, then continue inductively. Example 3.11 shows that this is impossible even for n=3, as the replacement of F_1 and H_1 in G by F_0 and H_0 respectively is not a 3-pasting scheme. In the next section, we explore conditions on an n-pasting scheme under which it is possible, as this is a useful technique in practice.

7 Unifying and evaluating n+1 composites

In this section, we shall show that 3-pasting schemes include all possible composites of a given set of 3-cells in a 3-category. We shall also discuss further the failure of this result for higher n. Finally, we shall indicate stronger conditions on n-pasting schemes that allow more flexibility in evaluating a composite, and the effect of such conditions upon our pasting result.

Proposition 7.1 Every composite of a set of 3-cells in a 3-category A is expressible as a 3-pasting composite of a labelled 3-pasting scheme.

Proof: We use induction on the number of non-trivial 3-cells in the composite. If the composite has no nontrivial 3-cells, then it is a composite of 2-cells. So it is a composite of a labelled 2-pasting scheme, hence of a labelled 3-pasting scheme. If the composite has a single non-trivial 3-cell, then it must be expressible without using vertical composition, so it is a composite of a labelled whisker and hence a composite of a labelled 3-pasting scheme. So it suffices to show that a k-composite of labelled 3-pasting schemes $((G_1, g_1), a_1)$ and $((G_2, g_2), a_2)$ is expressible as a labelled 3-pasting scheme for k=0 to 2. First suppose (G_1, g_1) and (G_2, g_2) are composed by identifying cod^0G with dom^0G_2. It is immediate that the result is a 3-pasting scheme by Definition 4.15. Second, if cod^1G_1 is identified with dom^1G_2, again by Definition 4.15, the result is immediate. Finally, for 2-composition, it suffices by induction and Corollary 4.6 to assume that (G_1, g_1) contains just one interior 3-cell, and we assume that $codG_1$ and $domG_2$ are identified. Again by Definition 4.15, the result is immediate. □

Example 3.11 illustrates why Proposition 7.1 fails in general for n=4. It is easy to write conditions on a 4-whisker which would allow it, subject to those conditions, to be appended to the top of a 4-pasting scheme and to have a new 4-pasting scheme as a result, but we can find nothing non-trivial. The difficulty is with domain replacement. It is easy to verify that if (G_1, g_1) and (G_2, g_2) are (n+1)-embeddings with $domG_2 = codG_1$ but otherwise disjoint, then their union is an (n+1)-embedding. Moreover, if (G_1, g_1) and (G_2, g_2) both satisfy the n-cell extension condition, then so does their union. Moreover, if they satisfy (*), then so does their union. The property that given any two (n+1)-cells F_1 and F_2, there exists an n-pasting scheme X from S to T such that precisely one of F_1 and F_2 is enclosed $B(X)$, also is preserved by composition. But none of these suffices to give domain replacement.

In fact, Example 3.11 suggests that it may be difficult to find conditions on labelled n-embeddings to include all composites, short of an assumption to the effect that the n-embedding can be decomposed into whiskers. For instance, Example 3.11 can be modified to show that one may have a 4-embedding with three 4-cells, such that one of them has the property that each of the others extends to a whisker that passes under its codomain, but which does not represent a composite.

Now, as promised at the end of Section 6, we study conditions on labelled pasting schemes that allow greater freedom in the way in which we may evaluate composites. Specifically, we should like to be able to evaluate them locally, by taking any n-ball in the pasting scheme and evaluating it, then substituting a single cell for it with the labelling determined by its composite. In general, that is impossible as illustrated by Example 3.11. However, there are reasonable conditions on a pasting scheme under which it is possible. For the case of n=1, the situation is trivial, so we shall restrict our attention to the case of n+1.

Definition 7.2 Given a (n+1)-embedding (G, g), a *directed (n+1)-ball* in (G, g) consists of a subset B of the set of (n+1)-cells of (G, g) such that, B together with the domains and codomains of its elements forms a spherical (n+1)-pasting scheme.

Definition 7.3 An (n+1)-embedding satisfies the *strong n-cell extension* condition if for every partition of the embedding into directed (n+1)-balls and every n-cell x not in the interior of any of the (n+1)-balls, x extends to an n-pasting scheme X in G with domain S and codomain T that respects the partition.

Observe that the strong n-cell extension condition implies the n-cell extension condition, since one may take the discrete partition. The converse is false, as illustrated by Example 3.11.

Definition 7.4 An (n+1)-embedding (G, g) satisfies the *strong domain replacement* condition if for every n-pasting scheme X in G from S to T,

1. if F is an (n+1)-cell in (G, g) with $domF$ a subpasting scheme of X, then replacing $domF$ by $codF$ in X gives another n-pasting scheme embedded in (n+1)-space
2. if B is a directed (n+1)-ball with $domB$ a subpasting scheme of X, then in X, replacing $domB$ by a single cell gives an n-pasting scheme

3. if B and B' are directed (n+1)-balls with empty intersection such that *domB* and *domB'* are both subpasting schemes of X, then in X, replacing both *domB* and *domB'* by single cells gives an n-pasting scheme.

By the same proof as for domain replacement in the presence of the n-cell extension condition, it follows that the asymmetry in the definition of strong domain replacement is illusory in the presence of the strong n-cell extension condition. Also, an acyclic 3-embedding satisfies strong domain replacement if it satisfies condition (*) because the replacement of any spherical subpasting scheme of a 2-pasting scheme by a single cell yields a 2-pasting scheme.

Definition 7.5 A *strong (n+1)-pasting scheme* is an (n+1)-embedding that satisfies the strong n-cell extension condition and the strong domain replacement condition.

Every strong (n+1)-pasting scheme is evidently an (n+1)-pasting scheme, but as Example 3.11 shows, the converse is in general false. However, every 2-pasting scheme is a strong 2-pasting scheme. Moreover, a 3-pasting scheme is strong if and only if it satisfies the strong 2-cell extension condition.

We now develop the extra topological machinery we require to support our more general notion of composite.

Lemma 7.6 Let (G, g) be an (n+1)-pasting scheme and let B be a directed (n+1)-ball in (G, g). Suppose further that *domB* extends to an n-pasting scheme X in G from S to T such that no n-cell of X lies in the interior of B. Then there exists an (n+1)-sequence of (G, g) for which, if F and F' are elements of B, every (n+1)-cell that lies between them in the (n+1)-sequence is also an element of B.

Proof: By the dual of Corollary 4.11, the (n+1)-cells of (G, g) not in $B(X)$ form a subpasting scheme of (G, g) with domain *domG* and codomain X. So they yield an (n+1)-sequence S_X. Since B is a spherical (n+1)-pasting scheme, it has an (n+1)-sequence S_B. Moreover, since X does not enter the interior of B, every (n+1)-cell of B lies inside $B(X)$. So we may extend S_X by S_B. Finally, by domain replacement, induction, and Corollary 4.11, the remaining (n+1)-cells of (G, g) may be appended to give an (n+1)-sequence of the whole of (G, g). □

Remark 7.7 By an argument similar to that of Proposition 4.3, it is routine to verify that given a directed (n+1)-ball B in an (n+1)-embedding (G, g), if *domB* extends to an n-pasting scheme X in G from S to T such that no n-cell of X lies in the interior of B, then (G, g) is "loop-free relative to" B, i.e. for any alternating sequence $F_0, x_0, \ldots, F_m$ of (n+1)-cells F_i and n-cells x_i of (G, g), with x_i an element of $outF_i \cap inF_{i+1}$, and with F_0 and F_m both in B, it follows that F_i is in B for every i. Just as in Proposition 4.9, the converse holds if (G, g) satisfies condition 1 of (strong) domain replacement. As suggested by the terminology, this new condition is closely related to Johnson's "loop-freeness" condition. More precisely, if (G, g) is the domain of a single (n+2)-cell, then our condition is a special case of Johnson's condition.

Proposition 7.8 If (G,g) is a strong (n+1)-pasting scheme and B is a directed (n+1)-ball in (G,g), then replacing B by a single (n+1)-cell in (G,g) gives a strong (n+1)-pasting scheme (G',g').

Proof: It is immediate that (G',g') is an (n+1)-embedding, that it satisfies the strong n-cell extension condition and conditions 2 and 3 of strong domain replacement. For condition 1 of strong domain replacement, let X be an n-pasting scheme in G' from S to T, and let F be an (n+1)-cell in (G',g') with $domF$ a subpasting scheme of X. Then, X is an n-pasting scheme in G. So, if F is an (n+1)-cell in (G,g), we are done. If not, then F must be our new (n+1)-cell. Then the result follows from Lemma 7.6 and domain replacement in (G,g) applied several times. □

Corollary 7.9 Given any partition of the (n+1)-cells of a strong (n+1)-pasting scheme (G,g) into directed (n+1)-balls, replacement of each directed (n+1)-ball by a single (n+1)-cell yields a strong (n+1)-pasting scheme.

Proof: Use Proposition 7.8 to replace one directed (n+1)-ball at a time. □

Corollary 7.9 gives us conditions under which we can evaluate a labelled (n+1)-pasting scheme by evaluating directed (n+1)-balls in any combination. However, for a particular purpose, we may only be interested in replacing a particular directed (n+1)-ball by a single (n+1)-cell. We can relax the above conditions a little in order to allow precisely that rather than the more general phenomenon.

Proposition 7.10 Let (G,g) be an (n+1)-pasting scheme and let B be a directed (n+1)-ball in (G,g). Suppose

1. $domB$ extends to an n-pasting scheme X from S to T in G such that no n-cell in X lies in the interior of B
2. for any n-pasting scheme X from S to T in G such that $domB$ is a subpasting scheme of X, the replacement of $domB$ in X by a single n-cell gives an n-pasting scheme
3. for any n-pasting scheme X from S to T in G such that $domB$ is a subpasting scheme of X and for any (n+1)-cell F of (G,g) not in B such that $domF$ is also a subpasting scheme of X, replacing both $domB$ and $domF$ by single cells gives an n-pasting scheme.

Then, in (G,g), replacing B by a single (n+1)-cell gives an (n+1)-pasting scheme.

Proof: The n-cell extension condition and condition 1 of domain replacement follow from Lemma 7.6. Everything else is immediate from the conditions. □

Corollary 7.11 Let (G,g) be a 3-pasting scheme and let B be a directed 3-ball in (G,g). Suppose moreover that $domB$ extends to a 2-pasting scheme X from S to T in G such that no 2-cell in X lies in the interior of B. Then, in (G,g), replacing B by a single 3-cell gives a 3-pasting scheme.

Remark 7.12 It is evident from Proposition 7.10 that if all (n+1)-pasting schemes that appear in an (n+2)-pasting scheme satisfy the conditions of the proposition, then part of domain replacement necessarily holds for that (n+2)-pasting scheme. Of course, one can strengthen the conditions of Proposition 7.10 mildly for (n+1)-pasting schemes to similarly dispense with condition 2 of domain replacement for an (n+2)-pasting scheme. However, even in the presence of condition (*), one still cannot simply discard domain replacement completely. Example 3.11 can be used to construct a counter-example.

Now, given the above topological machinery, we can define the notion of a weak composite of a labelled strong (n+1)-pasting scheme and prove that every such pasting scheme has a unique weak composite. We define the term weak composite inductively.

Definition 7.13 Given a labelled strong (n+1)-pasting scheme $((G,g),a)$ in an (n+1)-category A, a *weak composite* of $((G,g),a)$ is the value of

1. any (n+1)-pasting composite of $((G,g),a)$
2. any weak composite of the (n+1)-pasting composites determined by a partition of the (n+1)-cells of (G,g) into directed (n+1)-balls.

Observe that Corollary 7.9 has been used implicitly in the definition of weak composite: it is precisely Corollary 7.9 that allows us to speak of the weak composite in the second part of the definition, as we know that the partition of (G,g) yields a strong (n+1)-pasting scheme. The theorem follows directly from that.

Theorem 7.14 For every positive natural number n, every labelled strong (n+1)-pasting scheme $((G,g),a)$ in an (n+1)-category A has a unique weak composite.

Proof: By the (n+1)-categorical pasting theorem (Theorem 6.16), every labelled strong (n+1)-pasting scheme has an (n+1)-pasting composite. For the unicity, it suffices to show that a weak composite determined by taking a non-trivial partition of (G,g) into directed (n+1)-balls has the same value as some (n+1)-pasting composite. By induction, it suffices to assume that a non-trivial partition has only one non-trivial directed (n+1)-ball B. So take decompositions of B and of (G',g'), the (n+1)-pasting scheme given by replacing B in (G,g) by a single (n+1)-cell. Then, a decomposition of (G,g) is given by taking that of (G',g') and refining it by that of B. So, by the (n+1)-categorical pasting theorem, we are done. □

Corollary 7.15 Let $((G,g),a)$ be a labelled 3-pasting scheme that satisfies the strong 2-cell extension condition. Then $((G,g),a)$ has a unique weak composite.

Evidently, we can modify Theorem 7.14 mildly in order to capture the situation in which we wish only to evaluate a particular labelled directed (n+1)-ball, then to use that to take a composite.

Theorem 7.16 Let $((G,g),a)$ be a labelled (n+1)-pasting scheme in an (n+1)-category A and suppose (G,g) and a directed (n+1)-ball B satisfy the conditions of Proposition 7.10. Then the (n+1)-cell in A determined by evaluating B then evaluating the labelled (n+1)-pasting scheme given by replacing B in (G,g) by a single (n+1)-cell labelled by the (n+1)-pasting composite of B, is the (n+1)-pasting composite of $((G,g),a)$.

Proof: As for Theorem 7.14. □

Corollary 7.17 Let $((G,g),a)$ be a labelled 3-pasting scheme and let B be a directed 3-ball such that $domB$ extends to a 2-pasting scheme X from S to T with no 2-cell of X lying in the interior of B. Then if one evaluates B, then uses the result to evaluate $((G,g),a)$, the result is the composite of $((G,g),a)$.

8 Free n-categories

In [13], Street gives a definition of what it means for a given n-category to be free. His definition is not given in terms of the existence of a left adjoint. However, the following inductive procedure is implicit in his paper.

For each natural number n, it is possible to define an adjunction in the usual sense $F_n \dashv U_n : n\text{-}Cat \to n\text{-}Cmptd$, where $n\text{-}Cmptd$ is defined inductively as follows. Put $0\text{-}Cat = 0\text{-}Cmptd = Set$ with $F_0 = U_0 =$ the identity functor on the category of sets. A *weak (n+1)-computad* G consists of an object G_n of $n\text{-}Cmptd$, a set G_{n+1}, and two functions $dom, cod : G_{n+1} \to F_n(G_n)$ such that $dom\, dom = dom\, cod$ and $cod\, dom = cod\, cod$. A *morphism* of weak (n+1)-computads from (G_{n+1}, G_n, dom, cod) to (H_{n+1}, H_n, dom, cod) consists of a morphism $f_n : G_n \to H_n$ of weak n-computads and a function $f_{n+1} : G_{n+1} \to H_{n+1}$ such that $dom\, f_{n+1} = f_n\, dom$ and $cod\, f_{n+1} = f_n\, cod$. The category $(n+1)\text{-}Cmptd$ is the category of weak (n+1)-computads and morphisms of weak (n+1)-computads, with the evident composition. Given an (n+1)-category A, define the weak (n+1)-computad $U_{n+1}A = G$ by $G_n = U_n(A_n)$ where A_n is the underlying n-category of A, and G_{n+1} is the limit in the category of sets of the diagram

$$\begin{array}{ccccccc} F_nU_n(A_n) & & & A_{n+1} & & & F_nU_n(A_n) \\ & \searrow^{counit} & \swarrow_{dom} & & \searrow^{cod} & \swarrow_{counit} & \\ & A_n & & & & A_n & \end{array} \qquad (8.1)$$

where A_n is here identified with its set of n-cells.

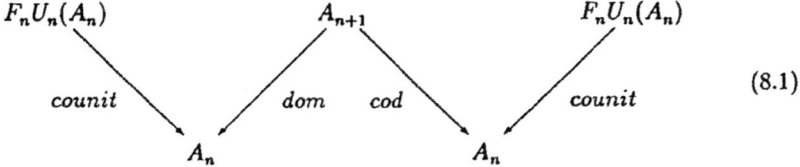

Each (n+1)-functor $f : A \to B$ induces a morphism $U_{n+1}f : U_{n+1}(A) \to U_{n+1}(B)$ in the evident way. The induction is complete once one establishes

Proposition 8.1 The functor U_{n+1} has a left adjoint F_{n+1}.

Proof: The proof is implicit in [13] and was discussed by Street at the International Category Theory meeting in Louvain-la-Neuve in 1987. The construction for the case n=1 appears in Street's paper [12]. □

Proposition 8.2 An n-category is free in Street's sense if and only if it is in the image of F_n.

Proof: This is implicit in the definition of free n-category in [13]. □

In fact, with mild conditions on the (n+1)-computad G, the equivalence relations needed in the construcion of $F_{n+1}(G)$ have normal forms determined by rewrite systems. This analysis is part of current work by Eilenberg and Street.

A little more can be said to strengthen Proposition 8.1. It is routine to verify that all of the data for the adjunction between $(n+1)$-Cat and $(n+1)$-$Cmptd$ commutes with the evident forgetful functors to n-Cat and n-$Cmptd$ as in the diagram

$$
\begin{array}{ccccccc}
(n+1)-Cat & \longrightarrow & n-Cat & \ldots & 1-Cat & \longrightarrow & 0-Cat \\
F_{n+1}\uparrow \dashv \downarrow U_{n+1} & & Fn\uparrow \dashv \downarrow U_n & & F_1\uparrow \dashv \downarrow U_1 & & F_0\uparrow \dashv \downarrow U_0 \\
(n+1)-Cmptd & \longrightarrow & n-Cmptd & \ldots & 1-Cmptd & \longrightarrow & 0-Cmptd
\end{array}
\qquad (8.2)
$$

So, assuming the existence of at least three strongly inaccessible cardinals, we may now define Cat_∞ to be the object of the large 2-category $[\omega^{op}, CAT]$ given by the top line of diagram (8.2). Similarly, we may define $Cmptd_\infty$ to that object of $[\omega^{op}, CAT]$ determined by the bottom line of diagram (8.2). So we may strengthen Proposition 8.1 to

Corollary 8.3 The collection $(F_n \dashv U_n | n < \omega)$ forms an adjuntion $F \dashv U$ in $[\omega^{op}, CAT]$.

Now we may define ∞-Cat to be the limit in CAT of the diagram Cat_∞. Similarly, define ∞-$Cmptd$ to be the limit in CAT of the diagram $Cmptd_\infty$. Then, taking the limits induces an adjunction $F_\infty \dashv U_\infty : \infty\text{-}Cat \to \infty\text{-}Cmptd$.

Our category ∞-Cat does not quite agree with Street's ω-Cat. Ours is the full subcategory of his determined by those of his ω-categories for which every element is an n-cell for some n. This agrees with the notation of Al-Agl and Steiner. Nevertheless, we can still deduce

Corollary 8.4 An ∞-category is free in Street's sense if and only if it is in the image of F_∞.

We now seek to compare our definition of n-computad with the above. First note that at several points we have implicitly used the notion of the underlying (n+1)-computad of an (n+1)-category, for instance in defining a labelling of an (n+1)-pasting scheme in Definition 6.1. The underlying (n+1)-computad is subtly different to the underlying weak (n+1)-computad: where the latter uses $F_n U_n(A_n)$ as in diagram (8.1), the former uses *diags* of the underlying computad of A_n. So although for each n, there is an evident fully faithful functor J_n : *n-Computad* $\rightarrow$ *n-Cmptd* expressing *n-Computad* as a full subcategory of *n-Cmptd* (and this lifts to a map in $[\omega^{op}, CAT]$), U_n does not factor through J_n in general. Our n-computads amount to those weak n-computads whose domains and codomains are given by pasting diagrams, and so the underlying n-computad and weak n-computad of an n-category differ accordingly. Since we do not capture all possible composites in general, J_n generally expresses *n-Computad* as a proper subcategory of *n-Cmptd*. However, since we do capture all composites for n=1 to 3, it follows that we have

Proposition 8.5 For n=1 to 4, the functor J_n is an equivalence of categories.

There is a strong case that what we have called a weak computad would be better called simply a computad, and what we have called a computad would be better called a strong computad. Our choice in this paper has been just for brevity. However, it should be noted that our computads do incorporate a large class of the more general weak computads. For instance, it follows from Johnson and Street's work that all of the simplexes are computads. Moreover, Examples 3.11 and 3.12 illustrate several pasting schemes that may form domains and codomains for higher dimensional computads such as those describing at least some of the higher level adjuncions. Finally, Johnson's "coherent situations" [4] may be discussed more directly in the geometrical terms of our computads.

9 Bibliography

[1] F.A.Al-Agl and R.Steiner, Nerves of multiple categories, preprint (1990)

[2] J.W.Gray, Formal Category Theory: Adjointness for 2-Categories, in Lecture Notes in Math 391 (Springer, Berlin, 1971)

[3] M.Johnson, Pasting diagrams in n-categories with applications to coherence theorems and categories of paths, Doctoral thesis, University of Sydney (1987)

[4] Michael Johnson, The combinatorics of n-categorical pasting, J. Pure Appl Algebra 62 (1989) 211-225

[5] Michael Johnson, Finding a representable family and the free construction on 1, International Category Theory Meeting 1990 (lecture)

[6] M.M.Kapranov and V.A.Voevodsky, Combinatorial-geometric aspects of polycategory theory: pasting schemes and higher Bruhat orders, preprint, Steklov Mathematical Institute (1990)

[7] G.M.Kelly and R.H.Street, Review of the elements of 2-categories, in Lecture Notes in Math 420 (Springer, Berlin, 1974) 75-103

[8] A.J.Power, A 2-categorical pasting theorem, J. Algebra 129 (1990) 439-445

[9] A.J.Power, An algebraic formulation for data refinement, in Lecture Notes in Comp. Science 442 (Springer, Berlin, 1990) 390-401

[10] A.J.Power, Coherence for bicategories with finite bilimits II (in preparation)

[11] M.B.Smyth and G.D.Plotkin, The category-theoretic solution to recursive domain equations, SIAM J. Computing 11 (1982) 761-783

[12] Ross Street, Limits indexed by category-valued 2-functors, J. Pure Appl Algebra 8 (1976) 149-181

[13] Ross Street, The algebra of oriented simplexes, J. Pure Appl. Algebra 49 (1987) 283-336

TOPOS-THEORETIC APPROACHES TO MODALITY

by

Gonzalo E. REYES and Houman ZOLFAGHARI

Département de mathématiques et de statistique

Université de Montréal

Dedicated to Max Kelly on the occasion of his 60th birthday.

Introduction

In the last few years, two topos-theoretic approaches to modal logic have been developed simultaneously but independently: one started by Reyes [9] and further developed in Lavendhomme, Lucas and Reyes [7], and the other due to Ghilardi and Meloni [2].

At first sight, these approaches appear quite unrelated and the formal systems to which they lead are different. It may come, therefore, as a surprise that both are particular cases of a more general one which we shall describe in detail in this paper.

There is a fundamental difference between modal operators in a topos such as $\Box$ (necessity) and other logical operators such as $\neg$ (negation): while $\neg$ is functorial in the sense that it commutes with pull-backs (and hence it defines a map $\neg : \Omega \to \Omega$), this is not so for $\Box$. Indeed, the only operator $\Box : \Omega \to \Omega$ such that $\Box p \subseteq p$ and $\Box T = T$ is the identity.

In Reyes [9], this difficulty is circumvented by considering a topos over a base topos and restricting the domain of application of modal operators to predicates of "constant" objects only, but keeping functoriality with respect to "constant" maps. Ghilardi and Meloni [2] on the other hand, restrict the domain of application as in Reyes [9] but relax functoriality to lax functoriality for constant maps. This feature of the approach of Ghilardi and Meloni [2] will be kept in our work. Indeed, it has to be kept in any context which aims to generalize their approach. In a sequel to this paper, we shall show that by imposing a restriction on the topos itself, we may define lax modal operators on predicates of *all* objects of the topos.

§1. The general context

We consider a topos $\mathcal{E}$ (thought of as a universe of "variable sets") defined over a topos $\mathcal{S}$ (thought of as a universe of "constant sets"). In other words, we have a geometric morphism (Δ,Γ), i.e., a diagram

$$\mathcal{S} \underset{\Gamma}{\overset{\Delta}{\rightleftarrows}} \mathcal{E}$$

such that $\Delta \dashv \Gamma$ and Δ preserves finite limits.

Since Δ is uniquely determined by Γ, we shall write $\mathcal{E} \xrightarrow{\Gamma} \mathcal{S}$ for the above geometric morphism, leaving Δ implicit, as is customary in topos theory.

From now on we make the following assumption on (Δ,Γ):

(*) Δ preserves infima of subobjects of $\mathcal{S}$

It is convenient to reformulate (*) in terms of the existence of adjoint functors. To do so, consider the diagrams, for each $S \in \mathcal{S}$,

(**) $$\mathrm{Sub}_{\mathcal{S}}(S) \underset{\Gamma_S}{\overset{\Delta_S}{\rightleftarrows}} \mathrm{Sub}_{\mathcal{E}}(\Delta S)\,.$$

where $\mathrm{Sub}_{\mathcal{S}}(S)$ is the ordered category of subobjects of S and Δ_S, Γ_S are defined as follows: given $P \hookrightarrow S$ apply Δ to obtain $\Delta P \hookrightarrow \Delta S$. Define $\Delta_S(P \hookrightarrow S) = \Delta P \hookrightarrow \Delta S$. On the other hand, given $K \hookrightarrow \Delta S$, apply Γ and take the pull-back

$$\begin{array}{ccc} \Gamma(K) & \hookrightarrow & \Gamma\Delta S \\ \uparrow \lrcorner & & \uparrow \eta_S \\ \Gamma_S(K) & \hookrightarrow & S \end{array}$$

where $\eta : \mathrm{Id} \to \Gamma\Delta$ is the unit of the adjunction $\Delta \dashv \Gamma$.

It is easy to check that $\Delta_S \dashv \Gamma_S$. Therefore our assumption (*) may be reformulated as

(*)' for each $S \in \mathcal{S}$, we have a diagram

$$\mathrm{Sub}_{\mathcal{S}}(S) \begin{array}{c} \overset{\Pi_S}{\longleftarrow} \\ \overset{\Delta_S}{\longrightarrow} \\ \underset{\Gamma_S}{\longleftarrow} \end{array} \mathrm{Sub}_{\mathcal{E}}(\Delta S)$$

with $\Pi_S \dashv \Delta_S \dashv \Gamma_S$.

From these adjunctions, we obtain at once explicit descriptions for Π_S and Γ_S.

1.1 Proposition. Let $\mathcal{E} \xrightarrow{\Gamma} \mathcal{S}$ be a geometric morphism satisfying (*)'. Then

$$\Pi_S(K) = \wedge\{P \hookrightarrow S \mid K \hookrightarrow \Delta P \hookrightarrow \Delta S\}$$

$$\Gamma_S(K) = \vee\{P \hookrightarrow S \mid \Delta P \hookrightarrow K \hookrightarrow \Delta S\}\,.$$

This proposition tells us that $\Pi_S(K)$ is the smallest subobject P of S such that K is contained in ΔP, whereas $\Gamma_S(K)$ is the largest subobject P of S such that K contains ΔP.

We notice the following difference between Π_S, Δ_S and Γ_S: whereas Δ_S and Γ_S are easily seen to be natural in S, this is not so for Π_S (unless we impose further conditions on the morphism (Δ,Γ)).

The naturality of Δ_S and Γ_S implies the existence of a diagram (in $\mathcal{S}$):

$$\Omega_{\mathcal{S}} \underset{\gamma}{\overset{\delta}{\rightleftarrows}} \Gamma(\Omega_{\mathcal{E}})$$

where $\delta \dashv \gamma$ and $\delta(T) = T$, $\delta(p \wedge q) = \delta(p) \wedge \delta(q)$.

Although this is well-known, let us elaborate a little: the domain category of Δ_S is $\mathrm{Sub}_{\mathcal{S}}(S) \simeq \mathcal{S}(S,\Omega_{\mathcal{S}})$, whereas the target category of Δ_S is $\mathrm{Sub}_{\mathcal{E}}(\Delta S) \simeq \mathcal{E}(\Delta S,\Omega_{\mathcal{E}}) \simeq \mathcal{S}(S,\Gamma(\Omega_{\mathcal{E}}))$. By identifying the domain and the codomain of Δ_S with $\mathcal{S}(S,\Omega_{\mathcal{S}})$ and $\mathcal{S}(S,\Gamma(\Omega_{\mathcal{E}}))$ respectively, we may consider Δ_S as a map which sends S-elements of $\Omega_{\mathcal{S}}$ into S-elements of $\Gamma(\Omega_{\mathcal{E}})$ in a way that is natural in S, i.e., $(\Delta_S)_{\mathcal{S}}$ defines a map $\Omega_{\mathcal{S}} \xrightarrow{\delta} \Gamma(\Omega_{\mathcal{E}})$ such that $\delta \circ p = \Delta_S(p)$ for all S-elements

$p : S \longrightarrow \Omega_{\mathcal{S}}$ of $\Omega_{\mathcal{S}}$. Similarly, $(\Gamma_S)_{\mathcal{S}}$ defines a map $\Gamma(\Omega_{\mathcal{E}}) \xrightarrow{\gamma} \Omega_{\mathcal{S}}$ such that $\gamma \circ K = \Gamma_S(K)$ for all S-elements $K : S \longrightarrow \Gamma(\Omega_{\mathcal{E}})$ of $\Gamma(\Omega_{\mathcal{E}})$.

On the other hand, $(\pi_S)_{\mathcal{S}}$ is only lax natural in S in the sense that for all $S \xrightarrow{f} T \in \mathcal{S}$

$$\Pi_S f^* \leq f^* \Pi_T .$$

This means that we have a cell as indicated in the diagram

$$\begin{array}{cccc}
S & \mathcal{S}(S,\Omega_{\mathcal{S}}) & \longleftarrow & \mathcal{S}(S,\Gamma(\Omega_{\mathcal{E}})) \\
{\scriptstyle f}\downarrow & {\scriptstyle f^*}\uparrow & \Swarrow & \uparrow{\scriptstyle f^*} \\
T & \mathcal{S}(T,\Omega_{\mathcal{S}}) & \longleftarrow & \mathcal{S}(T,\Gamma(\Omega_{\mathcal{E}}))
\end{array}$$

In fact, $K \leq \Delta_T \Pi_T K$ for all $K \in \mathcal{S}(T,\Gamma(\Omega_{\mathcal{E}}))$, since $\Pi_T \dashv \Delta_T$. Hence $f^*(K) \leq f^*\Delta_T\Pi_T K = \Delta_S f^* \Pi_T K$, by naturality of $(\Delta_S)_{\mathcal{S}}$. Since $\Pi_S \dashv \Delta_S$, this implies that $\Pi_S f^*(K) \leq f^*\Pi_T(K)$.

To express this lax functoriality of $(\Pi_S)_{\mathcal{S}}$ we shall write

$$\Gamma(\Omega_{\mathcal{E}}) \dashrightarrow \Omega_{\mathcal{S}} .$$

With this convention we may rewrite our main assumption on $\mathcal{E} \xrightarrow{\Gamma} \mathcal{S}$ as follows

(*)" There is a lax π such that

$$\Omega_{\mathcal{S}} \;\; \begin{array}{c} \xleftarrow{\;\pi\;} \text{(lax)} \\ \xrightarrow{\;\delta\;} \\ \xleftarrow[\;\gamma\;]{} \end{array} \;\; \Gamma(\Omega_{\mathcal{E}})$$

with $\pi \dashv \delta \vdash \gamma$.

Although several examples of our general context will be given later on, we mention 2 cases where our main assumption (*) (equivalently, (*)") is satisfied.

Example 1. Let $u : \mathbb{C} \longrightarrow \mathbb{D}$ be a functor. Then the induced geometric morphism $\mathrm{Set}^{\mathbb{D}^\circ} \longrightarrow \mathrm{Set}^{\mathbb{C}^\circ}$ is essential, i.e., $\Delta = u^*$ has a left adjoint $\Pi = u_!$.

This example can be generalized: any essential geometric morphism obviously satisfies (*)"

Example 2. An open geometric morphism $\mathcal{E} \xrightarrow{\Gamma} \mathcal{S}$, i.e., a geometric morphism such that there is a diagram (in $\mathcal{S}$)

$$\Omega_{\mathcal{S}} \;\; \begin{array}{c} \xleftarrow{\;\lambda\;} \\ \xrightarrow{\;\delta\;} \\ \xleftarrow[\;\gamma\;]{} \end{array} \;\; \Gamma\Omega_{\mathcal{E}}$$

where λ is an actual map such that $\lambda \dashv \delta \dashv \gamma$. Notice that λ satisfies the following Frobenius condition: $\lambda(k \wedge \delta(p)) = \lambda(k) \wedge p$.

§2. Modal operators

In this section we define modal operators $\Box$ ("necessity") and $\Diamond$ ("possibility") starting from a geometric morphism $\mathcal{E} \xrightarrow{\Gamma} \mathcal{S}$ satistying (*).

Recall that we can reformulate (*) as the existence of a lax π in the diagram

$$\Omega_{\mathcal{S}} \underset{\gamma}{\overset{\overset{\pi}{\longleftarrow}}{\underset{\longleftarrow}{\overset{\delta}{\longrightarrow}}}} \Gamma(\Omega_{\mathcal{E}})$$

with $\pi \dashv \delta \dashv \gamma$.

2.1 Definition. We define the *necessity operator* $\Box$ as follows $\Box = \delta_\gamma : \Gamma(\Omega_{\mathcal{E}}) \longrightarrow \Gamma(\Omega_{\mathcal{E}})$.

2.2 Proposition. The map $\Box$ has the following properties

$$\begin{aligned} \Box &\leq \mathrm{Id}_{\Gamma(\Omega_{\mathcal{E}})} \\ \Box^2 &= \Box \\ \Box T &= T \\ \Box(K \wedge L) &= \Box K \wedge \Box L \end{aligned}$$

Proof. Direct check. Alternatively, observe that $\Box = \delta\gamma$ is a lex cotriple.

Notice that we have not used our assumption (*) to define $\Box$, since it is definable from δ and γ alone.

From $\Box$ we can further define

$$\Box_S : \mathcal{S}(S, \Gamma(\Omega_{\mathcal{E}})) \longrightarrow \mathcal{S}(S, \Gamma(\Omega_{\mathcal{E}}))$$

by composition with $\Box$, i.e. $\Box_S(K) = \Box \circ K = \Delta_S \circ \Gamma_S \circ K$.

On the other hand, to define $\Diamond$ we need our assumption.

2.3 Definition. We define the *possibility operator* $\Diamond$ to be the lax $(\Diamond_S)_{\mathcal{S}}$ where, for each $S \in \mathcal{S}$,

$$\Diamond_S : \mathcal{S}(S, \Gamma(\Omega_{\mathcal{E}})) \longrightarrow \mathcal{S}(S, \Gamma(\Omega_{\mathcal{E}}))$$

is given by $\Diamond_S = \Delta_S \circ \Pi_S$.

2.4 Proposition. The couple $(\Diamond, \Box)$ has the following properties for all $S \in \mathcal{S}$

$$\begin{aligned} &\Box_S \leq \mathrm{Id}_{\Gamma(\Omega_{\mathcal{E}})} \leq \Diamond_S \\ &\Box_S^2 = \Box_S \,,\ \Diamond_S^2 = \Diamond_S \\ &\Diamond_S \dashv \Box_S \end{aligned}$$

Furthermore, for all $S \xrightarrow{f} T \in \mathcal{S}$,

$$\Box_S f^* = f^* \Box_T$$

$$\Diamond_S f^* \leq f^* \Diamond_T$$

Proof. Straight forward computation using the adjunctions $\Pi_S \dashv \Delta_S \dashv \Gamma_S$.

Using our convention once again, we may write $\Diamond : \Gamma(\Omega_{\mathcal{E}}) \dashrightarrow \Gamma(\Omega_{\mathcal{E}})$ to indicate that we have a family $(\Diamond_S)_{\mathcal{S}}$ which is lax natural in S . With this convention

$$\begin{cases} \Box = \delta\gamma \\ \Diamond = \delta\pi \end{cases}$$

Since we are interested in modal operators and $\Box : \Gamma(\Omega_{\mathcal{E}}) \dashrightarrow \Gamma(\Omega_{\mathcal{E}})$ always exists ($\Box = \delta\gamma$), it is natural to ask for a characterization of those geometric morphisms $\mathcal{E} \xrightarrow{\Gamma} \mathcal{S}$ for which there is a lax $\Diamond : \Gamma(\Omega_{\mathcal{E}}) \dashrightarrow \Gamma(\Omega_{\mathcal{E}})$ such that the couple $(\Diamond,\Box)$ has the properties listed in the previous proposition. Let us call these MAO couples for short.

The answer to this question justifies, to some extent, the choice of our context:

2.5 Proposition. Let $\mathcal{E} \xrightarrow{\Gamma} \mathcal{S}$ be a geometric morphism. Then the following are equivalent:

(1) There is a lax $\Diamond : \Gamma(\Omega_{\mathcal{E}}) \dashrightarrow \Gamma(\Omega_{\mathcal{E}})$ such that $(\Diamond,\Box)$ is a MAO couple.

(2) If

$$\begin{array}{ccc} \mathcal{E} & \xrightarrow{\Gamma} & \mathcal{S} \\ & {\scriptstyle\Gamma_o}\searrow \quad \nearrow{\scriptstyle\Gamma'_o} & \\ & \mathcal{S}_o & \end{array}$$

is the canonical surjection/inclusion factorization of Γ, then Γ_o satisfies (*)" , i.e., there is a diagram

$$\Omega_{\mathcal{S}_o} \underset{\gamma_o}{\overset{\pi_o \; (\leftarrow\text{-}\text{-})}{\underset{\longleftarrow}{\xrightarrow{\delta_o}}}} \Gamma_o(\Omega_{\mathcal{E}})$$

with a lax π_o such that $\pi_o \dashv \delta_o \dashv \gamma_o$. Furthermore, by identifying $\Gamma_o(\Omega_{\mathcal{E}})$ with $\Gamma(\Omega_{\mathcal{E}})$, $\Box = \delta_o\gamma_o$ and $\Diamond = \delta_o\pi_o$.

Proof. As we saw before, the geometric morphism $\mathcal{E} \xrightarrow{\Gamma} \mathcal{S}$ gives rise to a diagram

$$\Omega_{\mathcal{S}} \underset{\gamma}{\overset{\delta}{\rightleftarrows}} \Gamma(\Omega_{\mathcal{E}})$$

such that $\delta \dashv \gamma$, $\delta T = T$ and $\delta(p \wedge q) = \delta(p) \wedge \delta(q)$. We define $\Box = \delta\gamma$.

We now describe the surjection/inclusion factorization of Γ in terms of $\Box$ by using some elementary notions of the theory of locales (cf. [6]).

The fixpoints of $\Box$ constitute a sub locale Ω_o of the locale $\Gamma(\Omega_{\mathcal{E}})$, i.e., we have a diagram in $\mathcal{S}$

$$\Omega_o \underset{\gamma_o}{\overset{\delta_o}{\rightleftarrows}} \Gamma(\Omega_{\mathcal{E}})$$

such that δ_o is a locale morphism and $\delta_o \dashv \gamma_o$. Notice that this implies that $\gamma_o\delta_o = \mathrm{Id}_{\Omega_o}$. (This is easy to check directly; alternatively we can describe Ω_o as the Eilenberg-Moore category of coalgebras for the cotriple $\Box$).

Since $\Omega_{\mathcal{B}}$ is the initial locale in $\mathcal{B}$, there is a unique morphism of locales

$$\Omega_{\mathcal{B}} \xrightarrow{\delta'_o} \Omega_o \ .$$

We let γ'_o be its right adjoint.

For the same reason, the diagram

$$\begin{array}{ccc} \Omega_{\mathcal{B}} & \xrightarrow{\ \delta\ } & \Gamma(\Omega_{\mathcal{E}}) \\ & {\scriptstyle \delta'_o}\searrow \quad \nearrow{\scriptstyle \delta_o} & \\ & \Omega_o & \end{array}$$

is commutative, i.e., $\delta = \delta_o\delta'_o$. This implies that $\gamma = \gamma'_o\gamma_o$. We now claim that $\delta'_o\gamma'_o = \mathrm{Id}_{\Omega_o}$ and $\gamma'_o\delta'_o = \gamma\delta$. In fact, $\delta'_o\gamma'_o = (\gamma_o\delta_o)\delta'_o\gamma'_o(\gamma_o\delta_o) = \gamma_o(\delta_o\delta'_o)(\gamma'_o\gamma_o)\delta_o = \gamma_o\delta\gamma\delta_o = \gamma_o\Box\,\delta_o = \gamma_o\delta_o = \mathrm{Id}_{\Omega_o}$. On the other hand, $\gamma\delta = (\gamma'_o\gamma_o)(\delta_o\delta'_o) = \gamma'_o(\gamma_o\delta_o)\delta'_o = \gamma'_o\delta'_o$. To show that $\Box = \delta_o\gamma_o$, we simply compute $\delta_o\gamma_o(K) = \delta_o\Box(K) = \Box(K)$ for $K \in \Gamma(\Omega_{\mathcal{E}})$.

We have thus obtained a factorization at the level of locales, locale morphisms and their adjoints

$$\begin{array}{ccc} \Omega_{\mathcal{B}} & \underset{\gamma}{\overset{\delta}{\rightleftarrows}} & \Gamma(\Omega_{\mathcal{E}}) \\ & {\scriptstyle \gamma_o}\nwarrow\!\searrow{\scriptstyle \delta'_o} \quad {\scriptstyle \delta_o}\nearrow\!\swarrow{\scriptstyle \gamma_o} & \\ & \Omega_o & \end{array}$$

Defining $j = \gamma'_o\delta'_o = \gamma\delta : \Omega_{\mathcal{B}} \longrightarrow \Omega_{\mathcal{B}}$, we check that j is a topology with the following property

2.6 Lemma $X' \hookrightarrow X \in \mathcal{B}$ is j-dense iff $\Delta X' \xrightarrow{\sim} \Delta X \in \mathcal{E}$.

Proof (of lemma). We first observe the following connection between the classifying maps $\chi_{X'}$ of $X' \hookrightarrow X$ and $\chi_{\Delta X'}$ of $\Delta X' \hookrightarrow \Delta X$: $\delta o \chi_{X'} = t(\chi_{\Delta X'})$, where "t" stands for "the transpose of". Indeed, consider the diagram

$$\begin{array}{ccc} X' & \hookrightarrow & X \\ \downarrow & & \downarrow{\scriptstyle \chi_{X'}} \\ 1 & \underset{T}{\hookrightarrow} & \Omega_{\mathcal{B}} \\ & & \downarrow{\scriptstyle \delta} \\ & & \Gamma(\Omega_{\mathcal{E}}) \end{array}$$

By the very definition of δ, $t(\delta o \chi_{X'}) = \chi_{\Delta X'}$ and this proves that $\delta o \chi_{X'} = t(\chi_{\Delta X'})$.

To finish the proof of the lemma, we notice the following equivalences

$$\frac{\frac{\frac{T \le \gamma\delta(\chi_{X'})}{\delta T \le \delta(\chi_{X'})}}{t(\delta T) \le t(\delta\chi_{X'})}}{T_{\Delta X} \le \chi_{\Delta X'}} \qquad \text{(since } \delta \dashv \gamma)$$

2.7 Lemma. Each $\Gamma(E)$ is a j-sheaf.

Proof. Obvious from the previous lemma and the adjunction $\Delta \dashv \Gamma$.

We obtain thus the following surjection/inclusion factorization

$$\begin{array}{ccc} \mathcal{E} & \xrightarrow{\Gamma} & \mathcal{B} \\ & {\scriptstyle\Gamma_o}\searrow \quad \nearrow{\scriptstyle\Gamma'_o} & \\ & \mathcal{B}_o = Sh_j(\mathcal{B}) & \end{array}$$

Notice that Γ_o is a surjection, since Ω_o is the subobject classifier of $\mathcal{B}_o$ (because

$$\Omega_o \longrightarrow \Omega_{\mathcal{B}} \underset{Id}{\overset{j}{\rightrightarrows}} \Omega_{\mathcal{B}}$$

is an equalizer) and δ_o is mono.

End of Proof (of 2.5). (1) $\Rightarrow$ (2): define $\pi_o = \gamma_o \lozenge$ and check that $\pi_o \dashv \delta_o$. To do so, let

$$S \underset{K'}{\overset{K}{\rightrightarrows}} \Gamma(\Omega_{\mathcal{E}})$$

be two S-elements of $\Gamma(\Omega_{\mathcal{E}})$. We have the following equivalences

$$\frac{\frac{\frac{\frac{\frac{\pi_o(K) = \gamma_o\lozenge_S(K) \le K'}{\delta_o\gamma_o\lozenge_S(K) \le \delta_o(K')}}{\Box_S\lozenge_S(K) \le \delta_o(K')}}{\lozenge_S K \le \delta_o(K')}}{K \le \Box_S\delta_o(K')}}{K \le \delta_o(K')} \qquad \begin{array}{l}\text{(since } \delta_o \text{ mono)} \\ \\ \text{(since } \lozenge_S \dashv \Box_S)\end{array}$$

Finally, (2) $\Rightarrow$ (1) is clear.

§3. Functoriality of modal operators

The operators $\Box$, $\lozenge$ defined in §2 behave differently with respect to functoriality: while $\Box$ is functorial, $\lozenge$ is only lax functorial (2.4 Proposition).

In this section we investigate under which circumstances $\Diamond$ is functorial and how functoriality is related to well known properties of the geometric morphism $\mathcal{E} \xrightarrow{\Gamma} \mathcal{S}$ such as openness, Frobenius property, etc.

The first question is answered by the following

3.1 Theorem. Let $\Diamond$ be the possibility operator associated with a geometric morphism $\mathcal{E} \xrightarrow{\Gamma} \mathcal{S}$ satisfying (*) as in §2. Then $\Diamond$ is functorial iff in the surjection/inclusion factorization

$$\begin{array}{ccc} \mathcal{E} & \xrightarrow{\Gamma} & \mathcal{S} \\ & {\scriptstyle\Gamma_o}\searrow \quad \nearrow{\scriptstyle\Gamma'_o} & \\ & \mathcal{S}_o & \end{array}$$

Γ_o is open.

Proof. This follows from 2.5 Proposition. In fact, we have $\pi_o = \gamma_o \Diamond$ and $\Diamond = \delta_o \pi_o$. Therefore $\Diamond$ is natural iff π_o is natural. But the existence of such a π_o is precisely the definition of Γ_o to be open, since the further Frobenius property for openess, $\pi_o(k \wedge \delta(p)) = \pi_o(k) \wedge p$ is automatically satisfied in this case.

For (essential) morphisms $u_* : Set^{\mathbb{C}^o} \longrightarrow Set^{\mathbb{D}^o}$ derived from functors $u : \mathbb{C} \longrightarrow \mathbb{D}$ between small categories, Johnstone [5] gives the following characterization: u_* is open iff for all $C \in \mathbb{C}$, $D \in \mathbb{D}$ and all $D \xrightarrow{\alpha} u(C) \in \mathbb{D}$, there are morphisms $f : C' \longrightarrow C \in \mathbb{C}$ $\beta : u(C') \longrightarrow D \in \mathbb{D}$ and $\lambda : \Delta \longrightarrow u(C') \in \mathbb{D}$ such that

$$\beta\lambda = 1_D \quad \text{and} \quad \alpha\beta = u(f) .$$

For a preopen u_*, the caracterization is exactly the same except that we require only $\alpha = u(f)\lambda$ instead of $\alpha\beta = u(f)$.

On the other hand, it is easy to check that the surjection/inclusion factorization of u_* is

$$\begin{array}{ccc} Set^{\mathbb{C}^o} & \longrightarrow & Set^{\mathbb{D}^o} \\ & \searrow \quad \nearrow & \\ & Set^{\mathbb{E}^o} & \end{array}$$

where $\mathbb{E}$ is the full subcategory of $\mathbb{D}$ consisting of objects of the form $u(C)$ (cf. Johnstone [4]).

Putting these things together we obtain the following corollary of 3.1 theorem.

3.2 Corollary. Let $u : \mathbb{C} \longrightarrow \mathbb{D}$ be a functor between small categories and let $u_* : Set^{\mathbb{C}^o} \longrightarrow Set^{\mathbb{D}^o}$ be the induced (essential) geometric morphism. Then the associated operator $\Diamond$ is functorial iff we have the following condition: for all $C, C' \in \mathbb{C}$ and all $\alpha : u(C') \longrightarrow u(C) \in \mathbb{D}$ there are morphisms $f : C'' \longrightarrow C \in \mathbb{C}$, $\beta : u(C'') \longrightarrow u(C') \in \mathbb{D}$ and $\lambda : u(C') \longrightarrow u(C'') \in \mathbb{D}$ such that

$$\beta\lambda = 1_{u(C')} \quad \text{and} \quad \alpha\beta = u(f) .$$

Next we investigate the following properties:

(1) Γ is locally connected, i.e., Γ is essential and the left adjoint Π of Δ satisfies the "generalized Frobenius law": for all objects $F \in \mathcal{E}$, $X, Y \in \mathcal{S}$ and for all morphisms $X \longrightarrow Y \in \mathcal{S}$ and $F \longrightarrow \Delta Y \in \mathcal{E}$, the canonical morphism

$$\Phi_{F,X,Y} : \Pi(F\times_{\Delta Y}\Delta X) \longrightarrow \Pi(F)\times_Y X$$

is an isomorphism.

This canonical morphism corresponds to the pair of maps $\langle \Pi(p_1), \varepsilon_X \circ \Pi(p_2)\rangle$, where

$$\Pi(F\times_{\Delta Y}\Delta X) \xrightarrow{\Pi(p_1)} \Pi(F)$$

and

$$\Pi(F\times_{\Delta Y}\Delta X) \xrightarrow{\Pi(p_2)} \Pi(\Delta X) \xrightarrow{\varepsilon_X} \Pi(F)$$

p_1 and p_2 are the two projections and ε_X is the counit of the adjunction.

(2) Γ is essential and Π satisfies the "Frobenius law", i.e., for all $F \in \mathcal{E}$ and $X \in \mathcal{S}$, the canonical morphism

$$\psi_{F,X} : \Pi(F\times\Delta X) \longrightarrow \Pi(F) \times X$$

is an isomorphism.

Notice that in terms of Φ , $\psi_{F,X} = \Phi_{F,X,1}$.

(3) Γ is essential and all the $\Phi_{F,X,Y}$ are epis

(4) Γ is essential and all the $\psi_{F,X}$ are epis

(5) Γ is open

(6) Γ is preopen in the sense of Johnstone [5]: for all $X,Y \in \mathcal{S}$, the canonical map $\theta_{X,Y} : \Delta(X^Y) \longrightarrow \Delta X^{\Delta Y}$ is monic. (Recall that $\theta_{X,Y}$ is the exponential transpose of

$$\Delta(X^Y) \times \Delta Y \simeq \Delta(X^Y\times Y) \xrightarrow{\Delta(ev)} \Delta(X) .)$$

(7) The operator $\Diamond$ derived from Γ is functorial.

3.2 Theorem. Let $\Gamma : \mathcal{E} \longrightarrow \mathcal{S}$ be an essential geometric morphism. We have the following relations between properties (1) to (7):

$$\begin{array}{ccccccc} (1) & \Rightarrow & (3) & \Leftrightarrow & (5) & \Rightarrow & (7) \\ \Downarrow & & \Downarrow & & \Downarrow & & \\ (2) & \Rightarrow & (4) & \Leftrightarrow & (6) & & \end{array}$$

Proof. The implications (1) $\Rightarrow$ (3) , (2) $\Rightarrow$ (4) and (5) $\Rightarrow$ (7) are evident; (1) $\Rightarrow$ (2) and (3) $\Rightarrow$ (4) both follow from the fact that $\psi_{F,X} = \Phi_{F,X,1}$. It remains to prove the two equivalences. (4) $\Leftrightarrow$ (6):

We prove that

$$\begin{array}{ccc} \Delta(Y^X) \xrightarrow{\theta} \Delta Y^{\Delta X} & & \Pi(F\times\Delta X) \xrightarrow{\psi} \Pi F \times X \\ & \Leftrightarrow & \\ \text{is mono for all } X,Y & & \text{is epi for all } F,X . \end{array}$$

First let us calculate the transpose of ψ. Remember that $\psi = \langle \Pi(\pi_1), \varepsilon_X \circ \Pi(\pi_2) \rangle$ where $F \times \Delta X \xrightarrow{\pi_1} F$ and $F \times \Delta X \xrightarrow{\pi_2} \Delta X$ are the two projections. Transposition is obtained by applying Π and composing with the counity.

$\varepsilon_X \circ \Pi(\pi_2)$ is simply the transpose of π_2, and transposing $\eta_F \circ \pi_1$: $F \times \Delta X \xrightarrow{\pi_1} F \xrightarrow{\eta_F} \Delta\Pi F$, where F is the unity of the adjunction, we obtain

$$\varepsilon_{\Pi F} \circ \Pi(\eta_F) \circ \Pi(\pi_1) = \Pi(\pi_1)$$

because of the following commutative diagram given the adjunction

$$\begin{array}{ccc} \Pi F & \xrightarrow{\Pi(\eta_F)} & \Pi\Delta\Pi F \\ & {\scriptstyle Id} \searrow \quad \nearrow {\scriptstyle \varepsilon_{\Pi F}} & \\ & \Pi(F) & \end{array}$$

In conclusion, the transpose of ψ is the morphism

$$\langle \eta_p \circ \pi_1, \pi_2 \rangle = \eta_F \times \Delta X : F \times \Delta X \longrightarrow \Delta\Pi F \times \Delta X .$$

($\Rightarrow$): We start by remarking these natural correspondances:

$$\begin{array}{c} \Pi(F\times\Delta X) \xrightarrow{\Psi} \Pi F\times X \xrightarrow{\alpha} Y \\ \hline F \times \Delta X \xrightarrow{\eta_F\times\Delta X} \Delta\Pi F\times\Delta X \xrightarrow{\Delta\alpha} \Delta Y \\ \hline F \xrightarrow{\eta_F} \Delta\Pi F \xrightarrow{\tau(\Delta\alpha)} \Delta Y^{\Delta X} \\ \qquad {\scriptstyle \Delta\tau(\alpha)} \searrow \quad \nearrow {\scriptstyle \theta} \\ \Delta(Y^X) \end{array}$$

where $\tau(\Delta\alpha)$ and $\tau(\alpha)$ are the exponential transposes of $\Delta\alpha$ and α and the last factorization follows from the commutative diagram.

$$\begin{array}{ccc} \Pi F \times X & \xrightarrow{\alpha} & Y \\ & {\scriptstyle \tau(\alpha)\times X} \searrow \quad \nearrow {\scriptstyle ev} & \\ & Y^X \times X & \end{array}$$

given by the adjunction $(-) \times X \dashv (\ )^X$, on which we apply Δ and then take the transpose with respect to $(-) \times \Delta X \dashv (\ )^{\Delta X}$.

Now suppose that θ is mono, and let α,β be morphisms such that $\alpha\psi = \beta\varphi$. We have:

$$\begin{array}{c} \Pi(F\times\Delta X) \xrightarrow{\Psi} \Pi F\times X \overset{\alpha}{\underset{\beta}{\rightrightarrows}} Y \\ \hline F \xrightarrow{\eta_F} \Delta\Pi F \rightrightarrows \Delta Y^{\Delta X} \\ \qquad {\scriptstyle \Delta\tau(\alpha)} \searrow\!\!\searrow {\scriptstyle \Delta\tau(\beta)} \quad \nearrow {\scriptstyle \theta} \\ \Delta(Y^X) \end{array} \quad , \ \alpha\Psi = \beta\Psi$$

But θ is mono and

$$F \xrightarrow{\eta_F} \Delta\Pi F \overset{\Delta\tau(\alpha)}{\underset{\Delta\tau(\beta)}{\rightrightarrows}} \Delta(Y^X)$$

$$\overline{\Pi F \xrightarrow[\mathrm{Id}]{} \Pi F \overset{\tau(\alpha)}{\underset{\tau(\beta)}{\rightrightarrows}} Y^X}$$

therefore $\tau(\alpha) = \tau(\beta)$ and $\alpha = \beta$.

($\Leftarrow$) Suppose ψ is epi and let α, β be morphisms such that $\theta\alpha = \theta\beta$,

$$F \overset{\alpha}{\underset{\beta}{\rightrightarrows}} \Delta(Y^X) \xrightarrow{\theta} \Delta Y^{\Delta X}$$

Remark that transposing

$$\Pi F \xrightarrow{t(\alpha)} Y^X$$
$$\mathrm{Id} \searrow \quad \nearrow t(\alpha)$$
$$\Pi F$$

where $t(\alpha)$ is the transpose of α , we have a factorization

$$F \xrightarrow{\alpha} \Delta(Y^X)$$
$$\eta_F \searrow \quad \nearrow \Delta t(\alpha)$$
$$\Delta\Pi F$$

and we have the same thing for β .

We get the following correspondances.

$$F \overset{\alpha}{\underset{\beta}{\rightrightarrows}} \Delta(Y^X) \xrightarrow{\theta} \Delta Y^{\Delta X}$$
$$\eta_F \searrow \quad \overset{\Delta t(\alpha)}{\underset{\Delta t(\beta)}{\nearrow\!\!\!\nearrow}}$$
$$\overline{\Delta\Pi F}$$
$$F \times \Delta X \xrightarrow{\quad} \Delta(Y^X) \times \Delta X \xrightarrow{\Delta ev} \Delta Y$$
$$\eta_F \times \Delta X \searrow \quad \overset{\Delta t(\alpha) \times \Delta X}{\underset{\Delta t(\beta) \times \Delta X}{\nearrow\!\!\!\nearrow}}$$
$$\overline{\Delta\Pi F \times \Delta X}$$
$$\Pi(F \times \Delta X) \xrightarrow{\quad} Y^X \times \Delta X \xrightarrow{ev} Y$$
$$\psi \searrow \quad \overset{t(\alpha) \times X}{\underset{t(\beta) \times X}{\nearrow\!\!\!\nearrow}}$$
$$\Pi F \times X$$

ψ being epi, ev equalizes $t(\alpha) \times X$ and $t(\beta) \times X$. But the exponential transpose of this part of the diagram is

$$\Pi(F) \overset{t(\alpha)}{\underset{t(\beta)}{\rightrightarrows}} Y^X \xrightarrow{\mathrm{Id}} Y^X ,$$

therefore we have $t(\alpha) = t(\beta)$ and $\alpha = \beta$.

The equivalence (3) $\Leftrightarrow$ (5) follows from (4) $\Leftrightarrow$ (6) by using slice toposes, and the following result of Johnstone [5]: For an object Y of $\mathcal{S}$, let

$$\mathcal{E}/\Delta Y \underset{\Gamma/Y}{\overset{\Pi/Y}{\underset{\longrightarrow}{\overset{\longrightarrow}{\xleftarrow{\Delta/Y}}}}} \mathcal{S}/Y$$

be the induced morphism. Γ is open iff for all Y, Γ/Y is pre open. (Just notice that $\Delta/Y(X \longrightarrow Y) = \Delta X \longrightarrow \Delta Y$ and $\Pi/Y(F \longrightarrow \Delta Y) = \Pi F \longrightarrow Y$ which is the transpose of $F \longrightarrow \Delta F$). Γ/Y being pre open means that all the morphism

$$\Pi/Y((F \longrightarrow \Delta Y)\times(\Delta X \longrightarrow \Delta Y)) \xrightarrow{\Psi/Y} (\Pi F \longrightarrow Y) \times (X \longrightarrow Y)\ .$$

are epi.

We can verify that ψ/Y is just the morphism

$$\begin{array}{ccc} \Pi(F\times_{\Delta Y}\Delta X) & \xrightarrow{\Phi_{F,X,Y}} & \Pi F\times X \\ & \searrow \quad \nearrow & \\ & Y & \end{array}$$

in $\mathcal{S}/Y$. And ψ/Y is epi iff $\phi_{F,X,Y}$ is epi.

So Γ is open iff all the maps $\phi_{F,X,Y}$ are epi, which is precisely condition (3). We shall give counter examples to reverse arrows in the next section.

§4. Particular cases and examples

In this section we show that the approaches to modal logic mentioned in the introduction may be seen as particular cases of the one developed in this paper. Furthermore we show, by means of examples that some of the implications of 3.2 Theorem cannot be reversed.

The context of Reyes [9] is a locally connected geometric morphism $\mathcal{E} \xrightarrow{\Gamma} \mathcal{S}$, whereas the context of Lavendhomme, Lucas and Reyes [7] is an arbitrary geometric morphism $\mathcal{E} \xrightarrow{\Gamma} \mathcal{S}$ with $\mathcal{S}$ a Boolean topos. In the last case, Γ is automatically open (cf. [1]) and 3.2 Theorem show that (*) is satisfied and, furthermore, $\Diamond$ is actually functorial.

On the other hand, the context of Ghilardi and Meloni [2] is the essential geometric morphism $Set^{|\mathbb{C}|} \longrightarrow Set^{\mathbb{C}^{\circ}}$ derived from the obvious functor $|\mathbb{C}| \longrightarrow \mathbb{C}$. (Here $\mathbb{C}$ is a small category and $|\mathbb{C}|$ the discrete category of its objects). Since Δ has a left adjoint, Π_o, condition (*) is satisfied, but $\Diamond$ is not functorial, only lax functorial. We shall investigate functoriality of $\Diamond$ in this case in 4.2 Proposition.

In some sense, the approach of Lavendhomme, Lucas and Reyes [7] is at the opposite extreme from the one of Ghilardi and Meloni [2]. Whereas in the first, the base topos $\mathcal{S}$ is Boolean, in the second it is the topos $\mathcal{E}$ which is Boolean. The motivation for the first was to view $\mathcal{E}$ as a universe of variable sets and $\mathcal{S}$ as a universe of constant sets, providing a "standard of constancy" for change and modalities in $\mathcal{E}$. Modalities were defined for predicates of constant sets only and functoriality of modal operators resulted as a consequence. This is clearly not the motivation for the other approach.

We shall study these extremes in some detail.

Case A: $\mathcal{S}$ is Boolean.

In this case we have the following

4.1 Proposition. Let $\mathcal{E} \xrightarrow{\Gamma} \mathcal{S}$ be a geometric morphism with $\mathcal{S}$ Boolean. Then the resulting modal operators are endomaps of $\Gamma(\Omega_{\mathcal{E}})$ and have the following properties

$$\Box \leq \mathrm{Id}_{\Gamma(\Omega_{\mathcal{E}})} \leq \Diamond$$
$$\Box^2 = \Box\ ,\ \Diamond^2 = \Diamond$$
$$\Diamond \dashv \Box$$

Furthermore

$$\Box K \vee \neg\, \Box K = T$$
$$\Diamond K = \neg \Box \neg K$$

Proof. See Lavendhomme, Lucas and Reyes [7]. This system is called IBM, for Intuitionistic with Boolean modalities.

Example 1. The essential geometric morphism $\mathrm{Set}^I \xrightarrow{\Gamma} \mathrm{Set}$, where I is a set (or discrete small category), gives rise to the "possible worlds" semantics. See Reyes [9].

Example 2. The essential geometric morphism $\mathrm{Set}^{\mathbb{P}^o} \xrightarrow{\Gamma} \mathrm{Set}$, where $\mathbb{P}$ is a preordered set, gives rise to the "possible situations" semantics. See Reyes [9].

Example 3. The geometric morphism $\mathrm{Sh}(\mathbb{B}) \xrightarrow{\Gamma} \mathrm{Set}$, where $\mathbb{B}$ is a Beth tree, gives rise to the "Graft semantics". See Lavendhomme, Lucas and Reyes [7].

The interest of this example is the existence of a completeness theorem for IBM with respect to this "graft" semantics (see the above mentioned paper).

Case B: $\mathcal{E}$ is Boolean.

As we mentioned already, the basic example of this context is the one studied by Ghilardi and Meloni [2]: the essential geometric morphism

$$\mathrm{Set}^{|\mathbb{C}|} \begin{array}{c} \xrightarrow{u_!} \\ \xleftarrow{u^*} \\ \xrightarrow[u_*]{} \end{array} \mathrm{Set}^{\mathbb{C}^o}$$

where u is the obvious functor $u : |\mathbb{C}| \longrightarrow \mathbb{C}$.

We have a MAO couple $(\Diamond,\Box)$ on $\mathrm{Set}^{|\mathbb{C}|}$.

Objects of $\mathrm{Set}^{\mathbb{C}^o}$ are presheaves on $\mathbb{C}$ and those of $\mathrm{Set}^{|\mathbb{C}|}$ are just families indexed by objects of $\mathbb{C}$. The functor u^* applied to a presheaf X just forgets all the transition morphisms and "sees" X as a family. In the following discussion we will identify, by abuse of language, X and u^*X.

We have now the following nice characterization of the resulting modal operators (see Proposition 1.1): if X is a presheaf and F a subfamily of X, then

$\Box F$ is the largest subpresheaf of X contained in F.
$\Diamond F$ is the largest subpresheaf of X containing F.

So $\Box F$ and $\Diamond F$ are the two best "presheaf approximations" of the family F. We can go further and give explicit descriptions of $\Diamond$ and $\Box$ (cf. Ghilardi & Meloni [2]), by recalling that a subpresheaf is simply a subfamily closed under the action of the transition morphisms. So if F is a subfamily of the presheaf X, $\Diamond F$ is the closure under this action and $\Box F$ is the closed part:

$$(M)\qquad \begin{cases} x \in \lozenge F(C) \text{ iff for some } C \xrightarrow{f} C' \text{ in } \mathbb{C} \text{ and } y \in F(C'),\ x = X(f)(y) \\ x \in \square F(C) \text{ iff for all } C' \xrightarrow{f} C \text{ in } \mathbb{C},\ X(f)(x) \in F(C'). \end{cases}$$

Remark. We can think of $\lozenge F$ as the orbits of the elements of F under the action of all the transition morphisms of X. One interesting presheaf is h_A, the representable presheaf associated with an object A of $\mathbb{C}$. Then $\{1_A\}$ is a subfamily containing only 1_A, the identity of A, at stage A and nothing at other stages. (We will write families simply as sets, where each element belongs to its implicit stage). The transition morphisms $h_A(f)$ correspond to right composition by f, and we have by the above caracterization

$$\lozenge(\{1_A\}) = h_A .$$

Here is one nice and natural application of this idea. We know from general topos theory that a morphism $X \xrightarrow{f} Y$ in $Set^{\mathbb{C}^o}$ induces a diagram

$$\mathcal{P}(X) \underset{\forall_f}{\overset{\exists_f}{\rightleftarrows}} \mathcal{P}(Y) \quad (\text{with } f^* : \mathcal{P}(Y) \to \mathcal{P}(X))$$

where $\exists_f \dashv f^* \dashv \forall_f$ (as a functors of the lattices seen as ordered categories).

f^* is calculated "point by point" that is for $A \hookrightarrow Y$, and $C \in |\mathbb{C}|$, we do the computation in Set:

$$(f^*(A))(C) = f^*_C(A(C)) = f^{-1}_C(A(C)) .$$

But we cannot, in general, do the same for $\exists_f$ and $\forall_f$, the problem is that $\{\exists_{f_C}(A(C))\}_{C\in|\mathbb{C}|}$ and $\{\forall_{f_C}(A(C))\}_{C\in|\mathbb{C}|}$ are subfamilies of Y, but they need not be subpresheaves. The solution is simply to apply the modal operator, we just take:

$$\forall_f(A) = \square\{\forall_{f_C}(A(C))\}_{C\in|\mathbb{C}|}$$

and

$$\exists_f(A) = \lozenge\{\exists_{f_C}(A(C))\}_{C\in|\mathbb{C}|}$$

And then the adjunctions $\exists_f \dashv f^* \dashv \forall_f$ follow immediately from the adjunctions $\exists f_C \dashv f^*_C \dashv \forall f_C$ in Sets. In the situation where $\mathcal{E}$ is boolean, the functoriality of $\lozenge$ becomes a very strong condition, as the following result shows. (We concentrate just on the surjective part of the geometric morphism, which is the important one by 3.1 Theorem.)

4.2 Proposition. For $\mathcal{E} \xrightarrow{\Gamma} \mathcal{S}$ where Γ is surjective and $\mathcal{E}$ boolean, $\lozenge$ is functorial iff $\mathcal{S}$ is boolean.

Proof. $\lozenge$ functorial implies that Γ is open by 3.1 Theorem. Thus we have a morphism λ such that $\lambda \dashv \delta$ in the diagram

$$\Omega_{\mathcal{S}} \underset{\delta}{\overset{\lambda}{\rightleftarrows}} \Gamma(\Omega_{\mathcal{E}})$$

where $\Gamma(\Omega_{\mathcal{E}})$ is boolean because $\Omega_{\mathcal{E}}$

Then using the identity $\lambda(a \wedge \delta(b)) = \lambda(a) \wedge b$ in $\Omega_{\mathcal{B}}$ we show that δ preserves implications and therefore negations:

$$\begin{array}{c} C \leq \delta(a) \longrightarrow \delta(b) \\ \hline C \wedge \delta(a) \leq \delta(b) \\ \hline \lambda(C \wedge \delta(a)) \leq b \\ \hline \lambda C \wedge a \leq b \\ \hline \lambda C \leq a \longrightarrow b \\ \hline C \leq \delta(a \longrightarrow b)\,. \end{array}$$

For all a, we have $\delta(a \vee \neg a) = \delta(a) \vee \neg\, \delta(a) = 1$ because $\Gamma(\Omega_{\mathcal{E}})$ is boolean, so by injectivity of δ we have $a \vee \neg a = 1$, which complete the proof that $\mathcal{B}$ is boolean.

4.3 Corollary. For the morphism $Set^{|\mathbb{C}|} \xrightarrow{u_*} Set^{\mathbb{C}^o}$ of the last example, $\Diamond$ is functorial iff $\mathbb{C}$ is a groupoïd.

For the proof, we just use the following well-known lemma:

Lemma. $Set^{\mathbb{C}^o}$ is boolean iff $\mathbb{C}$ is a groupoïd.

Proof. Let $A \xrightarrow{f} B$ be a morphism in $\mathbb{C}$. We show that f has a left inverse.

We take its image $h_A \xrightarrow{h_f} h_B$, by the Yoneda embedding. Since h_B is generated by Id_B, we have this nice property: if $S \hookrightarrow h_B$, then $S = h_B$ or $S = \phi$.

The proof is easy, $Set^{\mathbb{C}^o}$ beeing boolean, the element Id_B of h_B must be either in S or in $\neg S$, but Id_B generates all h_B thus $S = h_B$ or $\neg S = h_B$.

Now take the image of h_f in h_B

$$\begin{array}{ccc} h_A & \xrightarrow{h_f} & h_B \\ & \searrow \quad \nearrow & \\ & \mathrm{Im}\, h_f & \end{array}$$

$\mathrm{Im}\, h_f$ is not empty (we have at least f as an element) so we must have $\mathrm{Im}\, h_f = h_B$. h_f is then surjective and we have $g : B \longrightarrow A$ such that $gf = 1_B$.

Each $f \in Mor(\mathbb{C})$ has a left inverse, so it has an inverse by a standard argument, therefore $\mathbb{C}$ is a groupoïd. In the case $Set^{|\mathbb{C}|} \longrightarrow Set^{\mathbb{C}^o}$, we even have a finer result, implicit in Ghilardi and Meloni [2].

4.4 Proposition. Let $\Diamond$ be the possibility operator associated with $Set^{|\mathbb{C}|} \longrightarrow Set^{\mathbb{C}^o}$. Then the following conditons are equivalent:

(i) $\mathbb{C}$ is a groupoïd

(ii) $\Diamond$ is functorial i.e. $\Diamond f^* = f^* \Diamond$ for all $X \xrightarrow{f} Y$ in $Set^{\mathbb{C}^o}$.

(iii) $\Diamond \pi^* = \pi^* \Diamond$ for all projections $X \times Y \xrightarrow{\pi} X$ in $Set^{\mathbb{C}^o}$.

(iv) $\Diamond d^* = d^* \Diamond$ for all the diagonals $X \xrightarrow{d} X \times X$ in $Set^{\mathbb{C}^o}$.

(v) All the transition morphisms of all objects X in $Set^{\mathbb{C}^o}$ are surjective.

(vi) All the transition morphisms of all objects X in $Set^{\mathbb{C}^o}$ are injective.

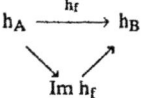

Proof. We already know (i) $\Leftrightarrow$ (ii) . The implications (ii) $\Rightarrow$ (iii) , (ii) $\Rightarrow$ (iv) and (i) $\Rightarrow$ (v) are immediate.

(v) $\Rightarrow$ (i) follows from the analysis of the representables. Let h_A be the one associated with A in $\mathbb{C}$, $h_A(B)$ is the set $\hom_{\mathbb{C}}[B,A]$ and the transition morphism $h_A(B) \xrightarrow{h_A(f)} h_A(C)$ for $C \xrightarrow{f} B$ in $\mathbb{C}$ is just right composition by f. Putting $C = A$ taking an inverse image of $1_A \in h_A(A)$ ($h_A(f)$ being surjective), we obtain a left inverse for f. Each f in $\mathbb{C}$ has a left inverse, therefore it is a groupoïd.

Next we show (iii) $\Rightarrow$ (i) and (iv) $\Rightarrow$ (vi) $\Rightarrow$ (i) and we will have all the equivalences.

(iii) $\Rightarrow$ (i): Suppose that $\Diamond$ commutes with pulling back along projections. We show that every morphism $B \xrightarrow{f} A$ in $\mathbb{C}$ has a left inverse, and so $\mathbb{C}$ is a groupoïd.

Let f be such a morphism. Take the first projection $h_A \times h_B \xrightarrow{\pi} h_A$.

$\{1_A\}$ is a subfamily of h_A , and as we argued earlier $\Diamond\{1_A\} = h_A$ thus $\pi^*\Diamond\{1_A\} = h_A \times h_B$. On the other hand: $\Diamond\pi^*(\{1_A\}) = \Diamond(\{1_A\}\times h_B)$, therefore we must have $h_A \times h_B = \Diamond(\{1_A\}\times h_B)$.

Now take $(f,1_B) \in (h_A\times h_B)(B)$. By the caracterization (M) there are morphisms g and k in $\mathbb{C}$ such that

$$\begin{aligned}(f,1_B) &= (h_A\times h_B)(k)((1_A,g))\\ &= (h_A(k)(1_A),h_B(k)(g))\\ &= (k,g\circ k)\end{aligned}$$

So $k = f$ and $g \circ f = 1_B$ wich gives the desired left inverse for f .

(iv) $\Rightarrow$ (vi) Let X be a presheaf and $f : B \longrightarrow A$ a morphism in $\mathbb{C}$. We show that $X(f)$ is injective.

Take the diagonal $d : X \longrightarrow X \times X$. We have $\Diamond d^* = d^*\Diamond$. Suppose that $X(f)(a) = X(f)(b) = c$ for some $a,b \in X(A)$. We have $(X\times X)(f)(a,b) = (X(f)(a),X(f)(b)) = (c,c) = d(c)$, therefore $d(c) = (c,c) \in \Diamond\{(a,b)\}$ because it is in the orbit of (a,b) .

So we have $c \in d^*\Diamond\{(a,b)\} = \Diamond d^*\{(a,b)\}$, and by (M) it must come from some element in $d^*\{(a,b)\}$ under the action of a transition morphism. But

$$d^*\{(a,b)\} = \{(x,x) | x\in X\} \cap \{(a,b)\}$$

and $d^*\{a,b\}$ beeing non empty, we must have $a = b$.

(vi) $\Rightarrow$ (i): Suppose all the transition morphisms are injective for all preshseaves. We show that every arrow $B \xrightarrow{f} A$ in $\mathbb{C}$ has a right inverse, therefore $\mathbb{C}$ is a groupoïd.

Let X be the colimit of the following diagram in $\mathrm{Set}^{\mathbb{C}^{o}}$

$$\begin{array}{ccc} h_B & \xrightarrow{h_f} & h_A \\ {\scriptstyle h_f}\downarrow & & \\ h_A & & \end{array}$$

$X(C)$ for $C \in \mathbb{C}$ consists of two copies of $h_A(C)$ quotiented by the congruence $\sim$ generated by $h_B(C)$ and the two versions of $h_f(C)$. More precisely, the elements are of the form (x,i) where $x \in h_A(C)$ and $i = 0$ or $i = 1$. The generators of the congruence are the basic relations $(h_f(y),i)\ R\ (h_f(y),j)$ where $y \in h_B(C)$ and $i,j \in \{0,1\}$, or equivalently $(f\circ y,i)\ R\ (f\circ y,j)$. Now take $C = A$. We have

$$X(f)(1_A,0) = (f,0) = (f,1) = X(f)(1_A,1)$$

and X(f) being injective, we must have $(1_A,0) \sim (1_A,1)$. But each non trivial congruence comes from a chain of basic relations. Therefore we must match $(1_A,0)$ with some $(f\circ y,i)$. So there is an $y \in h_B(A)$ such that $f \circ y = 1_A$ giving the desired right inverse.

Remark on connectivity properties

Another major difference between the approaches A and B is given by connectivity properties. A geometric mophism $\mathcal{E} \xrightarrow{\Gamma} \mathcal{S}$ is connected if $\Gamma\Delta = \mathrm{Id}$ (or equivalently $\Pi(1) = 1$ if it is essential).

In this case $Set^{\mathbb{C}^o} \xrightarrow{\Gamma} Set$, Γ is connected iff $\mathbb{C}$ is connected (the connected components of 1 are precisely the connected components of $\mathbb{C}$), whereas for $Set^{|\mathbb{C}|} \xrightarrow{\Gamma} Set^{\mathbb{C}^o}$, it is possible only for the trivial case:

Proposition. Let $u : Set^{|\mathbb{C}|} \xrightarrow{\Gamma} Set^{\mathbb{C}^o}$ be the geometric morphism induced by $|\mathbb{C}| \hookrightarrow \mathbb{C}$. Then the following are equivalent:

(1) Γ is connected

(2) $\mathbb{C} = |\mathbb{C}|$

(3) For all presheaf X, and A,B subfamilies of X,
$\Box(A\cup B) \cap \Diamond A \cap \Diamond B \subseteq \Diamond(A\cap B)$.

Proof. (2) $\Rightarrow$ (3) and (2) $\Rightarrow$ (1) are immediate. (1) $\Rightarrow$ (2): since (1) is equivalent to Δ being full and faithful. (3) $\Rightarrow$ (2): suppose we have $C' \xrightarrow{f} C$ in $\mathbb{C}$. We show that $C = C'$, and $f = \mathrm{Id}$.

Take $X = h_C$, $A = \{1_C\}$ and let B the complement of A. We have $\Diamond(A\cap B) = \varnothing$, $\Box(A\cup B) = h_C$, $\Diamond A = \Diamond\{1_C\} = h_C$ therefore the inequality forces $\Diamond B = \varnothing$, which implies $B = \varnothing$. Now $f \in h_C$, therefore $f \in A$ and so $f = 1_C$.

Examples and counter examples.

(a) For the following essential geometric morphism, we don't have Frobenius property, although it is open as we show by using our results on modality operators.

The functor $u : (\circ \rightrightarrows \circ) \longrightarrow (\circ \longrightarrow \circ)$, sending the two parallel arrows to the unique non-identity arrow of the right category, induces an essential morphism

$$Set^{\circ \rightrightarrows \circ} \begin{array}{c} \xrightarrow{u_!} \\ \xleftarrow{u^*} \\ \xrightarrow[u_*]{} \end{array} Set^{\circ \rightarrow \circ}$$

An object of $Set^{\circ \rightrightarrows \circ}$ is of the form $X = (X_1 \underset{\delta_1}{\overset{\delta_o}{\rightrightarrows}} X_o)$ where X_1 and X_o are sets and δ_o,δ_1 functions. We interpret X as a graph, X_1 and X_o being the sets of arrows and points, respectively, and δ_o,δ_1 the source and target functions. $Set^{\circ \rightrightarrows \circ}$ thus becomes the topos of (irreflexive) graphs. (cf. Lawvere [8]).

In $Set^{\circ \rightarrow \circ}$, an object Y consists of two sets Y_o, Y_1 and a function $\delta : Y_1 \longrightarrow Y_o$. In the same fashion, we will think of Y as a graph whose only arrows are loops, and δ is at the same time the source and target functions. Then the functor $\Delta = u^*$ is simply the "translation":

$$\Delta(Y_1 \xrightarrow{\delta} Y_o) = (Y_1 \underset{\delta}{\overset{\delta}{\rightrightarrows}} Y_o) \in Set^{\circ \rightrightarrows \circ}$$

The functor $\Gamma = u_*$ just extracts the loops out of all arrows of a graph, keeping all the vertices. The functor $\Pi = u_!$ collapses every arrow into a loop, identifying the source and target vertices of every arrow.

The Frobenius property does not hold here: take $X = (\emptyset \longrightarrow 1)$, the "generic vertex" and

$X = (1 \underset{i_1}{\overset{i_o}{\rightrightarrows}} 2)$, the "generic arrow". Then $Y \times \Delta X = (\emptyset \rightrightarrows 2)$ which is a graph with two vertices and no arrows. We have $\Pi(Y{\times}\Delta X) = (\emptyset \longrightarrow 2)$.

On the other hand $\Pi(Y) = (1 \longrightarrow 1)$, the "generic loop", which is also the point in this topos. Therefore $\Pi(Y) \times X = (\emptyset \longrightarrow 1)$.

For openness, we observe that u_* is surjective and by 3.1 theorem, it is open iff the induced possibility operator is functorial. But here $\Diamond$ is the identity operator: let X be an object of $Set^{\circ \rightarrow \circ}$ and $E \hookrightarrow \Delta X$ in $Set^{\circ \rightrightarrows \circ}$. ΔX is a graph consisting only of loops and vertices, so E which must be the same, is already of the form $\Delta S'$ where $S' \hookrightarrow X$. $\Diamond E$ being the smallest $\Delta S'$ where $S' \hookrightarrow X$ and $E \hookrightarrow \Delta S'$ must be E itself. We have $\Diamond = Id$ (and $\Box = Id$) which is functorial.

(b) Here we give an example where we have functoriality of $\Diamond$ but not openness. In fact the "Frobenius morphism" Ψ here is mono but not epi in general.

Let $\mathbb{C}$ be a small category, and Γ be the global sections functor in the geometric morphism:

$$Set^{\mathbb{C}^o} \underset{\Gamma}{\overset{\Delta}{\leftrightarrows}} Set$$

The functor Γ has a left adjoint B (the "codiscrete" or "chaotic" functor) if some representable of $Set^{\mathbb{C}^o}$ has a point (see Lawvere [8]). Suppose we have such a left adjoint, we get an essential morphism

$$Set \;\overset{\Delta}{\underset{B}{\overset{\longrightarrow}{\underset{\longrightarrow}{\overset{\Gamma}{\longleftarrow}}}}}\; Set^{\mathbb{C}^o}$$

The Frobenius morphism Ψ becomes

$$\Psi : \Delta S \times \Delta\Gamma X \simeq \Delta(S{\times}\Gamma X) \longrightarrow \Delta S \times X$$

which is not iso in general. In fact we have $\Psi = \Delta S \times \varepsilon_X$ and Ψ is iso iff the counity $\Delta\Gamma X \longrightarrow X$ is iso, which happens only if X is a sum of points.

Let us look at the particular example where $\mathbb{C} = \Delta_1$, the monoid with three elements: 1, δ_o and δ_1, and identities $\delta_i\delta_j = \delta_i$ for $i,j \in \{0,1\}$. This example is studied in Lawvere [8]. An object X of $Set^{\Delta_1^o}$ corresponds to a reflexive graph: a set X of arrows and two endofunctions $X(\delta_o)$ and $X(\delta_1)$, the source and target functions, respectively. Here the vertices are represented by reflexive arrows: i.e. arrows which are their own source and target. The set of vertices of X is precisely the set of points $\Gamma(X)$ and $\Delta\Gamma(X)$ is the discrete reflexive subgraph of the points of X. The counity $\varepsilon_x : \Delta\Gamma X \hookrightarrow X$ is just the inclusion as a subgraph. It is mono and therefore Ψ is mono. It is not epi except if X is discrete. So (Γ,B) is not open.

On the other hand, $\Diamond$ is functorial as we see either by checking that the image factorization of (Γ,B) is

$$\begin{array}{ccc} \mathrm{Set} & \longrightarrow & \mathrm{Set}^{\Delta_1^{\circ}} \\ & {\scriptstyle \mathrm{Id}}\searrow & \nearrow \\ & \mathrm{Set} & \end{array}$$

where the surjective part is trivially open, or by analyzing ΔS for $S \in \mathrm{Set}$ and remarking that is subgraphs come from subsets of S and therefore $\Diamond = \mathrm{Id}$ which is functorial.

(c) Here is an example where the modality operators are nontrival. Let $\mathbb{C}$ be the category

$$\begin{array}{ccc} \circ\, 0 & \xrightarrow{\alpha_o} & \circ\, 1 \\ {\scriptstyle \alpha_1}\downarrow & & \\ \circ\, 2 & & \end{array}$$

and $\mathbb{D} = (\circ \overset{\delta_o}{\underset{\delta_1}{\rightrightarrows}} \circ)$, and let $u : \mathbb{C} \longrightarrow \mathbb{D}$ be the functor which sends α_o to δ_o and α_1 to δ_1. As before we interpret an object $X = (X_1 \rightrightarrows X_o)$ of $\mathrm{Set}^{\mathbb{D}^o}$ as a graph. In $\mathrm{Set}^{\mathbb{C}^o}$ an object Y is of the type

$$\begin{array}{ccc} Y(0) & \xleftarrow{Y(\alpha_o)} & Y(1) \\ {\scriptstyle Y(\alpha_1)}\uparrow & & \\ Y(2) & & \end{array}$$

In the following discussion, we think of $Y(0)$ as the set of vertices of Y, and $Y(1)$ and $Y(2)$ as sets of loops, where we have two kind of them $x_1 \in Y(1)$ is a continuous loop and $x_2 \in Y(2)$ is a dotted one. The Y will be like this:

Y 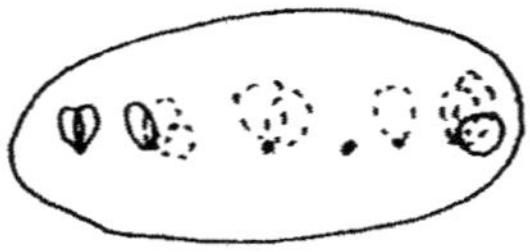

Now if X is a graph $\Delta X(0) = X(0)$ and $\Delta X(1) = \Delta X(2) = X(1)$ so ΔX is of the form

$$\begin{array}{ccc} X(0) & \xleftarrow{X(\delta_o)} & X(1) \\ {\scriptstyle X(\delta_1)}\uparrow & & \\ X(1) & & \end{array}$$

that is, every arrow of X gives rise to (or in fact splits into) two different loops, a continuous one at its source vertex, and a dotted one at the target vertex. So in ΔX, every dotted loop, coming from an arrow, has a continuous counterpart. For example take the following graph:

X 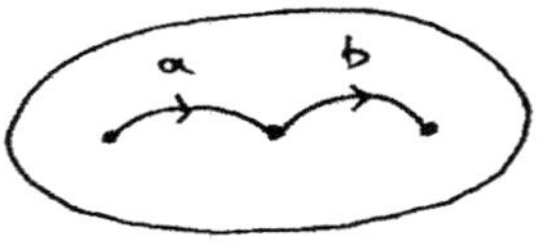

Then ΔX becomes:

ΔX 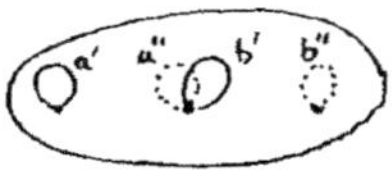

The arrow a has split into loops a' and a" and b into b' and b". Using the definition $\Delta X(1) = \Delta(X)(2) = X(1)$, we could have written $a' = a" = a$, but this somehow is not in the "generalized element" spirit of topos theory. In fact a' and a", being defined at different stages cannot be equal, not even compared (inside $Set^{\mathbb{C}^o}$).

Now if $E \hookrightarrow \Delta X$, $\Diamond E$ (and $\Box E$) must be of the form ΔS for some $S \hookrightarrow X$ therefore $\Diamond E$ is the counterpart completion of E and we have

$$a' \in \Diamond E(1) \Leftrightarrow a" \in \Diamond E(2).$$

In fact, this can be reversed: if $x_1 \in \Delta X(1)$ and $x_2 \in \Delta X(2)$, x_1 and x_2 come from the same arrow in X iff for all $E \hookrightarrow \Delta X$

$$x_1 \in \Diamond E(1) \Leftrightarrow x_2 \in \Diamond E(2).$$

Acknowledgments

We would like to express our debt of gratitude to Bill Lawvere. It is he who suggested that the approaches of Reyes [9] and Ghilardi and Meloni [2] are particular cases of a more general one. Furthermore, several of our examples were inspired by his work on toposes of generalized graphs.

The first author would like to thank the Natural Sciences and Engineering Research Council of Canada for its financial support.

References

[1] Barr, M. and R. Paré (1980), Molecular Toposes, J. Pure and Applied Algebra 17, 127-152.

[2] Ghilardi, S. and G.C. Meloni (1988), Modal and tense predicate logic: models in presheaves and categorical conceptualization, in Categorical Algebra and its Applications (Proceedings, Louvain-La-Neuve 1987), edited by F. Borceux, Lecture Notes in Mathematics 1348, Springer-Verlag.

[3] Hughes, G.E. and M.J. Cresswell (1968/1989), An Introduction to Modal Logic, Routledge, London, New York.

[4] Johnstone, P.T. (1977), Topos theory, Academic Press, London, New York, San Francisco.

[5] Johnstone, P.T. (1980), Open maps of toposes, Manuscripta Mathematica 31, 217-247.

[6] Joyal, A. and M. Tierney (1984), An extension of the Galois theory of Grothendieck, Memoirs AMS 309, American Mathematical Society, Providence, Rhode Island.

[7] Lavendhomme, R., Lucas, Th. and G.E. Reyes (1989), Formal systems for topos-theoretic modalities, Bull. Soc. math. Belgique (Série A), XLI Fascicule 2.

[8] Lawvere, F.W. (1986), Categories of spaces may not be generalized spaces as examplified by directed graphs, Rev. Colombiana Mat. 20.

[9] Reyes, G.E. ( ), A topos-theoretic approach to reference and modality, to appear in Notre Dame Journal of Formal Logic.

This paper is in final form and will not be published elsewhere.

NEGATIVE SETS HAVE EULER CHARACTERISTIC AND DIMENSION

Stephen H. Schanuel
Mathematics Department S.U.N.Y. at Buffalo
106 Diefendorf Hall, Buffalo, NY 14214-3093 (U.S.A.)

1. Where are the negative sets?

Though ill-posed, the question is suggestive; a good answer should complete the diagram

$$\begin{array}{ccc} \mathbb{S} & \hookrightarrow & \mathbb{E}\ ? \\ \downarrow & & \downarrow \\ \mathbb{N} & \hookrightarrow & \mathbb{Z} \end{array}$$

where $\mathbb{S}$ is the category of finite sets; we seek an enlargement $\mathbb{E}$, the isomorphism classes of which should give rise to all integers, rather than just natural numbers. Why is this desirable? The utility of the observation that natural numbers are the isomorphism classes of finite sets derives primarily from the fact that sets can carry structure. For instance, with Euler's function $\varphi(n)$ (the number of integers $0 \le x < n$ relatively prime to n), the equation $\varphi(mn) = \varphi(m)\varphi(n)$ for relatively prime m and n is but a pale reflection of the isomorphism of rings $\mathbb{Z}/mn \cong \mathbb{Z}/m \times \mathbb{Z}/n$. The isomorphism of rings induces an isomorphism of their groups of units, while the equation records only that these groups are isomorphic as sets.

What we seek, then, is a category $\mathbb{E}$ which would allow us to "lift" equations between integers to isomorphisms between objects, because the isomorphism may then preserve some structure relevant to the equation under consideration.

2. A "proof" that there are no negative sets.

We would hope to find $\mathbb{E}$ with finite coproducts and finite products, satisfying at least the distributive laws (that the canonical maps $0 \to A \times 0$ and $A \times B + A \times C \to A \times (B + C)$ are isomorphisms). But already with the coproduct, a difficulty presents itself: $A + B \cong 0$ implies $A \cong B \cong 0$, since to have exactly one map $A + B \to X$ is to have exactly one map $A \to X$ and one map $B \to X$. So the isomorphism classes of objects in a category with coproducts never constitute a nontrivial group under addition. The most we can expect is that the universal map from the set of (isomorphism classes of) objects of $\mathbb{E}$ to a cancellative monoid ($a + b = a + c$ implies $b = c$) will have $\mathbb{Z}$ as codomain.

To clarify our goal, then: $\mathbb{E} \supset \mathbb{S}$ should be a category satisfying distributive laws, and its "rig" of isomorphism classes should have the ring of integers as its reflection into cancellative rigs. A *rig* is a commutative "ring without negatives", that is, having two commutative monoid structures $(0, +)$ and $(1, \times)$ related by the distributive laws $0 = a0$ and $ab + ac = a(b + c)$. Examples abound, *e.g.* $\mathbb{N}$ and $\mathbb{N}/(1 + 1 \sim 1)$, whose modules are commutative monoids and, respectively, sup-semilattices. Other examples include the rig of isomorphism classes of vector bundles on a space, or of finitely generated projective modules over a commutative ring, under direct sum and tensor product. While it is customary to reflect rigs into rings by tensoring with $\mathbb{Z}$, it is by no means always desirable to ignore the extra information contained in the rig. (Steenrod remarked that much of his mathematics came from analyzing the information that others had deliberately discarded by performing such identifications; their "garbage", he called it.) Of most importance for us is the Burnside rig of isomorphism classes of objects in any distributive category (defined below).

3. Euler and counting.

Undeterred by the proof that there are no negative sets, Euler proceeded to find them, in his analysis of the formula $V - E + F = 2$ for the numbers of vertices, edges, and faces in suitable polyhedra. While some later accounts focus on this "Euler characteristic" as a topological invariant, we wish to emphasize instead the irrelevance of topology, and treat the Euler characteristic of a polyhedron rather as a finitely additive measure. Roughly, Betti numbers (ranks of homology groups) depend on how a space is pieced together, but Euler characteristic doesn't; if a space is a disjoint union of two parts, the Euler characteristics add.

Euler's analysis, which demonstrated that in counting suitably "finite" spaces one can get well-defined negative integers, was a revolutionary advance in the idea of cardinal number — perhaps even more important than Cantor's extension to infinite sets, if we judge by the number of areas in mathematics where the impact is pervasive. In any case, it leads us to the desired categories $\mathbb{E}$, which we now describe.

4. Polyhedra and semialgebraic sets.

By a *polyhedron*, (respectively, *semialgebraic set*), we mean a pair n, $P \subset \mathbb{R}^n$, where P is in the boolean algebra generated by subsets of the form $f(x_1, x_2, \ldots, x_n) > 0$, with $f = b + \Sigma a_i x_i$ (or, respectively, f a polynomial.) These are the objects of a category $\mathbb{P}$ (respectively, $\mathbb{SA}$); a map in the category from $P \subset \mathbb{R}^n$ to $Q \subset \mathbb{R}^m$ is any map of sets whose graph (in $\mathbb{R}^{m+n}$) is a polyhedron (respectively, a semialgebraic set.) We'll treat $\mathbb{P}$ in some detail, and just describe the corresponding facts for $\mathbb{SA}$.

A typical polyhedron in the plane might be the union of the open first quadrant and a line, with finitely many triangles, line segments, and points added or deleted. Any bijective map in $\mathbb{P}$, for example from $(0, 1) \cup \{2\} \cup (3, 4)$ to $(0, 2)$ by $f(t) = t$ for $t \in (0, 1)$, $f(2) = 1$, and $f(t) = t - 2$ for $t \in (3, 4)$, is invertible. Perhaps a better name for $\mathbb{P}$ would be $\mathbb{SL}$, for "semilinear", were it not for the usage requiring semilinear maps to be continuous.

The categories $\mathbb{S}$, $\mathbb{P}$, and $\mathbb{SA}$ are distributive, where $\mathbb{E}$ *distributive* means that $\mathbb{E}$ has finite limits, finite coproducts, and $\mathbb{E}^2 \to \mathbb{E}/(1+1)$ by $(A, B) \longmapsto [(A+B) \to (1+1)]$ is an equivalence. From this it follows that $\mathbb{E}$ satisfies the earlier distributive laws and that $\mathbb{E}/B$ is distributive for any object B in $\mathbb{E}$. (The terminology is not yet standard, with reason: Walters, Cockett, and others have shown that a weaker notion, not requiring all finite limits, is also useful in computer science and elsewhere. The strong notion we use here was suggested by lectures of Lawvere.) In addition these categories are *boolean:* every subobject is a summand; or equivalently, subobjects of P in $\mathbb{E}$ are classified by maps $P \to 1+1$. The full subcategory $\mathbb{P}_0 \subset \mathbb{P}$ of *bounded* polyhedra (those which are bounded in $\mathbb{R}^n$) shares all these properties. Our basic task is to calculate the Burnside rigs of these categories and to show their relationship to our original problem. We preface this with some general remarks on Burnside rigs.

5. On Burnside rigs of distributive categories.

The Burnside rig (of isomorphism classes of objects, added by coproduct and multiplied by product) of a distributive category has some special features, the first of which we have already seen.

1) If $a + b = 0$, then $a = b = 0$.

2) If $\Sigma a_i = \Sigma b_j$, then there exist c_{ij} such that $\Sigma_j c_{ij} = a_i$ and $\Sigma_i c_{ij} = b_j$.

3) If a is connected ($a \neq 0$, and $a = b + c$ implies $b = 0$ or $c = 0$), then a is cancellable ($a + x = a + y$ implies $x = y$).

4) 1 is cancellable (whether it is connected or not; in our examples it is connected).

5) If $ab = 1$, then $a = b = 1$.

Properties (1) and (3) follow from (2), which follows easily from the observation that coproduct decompositions $A = \Sigma_I A_i$ correspond to maps $A \to 1 + 1 + \ldots + 1$ (I terms). I don't know what additional properties characterize Burnside rigs of distributive categories.

6. The Burnside rig of bounded polyhedra: the open interval as "–1".

To calculate the Burnside rig $\mathcal{B}(\mathbb{P}_0)$ of the category of bounded polyhedra, it turns out that there is only one basic observation needed. The isomorphism class x of the open interval $(0, 1)$ satisfies $x = 2x + 1$, or perhaps better, $x = x + 1 + x$, because

$$(0, 1) \;=\; (0, {}^1\!/2) \cup \{{}^1\!/2\} \cup ({}^1\!/2, 1)$$

is a coproduct decomposition. (Recall that maps in our category are not required to be continuous.) Thus while $(0, 1)$ is not "minus one", it comes as close as it can: $0 = x + 1$ is impossible, but $x = 2x + 1$ can be achieved.

Hence the canonical map from the free rig on one generator, $\mathbb{N}[X]$, to $\mathcal{B}(\mathbb{P}_0)$, by $X \longmapsto x$, factors through $\mathbb{N}[X]/(X \sim 2X + 1)$:

$$\mathbb{N}[X]/(X \sim 2X + 1) \longrightarrow \mathcal{B}(\mathbb{P}_0),$$

and I claim this is an isomorphism. Surjectivity is easy, because every bounded polyhedron is a disjoint union of open simplices Δ_n°; and $\Delta_n^\circ \cong (0, 1)^n$. The heart of the matter is the injectivity of our map; and for this we need to introduce two invariants, Euler characteristic and dimension.

For any rig R (recall that all rigs are commutative), define the *Euler characteristic*

$$R \xrightarrow{\;\chi\;} E(R)$$

to be the universal map to a rig with additive cancellation. The description of this is well known: $E(R) = R/\!\sim$, where $r \sim s$ if and only if there is a t with $r + t = s + t$. Similarly, define the *dimension*

$$R \xrightarrow{\;\dim\;} D(R)$$

to be the universal map to a rig in which $1 + 1 = 1$ (and hence $x + x = x$). This seems less known: $D(R) = R/\!\sim$, where $r \sim s$ if and only if $r \leq s$ and $s \leq r$, where $r \leq s$ means that "a finite sum of copies of s can swallow r", *i.e.* there exist a natural number n, and $t \in R$, with $r + t = ns$.

Let $R = \mathbb{N}[X]/(X \sim 2X + 1)$; anticipating a bit, we will call this the *rig of geometric cardinalities*. Now, E(R) and D(R) are easy to calculate; we get $E(R) = \mathbb{Z}$, with $\chi(X) = -1$. Equally simple, if less familiar, is D(R): it is

$$D(R) = \{0 = d^{-\infty},\ 1 = d^0,\ d^1, d^2,\ \ldots\}$$

with $d^i d^j = d^{i+j}$ and $d^i + d^j = d^{\max(i,j)}$. The exponential notation is in keeping with the idea that multiplying polyhedra adds dimensions, while adding gives the maximum of the two dimensions.

To complete the proof that $R = \mathbb{N}[X]/(X \sim 2X + 1) \longrightarrow \mathcal{B}(\mathbb{P}_0)$ is an isomorphism, we need only define a rig homomorphism

$$(\overline{\chi}, \overline{\dim}):\ \mathcal{B}(\mathbb{P}_0) \longrightarrow \mathbb{Z} \times D(R),$$

check that the composite of this with $R \longrightarrow \mathcal{B}(\mathbb{P}_0)$ is $(\chi, \dim)$, and show that $(\chi, \dim)$ is injective. This last is a simple induction, after noting that $(\chi, \dim)(f(X)) = (f(-1), \text{degree } f)$. So the definitions of $\overline{\chi}$ and $\overline{\dim}$ need attention. One defines these, at an object P, by writing P as a disjoint union of atoms A in the boolean algebra given by a hyperplane decomposition of space, $P = \cup_{\text{atoms } A \subset P} A$, and then setting

$$\overline{\chi}(P) = \Sigma(-1)^{\text{gdim}(A)} \qquad \text{and} \qquad \overline{\dim}(P) = d^{\sup \text{gdim}(A)},$$

where gdim(A) is the ordinary geometric dimension of the atom A. It's easy to check that adding a hyperplane leaves these quantities unchanged, and then that they're isomorphism-invariant.

Summing up: a geometric cardinality can be identified with an equvalence class of polynomials with natural number coeficients — two such being equivalent if they have the same degree and the same value at -1 — or with an isomorphism class of bounded polyhedra. As we'll see shortly, it is also an isomorphism class of semialgebraic sets, or of finitely subanalytic sets, and is an equivalence class of constructible sets (the boolean closure of the class of algebraic sets in $\mathbb{C}^n$).

7. The Burnside rig of unbounded polyhedra.

The major themes of this paper can be understood without the corresponding (and somewhat more cumbersome) calculation for $\mathbb{P}$, the category of all polyhedra. Nevertheless, since $\mathbb{P}$ is of interest in connection with linear programming and related matters, we give a sketch of the necessary changes to convert the calculation for $\mathbb{P}_0$ to that for $\mathbb{P}$. There are two generators: $x = (0, 1)$, as before, and $y = (0, \infty)$. We easily get three relations: $x = 2x + 1$, $y = x + 1 + y$ because $(0, \infty) = (0, 1) \cup \{1\} \cup (1, \infty)$, and $y^2 = 2y^2 + y$ because $(0, \infty)^2 = \{(r, s) \mid r < s\} \cup \{(r,s) \mid r = s\} \cup \{(r, s) \mid r > s\}$. So the rig $\overline{R}$ presented by these maps to $\mathcal{B}(\mathbb{P})$:

$$\overline{R} = \mathbb{N}[X, Y]/(X \sim 2X + 1, Y \sim X + 1 + Y, Y^2 \sim 2Y^2 + Y) \longrightarrow \mathcal{B}(\mathbb{P}),$$

and we claim this is an isomorphism. As before, we calculate:

$$E(\overline{R}) = \mathbb{Z} \times \mathbb{Z}, \text{ with } \chi(X) = (-1, -1) \text{ and } \chi(Y) = (-1, 0).$$

(The first relation gives, after cancellation, $0 = X + 1$, so we get $\mathbb{Z}$ as subrig; then the second relation becomes vacuous, while the third gives that $Y + 1$ is idempotent in $E(\overline{R})$.) Again, $D(\overline{R})$ is less familiar. An element of $D(\overline{R})$ is a finitely generated (hence finite) order-ideal in the partially ordered set of monomials X^iY^j, ordered as a monoid with $1 < X < Y$; so $X^iY^j \leq X^pY^q$ means $j \leq q$ and $i+j \leq p+q$. These order ideals are multiplied by multiplying elementwise and down-closing, and are added by union. (Note that $D(\overline{R})$ could have been given by a similar description, using the poset $\{1 < X < X^2 < ....\}$.). It is worth noting that both $D(\overline{R})$ and $D(R)$ have multiplicative cancellation: for $a \neq 0$, $ab = ac$ implies $b = c$. This will show that any (bounded) polyhedron with cancellable (bounded) Euler characteristic is multiplicatively cancellable.

Checking the surjectivity of $\overline{R} \to \mathcal{B}(\mathbb{P})$ by $X \longmapsto (0, 1) = x$ and $Y \longmapsto (0, \infty) = y$ is harder than before, but not much. We must show that every polyhedron P is isomorphic to a sum of monomials $(0,1)^i \times (0, \infty)^j$. For this, we show that P can be decomposed into pieces each linearly (rather than just piecewise-linearly) isomorphic to $\Delta_i^\circ \times (0, \infty)^j$. This is done by induction on the geometric dimension of P, and P can be supposed to be an atom in a decomposition of $\mathbb{R}^n$ by hyperplanes; but it is important to first ensure that the family of hyperplanes includes at least n that are independent, *i.e.* the linear functionals f in the equations $f(x) = c$ are linearly independent. By induction, each face of the atom P can be suitably decomposed; and then one decomposes P by choosing any point p in P and taking the open *truncated cones* consisting of the points $tp + (1-t)x$ where $0 < t < 1$ and x ranges over any of the parts into which the faces of P have been decomposed. The truncated cone on $\Delta_i^\circ \times (0, \infty)^j$ is $\Delta_{i+1}^\circ \times (0, \infty)^j$. These cones do not exhaust P, but what's left is an infinite closed cone with vertex p; and decomposing its (bounded) intersection with a suitable hyperplane into open simplices cuts this cone into a sum of powers of y. The whole proof is thus quite parallel to the proof that bounded polyhedra can be decomposed into disjoint open simplices by decomposing the boundary, picking a point inside, and "coning"; the only new ingredient is that in the unbounded case there is still a cone left over, which one proves can also be decomposed as a sum of parts linearly isomorphic to monomials in y.

To map $\mathcal{B}(\mathbb{P})$ to $E(\overline{R}) \times D(\overline{R})$ turns out to be a bit easier than one might expect; the clue is that each atom in a decomposition of $\mathbb{R}^n$ (by at least n independent hyperplanes, as before) is in fact (polyhedrally isomorphic to) a monomial x^iy^j. The sum $i+j$ is just the geometric dimension of the atom, while j is that of the cone in $\mathbb{R}^n$ obtained by intersecting the closed half spaces given by the hyperplanes through the origin parallel to the faces of P. It's easier not to prove this at this stage; for now, it shows how to define the map from $\mathcal{B}(\mathbb{P})$ to $E(\overline{R}) \times D(\overline{R})$. The proof that this map is well-defined, *i.e.* unchanged by adding another hyperplane and isomorphism-invariant, is straightforward; and the rest goes just as before, with just a little more care in the inductive argument to show that $\overline{R} \to E(\overline{R}) \times D(\overline{R})$ is injective.

A geometric description of the "Euler characteristic" $\chi(P) = (m, n)$ and "dimension" $\dim(P) = F \subset \{X^iY^j\}$ is now not difficult. First, m is the "expected" Euler characteristic, since $x = (0, 1)$ and $y = (0, \infty)$ are "alike", except that there is no piecewise-linear isomorphism between them. (They are semi-algebraically isomorphic, by $t \longmapsto t^{-1} - 1$. This will be useful later.) Second, n is the "bounded Euler characteristic" of $P \subset \mathbb{R}^n$, *i.e.* $\chi(P \cap C)$ for any sufficiently large closed cube $C = [-B, B]^n$. The dimension of $P \subset \mathbb{R}^n$ is just the set of monomials $x^i y^j$ which are subobjects of P (linearly, if you want.) For example, for a geometrically 2-dimensional polyhedron, the possible dimensions, in increasing order, are the order ideals generated by x^2, by x^2 and y, by xy, and by y^2, exemplified by, respectively, a 2-simplex, the union of a 2-simplex and a ray, an infinite strip bounded by parallel lines, and the plane.

8. Related examples.

Modulo well-known facts, it is easy to check that the Burnside rig of the category $\mathbb{SA}$ of semialgebraic sets is the same as that for $\mathbb{P}_0$, the rig R of geometric cardinalities. The main ingredient is Hironaka's theorem of semialgebraic triangulability of semialgebraic sets.

Thus the inclusions of distributive categories

$$\mathbb{S} \hookrightarrow \mathbb{P}_0 \hookrightarrow \mathbb{P} \hookrightarrow \mathbb{SA}$$

(the first two of which are full) give, on passage to Burnside rigs,

$$\mathbb{N} \longrightarrow R \longrightarrow \overline{R} \longrightarrow R,$$

exhibiting, geometrically, R as a retract of $\overline{R}$ (by $y \longmapsto x$).

The geometric interpretation of the injectivity, for each of these rigs $A = \mathbb{N}, R$, or $\overline{R}$, of $A \to E(A) \times D(A)$, should be clear. For instance, for polyhedra, it says: if P and Q are polyhedra which are "cancellation equivalent" (*i.e.* $P + T \cong Q + T$ for some polyhedron T) and "comparable" (*i.e.* $P \leq nQ$ and $Q \leq mP$ for some natural numbers m, n), then $P \cong Q$. I do not know a proof of this, in any of the categories $\mathbb{P}_0$, $\mathbb{P}$, $\mathbb{SA}$, which does not use essentially the entire calculation sketched above! A similar remark applies to proving that $2P \cong 2Q$ implies $P \cong Q$. For these, it might be helpful if one could find a simpler characterization of those rigs A for which $A \to E(A) \times D(A)$ is injective; that is, those in which $a + t = b + t$ & $a + s = nb$ & $b + r = ma$ implies $a = b$.

One trivial generalization: everything said about $\mathbb{P}_0$ and $\mathbb{P}$ remains valid if the reals are replaced by any ordered field; for instance, the Burnside rig is unchanged. (One should note that not every order-convex subset of the line is a polyhedron; for instance with $\mathbb{Q}$ as the field, $(0, \sqrt{2})$ is not polyhedral, since it is not defined by finitely many inequalities with rational coefficients.) More interesting is the category $\mathbb{FA}$ of finitely subanalytic sets, which van den Dries has shown shares enough of the properties of $\mathbb{SA}$ that our calculation also gives R, the rig of geometric cardinalities, as $\mathcal{B}(\mathbb{FA})$. All our categories satisfy the *axiom of choice:* every epimorphism has a section; but $\mathbb{FA}$ and $\mathbb{SA}$ satisfy the stronger *generic triviality* theorem: every map $A \to B$ is isomorphic to a coproduct of product projections, $\sum B_i \times F_i \to \sum B_i$. This is false in $\mathbb{P}$, and even in $\mathbb{P}_0$, as the example of the map from the open triangle with vertices (0, 0), (0, 1), and (1, 1) to the interval (0, 1) by projection on the first coordinate demonstrates. However, any map $A \to B$ in $\mathbb{P}_0$ is a coproduct of maps $A_i \to B_i$ in which the isomorphism class of the fiber is constant; and the

Euler characteristic behaves as if such a map were a product: $\chi(A_i) = \chi(B_i)\chi(\text{fiber})$. This observation suggests a reduced Burnside rig in our next example.

Genuinely different is the category $\mathbb{CS}$ of *constructible sets:* an object is a subset of $\mathbb{C}^n$ in the boolean algebra generated by zero-sets of polynomials; a map is a function with a constructible graph. Essentially by construction, this category is distributive and boolean, though without axiom of choice; but its Burnside rig is complicated. A reduced Burnside rig $\mathcal{B}_{red}(\mathbb{CS})$ and rig homomorphism $\mathcal{B}(\mathbb{CS}) \to \mathcal{B}_{red}(\mathbb{CS})$ are defined to be universal for rig homomorphisms π with domain $\mathcal{B}(\mathbb{CS})$ satisfying: whenever $A \to B$ is in $\mathbb{CS}$ and $\pi(\text{fiber})$ is constant, $\pi(A) = \pi(B)\pi(\text{fiber})$. This gives rise to the rig of geometric cardinalities again! The generator X is the twice-punctured complex plane $\mathbb{C}\setminus\{0, 1\}$. To get the desired relation $\pi(X) = 2\pi(X) + 1$, consider $Y = \mathbb{C}\setminus\{0, 1, -1\}$, and note that $X = Y + 1$ while all fibers of the squaring map $Y \to X$ have two points. The proof that X generates is inductive, projecting the constructible set on a coordinate hyperplane; and the injectivity of the map from R to our reduced rig is proved by using the forgetful functor from $\mathbb{CS}$ to $\mathbb{SA}$, viewing $\mathbb{C}^n$ as $\mathbb{R}^{2n}$. (This calculation is related to work of van den Dries, Marker, and Martin, "Definable equivalence relations on algebraically closed fields", *J. Symbolic Logic* 54 (1989) 928-935.) It is suggestive that in this example as in the earlier ones, the "minus 1" object comes from the basic bipointed parameter object for homotopies: $\{0, 1\} \to \mathbb{C}$ for algebraic geometry, respectively $\{0, 1\} \to [0, 1]$ for topology, by deleting the two marked points.

Applications of these ideas to geometry will have to be treated on another occasion; some work by Beifang Chen on curvature measures along these lines will appear in *Advances in Mathematics.* Also postponed are the analysis of colimits and of the relation of a boolean distributive category to its "Gaeta topos", following ideas of Lawvere which have exerted a continuing influence on the shape of the work described here.

I wish to express my gratitude to the organizers of the conference, and particularly to Aurelio Carboni and many colleagues at the University of Milan, for their warm and generous hospitality.

This paper is in final form and will not be published elsewhere.

MODULAR CATEGORIES

Michel Thiébaud
Collège de Staël
1227 Carouge, Genève, Switzerland

Introduction

The notion of a modular category was defined by A. Carboni in [1], where he showed that those categories are precisely the slices of left exact additive categories, the so called affine categories.

We shall show here that, of the two conditions characterizing a modular category, the second one is in some sense the converse of the first, and that these two conditions are particular cases of a very general condition which can be imposed on an arbitrary adjunction. A careful study of this will give us two characterizations of a modular category. The first one is an elementary one and is very reminiscent of the characterization of a distributive category. The second one, which is to be compared with the characterization of [1], asks for the equivalence of slice categories and allows us to show that the fibration of pointed objects in a modular category is a constant (additive) fibration.

We are not assuming in this paper the existence of a terminal object in the definition of a modular category and so, in view of [1], an affine category will be a modular category with a terminal object.

1. Modular categories

The following definition is the definition of [1] but for the existence of a terminal object.

Definition 1.1. A *modular category* is a category $\mathbf{E}$ with pullbacks and finite coproducts such that :
(1) for every object U of $\mathbf{E}$ and for every map $f : X \to Z$ in $\mathbf{E}/U$ the canonical map

$$\begin{pmatrix}(i_X, f)\\ i_Y \times Z\end{pmatrix} : X + (Y \times Z) \to (X + Y) \times Z$$

is invertible for every object Y in $\mathbf{E}/U$;

(2) for every object U of $\mathbf{E}$ and for every map $f : X \to Y$ in $\mathbf{E}/U$ the square

$$\begin{array}{ccc} X & \xrightarrow{i_X} & 1+X \\ {\scriptstyle f}\downarrow & & \downarrow{\scriptstyle 1+f} \\ Y & \xrightarrow[i_Y]{} & 1+Y \end{array}$$

is a pullback, where 1 denotes the terminal object in $\mathbf{E}/U$.

Expressing those two conditions in $\mathbf{E}$ gives :

Definition 1.2. A *modular category* is a category $\mathbf{E}$ with pullbacks and finite coproducts such that :
(1) for every pair of commutative diagrams in $\mathbf{E}$:

$$\begin{array}{ccccc} W & \xrightarrow{k} & Z & \xleftarrow{f} & X \\ {\scriptstyle h}\downarrow & & \downarrow{\scriptstyle z} & \swarrow{\scriptstyle x} & \\ Y & \xrightarrow[y]{} & U & & \end{array} \qquad \begin{array}{ccccc} X+W & \xrightarrow{\binom{f}{k}} & Z & \xleftarrow{f} & X \\ {\scriptstyle X+h}\downarrow & & \downarrow{\scriptstyle z} & \swarrow{\scriptstyle x} & \\ X+Y & \xrightarrow[\binom{x}{y}]{} & U & & \end{array}$$

if the left-hand square is a pullback, then so is the right-hand square;
(2) for every object U and every map $f : X \to Y$ in $\mathbf{E}$ the square

$$\begin{array}{ccc} X & \xrightarrow{i_X} & U+X \\ {\scriptstyle f}\downarrow & & \downarrow{\scriptstyle U+f} \\ Y & \xrightarrow[i_Y]{} & U+Y \end{array}$$

is a pullback.

We shall now show that condition (2) in this definition is in fact the converse of condition (1) in the following sense :

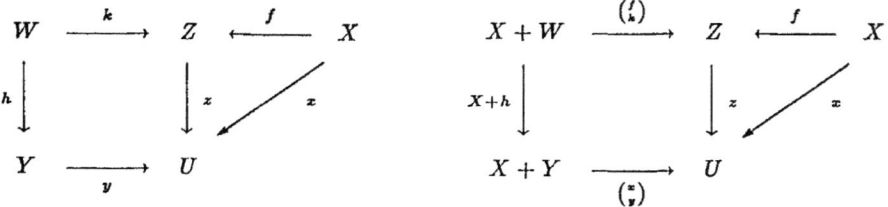

Lemma 1.3. *The condition (2) in the above definition is equivalent to the following condition :*
(2) If the right-hand square in (1) is a pullback then so is the left-hand square.*

Proof. (2) $\Rightarrow$ (2*) : Let the square

$$\begin{array}{ccc} X+W & \xrightarrow{\binom{f}{k}} & Z \\ {\scriptstyle X+h}\downarrow & & \downarrow{\scriptstyle z} \\ X+Y & \xrightarrow[\binom{x}{y}]{} & U \end{array}$$

be a pullback, and consider the following diagram :

$$\begin{array}{ccccc} W & \xrightarrow{i_W} & X+W & \xrightarrow{\binom{f}{k}} & Z \\ {\scriptstyle h}\downarrow & & \downarrow{\scriptstyle X+h} & & \downarrow{\scriptstyle z} \\ Y & \xrightarrow[i_Y]{} & X+Y & \xrightarrow[\binom{x}{y}]{} & U \end{array}$$

The condition (2), which says that the left-hand square is a pullback, implies that the composite square, which is the left-hand square in (1), is a pullback.
(2*) $\Rightarrow$ (2) : Apply condition (2*) to the diagram

$$\begin{array}{ccccc} X+W & = & X+W & \xleftarrow{i_X} & X \\ {\scriptstyle X+h}\downarrow & & \downarrow{\scriptstyle X+h} & \swarrow{\scriptstyle i_X} & \\ X+Y & = & X+Y & & \end{array}$$

□

So we have :

Proposition 1.4. *A category* **E** *with pullbacks and finite coproducts is modular iff it satisfies the following condition :*
(M) *For every pair of commutative diagrams as in* (1) *in Definition 1.2, the left-hand square is a pullback iff the right-hand square is a pullback.* □

Our aim now is to show that this rather peculiar condition (M) is in fact a very natural one to impose on a general adjunction, and we shall develop in the next section the necessary formalism for this.

2. Cartesian adjunctions

Let **A** and **B** be two categories with pullbacks and let

$$\mathbf{A} \underset{U}{\overset{F}{\rightleftarrows}} \mathbf{B}\ , \quad F \dashv U\ ,$$

be an adjunction between them, with unit η and counit ε. Let, for every object A of **A**,

$$\mathbf{A}/A \underset{U_A}{\overset{F_A}{\rightleftarrows}} \mathbf{B}/FA$$

be the induced adjunction between the corresponding slice categories, with

$$F_A(C \xrightarrow{a} A) = (FC \xrightarrow{Fa} FA)$$

$$U_A(D \xrightarrow{b} FA) = (P \xrightarrow{p} A)\ ,$$

where

$$\begin{array}{ccc} P & \xrightarrow{q} & UD \\ {\scriptstyle p}\downarrow & & \downarrow{\scriptstyle Ub} \\ A & \xrightarrow[\eta A]{} & UFA \end{array}$$

is a pullback.

We shall now consider pairs of commutative squares of the form

$$\begin{array}{ccc} C & \xrightarrow{g} & UD \\ {\scriptstyle a}\downarrow & (P) & \downarrow{\scriptstyle Ub} \\ A & \xrightarrow[f]{} & UB \end{array} \qquad\qquad \begin{array}{ccc} FC & \xrightarrow{\bar g} & D \\ {\scriptstyle Fa}\downarrow & (Q) & \downarrow{\scriptstyle b} \\ FA & \xrightarrow[\bar f]{} & B \end{array}$$

and, as a particular case, pairs of squares of the form

$$\begin{array}{ccc} C & \xrightarrow{g} & UD \\ {\scriptstyle a}\downarrow & (P^*) & \downarrow{\scriptstyle Ub} \\ A & \xrightarrow[\eta A]{} & UFA \end{array} \qquad\qquad \begin{array}{ccc} FC & \xrightarrow{\bar g} & D \\ {\scriptstyle Fa}\downarrow & (Q^*) & \downarrow{\scriptstyle b} \\ FA & = & FA \end{array}$$

where $f, \bar f$ and $g, \bar g$ correspond to each other via the adjunction $F \dashv U$. The left-hand squares are in $\mathbf{A}$, the right-hand squares in $\mathbf{B}$.

Proposition 2.1. *The following conditions are equivalent :*
(1) *For all pairs of squares as above, if (P) is a pullback then so is (Q).*
(2) *For all pairs of squares as above, if (P^*) is a pullback then so is (Q^*).*
(3) *For all A in $\mathbf{A}$ the counit of the adjunction $F_A \dashv U_A$ is an isomorphism.*
(4) *The counit ε of the adjunction $F \dashv U$ is cartesian (i.e. the squares expressing its naturality are pullbacks) and F preserves pullbacks of the form (P).*

Proof. (1) $\Rightarrow$ (2) is obvious.
(2) $\Rightarrow$ (1) : Suppose that (P) is a pullback and let

$$\begin{array}{ccc} K & \xrightarrow{q} & D \\ {\scriptstyle p}\downarrow & (*) & \downarrow{\scriptstyle b} \\ FA & \xrightarrow[\bar f]{} & B \end{array}$$

be a pullback. Applying U to it gives us a pullback through which (P) factors :

$$\begin{array}{ccccc} & & \overset{g}{\overbrace{\qquad\qquad\qquad\qquad}} & & \\ C & \xrightarrow{k} & UK & \xrightarrow{Uq} & UD \\ {\scriptstyle a}\downarrow & & {\scriptstyle Up}\downarrow & & \downarrow{\scriptstyle Ub} \\ A & \xrightarrow[\eta A]{} & UFA & \xrightarrow[U\bar{f}]{} & UB \\ & & \underset{f}{\underbrace{\qquad\qquad\qquad\qquad}} & & \end{array}$$

The left-hand square is then a pullback of the form (P^*). Applying condition (2) to it and composing with $(*)$ shows that (Q), which is the composite square in the following diagram, is a pullback :

$$\begin{array}{ccccc} FC & \xrightarrow{\bar{k}} & K & \xrightarrow{q} & D. \\ {\scriptstyle Fa}\downarrow & & {\scriptstyle p}\downarrow & (*) & \downarrow{\scriptstyle b} \\ FA & = & FA & \xrightarrow[\bar{f}]{} & B \end{array}$$

(2) $\Leftrightarrow$ (3) : Let $a : C \to A$ be an object in $\mathbf{A}/A$ and $b : D \to FA$ an object in $\mathbf{B}/FA$. Then (P^*) is a pullback iff the map from a to $U_A b$ induced by the pair of maps (a, g) is an isomorphism, and (Q^*) is a pullback iff the corresponding map $\bar{g}$ from $F_A a$ to b is an isomorphism. The result follows then from a general argument on adjunctions.
(1) $\Rightarrow$ (4) : Applying condition (1) to pullbacks of the form

$$\begin{array}{ccc} UD & = & UD \\ {\scriptstyle Ub}\downarrow & & \downarrow{\scriptstyle Ub} \\ UB & = & UB \end{array}$$

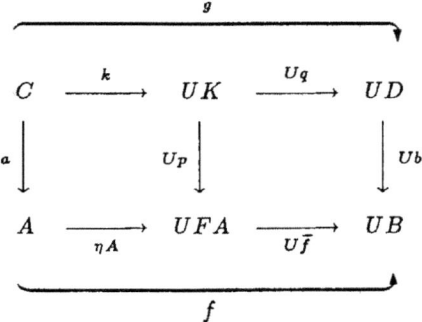

shows that ε is cartesian. Suppose now that (P) is a pullback. Applying F to it and composing with ε gives :

$$\begin{array}{ccccc} & & \overset{\bar{g}}{\overbrace{\hspace{12em}}} & & \\ FC & \xrightarrow{Fg} & FUD & \xrightarrow{\varepsilon D} & D \\ {\scriptstyle Fa}\downarrow & & {\scriptstyle FUB}\downarrow & & \downarrow{\scriptstyle b} \\ FA & \xrightarrow[Ff]{} & FUB & \xrightarrow[\varepsilon B]{} & B \\ & & \underset{\bar{f}}{\underbrace{\hspace{12em}}} & & \end{array}$$

The right-hand square is a pullback and condition (1) implies that the composite square is a pullback. And so is the left-hand square.
(4) $\Rightarrow$ (1) : In the above diagram, if the left-hand square and the right-hand square are pullbacks, then so is their composite, which is (Q). □

Proposition 2.2. *The following conditions are equivalent :*
(1) *For all pairs of squares (P), (Q), if (Q) is a pullback then so is (P).*
(2) *For all pairs of squares (P^*), (Q^*), if (Q^*) is a pullback then so is (P^*).*
(3) *For all A in $\mathbf{A}$ the unit of the adjunction $F_A \dashv U_A$ is an isomorphism.*
(4) *The unit η of the adjunction $F \dashv U$ is cartesian.*

Proof. (1) $\Rightarrow$ (2) is obvious.
(2) $\Rightarrow$ (4) : Applying condition (2) to pullbacks of the form

$$\begin{array}{ccc} FC & = & FC \\ {\scriptstyle Fa}\downarrow & & \downarrow{\scriptstyle Fa} \\ FA & = & FA \end{array}$$

shows that η is cartesian.
(4) $\Leftrightarrow$ (1) : Suppose that (Q) is a pullback. Applying U to it gives us a pullback. The following diagram then shows that η is cartesian iff (P), which is the composite square, is a pullback :

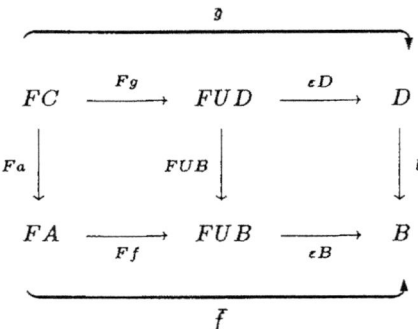

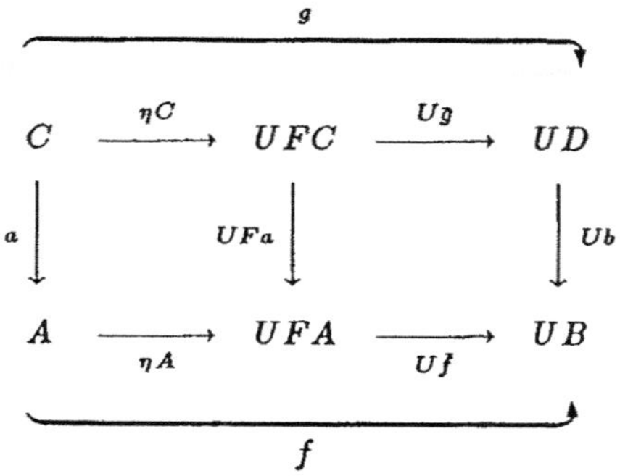

(2) $\Leftrightarrow$ (3) : This follows from a general argument on adjunctions, as in the proof of the equivalence (2) $\Leftrightarrow$ (3) in Proposition 2.1.

Putting together Proposition 2.1 and Proposition 2.2 and expressing the corresponding conditions more succintly, we have :

Proposition 2.3. *The following conditions are equivalent :*
(1) (P) *is a pullback iff* (Q) *is a pullback.*
(2) (P^*) *is a pullback iff* (Q^*) *is a pullback.*
(3) *For all* A *in* $\mathbf{A}$ *the adjunction* $F_A \dashv U_A$ *is an equivalence.*
(4) η *is cartesian,* ε *is cartesian, and* F *preserves all pullbacks.*

Proof. (1) $\Leftrightarrow$ (2), (1) $\Leftrightarrow$ (3) and (4) $\Rightarrow$ (1) follow directly from the two preceeding propositions.
(1) $\Rightarrow$ (4) : It suffices to show that condition (1) of Proposition 2.1 and condition (1) of Proposition 2.2 together imply that F preserves all pullbacks. Let

$$
\begin{array}{ccc}
C' & \xrightarrow{\ g\ } & C \\
{\scriptstyle a'}\downarrow & & \downarrow{\scriptstyle a} \\
A' & \xrightarrow[\ f\ ]{} & A
\end{array}
$$

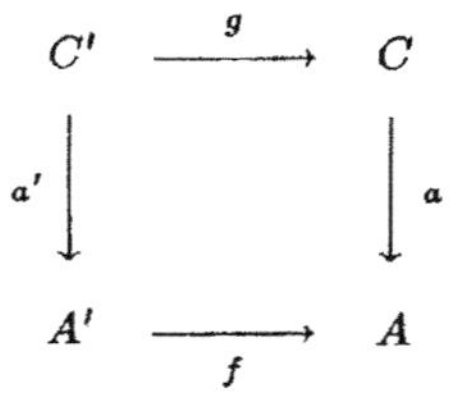

be a pullback. Composing with η, which is cartesian by the implication (1) $\Rightarrow$ (4) of

Proposition 2.2, and applying F gives a diagram

$$\begin{array}{ccccc} FC' & \xrightarrow{Fg} & FC & \xrightarrow{F\eta C} & FUFC \\ {\scriptstyle Fa'}\downarrow & & {\scriptstyle Fa}\downarrow & & \downarrow{\scriptstyle FUFa} \\ FA' & \xrightarrow[Ff]{} & FA & \xrightarrow[F\eta A]{} & FUFA \end{array}$$

in which, by the implication (1) $\Rightarrow$ (4) of Proposition 2.1, the right-hand square and the composite square are pullbacks, each one being the image by F of a pullback of the form (P). And so the left-hand square is a pullback. □

It should be mentioned here that the implication (4) $\Rightarrow$ (3) in the proposition above has also been proved in [3].

Remark 2.4. In each one of the last three propositions the equivalence (2) $\Leftrightarrow$ (3) is obviously true when the object A is fixed in both conditions (2) and (3).

Definition 2.5. An adjunction between two categories with pullbacks is called *cartesian* when it satisfies the equivalent conditions of Proposition 2.3.

Remark 2.6. The notion of cartesian adjunction can also be expressed in terms of fibrations as follows :

Let $d_1 : \mathbf{B}^2 \to \mathbf{B}$ be the canonical (or fundamental) fibration associated to $\mathbf{B}$ and let $F^*(\mathbf{B}) \to \mathbf{A}$ be the inverse image of d_1 along F, i.e. $F^*(\mathbf{B})$ is defined by the following pullback

$$\begin{array}{ccc} F^*(\mathbf{B}) & \longrightarrow & \mathbf{B}^2 \\ \downarrow & & \downarrow{\scriptstyle d_1} \\ \mathbf{A} & \xrightarrow[F]{} & \mathbf{B} \end{array}$$

Then the adjunction $F \dashv U$ is cartesian iff it induces an equivalence

$$\mathbf{A}^2 \rightleftarrows F^*(\mathbf{B})$$

of fibrations over $\mathbf{A}$.

Let us now assume furthermore that **A** and **B** have a terminal object (i.e. that **A** and **B** are left exact), and let us consider pairs of commutative squares of the form

$$\begin{array}{ccc} C & \xrightarrow{g} & UD \\ {\scriptstyle t}\downarrow & (P_1) & \downarrow{\scriptstyle Ub} \\ 1 & \xrightarrow[\eta 1]{} & UF1 \end{array} \qquad\qquad \begin{array}{ccc} FC & \xrightarrow{\bar{g}} & D \\ {\scriptstyle Ft}\downarrow & (Q_1) & \downarrow{\scriptstyle b} \\ F1 & = & F1 \end{array}$$

where t denotes the unique map from C to 1.

Proposition 2.7. *The following conditions are equivalent :*
(1) If (Q_1) is a pullback then so is (P_1).
(2) The unit of the adjunction $F_1 \dashv U_1$ is an isomorphism.
(3) η is cartesian.

Proof. (2) $\Leftrightarrow$ (3) : It is easy to see that η is cartesian iff all squares of the form

$$\begin{array}{ccc} C & \xrightarrow{\eta C} & UFC \\ {\scriptstyle t}\downarrow & & \downarrow{\scriptstyle UFt} \\ 1 & \xrightarrow[\eta 1]{} & UF1 \end{array}$$

are cartesian. The result follows then from the equivalence (2) $\Leftrightarrow$ (3) of Proposition 2.2 and Remark 2.4, for $A = 1$.
(1) $\Leftrightarrow$ (3) : Suppose that (Q_1) is a pullback. Applying U to it gives us a pullback. The following diagram then shows that η is cartesian iff (P_1), which is the composite square, is a pullback :

$$\begin{array}{ccccc} & & \xrightarrow{\qquad\qquad g \qquad\qquad} & & \\ C & \xrightarrow{\eta C} & UFC & \xrightarrow{U\bar{g}} & UD \\ {\scriptstyle t}\downarrow & & {\scriptstyle UFt}\downarrow & & \downarrow{\scriptstyle Ub} \\ 1 & \xrightarrow[\eta 1]{} & UF1 & = & UF1 \\ & & \xrightarrow[\qquad\qquad \eta 1 \qquad\qquad]{} & & \end{array}$$

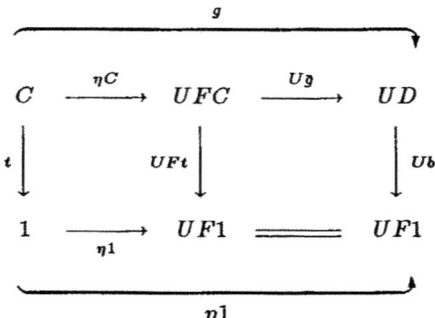

A similar proposition involving the counit of the adjunction $F_1 \dashv U_1$ does not exist, but we have :

Proposition 2.8. *The following conditions are equivalent :*
(1) (P_1) is a pullback iff (Q_1) is a pullback.
(2) The adjunction $F_1 \dashv U_1$ is an equivalence.
(3) The adjunction $F \dashv U$ is cartesian.

Proof. (1) $\Leftrightarrow$ (2) : By the equivalence (2) $\Leftrightarrow$ (3) of Proposition 2.3 and Remark 2.4, taking $A = 1$.
(3) $\Rightarrow$ (1) : By the implication (4) $\Rightarrow$ (3) of Proposition 2.3, taking A=1.
(1) $\Rightarrow$ (3) : By Proposition 2.7, (1) implies condition (4) of Proposition 2.2. We shall show now that (1) implies condition (2) of Proposition 2.1. Consider the following diagrams :

$$
\begin{array}{ccc}
C & \xrightarrow{g} & UD \\
{\scriptstyle a}\downarrow & & \downarrow{\scriptstyle Ub} \\
A & \xrightarrow{\eta A} & UFA \\
{\scriptstyle t}\downarrow & & \downarrow{\scriptstyle UFt} \\
1 & \xrightarrow{\eta 1} & UF1
\end{array}
\qquad\qquad
\begin{array}{ccc}
FC & \xrightarrow{\bar{g}} & D \\
{\scriptstyle Fa}\downarrow & & \downarrow{\scriptstyle b} \\
FA & = & FA \\
{\scriptstyle Ft}\downarrow & & \downarrow{\scriptstyle Ft} \\
F1 & = & F1
\end{array}
$$

First, (1) says that the lower square on the left is a pullback because the corresponding square on the right is one. Now, if the upper square on the left (which is of the form (P^*)) is a pullback, then the composite square on the left is a pullback and so is, by (1), the composite square on the right. It follows then that the upper square on the right (which is the corresponding square (Q^*)) is a pullback. (3) then follows from Proposition 2.3. $\square$

3. Characterization of modular categories

Let $\mathbf{E}$ be a category with pullbacks and finite coproducts. Let us denote by $\mathbf{E}^X$, for any object X of $\mathbf{E}$, the category of objects of $\mathbf{E}$ under X and let

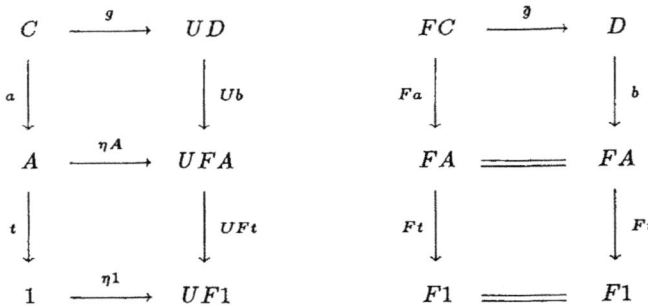

$$\mathbf{E} \underset{U^X}{\overset{F^X}{\rightleftarrows}} \mathbf{E}^X \ , \quad F^X \dashv U^X$$

be the corresponding adjunction exhibiting $\mathbf{E}^X$ as a category of algebras.

The crucial point now is to observe that, for a fixed X, the condition (M) of Proposition 1.4 is the condition (1) of Proposition 2.3 applied to the above adjunction.

With this observation and the propositions of the last section, we have :

Proposition 3.1. *A category* $\mathbf{E}$ *with pullbacks and finite coproducts is modular iff, given any commutative diagram of the form*

$$\begin{array}{ccccc} X & \xrightarrow{f} & Z & \xleftarrow{k} & W \\ \| & & \downarrow{\scriptstyle z} & & \downarrow{\scriptstyle h} \\ X & \xrightarrow[i_X]{} & X+Y & \xleftarrow[i_Y]{} & Y \end{array}$$

then the top row is a coproduct iff the right-hand square is a pullback.

Proof. This follows from Proposition 1.4 and the equivalence (1) $\Leftrightarrow$ (2) of Proposition 2.3. □

Assuming the existence of a terminal object we have :

Proposition 3.2. *A left exact category* $\mathbf{E}$ *with finite coproducts is modular (or affine) iff, given any commutative diagram of the form*

$$\begin{array}{ccccc} X & \xrightarrow{f} & Z & \xleftarrow{k} & W \\ \| & & \downarrow{\scriptstyle z} & & \downarrow{\scriptstyle t} \\ X & \xrightarrow[i_X]{} & X+1 & \xleftarrow[i_1]{} & 1 \end{array}$$

then the top row is a coproduct iff the right-hand square is a pullback.

Proof. It follows from Proposition 1.4, the equivalence (1) $\Leftrightarrow$ (3) of Proposition 2.8 and Proposition 2.3. $\square$

Remark 3.3. The last two propositions should be compared with the characterization of a distributive category, either as a category with pullbacks and finite coproducts such that, given any commutative diagram

$$\begin{array}{ccccc} V & \xrightarrow{j} & Z & \xleftarrow{k} & W \\ {\scriptstyle q}\downarrow & & \downarrow{\scriptstyle z} & & \downarrow{\scriptstyle p} \\ X & \xrightarrow[i_X]{} & X+Y & \xleftarrow[i_Y]{} & Y \end{array}$$

then the top row is a coproduct iff both squares are pullbacks or, assuming the existence of a terminal object, as a left exact category satisfying the same condition, where it suffices now to take those diagrams with the coproduct $1 \to 1+1 \leftarrow 1$ as the bottom row.

Proposition 3.4. *A category* $\mathbf{E}$ *with pullbacks and finite coproducts is modular iff, for all objects* X *and* Y *of* $\mathbf{E}$*, the adjunction*

$$\mathbf{E}/Y \underset{U_Y^X}{\overset{F_Y^X}{\rightleftarrows}} \mathbf{E}^X/(X \xrightarrow{i_X} X+1)$$

(induced, for any Y*, by the adjunction* $F^X \dashv U^X$*) is an equivalence.*

Proof. It follows from Proposition 1.4 and the equivalence (1) $\Leftrightarrow$ (3) of Proposition 2.3. $\square$

Assuming the existence of a terminal object we have :

Proposition 3.5. *A left exact category* $\mathbf{E}$ *with finite coproducts is modular (or affine) iff, for every object* X *of* $\mathbf{E}$*, the adjunction*

$$\mathbf{E} \underset{U_1^X}{\overset{F_1^X}{\rightleftarrows}} \mathbf{E}^X/(X \xrightarrow{i_X} X+1)$$

(where $\mathbf{E}$ *is identified with* $\mathbf{E}/1$*) is an equivalence.*

Proof. This follows from Proposition 1.4, the equivalence (2) $\Leftrightarrow$ (3) of Proposition 2.8 and Proposition 2.3. □

Remark that condition (2) of the characterization theorem of [1] is a particular case of the condition above, where X=1.

4. The fibration of pointed objects

Let **E** be a category with pullbacks, and let p : **Pt E** $\to$ **E** be the fibration of pointed objects in **E**. (An object in **Pt E** is a pair of maps

$$X \underset{f}{\overset{s}{\rightleftarrows}} Y$$

in **E** with $fs = 1_X$, a map in **Pt E** is a morphism of such diagrams, and $p(f,s) = X$.)

Let us denote by **Pt E** $[X]$ the fiber over X of **Pt E**.

We shall now prove that, for a modular category **E**, **Pt E** is equivalent to a constant fibration, and then that **Pt E** is additive.

Proposition 4.1. *If* **E** *is a modular category, then the universal map*

$$K \;:\; \mathbf{Pt\ E} \to \mathbf{E} \times \mathbf{Pt\ E}[0]$$

from the fibration **Pt E** *to a constant fibration is an equivalence.*

Proof. Identifying **Pt E** [X] with $\mathbf{E}^X/(X = X)$ and **Pt E** [0] with **E**/0, one checks that, over an object X of **E**, the functor K is the functor $U_0^X : \mathbf{E}^X/(X{=}X) \to \mathbf{E}/0$ which, by proposition 3.4, is an equivalence. □

Proposition 4.2. *If* **E** *is a modular category, then* **Pt E** *is an additive fibration.*

Proof. In view of Proposition 4.1, it suffices to show that **E**/0 is additive. First, the terminal object of **E**/0, the identity on 0, is also an initial object. Let now $x : X \to 0$ and $z : Z \to 0$ be two objects in **E**/0. Applying condition (1) of Definition 1.1 to the map $X \xrightarrow{x} 0 \to Z$ (denoted by 0) in **E**/0 and identifying $0 \times Z$ with Z and X+0 with X in **E**/0, shows (using a matrix notation) that the map

$$\begin{pmatrix} 1 & 0 \\ 0 & 1 \end{pmatrix} : X + Z \to X \times Z$$

is an isomorphism.

Applying the same condition to the map 1_X in $\mathbf{E}/0$ shows that the map

$$\begin{pmatrix} 1 & 1 \\ 0 & 1 \end{pmatrix} : X + X \to X \times X$$

is an isomorphism and this, by a standard argument (see [2] for example), complete the proof that $\mathbf{E}/0$ is additive. □

Assuming the existence of a terminal object we have, as in [1]

Proposition 4.3. *A modular category* $\mathbf{E}$ *with a terminal object is affine, i.e.* $\mathbf{E}$ *is equivalent to a slice of a left exact additive category.*

Proof. It follows from Proposition 3.5, by taking X=1, that $\mathbf{E}$ is equivalent to the category $\mathbf{E}^1/(1 \to 1 + 1)$. $\mathbf{E}^1$, denoted Pt(E) in [1], is left exact and, being the fiber over 1 of **Pt E**, is additive by Proposition 4.2. □

Aknowledgements

The author was in 1989/1990 on a study leave from the Collège de Staël in Geneva and spent that year in Sydney as a Visiting Fellow at Macquarie University, where the work reported on here was done.

The author wishes to thank the Département de l'instruction publique du canton de Genève for this opportunity, the Sydney Category Theory Research Group for inviting him to Sydney and its members for their hospitality.

References

[1] A. Carboni, Categories of affine spaces, J. Pure Appl. Algebra 61 (1989), 243-250.

[2] B. Mitchell, Theory of categories, Academic Press, 1965.

[3] P. Taylor, The trace factorization of stable functors (paper circulated at the International Category Theory Meeting '89, in Bangor).

This paper is in final form and will not be published elsewhere.

SOME CONSTRUCTIVE RESULTS RELATED TO COMPACTNESS AND THE (STRONG) HAUSDORFF PROPERTY FOR LOCALES

J.J.C. Vermeulen
Department of Mathematics, University of Cape Town
Private Bag, Rondebosch 7700, South Africa

0. Introduction

The categorical characterization of the Hausdorff property for topological spaces, viz. closedness of the diagonal, was for locales first considered by J.R. Isbell [1]. Despite being a natural and well-behaved property, it was, however, not immediately accepted as the "correct" notion of localic Hausdorffness, since an ordinary Hausdorff topological space need not have a closed *localic* diagonal: the standard and localic products of topological spaces are in general different. On the other hand, having a closed localic diagonal is sufficient for a topological space to be Hausdorff, and therefore the term "strongly Hausdorff" was used to refer to the former property.

In the constructive context of a topos, where locales are the "true" spaces, it is however evident that strong Hausdorffness is more fundamental than the various other definitions (e.g. [5], [7] and [9]) that have been proposed. In particular, with the strong definition, and the standard localic usage of the terms *closed, compact, locally compact* and *regular* (cf. [4]), the following familiar relationships hold in an arbitrary topos:

(I) *A regular locale is Hausdorff*;
(II) *A compact Hausdorff locale is locally compact*;
(III) *A locally compact Hausdorff locale is regular*;
(IV) *A compact sublocale of a Hausdorff locale is closed.*

Now, proofs of statements (I) and (implicitly) (II) may already be found in [1], that of (I) being essentially constructive, whereas (II) needed a choice principle: it was shown [1, Thm. 2.6] that all compact, (strongly) Hausdorff locales are spatial, and for such spaces the relevant localic and topological notions coincide. The latter also applies to statement (III); cf. [2] or [4, II 2.13], [4, VII 4.3, 4.5]. We are not aware of a published proof of statement (IV).

Our purpose here is to establish the constructive validity of (II) (Prop. 4.2), (III) (Prop. 4.5) and (IV) (Prop. 3.3). In accordance with the spirit of [1] and particularly [6], our language and style of proof have been chosen to exhibit locales as geometric objects rather than lattices. The results presented here have a close connection with the stability under pullback of perfect maps of locales [3], [10]; this is briefly discussed in the final section.

NOTATION. *Sets* are objects of an unspecified topos $\mathcal{S}$. *Spaces* are locales in $\mathcal{S}$; a space X is therefore completely determined by its frame $\mathcal{O}X$ of open subspaces, and a continuous map $X \xrightarrow{f} Y$ by the inverse-image frame homomorphism $f^-[-]\colon \mathcal{O}Y \to \mathcal{O}X$. If S is a subspace of X, $\neg S$ stands for the largest *open* subspace of X disjoint from S. If it exists, the (true) subspace-complement of S is denoted by $-S$; in particular, if S is closed, $-S = \neg S$. More generally, $-\bar{S} = \neg S$ where $\bar{S}$ is the closure of S. The interior of S is denoted by $\mathring{S}$. We say

0.1. Definition. A space X is *Hausdorff* if the diagonal $\Delta \equiv \langle \mathrm{id}, \mathrm{id} \rangle : X \hookrightarrow X \times X$ is closed.

Thus defined, the Hausdorff property is evidently inherited by subspaces and products. The interested reader might wish to consult standard references like [4] and [6] for the necessary background on locales; cf. also [3].

1. Some Machinery

We need a few results of a technical nature, involving binary products of spaces. Let X and Y be spaces. Then a base for the product space $X \times Y$ is provided by open "squares", that is, open subspaces of $X \times Y$ of the form $U \times V \simeq \pi_X^{-} U \wedge \pi_Y^{-} V$, where $U \in \mathcal{O}X$ and $V \in \mathcal{O}Y$ and π_X, π_Y are the projections. We say a binary relation $r\langle U, V\rangle$ between opens U of X and V of Y is *left-downstable, left-stable under suprema,* etc. if for each $V \in \mathcal{O}Y$, the corresponding closure property holds for the set $\{U \in \mathcal{O}X \mid r\langle U, V\rangle\}$ in $\mathcal{O}X$. By *(finitely) left-stable* we mean left-downstable and left-stable under (finite) suprema. The corresponding notions of right-stability are defined similarly. With the relation r we assosiate the mapping $j_r : \mathcal{O}X \to \mathcal{O}Y$ given by

$$j_r(P) = \bigvee \{Q \mid r\langle P, Q\rangle\}.$$

One checks that

1.1. Lemma. *The following are equivalent:*

i) *r is right-stable;*
ii) *For all $P \in \mathcal{O}X$ and $Q \in \mathcal{O}Y$, $Q \leq j_r(P) \iff r\langle P, Q\rangle$.* ∎

Lemma 1 says the assignment $r \mapsto j_r$ restricts to a bijection between right-stable relations, and functions $\mathcal{O}X \to \mathcal{O}Y$. For convenience, we say the relation r is *admissible* if i) r is left-downstable, ii) $r\langle 0, X\rangle$ holds and iii) $r\langle P \vee P', Q \wedge Q'\rangle$ holds whenever both $r\langle P, Q\rangle$ and $r\langle P', Q'\rangle$ do. An admissible relation r is finitely left-stable; the converse is true provided r is right-downstable. Expressed in terms of j_r,

1.2. Lemma. *If r is admissible, then j_r transforms finite joins into meets. The converse holds if r is right-stable.*

Thus, the mapping $r \mapsto j_r$ restricts to a bijection between finitely left-stable, right-stable relations, and functions $\mathcal{O}X \to \mathcal{O}Y$ which transform finite joins into meets.

Relations r which are both left- and right-stable, or *stable* for short, are the familiar "product ideals" of $\mathcal{O}X \times \mathcal{O}Y$, and correspond to mappings $j_r : \mathcal{O}X \to \mathcal{O}Y$ which transform all joins into meets (Galois-connections). These may be used to represent the opens of $X \times Y$, since:

1.3. Proposition [4, II 2.12]. *Let r be a relation between opens X and Y which is both left- and right-stable. Then, for all $U \in \mathcal{O}X$ and $V \in \mathcal{O}Y$,*

$$U \times V \leq \bigvee \{P \times Q \mid r\langle P, Q\rangle\} \iff r\langle U, V\rangle. \quad \blacksquare$$

To be useful in the sequel, we need to expand Proposition 1.3 slightly:

1.4. Lemma. *Let $r\langle P,Q\rangle$ and $s\langle P,Q\rangle$ be right-stable relations between opens P of X and Q of Y such that*

i) $r\langle P,Q\rangle \Rightarrow s\langle P,Q\rangle$ *for all* $P \in X$, $Q \in Y$;
ii) r *is finitely left-stable;*
iii) s *is left-downstable and left-stable under directed suprema.*

Then given any $U \in \mathcal{O}X$ and $V \in \mathcal{O}Y$, $U \times V \leq \bigvee\{P \times Q \mid r\langle P,Q\rangle\}$ only if $s\langle U,V\rangle$.

PROOF. Define a new right-stable relation s' by

$$s'\langle P,Q\rangle \iff \forall R \in \mathcal{O}X,\ S \leq Q \in \mathcal{O}Y\colon\ r\langle R,S\rangle \Rightarrow s\langle P \vee R, S\rangle.$$

Then it is straightforward to check that the hypotheses of the statement is satisfied with s' in the place of s; furthermore, $s' \Rightarrow s$, and

$$\forall P,\ R \in \mathcal{O}X,\ Q \in \mathcal{O}Y\colon\ s'\langle P,Q\rangle \text{ and } r\langle R,Q\rangle \Rightarrow s'\langle P \vee R, Q\rangle.$$

Now let s'' be the right-stable relation

$$s''\langle P,Q\rangle \iff \forall R \in \mathcal{O}X,\ S \leq Q \in \mathcal{O}Y\colon\ s'\langle R,S\rangle \Rightarrow s'\langle P \vee R, S\rangle.$$

Again, the hypotheses of the statement are satisfied with s'' in the place of s, and $s'' \Rightarrow s'$; also

$$\forall P,\ P' \in \mathcal{O}X,\ Q \in \mathcal{O}Y:\ s''\langle P,Q\rangle \text{ and } s''\langle P',Q\rangle \Rightarrow s''\langle P \vee P', Q\rangle.$$

But the last fact means s'' is finitely left-stable, hence stable, since $s''\langle 0,Q\rangle$ is implied by $r\langle 0,Q\rangle$, which is true for all $Q \in Y$. Suppose now that $U \times V \leq \bigvee\{P \times Q \mid r\langle P,Q\rangle\}$, $U \in \mathcal{O}X$ and $V \in \mathcal{O}Y$. Then $U \times V \leq \bigvee\{P \times Q \mid s''\langle P,Q\rangle\}$ and we may apply Proposition 1.3 for s'' to give $s''\langle U,V\rangle$; since $s'' \Rightarrow s' \Rightarrow s$, we conclude that $s\langle U,V\rangle$. ∎

2. Hausdorffness and regularity

Recall that a space X is *regular* if each open U of X is covered by the set $\{V \in \mathcal{O}X \mid \bar{V} \leq U\}$; V is said to be *rather below* U if $\bar{V} \leq U$. For completeness, we recall the proof of statement (I):

2.1. Proposition [1, 2.3.1], [4, III 1.3]. *Let X be a regular space. Then X is Hausdorff.*

PROOF. We want to show that $\bar{\Delta} \leq \Delta$. Now, since Δ is the equalizer of the projections $\pi_0, \pi_1\colon X \times X \to X$, it is (by symmetry) enough to prove that $\bar{\Delta} \wedge \pi_0^- U \leq \pi_1^- U$, that is, $\pi_0^- U \leq \neg\Delta \vee \pi_1^- U$ for all $U \in \mathcal{O}X$, where $\neg\Delta = -\Delta = \bigvee\{R \times S \mid R \wedge S \leq 0\}$. But if $\bar{P} \leq U$ for P open, then $X \leq \neg P \vee U$, i.e.

$$\begin{aligned}\pi_0^- P &= P \times X \\ &\leq P \times \neg P \vee P \times P \\ &\leq \neg\Delta \vee X \times U = \neg\Delta \vee \pi_1^- U.\end{aligned}$$

So, by regularity $\pi_0^- U \leq \bigvee\{\pi_0^- P \mid \bar{P} \leq U\} \leq \neg\Delta \vee \pi_1^- U$. ∎

Recall that a space X is *compact* if each cover of X contains a finite subset which still covers X. To prove statement (III), we first consider compact Hausdorff spaces.

2.2. Lemma. *Let $t\langle P,Q\rangle$ be an admissible relation between opens P of a compact space K and opens Q of a space X. Suppose $U \in \mathcal{O}X$ satisfies $K \times U \leq \bigvee\{P \times Q \mid t\langle P,Q\rangle\}$. Then $U \leq j_t(K) = \bigvee\{Q \mid t\langle K,Q\rangle\}$; in particular, if t is right-stable, then $t\langle K,U\rangle$ holds.*

PROOF. Define the relations r and s on $\mathcal{O}K \times \mathcal{O}X$ by

$$r\langle P,Q\rangle \iff Q \leq j_t(P)$$
$$s\langle P,Q\rangle \iff K \leq P \Rightarrow r\langle K,Q\rangle.$$

Then $r \Rightarrow s$, and both r and s are right-stable. Since j_t transforms finite joins into meets (Lemma 1.2), r is finitely left-stable. By compactness of K, s is left-stable under directed suprema; it is also clearly left-downstable. Since $t \Rightarrow r$, $K \times U \leq \bigvee\{P \times Q \mid r\langle P,Q\rangle\}$; we may therefore apply Lemma 1.4 to deduce that $s\langle K,U\rangle$ holds, or equivalently $r\langle K,U\rangle$. The final statement follows from Lemma 1.1. ∎

2.3. Proposition. *Let X be a compact Hausdorff space. Then X is regular.*

PROOF. Let $U \in \mathcal{O}X$ be given. Since X is Hausdorff, $\neg\Delta$ is the complement of Δ in $X \times X$, that is, $X \times X \leq \Delta \vee \neg\Delta$. Also, since $\Delta \wedge X \times U \leq U \times X$,

$$\begin{aligned}\Delta \wedge -U \times U &\leq \Delta \wedge X \times U \wedge -U \times X \\ &\leq U \times X \wedge -U \times X \\ &\leq (U \wedge -U) \times X \\ &\leq 0.\end{aligned}$$

It therefore follows that $-U \times U \leq \neg\Delta$, giving

$$X \times U \leq (U \times X) \vee \neg\Delta = \bigvee\{P \times Q \mid P \leq U \text{ or } P \wedge Q \leq 0\}. \qquad (\dagger)$$

Now, let t be the relation

$$t\langle P,Q\rangle \iff P \wedge \bar{Q} \leq U.$$

Then t is stable on the left and finitely stable on the right (closure preserves order and finite suprema). It follows that t is admissible. Further, $t\langle P,Q\rangle$ is clearly true if $P \leq U$ or $P \wedge Q \leq 0$. By (†), it follows that $X \times U \leq \bigvee\{P \times Q \mid t\langle P,Q\rangle\}$. We may therefore apply Lemma 2.2 and obtain $U \leq j_t(X) = \bigvee\{V \in \mathcal{O}X \mid \bar{V} \leq U\}$. ∎

We may remark that it has been shown [7] that Proposition 2.3 does in fact not hold for the notion of Hausdorffness considered in [5].

3. Compact subspaces of a Hausdorff space

Statement (IV) can be proved by a direct application of Lemma 2.2; however, conceptually it is useful to regard it as a consequence of a more general fact, viz. Proposition 3.2 below. Our proof of the latter uses the following ("Frobenius reciprocity"):

3.1. Lemma. *Let $f: X \to Y$ be a continuous map, S a subspace of X and U an open subspace of Y. Then*

$$f[S \wedge f^{-}[U]] = f[S] \wedge U.$$

PROOF. The diagram

$$\begin{array}{ccc} S \wedge f^{-}[U] & \hookrightarrow & S \\ \downarrow f & & \downarrow f \\ f[S] \wedge U & \hookrightarrow & f[S] \end{array}$$

is a pullback, and since $f[S] \wedge U$ is open in $f[S]$, the restriction of f (on the left) remains surjective. ∎

3.2. Proposition. *Let K and X be spaces, with K compact. Then the projection $K \times X \xrightarrow{\pi_X} X$ is closed.*

PROOF. Let C be a closed subspace of $K \times X$. Then, for any $U, V \in \mathcal{O}X$ we have

$$\begin{aligned} \pi_X[C] \wedge U \leq V &\iff C \wedge \pi_X^{-}[U] \leq \pi_X^{-}[V] \quad \text{using Lemma 3.1} \\ &\iff \pi_X^{-}[U] \leq \neg C \vee \pi_X^{-}[V] \quad \text{since } C \text{ is closed.} \qquad (*) \end{aligned}$$

To prove that $\pi_X[C]$ is closed, suppose $\overline{\pi_X[C]} \wedge U \leq V$ for some arbitrarily given $U, V \in \mathcal{O}X$; we have to deduce that $\pi_X[C] \wedge U \leq V$, that is, that $U \leq \neg\pi_X[C] \vee V$. Define the relation t on $\mathcal{O}K \times \mathcal{O}X$ by

$$t\langle P, Q\rangle \iff P \times Q \leq \neg C \text{ or } Q \leq V.$$

Then t is clearly left-downstable, and $t\langle 0, X\rangle$ holds. Further, if $t\langle P, Q\rangle$ and $t\langle P', Q'\rangle$ are true, then either $P \times Q \leq \neg C$ and $P' \times Q' \leq \neg C$, in which case $(P \vee P') \times (Q \wedge Q') \leq \neg C$, or $Q \wedge Q' \leq V$ (or both); in each case $t\langle P \vee P', Q \wedge Q'\rangle$ follows. Thus, t is admissible. Also, from $(*)$

$$\begin{aligned} K \times U &\leq \neg C \vee K \times V \\ &= \bigvee\{P \times Q \mid P \times Q \leq \neg C \text{ or } Q \leq V\} \\ &= \bigvee\{P \times Q \mid t\langle P, Q\rangle\}. \end{aligned}$$

Applying Lemma 2.2, we conclude that $U \leq j_t(K) = \neg\pi_X[C] \vee V$. ∎

Statement (IV) now follows by a standard argument:

3.3. Proposition. *Suppose X is Hausdorff. Then compact subspaces of X are closed.*

PROOF. Let $K \overset{i_K}{\hookrightarrow} X$ be a compact subspace of X. Then the embedding $\langle i_K, \mathrm{id}\rangle: K \hookrightarrow K \times X$ is the inverse image of Δ along $i_K \times \mathrm{id}: K \times X \to X \times X$, and consequently closed in $K \times X$ by Hausdorffness; since its image under π_X is just K, the result follows from Proposition 3.2. ∎

4. Locally compact Hausdorff spaces

Recall that a space X is said to be *locally compact* if the frame $\mathcal{O}X$ is a continuous lattice, that is, if each open U of X is covered by the set $\{V \in \mathcal{O}X \mid V \ll U\}$, where $V \ll U$ (V is *way below* U) means each open cover of U contains a finite subset covering V. Although this definition does not directly refer to compactness, for Hausdorff spaces it is equivalent to the familiar notion, as we shall see below.

First, to get from Proposition 2.3 to statement (II), we only need

4.1. Lemma. *Let X be a space, and U an open subspace of X with $U \ll X$. Then for all $V \in \mathcal{O}X$, $\bar{U} \leq V \Rightarrow U \ll V$.*

PROOF. Let $\bar{U} \leq V$. If V is is covered by $\mathcal{D} \subseteq \mathcal{O}X$, $\mathcal{D}$ directed, so is $\bar{U}$, giving $X \leq \neg U \vee \bigvee \mathcal{D} \leq \bigvee(\neg U \vee \mathcal{D})$ (where we have used the fact that $\mathcal{D}$ is inhabited). Since $U \ll X$, it follows that $U \leq \neg U \vee D$, that is, $U \leq D$, for some $D \in \mathcal{D}$. But this shows that $U \ll V$. ∎

4.2. Proposition. *A compact Hausdorff space is locally compact.*

PROOF. By Proposition 2.3, a compact Hausdorff space is regular, and by Lemma 4.1, the rather below relation implies the way below relation in compact spaces. ∎

To deduce statement (III), we need a variant of Lemma 2.2 to deal with local, rather than absolute compactness.

4.3. Lemma. *Let $t\langle P, Q\rangle$ be an admissible relation between opens P of a locally compact space X and opens Q of a space Y. Suppose $U \times V \leq \bigvee\{P \times Q \mid t\langle P, Q\rangle\}$, $U \in \mathcal{O}X$ and $V \in \mathcal{O}Y$. Then, for all $W \in \mathcal{O}X$, $W \ll U \Rightarrow V \leq j_t(W) = \bigvee\{Q \mid t\langle W, Q\rangle\}$; in particular, if t is right-stable, then $t\langle W, V\rangle$ holds.*

PROOF. Given an open $W \ll U$, define the relations r and s on $\mathcal{O}X \times \mathcal{O}Y$ by

$$r\langle P, Q\rangle \iff Q \leq j_t(P)$$
$$s\langle P, Q\rangle \iff W \ll P \Rightarrow r\langle W, Q\rangle.$$

Then $r \Rightarrow s$, since $W \ll P \Rightarrow W \leq P$. It is also clear that both r and s are right-stable. Since j_t transforms finite joins into meets (Lemma 1.2), r is finitely left-stable. It is immediate that s is left-downstable; s is also left-stable under directed suprema, since in locally compact spaces, the way below relation for opens satisfies the subdivision property: $U \ll V \Rightarrow U \ll W \ll V$ for some open W. The statement now follows by applying Lemma 1.4. ∎

4.4. Lemma. *Let X be a Hausdorff space. Then, if X is locally compact, the way below relation for opens implies the rather below relation. That is, for all $U, V \in \mathcal{O}X$, $U \ll V \Rightarrow \bar{U} \leq V$.*

PROOF. Suppose $U \ll V$. Since $-V \times V \leq \neg\Delta$, Hausdorffness gives

$$V \times X \leq (X \times V) \vee \neg\Delta = \bigvee\{P \times Q \mid Q \leq V \text{ or } P \wedge Q \leq 0\}. \qquad (\dagger\dagger)$$

Now let t be the finitely left-stable, right-stable (hence admissible) relation

$$t\langle P, Q\rangle \iff \bar{P} \wedge Q \leq V.$$

Then $t\langle P,Q\rangle$ holds if $Q \leq V$ or $P \wedge Q \leq 0$; by (††) it follows that $V \times X \leq \bigvee\{P \times Q \mid t\langle P,Q\rangle\}$. So, if X is locally compact, Lemma 4.3 may be applied to give $t\langle U,X\rangle$, that is, $\bar{U} \leq V$. ■

Lemma 4.4 immediately gives

4.5. Proposition. *A locally compact Hausdorff space is regular.* ■

We end by showing that for Hausdorff spaces, local compactness can be defined in terms of compact neighbourhoods.

4.6. Lemma. *Let X be a locally compact Hausdorff space, and U an open subspace of X. Then $U \ll X \iff \bar{U}$ is compact.*

PROOF. $\bar{U}$ compact $\Rightarrow U \ll X$ is immediate. For the converse, suppose $U \ll X$. Then, if $\bar{U}$ is covered by $\mathcal{D} \subseteq \mathcal{O}X$, $\mathcal{D}$ directed, we have $U \ll \bigvee \mathcal{D}$ by Lemma 4.1. Choose V open such that $U \ll V \ll \bigvee \mathcal{D}$, which is possible by the local compactness of X. Then $U \ll V \leq D$ for some $D \in \mathcal{D}$; but then $\bar{U} \leq D$ by Lemma 4.4. ■

4.7. Proposition. *The following are equivalent for a Hausdorff space X:*

i) *X is locally compact.*
ii) *X is covered by opens with compact closure.*
iii) *X is covered by the interiors of a set of compact subspaces.*

PROOF. The implication i)⇒ii) is given by Lemma 4.6. ii)⇒iii) is immediate. To show that iii)⇒i), let $\mathcal{K}$ be a set of compact subspaces of X such that $X \leq \bigvee\{\mathring{K} \mid K \in \mathcal{K}\}$. Then, given $U \in \mathcal{O}X$ and $K \in \mathcal{K}$, K is closed in X, by Proposition 3.3, and so the closure in K of any $V \in \mathcal{O}K$ coincides with its closure in X; if moreover $V \leq \mathring{K}$, then V is open in X, and $V \ll X$ by the compactness of K. Since K is regular (Proposition 2.3), it follows that

$$\begin{aligned} \mathring{K} \wedge U &\leq \bigvee\{V \in \mathcal{O}K \mid \bar{V} \leq \mathring{K} \wedge U \text{ in } K\} \\ &\leq \bigvee\{V \in \mathcal{O}X \mid V \ll X \text{ and } \bar{V} \leq U\} \\ &\leq \bigvee\{V \in \mathcal{O}X \mid V \ll U\}, \end{aligned}$$

where in the last step we have used Lemma 4.4. But then

$$U \leq \bigvee\{\mathring{K} \wedge U \mid K \in \mathcal{K}\} \leq \bigvee\{V \in \mathcal{O}X \mid V \ll U\}. \quad ■$$

Either property (ii) or (iii) above are sometimes referred to as *weak* local compactness. We remark that the link between (ii) and various other notions of Hausdorffness has been investigated in [8].

5. Applications to perfect maps

The constructive validity of the methods and results of the previous sections allow these to be (in particular) interpreted in spatial toposes relative to the chosen base topos $\mathcal{S}$. We recall [6] that the category of internal spaces in a topos Shv(X) of sheaves over a space X is simply the category Sp$/X$ of spaces over X. A (global) subspace in Shv(X) of a continuous map $E \xrightarrow{p} X$ is determined by an ordinary subspace S of E;

S is open (resp. closed) in E over X if and only if it is so in the ordinary sense in E. The map $E \xrightarrow{p} X$ is said to be *perfect* [3] if it is compact as a space in $\mathrm{Shv}(X)$; if p is moreover regular in $\mathrm{Shv}(X)$, it is said to be *proper*. Perfect (resp. proper) maps are closed under composition [3] and preserved under pullback; the latter may be deduced by interpreting the following particular case, which is fairly straightforwardly proved using Lemma 2.2 (cf. [10]), in the appropriate sheaf-category:

5.1. Proposition. *For spaces K and X, with K compact, the projection $K \times X \xrightarrow{\pi_X} X$ is perfect.* ∎

Conceptually, the stability properties of perfect maps may be regarded as the key fact behind the constructive (that is, base-independent) validity of the compactness related results that we have presented. For example, interpreting in $\mathrm{Shv}(X)$ the fact that the unique map from a compact space to the terminal space is closed, it follows that the projection $K \times X \xrightarrow{\pi_X} X$ is closed in $\mathrm{Shv}(X)$, which implies that it is a closed map in the ordinaray sense; this provides a conceptual proof of Proposition 3.2.

Generalizing openness from subspaces to arbitrary maps gives the notion of *open map*; perfect maps may be viewed as the corresponding extension to maps of the idea of closedness of a subspace (perfect embeddings into X are exactly closed embeddings, by interpreting in $\mathrm{Shv}(X)$ the fact that the compact subspaces of the terminal space are its closed subspaces). In this sense, the following is a generalization of statement (IV):

5.2. Proposition. *Any continuous map $f: K \to X$ from a compact space to a Hausdorff space is perfect.*

PROOF. The graph $\langle \mathrm{id}, f\rangle: K \hookrightarrow K \times X$ of f is the inverse image along $f \times \mathrm{id}: K \times X \to X \times X$ of the diagonal of X; it is therefore closed, hence a perfect embedding into $K \times X$. But this means $f = \pi_X \circ \langle \mathrm{id}, f\rangle$ is perfect, being the composite of perfect maps. ∎

Now, as an immediate corollary of the (constructive) regularity of compact Hausdorff spaces, we recover the description of a proper map (cf. [4 III 3.8]) in its familiar form:

5.3. Proposition. *A perfect map $E \xrightarrow{p} X$ is proper if and only if the diagonal $\Delta: E \hookrightarrow E \times_X E$ is closed.*

PROOF. $\Delta: E \hookrightarrow E \times_X E$ is the (ordinary) diagonal of $E \xrightarrow{p} X$ in Sp/X. The statement therefore follows by interpreting Proposition 2.3 in $\mathrm{Shv}(X)$. ∎

5.4. Corollary. *Any perfect map $E \xrightarrow{p} X$ with Hausdorff domain is proper.*

PROOF. If E has a closed diagonal, it has a closed diagonal over X. ∎

We are grateful to Bernhard Banaschewski for drawing our attention to an error in a preliminary draft of this paper.

REFERENCES

[1] J.R. Isbell. Atomless parts of spaces. *Math. Scand.* **31** (1972), 5–32.
[2] J.R. Isbell. Function spaces and adjoints. *Math. Scand.* **36** (1975), 317–39.
[3] P.T. Johnstone. Factorization and pullback theorems for localic geometric morphisms. Univ. Cath. de Louvain, Sém. de Math. Pure, Rapport no. 79.
[4] P.T. Johnstone. *Stone Spaces.* Cambridge studies in advanced Math. no. 3 (Cambridge University Press, 1982).
[5] P.T. Johnstone and Shu-Hao Sun. Weak products and Hausdorff locales. Springer Lecture Notes in Math. vol. 1348 (Springer-Verlag, 1988).
[6] A. Joyal and M. Tierney. An extension of the Galois theory of Grothendieck. *Memoirs Amer. Math Soc.* **309** (1984).
[7] J. Paseka and B. Šmarda. T_2-frames and almost compact frames. Preprint (?).
[8] J. Paseka and B. Šmarda. On some notions related to compactness for locales. *Acta Univ. Carolinae* (2) **29** (1988), 51–65.
[9] H. Simmons. The lattice-theoretic part of topological seperation properties. *Proc. Edin. Math. Soc.* (2) **21** (1978), 41–48
[10] J.J.C. Vermeulen. Pullback-stability properties of perfect maps of locales. Preprint.

This paper is in final form and will not be published elsewhere.

PART II

Table of content

AN INTRODUCTION TO TANNAKA DUALITY AND QUANTUM GROUPS

by
André JOYAL and Ross STREET
Université du Québec à Montréal, Canada and Macquarie University, Australia

Introduction

The goal of this paper is to give an account of classical Tannaka duality [C] in such a way as to be accessible to the general mathematical reader, and to provide a key for entry to more recent developments [SR, DM] and quantum groups [D1]. Expertise in neither representation theory nor category theory is assumed.

Naively speaking, Tannaka duality theory is the study of the interplay which exists between a group and the category of its representations. The early duality theorems of Tannaka-Krein [Ta, Kr] concentrate on the problem of reconstructing a compact group from the collection of its representations. In the abelian case, this problem amounts to reconstructing the group from its character group, and is the content of the Pontrjagin duality theorem. A good exposition of this theory can be found in the book by Chevalley [C]. In these early developments, there was little or no use of categorical concepts, partly because they did not exist at the time, and, partly because categorically concepts were not yet familiar to the mathematical community and it was possible to do without them [BtD].

To Grothendieck we owe the understanding that the process of Tannaka duality can be reversed. In his work to solve the Weil conjectures, he constructed the category of motives as the universal recipient of a Weil cohomology [Kl]. By using a fiber functor from his category of motives to vector spaces, he could construct a pro-algebraic group G. He also conjectured that the category of motives could be recaptured as the category of representations of G. This group is called the Grothendieck Galois group, since it is an extension of the Galois group of $\overline{Q}/Q$. The work spreading from these ideas can be found in [SR, DM].

An entirely different development came from mathematical physicists working on superselection principles in quantum field theory [DHR] where it was discovered that the superselection structure could be described in terms of a category whose objects are certain endomorphisms of the C^*-algebra of local observables, and whose arrows are intertwining operators. Reversing the duality process, they succeeded in constructing a compact group whose representations can be identified with their superselection category [DR].

Another impulse to the development of Tannaka duality comes from the theory of quantum groups. These new mathematical objects were discovered by Jimbo [J] and Drinfel'd [D1] in connection with the work of L.D. Faddeev and his collaborators on the quantum inverse scattering method. V.V. Lyubashenko [Ly] initiated the use of Tannaka duality in the construction of quantum groups. We should also mention S.L. Woronowicz [W] in the case of compact quantum groups. Recently, S. Majid [M3] has shown that one can use Tannaka-Krein duality for constructing the quasi-Hopf algebras introduced by Drinfel'd [D2] in connection with the solution of the Knizhnik-Zamolodchikov equation. Also see K.-H. Ulbrich [U].

Another motivation for studying Tannaka duality might come from the theory of angular momentum in Quantum Physics [BL1]. The Racah-Wigner algebra, the $9-j$ and $3-j$ symbols, and, the Racah and Wigner coefficients, all seem to be about the explicit description of the structures which exist on the category of representations of some compact groups like SU(2) or SU(3) [BL2]. Here the theory of orthogonal polynomials and special functions comes into play [AK]. Also, the q-analogues of the classical orthogonal polynomials show up as spherical functions on quantum groups [Ko].

Another essential aspect of the picture that should be mentioned is the connection between knot theory, Feynman diagrams, category theory, and quantum groups. It was discovered by Turaev [T] that the new invariants found by Jones [Jn1] could be constructed from Yang-Baxter operators. The method was formalised by showing that certain categories constructed geometrically from tangles of strings or ribbons could be given a simple algebraic presentation [FY, T, JS1]. It is remarkable that notation, introduced by Penrose [Pn] for calculating with ordinary tensors, has the right degree of generality to express correctly the calculations in any tensor category [JS2]. What is emerging here is a new symbiosis between algebra, geometry and physics, the consequences of which are not yet fully understood. A good review of the recent research on the whole subject is given by P. Cartier [Car].

After a brief review of Pontrjagin duality and Fourier transforms for locally compact abelian groups, we give a treatment in Section 1 of the classical Tannaka theory for compact (non-abelian) groups. In the compact case, representations must be considered, because characters are no longer sufficient to recapture the compact group. There is also a notion of Fourier transform applying, and we examine this in detail. Where possible we work with a general topological monoid M. A central object of our analysis is the algebra R(M) of *representative* complex-valued functions on M. In Section 2, we show that R(M) is a bialgebra and compute it in the case where M is the unitary group U(n).

Section 3 and Section 4 begin the modern treatment of Tannaka reconstruction, motivated at each stage by the example of a topological monoid. Instrumental here is the *Fourier cotransform* which can be seen as the continuous predual of the Fourier transform. In fact, the Fourier cotransform provides an isomorphism between the reconstructed object and R(M).

Section 5 is an introduction to Tannaka duality for homogeneous spaces [IS]. The proof of the duality theorem of Section 5 is independent of the proof of the one appearing in Section 1.

Sections 6 and 7 are devoted to the characterization of the category of comodules over a coalgebra. Sections 8 and 9 study extra structure which is possessed by the coalgebra $End^{\vee}(X)$ of Section 4.

Section 10 introduces the concept of braided tensor categories and Yang-Baxter operators at the appropriate level of generality. Section 11 is a brief description of the categorical axiomatization of the geometry of knots and tangles.

Section 12 is a too brief introduction to quantum groups. There are many important aspects of the theory of quantum groups which could be mentioned, such as the work of Lusztig [Lu], Rosso [R], and Deligne. Our goal here is a modest one (already longer than our editors expected), and we apologize to those whose work we have not mentioned.

This paper is in final form and will not be published elsewhere.

§1. Classical Tannaka duality.

Before describing Tannaka duality, we briefly recall Pontrjagin duality. Let G be a commutative locally-compact group. A *character* χ of G is a continuous homomorphism

$$\chi : G \longrightarrow T$$

where **T** is the multiplicative group of complex numbers of modulus 1. The characters form a group $G^\vee$ which is given the topology of uniform convergence on compact subsets of G. It turns out that $G^\vee$ is locally compact. There is a canonical pairing

$$\langle\ ,\ \rangle : G^\vee \times G \longrightarrow T,$$

and we obtain a canonical homomorphism

$$i : G \longrightarrow (G^\vee)^\vee.$$

Theorem 1 (Pontrjagin). *The canonical homomorphism* i *is an isomorphism of topological groups.*

The group G is compact if and only if the dual group $G^\vee$ is discrete. We have

$$\mathbf{T}^\vee = \mathbf{Z}, \quad \mathbf{Z}^\vee = \mathbf{T}, \quad \mathbf{R}^\vee = \mathbf{R}, \quad (\mathbf{Z}/n)^\vee = \mathbf{Z}/n.$$

Many groups are self dual, such as the additive groups of the local fields.

Pontrjagin duality goes hand-in-hand with the theory of Fourier transforms which we briefly describe. There is a positive measure dx on G called the Haar measure. It is the unique (up to scalar multiple) Borel measure which is invariant under translations. Using it, we can define the spaces $L^1(G)$ and $L^2(G)$ of integrable and square integrable functions. The Fourier transform

$$\mathcal{F} : L^1(G) \cap L^2(G) \longrightarrow L^2(G^\vee)$$

is defined as follows

$$(\mathcal{F}f)(s) = \int_G f(x) \langle s, x \rangle \, dx \ .$$

The set $L^1(G) \cap L^2(G)$ is a dense subspace of $L^2(G)$ and we have:

Theorem 2 (Plancherel). *With correct normalisation of the Haar measure* ds *on* $G^\vee$, *the mapping* $f \longmapsto \mathcal{F}f$ *extends uniquely to an isometry*

$$\mathcal{F} : L^2(G) \xrightarrow{\ \sim\ } L^2(G^\vee).$$

The inverse transformation $\mathcal{F}^{-1}$ is given by

$$\mathcal{F}^{-1}(g)(x) = \int_{G^\vee} g(s) \overline{\langle s, x \rangle} \, ds.$$

The measure ds on $G^\vee$ which produces the isometry is unique, and is called the measure associated to dx. When G is compact, we often choose dx so that the total mass of G is 1. In this case, the corresponding associated measure ds on the discrete group $G^\vee$ assigns mass 1 to singletons.

It is an open problem to formulate and prove a general duality theorem for non-commutative locally compact groups such as Lie groups. Even the case of simple algebraic groups is not well understood despite the enormous accumulating knowledge on their irreducible representations. However, when the group is compact, there is a good duality theory due to H. Peter, H. Weyl and T. Tannaka. In this case, the dual object $G^\vee$ is discrete and so belongs to the realm of algebra. In order to describe this theory, it will be convenient to introduce some of the needed concepts in a more general setting.

Let M be a topological monoid. *A (finite) representation of* M consists of a finite-dimensional complex vector space V together with a continuous homomorphism

$$\pi_V : M \longrightarrow \mathrm{End}(V)$$

into the monoid End(V) of linear endomorphisms of V. *Real representations* are defined using real vector spaces. We denote by $\mathcal{Rep}(M,\mathbf{C})$ [respectively, $\mathcal{Rep}(M,\mathbf{R})$] the category of all complex [respectively, real] finite representations of M. There is a forgetful functor

$$\mathcal{U} : \mathcal{Rep}(M,\mathbf{C}) \longrightarrow \mathcal{Vect}_{\mathbf{C}}$$

where $\mathcal{Vect}_{\mathbf{C}}$ denotes the category of all complex vector spaces. Recall that a natural transformation $u : \mathcal{U} \longrightarrow \mathcal{U}$ is a family of maps $u_V : V \longrightarrow V$ indexed by $V \in \mathcal{Rep}(M,\mathbf{C})$ such that the square

$$\begin{array}{ccc} V & \xrightarrow{\ u_V\ } & V \\ {\scriptstyle h}\downarrow & & \downarrow{\scriptstyle h} \\ W & \xrightarrow[\ u_W\ ]{} & W \end{array}$$

commutes for all morphisms $h : V \longrightarrow W$ of representations ("intertwining operators"). Clearly, each element $x \in M$ produces such a natural transformation $\pi(x) : \mathcal{U} \longrightarrow \mathcal{U}$ whose V component is the element $\pi_V(x) : V \longrightarrow V$.

There is a topology on the set $\mathrm{End}(\mathcal{U})$ of natural transformations from $\mathcal{U}$ to $\mathcal{U}$. It is the coarsest topology rendering all the projections $u \longmapsto u_V \in \mathrm{End}(V)$ continuous. Composition and addition of natural transformations turns $\mathrm{End}(\mathcal{U})$ into a topological algebra. There is a *conjugation operation*

$$\mathrm{End}(\mathcal{U}) \longrightarrow \mathrm{End}(\mathcal{U}), \qquad u \longmapsto \bar{u}\,,$$

given by

$$\bar{u}_V(x) = \overline{u_{\bar{V}}(\bar{x})}$$

where $\bar{V}$ denotes the conjugate representation of V (the elements of the vector space $\bar{V}$ are the same as those of V but the identity map $V \longrightarrow \bar{V}$, denoted $x \longmapsto \bar{x}$, is an anti-linear isomorphism).

We would like to characterize the natural transformations of the form $\pi(x)$. Recall that a natural transformation u is *tensor preserving* (or "monoidal") when we have

$$u_{V\otimes W} = u_V \otimes u_W \qquad \text{and} \qquad u_I = 1_I$$

where I denotes the trivial 1-dimensional representation of M. We say that u is *self-conjugate* when $u = \bar{u}$. The set of tensor preserving self-conjugate transformations is a closed subset of $\mathrm{End}(\mathcal{U})$ which is also closed under composition. We call this subset the *Tannaka monoid* of M and denote it by $\mathcal{T}(M)$. For any $x \in M$, the natural transformation $\pi(x) \in \mathrm{End}(\mathcal{U})$ belongs to $\mathcal{T}(M)$. We have a continuous homomorphism

$$\pi : M \longrightarrow \mathcal{T}(M).$$

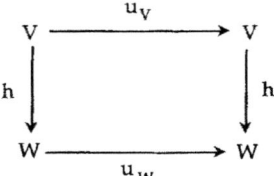

Proposition 3. *$\mathcal{T}(M)$ is a topological group if M is.*

Proof. Suppose that M is a topological group. Then any representation

$$\pi_V : M \longrightarrow GL(V)$$

has a dual, or "contragredient", representation

$$\pi_{V^\vee} : M \longrightarrow GL(V^\vee)$$

where $V^\vee$ is the dual vector space of V and

$$\pi_{V^\vee}(x) = {}^t\pi_V(x)^{-1}.$$

We have a morphism of representations $\varepsilon : V^\vee \otimes V \longrightarrow I = C$ defined by $\varepsilon(s \otimes x) = \langle s, x \rangle$. From the naturality of $u \in \mathcal{T}(M)$, we have the equality

$$\varepsilon\, u_{V^\vee \otimes V} = u_I\, \varepsilon .$$

Using the equalities

$$u_{V^\vee \otimes V} = u_{V^\vee} \otimes u_V \qquad \text{and} \qquad u_I = 1_I ,$$

we see that

$$\langle u_{V^\vee}(s), u_V(x) \rangle = \langle s, x \rangle.$$

This shows that $u_{V^\vee}$ is the contragredient transformation of u_V,

$$u_{V^\vee} \circ {}^t u_V = 1_{V^\vee} ,$$

and implies that u_V is invertible.$_{\text{qed}}$

When $M = G$ is a compact group, the algebra $\mathrm{End}(\mathfrak{U})$ has a particularly simple structure. Let $G^\vee$ be the set of isomorphism classes of irreducible representations of G. Let us choose a representation (V_λ, π_λ) in each such class λ. For any $u \in \mathrm{End}(\mathfrak{U})$, let us write u_λ for the V_λ component of u, and put

$$q(u) = (u_\lambda \mid \lambda \in G^\vee).$$

Proposition 4. *The map*

$$q : \mathrm{End}(\mathfrak{U}) \longrightarrow \prod_{\lambda \in G^\vee} \mathrm{End}(V_\lambda)$$

is an isomorphism of topological algebras.

Proof. Using naturality, we have the commutative diagram

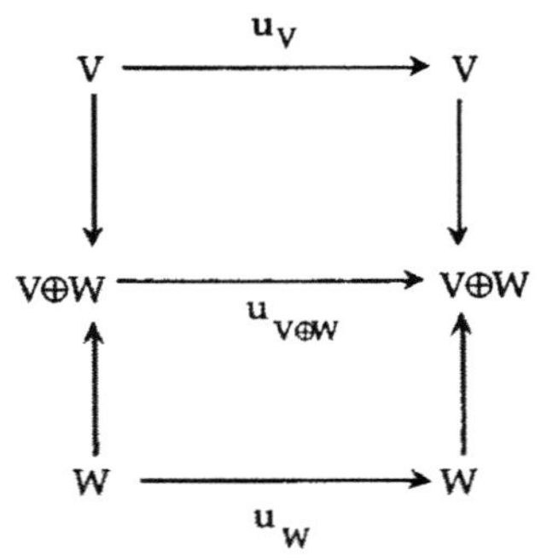

for any $V, W \in \mathcal{Rep}(G, \mathbf{C})$, which shows that

$$u_{V\oplus W} = u_V \oplus u_W .$$

This implies that u is entirely determined by $q(u)$ since any representation of G decomposes as a direct sum of irreducibles.

To prove that q is surjective, let $t = (t_\lambda \mid \lambda \in G^\vee)$ be a family of elements $t_\lambda \in \mathrm{End}(V_\lambda)$. We want to construct an element $u \in \mathrm{End}(\mathcal{U})$ so that $u_\lambda = t_\lambda$ for every $\lambda \in G^\vee$. Let $S \in \mathcal{Rep}(G, \mathbf{C})$. There is a unique decomposition of S as a direct sum of isotypical components

$$S = \sum_{\lambda \in G^\vee} S_\lambda .$$

Moreover, for each $\lambda \in G^\vee$, the canonical map

$$\psi_\lambda : V_\lambda \otimes \mathrm{Hom}_G(V_\lambda, S_\lambda) \longrightarrow S_\lambda$$

is an isomorphism of G-modules. We put

$$u_{S_\lambda} = \psi_\lambda \circ (t_\lambda \otimes 1) \circ \psi_\lambda^{-1}$$

and

$$u_S = \sum_{\lambda \in G^\vee} u_{S_\lambda} .$$

It is not difficult to verify that the family

$$u = (u_S \mid S \in \mathcal{Rep}(G, \mathbf{C}))$$

defines a natural transformation $u : \mathcal{U} \longrightarrow \mathcal{U}$ having the required property. The continuity of q^{-1} is a consequence of the fact that the topology on $\mathrm{End}(\mathcal{U})$ is the coarsest rendering continuous all the projections $u \longmapsto u_\lambda$, $\lambda \in G^\vee$. **qed**

Remark. The proof of Proposition 4 is based on the fact that the category $\mathcal{Rep}(G, \mathbf{C})$ is the closure under direct sums of its subcategory of irreducible representations.

Proposition 5. *$\mathcal{T}(G)$ is compact if G is a compact group.*

Proof. Any representation V of a compact group admits a positive definite invariant hermitian form $g : V \otimes V \longrightarrow \mathbf{C}$. We can view g as a C-linear pairing

$$h : \overline{V} \otimes V \longrightarrow \mathbf{C}, \qquad h(\overline{x}, y) = g(x, y).$$

For any $u \in \mathrm{End}(\mathcal{U})$, we have

$$h \circ u_{\overline{V} \otimes V} = u_I \circ h.$$

But, if u is tensor preserving and self conjugate, we have

$$u_{\overline{V} \otimes V} = u_{\overline{V}} \otimes u_V, \quad u_I = 1, \quad u_{\overline{V}}(\overline{x}) = \overline{u_V(x)}.$$

This means that

$$h(\overline{u_V(x)}, u_V(y)) = h(\overline{x}, y) ;$$

that is,

$$g(u_V(x), u_V(y)) = g(x, y),$$

which means that u_V belongs to the unitary group $U(V, g)$. It follows from this that $\mathcal{T}(G)$

is a closed subgroup of a product of compact groups.$_{\text{qed}}$

To better understand the structure of $\mathcal{T}(G)$ we need to take account of the $*$–involution

$$(\ )^*: \mathrm{End}(\mathcal{U}) \longrightarrow \mathrm{End}(\mathcal{U}).$$

Its definition can be given in terms of two other involutions:

$$u^* = (\bar{u})^\vee = \overline{(u^\vee)}$$

where, by definition, $\bar{u}$ is the conjugation operation considered earlier and

$$(u^\vee)_V = (u_{V^\vee})^\vee .$$

A better description, valid only for compact groups, is to say that it transports across the isomorphism q of Proposition 4 to the *canonical* C*–algebra structure on each $\mathrm{End}(V_\lambda)$ for $\lambda \in G^\vee$: the adjoint h^* of an element $h \in \mathrm{End}(V_\lambda)$ is defined by the equation

$$g(h^*(x), y) = g(x, h(y))$$

and does not depend on the choice of g since g is unique up to a scalar multiple (by Schur's Lemma applied to the irreducible representation V_λ).

Remark. The algebra $\mathrm{End}(V_\lambda)$ does not depend on the choice of V_λ in the class λ. For, it follows from Schur's Lemma that the algebras $\mathrm{End}(V_1)$ and $\mathrm{End}(V_2)$ are *canonically* isomorphic for any two isomorphic irreducible representations V_1, V_2.

An element $u \in \mathrm{End}(\mathcal{U})$ is *unitary* if $u^*u = uu^* = 1$. The group of unitary elements is isomorphic to the product

$$\prod_{\lambda \in G^\vee} U(d_\lambda)$$

where $U(d_\lambda) \subset \mathrm{End}(V_\lambda)$ is a unitary group of dimension $d_\lambda = \dim V_\lambda$. We have proved that $\mathcal{T}(G)$ is a closed subgroup of this product.

The theory of Fourier transforms on compact groups can now be described. The *Fourier transform* of a continuous map $f : G \longrightarrow \mathbb{C}$ is an element $\mathcal{F}f \in \mathrm{End}(\mathcal{U})$. It is defined by the integral

$$(\mathcal{F}f)_V = \int_G f(x)\, \pi_V(x)\, dx .$$

It is easy to see that

$$\mathcal{F}(f * g) = (\mathcal{F}f)(\mathcal{F}g),$$

where $f * g$ is the convolution of f and g given by

$$(f * g)(x) = \int_G f(x\, y^{-1})\, g(y)\, dy .$$

Also, we have

$$\mathcal{F}f^* = (\mathcal{F}f)^*$$

where f^* is defined by $f^*(x) = \bar{f}(x^{-1})$.

It follows from Proposition 4 that $\mathcal{F}f$ is entirely determined by its effect on irreducible representations

$$(\mathcal{F}f)(\lambda) = \int_G f(x)\, \pi_\lambda(x)\, dx ,$$

so that $\mathcal{F}$ defines a map

$$\mathcal{F}: \mathcal{C}(G, \mathbf{C}) \longrightarrow \prod_{\lambda \in G^\vee} \mathrm{End}(V_\lambda)$$

on the algebra $\mathcal{C}(G, \mathbf{C})$ of complex-valued continuous functions on G. We would like to describe an inverse to $\mathcal{F}$. It should be noted that $\mathcal{F}$ cannot be surjective because we have

$$\| \mathcal{F}f(\lambda)\| \leq \| f \|_\infty$$

where the norm on the C*-algebra $\mathrm{End}(V_\lambda)$ is used on the left hand side. This shows that $\mathcal{F}$ lands in the set of elements of bounded norm

$$\| u \| = \sup_{\lambda \in G^\vee} \| u_\lambda \| < +\infty .$$

This set forms a genuine C*-algebra

$$\prod_{\lambda \in G^\vee}^{\mathrm{bded}} \mathrm{End}(V_\lambda) \subset \prod_{\lambda \in G^\vee} \mathrm{End}(V_\lambda) .$$

We begin by describing a partial inverse $\mathcal{F}^{-1}$ to $\mathcal{F}$:

$$\mathcal{F}^{-1}: \sum_{\lambda \in G^\vee} \mathrm{End}(V_\lambda) \longrightarrow \mathcal{C}(G, \mathbf{C})$$

$$(\mathcal{F}^{-1}g)(x) = \sum_{\lambda \in G^\vee} \mathrm{Tr}(g(\lambda)\, \pi_\lambda(x)^{-1})\, d_\lambda .$$

But, to prove anything about this inverse, we need the orthogonality relations.

Lemma 6. *If* $\lambda, \delta \in G^\vee$ *and* $A \in \mathrm{Hom}(V_\delta, V_\lambda)$ *then*

$$\int_G \pi_\lambda(x) A \pi_\delta(x)^{-1} dx = \begin{cases} \frac{\mathrm{Tr}(A)}{d_\lambda}\mathrm{id} & \text{if } \lambda = \delta \\ 0 & \text{otherwise} . \end{cases}$$

Proof. This is a classical lemma, but we give the proof for completeness. The left hand side is an operator obtained by averaging the function $x \longmapsto \pi_\lambda(x)\, A\, \pi_\delta(x)^{-1}$ over the group. It is therefore an invariant map from V_δ to V_λ. By Schur's lemma, it must be equal to 0 if $\lambda \neq \delta$ and a scalar multiple of the identity if $\lambda = \delta$. To check the equality, we just need to take the trace on both sides.**qed**

Lemma 7. *If* $\lambda, \delta \in G^\vee$ *and* $v \in V_\lambda$, $w \in V_\delta$ *then*

$$\int_G \pi_\lambda(x)(v) \otimes \pi_\delta(x)^{-1}(w)\, dx = \begin{cases} \frac{w \otimes v}{d_\lambda} & \text{if } \lambda = \delta \\ 0 & \text{otherwise} . \end{cases}$$

Proof. It is enough to prove that, for any linear form $\phi: V_\lambda \longrightarrow \mathbf{C}$, we have

$$\int_G \pi_\lambda(x)(v)\, \phi(\pi_\delta(x)^{-1}(w))\, dx = \begin{cases} \frac{w\phi(v)}{d_\lambda} & \text{if } \lambda = \delta \\ 0 & \text{otherwise} . \end{cases}$$

This equality is a special case of Lemma 6 applied to the operator A given by $A(y) = \phi(y)\, v$.**qed**

Proposition 8 (Orthogonality Relations). *If* $\lambda, \delta \in G^\vee$ *and* $A \in \mathrm{End}(V_\lambda)$, $B \in \mathrm{End}(V_\delta)$ *then*

$$\text{(i)} \int_G \mathrm{Tr}(A\,\pi_\lambda(x))\,\mathrm{Tr}(B\,\pi_\delta(x^{-1}))\,dx = \begin{cases} \frac{\mathrm{Tr}(AB)}{d_\lambda} & \text{if } \lambda = \delta \\ 0 & \text{otherwise} . \end{cases}$$

$$\text{(ii)} \int_G \mathrm{Tr}(A\,\pi_\lambda(x))\,\overline{\mathrm{Tr}(B\,\pi_\delta(x))}\,dx = \begin{cases} \frac{\mathrm{Tr}(AB^*)}{d_\lambda} & \text{if } \lambda = \delta \\ 0 & \text{otherwise} . \end{cases}$$

Proof. It suffices to verify relation (i) when both A and B are of rank 1, since the identity is bilinear in A, B and the rank 1 operators span all the operators. Put $A(x) = \phi(x)\,v$ and $B(x) = \psi(x)\,v$. The result follows from application of the linear form $\phi \otimes \psi : V_\lambda \otimes V_\delta \longrightarrow \mathbf{C}$ to the equality in Lemma 7. The second relation follows from the first and

$$\overline{\mathrm{Tr}(B\pi_\delta(x))} = \mathrm{Tr}(B^*\pi_\delta(x)^*) = \mathrm{Tr}(B^*\pi_\delta(x)^{-1}) \text{ .}_{\text{qed}}$$

Corollary 9. *If* $\lambda, \delta \in G^\vee$ *and* $A \in \mathrm{End}(V_\lambda)$ *then the following identity holds:*

$$d_\lambda \int_G \mathrm{Tr}(A\,\pi_\lambda(x))\,\pi_\delta(x^{-1})\,dx = \begin{cases} A & \text{for } \lambda = \delta \\ 0 & \text{otherwise} . \end{cases}$$

Proof. Since the pairing $\langle A,B\rangle = \mathrm{Tr}(AB)$ is exact, it suffices to check the equation after applying $\mathrm{Tr}(B\,-)$ on both sides. But then the orthogonality relations give the result.$_{\text{qed}}$

On $\mathcal{C}(G, \mathbf{C})$ we introduce the inner product

$$\langle f, g\rangle = \int_G \bar{f}(x)g(x)dx .$$

Note that we have

$$\langle f, g\rangle = \varepsilon(f^* * g)$$

where $\varepsilon : \mathcal{C}(G, \mathbf{C}) \longrightarrow \mathbf{C}$ is the functional "evaluate at $e \in G$" : $\varepsilon(f) = f(e)$.

On $\sum_{\lambda \in G^\vee} \mathrm{End}(V_\lambda)$ we introduce the inner product

$$\langle g, h\rangle = \sum_{\lambda \in G^\vee} \mathrm{Tr}(g^*(\lambda)\,h(\lambda))d_\lambda .$$

Proposition 10. *The following equalities hold:*

(a) $\mathcal{F}\mathcal{F}^{-1} g = g$,

(b) $\mathcal{F}^{-1}(g\,h) = (\mathcal{F}^{-1} g) * (\mathcal{F}^{-1} h)$,

(c) $\mathcal{F}^{-1}(g^*) = (\mathcal{F}^{-1} g)^*$,

(d) $\langle \mathcal{F}^{-1} g, \mathcal{F}^{-1} h\rangle = \langle g, h\rangle$.

Proof. We have

$$(\mathcal{F}(\mathcal{F}^{-1}g))(\delta) = \sum_{\lambda \in G^\vee} \int_G d_\lambda\,\mathrm{Tr}(g(\lambda)\pi_\lambda(x)^{-1})\,\pi_\delta(x)\,dx .$$

But, by Corollary 9, we have

$$d_\lambda \int_G \mathrm{Tr}(g(\lambda)\pi_\lambda(x^{-1}))\pi_\delta(x)dx = \begin{cases} g(\lambda) & \text{for } \lambda = \delta \\ 0 & \text{otherwise} \end{cases} .$$

The equality $\mathcal{F}\mathcal{F}^{-1} g = g$ follows. Similarly for (b). The equality (c) is trivial. To prove the last, we note that

$$\varepsilon(\mathcal{F}^{-1}g) = \sum_{\lambda \in G^\vee} \mathrm{Tr}(g(\lambda))d_\lambda ,$$

and therefore

$$\begin{aligned} \langle \mathcal{F}^{-1}g, \mathcal{F}^{-1}h \rangle &= \varepsilon((\mathcal{F}^{-1}g)^* * (\mathcal{F}^{-1}h)) \\ &= \varepsilon\, \mathcal{F}^{-1}(g^* h) \\ &= \sum_{\lambda \in G^\vee} \mathrm{Tr}(g^*(\lambda)h(\lambda))d_\lambda \\ &= \langle g, h \rangle . \text{qed} \end{aligned}$$

The domain of $\mathcal{F}^{-1}$ can be completed by using the norm

$$\| u \|^2 = \langle u, u \rangle .$$

Its completion is a Hilbert space isomorphic to the Hilbert sum

$$\sum_{\lambda \in G^\vee}^{\text{hilbert}} \mathrm{End}(V_\lambda)$$

where $\mathrm{End}(V_\lambda)$ is given the Hilbert space metric defined by

$$\| A \|^2 = d_\lambda \mathrm{Tr}(A^* A).$$

The continuous extension of $\mathcal{F}^{-1}$ defines an isometric embedding

$$\mathcal{F}^{-1} : \sum_{\lambda \in G^\vee}^{\text{hilbert}} \mathrm{End}(V_\lambda) \longrightarrow L^2(G) .$$

Definition. Let M be a topological monoid. A function $f : M \longrightarrow \mathbf{C}$ is said to be *representative* of $(V, \pi_V) \in \mathcal{R}ep(M,\mathbf{C})$ when there is a linear form

$$\phi : \mathrm{End}(V) \longrightarrow \mathbf{C}$$

such that $f = \phi \circ \pi_V$.

Equivalently, f is representative of (V, π_V) when there is some $A \in \mathrm{End}(V)$ such that $f(x) = \mathrm{Tr}(A\, \pi_V(x))$ for every $x \in M$. This means that f is a linear combination of the coefficients $\pi_{ij}(x)$ of π_V in some basis of V. We denote by $R(V, \pi_V)$, or $R(V)$, the set of functions that are representative of (V, π_V). Let $R(M)$ be the set of functions on M which are representative of some representation (V, π_V).

Proposition 11. $R(M)$ *is a subalgebra of* $\mathcal{C}(M, \mathbf{C})$ *which is closed under conjugation* $f \mapsto \bar{f}$.

Proof. The result follows from the easy observations below.

$$R(V_1 \oplus V_2) = R(V_1) + R(V_2), \quad R(V_1 \otimes V_2) \subseteq R(V_1)R(V_2)$$

$$R(\bar{V}) = \overline{R(V)}$$

$$R(I) = \{\text{constant functions}\} \text{ qed}$$

Theorem 12 (Peter-Weyl). *For any compact group* G *and any* $x \in G$, $x \neq e$, *there exists a finite dimensional representation* (V, π_V) *such that* $\pi_V(x) \neq \mathrm{id}$.

Corollary 13. *On a compact group any continuous function can be uniformly approximated by representative functions.*

Proof. Let $x, y \in G$, $x \neq y$. Theorem 12 says that there exists a representation π_V such that $\pi_V(x y^{-1}) \neq \mathrm{id}$; that is, $\pi_V(x) \neq \pi_V(y)$. But there exists a linear form $\phi : \mathrm{End}(V) \longrightarrow \mathbf{C}$ such that

$$\phi(\pi_V(x)) \neq \phi(\pi_V(y)).$$

This shows that the subalgebra R(G) separates points; and since R(G) is closed under conjugation, the result follows from the Stone-Weierstrass theorem.$_{\mathbf{qed}}$

Theorem 14 (Plancherel theorem for compact groups). *The Fourier transform* $\mathcal{F}$ *can be extended continuously to an isometry*

$$\mathcal{F} : L^2(G) \xrightarrow{\ \simeq\ } \sum_{\lambda \in \check{G}}^{\mathrm{hilbert}} \mathrm{End}(V_\lambda) .$$

Proof. We have already defined an isometric embedding

$$\mathcal{F}^{-1} : \sum_{\lambda \in \check{G}}^{\mathrm{hilbert}} \mathrm{End}(V_\lambda) \longrightarrow L^2(G) .$$

The subspace $\mathrm{Im}(\mathcal{F}^{-1})$ is closed since it is complete. But it contains the dense subspace R(G); so $\mathrm{Im}(\mathcal{F}^{-1}) = L^2(G)$, and the theorem is proved.$_{\mathbf{qed}}$

The group G acts on the left and right of $L^2(G)$ via the equalities:

$$(x \bullet f)(y) = f(x^{-1} y), \qquad (f \bullet x)(y) = f(y x^{-1}).$$

Also, G acts on both sides of each of the Hilbert spaces $\mathrm{End}(V_\lambda)$:

$$x \bullet A = \pi_\lambda(x) A, \qquad A \bullet x = A \pi_\lambda(x) .$$

The Fourier transform respects these actions:

$$\mathcal{F}(x \bullet f) = x \bullet (\mathcal{F}f), \qquad \mathcal{F}(f \bullet x) = (\mathcal{F}f) \bullet x .$$

A function $f \in \mathcal{C}(G, \mathbf{C})$ is a *class* function if it is constant on conjugacy classes of G; or equivalently, if it is invariant under conjugation:

$$f = x \bullet f \bullet x^{-1}.$$

The class functions are exactly the members of the *centre* of the algebra $\mathcal{C}(G, \mathbf{C})$ with respect to the convolution product.

Proposition 15. *An element* $u \in \mathrm{End}(\mathcal{U})$ *is central if and only if, for every* $V \in \mathcal{R}ep(G, \mathbf{C})$, *the linear map* $u_V : V \longrightarrow V$ *is a G-homomorphism.*

Proof. If u is central then, for every $x \in G$, we have $u\,\pi(x) = \pi(x)\,u$, and so

$$u_V \pi_V(x) = \pi_V(x) u_V$$

for every V. Conversely, if u_V is a G-homomorphism then, for every $w \in \mathrm{End}(\mathcal{U})$, we have the commutative square

$$\begin{array}{ccc} V & \xrightarrow{u_V} & V \\ {\scriptstyle w_V}\downarrow & & \downarrow{\scriptstyle w_V} \\ V & \xrightarrow[u_V]{} & V \end{array}$$

showing that $uw = wu$; so u is central.$_{\text{qed}}$

We would like to restrict the Fourier transform to its central part. If $f \in \mathcal{C}(G, \mathbf{C})$ is a class function then $x \bullet f = f \bullet x$, and therefore

$$\pi(x)(\mathcal{F}f) = (\mathcal{F}f)\,\pi(x),$$

which shows that $\mathcal{F}f$ is central.

For any $\lambda \in G^{\vee}$, the map $(\mathcal{F}f)(\lambda) : V_\lambda \longrightarrow V_\lambda$ is a G-homomorphism. According to Schur's lemma, it must be a scalar multiple of the identity. Denoting this scalar by $(Tf)(\lambda)$, we have

$$(Tf)(\lambda) = \frac{1}{d_\lambda}\,\mathrm{Tr}(Tf)(\lambda) = \frac{1}{d_\lambda}\int_G f(x)\,\chi_\lambda(x)\,dx$$

where $\chi_\lambda(x) = \mathrm{Tr}(\pi_\lambda(x))$ is the character of the irreducible representation V_λ. From the relation $(\mathcal{F}f)(\lambda) = (Tf)(\lambda)\,\mathrm{id}_\lambda$, we deduce the relations

$$\begin{aligned} T(f * g) &= (Tf)(Tg), \\ T(f^*) &= \overline{Tf}. \end{aligned}$$

The centre of each algebra $\mathrm{End}(V_\lambda)$ is $\mathbf{C} \bullet \mathrm{id}_\lambda$. According to Propositions 4 and 15, the centre of the algebra $\mathrm{End}(\mathcal{U})$ is equal to the product

$$\prod_{\lambda \in G^{\vee}} \mathbf{C} \cdot \mathrm{id}_\lambda \;\cong\; \mathbf{C}^{G^{\vee}}.$$

We define the inverse Fourier transform of any function $g : G^{\vee} \longrightarrow \mathbf{C}$ of finite support by

$$(T^{-1}g)(x) = \sum_{\lambda \in G^{\vee}} g(\lambda)\overline{\chi_\lambda(x)}\,d_\lambda .$$

It is a consequence of Proposition 10 that:

(a) $T\,T^{-1}\,g = g$,

(b) $T^{-1}(g\,h) = (T^{-1}g) * (T^{-1}h)$,

(c) $T^{-1}(g^*) = (T^{-1}g)^*$,

(d) $\langle T^{-1}g, T^{-1}h\rangle = \langle g, h\rangle$.

In this last equality, the inner product on the right hand side is defined by

$$\langle g, h\rangle = \sum_{\lambda \in G^{\vee}} \overline{g}(\lambda)\,h(\lambda)\,d_\lambda^2 .$$

The *spectral measure* d on $G^{\vee}$ is the one assigning weight $d_\lambda^{\,2}$ to the singletons $\{\lambda\}$. The Hilbert space $L^2(G^{\vee})$ is the space of square summable functions with respect to the spectral measure. We write $C(G)$ for the space of conjugacy classes of G. It is the orbit space of G acting on itself by conjugation. There is a canonical measure on $C(G)$ obtained by taking the image of the Haar measure along the projection $G \longrightarrow C(G)$. Obviously,

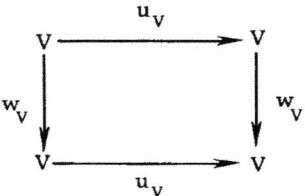

$L^2(C(G))$ is isomorphic to the subspace of $L^2(G)$ consisting of square integrable class functions.

Theorem 16. *The Fourier transforms* T *and* T^{-1} *have continuous extensions to mutually inverse isometries*

$$L^2(C(G)) \xleftrightarrow{\ \simeq\ } L^2(G^\vee).$$

Proof. This is just a description of the restriction of the Fourier transforms $\mathcal{F}$ and $\mathcal{F}^{-1}$ to the central parts.$_{\mathbf{qed}}$

A collection $\mathcal{X}$ of representations of G is *closed* when it contains
(i) π_1 if π_1 is isomorphic to some $\pi_2 \in \mathcal{X}$,
(ii) π_1 if π_1 is a subrepresentation of $\pi_2 \in \mathcal{X}$,
(iii) $\pi_1 \oplus \pi_2$ if $\pi_1, \pi_2 \in \mathcal{X}$,
(iv) $\pi_1 \otimes \pi_2$ if $\pi_1, \pi_2 \in \mathcal{X}$,
(v) $\bar{\pi}_1$ if $\pi_1 \in \mathcal{X}$,
(vi) I.

The set of representative functions of the members of $\mathcal{X}$ is a subalgebra $R(\mathcal{X})$ of the algebra R(G).

Lemma 17. *Let* $\mathcal{X}$ *be a closed collection of representations of* G. *Suppose that, for every* $x \in G$, $x \neq e$, *there exists a representation* $\pi_V \in \mathcal{X}$ *such that* $\pi_V(x) \neq e$. *Then* $\mathcal{X} = \mathcal{R}ep(G,C)$.

Proof. If $\mathcal{X} \neq \mathcal{R}ep(G,C)$, there will be an element $\lambda \in G^\vee$ such that $\pi_\lambda \notin \mathcal{X}$. The orthogonality relations then imply that the representative functions of π_λ are orthogonal to the elements of $R(\mathcal{X})$. In particular, χ_λ is orthogonal to $R(\mathcal{X})$. On the other hand, the hypothesis implies that the subalgebra $R(\mathcal{X})$ separates the points of G and is closed under conjugation. According to the Stone-Weierstrass theorem, $R(\mathcal{X})$ is dense in $\mathcal{C}(G, C)$. This is a contradiction.$_{\mathbf{qed}}$

We would like to prove that, for any compact group G, the map

$$\pi : G \longrightarrow \mathcal{T}(G)$$

induces, via restriction, an isomorphism of algebras

$$\pi^* : R(\mathcal{T}(G)) \longrightarrow R(G).$$

We shall obtain this from the next result.

Lemma 18. *The restriction functor*

$$\pi^* : \mathcal{R}ep(\mathcal{T}(G), C) \longrightarrow \mathcal{R}ep(G, C)$$

is an equivalence of categories.

Proof. We define an extension functor

$$e : \mathcal{R}ep(G, C) \longrightarrow \mathcal{R}ep(\mathcal{T}(G), C)$$

as follows. For any $V \in \mathcal{R}ep(G, C)$, the map $u \longmapsto u_V$ is a representation of $\mathcal{T}(G)$ on V. The commutative triangle

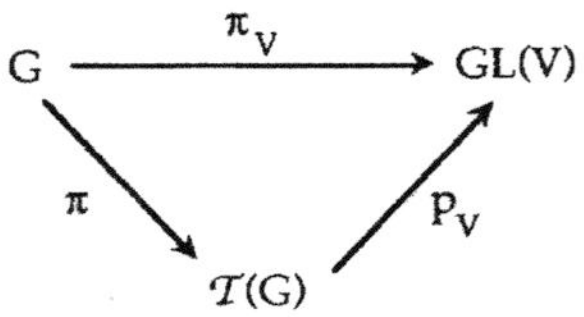

shows that $\pi^*(p_V) = \pi_V$. We put $e(V, \pi_V) = (V, p_V)$. For any morphism $h : (V, \pi_V) \longrightarrow (W, \pi_W)$ in $\mathcal{Rep}(G, \mathbf{C})$ and any $u \in \mathcal{T}(G)$, we have $h \circ u_V = u_W \circ h$. This shows that h is also a morphism $h : (V, p_V) \longrightarrow (W, p_W)$ in $\mathcal{Rep}(\mathcal{T}(G), \mathbf{C})$. It is easy to see that e preserves direct sums, tensor products and conjugate representations. Moreover, from the same commutative triangle (above), if (V, π_V) is irreducible then $e(V, \pi_V)$ is irreducible. From this it follows that the image of the functor e is a full subcategory whose objects constitute a closed collection $\mathcal{X}$ of representations of $\mathcal{T}(G)$. We prove that $\mathcal{X}$ separates the elements of $\mathcal{T}(G)$. Let $u \in \mathcal{T}(G)$, $u \neq e$. There exists a representation V of G such that $u_V \neq id_V$. This means that $p_V(u) \neq id_V$ and thus p_V separates u from e. By Lemma 17, the collection $\mathcal{X}$ is all of $\mathcal{Rep}(\mathcal{T}(G), \mathbf{C})$, and so e is an equivalence of categories. Finally, π^* is an equivalence since $\pi^* \circ e = id$ .qed

Lemma 19. *The restriction map*

$$\pi^* : R(\mathcal{T}(G)) \longrightarrow R(G)$$

is an isomorphism of algebras.

Proof. We define an inverse $e : R(G) \longrightarrow R(\mathcal{T}(G))$ in the following manner. Any $f \in R(G)$ has a unique representation in the form

$$f(x) = \sum_{\lambda \in G^\vee} Tr(g(\lambda)\, \pi_\lambda(x))\, d_\lambda$$

where $g \in \sum_{\lambda \in G^\vee} End(V_\lambda)$. (In fact, $g = \mathcal{F}f$.) We define

$$e(f)(u) = \sum_{\lambda \in G^\vee} Tr(g(\lambda)\, p_\lambda(u))\, d_\lambda .$$

According to Lemma 18, we have $G^\vee = \mathcal{T}(G)^\vee$ from which the bijectivity of e follows. Finally, π^* is bijective since $\pi^* \circ e = id$ .qed

Theorem 20 (Tannaka-Krein). *For any compact group* G, *the canonical map*

$$\pi : G \longrightarrow \mathcal{T}(G)$$

is an isomorphism.

Proof. We first prove the injectivity of π. According to Peter-Weyl (Theorem 12), for any $x \in G$, $x \neq e$, there is a representation (V, π_V) such that $\pi_V(x) \neq id$. But then $\pi(x) \neq e$ since $p_V(\pi(x)) = \pi_V$. To prove the surjectivity of π, we first prove

$$(*) \qquad \int_{\mathcal{T}(G)} f(u)\, du = \int_G f(\pi(x))\, dx$$

for any $f \in \mathcal{C}(\mathcal{T}(G), \mathbf{C})$. It is enough to prove this equality when $f \in R(\mathcal{T}(G))$ since

$R(\mathcal{T}(G))$ is dense in $\mathcal{C}(\mathcal{T}(G), \mathbf{C})$ by Corollary 13. Using Lemma 19, we see that a function f $\in R(\mathcal{T}(G))$ has a unique representation

$$f(u) = \sum_{\lambda \in G^{\vee}} \mathrm{Tr}(g(\lambda)\, p_{\lambda}(u))\, d_{\lambda}\,,$$

so that

$$f(\pi(x)) = \sum_{\lambda \in G^{\vee}} \mathrm{Tr}(g(\lambda)\, \pi_{\lambda}(x))\, d_{\lambda}\,.$$

But the orthogonality relations give

$$\int_G \mathrm{Tr}(g(\lambda)\pi_{\lambda}(u))\, du = \begin{cases} g(I) & \text{for } \lambda = I \\ 0 & \text{otherwise}\end{cases}.$$

Similarly,

$$\int_{\mathcal{T}(G)} \mathrm{Tr}(g(\lambda)\, p_{\lambda}(u))\, du = \begin{cases} g(I) & \text{for } \lambda = I \\ 0 & \text{otherwise}\end{cases}.$$

Equality (*) follows.

To prove the surjectivity, let us suppose that $\mathrm{Im}(\pi) \neq \mathcal{T}(G)$. Let f be a positive function whose support is contained in the complement of the closed subset $\mathrm{Im}(\pi)$. Then we have

$$\int_{\mathcal{T}(G)} f(u)\, du > 0 \quad \text{and} \quad \int_G f(\pi(x))\, dx = 0$$

which is a contradiction.$_{\mathbf{qed}}$

§2. The bialgebra of representative functions.

This Section provides a more complete description of the algebra of representative functions on a compact group and, more generally, on a topological monoid. We shall see that $R(M)$ is a bialgebra, and we shall compute it in some special cases. We begin by giving a simple characterization of representative functions on locally compact monoids.

Let M be a locally compact monoid. For each $y \in M$, consider the translation operators $\lambda_y(x) = yx$ and $\rho_y(x) = xy$. We enrich $\mathcal{C}(M, \mathbf{C})$ with the topology of uniform convergence on compact subsets of M. The maps

$$(y, f) \longmapsto f \circ \rho_y \qquad \text{and} \qquad (f, y) \longmapsto f \circ \lambda_y$$

define continuous left and right actions of M on $\mathcal{C}(M, \mathbf{C})$.

Proposition 1. *A continuous complex-valued function* f *on a locally compact monoid* M *is representative if and only if the linear subspace of* $\mathcal{C}(M, \mathbf{C})$ *spanned by any one of the three subsets*

$$\{ f \circ \lambda_y \mid y \in M \}, \qquad \{ f \circ \rho_y \mid y \in M \}, \qquad \{ f \circ \lambda_y \circ \rho_z \mid y, z \in M \}$$

is finite dimensional.

Proof. Let $\pi_V : M \longrightarrow \mathrm{End}(V)$ be a representation of M. The set $R(V, \pi_V)$ of representative functions for π_V is closed under left and right translation; for, if $f(x) = \mathrm{Tr}(A\, \pi_V(x))$, we have

$$f \circ \lambda_y \circ \rho_z(x) = \mathrm{Tr}(A\, \pi_V(yxz)) = \mathrm{Tr}(\pi_V(z)\, A\, \pi_V(y)\, \pi_V(x)).$$

Conversely, let us assume, for example, that $\{ f \circ \rho_y \mid y \in M \}$ generates a finite dimensional vector subspace V. The continuous left action of M on $\mathcal{C}(M, \mathbf{C})$ restricts to a continuous left action of M on V. We obtain in this way a continuous representation $\pi_V : M \longrightarrow \mathrm{End}(V)$ of M. Let $\varepsilon : V \longrightarrow \mathbf{C}$ be the linear form $\varepsilon(h) = h(e)$ where $e \in M$ is the unit. We have

$$f(x) = \varepsilon(f \circ \rho_x) = \varepsilon(\pi_V(x)(f)) = \mathrm{Tr}(A\, \pi_V(x))$$

where $A \in \mathrm{End}(V)$ is the rank one linear function $h \longmapsto \varepsilon(h)\, f$ .qed

For any topological monoid M, there is a bialgebra structure on R(M) which we now describe. Let

$$\Delta : \mathcal{C}(M, \mathbf{C}) \longrightarrow \mathcal{C}(M \times M, \mathbf{C})$$

be the algebra homomorphism defined by

$$(\Delta f)(x, y) = f(xy).$$

We have canonical inclusions

$$R(M) \otimes R(M) \subset \mathcal{C}(M, \mathbf{C}) \otimes \mathcal{C}(M, \mathbf{C}) \subset \mathcal{C}(M \times M, \mathbf{C}).$$

Lemma 2. *For all* $f \in R(M)$,

$$\Delta f \in R(M) \otimes R(M).$$

Proof. It is enough to prove the result when $f(x) = \pi_{ij}(x)$ where $\pi = (\pi_{ij})$ is some representation of M. In this case we have

$$\pi_{ij}(xy) = \Sigma_k\, \pi_{ik}(x)\, \pi_{kj}(y);$$

that is,

$$\Delta\pi_{ij} = \Sigma_k\, \pi_{ik} \otimes \pi_{kj} \text{ .qed}$$

Hence Δ gives a comultiplication for R(M). The counit $\varepsilon : R(M) \longrightarrow \mathbf{C}$ of the coalgebra is given by $\varepsilon(f) = f(e)$.

When the monoid M is a topological group, the bialgebra R(M) becomes a Hopf algebra. The antipode $\nu : R(M) \longrightarrow R(M)$ is defined by

$$\nu(f)(x) = f(x^{-1}).$$

To see that $\nu(f) \in R(M)$, suppose that $f(x) = \mathrm{Tr}(\pi(x)A)$. Then we have

$$f(x^{-1}) = \mathrm{Tr}(\pi(x^{-1})A) = \mathrm{Tr}({}^tA\, {}^t\pi(x^{-1})) = \mathrm{Tr}(\pi^\vee(x)\, {}^tA)$$

where $\pi^\vee(x) = {}^t\pi(x^{-1})$ is the contragredient representation of π.

The algebra R(G) is easy to describe when G is a compact abelian group. In this case, R(G) is the linear subspace of $\mathcal{C}(G, \mathbf{C})$ spanned by the set $G^\vee$ of characters of G. However, the orthogonality relations imply that $G^\vee$ is a linearly independent subset of $\mathcal{C}(G, \mathbf{C})$. This shows that R(G) is isomorphic to the enveloping algebra $\mathbf{C}[G^\vee]$ (usually called the group algebra) of the dual group $G^\vee$. For any $\chi \in G^\vee$, the relation

$$\chi(xy) = \chi(x)\,\chi(y)$$

implies the relation

$$\Delta\chi = \chi \otimes \chi\,.$$

This shows that the coalgebra structure $\Delta : \mathbf{C}[G^\vee] \longrightarrow \mathbf{C}[G^\vee] \otimes \mathbf{C}[G^\vee]$ is simply the linear extension of the diagonal map $\Delta : G^\vee \longrightarrow G^\vee \times G^\vee$. The antipode ν is the linear extension of the inverse operation $G^\vee \longrightarrow G^\vee$. It is also worthwhile to identify the operation corresponding to conjugation $f \longmapsto \bar{f}$ in R(G): it is the *antilinear* extension of the inverse operation.

If G is the circle group T then R(G) is the ring of finite Fourier series:

$$R(T) = \mathbf{C}[\mathbf{Z}] = \mathbf{C}[z, z^{-1}]$$

where $z = e^{i\theta}$ and $\bar{z} = z^{-1}$.

When G is not abelian, the following result is useful in identifying R(G). For any representation π_V of G, let $\det(\pi_V)$ be the one-dimensional representation obtained as the composite of $\pi_V : G \longrightarrow GL(V)$ and $\det : GL(V) \longrightarrow GL(1)$.

Proposition 3. *Let* G *be a compact group, and let* π_V *be a faithful representation of* G. *The algebra* R(G) *is generated by the coefficients of* π_V *together with* $\det(\pi_V)^{-1}$.

Proof. Let A be the subalgebra of R(G) generated by the coefficients of π_V and $\det(\pi_V)^{-1}$. Let us verify first that A is closed under conjugation. But we have

$$\overline{\det(\pi_V)^{-1}} = \det(\pi_V),$$

and also

$$\overline{R(\pi_V)} = R(\pi_{\overline{V}}) = R(\pi_{V^\vee}).$$

It is therefore sufficient to verify that the entries of the contragredient representation are contained in A. This is true because of the familiar formula expressing the entries of an inverse matrix as cofactors divided by the determinant.

To finish the proof, let $\mathcal{X}$ be the collection of representations whose coefficients belong to A. Then $\mathcal{X}$ is a closed collection, and, since π_V is faithful, we can apply Lemma 16. This gives $\mathcal{X} = \mathcal{R}ep(G, \mathbf{C})$ so that $A = R(G)$.qed

We would like to apply Proposition 3 to compact Lie groups, but to do this we need the following result.

Proposition 4. *Each compact Lie group* G *admits a faithful representation.*

Proof. We shall use the fact that the poset of closed submanifolds of a given compact manifold is artinian. This means that any non-empty collection of submanifolds contains a minimal element; or, equivalently, that any decreasing sequence of submanifolds must stop. For any $V \in \mathcal{R}ep(G, \mathbf{C})$, the kernel of π_V is a closed submanifold of G. Let $\ker(\pi_W)$ be a minimal element in this collection. According to the Peter-Weyl Theorem, for any $x \neq e$, there exists $V \in \mathcal{R}ep(G, \mathbf{C})$ such that $x \notin \ker(\pi_V)$. But we have

$$\ker(\pi_V \oplus \pi_W) = \ker(\pi_W) \cap \ker(\pi_V),$$

so that $\ker(\pi_W) \subseteq \ker(\pi_V)$, and so $x \notin \ker(\pi_W)$. This proves that $\ker(\pi_W) = \{e\}$.qed

The last two Propositions have the following immediate consequence.

Corollary 5. *The algebra of representative functions on a compact Lie group is finitely generated.*

We shall now give a complete description of R(G) for the compact Lie group G = U(n) of unitary $n \times n$ complex matrices. Let $\mathbf{C}[(z_{ij})]$ be the ring of polynomials in the n^2 indeterminates z_{ij} $(1 \leq i, j \leq n)$ and let $d = \det(z_{ij})$. If we map z_{ij} into the coefficient ρ_{ij} of the standard representation

$$\rho : U(n) \hookrightarrow \mathrm{End}(\mathbf{C}^{n\times n}),$$

we obtain a homomorphism

$$\iota : \mathbf{C}[(z_{ij}), d^{-1}] \longrightarrow R(U(n))$$

sending d^{-1} to $\det(\rho)^{-1}$.

Proposition 6. *The homomorphism* ι *is an algebra isomorphism, so that*

$$R(U(n)) \cong \mathbf{C}[(z_{ij}), d^{-1}].$$

Proof. Proposition 3 implies that ι is surjective. To show that ι is injective, let us first remark that, for any $P(z) \in \mathbf{C}[(z_{ij}), d^{-1}]$, the function $\iota(P(z))$ on $U(n)$ is exactly the function $u \longmapsto P(u)$. Therefore, if $\iota(P(z)) = 0$ then $P(u) = 0$ for every $u \in U(n)$. We have to show that this implies that $P = 0$. This property is sometimes formulated by saying that $U(n)$ is Zariski dense in $GL(n, \mathbf{C})$ (it is the famous "unitarian trick" of Hermann Weyl which he used to reduce the theory of rational representations of $GL(n, \mathbf{C})$ to the easier case of continuous representations of $U(n)$). Although this is very classical, we give a brief proof. The *Cauchy transformation* $Z = C(A)$ of a matrix $A \in \mathrm{End}(\mathbf{C}^n)$ is defined by the formula

$$Z = (I + A)(I - A)^{-1}.$$

Its inverse is given by

$$A = (Z - I)(Z + I)^{-1}.$$

The Cauchy transformation is a birational correspondence which is well defined between a neighbourhood of $A = 0$ and a neighbourhood of $Z = I$ in $\mathbf{C}^{n\times n}$. When A is anti-Hermitian, Z is unitary, and conversely. As well as the Cauchy transformation, we shall use the invertible linear transformation $A = L(B)$ defined by

$$A = (B - {}^tB) + i(B + {}^tB),$$

$$B = \frac{A - {}^tA}{2} + \frac{A + {}^tA}{2i}.$$

When B is real, A is anti-Hermitian, and conversely. By hypothesis, the function

$$Q(B) = P(C(L(B)))$$

is a rational function of B which vanishes identically when B is real in some neighbourhood of 0. This clearly implies that $Q = 0$ identically; and therefore $P = 0$ identically since

$$P(Z) = Q(L^{-1}(C^{-1}(Z))) \,.\text{qed}$$

The Hopf algebra structure on $\mathbf{C}[(z_{ij}), d^{-1}]$ which corresponds to the one on $R(U(n))$ is easy to identify. The comultiplication is given by

$$\Delta z_{ij} = \Sigma_k z_{ik} \otimes z_{kj}$$

and the counit by

$$\varepsilon z_{ij} = \delta_{ij} \text{ (the Kronecker delta)}.$$

The antipode is given by

$$\nu z = z^{-1}$$

where $z = (z_{ij})$ and $\nu z = (\nu z_{ij})$. With this notation, we have

$$\Delta z = z \otimes z$$

where $\Delta z = (\Delta z_{ij})$ and $z \otimes z = (\Sigma_k z_{ik} \otimes z_{kj})$.

We should also identify the conjugation operation $(\overline{\ })$ on $\mathbf{C}[(z_{ij}), d^{-1}]$ which corresponds to the conjugation $f \longmapsto \bar{f}$ on $R(U(n))$. It is easy to see that $(\bar{\ })$ is the unique antilinear ring homomorphism such that

$$\bar{z} = {}^t z^{-1}$$

where $\bar{z} = (\bar{z}_{ij})$.

For any commutative C-algebra A, we define its *spectrum* $\mathrm{Spec}(A)$ as the set of all algebra homomorphisms $\chi : A \longrightarrow C$. We give $\mathrm{Spec}(A)$ the topology of pointwise convergence. When A is finitely generated, say $A = C[a_1, \ldots, a_n]$, the mapping $\chi \longmapsto (\chi(a_1), \ldots, \chi(a_n))$ is a homomorphism between $\mathrm{Spec}(A)$ and a closed algebraic subset of C^n. When A is enriched with an antilinear involution $a \longmapsto \bar{a}$, we can define a conjugation operation on $\mathrm{Spec}(A)$ by putting, for all $a \in A$,

$$\bar{\chi}(a) = \overline{\chi(\bar{a})}\ .$$

The *real spectrum* $\mathrm{Spec}_{\mathbf{R}}(A)$ of A is the set of $\chi \in \mathrm{Spec}(A)$ such that $\bar{\chi} = \chi$.

The spectrum of a finitely generated commutative Hopf algebra is a complex algebraic group. If the Hopf algebra is enriched with an antilinear involution (respecting the Hopf algebra structure) then its real spectrum is a real algebraic group. In the example above we see that the spectrum of $R(U(n))$ is the algebraic group $GL(n, C)$ and its real spectrum is $U(n)$. This is a special case of a more general result. For any compact group G, consider the canonical map

$$G \longrightarrow \mathrm{Spec}_{\mathbf{R}}(R(G))$$

sending $x \in G$ into the homomorphism χ_x given by $\chi_x(f) = f(x)$.

Theorem 7. *For any compact group* G, *there is a canonical homeomorphism*

$$G \xrightarrow{\ \approx\ } \mathrm{Spec}_{\mathbf{R}}(R(G))\ .$$

The proof of the above Theorem will be delayed until the next Section. It follows from Tannaka reconstruction. We note immediately this striking consequence:

Corollary 8. *Every compact Lie group is (real) algebraic.*

Proof. If G is a compact Lie group then, according to Corollary 5, the Hopf algebra $R(G)$ is finitely generated. So $\mathrm{Spec}(R(G))$ is therefore a complex algebraic group whose real part is equal to G.**qed**

To end this Section, we prove the following result needed in Section 8. Let M and M' be topological monoids. We have an obvious canonical map

$$i : R(M) \otimes R(M') \longrightarrow R(M \times M').$$

Proposition 9. *The canonical map* i *(above) is an isomorphism.*

Proof. Clearly i is injective. We prove it surjective. It suffices to prove that the matrix entries π_{rs} of any representation π_V of $M \times M'$ belongs to the image of i. We have

$$\pi_{rs}(x, y) = \Sigma_t\, \pi_{rt}(x, e)\, \pi_{ts}(e, y).$$

This shows that

$$\pi_{rs} = i\,(\Sigma_t\, \pi^1{}_{rt} \otimes \pi^2{}_{ts})$$

where $\pi^1{}_V(x) = \pi_V(x, e)$ and $\pi^2{}_V(y) = \pi_V(e, y)$.**qed**

§3. The Fourier cotransform.

In this Section we shall describe a transformation which has the Fourier transform as its dual. It is more fundamental than the Fourier transform in the sense that it behaves better algebraically. Starting with a category $\mathcal{C}$ and a functor $X : \mathcal{C} \longrightarrow \mathcal{V}ect_{\mathbb{C}}$ with values in finite dimensional vector spaces, we shall construct a certain vector space $\mathrm{End}^{\vee}(X)$ whose dual is the algebra $\mathrm{End}(X) = \mathrm{Hom}(X, X)$ of natural transformations from X to X. When $\mathcal{C} = \mathcal{R}ep(M, \mathbb{C})$ and $X = \mathcal{U}$ is the forgetful functor, the Fourier cotransform defines an isomorphism

$$\mathcal{F}^{\vee} : \mathrm{End}^{\vee}(\mathcal{U}) \xrightarrow{\sim} R(M)$$

whose transpose is the (generalised) Fourier transform. In Section 4 we shall show that $\mathrm{End}^{\vee}(\mathcal{U})$ supports a natural coalgebra structure and that $\mathcal{F}^{\vee}$ is an isomorphism of coalgebras.

Let $\mathcal{C}$ be a small category. For any pair of functors $X, Y : \mathcal{C} \longrightarrow \mathcal{V}ect_{\mathbb{C}}$, let $\mathrm{Hom}(X, Y)$ be the set of natural transformations between X and Y. Now $\mathrm{Hom}(X, Y)$ is a vector space which can be described as follows. For any arrow $f : A \longrightarrow B$ in $\mathcal{C}$, let

$$p_f \, , \, q_f \; : \; \prod_{C \in \mathcal{C}} \mathrm{Hom}(X(C), Y(C)) \longrightarrow \mathrm{Hom}(X(A), Y(B))$$

be the maps out of the product whose values at $u = (u_C \mid C \in \mathcal{C})$ are given by

$$p_f(u) = Y(f)\, u_A \, , \qquad q_f(u) = u_B \, X(f) \, .$$

Then clearly, $\mathrm{Hom}(X, Y)$ is the common equalizer of the pairs (p_f, q_f):

$$u \in \mathrm{Hom}(X, Y) \quad \Leftrightarrow \quad \text{for all } f \in \mathcal{C}, \; p_f(u) = q_f(u) \, .$$

This construction can be dualised when $X(C)$ and $Y(C)$ are finite dimensional for every $C \in \mathcal{C}$ (actually we just need $Y(C)$ to be finite dimensional for every $C \in \mathcal{C}$, but here we shall not consider this more general situation). More precisely, we want to construct a vector space $\mathrm{Hom}^{\vee}(X, Y)$ whose dual will be $\mathrm{Hom}(X, Y)$ (it is therefore suggestive to think of $\mathrm{Hom}^{\vee}(X, Y)$ as a *predual* of $\mathrm{Hom}(X, Y)$). By definition, $\mathrm{Hom}^{\vee}(X, Y)$ is the common coequalizer of the maps

$${}^t p_f \, , \, {}^t q_f \; : \; \mathrm{Hom}(X(A), Y(B))^* \longrightarrow \sum_{C \in \mathcal{C}} \mathrm{Hom}(X(C), Y(C))^*$$

into the direct sum. But, for any pair of finite dimensional vector spaces V, W, we have canonical isomorphisms

$$\mathrm{Hom}(V, W)^* \cong (V^* \otimes W)^* \cong W^* \otimes V \cong \mathrm{Hom}(W, V) \, .$$

The pairing between $\mathrm{Hom}(V, W)$ and $\mathrm{Hom}(W, V)$ is explicitly given by

$$\langle S, T \rangle = \mathrm{Tr}(ST) = \mathrm{Tr}(TS) \, .$$

Taking this duality into consideration, we see that $\mathrm{Hom}^{\vee}(X, Y)$ can be defined as the common coequalizer of the maps

$$i_f \, , \, j_f \; : \; \mathrm{Hom}(Y(B), X(A)) \longrightarrow \sum_{C \in \mathcal{C}} \mathrm{Hom}(Y(C), X(C))$$

where, for any $S \in \mathrm{Hom}(Y(B), X(A))$,

$$i_f(S) = (S \circ Y(f), A), \qquad j_f(S) = (X(f) \circ S, B).$$

In this description, the second component of a pair $(S \circ Y(f), A)$ indicates to which component of the direct sum this element belongs. A handier description of $\mathrm{Hom}^\vee(X, Y)$ is the following. For any $C \in \mathcal{C}$ and any $S \in \mathrm{Hom}(Y(C), X(C))$, let us write $[S]$ for the image of S under the canonical map

$$\mathrm{Hom}(Y(C), X(C)) \longrightarrow \mathrm{Hom}^\vee(X, Y).$$

We have that $\mathrm{Hom}^\vee(X, Y)$ is generated as a vector space by the symbols $[S]$ subject to the relations

(i) $[\alpha S + \beta T] = \alpha[S] + \beta[T]$ for $S, T \in \mathrm{Hom}(Y(C), X(C))$,

(ii) $[S \circ Y(f)] = [X(f) \circ S]$ for $f : A \longrightarrow B$ and $S \in \mathrm{Hom}(Y(B), X(A))$.

It is clear by construction that $\mathrm{Hom}(X, Y)$ is the linear dual of $\mathrm{Hom}^\vee(X, Y)$. The explicit pairing

$$\mathrm{Hom}(X, Y) \otimes \mathrm{Hom}^\vee(X, Y) \longrightarrow \mathbf{C}, \qquad \langle u, [S] \rangle = \mathrm{Tr}(u_C \circ S)$$

where $S \in \mathrm{Hom}(Y(C), X(C))$.

Before continuing the general study of $\mathrm{Hom}^\vee(X, Y)$, let us compute $\mathrm{Hom}^\vee(X, X) = \mathrm{End}^\vee(X)$ in the special case where $\mathcal{C} = \mathcal{R}ep(M, \mathbf{C})$ and $X = \mathcal{U}$ is the forgetful functor. We define the *Fourier cotransform*

$$\mathcal{F}^\vee : \mathrm{End}^\vee(\mathcal{U}) \longrightarrow R(M),$$

as follows:

$$\mathcal{F}^\vee(z)(x) = \langle \pi(x), z \rangle.$$

More explicitly, if $z = [A]$ where $A \in \mathrm{End}(V)$ and $\pi_V : M \longrightarrow \mathrm{End}(V)$ then

$$\mathcal{F}^\vee([A])(x) = \langle \pi(x), [A] \rangle = \mathrm{Tr}(\pi_V(x)\, A).$$

Theorem 1. *For any topological monoid* M, *the Fourier cotransform is an isomorphism*

$$\mathcal{F}^\vee : \mathrm{End}^\vee(\mathcal{U}) \xrightarrow{\ \simeq\ } R(M).$$

To prove Theorem 1, we shall need two Lemmas.

Lemma 2. *If* $V_1, V_2 \in \mathcal{R}ep(M, \mathbf{C})$, *and if* $A \in \mathrm{End}(V_1 \oplus V_2)$ *is written as*

$$A = \begin{pmatrix} A_{11} & A_{12} \\ A_{21} & A_{22} \end{pmatrix},$$

then

$$[A] = [A_{11}] + [A_{22}].$$

Proof. Let $j_i : V_i \longrightarrow V_1 \oplus V_2$ and $p_i : V_1 \oplus V_2 \longrightarrow V_i$ $(i = 1, 2)$ be the canonical inclusions and projections. We have

$$\begin{aligned} [A] &= [A(j_1 p_1 + j_2 p_2)] \\ &= [A\, j_1 p_1] + [A\, j_2 p_2] \\ &= [p_1 A\, j_1] + [p_2 A\, j_2] \\ &= [A_{11}] + [A_{22}]. \text{ qed} \end{aligned}$$

Lemma 3. *Any* $z \in \mathrm{End}^\vee(\mathcal{U})$ *can be represented as* $z = [\phi \otimes v]$ *for some* $v \in V$ *and* $\phi \in$

V^* *where* $\pi_V : M \longrightarrow \mathrm{End}(V)$. *Moreover, this can be done in such a way that* v *generates* V *as an* M*-module.*

Proof. Each element $z \in \mathrm{End}^\vee(\mathcal{U})$ can be represented as $z = [A]$ for some $A \in \mathrm{End}(W)$ and $W \in \mathcal{R}ep(M, \mathbf{C})$. But we can express A as a sum of matrices of rank one:

$$A = \sum_{i \in I} \phi_i \otimes v_i \ ;$$

so that,

$$z = \sum_{i \in I} [\phi_i \otimes v_i] \,.$$

Let V be the representation obtained as the direct sum of I copies of W, and let

$$v = (\, v_i \,|\, i \in I\,)\,, \qquad \phi = \sum_{i \in I} \phi_i p_i$$

where $p_i : V \longrightarrow W$ is the i-th projection. Using Lemma 2, we see that

$$z = [\,\phi \otimes v\,]\,.$$

To prove the last statement, let V' be the submodule of V generated by v. The inclusion $j : V' \longrightarrow V$ is an arrow in the category $\mathcal{R}ep(M, \mathbf{C})$. Let B be the composite

$$V \xrightarrow{\ \phi\ } \mathbf{C} \xrightarrow{\ v\ } V' \,.$$

Then we have

$$[\phi \otimes v] = [\,j B\,] = [\,B j\,] = [\,(\phi | V') \otimes v\,]._{\mathrm{qed}}$$

Proof of Theorem 1. The surjectivity of $\mathcal{F}^\vee$ is immediate from the definition of representative function. Suppose that $\mathcal{F}^\vee(z) = 0$. According to Lemma 3, we can suppose that $z = [\,\phi \otimes v\,]$ and that v generates the M-module V. Then we have

$$\phi\,(\,\pi_V(x)\,(v)\,) = \mathcal{F}^\vee(z)\,(x) = 0$$

for every $x \in M$. This shows that the submodule generated by v is contained in $\ker(\phi)$. Therefore $\phi = 0$ and $z = 0._{\mathrm{qed}}$

At this point it might be worthwhile to explain why we call $\mathcal{F}^\vee$ the "Fourier cotransform". Suppose G is a compact group. The continuous dual of the Banach space $\mathcal{C}(G, \mathbf{C})$ is the space $\mathcal{M}(G, \mathbf{C})$ of bounded measures on G. The Fourier transform of $\mu \in \mathcal{M}(G, \mathbf{C})$ is the element $\mathcal{F}\mu \in \mathrm{End}(\mathcal{U})$ given by

$$(\mathcal{F}\mu)_V = \int_G \pi_V(x)\, d\mu(x)\,.$$

The Fourier transform of a function $f \in \mathcal{C}(G, \mathbf{C})$ is simply the Fourier transform of $f\,dx$ where dx denotes the Haar measure on G.

Proposition 4. *If* $z \in \mathrm{End}^\vee(\mathcal{U})$ and $\mu \in \mathcal{M}(G, \mathbf{C})$ *then*

$$\langle\, \mathcal{F}\mu\,, z \,\rangle = \langle\, \mu\,, \mathcal{F}^\vee z \,\rangle.$$

Proof. If $z = [\,A\,]$ where $A \in \mathrm{End}(V)$ and $\pi_V : G \longrightarrow \mathrm{End}(V)$ then we have

$$\langle\, \mathcal{F}\mu\,, [\,A\,]\,\rangle = \mathrm{Tr}\,(\,(\mathcal{F}\mu)_V\, A\,)$$

$$= \operatorname{Tr}\Big(\int_G \pi_V(x)\, A\, d\mu(x)\Big)$$

$$= \int_G \operatorname{Tr}(\pi_V(x)\, A)\, d\mu(x)$$

$$= \int_G \mathcal{F}^\vee([A])(x)\, d\mu(x)$$

$$= \langle \mu, \mathcal{F}^\vee([A]) \rangle. \quad \text{q ed}$$

The above Proposition 4 shows that the Fourier transform

$$\mathcal{F} : \mathcal{M}(G, \mathbf{C}) \longrightarrow \operatorname{End}(\mathfrak{U})$$

is the continuous dual of the linear map

$$\mathcal{F}^\vee : \operatorname{End}^\vee(\mathfrak{U}) \longrightarrow R(G) \subset \mathcal{C}(G, \mathbf{C}).$$

At this point, it becomes possible to extend the domain of the Fourier transform. We have a canonical inclusion

$$\mathcal{M}(G, \mathbf{C}) \subset R(G)^*$$

since $R(G)$ is dense in the Banach space $\mathcal{C}(G, \mathbf{C})$ (Section 1 Corollary 13). We shall extend $\mathcal{F}$ to the full linear dual of $R(G)$:

$$\mathcal{F} : R(G)^* \longrightarrow \operatorname{End}(\mathfrak{U}).$$

Let us write $\langle h, f \rangle$ for the evaluation pairing between $R(G)^*$ and $R(G)$. The Fourier transform of $h \in R(G)^*$ can be defined by the formula

$$(\mathcal{F}h)_V = \langle h, \pi_V \rangle$$

by which we mean that, if (π_{ij}) is the matrix of π_V for some basis of V then $(\langle h, \pi_{ij} \rangle)$ is the matrix of $(\mathcal{F}h)_V$ for the same basis.

One might think of the elements of $R(G)^*$ as generalised distributions on G. More precisely, when G is a compact Lie group, it can be proved that $R(G)$ is a dense subspace of the space $\mathcal{C}^\infty(G, \mathbf{C})$ of smooth functions with the smooth topology [Sc]. The continuous dual of $\mathcal{C}^\infty(G, \mathbf{C})$ is the space $\mathcal{D}(G, \mathbf{C})$ of distributions on G. We thus have a canonical inclusion

$$\mathcal{D}(G, \mathbf{C}) \subset R(G)^*,$$

showing that the Fourier transform is defined on distributions. Recall that there is an algebra structure on $\mathcal{D}(G, \mathbf{C})$ given by the convolution product of distributions. There is also an algebra structure on $R(G)^*$ (and more generally on each $R(M)^*$) which is the dual of the coalgebra structure on $R(G)$. It is easy to verify that this product on $R(G)^*$ extends the convolution product of distributions. In particular, the universal enveloping algebra $U(\mathfrak{g})$ of the Lie algebra $\mathfrak{g}$ of the group G is a subalgebra of $R(G)^*$, since it is equal to the subalgebra of $\mathcal{D}(G, \mathbf{C})$ consisting of distributions whose supports are concentrated at the unit element $e \in G$. The Lie algebra $\mathfrak{g} \subset U(\mathfrak{g}) \subset R(G)^*$ corresponds to the ε-derivations

$$D : R(G) \longrightarrow \mathbf{C};$$

that is, the linear maps such that

$$D(fg) = D(f)\,\varepsilon(g) + \varepsilon(f)\,D(g)$$

where $\varepsilon : R(G) \longrightarrow \mathbf{C}$ is evaluation at $e \in G$.

In defining the generalised Fourier transform there was no need to restrict to compact Lie groups. Quite generally, we have:

Proposition 5. *For any topological monoid* M, *the (generalised) Fourier transform is an isomorphism of topological algebras*

$$\mathcal{F} : R(M)^* \xrightarrow{\sim} \mathrm{End}(\mathfrak{U}).$$

Proof. As in the proof of Proposition 4, it can be shown that, for any $z \in \mathrm{End}^\vee(\mathfrak{U})$ and $h \in R(M)^*$, we have

$$\langle \mathcal{F}h, z \rangle = \langle h, \mathcal{F}^\vee z \rangle.$$

This proves that $\mathcal{F}$ is an isomorphism since it is the transpose of $\mathcal{F}^\vee$ and we have Theorem 1. Also this shows that $\mathcal{F}$ is bicontinuous since the topology on both sides is the usual topology for linear duals (namely, pointwise convergence for linear functionals). To finish the proof, we must verify that $\mathcal{F}$ is an algebra homomorphism. By definition of the convolution product, we have, for all $h, k \in R(M)^*$ and $f \in R(M)$,

$$\langle h * k, f \rangle = \sum_{r=1}^{n} \langle h, f_r \rangle \langle k, g_r \rangle$$

where

$$\Delta f = \sum_{r=1}^{n} f_r \otimes g_r \ .$$

If we apply this formula to the entries π_{ij} of the matrix of a representation π_V of M (relative to some basis of V), we obtain

$$\langle h * k, \pi_{ij} \rangle = \sum_{r=1}^{n} \langle h, \pi_{ir} \rangle \langle k, \pi_{rj} \rangle ,$$

which means exactly that

$$\mathcal{F}(h * k)_V = (\mathcal{F}h)_V (\mathcal{F}k)_V \ . \text{qed}$$

When G is a compact group, we can combine the last Proposition 5 with Section 1 Proposition 4 to obtain:

Corollary 6. *For any compact group* G, *the Fourier transform defines an isomorphism of topological algebras*

$$\mathcal{F} : R(G)^* \xrightarrow{\sim} \prod_{\lambda \in G^\vee} \mathrm{End}(V_\lambda) \ .$$

We now return to the more general situation of a topological monoid M. Recall that we have a canonical map

$$j : M \longrightarrow R(M)^*$$

defined by $\langle j(x), f \rangle = f(x)$. We might say that $j(x)$ is the Dirac measure concentrated at $x \in M$. We have a commutative triangle

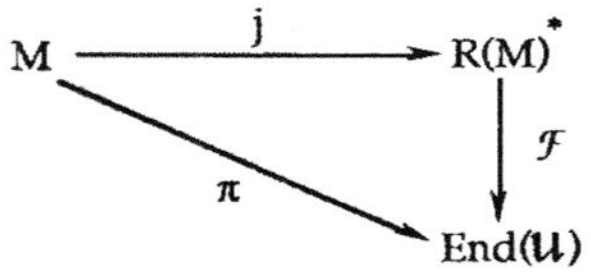

since, for every $V \in \mathcal{R}ep(M,C)$,

$$\mathcal{F}(j(x))_V = \langle j(x), \pi_V \rangle = \pi_V(x) = \pi(x)_V .$$

Let $\mathrm{End}^{\otimes}(\mathcal{U})$ denote the submonoid of $\mathrm{End}(\mathcal{U})$ consisting of the tensor-preserving natural transformations (the product is composition). The Tannaka monoid $\mathcal{T}(M)$ of M is the submonoid of $\mathrm{End}^{\otimes}(\mathcal{U})$ consisting of the self-conjugate natural transformations.

Proposition 7. *The Fourier transform* $\mathcal{F}$ *induces the following isomorphisms of topological monoids:*

$$\mathrm{Spec}(R(M)) \xrightarrow{\sim} \mathrm{End}^{\otimes}(\mathcal{U}), \qquad \mathrm{Spec}_{\mathbf{R}}(R(M)) \xrightarrow{\sim} \mathcal{T}(M).$$

Moreover, the following triangle commutes.

$$\begin{array}{ccc} M & \xrightarrow{\ j\ } & \mathrm{Spec}_{\mathbf{R}}(R(M)) \\ & \searrow^{\pi} & \downarrow \wr \\ & & \mathcal{T}(M) \end{array}$$

Proof. For any representations $V, W \in \mathcal{R}ep(M,C)$, let (π_{ij}) and (ρ_{rs}) be the matrices of π_V and π_W in some basis of V and W, respectively. If we express the equality

$$(\mathcal{F}k)_{V\otimes W} = (\mathcal{F}k)_V \otimes (\mathcal{F}k)_W$$

in matrix form, we obtain the equality

$$\langle k, \pi_{ij}\, \rho_{rs} \rangle = \langle k, \pi_{ij} \rangle \langle k, \rho_{rs} \rangle$$

for all i, j, r, s. This proves that k is an algebra homomorphism if and only if $\mathcal{F}k \in \mathrm{End}^{\otimes}(\mathcal{U})$ (we are using here the fact that $R(M)$ is the linear span of the matrix entries of the representations of M). This establishes the first bijection. The second bijection is a consequence of the formula

$$\mathcal{F}\bar{k} = \overline{\mathcal{F}k}$$

where the conjugate $\bar{k}$ of an element $k \in R(M)^*$ is defined by

$$\langle \bar{k}, f \rangle = \overline{\langle k, \bar{f} \rangle}.$$

Bicontinuity of the bijections follows from that of the Fourier transform.$_{\mathbf{qed}}$

Corollary 8. *For any compact group* G, *there is a commutative triangle of compact group isomorphisms:*

$$\begin{array}{ccc} G & \xrightarrow[\ j\ ]{\sim} & \mathrm{Spec}_{\mathbf{R}}(R(G)) \\ & {}_{\pi}\searrow \wr \quad \wr \swarrow_{\mathcal{F}} & \\ & \mathcal{T}(G) & \end{array}$$

Proof. According to Tannaka-Krein (Theorem 20), π is an isomorphism. Hence, $j = \mathcal{F}^{-1} \circ \pi$ is also an isomorphism.$_{\mathbf{qed}}$

§4. The coalgebra $\mathrm{End}^{\vee}(X)$.

In this Section, we shall introduce a coalgebra structure on $\mathrm{End}^{\vee}(X)$ for any functor

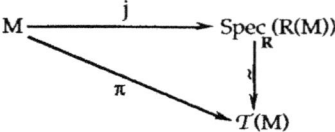

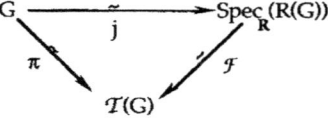

$X : \mathcal{C} \longrightarrow \mathcal{V}ect_{\mathbb{C}}$ whose values are finite dimensional vector spaces. More generally, under the same hypotheses on functors Y, Z, we shall describe a map

$$\Delta : \mathrm{Hom}^{\vee}(X, Z) \longrightarrow \mathrm{Hom}^{\vee}(Y, Z) \otimes \mathrm{Hom}^{\vee}(X, Y)$$

which dualises the usual composition map. We shall also show that there is a coaction

$$\gamma : X \longrightarrow Y \otimes \mathrm{Hom}^{\vee}(X, Y)$$

dualising the usual evaluation action

$$\mathrm{Hom}(X, Y) \otimes X \longrightarrow Y.$$

We begin by giving another description of $\mathrm{Hom}^{\vee}(X, Y)$ useful for many purposes. It is based on the concept of *tensor product* [F] of a contravariant functor $S : \mathcal{C}^{op} \longrightarrow \mathcal{V}ect_{\mathbb{C}}$ with a covariant functor $T : \mathcal{C} \longrightarrow \mathcal{V}ect_{\mathbb{C}}$. The behaviour of this tensor product is similar to that of the tensor product of right and left modules. To stress this analogy, we write *on the right* the action of $f : A \longrightarrow B$ on an element $x \in S(B)$:

$$S(f)(x) = x \cdot f$$

and similarly, we shall write *on the left* the action of f on $y \in T(A)$:

$$T(f)(y) = f \cdot y.$$

The required tensor product is based on the notion of *bilinear pairing*

$$q : S \times T \longrightarrow V$$

where V is a vector space. This q is a family $(q_C \mid C \in \mathcal{C})$ of bilinear pairings

$$q_C : S(C) \times T(C) \longrightarrow V$$

which respect the actions of S and T. More precisely, it is required that we have

$$q_A(x \cdot f, y) = q_B(x, f \cdot y)$$

for all $f : A \longrightarrow B$ in $\mathcal{C}$, $x \in S(B)$, $y \in T(A)$. The universal recipient of such a pairing is a vector space called the *tensor product* over $\mathcal{C}$ of S with T, and is denoted by $S \otimes_{\mathcal{C}} T$. We now formulate one of its fundamental properties. Note that a family $(q_C \mid C \in \mathcal{C})$ of bilinear pairings

$$q_C : S(C) \times T(C) \longrightarrow V$$

corresponds to a family $(q'_C \mid C \in \mathcal{C})$ of linear maps

$$q'_C : T(C) \longrightarrow \mathrm{Hom}(S(C), V).$$

Let us denote by $\mathrm{Hom}(S, V)$ the *covariant* functor whose value at $C \in \mathcal{C}$ is the vector space $\mathrm{Hom}(S(C), V)$.

Proposition 1. *The correspondence described above determines a bijection between linear maps* $S \otimes_{\mathcal{C}} T \longrightarrow V$ *and natural transformations* $T \longrightarrow \mathrm{Hom}(S, V)$.

Suppose now that the functors $X, Y : \mathcal{C} \longrightarrow \mathcal{V}ect_{\mathbb{C}}$ take their values in finite dimensional vector spaces. For any object $C \in \mathcal{C}$, the map

$$[\ ,\ ] : Y(C)^* \otimes X(C) \longrightarrow \mathrm{Hom}^{\vee}(X, Y), \qquad [\phi, x] = [\phi \otimes x]$$

is a bilinear pairing between the contravariant functor Y^* and the covariant functor X since we have, for all $f : A \longrightarrow B$ in $\mathcal{C}$, $x \in X(A)$, $\phi \in Y(B)^*$,

$$[\phi\, Y(f) \otimes x] = [\phi \otimes X(f) x].$$

Using this pairing, we obtain a canonical map

$$Y^* \otimes_{\mathcal{C}} X \longrightarrow \mathrm{Hom}^{\vee}(X, Y).$$

The verification of the next result is left to the reader.

Proposition 2. *The canonical map described above is an isomorphism*

$$Y^* \otimes_C X \xrightarrow{\ \simeq\ } \mathrm{Hom}^\vee(X, Y).$$

We can now apply Proposition 1 to this situation using the canonical isomorphism

$$\mathrm{Hom}_C(Y^*, V) \xrightarrow{\ \simeq\ } Y \otimes V.$$

We see that there is a bijection between the linear maps $\mathrm{Hom}^\vee(X, Y) \longrightarrow V$ and the linear maps $X \longrightarrow Y \otimes V$. To describe this bijection explicitly, we need the natural transformation

$$\gamma : X \longrightarrow Y \otimes \mathrm{Hom}^\vee(X, Y)$$

defined as follows. For any $C \in C$ and any basis $e_1, \dots, e_n$ of $Y(C)$, we put

$$\gamma_C(x) = \sum_{i=1}^{n} e_i \otimes [e_i^* \otimes x].$$

We leave to the reader the verification that γ is natural.

Proposition 3. *For any natural transformation* $n : X \longrightarrow Y \otimes V$, *there exists a unique linear map* $\tilde{n} : \mathrm{Hom}^\vee(X, Y) \longrightarrow V$ *such that the following triangle commutes.*

$$\begin{array}{ccc} X & \xrightarrow{\ \gamma\ } & Y \otimes \mathrm{Hom}^\vee(X, Y) \\ & \searrow^{n} & \downarrow{\scriptstyle 1 \otimes \tilde{n}} \\ & & Y \otimes V \end{array}$$

We can now describe the map

$$\Delta : \mathrm{Hom}^\vee(X, Z) \longrightarrow \mathrm{Hom}^\vee(Y, Z) \otimes \mathrm{Hom}^\vee(X, Y)$$

whose dual is the usual composition map

$$\mathrm{Hom}(Y, Z) \otimes \mathrm{Hom}(X, Y) \longrightarrow \mathrm{Hom}(X, Z).$$

We have $\Delta = \tilde{n}$ where n is the composite

$$X \xrightarrow{\ \gamma\ } Y \otimes \mathrm{Hom}^\vee(X, Y) \xrightarrow{\ \gamma \otimes 1\ } Z \otimes \mathrm{Hom}^\vee(Y, Z) \otimes \mathrm{Hom}^\vee(X, Y).$$

A short calculation gives

$$\Delta[\phi \otimes x] = \sum_{i=1}^{n} [\phi \otimes y_i] \otimes [y_i^* \otimes x]$$

where $x \in X(C)$, $\phi \in Z(C)^*$ and $\{y_1, \dots, y_n\}$, $\{y_1^*, \dots, y_n^*\}$ are dual bases of $Y(C)$, $Y(C)^*$. In particular, Δ defines a coalgebra structure on $\mathrm{End}^\vee(X) = \mathrm{Hom}^\vee(X, X)$:

$$\Delta : \mathrm{End}^\vee(X) \longrightarrow \mathrm{End}^\vee(X) \otimes \mathrm{End}^\vee(X).$$

To compute Δ explicitly, recall that, for any vector space V, there is a coalgebra structure on $\mathrm{End}(V)$ since $\mathrm{End}(V)$ is self dual and is itself an algebra. If $e_1, \dots, e_n$ is a basis of V then the coalgebra structure $\delta : \mathrm{End}(V) \longrightarrow \mathrm{End}(V) \otimes \mathrm{End}(V)$ is given by

$$\delta(e_{ij}) = \sum_{k=1}^{n} e_{ik} \otimes e_{kj}$$

where $e_{ij} = e_i^* \otimes e_j$; the counit is the trace map $\mathrm{Tr} : \mathrm{End}(V) \longrightarrow \mathbf{C}$. The canonical map

$$\sum_{C \in C} \mathrm{End}(X(C)) \longrightarrow \mathrm{End}^\vee(X)$$

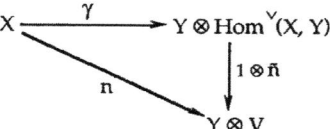

expresses the coalgebra $End^\vee(X)$ as a quotient of the direct sum of coalgebras. This implies that we have

$$\Delta[e_{ij}] = \sum_{k=1}^{n} [e_{ik}] \otimes [e_{kj}] ,$$

and that the counit $\varepsilon : End^\vee(X) \longrightarrow \mathbb{C}$ is given by $\varepsilon[A] = \mathrm{Tr}\, A$.

The following result supplements Section 3 Theorem 1.

Proposition 4. *For any topological monoid* M, *the Fourier cotransform*

$$\mathcal{F}^\vee : End^\vee(\mathcal{U}) \longrightarrow R(M)$$

is an isomorphism of coalgebras.

Proof. Let $V \in \mathcal{R}ep(M, \mathbb{C})$ and take $A \in End(V)$. The calculation

$$\varepsilon\, \mathcal{F}^\vee[A] = \mathrm{Tr}(\pi_V(e)A) = \mathrm{Tr}(A) = \varepsilon[A]$$

shows that $\mathcal{F}^\vee$ preserves counits. If $e_1, \ldots, e_n$ is a basis of V then we have

$$\mathcal{F}^\vee([e_{ij}])(x) = \mathrm{Tr}(\pi_V(x)\, e_{ij}) = e_i^*(\pi_V(x) e_j) = \pi_{ij}(x).$$

This shows that $\mathcal{F}^\vee$ preserves Δ since

$$\Delta\pi_{ij} = \sum_{k=1}^{n} \pi_{ik} \otimes \pi_{kj}, \quad \Delta[e_{ij}] = \sum_{k=1}^{n} [e_{ik}] \otimes [e_{kj}],$$

and $End^\vee(\mathcal{U})$ is the linear span of the elements $[e_{ij}]$ as V runs over the representations of M.**qed**

When G is a compact group, we have a canonical map of coalgebras

$$i : \sum_{\lambda \in G^\vee} End(V_\lambda) \longrightarrow End^\vee(\mathcal{U}).$$

It is an isomorphism since ${}^t i$ is the isomorphism q of Section 1 Proposition 4. We shall use this map to identify $End^\vee(\mathcal{U})$ with $\Sigma_\lambda End(V_\lambda)$. With this convention, the following result is the dual of Section 3 Corollary 6.

Corollary 5. *For any compact group* G, *the Fourier cotransform provides an isomorphism of coalgebras:*

$$\mathcal{F}^\vee : \sum_{\lambda \in G^\vee} End(V_\lambda) \xrightarrow{\sim} R(G) .$$

It is of interest to see the description of the inverse cotransform

$$\mathcal{F}^{\vee -1} : R(G) \longrightarrow \sum_{\lambda \in G^\vee} End(V_\lambda) .$$

The orthogonality relations imply that, for any $f \in R(G)$, we have

$$\mathcal{F}^{\vee -1}(f)(\lambda) = d_\lambda \int_G \pi_\lambda(x)^{-1} f(x)\, dx .$$

Recall that a (right) comodule over a coalgebra E is a vector space V equipped with a coaction

$$\alpha : V \longrightarrow V \otimes E$$

which is associative and unitary; that is, the following diagrams commute.

$$\begin{array}{ccc} V & \xrightarrow{\alpha} & V \otimes E \\ {\scriptstyle\alpha}\downarrow & & \downarrow{\scriptstyle\alpha \otimes E} \\ V \otimes E & \xrightarrow[V \otimes \delta]{} & V \otimes E \otimes E \end{array} \qquad \begin{array}{ccc} V & \xrightarrow{\alpha} & V \otimes E \\ & \searrow & \downarrow{\scriptstyle V \otimes \varepsilon} \\ & & V \end{array}$$

When V is finite dimensional, there is a bijection between comodule structures $\alpha : V \longrightarrow V \otimes E$ and coalgebra maps

$$\tilde{\alpha} : \mathrm{End}(V) \longrightarrow E .$$

If $e_1, \ldots, e_n$ is a basis for V then we have

$$\alpha(e_i) = \sum_{j=1}^{n} e_j \otimes \alpha_{ij}$$

where (α_{ij}) is a matrix with entries in E. Putting $e_{ij} = e_i^* \otimes e_j$, we then see that the coalgebra map determined by α is given by the equations

$$\tilde{\alpha}(e_{ij}) = \alpha_{ij} .$$

We have the formulas

$$\tilde{\alpha}(\phi \otimes v) = (\phi \otimes 1)\alpha(v), \qquad \alpha(v) = \sum_{i=1}^{n} e_i \otimes \tilde{\alpha}(e_i^* \otimes v)$$

valid for all $v \in V$ and $\phi \in V^*$. Expressed in terms of the matrix (α_{ij}), the associativity and unitarity conditions are very simply:

$$\delta\alpha = \alpha \underline{\otimes} \alpha \qquad \text{and} \qquad \varepsilon\alpha = \mathrm{id}$$

where

$$\delta(\alpha_{ij}) = (\delta\alpha_{ij}), \quad (\alpha_{ij}) \underline{\otimes} (\alpha_{ij}) = (\Sigma_k \alpha_{ik} \otimes \alpha_{kj}), \quad \varepsilon(\alpha_{ij}) = (\varepsilon\alpha_{ij}), \quad \mathrm{id} = (\delta_{ij}).$$

The vector space V is canonically a comodule over the coalgebra $\mathrm{End}(V)$ via the coaction

$$c : V \longrightarrow V \otimes \mathrm{End}(V), \qquad c(x) = \Sigma_i e_i \otimes (e_i^* \otimes x);$$

the corresponding coalgebra map

$$\tilde{c} : \mathrm{End}(V) \longrightarrow \mathrm{End}(V)$$

is the identity map of $\mathrm{End}(V)$.

Warning. The dual E^* of a coalgebra is naturally an algebra. However, the reader should note that a right comodule over E has a natural structure of a *left* E^*-module (but not of a *right* E^*-module). If $\alpha : V \longrightarrow V \otimes E$ is the coaction then, for all $\phi \in V^*$ and $v \in V$, we have

$$\phi \bullet v = (1 \otimes \phi)\, \alpha(v).$$

A *morphism* $f : (V, \alpha) \longrightarrow (W, \beta)$ of E-comodules is a linear mapping $f : V \longrightarrow W$ such that the square below commutes. The category of (right) E-comodules will be designated by *Comod*E, and its full subcategory of finite dimensional comodules will be denoted by $Comod_f$E.

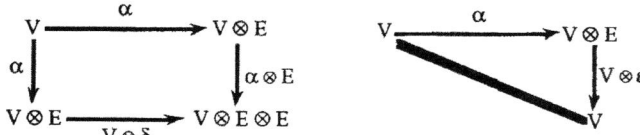

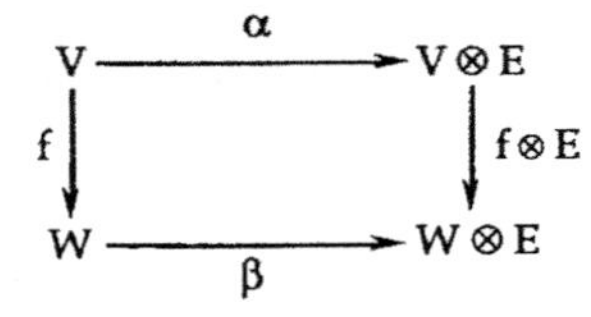

It is instructive to compute the finite dimensional comodules over the coalgebra R(M) for any topological monoid M. First, for any $V \in \mathcal{Rep}(M, C)$, if we compose the map $\gamma_V : V \longrightarrow V \otimes \mathrm{End}^{\vee}(\mathcal{U})$ with the Fourier cotransform, we obtain a map $c_V : V \longrightarrow V \otimes R(M)$ which is a coaction. To see this, pick a basis $e_1, \ldots, e_n$ of V. Then a short computation gives

$$c_V(e_i) = \sum_{j=1}^{n} e_j \otimes \pi_{ij}$$

where (π_{ij}) is the matrix of π_V in the basis $e_1, \ldots, e_n$. The identities

$$\pi_{ij}(xy) = \sum_{k=1}^{n} \pi_{ik}(x)\, \pi_{kj}(y)\,, \qquad \pi_{ij}(e) = \delta_{ij}$$

mean that $\delta(\pi) = \pi \underline{\otimes} \pi$, $\varepsilon(\pi) = \mathrm{id}$, and therefore that c_V is a coaction. We have defined a functor

$$\mathcal{U}^{\sim} : \mathcal{Rep}(M, C) \longrightarrow \mathcal{Comod}_f R(M).$$

Proposition 6. *For any topological monoid* M, *the functor* $\mathcal{U}^{\sim}$ *is an equivalence of categories.*

Proof. It suffices to describe an inverse $\mathcal{U}^{\sim -1}$. Let (V, α) be a finite dimensional comodule over R(M). We have

$$\alpha(e_i) = \sum_{j=1}^{n} e_j \otimes \alpha_{ij}$$

where $e_1, \ldots, e_n$ is a basis for V. The relations $\delta(\alpha) = \alpha \underline{\otimes} \alpha$, $\varepsilon(\alpha) = \mathrm{id}$ mean that the matrix (α_{ij}) determines a representation π of M which is continuous since the functions α_{ij} are. We put $\mathcal{U}^{\sim -1}(V, \alpha) = (V, \pi)$.qed

Let C be a category. An E-*comodule structure* on a functor $X : C \longrightarrow \mathcal{Vect}_C$ is a natural transformation

$$\alpha : X \longrightarrow X \otimes E$$

such that α_C is a comodule structure on X(C) for all $C \in C$. Using α, we obtain a functor

$$(X, \alpha) : C \longrightarrow \mathcal{Comod}(E)$$

by putting $(X, \alpha)(C) = (X(C), \alpha_C)$. This functor fits into a commutative triangle

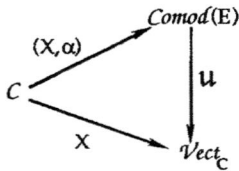

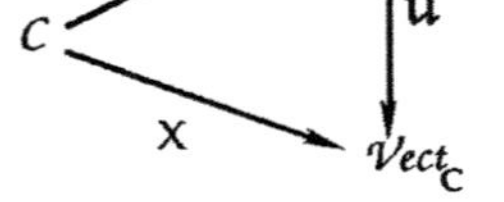

where $\mathcal{U}$ is the forgetful functor. What is more, it is clear that any functor $X^{\sim} : \mathcal{C} \longrightarrow \mathit{Comod}(E)$ such that $\mathcal{U}X^{\sim} = X$ is of the form (X, α) for a unique comodule structure α on X. When X takes its values in finite dimensional vector spaces, the natural transformation $\gamma : X \longrightarrow X \otimes \mathrm{End}^{\vee}(X)$ defines an $\mathrm{End}^{\vee}(X)$-comodule structure on X. This follows from the commutativity of the triangles

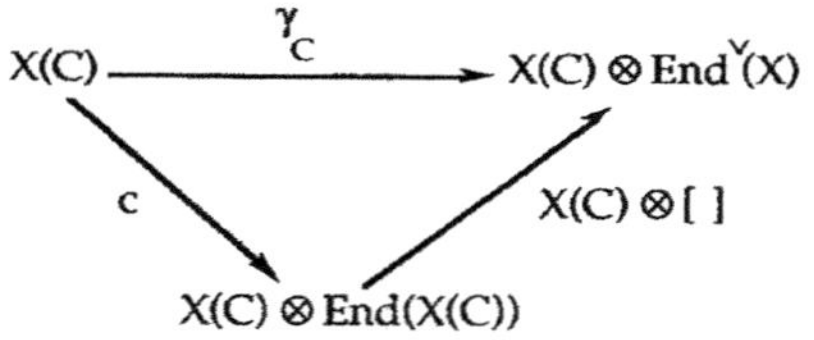

where c is the $\mathrm{End}(X(C))$-comodule structure on $X(C)$ and where $[\ \ ] : \mathrm{End}(X(C)) \longrightarrow \mathrm{End}^{\vee}(X)$ is the canonical morphism of coalgebras.

Proposition 7. *Let* $X : \mathcal{C} \longrightarrow \mathcal{V}ect_{\mathbb{C}}$ *be a functor whose values are finite dimensional vector spaces. For any coalgebra* E *and any comodule structure* $n : X \longrightarrow X \otimes E$*, there is precisely one morphism* $\tilde{n} : \mathrm{End}^{\vee}(X) \longrightarrow E$ *of coalgebras such that the following triangle commutes.*

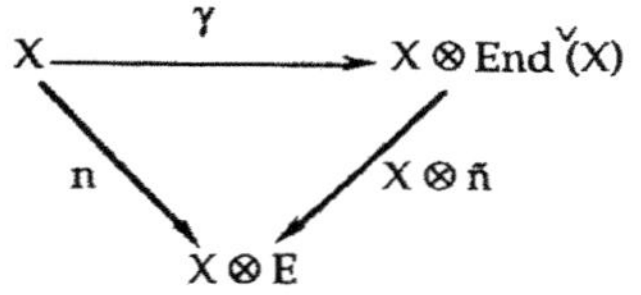

Proof. To prove this Proposition, it suffices to verify that the linear map $\tilde{n}$, whose existence and uniqueness are assured by Proposition 3, is a map of coalgebras. The verification is left to the reader.$_{\mathbf{qed}}$

§5. Tannaka duality for homogeneous spaces.

In this part we show how Tannaka duality can be used to obtain results on homogeneous spaces. The proof given here is independent of the proof in Section 1. As in Section 2, the results proved here can be used to show that any homogeneous space over a compact Lie group is a (real) algebraic variety. This is the basis for the construction of quantum homogeneous spaces such as quantum spheres [Pd].

Let G be a compact group. By a *(left) G-space* we mean a topological space X equipped with a continuous (left) action $G \times X \longrightarrow X$. A G-space is *homogeneous* when it is non-empty and the action is transitive. For each closed subgroup $H \subset G$, the space G/H of orbits for the action of H on the right of G is a homogeneous G-space. If $K \subset G$ is another subgroup, we shall write $K \backslash G/H$ for the space of orbits for the action of K on the left of G/H.

Let X be a homogeneous G-space. For any representation V of G, we shall write V^X for the vector space $\mathrm{Hom}_G(X, V)$ of G-invariant maps from X to V. If we choose a basepoint $x_0 \in X$ and H is the stabilizer of x_0, then the map $g \longmapsto g x_0$ induces an isomorphism between G/H and X, while the map $a \longmapsto a(x_0)$ defines an isomorphism between V^X and the subspace $V^H \subset V$ consisting of the H-invariant vectors. For all $V \in \mathcal{Rep}(G,C)$, putting

$$\mathcal{U}^X(V) = V^X \quad \text{and} \quad \mathcal{U}^H(V) = V^H,$$

we obtain two isomorphic functors

$$\mathcal{U}^X,\ \mathcal{U}^H : \mathcal{Rep}(G,C) \longrightarrow \mathcal{Vect}_C .$$

Let Y be another homogeneous G-space. We define the *Fourier transform* of a G-invariant function $f \in \mathcal{C}(Y\times X, C)$ to be the natural transformation

$$\mathcal{F}f : \mathcal{U}^X \longrightarrow \mathcal{U}^Y$$

whose value at $V \in \mathcal{Rep}(G,C)$ is the linear map

$$(\mathcal{F}f)_V : V^X \longrightarrow V^Y$$

given by the formula

$$(\mathcal{F}f)_V(a)(y) = \int_X f(y,x)\,a(x)\,dx$$

where dx is the normalised Haar measure on X. As in Section 1, it can be proved that the projection

$$\mathrm{Hom}(\mathcal{U}^X, \mathcal{U}^Y) \longrightarrow \prod_{\lambda \in G^\vee} \mathrm{Hom}\,(V_\lambda^X, V_\lambda^Y)$$

is an isomorphism, so that we can view the Fourier transform as a map

$$\mathcal{F} : \mathcal{C}(Y \times X, C)^G \longrightarrow \prod_{\lambda \in G^\vee} \mathrm{Hom}\,(V_\lambda^X, V_\lambda^Y).$$

We now want to show that $\mathcal{F}$ defines an isometry of Hilbert spaces. Let $L^2(Y\times X)$ denote the space of square integrable functions on $Y\times X$ with respect to the product $dx\,dy$ of the normalised Haar measures on X and Y. Also, let $L^2(Y\times X/G)$ denote the space of square integrable functions on $Y\times X/G$ with respect to the projection onto $Y\times X/G$ of the product measure $dx\,dy$. The space $L^2(Y\times X/G)$ is isometric to the subspace $L^2(Y\times X)^G$ of G-invariant elements of $L^2(Y\times X)$ in which $\mathcal{C}(Y\times X, C)^G$ is dense. On the other hand, for any $\lambda \in G^\vee$, there is a canonical Hilbert space structure on $\mathrm{Hom}(V_\lambda{}^X, V_\lambda{}^Y)$ that we shall now describe. More generally, any G-invariant inner product on X induces an inner product on V^X given by

$$\langle a, b\rangle = \langle a(x_0), b(x_0)\rangle$$

where the value of the right hand side does not depend on the choice of the basepoint $x_0 \in X$ since we have

$$\langle a(\sigma x_0), b(\sigma x_0)\rangle = \langle \sigma a(x_0), \sigma b(x_0)\rangle = \langle a(x_0), b(x_0)\rangle .$$

Using inner products on $V_\lambda{}^X$ and $V_\lambda{}^X$ which are induced by some G-invariant metric on V_λ, we can define the adjoint $f^* : V_\lambda{}^Y \longrightarrow V_\lambda{}^X$ of any map $f : V_\lambda{}^X \longrightarrow V_\lambda{}^Y$. This adjoint does not depend on the particular choice of the G-invariant metric on V_λ. On the vector space $\mathrm{Hom}(V_\lambda{}^X, V_\lambda{}^Y)$ we shall use the inner product

$$\langle f, g\rangle = d_\lambda \mathrm{Tr}(f^* g)$$

where d_λ denotes (as before) the dimension of V_λ.

Theorem 1. *The Fourier transform extends continuously to an isometry*

$$\mathcal{F} : L^2(X \times Y / G) \xrightarrow{\sim} \sum_{\lambda \in G^\vee}^{\text{hilbert}} \mathrm{Hom}(V_\lambda^X, V_\lambda^X) \ .$$

Proof. We can suppose that $X = G/H$ and $Y = G/K$. Let $p_H : V \longrightarrow V^H$ be the averaging operator defined for any $V \in \mathcal{R}ep(G, C)$ by the formula

$$p_H(v) = \int_H hv\,dh$$

where dh is the normalised Haar measure on H. If $i_K : V^K \hookrightarrow V$ denotes the inclusion, it is easy to see that the map $f \longmapsto i_K\, f\, p_H$ defines an isometric embedding of $\mathrm{Hom}(V_\lambda{}^H, V_\lambda{}^K)$ into $\mathrm{Hom}(V_\lambda, V_\lambda)$. We obtain in this way an isometric embedding

$$\alpha : \sum_{\lambda \in G^\vee}^{\text{hilbert}} \mathrm{Hom}(V_\lambda^H, V_\lambda^K) \longrightarrow \sum_{\lambda \in G^\vee}^{\text{hilbert}} \mathrm{Hom}(V_\lambda, V_\lambda)$$

whose image consists of the elements which are fixed under the actions of K on the left and H on the right. On the other hand, the map $g \longmapsto (y_0\,, g\,x_0)$ induces a measure preserving homeomorphism between $K\backslash G/H$ and $Y \times X / G$ (the measure on $K\backslash G/H$ is obtained by projecting the Haar measure on G). This defines an isometric embedding

$$\beta : L^2(Y \times X / G) \longrightarrow L^2(G)$$

whose image consists of the elements fixed by K (on the left) and by H (on the right). The Theorem then follows from the relation

$$\mathcal{F} \circ \beta = \alpha \circ \mathcal{F}$$

and the fact that the Fourier transform of Section 1 Theorem 14 respects the actions of G on the left and on the right on both its domain and codomain.$_{\text{qed}}$

When $Y = G$ acting on itself by left translation, we have $Y \times X / G \cong X$. We can define the Fourier transform of a function $f \in \mathcal{C}(X, C)$ to be the natural transformation

$$\mathcal{F}f : \mathcal{U}^X \longrightarrow \mathcal{U}$$

whose value at $V \in \mathcal{R}ep(G, C)$ is given by the formula

$$(\mathcal{F}f)_V(a) = \int_X f(x)\,a(x)\,dx\ .$$

Corollary 2. *The Fourier transform can be extended continuously to an isometry*

$$\mathcal{F} : L^2(X/G) \xrightarrow{\sim} \sum_{\lambda \in G^\vee}^{\text{hilbert}} \mathrm{Hom}(V_\lambda^X, V_\lambda)\ .$$

At this point we want to have similar results for the Fourier transform of generalised distributions on X. We shall let $R(X)$ denote the set of elements of $\mathcal{C}(X, C)$ whose orbits generate a finite dimensional subspace of $\mathcal{C}(X, C)$. According to Section 2 Proposition 1, this $R(G)$ is the set of representative functions on G (as before). When $X = G/H$, we have $R(G/H) = R(G)^H$. Let us see that, for all $V \in \mathcal{R}ep(G, C)$, $a \in V^X$, and $\phi \in V^*$, the function $x \longmapsto \phi(a(x))$ belongs to $R(X)$. But, if $x_0 \in X$, then the function

$$g \longmapsto \phi(a(g\,x_0)) = \phi(\pi_V(g)\,a(x_0))$$

belongs to $R(G)$ and is invariant under right translation of the stabilizer H of x_0; this proves the claim. We now define the Fourier transform $\mathcal{F}h$ of an element $h \in R(X)^*$. For $V \in \mathcal{R}ep(G, C)$, let $e_1, \ldots, e_n$ be a basis for V. Every element $a \in V^X$ can be written as

$$a(x) = \sum_{i=1}^{n} a_i(x)\, e_i$$

where $a_i(x) \in R(X)$ since $a_i(x) = e_i^*(a(x))$. We put

$$(\mathcal{F}h)_V(a) \;=\; \sum_{i=1}^{n} \langle h, a_i \rangle e_i \,.$$

Proposition 3. *The Fourier transform defines an isomorphism of topological vector spaces*

$$\mathcal{F} : R(X)^* \xrightarrow{\;\sim\;} \prod_{\lambda \in G^\vee} \mathrm{Hom}(V_\lambda^X, V_\lambda)\,.$$

Proof. We can suppose that $X = G/H$ and consider the averaging operation

$$p_H : R(G) \longrightarrow R(G/H)$$

defined by

$$p_H(f) = \int_H f(xh)\, dh\,.$$

Composition with p_H determines an embedding

$$\beta : R(G/H)^* \longrightarrow R(G)^*$$

whose image is the set of (generalized) distributions which are invariant under right translation by the elements of H. Similarly, if we compose with the averaging operators $p_H : V_\lambda \longrightarrow V_\lambda{}^H$, we obtain a map

$$\alpha : \prod_{\lambda \in G^\vee} \mathrm{Hom}(V_\lambda^H, V_\lambda) \longrightarrow \prod_{\lambda \in G^\vee} \mathrm{Hom}(V_\lambda, V_\lambda)$$

whose image consists of the elements which are invariant under right translation by elements of H. The result then follows from the identity

$$\mathcal{F} \circ \beta = \alpha \circ \mathcal{F}$$

and Section 3 Corollary 6.$_{\mathbf{qed}}$

The next thing to do is to define the Fourier transform of G-invariant generalized distributions on $Y \times X$. We shall do this without choosing basepoints in X and Y.

Definition. Let G be a compact group acting continuously on a topological vector space V. An element $v \in V$ is *G-finite* when its orbit generates a finite dimensional subspace of V.

By definition, $R(X)$ is the set of G-finite elements of $\mathcal{C}(X, \mathbf{C})$. Let us see that $R(X)$ is also equal to the G-finite elements of $R(X)^*$. To do this, let $j : R(X) \longrightarrow R(X)^*$ be the map which associates to a function $f \in R(X)$ the measure $f(x)dx$ on X. For any $g \in R(X)$, we have

$$\langle j(f), g \rangle = \int_X f(x)\, g(x)\, dx\,.$$

The image of j consists of G-finite elements since j is G-equivariant and every element of $R(X)$ is G-finite. Let $R(X)^\circ$ denote the set of G-finite elements of $R(X)^*$.

Proposition 4. *The mapping* j *determines an isomorphism*

$$R(X) \xrightarrow{\;\sim\;} R(X)^\circ.$$

Moreover, $R(X)^\circ$ *is dense in* $R(X)^*$.

Proof. We shall prove the result when $X = G$; the general result is proved similarly. If we use the isomorphism

$$\mathcal{F} : R(G)^* \xrightarrow{\simeq} \prod_{\lambda \in G^\vee} \mathrm{Hom}(V_\lambda, V_\lambda).$$

This reduces the problem to proving that, if an element

$$g \in \prod_{\lambda \in G^\vee} \mathrm{End}(V_\lambda)$$

is G-finite (for the left action of G) then all except a finite number of its components are zero. For each $\lambda \in G^\vee$, let $p_\lambda = (p_\lambda(\delta) \mid \delta \in G^\vee)$ be the projection operator, where

$$p_\lambda(\delta) = \begin{cases} \mathrm{id} & \text{if } \delta = \lambda \\ 0 & \text{otherwise} \end{cases}.$$

Then the orthogonality relations imply that $p_\lambda = \mathcal{F}(d_\lambda \bar{\chi}_\lambda)$ where χ_λ is the character of V_λ. We have the formula

$$p_\lambda g = d_\lambda \int_G \overline{\chi_\lambda}(x)\, \pi(x) g \, dx$$

which shows that, if g is G-finite, then all its components $p_\lambda g$ belong to the finite dimensional subspace generated by the orbit Gg. This proves that $p_\lambda g = 0$ except for a finite number of $\lambda \in G^\vee$. qed

Let $R(Y \times X)$ be the set of $G \times G$-finite elements of $\mathcal{C}(Y \times X, \mathbf{C})$. We have a canonical isomorphism

$$R(Y \times X) \cong R(X) \otimes R(Y).$$

For all elements $h \in R(X)^{*Y}$, the pairing

$$(g, f) \longmapsto \int_Y g(y) \langle h(y), f \rangle \, dy$$

is a G-invariant bilinear form $\rho(h)$ on $R(Y) \times R(X)$. Let us say that a generalized distribution $t \in R(Y \times X)^*$ is *right regular* when it is of the form $\rho(h)$ for some $h \in \mathcal{C}(Y, R(X)^*)$.

Proposition 5. *Every G-invariant generalized distribution is regular. More precisely, the map* ρ *defines an isomorphism*

$$\mathrm{Hom}_G(Y, R(X)^*) \cong \mathrm{Hom}_G(R(Y \times X), \mathbf{C}).$$

Proof. We describe an inverse to ρ. We compose the canonical isomorphisms

$$\mathrm{Hom}_G(R(Y \times X), \mathbf{C}) \cong \mathrm{Hom}_G(R(Y) \otimes R(X), \mathbf{C}) \cong \mathrm{Hom}_G(R(X), R(Y)^*)$$
$$\cong \mathrm{Hom}_G(R(X), R(Y)^\circ) \cong \mathrm{Hom}_G(R(X), R(Y))$$

with the isomorphism

$$k : \mathrm{Hom}_G(R(X), R(Y)) \longrightarrow \mathrm{Hom}_G(Y, R(X)^*)$$

defined as follows. For any G-invariant operator $u : R(X) \longrightarrow R(Y)$, we put

$$\langle k(u)(y), f \rangle = u(f)(y).$$

We leave the reader to check that k is bijective and we have defined an inverse to ρ. qed

We can now describe the Fourier transform of $t \in R(Y \times X)^{*G}$. According to Proposition 5, t is right regular, so that $t = \rho(h)$ for some $h \in \mathrm{Hom}_G(Y, R(X)^*)$. For all

$V \in \mathcal{R}ep(G,C)$, we want to define the Fourier transform

$$(\mathcal{F}t)_V : V^X \longrightarrow V^Y.$$

Let $e_1, \ldots, e_n$ be a basis for V. For all $a \in V^X$, we can write $a = \sum_i a_i e_i$ where $a_i \in R(X)$. Then we have

$$(\mathcal{F}t)_V(a)(y) = \sum_{i=1}^{n} \langle h(y), a_i \rangle e_i .$$

Proposition 6. *The Fourier transform defines an isomorphism of topological vector spaces*

$$R(X \times Y)^{*G} \cong \prod_{\lambda \in \check{G}} \mathrm{Hom}(V_\lambda^X, V_\lambda^Y) .$$

Proof. If we exponentiate the Fourier transform of Proposition 3 by Y, we obtain an isomorphism

$$R(X)^{*Y} \cong \prod_{\lambda \in \check{G}} \mathrm{Hom}(V_\lambda^X, V_\lambda^Y) .$$

Composing this with the isomorphism of Proposition 5, we have the Fourier transform as defined above.$_{\mathbf{qed}}$

To complete these results, we should describe the Fourier *cotransform*

$$\mathcal{F}^\vee : \mathrm{Hom}^\vee(\mathcal{U}^X, \mathcal{U}^Y) \longrightarrow R(Y \times X)^G.$$

We shall suppose that $Y = G$ and postpone the general case. We want to describe the cotransform

$$\mathcal{F}^\vee : \mathrm{Hom}^\vee(\mathcal{U}^X, \mathcal{U}) \longrightarrow R(X).$$

For all $V \in \mathcal{R}ep(G,C)$, all $\phi \in V^*$ and all $\phi \in V^X$, we put

$$\mathcal{F}^\vee([\phi \otimes a]) = \phi \circ a .$$

Proposition 7. *The Fourier cotransform* $\mathcal{F}^\vee$ *is an isomorphism*

$$\mathrm{Hom}^\vee(\mathcal{U}^X, \mathcal{U}) \xrightarrow{\ \simeq\ } R(X).$$

We now want to use these results to study the spectrum of the algebra $R(X)$. Let us first remark that, for all $V, W \in \mathcal{R}ep(G,C)$, we have a canonical pairing

$$\otimes : V^X \times W^X \longrightarrow (V \otimes W)^X \qquad (a,b) \longmapsto a \otimes b$$

defined by $(a \otimes b)(x) = a(x) \otimes b(x)$. We also have a unit element $1 \in \mathbf{C}^X = \mathbf{C}$.

Definition. A natural transformation $u : \mathcal{U}^X \longrightarrow \mathcal{U}$ is *tensor preserving* when, for all V, $W \in \mathcal{R}ep(G,C)$, $a \in V^X$, $b \in W^X$, the following equations hold:

$$u_{V \otimes W}(a \otimes b) = u_V(a) \otimes u_W(b) \quad \text{and} \quad u_I(1) = 1.$$

Let $\mathrm{Hom}^\otimes(\mathcal{U}^X, \mathcal{U})$ denote the set of tensor-preserving natural transformations. There is also the notion of *self-conjugate* natural transformation $u : \mathcal{U}^X \longrightarrow \mathcal{U}$; this means that

$$u_{\overline{V}}(\overline{a}) = \overline{u_V(a)}$$

for all $V \in \mathcal{R}ep(G,C)$ and all $a \in V^X$. Let $\mathcal{T}(X)$ denote the set of self-conjugate tensor-preserving natural transformations $\mathcal{U}^X \longrightarrow \mathcal{U}$.

For all $x \in X$, we have an element $\pi(x) \in \mathcal{T}(X)$ defined by $\pi(x)_V(a) = a(x)$ for $a \in V^X$. There is also the canonical map $j : X \longrightarrow R(X)^*$ defined by $\langle j(x), f \rangle = f(x)$.

Proposition 8. *The Fourier transform* $\mathcal{F}$ *induces the following isomorphisms of topological spaces:*

$$\mathrm{Spec}(R(X)) \xrightarrow{\ \simeq\ } \mathrm{Hom}^{\otimes}(\mathcal{U}^X, \mathcal{U})$$
$$\mathrm{Spec}_{\mathbb{R}}(R(X)) \xrightarrow{\ \simeq\ } \mathcal{T}(X).$$

Moreover, the following triangle commutes.

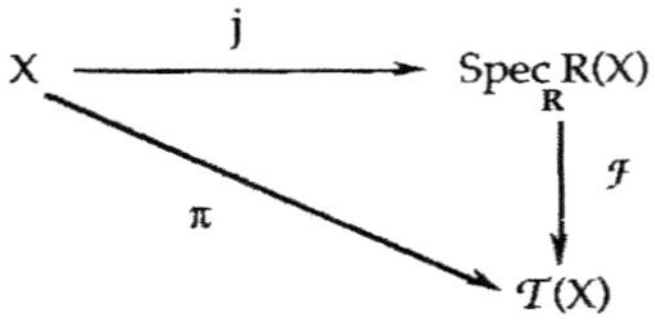

Proof. Let $h \in R(X)^*$ and let $V, W \in \mathcal{R}ep(G, \mathbb{C})$. Let us choose a basis $e_1, \ldots, e_n$ for V and a basis $f_1, \ldots, f_n$ for W. For any $a \in V^X$ and $b \in W^X$, we have $a = \sum_i a_i e_i$ and $b = \sum_j b_j f_j$. The equation

$$(\mathcal{F}h)_{V \otimes W}(a \otimes b) \;=\; (\mathcal{F}h)_V(a) \otimes (\mathcal{F}h)_W(b)$$

means exactly that we have

$$\langle h, a_i b_j \rangle = \langle h, a_i \rangle \langle h, b_j \rangle.$$

Similarly, the self-conjugacy condition for $\mathcal{F}h$ means exactly that we have

$$\langle h, \overline{a_i} \rangle = \overline{\langle h, a_i \rangle}\,.$$

This proves the Proposition since the coefficients a_i linearly generate $R(X)$.$_{\mathbf{qed}}$

Let $V \in \mathcal{R}ep(G, \mathbb{C})$ and let $g : V \otimes V \longrightarrow \mathbb{C}$ be a positive definite invariant hermitian form. Let us see that, for any $u \in \mathcal{T}(X)$, the map $u_V : V^X \longrightarrow V$ is an isometric embedding, where the metric on V^X is the one induced from g on V. We can view g as a $\mathbb{C}$-linear pairing $h : \overline{V} \otimes V \longrightarrow \mathbb{C}$ and we can repeat *mutatis mutandis* the proof of Section 1 Proposition 5. We obtain that

$$g(u_V(a), u_V(b)) = g(a, b),$$

which means that u_V is an isometric embedding. The collection of these embeddings is compact, since it is a homogeneous space over the unitary group $U(V, g)$. It follows that $\mathcal{T}(X)$ is a closed subspace of a product of compact spaces. This proves:

Proposition 9. $\mathcal{T}(X)$ *is compact.*

We next want to prove that $X \xrightarrow{\ \simeq\ } \mathcal{T}(X)$. For this we need some preliminaries. Suppose that G acts continuously on some compact space S. Let $A \subset \mathcal{C}(S, \mathbb{C})$ be a subalgebra which separates points of S and which is closed under conjugation.

Lemma 10. *If* A *is stable under the action of* G *and every element of* A *is* G*-finite then* A^G *separates the orbits of* G *in* S.

Proof. Consider the averaging operator $E : \mathcal{C}(S, \mathbb{C}) \longrightarrow \mathcal{C}(S, \mathbb{C})$ defined by

$$E(f)(x) = \int_G f(g\,x)dg\ .$$

If $f \in A$ then $E(f) \in A$ since $E(f)$ belongs to the convex hull of the orbit Gf whose linear span is a finite dimensional subspace of A. Let $p : S \longrightarrow S/G$ be the projection, and let c, $c' \in S/G$ be any two distinct orbits of G in S. The space S/G is compact and so there is a continuous function $\phi : S/G \longrightarrow \mathbf{R}$ such that $\phi(c) = 0$ and $\phi(c') = 1$. It follows from the Stone-Weierstrass Theorem that A is dense in $\mathcal{C}(S,C)$; so we can find $f \in \mathcal{C}(S,C)$ such that

$$\|f - \phi \circ p\| \le \tfrac{1}{4}$$

where we use the uniform norm. We have

$$\|E(f) - \phi \circ p\| = \|E(f - \phi \circ p)\| \le \|f - \phi \circ p\| \le \tfrac{1}{4}\ .$$

This implies that, for all $x \in c$ and $y \in c'$,

$$|E(f)(x) - E(f)(y)| \ge \tfrac{1}{2}\ .$$

So we have found an element $E(f) \in A^G$ which separates the orbits c, c'.$_{\text{qed}}$

Theorem 11. *The maps* π, j *provide two homeomorphisms*

$$\pi : X \xrightarrow{\ \simeq\ } \mathcal{T}(X) \quad \textit{and} \quad j : X \xrightarrow{\ \simeq\ } \mathrm{Spec}_{\mathbf{R}}(R(X)).$$

Proof. To prove that j is injective is equivalent to proving that $R(X)$ separates the points of X. But this follows from Lemma 10 since we have $X \cong G/H$, $R(X) = R(G)^H$ and $R(G)$ separates the points of G (Section 1 Corollary 13). Let $S = \mathrm{Spec}_{\mathbf{R}}(R(X))$. Then S is compact since $\mathcal{T}(X)$ is (Proposition 9) and Fourier transform gives a homeomorphism $S \cong \mathcal{T}(X)$. The group G acts on S since it acts on $R(X)$. This action is continuous since $S \subset R(X)^*$ and G acts continuously on $R(X)^*$. The algebra $A = R(X)$ becomes a subalgebra of $\mathcal{C}(S,C)$ if we put $f(\chi) = \chi(f)$ for all $f \in R(X)$ and $\chi \in S$. Clearly, A separates the points of S and the hypotheses of Lemma 10 are satisfied. This proves that A^G separates the points of S/G. But we have $A^G = R(X)^G = \mathbf{C}$, and so S/G must reduce to a singleton. This implies that $j : X \longrightarrow S$ is surjective since its image $j(X)$ is an orbit of G in S. We have proved that j is a homeomorphism. It follows that π is a homeomorphism, since $\pi = \mathcal{F} \circ j$.$_{\text{qed}}$

§6. Minimal models.

A *model* of a representative function f on a topological monoid M is a pair $(\phi, v) \in V^* \times V$ such that $f(x) = \phi(\pi_V(x)v)$ where V is some representation of M. Such a pair is far from being unique. A model is *minimal* if $\dim V$ is minimal. It is easy to see that a given representative function has a minimal model which is unique up to a (unique) isomorphism. In this part, we shall extend the concept of minimal model by giving a description of the elements of the tensor product of any two functors which satisfy some exactness conditions. This description is the key technical tool for proving the Representation Theorem of Section 7.

Recall that a category $\mathcal{C}$ is **C**-*additive* when each of its homsets $\mathcal{C}(A,B)$ has the structure of a complex vector space and composition is **C**-bilinear. Recall also that a functor $F : \mathcal{C} \longrightarrow \mathcal{D}$ between **C**-additive categories is **C**-*additive* when the maps

$$F : \mathcal{C}(A,B) \longrightarrow \mathcal{D}(FA,FB)$$

are C-linear.

Let $\mathcal{C}$ be an abelian C-additive category. We shall suppose that $\mathcal{C}$ is artinian, meaning that each decreasing sequence of subobjects in $\mathcal{C}$ terminates. Let $T : \mathcal{C} \longrightarrow \mathcal{V}ect_{\mathbb{C}}$ be a left exact C-additive functor. We shall need the category $\mathcal{el}(T)$ of *elements* of T. An object of $\mathcal{el}(T)$ is a pair (A, x) consisting of $A \in \mathcal{C}$ and $x \in T(A)$. A arrow $f : (A, x) \longrightarrow (B, y)$ in $\mathcal{el}(T)$ is an arrow $f : A \longrightarrow B$ such that $T(f)(x) = y$. We shall say that $x \in T(A)$ is *contained* in a subobject $B \rightarrowtail A$ when x is in the image of the map $T(B) \rightarrowtail T(A)$. From the left exactness of T we see that, if x is contained in the subobjects $B \rightarrowtail A$ and $B' \rightarrowtail A$, then it is contained in their intersection $B \wedge B' \rightarrowtail A$. Using the artinian property of the poset of subobjects of A, it follows that x is contained in a smallest subobject Span(x) of A. When Span(x) = A, we say that x *generates* A. We also need the full subcategory Span(T) of $\mathcal{el}(T)$ whose objects are the pairs (A, x) in which A = Span(x). Observe that, between any two objects of Span(T), there is at most one arrow. To see this, suppose $f, g : (A, x) \longrightarrow (B, y)$; so x is contained in $\mathrm{Ker}(f - g)$ since $T(f)(x) = y = T(g)(x)$. Therefore, $f = g$ since A = Span(x). Note also that, if $f : (A,x) \longrightarrow (B, y)$ is an arrow in Span(T), the map $f : A \longrightarrow B$ is epimorphic. This is because $T(f)(x)$ is contained in the image of f and therefore $\mathrm{Im}(f) = B$ since $T(f)(x) = y$ is generating.

We can now give a first description of the elements of the tensor product $S \otimes_{\mathcal{C}} T$ for any contravariant C-additive functor S. We suppose that the abelian category $\mathcal{C}$ is artinian and that the covariant functor T is left exact. We shall write $[\phi \otimes x]$ for the image of a pair (ϕ, x) by the canonical map

$$S(A) \times T(A) \longrightarrow S \otimes_{\mathcal{C}} T.$$

Proposition 1. *Under the above hypotheses, any element of* $S \otimes_{\mathcal{C}} T$ *is of the form* $[\phi \otimes x]$ *for some* $(\phi, x) \in S(A) \times T(A)$ *for which* A = Span(x). *Moreover, if* $(\phi, x) \in S(A) \times T(A)$ *and* $(\psi, y) \in S(B) \times T(B)$, *then the equality* $[\phi \otimes x] = [\psi \otimes y]$ *holds if and only if there exists an object* $(C, z) \in$ Span(T) *and arrows*

$$(A, x) \xleftarrow{\ f\ } (C, z) \xrightarrow{\ g\ } (B, y)$$

such that $S(f)(\phi) = S(g)(\psi)$.

Proof. We shall use some standard results from category theory [ML]. Since T is left exact, the category $\mathcal{el}(T)$ has finite limits and is therefore filtered (or codirected). If $p : \mathcal{el}(T) \longrightarrow \mathcal{C}$ is the projection functor and $y : \mathcal{C}^{op} \longrightarrow \mathcal{V}ect_{\mathbb{C}}{}^{\mathcal{C}}$ is the Yoneda embedding then we have the canonical isomorphism

$$T \xleftarrow{\ \simeq\ } \varinjlim \left(\mathcal{el}(T)^{op} \xrightarrow{\ p\ } \mathcal{C}^{op} \xrightarrow{\ y\ } \mathcal{V}ect_{\mathbb{C}}{}^{\mathcal{C}} \right).$$

Tensoring this isomorphism with S and using the isomorphisms

$$S \otimes_{\mathcal{C}} y(A) \cong S(A),$$

we obtain a canonical isomorphism

$$S \otimes_{\mathcal{C}} T \xleftarrow{\ \simeq\ } \varinjlim \left(\mathcal{el}(T)^{op} \xrightarrow{\ p\ } \mathcal{C}^{op} \xrightarrow{\ S\ } \mathcal{V}ect_{\mathbb{C}} \right).$$

The result is then a consequence of the standard description of colimits of directed diagrams, and the fact that Span(T) is initial in $\mathcal{el}(T)$.$_{\text{qed}}$

To obtain a more complete description of the elements of $S \otimes_{\mathcal{C}} T$, we shall make additional assumptions on the category $\mathcal{C}$ and the functor S. More precisely, we shall suppose that $\mathcal{C}$ is *noetherian*, meaning that each increasing sequence of subobjects terminates; or equivalently, that $\mathcal{C}^{op}$ is artinian. We shall also suppose that S is *left exact*, meaning that S transforms right exact sequences in $\mathcal{C}$ into left exact sequences of linear maps. As before, for any $\phi \in S(A)$, there is a smallest quotient object $\mathrm{Cospan}(\phi)$ *supporting* ϕ (we adopt this terminology to avoid confusion with the previous case of a covariant functor T). If $\mathrm{Cospan}(\phi) = A$, we say that ϕ *cogenerates* A. We shall say that a pair $(\phi, x) \in S(A) \times T(A)$ is a *model* of an element $z \in S \otimes_{\mathcal{C}} T$ when $z \in [\phi \otimes x]$. A model (ϕ, x) is *minimal* when $\mathrm{Cospan}(\phi) = A = \mathrm{Span}(x)$. An *isomorphism* between $(\phi, x) \in S(A) \times T(A)$ and $(\psi, y) \in S(B) \times T(B)$ is an invertible map $f : A \longrightarrow B$ such that $S(f^{-1})(\phi) = \psi$ and $T(f)(x) = y$. In this case we have

$$[\phi \otimes x] = [S(f)(\psi) \otimes x] = [\psi \otimes T(f)(x)] = [\psi \otimes y].$$

If two minimal models (ϕ, x), (ψ, y) are isomorphic, the isomorphism is unique since x and y are generating.

Theorem 2. *Suppose that the abelian* $\mathbb{C}$*-linear category* $\mathcal{C}$ *is artinian and noetherian, and that the functors* S *and* T *are left exact. Then every element of* $S \otimes_{\mathcal{C}} T$ *has a minimal model which is unique up to a unique isomorphism. Moreover, for any* $(\phi, x) \in S(A) \times T(A)$, *there is a minimal pair* $(\phi', x') \in S(A') \times T(A')$ *such that* $[\phi' \otimes x'] = [\phi \otimes x]$ *where* A' *is a subquotient of* A.

Proof. We begin by proving the last statement. Let

$$\mathrm{Span}(x) \xrightarrow{\;q\;} A' \xrightarrow{\;j\;} \mathrm{Cospan}(\phi)$$

be the image factorization of the composite

$$\mathrm{Span}(x) \xrightarrow{\;i\;} A \xrightarrow{\;p\;} \mathrm{Cospan}(\phi).$$

By definition, there exist x_1 and ϕ_1 such that $S(p)(\phi_1) = \phi$ and $T(i)(x_1) = x$. If we put $\phi' = S(j)(\phi_1)$ and $x' = T(q)(x_1)$ then we have

$$\begin{aligned}[\phi' \otimes x'] &= [\phi' \otimes T(q)(x_1)] = [S(q)(\phi') \otimes x_1] = [S(jq)(\phi_1) \otimes x_1] \\ &= [S(pi)(\phi_1) \otimes x_1] = [S(i)(\phi) \otimes x_1] = [\phi \otimes T(i)(x_1)] = [\phi \otimes x].\end{aligned}$$

Moreover, x' generates A' since x_1 generates Span(x). Similarly, ϕ' cogenerates A'. This proves that the pair (ϕ', x') is a minimal model of $[\phi \otimes x]$. This proves the existence of a minimal model for each element of $S \otimes_{\mathcal{C}} T$ since all elements of $S \otimes_{\mathcal{C}} T$ are of the form $[\phi \otimes x]$ by Proposition 1. To prove the uniqueness, suppose that $(\phi, x) \in S(A) \times T(A)$ and $(\psi, y) \in S(B) \times T(B)$ are both minimal models of z. Let (C, z), f, g be as in Proposition 1. The map f is epimorphic since $T(f)(z) = x$ and z, x are both generating. If $\sigma = S(f)(\phi)$ then $\mathrm{Cospan}(\sigma) \cong A$ since f is epimorphic and ϕ is cogenerating. Similarly, g is epimorphic and $\mathrm{Cospan}(\sigma) \cong B$ since ψ is cogenerating. This proves that we have an isomorphism $i : A \xrightarrow{\sim} B$ such that $i f = g$ and $S(i)(\psi) = \phi$. We have also

$$T(i)(x) = T(i)T(f)(z) = T(i\,f)(z) = T(g)(z) = y.$$

Hence i is an isomorphism between the pairs (ϕ, x) and (ψ, y). We have already remarked that such an isomorphism must be unique.$_{\text{qed}}$

Corollary 3. *For all* $(\phi,x) \in S(A) \times T(A)$, *the equality* $[\phi \otimes x] = 0$ *holds if and only if there exists a monomorphism* $j : B \rightarrowtail A$ *such that* $x \in \mathrm{Im}(T(j))$ *and* $S(j)(\phi) = 0$.

Proof. Let $(\phi', x') \in S(A') \times T(A')$ be a minimal model of $[\phi \otimes x]$ where A' is the image of the map $\mathrm{Span}(x) \rightarrowtail A \twoheadrightarrow \mathrm{Cospan}(\phi)$ as in the proof of the Theorem. If $[\phi' \otimes x'] = 0$ then $A' = 0$ by uniqueness of the minimal model. This shows that the composite $\mathrm{Span}(x) \rightarrowtail A \twoheadrightarrow \mathrm{Cospan}(\phi)$ is zero, and the result follows with $B = \mathrm{Span}(x)$.$_{\text{qed}}$

To end this Section, we shall prove that, for any coalgebra E, we have an isomorphism $E \cong \mathrm{End}^{\vee}(U)$ where $U : \mathit{Comod}_f E \longrightarrow \mathit{Vect}_C$ is the forgetful functor. We have an obvious coaction $\alpha : U \longrightarrow U \otimes E$ obtained by putting together all the coactions $\alpha_V : V \longrightarrow V \otimes E$ where $(V, \alpha) \in \mathit{Comod}_f E$. Using Section 4 Proposition 5, we obtain a morphism of coalgebras

$$\tilde{\alpha} : \mathrm{End}^{\vee}(U) \longrightarrow E\,.$$

In order to prove this morphism is invertible, we need the following Lemma whose proof is left to the reader.

Lemma 4. *Let* (V, α) *be a finite dimensional comodule over a coalgebra* E. *For each* $\phi \in V^*$, *the comodule* $\mathrm{Cospan}(\phi)$ *is obtained by taking the image factorization of the map*

$$V \xrightarrow{\ \alpha\ } V \otimes E \xrightarrow{\ \phi \otimes 1\ } E\,.$$

Proposition 5. *For each coalgebra* E, *the map*

$$\tilde{\alpha} : \mathrm{End}^{\vee}(U) \longrightarrow E$$

is an isomorphism of coalgebras.

Proof. We easily see that, for all $x \in V$, $\phi \in V^*$ and $(V, \alpha) \in \mathit{Comod}_f E$,

$$\tilde{\alpha}\,[\phi \otimes x] = (\phi \otimes 1)\,\alpha_V(x).$$

To prove surjectivity, take $x \in E$. There is a finite dimensional submodule $V \subset E$ such that $x \in V$. Let ϕ be the restriction of the counit $\varepsilon : E \longrightarrow C$ to the subspace V. Then we have

$$\tilde{\alpha}\,[\phi \otimes x] = (\phi \otimes 1)\delta(x) = (\varepsilon \otimes 1)\,\delta(x) = x\,.$$

To prove injectivity, take $z \in \mathrm{End}^{\vee}(U)$ in the kernel, and let $(\phi, x) \in V^* \times V$ be a minimal model of z. We have

$$(\phi \otimes 1)\,\alpha_V(x) = \tilde{\alpha}\,(z) = 0\,.$$

But Lemma 4 shows that the map $(\phi \otimes 1)\,\alpha_V$ is injective since ϕ is cogenerating. This shows that $x = 0$ and hence $z = 0$.$_{\text{qed}}$

§7. The representation theorem.

In this Section, we shall prove a representation theorem for an abelian category $\mathcal{C}$ equipped with an exact faithful functor U with values in finite dimensional vector spaces. We prove that $\mathcal{C}$ is equivalent to the category of finite dimensional E-comodules, where E is the coalgebra $End^{\vee}(U)$ constructed in Section 4. This result is basic in the theory of Tannakian categories in which the category comes first, or is given, and the group comes second, or is constructed. There are many examples of Tannakian categories in nature, especially from algebraic geometry in Grothendieck's theory of motives [DM]. The representation theorem proved here is not needed in the rest of this paper.

We first recall a few properties of the category of comodules over a coalgebra C. For any vector space V, the tensor product $V \otimes C$ has a comodule structure obtained by left comultiplication

$$V \otimes \delta : V \otimes C \longrightarrow V \otimes C \otimes C.$$

It is the cofree comodule over the vector space V. If $\alpha : V \longrightarrow V \otimes C$ is a comodule structure on V then α is also a morphism of comodules from V to the cofree comodule $V \otimes C$. The set of subcomodules of V is closed under intersections and sums. If $f : V \longrightarrow W$ is a morphism of comodules then the direct and inverse images under f of subcomodules are subcomodules. For completeness we include a proof of the following classical result.

Proposition 1. *Every comodule is the directed union of its finite dimensional subcomodules. Every coalgebra is the directed union of its finite dimensional subcoalgebras.*

Proof. Let $E \subset V$ be a finite dimensional subspace of the comodule (V, α). There exists a finite dimensional subspace $F \subset V$ such that $\alpha(E) \subset F \otimes C$. We have $E \subset \alpha^{-1}(F \otimes C)$ and $\alpha^{-1}(F \otimes C)$ is a submodule since it is the inverse image of the subcomodule $F \otimes C$ of the cofree comodule $V \otimes C$. Moreover, we have $\alpha^{-1}(F \otimes C) \subset F$ since

$$F \otimes C \subset (1 \otimes \varepsilon)^{-1}(F) \quad \text{and} \quad \alpha^{-1}(1 \otimes \varepsilon)^{-1}(F) = ((1 \otimes \varepsilon)\alpha)^{-1}(F) = F.$$

This proves the first sentence of the Proposition, and the second sentence follows once we add this remark: the subcomodules of C, as a comodule over $C^{op} \otimes C$ via left and right comultiplication, are exactly the subcoalgebras of C (where C^{op} is the opposite coalgebra of C).$_{\text{qed}}$

We need to prove some results on the category $\mathit{Comod}_f C$ of finite dimensional comodules over a coalgebra C.

Definition. A full subcategory of an abelian category is *replete* if it is closed under finite direct sums, subobjects and quotients.

In this definition it is explicitly assumed that any object isomorphic to an object of the replete subcategory also belongs to it. For any subcoalgebra $C' \subset C$, we have an inclusion $\mathit{Comod}_f C' \subset \mathit{Comod}_f C$ since every comodule over C' is naturally a comodule over C. Note that $\mathit{Comod}_f C'$ is a replete subcategory of $\mathit{Comod}_f C$.

Proposition 2. *For any coalgebra* C, *the assignment* $C' \longmapsto \mathit{Comod}_f C'$ *defines a bijection between the subcoalgebras of* C *and the replete subcategories of* $\mathit{Comod}_f C$.

Proof. We shall produce the inverse assignment. For any C-comodule (V,α), let $\mathrm{Im}(V)$ be the image of the coalgebra morphism

$$\tilde{\alpha} : \mathrm{End}(V) \longrightarrow C.$$

So $\mathrm{Im}(V,\alpha)$ is a subcoalgebra of C. For any replete subcategory $\mathcal{D} \subset \mathit{Comod}_f C$, let $\mathrm{Im}(\mathcal{D})$ be the sums of the subcoalgebras $\mathrm{Im}(V,\alpha)$ as (V,α) runs over the objects of $\mathcal{D}$. For any subcoalgebra $C' \subset C$, we obviously have $\mathrm{Im}(\mathcal{D}) \subset C'$ if and only if $\mathcal{D} \subset \mathit{Comod}_f C'$. This shows that

$$\mathrm{Im}(\mathit{Comod}_f C') \subset C' \quad \text{and} \quad \mathcal{D} \subset \mathit{Comod}_f(\mathrm{Im}(\mathcal{D})).$$

To prove that $C' \subset \mathrm{Im}(\mathit{Comod}_f C')$ we can suppose that $C' = C$. Using Proposition 1, it will be enough to show that every finite dimensional subcoalgebra V of C is contained in $\mathrm{Im}(\mathit{Comod}_f C)$. Now the map

$$\tilde{\delta} : \mathrm{End}(V) \longrightarrow V$$

corresponding to $\delta : V \longrightarrow V \otimes V$ is surjective since

$$\tilde{\delta}(\varepsilon \otimes v) = (\varepsilon \otimes 1)\,\delta(v) = v.$$

This shows that $\mathrm{Im}(V,\alpha) = V$ and hence $\mathrm{Im}(\mathit{Comod}_f C) \supset V$. To prove that $\mathit{Comod}_f(\mathrm{Im}(\mathcal{D})) \subset \mathcal{D}$, we can suppose that $\mathrm{Im}(\mathcal{D}) = C$. Then our task is to prove that $\mathcal{D} \subset \mathit{Comod}_f C$. Let us first remark that, for any comodule (V,α), the subcoalgebra $\mathrm{Im}(V,\alpha)$ is generated by the images of the maps

$$V \xrightarrow{\ \alpha\ } V \otimes C \xrightarrow{\ \phi \otimes 1\ } C,$$

where ϕ runs over V^*. This is a direct consequence of the identity

$$\tilde{\alpha}(\phi \otimes x) = (\phi \otimes 1)\,\alpha(x).$$

If $V \in \mathcal{D}$ then the images of the maps $(\phi \otimes 1)\,\alpha$ belongs to $\mathcal{D}$ since $\mathcal{D}$ is replete. Therefore $\mathrm{Im}(V,\alpha)$ belongs to $\mathcal{D}$. By hypothesis, C is generated by the $\mathrm{Im}(V,\alpha)$ for $(V,\alpha) \in \mathcal{D}$. This shows that every finite dimensional submodule of C belongs to $\mathcal{D}$. Let us prove that each $(W,\beta) \in \mathit{Comod}_f C$ belongs to $\mathcal{D}$. If $\phi_1, \ldots, \phi_n$ is a basis for W^* then the images W_i of the maps

$$W \xrightarrow{\ \beta\ } W \otimes C \xrightarrow{\ \phi_i \otimes 1\ } C$$

belongs to $\mathcal{D}$ since they are subcomodules of C. Combining the maps $W \longrightarrow W_i$ together, we obtain a map from W into the direct sum of the W_i. This map is easily seen to be monomorphic. So $W \in \mathcal{D}$. **qed**

Suppose now that $\mathcal{C}$ is a C-additive category and $U : \mathcal{C} \longrightarrow \mathcal{V}ect_C$ is a C-additive functor with values in finite dimensional spaces. In Section 4 we saw that the natural transformation $\gamma : U \longrightarrow U \otimes \mathrm{End}^\vee(U)$ determines a lifting

$$U^{\sim} : \mathcal{C} \longrightarrow \mathit{Comod}_f \mathrm{End}^\vee(U)$$

of U up into the category of $\mathrm{End}^\vee(U)$-comodules.

Theorem 3. *If $\mathcal{C}$ is abelian and* U *is exact and faithful then* $U^{\sim}$ *is an equivalence of categories.*

Proof. Obviously $U^{\sim}$ is faithful. Before proving that $U^{\sim}$ is full, let us see that, for any object $A \in \mathcal{C}$ and any subcomodule $E \subset U^{\sim}(A)$, there exists a subobject $j : B \rightarrowtail A$ such that $E = \mathrm{Im}U(j)$. Let $e_1, \ldots, e_n$ be a basis for the vector space U(A) chosen in such a way that $e_1, \ldots, e_k$ generate E. For any $x \in U(A)$, the coaction $\alpha : U(A) \longrightarrow U(A) \otimes \mathrm{End}^{\vee}(U)$ is given by the formula

$$\alpha(x) = \sum_{i=1}^{n} e_i \otimes [e_i^* \otimes x].$$

The hypothesis $\alpha(E) \subset E \otimes \mathrm{End}^{\vee}(U)$ means that, for all $i \le k$ and $j > k$, we have

$$[e_j^* \otimes e_i] = 0 .$$

We can now apply Section 6 Corollary 3 since $\mathrm{End}^{\vee}(U) = U^* \otimes_{\mathcal{C}} U$ and both U and U^* are exact ($\mathcal{C}$ is both artinian and noetherian since U is faithful). This means that, for all $i \le k$ and $j > k$, there is a subobject $B_{ij} \hookrightarrow A$ such that $e_i \in U(B_{ij})$ and $e_j^*(U(B_{ij})) = 0$. If we put

$$B_i = \bigcap_{j>k} B_{ij} \qquad \text{and} \qquad B = \bigcup_{i \le k} B_i$$

then we have $e_i \in U(B)$ for every $i \le k$, and $e_j^*(U(B)) = 0$ for every $j > k$. This shows that $E \subset U(B)$, and $U(B) \subset E$ since

$$E = \bigcap_{j>k} \mathrm{Ker}(e_j^*) .$$

To prove that $U^{\sim}$ is full, let $f : U^{\sim}(A) \longrightarrow U^{\sim}(B)$ be a morphism of comodules. Then the graph Γ_f of f is a subcomodule of $U^{\sim}(A) \oplus U^{\sim}(B) \cong U^{\sim}(A \oplus B)$. Therefore there is a subcomodule $C \subset A \oplus B$ whose image can be identified with Γ_f. We claim that C is the graph of an arrow $u : A \longrightarrow B$ whose image under U is f. To see this, let i be the composite of the inclusion $C \hookrightarrow A \oplus B$ and the first projection $p_1 : A \oplus B \longrightarrow A$. Then U(i) is the isomorphism $\Gamma_f \subset U(A) \oplus U(B) \longrightarrow U(A)$. Since U is exact and faithful, it follows that Ker(i) = Coker(i) = 0; so i is an isomorphism. The arrow u is defined as the composite

$$A \xrightarrow{i^{-1}} C \hookrightarrow A \oplus B \xrightarrow{p_2} B .$$

Obviously U(u) = f since U transforms this description of u into the description of f.

To finish the proof, we have to show that every object of $Comod_f\mathrm{End}^{\vee}(U)$ is isomorphic to a comodule in the image of $U^{\sim}$. If we let $\mathcal{D}$ be the full subcategory consisting of the comodules isomorphic to those in the image of $U^{\sim}$, we have to prove that $\mathcal{D} = Comod_f\mathrm{End}^{\vee}(U)$. But we have seen that the image of $U^{\sim}$, and therefore $\mathcal{D}$, is closed under subobjects. A similar argument shows that it is closed under quotients. Thus, $\mathcal{D}$ is a replete subcategory, and, from Proposition 2, we have $\mathcal{D} = Comod_f C'$ for some subcoalgebra $C' \subset \mathrm{End}^{\vee}(U)$. This implies that, for all $A \in \mathcal{C}$, the vector space U(A) has the structure of a C'-comodule $\gamma'_A : U(A) \longrightarrow U(A) \otimes C'$ which 'lifts' the coaction of $\mathrm{End}^{\vee}(U)$. More precisely, we have the commutative triangle of natural transformations below.

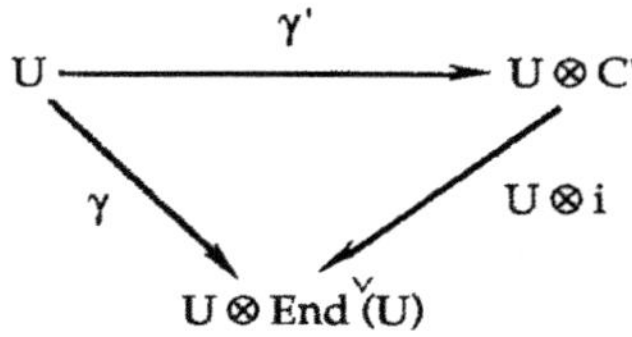

Using Section 4 Proposition 5, we have a unique coalgebra map

$$\tilde{\gamma}' : \mathrm{End}^{\vee}(U) \longrightarrow C'$$

such that the following triangle commutes.

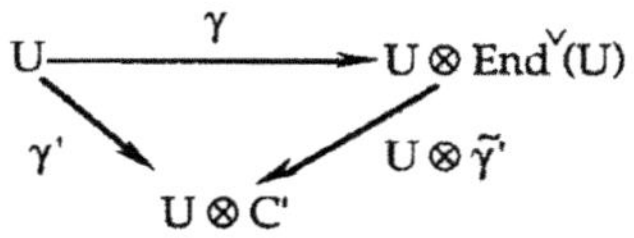

But then we have

$$i\,\tilde{\gamma}' = \mathrm{id}$$

by the uniqueness property of Section 4 Proposition 5. This proves that i is surjective, and therefore $C' = \mathrm{End}^{\vee}(U)$ since i is an inclusion.$_{\text{qed}}$

For any coalgebra C, the dual algebra C^* acts on the left of any (right) C-comodule. For any $V \in \mathit{Comod}(C)$, $\phi \in C^*$ and $x \in C$, we write $\phi\rfloor x$ for the action of ϕ on x. We have

$$\phi\rfloor x = (1_V \otimes \phi)\, \alpha_V(x) .$$

Each element $\phi \in C^*$ defines a natural transformation

$$\phi\rfloor(\ ) : \mathcal{U} \longrightarrow \mathcal{U}$$

where $\mathcal{U} : \mathit{Comod}(C) \longrightarrow \mathit{Vect}_C$ is the forgetful functor.

Proposition 4. *For any coalgebra* C, *the map* $\phi \longmapsto \phi\rfloor(\ )$ *is an isomorphism of algebras*

$$C^* \xrightarrow{\ \sim\ } \mathrm{Hom}(\mathcal{U}, \mathcal{U}) .$$

Proof. Using the fact that any comodule is a direct colimit of finite dimensional comodules, we can replace the category $\mathit{Comod}(C)$ by its subcategory $\mathit{Comod}_f(C)$. In this case, we have defined a coalgebra isomorphism

$$\tilde{\alpha} : \mathrm{End}^{\vee}(\mathcal{U}) \longrightarrow C .$$

The result will follow if we verify that the natural transformation $u : \mathcal{U} \longrightarrow \mathcal{U}$, corresponding to the composite of ϕ with the above coalgebra isomorphism, is equal to $\phi\rfloor(\ )$. Let $V \in \mathit{Comod}_f(C)$, $\psi \in V^*$ and $x \in V$. By definition of u, we have

$$\langle u, [\psi \otimes x] \rangle = \phi(\ \tilde{\alpha}\,(\psi \otimes x)) = \phi((\psi \otimes 1_C)\ \alpha_V(x))$$
$$= \psi((1_V \otimes \phi)\, \alpha_V(x)) = \psi(\phi\rfloor x) = \langle \phi\rfloor(\,) , [\psi \otimes x] \rangle$$

which shows that $u = \phi\rfloor(\ )$.$_{\text{qed}}$

§8. The bialgebra $\mathrm{End}^\vee(X)$ and tensor categories.

The tensor product $M\otimes N$ of two modules over an algebra A is an $A\otimes A$-module, but it is not in general an A-module. However, when A is a bialgebra, we can define an A-module structure on $M\otimes N$ by restricting the $A\otimes A$-module structure along the diagonal (= comultiplication) map $\delta : A \longrightarrow A\otimes A$. Similarly, the tensor product $M\otimes N$ of two comodules over a coalgebra C is a $C\otimes C$-comodule, and, when C is a bialgebra, we can corestrict this comodule structure along the multiplication map $\mu : C\otimes C \longrightarrow C$ to obtain a C-comodule structure on $M\otimes N$. The category *Comod*(C) of comodules over a bialgebra C is therefore a tensor category (also called a monoidal category). The main purpose of this section is to reverse this process: starting with a pair $(\mathcal{C}, X)$ consisting of a tensor category $\mathcal{C}$ and a tensor-product-preserving functor $X : \mathcal{C} \longrightarrow \mathcal{V}ect_f$, we shall show that the coalgebra $\mathrm{End}^\vee(X)$ can be enriched with the structure of a bialgebra.

In what follows, we let $\mathcal{V}ect_f$ denote the category of finite dimensional vector spaces. In Section 4 we saw how to construct a coalgebra $\mathrm{End}^\vee(X)$ from a pair $(\mathcal{C}, X)$ where $\mathcal{C}$ is a category and $X : \mathcal{C} \longrightarrow \mathcal{V}ect_f$ is a functor. It is easy to see in addition that a commutative triangle of functors

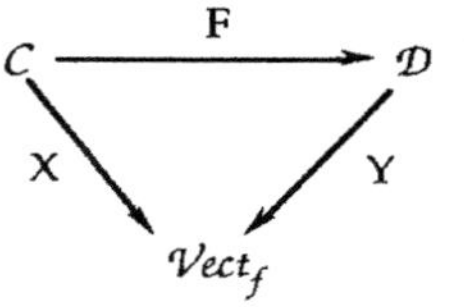

gives rise to a map of coalgebras $\mathrm{End}^\vee(X) \longrightarrow \mathrm{End}^\vee(Y)$ which we might call the *corestriction* along F (it is predual to the usual restriction map $\mathrm{End}(Y) \longrightarrow \mathrm{End}(X)$). When the functor F is an equivalence of categories, the corestriction map along F is an isomorphism of coalgebras. It is useful to formalise this process by introducing the category $\mathcal{C}at/\mathcal{V}ect_f$ of categories over $\mathcal{V}ect_f$. An object of $\mathcal{C}at/\mathcal{V}ect_f$ is a pair $(\mathcal{C}, X)$ where $\mathcal{C}$ is a (small) category and $X : \mathcal{C} \longrightarrow \mathcal{V}ect_f$ is a functor. A morphism $(F, \alpha) : (\mathcal{C}, X) \longrightarrow (\mathcal{D}, Y)$ consists of a functor $F : \mathcal{C} \longrightarrow \mathcal{D}$ and a natural isomorphism $\alpha : X \xrightarrow{\sim} YF$. Composition of morphisms is the obvious one. We have a covariant functor

$$\mathrm{End}^\vee : \mathcal{C}at/\mathcal{V}ect_f \longrightarrow \mathcal{C}oalg$$

with values in the category $\mathcal{C}oalg$ of coalgebras. We define the *(external) tensor product* of $X : \mathcal{C} \longrightarrow \mathcal{V}ect_f$ with $Y : \mathcal{D} \longrightarrow \mathcal{V}ect_f$ to be the functor

$$X \underline{\otimes} Y : \mathcal{C} \times \mathcal{D} \longrightarrow \mathcal{V}ect_f$$

where, for $(A, B) \in \mathcal{C} \times \mathcal{D}$,

$$(X \underline{\otimes} Y)(A, B) = X(A) \otimes Y(B).$$

Proposition 1. *There is a canonical isomorphism*

$$\theta : \mathrm{End}^\vee(X) \otimes \mathrm{End}^\vee(Y) \xrightarrow{\sim} \mathrm{End}^\vee(X \underline{\otimes} Y).$$

Proof. For any $(A, B) \in \mathcal{C} \times \mathcal{D}$, $S \in \mathrm{End}(X(A))$, $T \in \mathrm{End}(Y(B))$, we have $S\otimes T \in \mathrm{End}(X(A)\otimes Y(B))$. We put $\theta([S]\otimes[T]) = [S\otimes T]$. The best way to prove that θ is well defined and an isomorphism is to see that it is a special case of the following canonical isomorphism between tensor products of functors

$$(H \otimes_{\mathcal{C}} X) \otimes (K \otimes_{\mathcal{D}} Y) \xrightarrow{\sim} (H \underline{\otimes} K) \otimes_{\mathcal{C}\times\mathcal{D}} (X \underline{\otimes} Y)$$

where H, K are contravariant functors on $\mathcal{C}, \mathcal{D}$, respectively. When $H = X^*$ and $K = Y^*$, we have $H \otimes_{\mathcal{C}} X = \mathrm{End}^{\vee}(X)$ and $K \otimes_{\mathcal{D}} Y = \mathrm{End}^{\vee}(Y)$.$_{\text{qed}}$

Recall [ML] that a *tensor* (or "monoidal") *category* $\mathcal{C} = (\mathcal{C}, \otimes, I, a, \ell, r)$ consists of a category $\mathcal{C}$, a functor $\otimes : \mathcal{C} \times \mathcal{C} \longrightarrow \mathcal{C}$ (called the *tensor product*), an object $I \in \mathcal{C}$ (called the *unit object*) and natural isomorphisms

$$a = a_{A,B,C} : (A \otimes B) \otimes C \xrightarrow{\sim} A \otimes (B \otimes C),$$

$$\ell = \ell_A : I \otimes A \xrightarrow{\sim} A, \qquad r = r_A : A \otimes I \xrightarrow{\sim} A$$

(called the *associativity, left unit, right unit constraints*, respectively), such that, for all objects $A, B, C, D \in \mathcal{C}$, the following two *coherence* conditions hold:

$$a_{A,B,C\otimes D} \circ a_{A\otimes B,C,D} = (A \otimes a_{B,C,D}) \circ a_{A,B\otimes C,D} \circ (a_{A,B,C} \otimes D) \quad \text{and}$$

$$(A \otimes \ell_B) \circ a_{A,I,B} = r \otimes B.$$

It follows [ML] that all objects obtained by computing the tensor product of a sequence $A_1 \otimes A_2 \otimes \ldots \otimes A_m$ by bracketing it differently and by cancelling units are coherently isomorphic to each other.

A tensor category is called *strict* when all the constraints $a_{A,B,C}$, ℓ_A, r_A are identity arrows. Each tensor category $\mathcal{C}$ is equivalent to a *strict* tensor category $\mathrm{st}(\mathcal{C})$. The objects of $\mathrm{st}(\mathcal{C})$ are words $w = A_1 A_2 \ldots A_m$ in objects of $\mathcal{C}$. An arrow $f : w \longrightarrow w'$ is an arrow $f : [w] \longrightarrow [w']$ in $\mathcal{C}$ where we define

$$[\emptyset] = I, \qquad [A] = A, \qquad \text{and}$$

$$[A_1 A_2 \ldots A_{m+1}] = [A_1 A_2 \ldots A_m] \otimes A_{m+1}.$$

The tensor $\otimes$ for $\mathrm{st}(\mathcal{C})$ is given by $v \otimes w = vw$ and by commutativity of the following square

$$\begin{array}{ccc} [v] \otimes [w] & \xrightarrow{\; f \otimes g \;} & [v'] \otimes [w'] \\ \downarrow\wr & & \downarrow\wr \\ [vw] & \xrightarrow[\; f \otimes g \;]{} & [v'w'] \end{array} .$$

An example of a tensor category is the category $\mathit{Cat}/\mathcal{V}ect_f$ with the external tensor product described above. The unit object I in $\mathit{Cat}/\mathcal{V}ect_f$ is the functor $\mathbf{C} : \mathbf{1} \longrightarrow \mathcal{V}ect_f$ where $\mathbf{1}$ is the category with a single object $*$ and a single arrow (the identity of $*$) and $\mathbf{C}$ denotes the functor assigning to $*$ the one-dimensional vector space $\mathbf{C} \in \mathcal{V}ect_f$.

Let $\mathcal{C}, \mathcal{D}$ denote tensor categories. Recall [ML] that a *tensor* (or "strong monoidal") *functor* $F = (F, \phi, \phi_0) : \mathcal{C} \longrightarrow \mathcal{D}$ consists of a functor $F : \mathcal{C} \longrightarrow \mathcal{D}$, a natural isomorphism

$$\phi = \phi_{A,B} : FA \otimes FB \xrightarrow{\sim} F(A \otimes B),$$

and an isomorphism $\phi_0 : I \xrightarrow{\sim} FI$, such that the following three equations hold (where we write as if $\mathcal{C}, \mathcal{D}$ were strict):

$$\phi_{A\otimes B,C} \circ (\phi_{A,B} \otimes FC) = \phi_{A,B\otimes C} \circ (FA \otimes \phi_{B,C}),$$

$$\phi_0 \otimes FA = \phi_{I,A} \circ F(\phi_0 \otimes A) \quad \text{and} \quad FA \otimes \phi_0 = \phi_{A,I} \circ F(A \otimes \phi_0).$$

The tensor functor is strict when all the isomorphisms $\phi_{A,B}$, ϕ_0 are identities. One example of a tensor functor is the equivalence $C \xrightarrow{\sim} st(C)$ taking A to [A]. Another example is the functor $End^\vee : Cat/Vect_f \longrightarrow Coalg$.

Recall that a *monoid* $M = (M, \mu, \eta)$ in a tensor category C consists of an object $M \in C$ and arrows $\mu : M \otimes M \longrightarrow M, \eta : I \longrightarrow M$ such that the following diagrams commute.

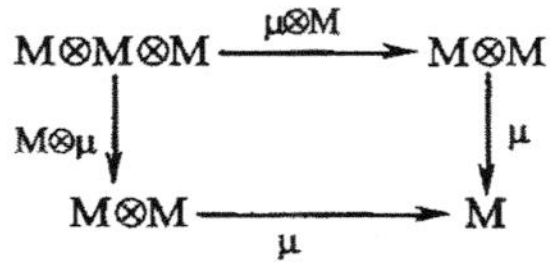

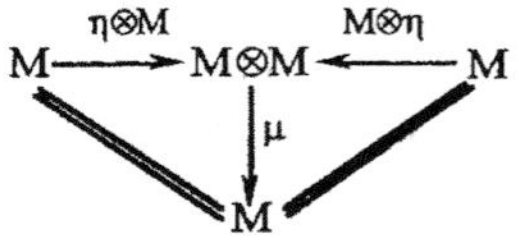

A *comonoid* is a monoid in C^{op}. For example, algebras are monoids in the category of vector spaces, coalgebras are comonoids in the same category, and bialgebras are monoids in the category of coalgebras. Monoids in Cat (where the tensor product is cartesian product) are strict tensor categories. Monoids in $Cat/Vect_f$ are the pairs (C, X) for which C is a strict tensor category and $X : C \longrightarrow Vect_f$ is a tensor functor (not necessarily strict). It follows from Proposition 1 that the coalgebra $End^\vee(X)$ corresponding to such a pair has the structure of a bialgebra since it enherits a monoid structure in the category of coalgebras. More generally, if we have a pair (C, X) where C is a tensor category, not necessarily strict, and where X is a tensor functor, we also have a bialgebra structure on $End^\vee(X)$. To see this, we can replace the pair (C, X) by a pair (C', X') where $C' = st(C)$ is a strict tensor category equivalent to C. We use the fact that the corestriction map $End^\vee(X')$ $\longrightarrow End^\vee(X)$ along an equivalence $C' \longrightarrow C$ is an isomorphism of coalgebras. However, we shall directly describe the algebra structure on $End^\vee(X)$ without recourse to (C', X'). For any $A, B \in C$, $S \in End(X(A))$ and $T \in End(X(B))$, let us write $S \otimes T$ for the dotted arrow in the square

$$
\begin{array}{ccc}
X(A \otimes B) & - - - - \longrightarrow & X(A \otimes B) \\
\downarrow\wr & & \wr\downarrow \\
X(A) \otimes X(B) & \xrightarrow{\quad S \otimes T \quad} & X(A) \otimes X(B)
\end{array}
$$

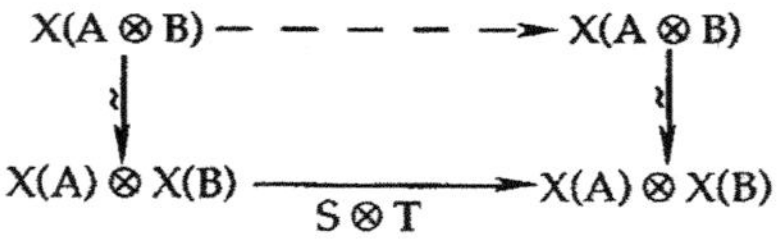

Also, let us write $1 \in X(I)$ for the element corresponding to $1 \in C$ under the isomorphism $C \cong X(I)$. These notational abuses are harmless, not only because the context will dissipate the ambiguity, but also because tensor functors satisfy a coherence theorem [Le]. We can now specify the algebra structure on $End^\vee(X)$. The product of the elements [S] and [T] of $End^\vee(X)$ is given by the simple formula

$$[S][T] = [S \otimes T].$$

The unit element of $End^\vee(X)$ is equal to [1].

When $C = Rep(M, C)$ and X is the forgetful functor U, we obtain a bialgebra structure on $End^\vee(U)$. The meaning of the bialgebra structure is elucidated by the next result.

Proposition 2. *For any topological monoid M, the Fourier cotransform*
$$\mathcal{F}^\vee : End^\vee(U) \longrightarrow R(M)$$
is an isomorphism of bialgebras.

Proof. It remains to verify that $\mathcal{F}^{\vee}$ is a homomorphism of algebras. For all $V, W \in \mathcal{R}ep(M,C)$ and all $A \in End(V)$, $B \in End(W)$, we have

$$\mathcal{F}^{\vee}([A][B]) = \mathcal{F}^{\vee}([A\otimes B]) = Tr(\pi_{V\otimes W}\circ(A\otimes B)) = Tr((\pi_V\otimes\pi_W)\circ(A\otimes B))$$
$$= Tr(\pi_V A\otimes\pi_W B) = Tr(\pi_V A)\,Tr(\pi_W B) = \mathcal{F}^{\vee}([A])\,\mathcal{F}^{\vee}([B]) \text{ and}$$
$$\mathcal{F}^{\vee}(1) = Tr(\pi_I) = 1\,._{\text{qed}}$$

The next thing we shall do is to characterize the algebra structure on $End^{\vee}(X)$ by a universal property. More precisely, for all algebras A, we shall prove that the correspondence $n \mapsto \tilde{n}$ is a bijection between tensor-preserving natural transformations $X \longrightarrow X\otimes A$ and algebra homomorphisms $End^{\vee}(X) \longrightarrow A$. Our first task is to define the former. A *coaction* of the algebra A on a vector space V is a linear map $\alpha : V \longrightarrow V\otimes A$, or equivalently, a linear map

$$\tilde{\alpha} : End(V) \longrightarrow A\,.$$

We define the *trace* of α as the value of the last linear map at the identity endomorphism of V. If $e_1, \ldots, e_n$ is a basis of V and

$$\alpha(e_i) = \sum_{j=1}^{n} e_j \otimes \alpha_{ij}$$

then

$$Tr(\alpha) = \sum_{i=1}^{n} \alpha_{ii}\,.$$

We have the formula

$$\tilde{\alpha}(S) = Tr(\alpha S)$$

which is valid for all $S \in End(V)$. The *tensor product* $\alpha \underline{\otimes} \beta$ of coactions $\alpha : V \longrightarrow V\otimes A$, $\beta : W \longrightarrow W\otimes A$ is defined to be the composite

$$V\otimes W \xrightarrow{\alpha\otimes\beta} V\otimes A\otimes W\otimes A \xrightarrow{\sim} V\otimes W\otimes A\otimes A \xrightarrow{V\otimes W\otimes\mu} V\otimes W\otimes A$$

where $\mu : A\otimes A \longrightarrow A$ is the multiplication of the algebra A, and the middle isomorphism in the composite uses the symmetry map $A\otimes W \xrightarrow{\sim} W\otimes A$. If

$$\beta(f_r) = \sum_{s=1}^{m} f_s \otimes \beta_{rs}$$

gives a matrix for β then

$$(\alpha\underline{\otimes}\beta)(e_i\otimes f_r) = \sum_{j,\,s=1}^{n,\,m} (e_j\otimes f_s)\otimes(\alpha_{ij}\beta_{rs})$$

gives a matrix for $\alpha\underline{\otimes}\beta$.

We shall say that a natural transformation $u : X \longrightarrow X\otimes A$ is *tensor preserving* when, for all $C, D \in \mathcal{C}$, we have

$$u_{C\otimes D} = u_C \underline{\otimes} u_D \quad \text{and} \quad u_I = 1\,.$$

Let $\mathcal{C}$ be a tensor category and $X : \mathcal{C} \longrightarrow \mathcal{V}ect_f$ be a tensor functor. In Section 4 Proposition 3, we defined a natural transformation $\gamma : X \longrightarrow X\otimes End^{\vee}(X)$.

Proposition 3. *The natural transformation* γ *is tensor preserving. Moreover, for all algebras* A *and all tensor-preserving natural transformations* $n : X \longrightarrow X \otimes A$, *there is precisely one algebra homomorphism* $\tilde{n} : End^{\vee}(X) \longrightarrow A$ *such that the following triangle commutes.*

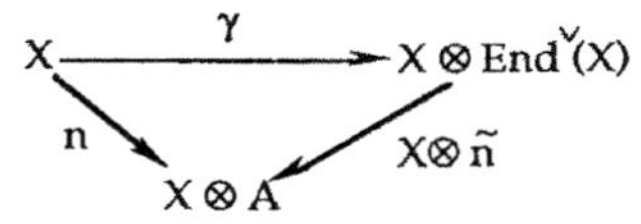

Proof. To prove γ tensor preserving, take $C, D \in \mathcal{C}$. Choose a basis $e_1, \ldots, e_m$ of X(C) and a basis $f_1, \ldots, f_n$ of X(D). By definition of $\gamma_C \underline{\otimes} \gamma_D$, we have the equality

$$(\gamma_C \underline{\otimes} \gamma_D)(x \otimes y) = \sum_{i,j} e_i \otimes f_j \otimes [e_i^* \otimes x][f_i^* \otimes y].$$

On the other hand, we have

$$\gamma_{C \otimes D}(x \otimes y) = \sum_{i,j} e_i \otimes f_j \otimes [e_i^* \otimes f_i^* \otimes x \otimes y].$$

The equality $\gamma_C \underline{\otimes} \gamma_D = \gamma_{C \otimes D}$ is now a consequence of the identity

$$[\phi \otimes x][\psi \otimes y] = [\phi \otimes \psi \otimes x \otimes y]$$

which holds in $End^{\vee}(X)$. To prove the rest of the Proposition, let $n : X \longrightarrow X \otimes A$ be a tensor-preserving natural transformation. According to Section 4 Proposition 3, there exists precisely one linear map $\tilde{n} : End^{\vee}(X) \longrightarrow A$ such that the triangle of the Proposition commutes. A straightforward computation shows that we have $\tilde{n}([S]) = Tr(n_C S)$, where we are using the trace introduced above. Using this we have

$$\tilde{n}([S][T]) = \tilde{n}([S \otimes T]) = Tr(n_{C \otimes D}(S \otimes T)) = Tr(n_C \otimes n_D \circ (S \otimes T))$$

$$= Tr(n_C S \otimes n_D T) = Tr(n_C S)\, Tr(n_D T) = \tilde{n}([S])\, \tilde{n}([T]) \,. \quad \text{qed}$$

When A is a bialgebra, a coaction $\alpha : V \longrightarrow V \otimes A$ defines a comodule structure on V if and only if it is associative and unitary. Also, the tensor product of two coactions α, β is a comodule structure if both of α, β are. The category $\mathcal{C}omod_f(A)$ of finite dimensional comodules is a tensor category. Clearly, if a natural transformation $n : X \longrightarrow X \otimes A$ is tensor preserving and defines a comodule structure on X then we obtain a functor

$$X' = (X, n) : \mathcal{C} \longrightarrow \mathcal{C}omod_f(A)$$

which is tensor preserving and renders commutative the following triangle, where $\mathcal{U}$ is the forgetful functor.

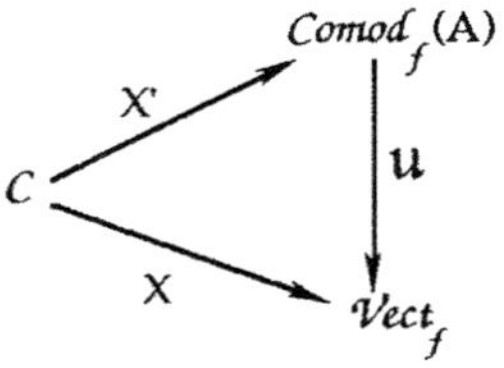

Proposition 4. *Let* $\mathcal{C}$ *be a tensor category and* $X : \mathcal{C} \longrightarrow \mathcal{V}ect_f$ *be a tensor functor. For all bialgebras* A, *there is a bijection between tensor-preserving functors* $X' : \mathcal{C} \longrightarrow \mathcal{C}omod_f(A)$ *such that* $\mathcal{U}X' = X$ *(i.e.* X' lifts X *) and bialgebra homomorphisms* $\mathrm{End}^{\vee}(X) \longrightarrow A$.

Proof. Just combine Proposition 4 with Section 4 Proposition 7.$_{\text{qed}}$

§9. Duality and Hopf algebras.

We begin by recalling the basic concepts of duality theory in a tensor category $\mathcal{C}$. Given $A, B \in \mathcal{C}$, we shall say that a pair of maps $\eta : I \longrightarrow B \otimes A$, $\varepsilon : A \otimes B \longrightarrow I$ form an *adjunction* between A and B if the following two composites are identities:

$$A \xrightarrow{A \otimes \eta} A \otimes B \otimes A \xrightarrow{\varepsilon \otimes A} A\ , \quad B \xrightarrow{\eta \otimes B} B \otimes A \otimes B \xrightarrow{B \otimes \varepsilon} B\ .$$

We call η the *unit* and ε the *counit* of the adjunction. We say that A (respectively, B) is *left adjoint* or *left dual* to B (respectively *right adjoint* or *right dual* to A). We also write $(\eta, \varepsilon) : A \dashv B$ to indicate that the pair (η, ε) is an adjunction between A and B.

The unit and counit of an adjunction determine each other. More precisely, let us say that a map $\varepsilon : A \otimes B \longrightarrow I$ (a 'pairing') is *exact* when, for all objects X, Y, the function

$$\varepsilon^{\#} : \mathcal{C}(X, B \otimes Y) \longrightarrow \mathcal{C}(A \otimes X, Y)\ , \qquad f \longmapsto (\varepsilon \otimes Y)(A \otimes f)$$

is bijective.

Proposition 1. *A pairing* $\varepsilon : A \otimes B \longrightarrow I$ *is exact if and only if there exists a map* $\eta : I \longrightarrow B \otimes A$ *such that the pair* (η, ε) *is an adjunction between* A *and* B.

Proof. If ε is exact then take η to be the unique map such that $\varepsilon^{\#}(\eta)$ is the canonical isomorphism $A \otimes I \xrightarrow{\sim} A$. Conversely, if (η, ε) is an adjunction then the function $h \longmapsto (B \otimes h)(\eta \otimes X)$ is an inverse for $\varepsilon^{\#}$.$_{\text{qed}}$

In the category of finite dimensional vector spaces, a pairing $\varepsilon : V \otimes W \longrightarrow \mathbb{C}$ is exact if and only if the corresponding map $x \longmapsto \varepsilon(x, -)$ from V to W^* is an isomorphism. When the pairing is exact, we can describe $\eta : \mathbb{C} \longrightarrow W \otimes V$ by giving the value $\eta(1) \in W \otimes V$. To any basis $e_1, \ldots, e_n$ of V there corresponds a dual basis $f_1, \ldots, f_n$ of W such that $\varepsilon(e_i, f_j) = \delta_{ij}$. We have

$$\eta(1) = f_1 \otimes e_1 + \ldots + f_n \otimes e_n\ .$$

In a tensor category $\mathcal{C}$, let $\varepsilon : A \otimes B \longrightarrow I$ and $\varepsilon' : A' \otimes B' \longrightarrow I$ be two exact pairings. We shall say that a map $f : A \longrightarrow A'$ is *left adjoint* to $g : B' \longrightarrow B$ (or that g is *right adjoint* to f) when we have

$$\varepsilon'(f \otimes A) = \varepsilon(B \otimes g)\ .$$

For all f, there is a unique right adjoint g given by

$$g = (\eta' \otimes B)(B' \otimes f \otimes B)(B' \otimes \varepsilon)\ .$$

Similarly, for all g, there is a unique left adjoint f given by

$$f = (A' \otimes \eta)(A' \otimes g \otimes A)(\varepsilon' \otimes A)\ .$$

Applying this to the case where A = A', we see that two right adjoints B, B' of A are canonically isomorphic. Similarly for left adjoints.

Definition. A tensor category $\mathcal{C}$ is *autonomous* when every object of $\mathcal{C}$ has both a left and a right adjoint.

When $\mathcal{C}$ is autonomous, we can choose, for each $C \in \mathcal{C}$, a pair of adjunctions

$$(\eta_C, \varepsilon_C) : C^{\ell} \dashv C \quad \text{and} \quad (\eta'_C, \varepsilon'_C) : C \dashv C^{r}.$$

We obtain in this way a pair of contravariant functors

$$(\ )^{\ell} : \mathcal{C}^{op} \longrightarrow \mathcal{C} \quad \text{and} \quad (\ )^{r} : \mathcal{C}^{op} \longrightarrow \mathcal{C}.$$

Obviously, for all $C \in \mathcal{C}$, we have canonical isomorphisms

$$(C^{r})^{\ell} \cong C \cong (C^{\ell})^{r}$$

making the functors $(\ )^{\ell}$, $(\ )^{r}$ mutually quasi-inverse (i.e. they give an equivalence of categories).

It is instructive to work out an example of an autonomous tensor category where right and left adjoint are different. For any algebra A, let us write $\mathcal{C}o_f(A)$ for the category whose objects are the coactions $\alpha : V \longrightarrow V \otimes A$ on finite dimensional vector spaces. In Section 8 we defined a tensor product of coactions, and so this category becomes a tensor category. We first identify the adjunctions within this category. Let (V,α), $(W,\beta) \in \mathcal{C}o_f(A)$ and let

$$(\eta, \varepsilon) : (V,\alpha) \dashv (W,\beta)$$

be an exact pairing. Clearly the pairing (η, ε) defines an exact pairing between the vector spaces V and W (since the forgetful functor $\mathcal{C}o_f(A) \longrightarrow \mathcal{V}ect_f$ preserves tensor product). Let $e_1, \ldots, e_n$ be a basis of V and let $f_1, \ldots, f_n$ be a dual basis. We have

$$\alpha(e_i) = \sum_j e_j \otimes \alpha_{ji} \quad , \quad \beta(f_i) = \sum_j f_j \otimes \beta_{ji} \quad ,$$

$$\varepsilon(e_i \otimes f_j) = \delta_{ij} \quad , \quad \eta(1) = \sum_i f_i \otimes e_i \quad .$$

Expressing that ε is a morphism in $\mathcal{C}o_f(A)$, we obtain

$$\sum_k \alpha_{ki} \beta_{kj} = \delta_{ij} \ .$$

Similarly, expressing that η is a morphism, we obtain

$$\sum_k \beta_{ik} \alpha_{jk} = \delta_{ij} \ .$$

If $\alpha = (\alpha_{ij})$ and $\beta = (\beta_{ij})$, these equalities can be formulated as the matrix equations $({}^t\alpha)\beta = \beta\,({}^t\alpha) = \mathrm{id}$. In other words, the *right adjoint* α^{r} of the matrix α is equal to $({}^t\alpha)^{-1}$. Similarly, we obtain that the *left adjoint* α^{ℓ} of α is the matrix ${}^t(\alpha^{-1})$. If the algebra A is not commutative, there is in general no relationship between ${}^t(\alpha^{-1})$ and $({}^t\alpha)^{-1}$. One might exist and not the other. We can inductively define

$$\alpha^{(0)} = \alpha, \qquad \alpha^{(n+1)} = (\alpha^{(n)})^{\ell} \text{ for } n \geq 0, \qquad \text{and} \qquad \alpha^{(n-1)} = (\alpha^{(n)})^{r} \text{ for } n \leq 0.$$

Let us say that a matrix α is *totally invertible* when $\alpha^{(n)}$ exists for all $n \in \mathbf{Z}$. Clearly the coactions with totally invertible matrices form an example of an autonomous category for which left and right duals do not coincide in general.

We now give a brief review of the basic theory of Hopf algebras. Recall that, for any coalgebra C and any algebra A, the *convolution product* defines an algebra structure on the vector space Hom(C, A), where the convolution $\phi * \psi$ of $\phi : C \longrightarrow A$ with $\psi : C \longrightarrow A$

is the composite

$$C \xrightarrow{\delta} C \otimes C \xrightarrow{\phi \otimes \psi} A \otimes A \xrightarrow{\mu} A .$$

The unit of $\mathrm{Hom}(C, A)$ is the composite

$$C \xrightarrow{\varepsilon} C \xrightarrow{\eta} A$$

where $\eta(\lambda) = \lambda.1$. If $f : C' \longrightarrow C$ is a morphism of coalgebras and $g : A \longrightarrow A'$ is a morphism of algebras then the assignment $\phi \longmapsto g\phi f$ is a morphism of algebras

$$\mathrm{Hom}(C, A) \longrightarrow \mathrm{Hom}(C', A').$$

When $C = A = H$ is a bialgebra, we obtain an algebra structure on $\mathrm{Hom}(H, H)$. An *antipode* ν on a bialgebra H is a two-sided inverse for the identity map $1_H : H \longrightarrow H$ with respect to the convolution product. More explicitly, this means that the following two diagrams commute.

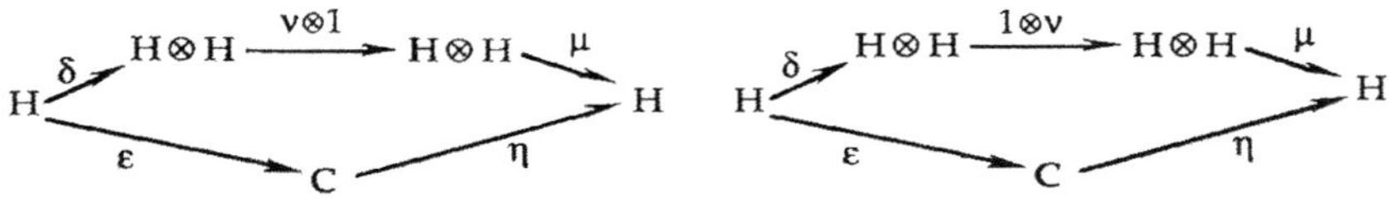

An antipode, when it exists, is unique. A bialgebra with an antipode is a *Hopf algebra.*

The translation of all the axioms on a bialgebra into pictorial notation [JS2] leads to the following diagrams.

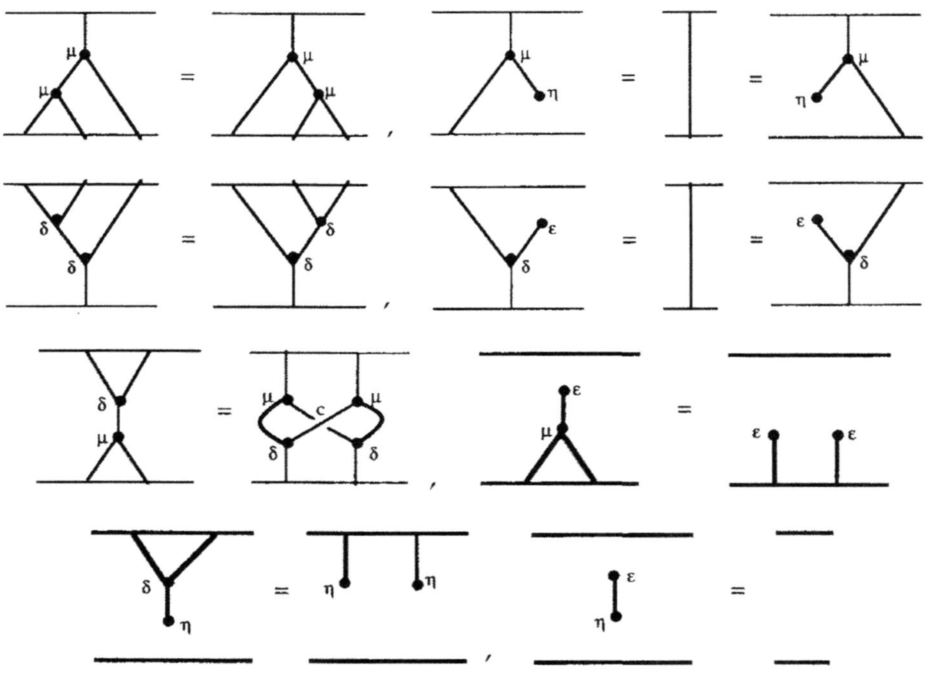

Now we add to these the two axioms for an antipode ν .

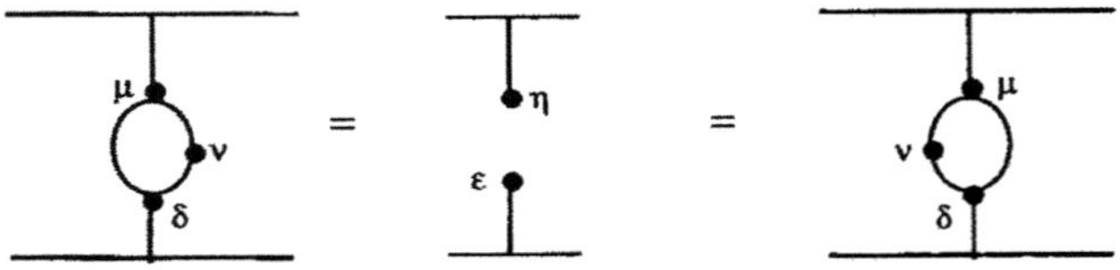

Proposition 2. *In any Hopf algebra* H, *the antipode* $\nu : H \longrightarrow H$ *is an anti-endomorphism of the algebra structure and of the coalgebra structure.*

Proof. We will compute the convolution inverse of the multiplication $\mu : H \otimes H \longrightarrow H$ in the algebra $\mathrm{Hom}(H \otimes H, H)$. First, the map $\phi \longmapsto \phi\mu$ is an algebra morphism

$$\mathrm{Hom}(H, H) \longrightarrow \mathrm{Hom}(H \otimes H, H)$$

since $\mu : H \otimes H \longrightarrow H$ is a coalgebra morphism. This proves that $\nu\mu$ is the convolution inverse of $1_H\, \mu = \mu$. Secondly, we shall verify, by a direct pictorial computation, that the map

$$\mu' = (H \otimes H \xrightarrow{c} H \otimes H \xrightarrow{\nu \otimes \nu} H \otimes H \xrightarrow{\mu} H)$$

is a convolution inverse to μ. The diagram for μ' is as follows.

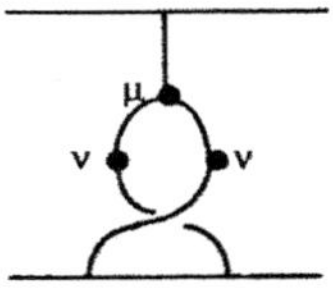

Starting with the diagram which expresses the product $\mu' * \mu$, we obtain the following sequence of diagrams having the same value:

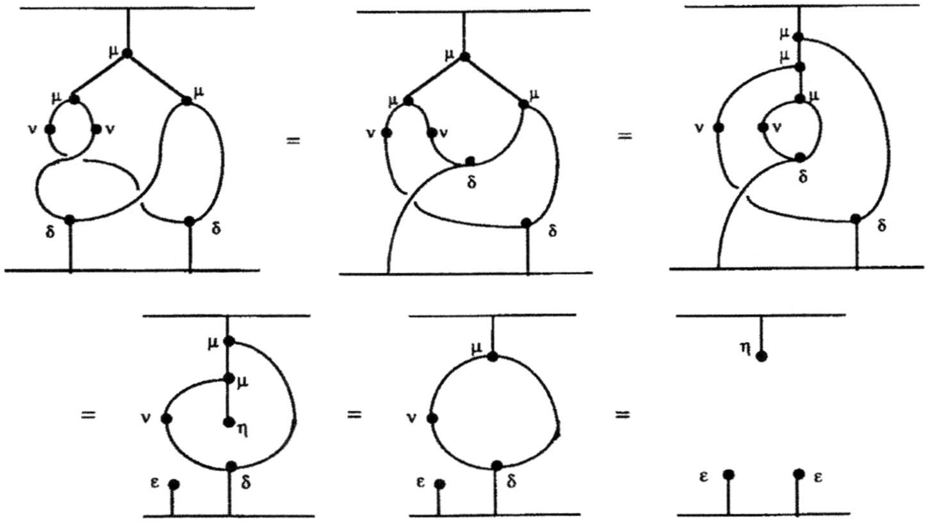

The last picture is the unit of the algebra $\mathrm{Hom}(H\otimes H, H)$, so we have proved that μ' is the convolution inverse of μ. By uniqueness of convolution inverse, we obtain $v\,\mu = \mu'$, which means that v is an anti-endomorphism of the algebra structure of H. Using a similar argument (invert the diagrams and replace ε, μ by η, δ), we obtain that v is an anti-endomorphism of the coalgebra structure.$_{\text{qed}}$

Now consider the bialgebra $H' = (H, \delta, \mu')$ obtained by reversing the multiplication μ: that is, $\mu'(x\otimes y) = \mu(y\otimes x)$.

Proposition 3. *For any Hopf algebra* H, *an antipode* v' *exists for* H' *if and only if the antipode* v *of* H *is bijective. In this case,* $v' = v^{-1}$.

Proof. If v is bijective, composition with v^{-1} defines an algebra morphism $\mathrm{Hom}(H,H) \longrightarrow \mathrm{Hom}(H,H')$ since $v^{-1} : H \longrightarrow H'$ is an algebra morphism. Therefore $1_H = v^{-1}v$ and $v^{-1} = v^{-1}1_H$ are mutually inverse in $\mathrm{Hom}(H, H')$. This shows that v^{-1} is an antipode for H'. Conversely, if v' exists, composition with v' gives an algebra morphism $\mathrm{Hom}(H,H) \longrightarrow \mathrm{Hom}(H,H')$, and therefore $v' = v'\,1_H$ and $v'v$ are mutually inverse in $\mathrm{Hom}(H, H')$. By uniqueness of inverses, $v'v = 1_H$. A similar argument gives $v\,v' = 1_H$ which proves the result.$_{\text{qed}}$

Recall that the category $\mathit{Comod}_f(H)$ of (right) comodules over a bialgebra H is a tensor category.

Proposition 4. *Let* H *be a bialgebra. Then* H *has an antipode if and only if every finite dimensional (right)* H*-comodule has a left dual.*

Proof. Let $\alpha : V \longrightarrow V\otimes H$ be a finite dimensional comodule. If $e_1, \ldots, e_n$ is a basis of V then we have

$$\alpha(e_i) = \sum_{j=1}^{n} e_i \otimes \alpha_{ji}\,, \quad \delta(\alpha_{ij}) = \sum_{k=1}^{n} \alpha_{ik}\otimes \alpha_{kj}\,, \quad \varepsilon(\alpha_{ij}) = \delta_{ij}\,.$$

It follows that

$$\sum_{k=1}^{n} v(\alpha_{ik})\,\alpha_{kj} = \mu(v\otimes 1_H)\,\delta(\alpha_{ij}) = \mu(\varepsilon(\alpha_{ij})) = \delta_{ij}\,,$$

and similarly that

$$\sum_{k=1}^{n} \alpha_{ik}\, v(\alpha_{kj}) = \delta_{ij}\,.$$

Hence $\alpha = (\alpha_{ij})$ is an invertible matrix. If we put

$$\beta = (\beta_{ij}) = (v\,\alpha_{ji}) = {}^{t}(\alpha^{-1})$$

then we have a coaction $\beta : V^* \longrightarrow V^*\otimes H$ defined as

$$\beta(e_i^*) = \sum_{j=1}^{n} e_i^*\otimes \beta_{ji}\,.$$

we have seen earlier in this Section that such a coaction is *left dual* to α in the category

$Co_f(H)$. It remains to see that β provides a comodule structure. But we have

$$\delta\,\beta_{ij} = \delta\, v\,\alpha_{ji} = c\,(v\otimes v)\,\delta\,\alpha_{ji} = c\sum_k v\,(\alpha_{jk})\otimes v\,(\alpha_{ki})$$
$$= \sum_k v\,(\alpha_{ki})\otimes v\,(\alpha_{jk}) = \sum_k \beta_{ik}\otimes\beta_{kj}\,,$$
$$\varepsilon\,\beta_{ij} = \varepsilon\, v\,\alpha_{ji} = \varepsilon\,\alpha_{ji} = \delta_{ji} = \delta_{ij}\,.$$

The rest of this Proposition follows from the next Proposition 5.$_{\text{qed}}$

Proposition 5. *Let C be a tensor category and $X : C \longrightarrow \mathcal{V}ect_f$ be a tensor functor. If every object of C has a left dual then the coalgebra* $\mathrm{End}^{\vee}(X)$ *has an antipode.*

Proof. For all objects $C\in C$, we have an exact pairing $\varepsilon : C^{\ell}\otimes C \longrightarrow I$. The vector space $X(C^{\ell})$ is then a dual $X(C)^*$ of $X(C)$, since tensor functors preserve dual pairs. Therefore, for all $A\in\mathrm{End}(X(C))$, there is a transposed endomorphism ${}^tA\in\mathrm{End}(X(C^{\ell}))$. Let us put

$$v\,([A]) = [{}^tA]\,.$$

Then v is easily seen to be a well defined linear endomorphism of $\mathrm{End}^{\vee}(X)$. It remains to prove that v is an antipode. Let $e_1, \ldots, e_n$ be a basis of $X(C)$ and let $e_1^*, \ldots, e_n^*$ be the dual basis of $X(C^{\ell})$ (the pairing is $X(\varepsilon)$). Then, for all $\phi\in X(C^{\ell})$ and $x\in X(C)$, we have

$$\delta\,[\phi\otimes x] = \sum_i\,[\phi\otimes e_i]\otimes[e_i^*\otimes x] \quad\text{and}\quad v\,[\phi\otimes x] = [x\otimes\phi]\,.$$

So that we have

$$\mu\,(v\otimes 1)\,\delta\,[\phi\otimes x] = \sum_i\,[e_i\otimes\phi]\,[e_i^*\otimes x] = \sum_i\,[(e_i\otimes e_i^*)\otimes(\phi\otimes x)] = [\,t\otimes(\phi\otimes x)]$$

where $t = \sum_i e_i\otimes e_i^*$ is the linear form $X(\varepsilon)$ on the space $X(C^{\ell})\otimes X(C)$ (so that t is also just the trace map on $X(C)^*\otimes X(C)$). If $1 : C \longrightarrow C$ denotes the identity form, we have

$$[\,t\otimes(\phi\otimes x)] = [\,1\,X(\varepsilon)\otimes(\phi\otimes x)] = [1\otimes X(\varepsilon)\,(\phi\otimes x)] = [\phi(x)] = \eta_0\varepsilon_0\,([\phi\otimes x])$$

where η_0, ε_0 here denote the unit and counit of H.$_{\text{qed}}$

Definition. A Hopf algebra is *autonomous* when the antipode is a bijective map.

Proposition 6. *Let H be a bialgebra. The category $Comod_f(H)$ is autonomous if and only if H is an autonomous Hopf algebra.*

Proof. Let H' be the bialgebra obtained by reversing the multiplication of H. Let $Comod_f(H)'$ be the tensor category obtained by reversing the tensor product on $Comod_f(H)$, so that $V\otimes' W = W\otimes V$. Then the map c given by $c(x\otimes y) = y\otimes x$ is an isomorphism between $V\otimes' W$ and $V\otimes W$ computed as H'-comodules. This shows that we have a canonical isomorphism of tensor categories between $Comod_f(H')$ and $Comod_f(H)'$. A right dual in $Comod_f(H)$ is a left dual in $Comod_f(H)'$. The result now follows from Propositions 3, 4, 5.$_{\text{qed}}$

§10. Braidings and Yang-Baxter operators.

Recall [JS1] that a *braiding* for a tensor category $\mathcal{V}$ consists of a natural family of isomorphisms

$$c = c_{A,B} : A \otimes B \xrightarrow{\sim} B \otimes A$$

in $\mathcal{V}$ such that the following two diagrams commute (where the unnamed arrows are associativity constraints).

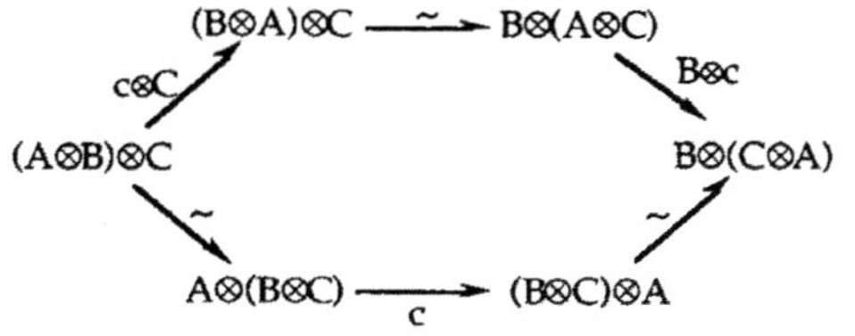

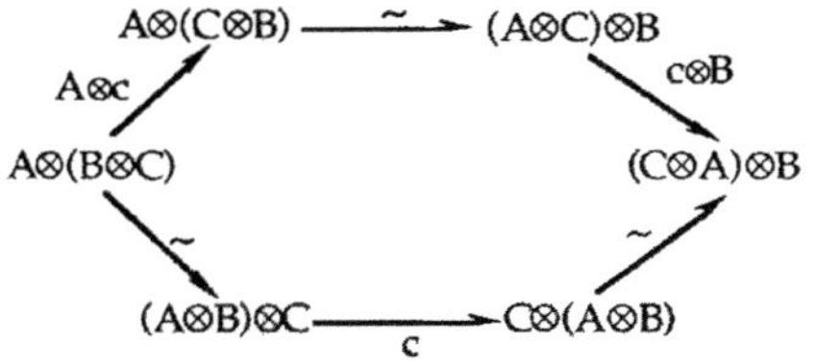

It follows from these axioms that $c_{A,I} : A \otimes I \xrightarrow{\sim} I \otimes A$ is equal to the canonical isomorphism $A \otimes I \xrightarrow{\sim} A \xrightarrow{\sim} I \otimes A$. Similarly for $c_{I,A} : I \otimes A \xrightarrow{\sim} A \otimes I$.

If c is a braiding then so is c' given by $c'_{A,B} = (c_{B,A})^{-1}$. A *symmetry* is a braiding for which $c = c'$.

A *braided tensor category* is a pair $(\mathcal{V}, c)$ consisting of a tensor category $\mathcal{V}$ and a braiding c.

Example 1. Let $\mathbf{B}_n$ be the Artin braid group. A presentation for $\mathbf{B}_n$ is given by the generators $s_1, \ldots, s_{n-1}$ and the relations

(A1) $s_i s_{i+1} s_i = s_{i+1} s_i s_{i+1}$ for $1 \le i \le n-2$,

(A2) $s_i s_j = s_j s_i$ for $1 \le i < j-1 \le n-2$.

The *braid category* $\mathbf{B}$ is the disjoint union of the $\mathbf{B}_n$. More explicitly, the objects of $\mathbf{B}$ are the natural numbers $0, 1, 2, \ldots,$ the homsets are given by

$$\mathbf{B}(m, n) = \begin{cases} \mathbf{B}_n & \text{when } m = n \\ \varnothing & \text{otherwise} \end{cases},$$

and composition is the multiplication of the braid groups. The category $\mathbf{B}$ is equipped with a strict tensor structure defined by *addition of braids*

$$\oplus : \mathbf{B}_m \times \mathbf{B}_n \longrightarrow \mathbf{B}_{m+n}$$

which is algebraically described by

$$s_i \oplus s_j = s_i\, s_{m+j}\,.$$

A braiding for $\mathbf{B}$ is given by the elements

$$c = c_{m,n} : m + n \longrightarrow n + m$$

illustrated by the following figure.

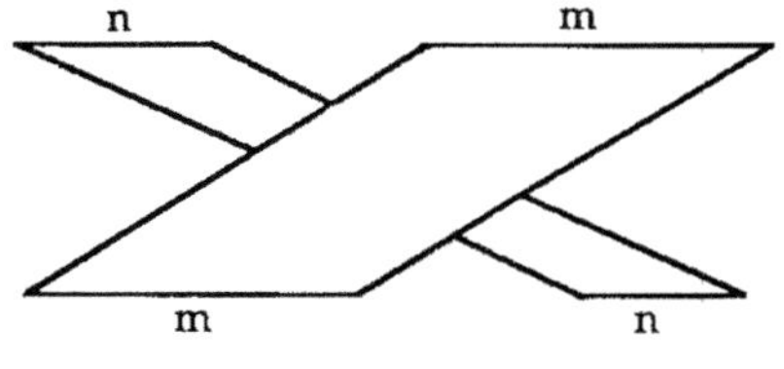

Theorem 1 [JS1]. $\mathbf{B}$ *is the free braided strict tensor category on one generating object.*

Example 2 [FY]. Let G be an arbitrary fixed (discrete) group. A *crossed G-set* is a G-set together with a function

$$| \ | : X \longrightarrow G$$

satisfying the condition $|gx| = g\,|x|\,g^{-1}$. A ***morphism*** $f : X \longrightarrow Y$ of crossed G-sets is a function satisfying $f(gx) = g\,f(x)$ and $|f(x)| = |x|$. We have a category $\mathcal{C}r(G)$ of crossed G-sets. This becomes a tensor category on taking the tensor product of crossed G-sets X, Y to be their cartesian product together with $|(x,y)| = |x||y|$. A braiding $c = c_{X,Y} : X \otimes Y \xrightarrow{\sim} Y \otimes X$ for the tensor category $\mathcal{C}r(G)$ is given by

$$c(x,y) = (|x|\,y, x).$$

Example 3. Let R be a commutative ring. The category $\mathbf{Z}\mathcal{M}od(R)$ of Z-graded R-modules has a well-known tensor product:

$$(A \otimes B)_n = \sum_{p+q=n} A_p \underset{R}{\otimes} B_q .$$

For any invertible element $k \in R$, we can define a braiding via the formula

$$c(x \otimes y) = k^{pq}\, y \otimes x$$

for $x \in A_p$ and $y \in B_q$. When $k = -1$ we get the usual (anti-)symmetry on graded modules.

Example 4. The character group of the circle group $T = U(1)$ is isomorphic to $\mathbf{Z}$. This shows that any representation $V \in \mathcal{R}ep(T, C)$ splits as a direct sum of isotypical components

$$V = \sum_{k \in Z} V_k$$

where the action of $z \in T$ on $x \in V_k$ is equal to $z^k x$. The tensor category $\mathcal{R}ep(T, C)$ is actually isomorphic to the category of Z-graded C-modules. This shows that, for any non-zero complex number $k \in C$, we can define a braiding on $\mathcal{R}ep(T, C)$ via the formula in the previous example.

Example 5. The centre of the unitary group U(n) is a one-dimensional torus $T = \{ zI : z \in U(1) \}$. If we restrict to T the action of U(n) on $V \in \mathcal{R}ep(U(n), C)$, we obtain a splitting

$$V = \sum_{k \in Z} V_k$$

where the V_k are stable under the action of U(n) since T is in the centre of U(n). This shows that we can transfer the braiding of Example 4 to the category $\mathcal{R}ep(U(n), C)$.

Example 6. The centre of the group SU(n) is the group $\mu_n \subset \Pi$ of n-th roots of unity. The dual group $\mu_n^{\vee}$ is isomorphic to Z/n. For any $V \in \mathcal{R}ep(SU(n), C)$, we obtain a splitting

$$V = \sum_{k \in Z/n} V_k .$$

If $k \in C$ is an n-th root of unity, there is a braiding defined by

$$c(x \otimes y) = k^{pq}\, y \otimes x$$

where $p, q \in Z/n$ are the degrees of x and y. When $n = 2$, the braiding is a symmetry. If we choose $k = -1$, the odd degree representations behave differently from the even ones under permutation symmetry. In theoretical physics, this is the mathematical structure which distinguishes fermions (odd representations) from bosons (even representations). When $n = 3$, we obtain a three-fold classification of the irreducible representations of

SU(3). If k is a primitive cubic root of unity, the braiding is not a symmetry and the braid group takes the place of the symmetric group as acting on tensor powers of $V^{\otimes n}$. In this case however, the operator

$$c = c_{V,V} : V \otimes V \xrightarrow{\sim} V \otimes V$$

satisfies the quadratic equation

$$c^2 + c + 1 = 0,$$

so that we are not too far away from the symmetric case (characterized by the equation $c^2 = 1$). The group SU(3) is the one used by theoretical physicists for the theory of quarks. Is there any physical significance to the above braiding?

Example 7. Example 3 can be generalized as follows. For any abelian group A, let $A\mathcal{Mod}(R)$ be the category of A-graded R-modules. Let $k : A \times A \longrightarrow R^{\times}$ be a pairing of the (additive) group A into the (multiplicative) group $R^{\times}$ of invertible elements of R. We can define a braiding via the formula

$$c(x \otimes y) = k(p,q)\, y \otimes x$$

where x, y are homogeneous of degree p, q, respectively.

Example 8. Examples 4, 5 and 6 can be generalized by taking a compact group A whose centre contains the dual group $A^{\vee}$ and where the pairing is $k : A \times A \longrightarrow C^{\times}$.

Example 9. For any bialgebra A, the category $\mathcal{Mod}(A)$ of (left) A-modules is a tensor category. A braiding c on $\mathcal{Mod}(A)$ is completely determined by the element $\gamma = c_{A,A}(1 \otimes 1)$. To see this, let M, N be two A-modules. For any $x \in M, y \in N$, we have the commutative diagram below, in which $x : A \longrightarrow M$, $y : A \longrightarrow N$ send $1 \in A$ to $x \in M, y \in N$, respectively.

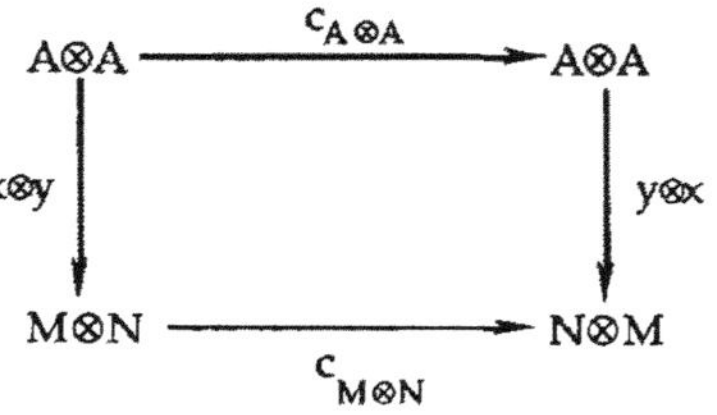

This proves that

$$c_{M,N}(x \otimes y) = \gamma(y \otimes x)$$

where the right-hand $A \otimes A$-module structure on $N \otimes M$ is used. Of course, this element $\gamma \in A \otimes A$ must satisfy certain conditions. The first is that the assignment $x \otimes y \longmapsto \gamma(y \otimes x)$ should be A-linear. To express this condition, let us use the notation $\xi_{ijk\ldots}$ for the image of ξ under the *usual* canonical isomorphism

$$\sigma_{ijk\ldots} : M_1 \otimes M_2 \otimes M_3 \otimes \ldots \xrightarrow{\sim} M_i \otimes M_j \otimes M_k \otimes \ldots$$

induced by the permutation $1\,2\,3 \ldots \longmapsto i\,j\,k \ldots$. The formula for the braiding is then

$$c(x \otimes y) = \gamma \,.\, (x \otimes y)_{21}.$$

The A-linearity of c then amounts to the equation

$$\gamma \,.\, (\delta(a) \,.\, (x \otimes y))_{21} = \delta(a) \,.\, \gamma \,.\, (x \otimes y).$$

This equation is valid for all $a \in A$ and all x, y if and only if, for all $a \in A$,

$$\gamma \,.\, \delta(a)_{21} = \delta(a) \,.\, \gamma.$$

The reader might enjoy proving the following result: the braidings on $\mathcal{Mod}(A)$ are in bijection with the invertible elements $\gamma \in A \otimes A$ which satisfy the two equations below in addition to the one above.

$$(1_A \otimes \delta)(\gamma) = (1 \otimes \gamma) . (\gamma \otimes 1)_{132} , \quad (\delta \otimes 1_A)(\gamma) = (\gamma \otimes 1) . (1 \otimes \gamma)_{213} .$$

These equations can be written more elegantly as

$$(1_A \otimes \delta)(\gamma) = s_{23}(\gamma) s_{13}(\gamma), \quad (\delta \otimes 1_A)(\gamma) = s_{12}(\gamma) s_{13}(\gamma)$$

where we are using the insertion operators $s_{ij} : A \otimes A \longrightarrow A \otimes A \otimes A$ defined by

$$s_{12}(x \otimes y) = x \otimes y \otimes 1, \quad s_{23}(x \otimes y) = 1 \otimes x \otimes y, \quad s_{13}(x \otimes y) = x \otimes 1 \otimes y .$$

Translated into diagrammatic notation, the three equations for γ are as follows, where we put $\mu'(x \otimes y) = y \otimes x$ and $\delta' = \delta_{21}$.

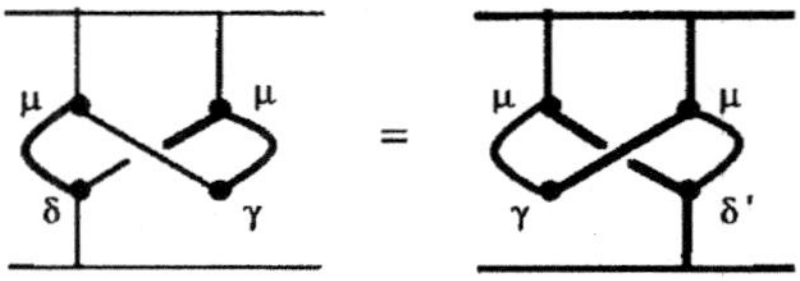

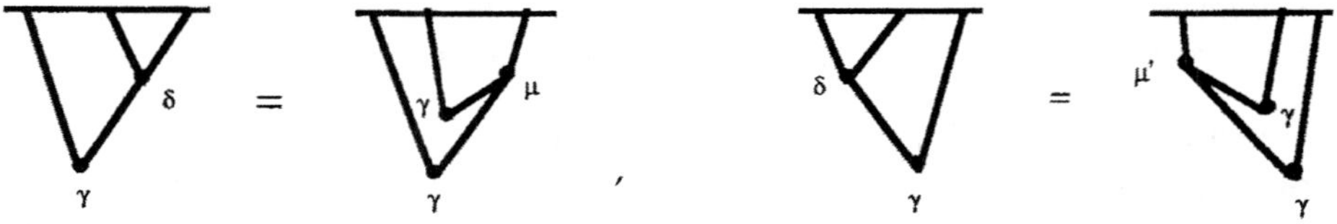

To accommodate our next example, we need the following result which holds for any coalgebra C. Let $n \geq 1$ be an integer and let $h \in (C^{\otimes n})^*$ be a linear form. For any n-sequence $(V_1, \ldots, V_n)$ of C-comodules, we can define an operator

$$\phi\rfloor : V_1 \otimes \ldots \otimes V_n \longrightarrow V_1 \otimes \ldots \otimes V_n$$

by using the fact that $V_1 \otimes \ldots \otimes V_n$ is a comodule over the coalgebra $C^{\otimes n} = C \otimes C \otimes \ldots \otimes C$. More precisely, for all $x_1 \in V_1, \ldots, x_n \in V_n$, we have

$$\phi\rfloor (x_1 \otimes \ldots \otimes x_n) = (1 \otimes \phi) \, \alpha(x_1 \otimes \ldots \otimes x_n)$$

where we have used the $C^{\otimes n}$-comodule structure

$$\alpha : V_1 \otimes \ldots \otimes V_n \longrightarrow C^{\otimes n} \otimes (V_1 \otimes \ldots \otimes V_n).$$

If $\mathcal{U}^{\otimes n} : Comod(C)^n \longrightarrow \mathcal{V}ect_C$ denotes the functor

$$(V_1, \ldots, V_n) \longmapsto V_1 \otimes \ldots \otimes V_n$$

then we have defined a natural transformation

$$\phi\rfloor( ) : \mathcal{U}^{\otimes n} \longrightarrow \mathcal{U}^{\otimes n} .$$

Proposition 2. *The assignment* $\phi \longmapsto \phi\rfloor( )$ *is an algebra isomorphism*

$$(C^{\otimes n})^* \xrightarrow{\sim} Hom(\mathcal{U}^{\otimes n}, \mathcal{U}^{\otimes n}).$$

Proof. This is a consequence of Section 7 Proposition 4 and Section 8 Proposition 1. qed

For any bialgebra A, the category $Comod(A)$ of (right) comodules is a tensor category. If c is a braiding on $Comod(A)$, we obtain a linear form

$$\gamma = (A\otimes A \xrightarrow{c_{A,A}} A\otimes A \xrightarrow{\varepsilon\otimes\varepsilon} C).$$

To state the next Proposition we shall use the insertion operators $s_{ij} : (A\otimes A)^* \longrightarrow (A\otimes A\otimes A)^*$ defined by

$$s_{12} = {}^t(A\otimes A\otimes\varepsilon),\quad s_{23} = {}^t(\varepsilon\otimes A\otimes A),\quad s_{13} = {}^t(A\otimes\varepsilon\otimes A).$$

Proposition 3. *The assignment* $c \longmapsto \gamma$ *described above is a bijection between braidings* c *on* Comod(A) *and linear forms* $\gamma\in(A\otimes A)^*$ *which are invertible for the convolution product* $*$ *and satisfy the following identities:*

$$\mu' * \gamma = \gamma * \mu,\quad \gamma(\mu\otimes 1_A) = s_{13}(\gamma) * s_{12}(\gamma),\quad \gamma(1_A\otimes\mu) = s_{13}(\gamma) * s_{23}(\gamma).$$

Proof (Sketch). The braiding is obtained from γ by the formula

$$c(x\otimes y) = (\gamma\rfloor x\otimes y)_{21} = \gamma_{21}\rfloor y\otimes x.$$

The situation is then entirely dual to that of Example 9. We can obtain these equations by rotating the pictures in that Example through 180°.qed

Recall [J, D] that a *Yang-Baxter operator* on a vector space V is a linear isomorphism

$$R : V\otimes V \xrightarrow{\sim} V\otimes V$$

such that the following hexagon commutes.

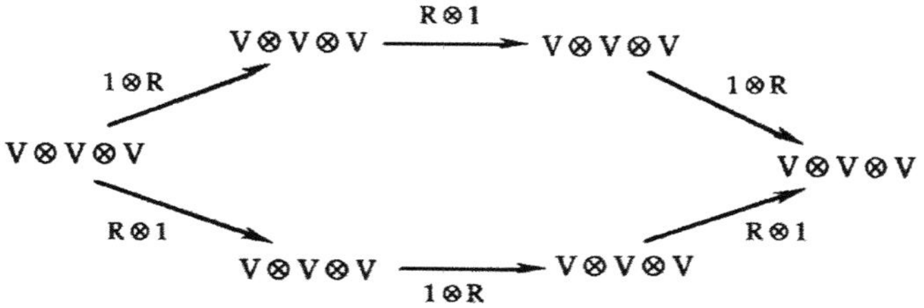

The equation

$$(R\otimes 1)(1\otimes R)(R\otimes 1) = (1\otimes R)(R\otimes 1)(1\otimes R)$$

is called the Yang-Baxter equation. The translation into pictures is as below.

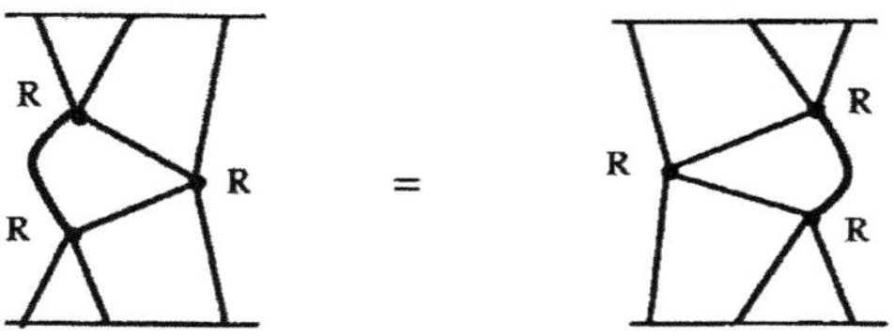

The rule of the game is to replace (whenever possible) these planar diagrams by 3-dimensional ones in which crossings replace the nodes labelled by R or R^{-1}, as indicated in the following picture. The Yang-Baxter equation is then depicted as the equality shown after that.

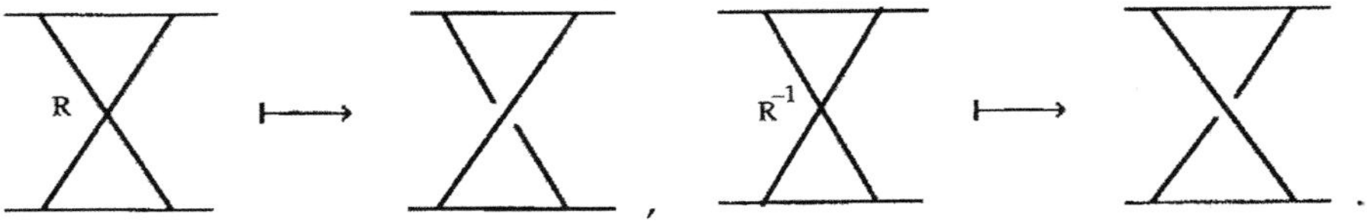

An example of a Yang-Baxter operator [J, T, JS3] is the following. Let $e_1, \dots, e_n$ be a basis for V, and let $q \in \mathbb{C}$ be a non-zero complex number. We define $R = R_q : V \otimes V \xrightarrow{\sim} V \otimes V$ as follows:

$$R(e_i \otimes e_j) = \begin{cases} e_j \otimes e_i & \text{for } i > j \\ e_j \otimes e_i + (q - q^{-1}) e_i \otimes e_j & \text{for } i < j \\ q e_i \otimes e_i & \text{for } i = j \ . \end{cases}$$

The operator R satisfies the equation

$$(R - q)\,(R + q^{-1}) = 0.$$

Moreover, the inverse of R is given by

$$R^{-1}(e_i \otimes e_j) = \begin{cases} e_j \otimes e_i & \text{for } i < j \\ e_j \otimes e_i + (q^{-1} - q) e_i \otimes e_j & \text{for } i > j \\ q^{-1} e_i \otimes e_i & \text{for } i = j \ . \end{cases}$$

One can check directly that this R is a Yang-Baxter operator.

Given any Yang-Baxter operator R on V, we can define, for every $n \geq 0$, a representation π_R of the braid group $\mathbf{B}_n$ in the general linear group $GL(V^{\otimes n})$ by putting

$$\pi_R(s_i) = \overbrace{1 \otimes \dots \otimes 1}^{i-1} \otimes R \otimes 1 \otimes \dots \otimes 1$$

for each generator s_i of $\mathbf{B}_n$. Putting together these representations, we obtain a tensor functor

$$\pi_R : \mathbf{B} \longrightarrow \mathcal{V}ect_{\mathbb{C}} \ .$$

Proposition 4. *The correspondence* $R \mapsto \pi_R$ *is a bijection between the Yang-Baxter operators on* V *and the (isomorphism classes of) tensor functors* $\pi : \mathbf{B} \longrightarrow \mathcal{V}ect_{\mathbb{C}}$ *such that* $\pi(1) = V$.

More generally, let $T : \mathcal{A} \longrightarrow \mathcal{V}$ be a functor from a category $\mathcal{A}$ to a tensor category $\mathcal{V}$.

Definition [JS3]. A Yang-Baxter operator on T is a natural family of isomorphisms

$$y = y_{A,B} \; : \; TA \otimes TB \xrightarrow{\sim} TB \otimes TA$$

such that the following hexagon commutes.

$$\begin{array}{ccccccc} & & TA\otimes TC\otimes TB & \xrightarrow{y\otimes 1} & TC\otimes TA\otimes TB & & \\ & \nearrow^{1\otimes y} & & & & \searrow^{1\otimes y} & \\ TA\otimes TB\otimes TC & & & & & & TC\otimes TB\otimes TA \\ & \searrow_{y\otimes 1} & & & & \nearrow_{y\otimes 1} & \\ & & TB\otimes TA\otimes TC & \xrightarrow[1\otimes y]{} & TB\otimes TC\otimes TA & & \end{array}$$

Any functor $T : \mathcal{A} \longrightarrow \mathcal{V}$ into a braided tensor category $\mathcal{V}$ comes equipped with a Yang-Baxter operator obtained from the braiding of $\mathcal{V}$:

$$y_{A,B} = c_{TA,TB} \; : \; TA \otimes TB \xrightarrow{\sim} TB \otimes TA .$$

The importance of Yang-Baxter operators is partly explained by the following considerations. For any category $\mathcal{A}$, there is a braided tensor category $\mathbf{B}\!\int\!\mathcal{A}$ of braids having their strings labelled by arrows of $\mathcal{A}$. (The notation $\mathbf{B}\!\int\!\mathcal{A}$ is intended to indicate that it is a wreath product in a generalized sense [K, JS3].) The objects of $\mathbf{B}\!\int\!\mathcal{A}$ are finite sequences of objects of $\mathcal{A}$. An arrow

$$(\alpha, f_1, \ldots, f_n) : (A_1, \ldots, A_n) \longrightarrow (B_1, \ldots, B_n)$$

consists of $\alpha \in \mathbf{B}_n$ and $f_i \in \mathcal{A}(A_i, B_{\alpha(i)})$ where $i \longmapsto \alpha(i)$ is the permutation defined by α. Composition of labelled braids is performed by composing the label on each string of the composite braid. The operation of addition of braids extends in the obvious way to labelled braids $\mathbf{B}\!\int\!\mathcal{A} \times \mathbf{B}\!\int\!\mathcal{A} \longrightarrow \mathbf{B}\!\int\!\mathcal{A}$, yielding a tensor structure on $\mathbf{B}\!\int\!\mathcal{A}$. There is an obvious braiding on $\mathbf{B}\!\int\!\mathcal{A}$ obtained from the braiding on $\mathbf{B}$. We have an inclusion functor

$$i : \mathcal{A} \longrightarrow \mathbf{B}\!\int\!\mathcal{A}$$

identifying $\mathcal{A}$ with the labelled braids with a single string. The braiding on $\mathbf{B}\!\int\!\mathcal{A}$ defines a (formal) Yang-Baxter operator $\mathbf{z}$ on the functor i. The next Proposition explains the sense in which this $\mathbf{z}$ is universal.

Proposition 5 [JS3]. *The braided tensor category* $\mathbf{B}\!\int\!\mathcal{A}$ *is free on* $\mathcal{A}$. *Moreover, for any tensor category* $\mathcal{V}$ *and any pair* $(\mathbf{y}, T)$, *where* $\mathbf{y}$ *is a Yang-Baxter operator on* $T : \mathcal{A} \longrightarrow \mathcal{V}$, *there exists a unique (up to a unique isomorphism) tensor functor* $T' : \mathbf{B}\!\int\!\mathcal{A} \longrightarrow \mathcal{V}$ *such that* $T'(\mathbf{z}) = \mathbf{y}$ *and the following triangle commutes.*

$$\begin{array}{ccc} \mathcal{A} & \xrightarrow{i} & \mathbf{B}\!\int\!\mathcal{A} \\ & {}_{T}\searrow & \downarrow{T'} \\ & & \mathcal{V} \end{array}$$

Example 10. For any algebra A, a Yang-Baxter operator on the forgetful functor

$$U : \mathcal{M}od(A) \longrightarrow \mathcal{V}ect_{\mathbb{C}}$$

is completely determined by the element

$$\gamma = y_{A,A}(1 \otimes 1) \in A \otimes A .$$

The operator is given by

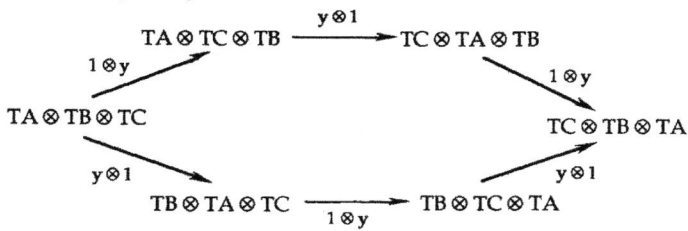

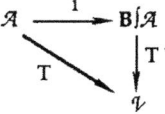

$$y_{M,N}(x \otimes y) = \gamma(y \otimes x).$$

Apart from invertibility, the only condition on γ is the equation

$$s_{23}(\gamma)\, s_{13}(\gamma)\, s_{12}(\gamma) = s_{12}(\gamma)\, s_{13}(\gamma)\, s_{23}(\gamma)$$

where $s_{ij} : A\otimes A \longrightarrow A\otimes A\otimes A$ are the insertion operators. We shall say that an invertible element $\gamma \in A\otimes A$ satisfying these equations is a *Yang-Baxter element* of the algebra A. It should be distinguished from the operator y that it defines. More precisely, when A = End(V) where V is a finite dimensional vector space, a Yang-Baxter element $\gamma \in$ End(V)$\otimes$End(V) $\cong$ End(V$\otimes$V) defines a Yang-Baxter operator $R = \gamma \circ c$ where $c : V\otimes V \longrightarrow V\otimes V$ is the usual symmetry operator.

Example 11. For any coalgebra C, a Yang-Baxter operator y on the forgetful functor

$$\mathcal{U} : \mathcal{C}omod(C) \longrightarrow \mathcal{V}ect_C$$

is determined by the linear form

$$\gamma = (C \otimes C \xrightarrow{y_{C,C}} C \otimes C \xrightarrow{\varepsilon\otimes\varepsilon} C).$$

We have the formula

$$y(x \otimes y) = (\gamma\rfloor x \otimes y)_{21} = \gamma_{21}\rfloor y \otimes x .$$

The linear form γ is invertible in the algebra $(C\otimes C)^*$ and satisfies the following equation in the algebra $(C\otimes C\otimes C)^*$:

$$s_{12}(\gamma) * s_{13}(\gamma) * s_{23}(\gamma) = s_{23}(\gamma) * s_{13}(\gamma) * s_{12}(\gamma)$$

where the s_{ij} are the insertion operators defined before Proposition 3.

Definition [JS3]. Suppose $\mathcal{V}$ is a braided tensor category. A (full) *twist* for $\mathcal{V}$ is a natural family of isomorphisms

$$\theta = \theta_A : A \xrightarrow{\sim} A$$

such that $\theta_I = 1$ and the following diagram (T) commutes.

$$\begin{array}{ccc} A\otimes B & \xrightarrow{c_{A,B}} & B\otimes A \\ {\scriptstyle \theta_{A\otimes B}}\downarrow & & \downarrow{\scriptstyle \theta_B\otimes\theta_A} \\ A\otimes B & \xleftarrow[c_{B,A}]{} & B\otimes A \end{array}$$

A tensor category equipped with a braiding and a twist is called a *balanced* (or *ribbon*) tensor category.

For any braided bialgebra (A,γ) (Example 9), the twists on $\mathcal{M}od(A)$ are in bijection with invertible central elements $\tau \in A$ satisfying the equations

$$\varepsilon(\tau) = 1 \quad \text{and} \quad \delta(\tau)_{21} = \gamma_{21}(\tau\otimes\tau)\,\gamma .$$

We have $\tau = \theta_A(1)$ and $\theta(x) = \tau x$.

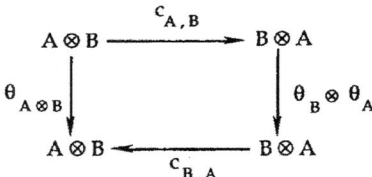

Similarly, for any cobraided bialgebra (A,γ), the twists on $\mathcal{C}omod(A)$ are in bijection with the invertible central elements $\tau \in A^*$ satisfying the equations

$$\tau(1) = 1 \quad \text{and} \quad \tau \circ \mu = \gamma * (\tau \otimes \tau) * \gamma_{21} \; .$$

Definition [JS3, Sh]. A tensor category is said to be *tortile* when it is balanced and each object A has a left dual A^* satisfying

$$\theta_{A^*} = \theta_A{}^* .$$

Definition. A *tortile* Hopf algebra (H, γ, τ) is a Hopf algebra H equipped with a braiding γ and a twist τ such that $\nu(\tau) = \tau$ where ν is the antipode.

Definition. A *cotortile* Hopf algebra (H, γ, τ) is a Hopf algebra equipped with a cobraiding γ and a cotwist τ such that $\tau \circ \nu = \nu$.

The category of comodules over a cotortile Hopf algebra is a tortile tensor category.

Proposition 6. *Every tortile tensor category is autonomous.*

Proof. In any braided tensor category, if $(\eta, \varepsilon) : A^* \dashv A$ is an adjunction then $(\eta_1, \varepsilon_1) : A \dashv A^*$ is also an adjunction where $\varepsilon_1 = \varepsilon \circ c_{A, A^*}$ and $\eta_1 = (c_{A^*, A})^{-1} \circ \eta$. To see this, we can use the following abstract argument. On any tensor category $(\mathcal{C}, \otimes)$ there is a *reverse* tensor product $C \otimes' D = D \otimes C$. Clearly, if $(\eta, \varepsilon) : A^* \dashv A$ in $(\mathcal{C}, \otimes)$ then $(\eta, \varepsilon) : A \dashv A^*$ in $(\mathcal{C}, \otimes')$. When $\mathcal{C}$ is braided, we have a natural isomorphism

$$c = c_{C,D} : C \otimes D \xrightarrow{\sim} C \otimes' D$$

which is *coherent* : it is an isomorphism of the two tensor structures [JS3, JS4]. Using c we can transport the adjunction $(\eta, \varepsilon) : A \dashv A^*$ in $(\mathcal{C}, \otimes')$ to an adjunction $(\eta_1, \varepsilon_1) : A \dashv A^*$ in $(\mathcal{C}, \otimes)$. That the formulas for η_1, ε_1 are as claimed is now clear.$_{\text{qed}}$

Proposition 7. *In a tortile tensor category, the square of the twist θ_A is given by*

$$(\theta_A)^2 =$$

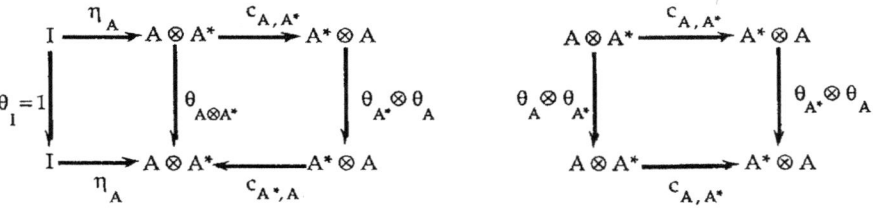

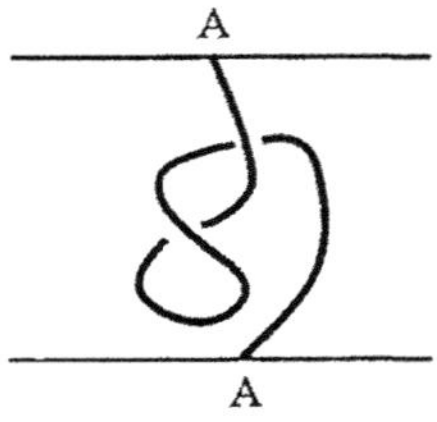

Proof. We have the following commutative diagrams:

$$\begin{array}{ccccc}
I & \xrightarrow{\eta_A} & A \otimes A^* & \xrightarrow{c_{A,A^*}} & A^* \otimes A \\
\downarrow{\scriptstyle \theta_I = 1} & & \downarrow{\scriptstyle \theta_{A \otimes A^*}} & & \downarrow{\scriptstyle \theta_{A^*} \otimes \theta_A} \\
I & \xrightarrow[\eta_A]{} & A \otimes A^* & \xleftarrow[c_{A^*,A}]{} & A^* \otimes A
\end{array}
\qquad
\begin{array}{ccc}
A \otimes A^* & \xrightarrow{c_{A,A^*}} & A^* \otimes A \\
\downarrow{\scriptstyle \theta_A \otimes \theta_{A^*}} & & \downarrow{\scriptstyle \theta_{A^*} \otimes \theta_A} \\
A \otimes A^* & \xrightarrow[c_{A,A^*}]{} & A^* \otimes A
\end{array}$$

which show that

$$c_{A,A^*}^{-1}\ c_{A^*,A}^{-1}\ \eta_A \;=\; (\theta_A \otimes \theta_{A^*})\ \eta_A$$

and therefore, tensoring this equality on the right with A and composing on the left with $A \otimes \varepsilon_A$, we have

$$(A \otimes \varepsilon_A)\,((c_{A,A^*}^{-1}\ c_{A^*,A}^{-1}\ \eta_A) \otimes A) \;=\; (A \otimes \varepsilon_A)\,(\theta_A \otimes \theta_{A^*} \otimes A)\,(\eta_A \otimes A)\,.$$

The left-hand side is equal to the value of the picture in the Proposition, so it remains to show that the right-hand side is equal to $(\theta_A)^2$. But we have

$$\varepsilon\,(\theta_{A^*} \otimes A) \;=\; \varepsilon\,(A \otimes \theta_A)$$

since $\theta_{A^*} = \theta_A{}^*$. The following sequence of pictures finishes the proof.

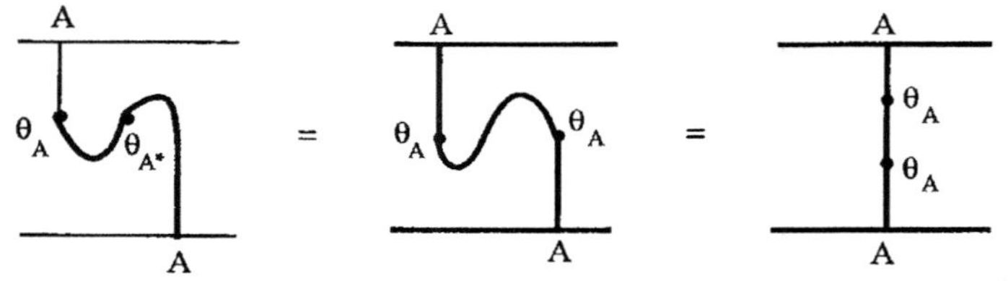

qed

The adjunction $(\eta_1, \varepsilon_1) : A \dashv A^*$ is not the appropriate one in a tortile tensor category. The reason is that, if (η_2, ε_2) denotes the pair obtained by a twofold application of the assignment $(\eta, \varepsilon) \longmapsto (\eta_1, \varepsilon_1)$, then we do not have $(\eta_2, \varepsilon_2) = (\eta, \varepsilon)$. Let us analyse the situation. For these pictures we shall use ribbons rather than strings. The twist θ will be represented by a full right hand screw turn of the ribbon. First, the pictures for ε and η are:

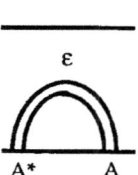

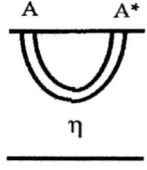

The adjunction equations are:

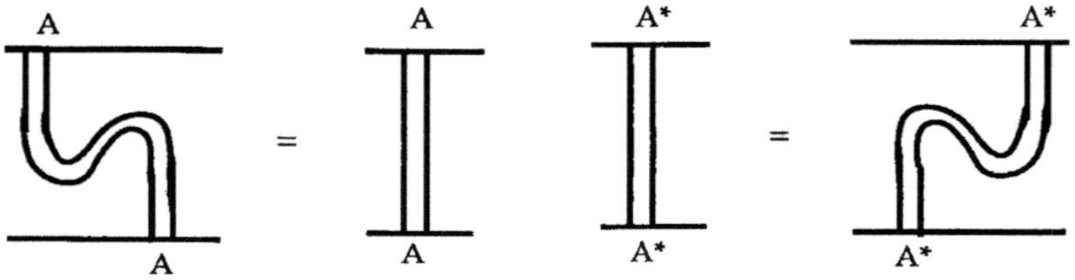

The pictures for ε_1, η_1 and the adjunction identities for them are shown below. By viewing the adjunction diagrams in 3-dimensional space, we can move the twisted ribbon on the left of these pictures and put it in the untwisted position on the right (the motion, technically called an isotopy, is restricted to a left-right sliding of the attached parts at the top and bottom).

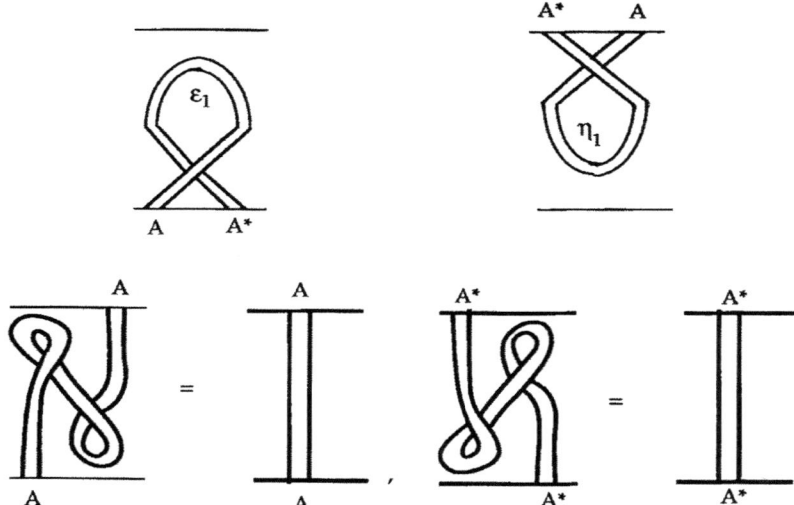

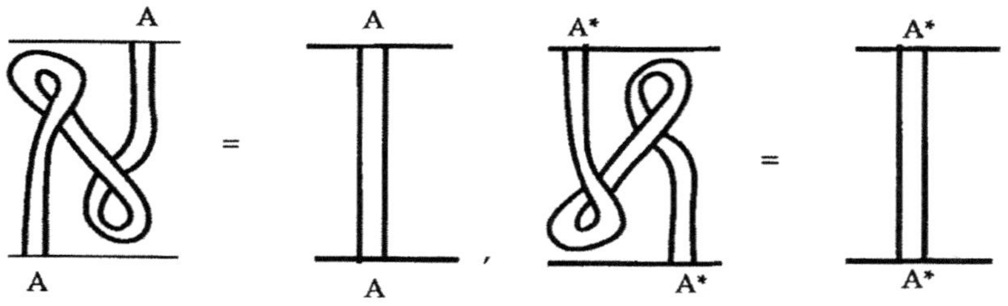

We would like to eliminate the looping in the pair (η_1, ε_1). One way to do this is to cancel the looping by a twist. If we put

$\varepsilon' = \varepsilon_1 \circ (\theta_A \otimes A^*) = \varepsilon \circ c_{A^*, A} \circ (\theta_A \otimes A^*)$, $\eta' = (A^* \otimes \theta_A^{-1}) \circ \eta_1 = (A^* \otimes \theta_A^{-1}) \circ \eta \circ (c_{A^*, A})^{-1}$

then the pictures for ε' and η' are:

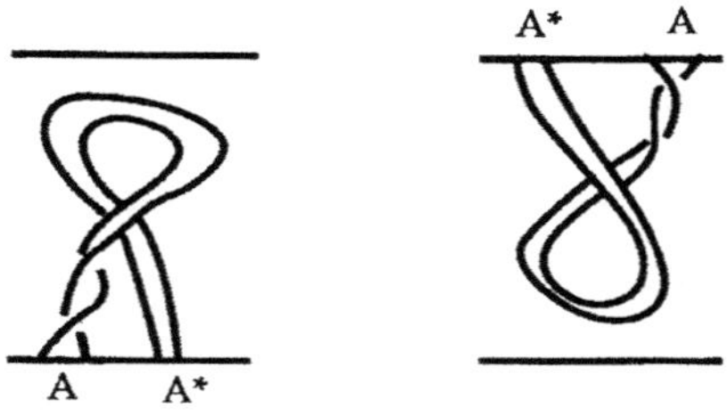

which should be redesigned to look like:

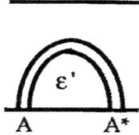

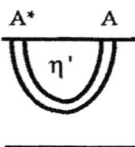

A formal verification that this is correct is as follows. If (η'', ε'') is the pair obtained by a two-fold application of the transformation $(\eta, \varepsilon) \mapsto (\eta', \varepsilon')$ then we have $(\eta'', \varepsilon'') = (\eta, \varepsilon)$. To see this, look at the picture of η'' and compute (some steps of this calculation are missing and we invite the reader to fill in the gaps):

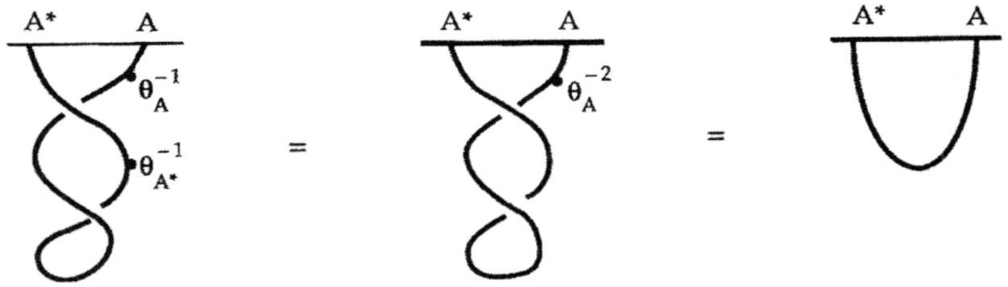

Suppose we have two adjunctions

$$(\eta_A, \varepsilon_A) : A^* \dashv A, \qquad (\eta_B, \varepsilon_B) : B^* \dashv B$$

in a tensor category. The *mate* of a map $f : X \otimes A \longrightarrow B \otimes Y$ is the map $f^{\circledast} : B^* \otimes X \longrightarrow Y \otimes A^*$ described by the diagram

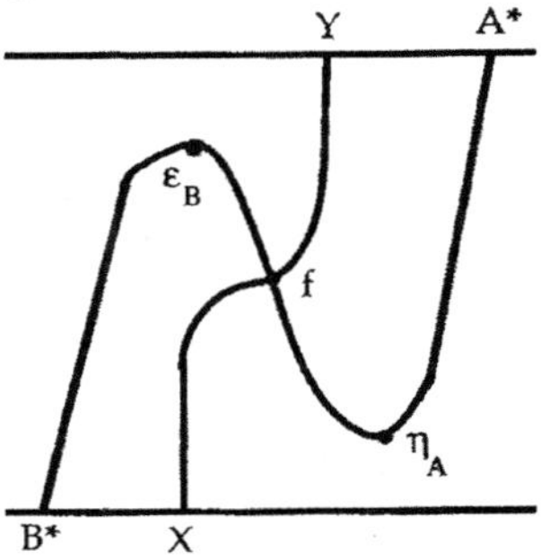

Equationally, this means that

$$f^{\circledast} = (\varepsilon_B \otimes Y \otimes A^*)(B^* \otimes f \otimes A^*)(B^* \otimes X \otimes \eta_A).$$

Proposition 8. *In a braided tensor category, the mate of* $c_{A,B} : A \otimes B \longrightarrow B \otimes A$ *is* $(c_{A,B^*})^{-1} : B^* \otimes A \longrightarrow A \otimes B^*$.

Proof. It suffices to prove that $(c_{A,B^*})(c_{A,B})^{\circledast} = 1$, which is done by the following sequence of diagrams.

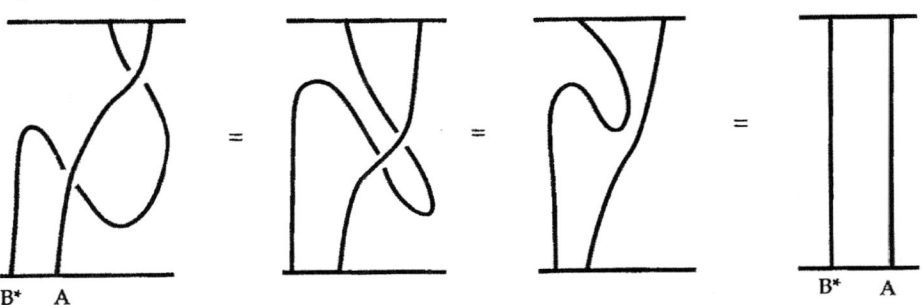

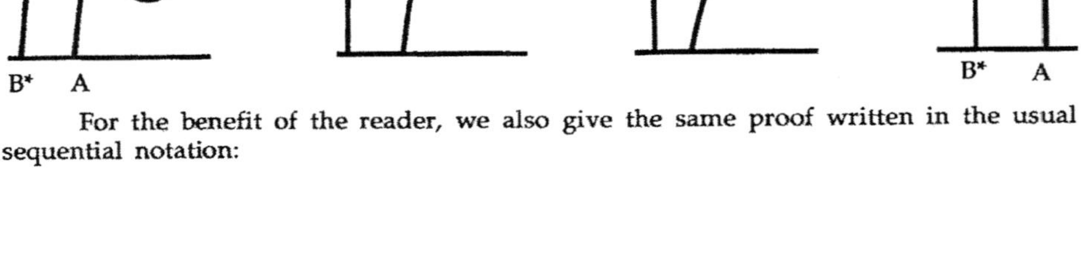

For the benefit of the reader, we also give the same proof written in the usual sequential notation:

$$
\begin{aligned}
(c_{A,B^*})(c_{A,B})^{@} &= (c_{A,B^*})(\varepsilon_B \otimes A \otimes B^*)(B^* \otimes c_{A,B} \otimes B^*)(B^* \otimes A \otimes \eta_B) \\
&= (\varepsilon_B \otimes c_{A,B^*})(B^* \otimes c_{A,B} \otimes B^*)(B^* \otimes A \otimes \eta_B) \\
&= (\varepsilon_B \otimes B^* \otimes A)(B^* \otimes B \otimes c_{A,B^*})(B^* \otimes c_{A,B} \otimes B^*)(B^* \otimes A \otimes \eta_B) \\
&= (\varepsilon_B \otimes B^* \otimes A)(B^* \otimes ((B \otimes c_{A,B^*}) \circ (c_{A,B} \otimes B^*))(B^* \otimes A \otimes \eta_B) \\
&= (\varepsilon_B \otimes B^* \otimes A)(B^* \otimes (c_{A,B \otimes B^*}))(B^* \otimes A \otimes \eta_B) \\
&= (\varepsilon_B \otimes B^* \otimes A)(B^* \otimes (c_{A,B \otimes B^*}(A \otimes \eta_B))) \\
&= (\varepsilon_B \otimes B^* \otimes A)(B^* \otimes ((\eta_B \otimes A) \circ c_{A,I})) \\
&= ((\varepsilon_B \otimes B^*) \circ (B^* \otimes \eta_B) \otimes A \\
&= B^* \otimes A \ \cdot_{\text{qed}}
\end{aligned}
$$

Corollary 9. *If* $(\eta, \varepsilon) : A^* \dashv A$ *in a braided tensor category then the following equations hold:*

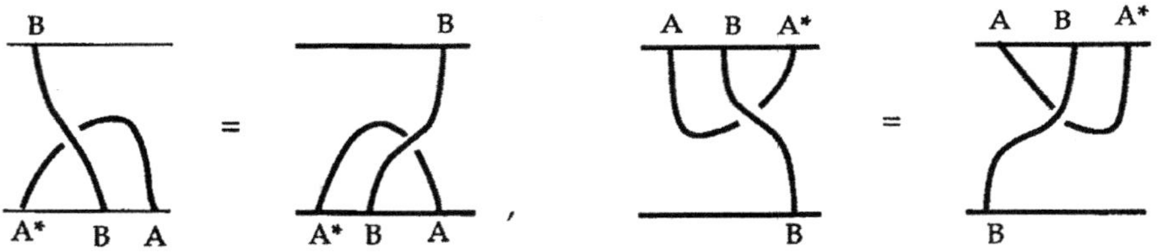

Proof. According to Proposition 8, we have

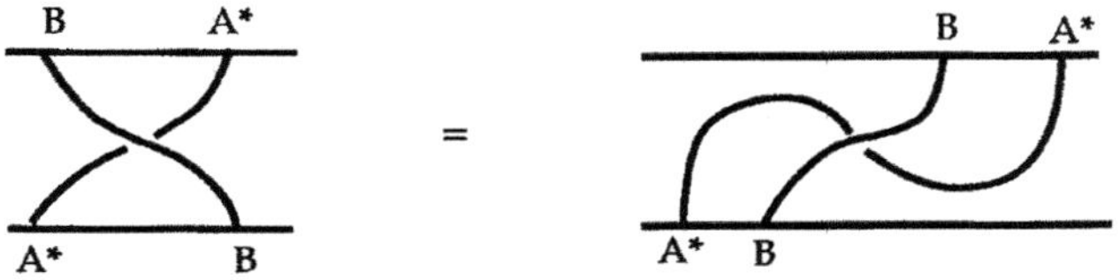

which can be composed on the top with $B \otimes \varepsilon$ to yield

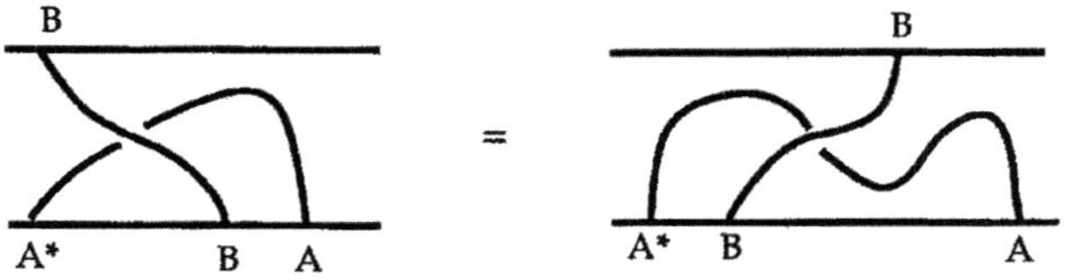

which proves the result.$_{\text{qed}}$

Definition [JS3, JS4] Let $T : \mathcal{A} \longrightarrow \mathcal{V}$ be a functor from a category $\mathcal{A}$ to a tensor category $\mathcal{V}$. A Yang-Baxter operator y on T is called *dualisable* when, for all $A \in \mathcal{A}$, the object TA has a left dual $(TA)^*$ and, for all $A, B \in \mathcal{A}$, the mates of $y_{A,B}$, $(y_{B,A})^{-1} : TA \otimes TB \longrightarrow TB \otimes TA$ are invertible.

It was shown by Proposition 8 that a braiding on a tensor category is a dualisable Yang-Baxter operator if every object has a left dual.

A dualisable Yang-Baxter operator y on a functor $T : \mathcal{A} \longrightarrow \mathcal{V}$ can be extended by duality to a Yang-Baxter operator y' on a functor $T' : \mathcal{A}' \longrightarrow \mathcal{V}$ where $\mathcal{A}'$ is the disjoint union $\mathcal{A} + \mathcal{A}^{op}$ of the category $\mathcal{A}$ and its opposite $\mathcal{A}^{op}$. To avoid ambiguities, we shall write A° and f° for the object and arrow in $\mathcal{A}^{op}$ corresponding to A and f in $\mathcal{A}$. The extension of T is given as follows:

$$T'(A) = T(A),\quad T'(A^{\circ}) = T(A)^{*},\quad T'(f) = T(f),\quad T'(f^{\circ}) = T(f)^{*}.$$

The extension of y is given as follows:

$$y'_{A,B} = y_{A,B},\quad y'_{A,B^{\circ}} = (y_{A,B^{\circledast}})^{-1},\quad y'_{A^{\circ},B} = (y_{B,A}{}^{-1})^{\circledast},\quad y'_{A^{\circ},B^{\circ}} = (y_{B,A})^{*}.$$

Proposition 10 [JS3] *The extension y' of a dualisable Yang-Baxter operator is a Yang-Baxter operator.*

The example of a Yang-Baxter operator $R = R_q : V \otimes V \longrightarrow V \otimes V$ previously given, on a finite dimensional vector space V and involving a non-zero $q \in \mathbb{C}$, is dualisable. A tedious straightforward calculation gives the following formulas, where we write $y_{V,V}$ for R and y_{V,V^*} for $y_{V,V^{\circ}}$:

$$y_{V,V}(e_i \otimes e_j) = \begin{cases} e_j \otimes e_i & \text{for } i > j \\ e_j \otimes e_i + (q - q^{-1}) e_i \otimes e_j & \text{for } i < j \\ q\, e_i \otimes e_i & \text{for } i = j \end{cases}$$

$$y_{V,V^*}(e_i \otimes e_j^*) = \begin{cases} e_j^* \otimes e_i & \text{for } i \neq j \\ q^{-1} e_i^* \otimes e_i + \sum_{k>i} q^{2(i-k)} e_k^* \otimes e_k & \text{for } i = j \end{cases}$$

$$y_{V^*,V}(e_i^* \otimes e_j) = \begin{cases} e_j \otimes e_i^* & \text{for } i \neq j \\ q^{-1} e_i \otimes e_i^* + \sum_{k<i} (q^{-1} - q) e_k \otimes e_k^* & \text{for } i = j \end{cases}$$

$$y_{V^*,V^*}(e_i^* \otimes e_j^*) = \begin{cases} e_j^* \otimes e_i^* & \text{for } i < j \\ e_j^* \otimes e_i^* + (q - q^{-1}) e_i^* \otimes e_j^* & \text{for } i > j \\ q\, e_i^* \otimes e_i^* & \text{for } i = j\ . \end{cases}$$

When $R : V \otimes V \longrightarrow V \otimes V$ is a dualisable Yang-Baxter operator on a finite dimensional vector space, we can use the extension R' to define a Yang-Baxter operator on $V \oplus V^*$. The vector space $V \oplus V^*$ is equipped with a non-degenerate symmetric pairing

$$\langle x \oplus \phi \mid y \oplus \psi \rangle = \phi(y) + \psi(x)$$

and also with a non-degenerate simplectic form

$$\omega(x \oplus \phi \mid y \oplus \psi) = \phi(y) - \psi(x).$$

Definition. Let R be a Yang-Baxter operator on an object Z in a tensor category, and let $\varepsilon : Z \otimes Z \longrightarrow I$ be an exact pairing. We say that R *respects* ε when we have the equations (in which the crossings are labelled by R and R^{-1} according to the convention previously explained):

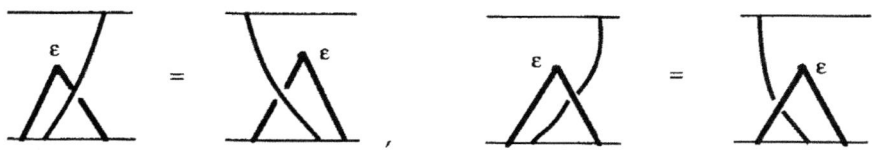

Proposition 11. *If* $(\eta, \varepsilon) : Z \dashv Z$ *and if* R *respects* ε *then the equations below hold:*

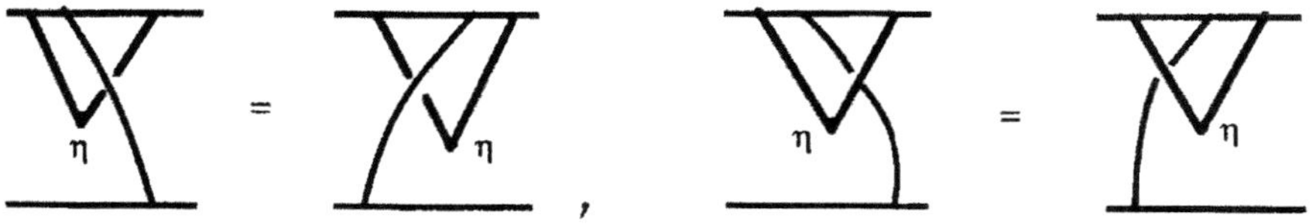

Proof. Exercise for the reader.$_{\text{qed}}$

Proposition 12. *A Yang-Baxter operator* $R : Z \otimes Z \longrightarrow Z \otimes Z$ *which respects an exact pairing is dualisable.*

Proof. The mate of R is equal to R^{-1} since we have:

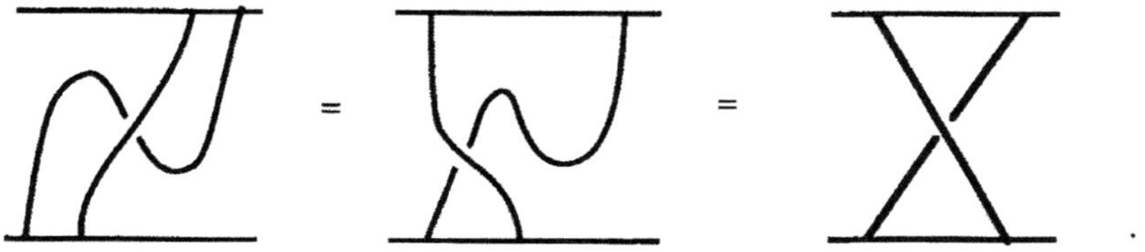

Similarly, the mate of R^{-1} is equal to R. This proves that these mates are invertible.$_{\text{qed}}$

Proposition 13. *If* R *is a dualisable Yang-Baxter operator on a finite dimensional vector space* V *then its extension* R' *to* $V \oplus V^*$ *respects both the canonical symmetric and the canonical symplectic pairings on* $V \oplus V^*$.

Proof. Exercise for the reader.$_{\text{qed}}$

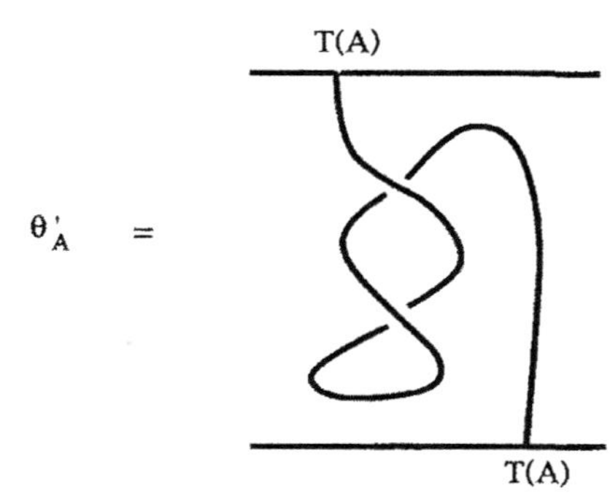

Suppose now that y is an arbitrary dualisable Yang-Baxter operator on a functor T $: \mathcal{A} \longrightarrow \mathcal{V}$. The picture above defines a canonical natural transformation

$$\theta' = \theta'_A : T(A) \longrightarrow T(A)$$

called the *double twist*. In the picture we use the extended Yang-Baxter operator y' to label the crossings.

Proposition 14. *For any dualisable Yang-Baxter operator* (y, T), *the double twist* θ' *is a natural isomorphism* $\theta' : T \xrightarrow{\sim} T$. *Moreover, the following equations hold:*

$$y(\theta' \otimes T) = (T \otimes \theta')\, y, \qquad y(T \otimes \theta') = (\theta' \otimes T)\, y.$$

Proof. The picture for the inverse of θ' is obtained from that for θ' by rotating through 180° and changing all the crossings; the rest is left to the reader.$_{\text{qed}}$

Definition [JS3, JS4] Let y be a Yang-Baxter operator on a functor $T : : \mathcal{A} \longrightarrow \mathcal{V}$. A *twist* on y is a natural isomorphism $\theta : T \xrightarrow{\sim} T$ such that $y(\theta \otimes T) = (T \otimes \theta)\, y$, $y(T \otimes \theta) = (\theta \otimes T)\, y$. A *tortile* Yang-Baxter operator is a pair (y, θ) where y is a dualisable Yang-Baxter operator and θ is a a twist on y such that $\theta^2 = \theta'$ where θ' is the double twist defined by y.

Example 11. In a tortile tensor category, the pair (c, θ) is a tortile Yang-Baxter operator since we have proved that $\theta^2 = \theta'$ (Proposition 7).

Example 12. A short calculation gives that the double twist on the operator R_q is equal to the map $x \longmapsto q^{2n} x$ on V of dimension n. If we put $\theta(x) = q^n x$, we obtain a tortile Yang-Baxter operator (R_q, θ).

§11. Knot invariants.

In this Section we provide a brief introduction to the method used by N. Yu. Reshetikhin and V.G. Turaev for obtaining knot invariants. We describe how Yang-Baxter operators can be used to produce tensor functors from the category of tangles of ribbons to vector spaces. If we apply the Tannaka duality machinery to these functors, we obtain quantum groups. This will be the subject of Section 12.

Let **P** be a Euclidean plane. A *geometric tangle* T is a compact 1-dimensional oriented submanifold of $[0,1] \times \mathbf{P}$ which is tamely embedded and whose boundary ∂T is equal to $T \cap \partial([0,1] \times \mathbf{P})$. We suppose that T meets $\partial([0,1] \times \mathbf{P})$ transversally. The *target* of T is the subset $\partial T \cap (\{1\} \times \mathbf{P})$ as an oriented submanifold. The *source* of T is the subset $\partial T \cap (\{0\} \times \mathbf{P})$, but with orientation reversed. A geometric tangle can be pictured as shown at the top of the next page.

A *tangle* is an isotopy class of geometric tangles where the isotopies keep the boundaries fixed. The source and target of a tangle are regarded as signed subsets of **P**. Let 1, 2, 3, . . . denote equally spaced collinear points in the plane **P**.

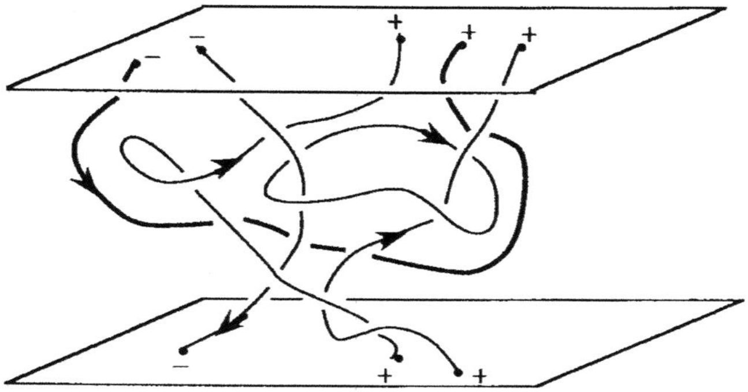

Now we can define the *autonomous braided tensor category* $\mathcal{T}$ *of tangles* [FY, Y]. The objects are functions $A : \{1, 2, 3, \ldots, n\} \longrightarrow \{+, -\}$ for $n \geq 0$, called *signed sets*. The arrows are the tangles which have these signed sets as sources and targets. Composition and tensor are as for braids. The braiding is illustrated by the following figure.

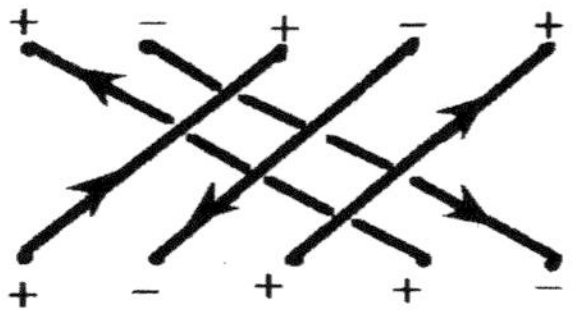

$c_{(+,-,+),(+,-)}$

The left dual A^* of a signed set A is given by reversing the order and the signs of the points. The arrows η_A and ε_A are illustrated in the following figure:

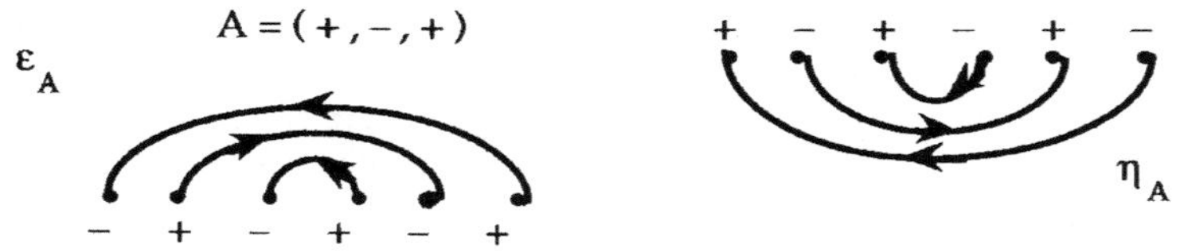

Let S^1 be the unit circle in the tangent space of the Euclidean plane $\mathbf{P}$. A *framing* on a geometric tangle T is a continuous function $f : T \longrightarrow S^1$. If $\varepsilon > 0$ is small enough then the set

$$T_\varepsilon = \{x + \alpha f(x) : x \in T,\ 0 \leq \alpha \leq \varepsilon\}$$

is an embedded surface in $\mathbf{P} \times [0,1]$ which is called a *tangle of ribbons*. We should think of the pair (T, f) as a tangle of ribbons with arbitrary small width. The *source* of (T, f) is the source of T equipped with the framing induced by f. Similarly for the target. The direction of the line on which $1, 2, 3, \ldots$ are listed will be called the *eastward* framing. The opposite direction is the *westward* framing. We can now define the *tortile tensor category* $\mathcal{T}^\sim$ of tangles on ribbons. The *objects* of $\mathcal{T}^\sim$ are the signed subsets of $1, 2, 3, \ldots$. We suppose that the framing of a positive element of a signed subset is eastward, while the framing of a negative element is westward. The *arrows* are the (isotopy classes of) tangles of ribbons having these framed signed subsets as sources and targets. Composition and tensor are as in $\mathcal{T}$. The braiding is as in $\mathcal{T}$. The units and counits of an adjunction

$(\eta_A, \varepsilon_A) : A^* \dashv A$ are illustrated in the following figure for $A = (+, -, -+)$.

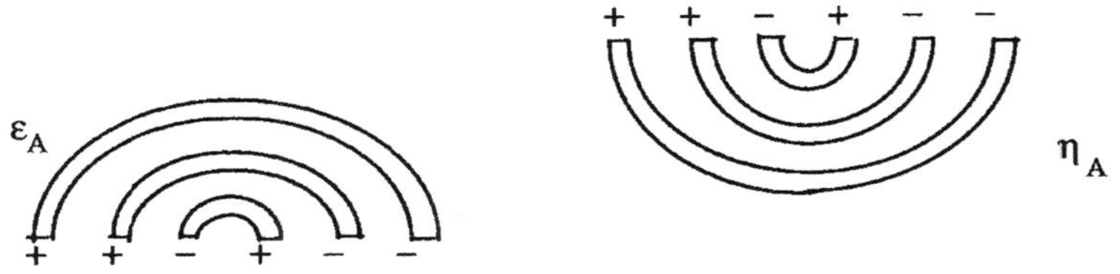

The twist θ is illustrated by the picture:

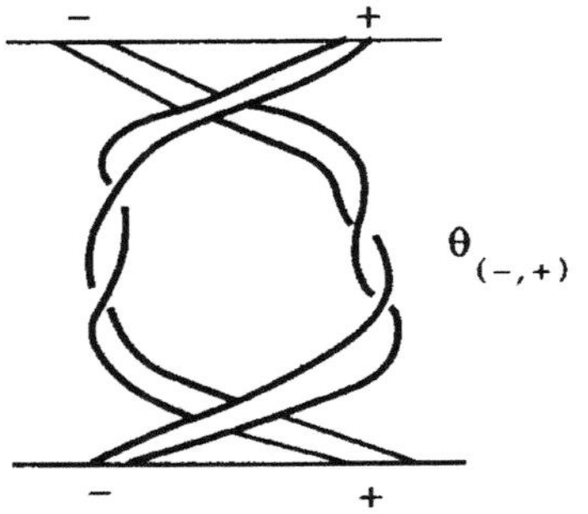

Theorem 1 [Sh] *The category* $\mathbf{T}^{\sim}$ *of tangles on ribbons is the free tortile tensor category generated by a single object.*

This means that, given any tortile tensor category $\mathcal{V}$ and any object $V \in \mathcal{V}$, there exists a tensor functor $F : \mathbf{T}^{\sim} \longrightarrow \mathcal{V}$, which preserves the braidings and the twists, such that $F(+) = V$; moreover, F is unique up to a unique isomorphism of functors. In this theorem, the functor F is entirely determined by the *object* V if everything else is kept fixed. For example, if $\mathcal{V} = \mathcal{V}ect_{f\mathbb{C}}$ is the category of finite dimensional vector spaces, then the functor F depends only (up to isomorphism) on the dimension of the vector space $V = F(+)$. We obtain, in this way, poor invariants of tangles and knots. To obtain better invariants, we need to complement this theorem by another one.

Recall that a tortile Yang-Baxter operator in a tensor category $\mathcal{C}$ is a triple (V, R, θ) where $R : V \otimes V \xrightarrow{\sim} V \otimes V$ is a dualisable Yang-Baxter operator and $\theta : V \xrightarrow{\sim} V$ is a twist on R such that $\theta^2 = \theta'$, where θ' is the canonical double twist defined by R (as defined before Section 10 Proposition 14).

Theorem 2 [T, JS4] *The category* $\mathbf{T}^{\sim}$ *of tangles on ribbons is free on a tortile Yang-Baxter operator.*

This means that, in order to define a tensor functor $F : \mathbf{T}^{\sim} \longrightarrow \mathcal{C}$, it suffices to select a tortile Yang-Baxter operator (V, R, θ) in $\mathcal{C}$. The tensor functor F is the only (up to a unique isomorphism) one for which $V = F(+)$, $R = F(c_{+,+})$ and $\theta = F(\theta_+)$.

This is the method used by Turaev to obtain knot invariants like the Jones polynomial. In this case $\mathcal{C}$ is the category of vector spaces. Using the tortile Yang-Baxter operator (R_q, θ) defined in Section 10, we can associate a number $P(q) = F(K)$ to any framed knot K (considered as a morphism $K : I \longrightarrow I$ in the tensor category $\mathcal{T}^{\sim}$). This number depends on q, and is, in fact, a Laurent polynomial in q.

§12. Quantum groups.

Let R be a Yang-Baxter operator on a finite dimensional vector space V. We saw in Section 10 how R can be used to define a tensor preserving functor

$$\pi : \mathbf{B} \longrightarrow \mathcal{V}ect_{f\mathbb{C}}$$

such that $\pi(1) = V$ and $\pi(c_{1,1}) = R$. If we apply the Tannaka duality machinery to the functor π, we obtain a bialgebra $\mathrm{End}^{\vee}(\pi)$ that we shall denote by $\mathcal{O}_R(\mathrm{End}(V))$. This notation is borrowed from algebraic geometry where $\mathcal{O}(X)$ usually denotes the ring of regular functions on an affine algebraic variety X. When $X = G$ is a compact Lie group, $\mathcal{O}(G)$ is the ring of representative functions on G. If R is the usual symmetry operator $x \otimes y \longmapsto y \otimes x$ then $\mathcal{O}_R(\mathrm{End}(V)) = S(\mathrm{End}(V)) = S(\mathrm{End}(V)^*)$ is the symmetric algebra on $\mathrm{End}(V)^*$, or equivalently, the algebra of polynomial functions on End(V). For a general R, the algebra $\mathcal{O}_R(\mathrm{End}(V))$ isnot commutative, and should be thought of as the algebra of regular functions on a "non-commutative" geometric object. Let us describe the algebra $\mathcal{O}_R(\mathrm{End}(V))$ by generators and relations.

Let $e_1, \ldots, e_n$ be a basis for V and let $x_i^j = [e_j^* \otimes e_i]$ be the images of the elements $e_j^* \otimes e_i$ under the canonical map

$$[\ ] : \mathrm{End}(V) \longrightarrow \mathcal{O}_R(\mathrm{End}(V)).$$

With respect to the basis $\{e_i \otimes e_j\}$ of $V \otimes V$, we have

$$R(e_i \otimes e_j) = \sum_{r,s} e_r \otimes e_s R_{ij}^{rs} .$$

Proposition 1. *A presentation of the algebra* $\mathcal{O}_R(\mathrm{End}(V))$ *is provided by the generators* x_i^j *for* $1 \le i, j \le n$ *and the relations*

$$\sum_{k,r} R_{kr}^{st} x_i^k x_j^r = \sum_{k,r} R_{ij}^{kr} x_k^s x_r^t .$$

Before giving a proof, we shall analyse the meaning of the relations in the presentation. For an algebra A, we shall say that a coaction $\alpha : V \longrightarrow V \otimes A$ *respects* the Yang-Baxter operator R when the following square commutes.

$$\begin{array}{ccc} V \otimes V & \xrightarrow{\alpha \underline{\otimes} \alpha} & V \otimes V \otimes A \\ {\scriptstyle R}\downarrow & & \downarrow{\scriptstyle R \otimes A} \\ V \otimes V & \xrightarrow[\alpha \underline{\otimes} \alpha]{} & V \otimes V \otimes A \end{array}$$

This condition is expressed by the equations

$$\sum_{k,r} R_{kr}^{st} \alpha_i^k \alpha_j^r = \sum_{k,r} R_{ij}^{kr} \alpha_k^s \alpha_r^t$$

where $\alpha(e_i) = \sum_i e_i \otimes \alpha_i^j$.

The canonical coaction (Section 8 Proposition 3)

$$\gamma_1 : V \longrightarrow V \otimes \mathrm{End}^{\vee}(\pi)$$

respects the Yang-Baxter operator R since, by the naturality of γ, the square

$$\begin{array}{ccc} V\otimes V & \xrightarrow{\gamma_2} & V\otimes V\otimes \mathrm{End}^{\vee}(\pi) \\ {\scriptstyle \pi(c_{1,1})}\downarrow & & \downarrow{\scriptstyle \pi(c_{1,1})\otimes 1} \\ V\otimes V & \xrightarrow[\gamma_2]{} & V\otimes V\otimes \mathrm{End}^{\vee}(\pi) \end{array}$$

commutes, and we have (Section 10 Proposition 4)

$$\pi\,(c_{1,1}) = R, \qquad \gamma_2 = \gamma_1 \underline{\otimes} \gamma_1 .$$

This implies that the relations in the purported presentation of $O_R(\mathrm{End}(V)) = \mathrm{End}^{\vee}(\pi)$ are satisfied since

$$\gamma_1(e_i) = \sum_j e_i \otimes [e_j^* \otimes e_i] = \sum_j e_j \otimes x_i^j .$$

Proposition 1 is now an immediate consequence of Proposition 2 below.

Proposition 2. *For any algebra* A *and any coaction* $\alpha : V \longrightarrow V \otimes A$ *respecting* R, *there is a unique map of algebras* $\alpha^{\sim} : O_R(\mathrm{End}(V)) \longrightarrow A$ *such that the triangle below commutes.*

$$\begin{array}{ccc} V & \xrightarrow{\gamma_1} & V\otimes O_R(\mathrm{End}(V)) \\ & {\scriptstyle \alpha}\searrow & \downarrow{\scriptstyle V\otimes\alpha^{\sim}} \\ & & V\otimes A \end{array}$$

Proof. Let $\alpha : V \longrightarrow V \otimes A$ be a coaction respecting R. Then R is a Yang-Baxter operator on the object (V, α) of the category $\mathcal{Co}_f(A)$ (see Section 10). Using Section 10 Proposition 4, we obtain a tensor functor $\pi^{\sim} : \mathbf{B} \longrightarrow \mathcal{Co}_f(A)$, or equivalently, a tensor-preserving natural transformation $\pi \longrightarrow \pi \otimes A$. Using Section 8 Proposition 3, we obtain a map of algebras $\mathrm{End}^{\vee}(\pi) \longrightarrow A$. The rest of the proof is left to the reader.$_{\mathbf{qed}}$

The coalgebra structure on $O_R(\mathrm{End}(V))$ is the usual one. On the generators x_i^j we have

$$\Delta x_i^j = \textstyle\sum_k x_k^j\, x_i^k , \qquad \varepsilon\,(x_i^j) = \delta_i^j .$$

If R is the usual symmetry operator $x \otimes y \longmapsto y \otimes x$, the relations reduce to

$$x_i^t\, x_j^s = x_j^s\, x_i^t$$

so that $O_R(\mathrm{End}(V)) = S(\mathrm{End}(V))$.

When R is the Yang-Baxter operator $x \otimes y \longmapsto (-1)^{pq} y \otimes x$ on a $\mathbf{Z}/2$-graded vector space $V = V_0 \oplus V_1$, we have a decomposition

$$\mathrm{End}(V) = \mathrm{End}(V_0) \oplus \mathrm{End}(V_1) \oplus \mathrm{Hom}(V_0, V_1) \oplus \mathrm{Hom}(V_1, V_0)$$

giving rise to a decomposition

$$O_R(\mathrm{End}(V)) = S(\mathrm{End}(V_0) \oplus \mathrm{End}(V_1)) \otimes \Lambda(\mathrm{Hom}(V_0, V_1) \oplus \mathrm{Hom}(V_1, V_0))$$

where Λ indicates an exterior algebra.

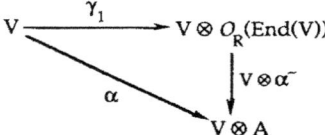

When $R = R_q$ (see just before Proposition 4 in Section 10), we have the following presentation for $\mathcal{O}_R(\mathrm{End}(V)) = \mathcal{O}_q(\mathrm{End}(V))$:

$$x_j^k x_i^r = \begin{cases} x_i^r x_j^k & \text{for } i<j \text{ and } k<r \\ x_i^r x_j^k + (q - q^{-1}) x_j^r x_i^k & \text{for } i<j \text{ and } r<k \\ q\, x_i^k x_j^k & \text{for } i<j \text{ and } k=r \\ q\, x_i^k x_i^r & \text{for } i=j \text{ and } k<r \end{cases}$$

When R is a dualisable Yang-Baxter operator we shall describe a quantum group $\mathcal{O}_R(\mathrm{GL}(V))$. For this we can follow a method similar to the one just used to describe the bialgebra $\mathcal{O}_R(\mathrm{End}(V))$. First, let **B'** be the free tensor category generated by the quintuple $(D^*, D, \eta, \varepsilon)$ where $(\eta, \varepsilon) : D^* \dashv D$ and $R : D\otimes D \longrightarrow D\otimes D$ is a dualisable Yang-Baxter operator on D. It is possible to give a geometric model for **B'** like the one described in Section 11 (but, in this case, we use tangles of strings such that the front projection defines an immersion into the yz-plane). However, no explicit description of **B'** is necessary. It can be proved that **B'** is braided and autonomous (it is in fact the free autonomous braided tensor category on a single object [FY]). Using the dualisable Yang-Baxter operator R, we can define a tensor functor

$$\pi' : \mathbf{B'} \longrightarrow \mathcal{V}ect_f$$

such that $\pi'((+)) = V$ and $\pi'(c_{+,+}) = R$. We put

$$\mathcal{O}_R(\mathrm{GL}(V)) = \mathrm{End}^\vee(\pi'),$$

which is a Hopf algebra since **B'** is autonomous. It is also cobraided since **B'** is braided. We shall give a presentation of $\mathcal{O}_R(\mathrm{GL}(V))$ in terms of $\mathcal{O}_R(\mathrm{End}(V))$. Recall that an action $\alpha : V \longrightarrow V\otimes A$ is (left) dualisable if its matrix $\alpha = (\alpha_{ij})$, with respect to some basis of V, is invertible. The canonical coaction

$$\gamma_+ : V \longrightarrow V \otimes \mathcal{O}_R(\mathrm{GL}(V))$$

is left dualisable with

$$\gamma_- : V^* \longrightarrow V^* \otimes \mathcal{O}_R(\mathrm{GL}(V))$$

as left dual. We have the following result whose proof is left to the reader.

Proposition 3. *Let* R *be a dualisable Yang-Baxter operator on* V*. For each algebra* A *and each dualisable coaction* $\alpha : V \longrightarrow V\otimes A$ *respecting* R*, there is a unique algebra map* $\alpha^\sim$ *:* $\mathcal{O}_R(\mathrm{GL}(V)) \longrightarrow A$ *such that the following triangle commutes.*

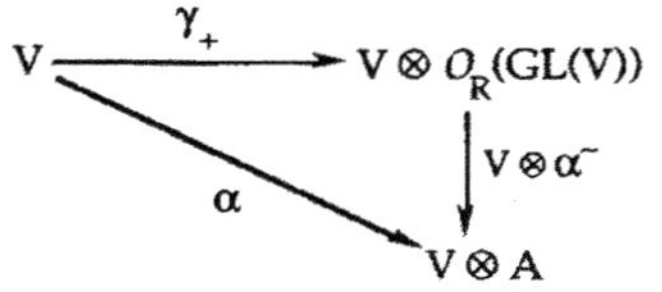

Let us denote by $\mathcal{O}_R(\mathrm{End}(V))[x^{-1}]$ the algebra obtained from $\mathcal{O}_R(\mathrm{End}(V))$ by adjoining the entries of an inverse x^{-1} of the matrix $x = (x_i^j)$ together with the relation $x^{-1} x = x x^{-1} = \mathrm{id}$.

Corollary 4. *There is a canonical isomorphism*

$$\mathcal{O}_R(GL(V)) \cong \mathcal{O}_R(End(V))[x^{-1}].$$

When $R = R_q$, the description of $\mathcal{O}_q(GL(V))$ is even simpler. The element

$$d = \sum_\sigma (-q)^{|\sigma|} x_1^{\sigma(1)} \ldots x_m^{\sigma(m)}$$

turns out to be in the centre of the algebra $\mathcal{O}_q(End(V))$. According to [FRT], it suffices to invert d in order to obtain $\mathcal{O}_q(GL(V))$; that is, we have

$$\mathcal{O}_q(GL(V)) = \mathcal{O}_q(End(V))[d^{-1}].$$

If we force d to be 1, we obtain the quantum group

$$\mathcal{O}_q(SL(V)) = \mathcal{O}_q(End(V))/d{=}1.$$

A lot of other quantum groups can be obtained as subgroups of $\mathcal{O}_R(GL(V))$. For example, if $\varepsilon : V \otimes V \longrightarrow C$ is a pairing respected by the Yang-Baxter operator R then there will be a quantum subgroup

$$\mathcal{O}_R(GL(V)) \twoheadrightarrow \mathcal{O}_R(O(V, \varepsilon))$$

of orthogonal transformations preserving ε.

Also, the reader might enjoy reading [W] on the compact quantum group $\mathcal{O}_q(SU(n))$. Finally, we encourage the reader to study quantum Grassmannians and quantum spheres [Pd].

REFERENCES

[AK] R.A. Askey, T.H. Koornwinder and W. Schempp, *Special Functions: Group Theoretical Aspects and Applications*, Mathematics and its Applications (D. Reidel, 1984).

[BL1] L.C. Biedenharm and J.D. Louck, *Angular Momentum in Quantum Physics*, Encyclopedia of Mathematics and its Applications 8 (Addison-Wesley, 1981).

[BL2] L.C. Biedenharm and J.D. Louck, *The Racah-Wigner Algebra in Quantum Theory*, Encyclopedia of Mathematics and its Applications 9 (Addison-Wesley, 1982).

[BtD] T. Bröcker and T. tom Dieck, *Representations of Compact Lie Groups*, Graduate Texts in Math. 98 (Springer-Verlag, Berlin 1985).

[Car] P. Cartier, Developpement recents sur les groupes de tresses. Applications à la topologie et à l'algèbre, *Seminaire Bourbaki 42ie annee, 1989-90, nº 716.*

[C] C. Chevalley, *Theory of Lie Groups*, (Princeton University Press, 1946).

[DM] P. Deligne and J.S. Milne, Tannakian categories, *Hodge Cocycles Motives and Shimura Varieties*, Lecture Notes in Math. 900 (Springer-Verlag, Berlin 1982) 101-228.

[DHR] S. Doplicher, R. Haag, J.E. Roberts, Local observables and particle statistics II, *Comm. Math. Phys. 35* (1974) 49-85.

[DR] S. Doplicher and J.E. Roberts, A new duality theory for compact groups, *Inventiones Math. 98* (1989) 157-218.

[D1] V.G. Drinfel'd, Quantum groups, *Proceedings of the International Congress of Mathematicians at Berkeley, California, U.S.A. 1986* (1987) 798-820.

[D2] V.G. Drinfel'd, Quasi-Hopf algebras and Knizhnik-Zamolodchikov equations, *Acad. Sci. Ukr.* (Preprint, ITP-89-43E, 1989)

[FRT] L.D. Faddeev, N.Yu. Reshetikhin and L.A. Takhtajan, Quantization of Lie algebras and Lie groups, *LOMI Preprints* (Leningrad 1987); *Algebra i Analiz 1:1* (1989, in Russian).

[F] P. Freyd, *Abelian Categories*, (Harper & Row, New York 1964).

[FY] P. Freyd and D. Yetter, Braided compact closed categories with applications to low dimensional topology, *Advances in Math. 77* (1989) 156-182.

[IS] N. Iwahori and M. Sugiura, A duality theorem for homogeneous compact manifolds of compact Lie groups, *Osaka J. Math.* 3 (1966) 139-153.

[J] M. Jimbo, A q-difference analog of U(g) and the Yang-Baxter equation, *Lett. Math. Phys.10* (1985) 63-69.

[Jn1] V. Jones, A polynomial invariant for knots via von Neumann algebras, *Bulletin American Math. Soc. 12* (1985) 103-111.

[Jn2] V. Jones, Index for subfactors, *Inventiones Math. 72* (1983) 1-25.

[JS1] A. Joyal and R. Street, *Braided monoidal categories*, Macquarie Math. Reports #850067(Dec. 1985); Revised #860081 (Nov. 1986).

[JS2] A. Joyal and R. Street, The geometry of tensor calculus I, *Advances in Math.* (to appear).

[JS3] A. Joyal and R. Street, Braided tensor categories, *Advances in Math.* (to appear).

[JS4] A. Joyal and R. Street, Tortile Yang-Baxter operators in tensor categories, *J. Pure Appl. Algebra 71* (1991) 43-51.

[K] G.M. Kelly, On clubs and doctrines, *Category Seminar Sydney 1972-73*, Lecture Notes in Math. 420 (Springer-Verlag, Berlin 1974) 181-256.

[Kl] S. Kleinman, Motives, *Proceedings of the Fifth Nordic Summer School, Oslo 1970* (Wolters-Noordhoff, Holland 1972).

[Ko] T.H. Koornwinder, *Orthogonal polynomials in connection with quantum groups.*

[Kr] M.G. Krein, A principle of duality for a bicompact group and square block algebra, *Dokl. Akad. Nauk. SSSR 69* (1949) 725-728.

[Le] G. Lewis, Coherence for a closed functor, *Coherence in Categories*, Lecture Notes in Math. 281 (Springer-Verlag, Berlin 1972) 148-195.

[Lu] G. Lusztig, Quantum deformations of certain simple modules over enveloping algebras, *Advances in Math. 70* (1988) 237-249.

[Ly] V.V. Lyubashenko, Hopf algebras and vector symmetries, *Uspekhi Mat. Nauk. 41 #5* (1986) 185-186.

[ML] S. Mac Lane, *Categories for the Working Mathematician*, Graduate Texts in Math. 5 (Springer-Verlag, Berlin 1971).

[Mn] Yu. I. Manin, *Quantum Groups and Non-Commutative Geometry*, Les publications du Centre de Recherches Mathématiques (Université de Montréal, 4e trimestre 1988).

[M1] S. Majid, Representations, duals and quantum doubles of monoidal categories, *Supl. Rend. Circ. Mat. Palermo* (to appear).

[M2] S. Majid, Braided groups, Preprint, DAMTP/90-42 (Cambridge 1990).

[M3] S. Majid, Tannaka-Krein theorem for quasiHopf algebras and other results, *Contemp. Math.* (to appear).

[P] B. Pareigis, A non-commutative non-cocommutative Hopf algebra in nature, *J. Algebra 70* (1981) 356-374.

[Pn] R. Penrose, Applications of negative dimensional tensors, *Combinatorial Mathematics and its Applications,* Edited by D.J.A. Welsh (Academic Press, 1971) 221-244.

[Pd] P. Podles, Quantum spheres, *Lett. Math. Phys.* (1987)193-202.

[RT] N.Yu. Reshetikhin and V.G. Turaev, Ribbon graphs and their invariants derived from quantum groups, *Comm. Math. Phys. 127(1)* (1990) 1-26.

[R] M. Rosso, *C.R. Acad. Sc. Paris 305, Série I* (1987) 587-590.

[SR] N. Saavedra Rivano, *Catégories Tannakiennes,* Lecture Notes in Math. 265 (Springer-Verlag, Berlin 1972).

[Sc] L. Schwartz, *Théorie des Distributions,* (Hermann, Paris 1966).

[Sh] M.C. Shum, *Tortile Tensor Categories,* PhD Thesis (Macquarie University, November 1989); Macquarie Math. Reports #900047(Apr. 1990).

[Sw] M.E. Sweedler, *Hopf Algebras,* Mathematical Lecture Notes Series (Benjamin, 1969).

[Ta] T. Tannaka, Über den Dualitätssatz der nichtkommutativen topologischen Gruppen, *Tôhoku Math. J. 45* (1939) 1-12.

[T] V.G. Turaev, The Yang-Baxter equation and invariants of links, *Invent. Math. 92* (1988) 527-553.

[U] K.-H. Ulbrich, On Hopf algebras and rigid monoidal categories, *Israel J. Math 70* (to appear).

[W] S.L. Woronowicz, Tannaka-Krein duality for compact matrix pseudogroups. Twisted SU(N) groups, *Inventiones Math. 93* (1988) 35-76.

[Y1] D.N. Yetter, Quantum groups and representations of monoidal categories, *Math. Proc. Camb. Phil. Soc.* (to appear).

[Y2] D.N. Yetter, Framed tangles and a theorem of Deligne on braided deformations of Tannakian categories, Preprint.

List of Participants

J. ADAMEK; *Prague, CZECHOSLOVAKIA*
J. BENABOU; *Paris, FRANCE*
D.B. BENSON; *Pullman, USA*
R. BETTI; *Turin, ITALY*
L. BIONDO; *Milan, ITALY*
F. BORCEUX; *Louvain-la-Neuve, BELGIUM*
G. VAN DEN BOSSCHE; *Louvain, BELGIUM*
D. BOURN; *Amiens, FRANCE*
R.D. BRANDT; *Hannover, GERMANY*
R. BROWN; *Bangor, UK*
M. BULLEJOS; *Granada, SPAIN*
M. BUNGE; *Montréal, CANADA*
A. BURRONI; *Paris, FRANCE*
F. CAGLIARI; *Bologna, ITALY*
A. CARBONI; *Milan, ITALY*
P.C. CARRASCO; *Granada, SPAIN*
R. CRUCIANI; *Rome, ITALY*
T. DATUASHVILI; *Tbilisi, GEORGIA*
R. DAWSON; *Halifax, CANADA*
Y. DIERS; *Valenciennes, FRANCE*
A. DOLD; *Heidelberg, GERMANY*
J. DONCEL; *La Coruña, SPAIN*
P. DUPONT; *Grez-Doiceau, BELGIUM*
J. DUSKIN; *Buffalo, USA*
S. EILENBERG; *New York, USA*
E. FARO; *Buffalo, USA*
T.F. FOX; *Montréal, CANADA*
A. FREI; *Vancouver, CANADA*
J.L. FREIRE; *Santiago de Compostela, SPAIN*
G. FRENCH; *Brighton, UK*
P.J. FREYD; *Philadelphia, USA*
F. GAGO; *Santiago de Compostela, SPAIN*
A. GARZON; *Granada, SPAIN*
M. GEHRKE; *Holte, DENMARK*
S. GHILARDI; *Milan, ITALY*
M. GRANDIS; *Genova, ITALY*
J.W. GRAY; *Urbana, USA*
M. HARADA; *Osaka, JAPAN*
A. HELLER; *New York, USA*
H.J. HOENKE; *Berlin, GERMANY*
J.M.E. HYLAND; *Cambridge, UK*
HÉBERT; *Franceville, GABON*
J. ISBELL; *Buffalo, USA*
B. JACOBS; *Nijmegen, HOLLAND*
G. JANELIDZE; *Tbilisi, GEORGIA*
C.B. JAY; *Edinburgh, UK*
M. JIBLADZE; *Tbilisi, GEORGIA*
M. JOHNSON; *Sydney, AUSTRALIA*
P.T. JOHNSTONE; *Cambridge, UK*
A. JOYAL; *Montréal, CANADA*
JUGNIA; *Lovain-la-Neuve, BELGIUM*
S. KASANGIAN; *Milan, ITALY*
G.M. KELLY; *Sydney, AUSTRALIA*
A. KOCK; *Aarhus, DENMARK*
M. KOROSTENSKI; *Johannesburg, SOUTH AFRICA*
A. LABELLA; *Rome, ITALY*
P. LAMBAN; *Logroño, SPAIN*
J. LAMBEK; *Montréal, CANADA*
D.M. LATCH; *Raleigh, USA*
F.W. LAWVERE; *Buffalo, USA*
F.E.J. LINTON; *Middletown, USA*
M.A. LOPEZ; *Santiago de Compostela, SPAIN*
R. LOZANOV; *Sofia, BULGARIA*
N. LUBIKULU; *Louvain-la-Neuve, BELGIUM*
S. MACLANE; *Chicago, USA*
J.L. MACDONALD; *Vancouver, CANADA*
H. MAHMOOD; *Edinburgh, UK*
S. MANTOVANI; *Turin, ITALY*
H. MARCUM; *Newark, USA*
G. MASCARI; *Rome, ITALY*
V. MATIJEVIC; *Split, YUGOSLAVIA*
G.C. MELONI; *Milan, ITALY*
I. MOERDIJK; *Utrecht, NETHERLANDS*
P.S. MULRY; *Hamilton, USA*
C.J. MULVEY; *Brighton, UK*
G. NASSOPOULOS; *Athens, GREECE*
M.L. NELSON; *Madison, USA*
S. NIEFIELD; *Schenectady, USA*
Z. OMIADZE; *Tbilisi, GEORGIA*
A. PACHKORIA; *Tbilisi, GEORGIA*
B. PACHUASHVILI; *Tbilisi, GEORGIA*
R. PARÉ; *Halifax, CANADA*
D. PAVLOVIC; *Utrecht, NETHERLANDS*
M.C. PEDICCHIO; *Trieste, ITALY*
J.W. PELLETIER; *North York, CANADA*
M. PFENNIGER; *Zurich, SWITZERLAND*
W. PHOA; *Cambridge, UK*
R.A. PICCININI; *St John's, CANADA*
T. PIRASHVILI; *Tbilisi, GEORGIA*
A.M. PITTS; *Cambridge, UK*

V. POLLARA; *Passau, GERMANY*
T. PORTER; *Bangor, UK*
J. PRADINES; *Touluose, FRANCE*
J.P.G. PRETORIUS; *Cambridge, UK*
G. RICHTER; *Bielefeld, GERMANY*
J.R. ROISIN; *Louvain-la-Neuve, BELGIUM*
R. ROSEBRUGH; *Sackville, CANADA*
J. ROSICKY; *Brno, CZECHOSLOVAKIA*
G. ROSOLINI; *Parma, ITALY*
S.H. SCHANUEL; *Buffalo, USA*
D. SCHUMACHER; *Wolfwille, CANADA*
R. SEELY; *Montréal, CANADA*
R. SIGAL; *New Haven, USA*
R.H. STREET; *Sydney, AUSTRALIA*
T. STREICHER; *Passau, GERMANY*
F. SUCCI; *Rome, ITALY*
J. TAVAKOLI; *Teheran, IRAN*
P. TAYLOR; *London, UK*
M. THIÉBAUD; *Thonex, SWITZERLAND*
W. THOLEN; *North York, CANADA*
M. TIERNEY; *Montréal, CANADA*
V.V. TOPENCHAROV; *Sofia, BULGARIA*
A. TOZZI; *L'Aquila, ITALY*
V. TRNKOVA; *Praha, CZECHOSLOVAKIA*
N. UGLESIC; *Split, YUOGOSLAVIA*
M. URIDIA; *Tbilisi, GEORGIA*
J.J.C. VERMEULEN; *Cape Town, SOUTH AFRICA*
C. VICENTINI; *Louvain-la-Neuve, BELGIUM*
S. VIGNA; *Milan, ITALY*
R.F.C. WALTERS; *Sydney, AUSTRALIA*
R.J. WOOD; *Halifax, CANADA*
M. WRIGHT; *Middlesex, UK*
LA MONTE YARROL; *Chicago, USA*
M. ZAWADOSKI; *Montréal, CANADA*
K. ZIMMZERMANN; *Schenectday, USA*

GPSR Compliance

The European Union's (EU) General Product Safety Regulation (GPSR) is a set of rules that requires consumer products to be safe and our obligations to ensure this.

If you have any concerns about our products, you can contact us on ProductSafety@springernature.com

In case Publisher is established outside the EU, the EU authorized representative is:

Springer Nature Customer Service Center GmbH
Europaplatz 3
69115 Heidelberg, Germany

Batch number: 09151036

Printed by Printforce, the Netherlands